Approximate Approximations

Mathematical
Surveys
and
Monographs

Volume 141

Approximate Approximations

Vladimir Maz'ya
Gunther Schmidt

American Mathematical Society

2000 *Mathematics Subject Classification.* Primary 41A30.

For additional information and updates on this book, visit
www.ams.org/bookpages/surv-141

Library of Congress Cataloging-in-Publication Data

Maz′ia, V. G.
Approximate approximations / Vladimir Maz′ya, Gunther Schmidt.
p. cm. — (Mathematical surveys and monographs ; v. 141)
Includes bibliographical references and index.
ISBN 978-0-8218-4203-4 (alk. paper)
1. Approximation theory. I. Schmidt, Gunther, 1952– II. Title.

QA221.M37 2007
511′.4—dc22 2007060769

∞ The paper used in this book is acid-free and falls within the guidelines
established to ensure permanence and durability.
Visit the AMS home page at `http://www.ams.org/`

10 9 8 7 6 5 4 3 2 1   12 11 10 09 08 07

Contents

Preface

• **General idea and motivation**. In this book, we discuss realizations and applications of a new concept of approximation procedures, called *approximate approximations*. Most of these procedures, which include approximate quasi-interpolation, interpolation, least square approximation, cubature of integral operators, and wavelet approximations, have one common feature. They are accurate without being convergent in a rigorous sense. In numerical mathematics, such a situation is not exceptional. For instance, non-convergent algorithms are natural in solving overdetermined ill-posed problems. However, for the approximation processes mentioned above, convergence is required.

Needless to say, the engineers and researchers who use numerical methods for solving applied problems do not need the convergence of the method. In fact, they need results which are exact within a prescribed accuracy, determined mainly by the tolerance of measurements and other physical parameters, and always by the precision of the computing system. Their attitude, supported by common sense, was a powerful motivation for the development of our theory.

The lack of convergence in approximate approximations is compensated for, first of all, by the flexibility in the choice of basis functions and by the simplicity of the multi-dimensional generalization. Another, and probably the most important, advantage is the possibility of obtaining explicit formulas for values of various integral and pseudodifferential operators of mathematical physics applied to the basis functions.

The concept of approximate approximations and first related results were published by the first author in [**62**] – [**64**]. Later on, various aspects of a general theory of these approximations were systematically investigated in several joint papers by the authors ([**66**] – [**70**]). The present book is essentially based on the papers just mentioned and on our recent unpublished results. We also report on computational algorithms of the approximate approximations developed together with V. Karlin, T. Ivanov, W. Wendland, F. Lanzara, A. B. Movchan, *et al.*

The theory under consideration is at the very beginning of its development and we wrote this book with the hope of attracting new researchers to this area.

• **Approximate quasi-interpolation**. To give an impression of what we have in mind, recall, for example, that a typical error estimate of spline interpolation $M_h u$ on a uniform grid with size h, for a function $u \in C^N$, has the form

$$\|u - M_h u\|_C \leq c h^N \|u\|_{C^N}$$

with some integer N and a constant c independent of u and h. Here C and C^N are the spaces of continuous and N-times continuously differentiable functions.

In contrast to this situation, we fix $\varepsilon > 0$ and construct an *approximate quasi-interpolant* $M_{h,\varepsilon}u$ using the translates of a more or less arbitrary function η_ε instead of piecewise polynomials

$$M_{h,\varepsilon}u(x) = \sum_m u(hm)\eta_\varepsilon(x/h - m)\ .$$

One can show that

$$\|u - M_{h,\varepsilon}u\|_C \leq c_1(u)h^N + c_2(u)\,\varepsilon\ .$$

Thus, the error consists of a part converging with order N to zero as $h \to 0$ and a non-convergent part $c_2(u)\varepsilon$ called the *saturation error*. Thus, the procedure provides good approximations up to some prescribed error level, but it does not converge as $h \to 0$.

The approximate quasi-interpolation procedure can be extended to the approximation of functions on domains and manifolds with non-uniformly distributed nodes.

• **Cubature formulas**. The numerical treatment of potentials and other integral operators with singular kernels arises as a computational task in different fields. Since standard cubature methods are very time-consuming, there is ongoing research to develop new effective algorithms like panel clustering, multipole expansions, or wavelet compression based on piecewise polynomial approximations of the density. The effective treatment of integral operators is also one of the main applications of approximate approximation.

The richness of the class of generating functions η makes it easier to find approximations for which the action of a given pseudodifferential operator can be effectively determined. For example, suppose one has to evaluate the convolution with a singular radial kernel as in the case of many potentials in mathematical physics. If the density is replaced by a quasi-interpolant with radial η, then after passing to spherical coordinates, the convolution is approximated by one-dimensional integrals. For many important integral operators $\mathcal{K}$ one can choose η even such that $\mathcal{K}\eta$ is analytically known, which results in semi-analytic cubature formulas for these operators. The special structure of the quasi-interpolation error gives rise to an interesting effect. Since the saturation error is a fast oscillating function and converges weakly to zero, the cubature formulas for potentials converge even in the rigorous sense, although there is no convergence for their densities.

• **Approximate wavelets**. Another example of approximate approximations is the notion of *approximate wavelet* decompositions for spaces generated by smooth functions satisfying refinement equations with a small error. It appears that those *approximate refinement equations* are satisfied by a broad class of scaling functions. This relation allows one to perform an approximate multi-resolution analysis of spaces generated by those functions. Therefore a wavelet basis can be constructed in which elements of fine scale spaces are representable within a given tolerance. The approximate wavelets provide most of the properties utilized in wavelet-based numerical methods and possess additionally simple analytic representations. Therefore the sparse approximation of important integral operators in the new basis can be computed using special functions or simple quadrature. One can give explicit formulas for harmonic and diffraction potentials whose densities are approximate wavelets.

• **Applications in mathematical physics**. The capability of approximate approximations to treat multi-dimensional integral operators enables one to develop new efficient numerical and semi-analytic methods for solving various problems in mathematical physics. First of all, this tool can be effectively used as an underlying approximation method in numerical algorithms for solving problems with integro-differential equations. Another very important application of approximate approximations is in the large field of integral equation methods for solving initial and boundary value problems for partial differential equations.

• **Structure of the book**. We describe briefly the contents of the book. More details are given in the introduction of each chapter. Most of the references to the literature are collected in Notes at the end of Chapters 2–13.

In Chapters 1 and 2 we analyze the approximate quasi-interpolation on uniform lattices. We start with simplest examples of second- and higher-order quasi-interpolants in both the one-dimensional and multi-dimensional cases. Then we turn to pointwise and integral error estimates for quasi-interpolation of functions given on the whole space. We formulate conditions on the generating functions η of quasi-interpolation formulas which ensure the smallness of saturation errors and the convergence with a given order up to the saturation bound.

A variety of basis functions and algorithms for their construction are the subject of Chapter 3. We provide examples giving rise to new classes of simple multivariate quasi-interpolation formulas which behave in numerical computations like high-order approximations.

Chapters 4 and 5 are dedicated to semi-analytic cubature formulas for numerous integral and pseudodifferential operators of mathematical physics, in particular for harmonic, elastic, and diffraction potentials. In Chapter 6 we obtain approximations of the inverse operator of the Cauchy problem for the heat, wave, and plate equations. There we also give formulas for the value of integral operators applied to more general basis functions.

The Gaussian functions possess remarkable approximation properties. Chapter 7 is devoted to quasi-interpolation and interpolation with these basis functions.

In Chapter 8 we perform approximate multi-resolution analysis for spaces generated by functions of the Schwartz class and introduce approximate wavelets. For the example of the Gaussian kernel we give simple analytic formulas of such wavelets first in the one-dimensional case and then in the case of many dimensions. We obtain quadratures of Newton and diffraction potentials acting on these wavelets.

In Chapter 9 the method of cubature of potentials is extended to the computation of these potentials over a bounded domain. Here we use mesh refinement towards the boundary of the domain and construct special boundary layer approximations. Our algorithm relies heavily on approximate refinement equations which, as was mentioned, play a crucial role in the construction of approximate wavelets also.

The approximate quasi-interpolation is extended in Chapter 10 to the approximation of functions on non-cubic grids and on domains and manifolds with non-uniformly distributed nodes.

In Chapter 11 we study approximate quasi-interpolation of scattered data. We show that simple modifications of basis functions provide an approximate partition of unity which allows the construction of high-order approximate quasi-interpolants on scattered centers.

Finally, in Chapters 12 and 13, we treat applications of approximate approximations to numerical algorithms of solving linear and non-linear pseudodifferential equations of mathematical physics. To be more specific, in Chapter 12 we apply the cubature methods developed in Chapter 4 to the solution of Lippmann-Schwinger type equations of scattering theory. We describe the boundary point method, the application of approximate approximations to the solution of boundary integral equations. The same chapter contains formulas for the harmonic single layer potential acting on basis functions given on a surface. In Chapter 13, we describe applications to non-linear evolution equations with local and non-local operators, including the Navier-Stokes, Joseph, Benjamin-Ono, and Sivashinsky equations.

• **Readership**. The book is intended for graduate students and researchers interested in applied approximation theory and numerical methods for solving problems of mathematical physics. No special knowledge is required to read this book, except for conventional university courses on functional analysis and numerical methods.

•**Acknowledgments**. The authors would like to thank F. Lanzara, V. Karlin, and T. Ivanov for the help in obtaining some of the numerical results presented in the book.

This research was made possible by support from DAAD and Svenska institutet (grant 313/S-PPP 4/96) and INTAS (grant 97-30551). The first author was partially supported by the NSF grant DMS 0500029. The second author was partially supported within the DFG Priority Program 1095. Both authors want to thank the Department of Mathematics of the University of Rome "La Sapienza" for the warm hospitality, and the second author gratefully acknowledges the hospitality of the Department of Mathematical Sciences at the University of Liverpool.

V. Maz′ya and G. Schmidt

CHAPTER 1

Quasi-interpolation

1.1. Introduction

1.1.1. Exercise for a freshman. Suppose we are given the task of drawing the graph of the function

$$f(x) = \sum_{m=-\infty}^{\infty} \mathrm{e}^{-(x-m)^2/2}$$

obtained by summation of shifted Gaussians, which are depicted in Fig. 1.1.

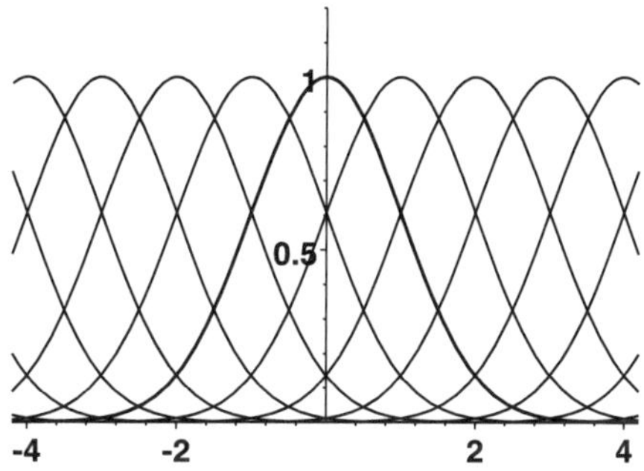

FIGURE 1.1

The function $f(x)$ is, of course, bounded, positive, and smooth. Moreover, $f(x+1) = f(x)$, i.e., it is periodic with period 1. So we expect that the graph of f should look like a nice periodic wavy curve. However, it is quite astonishing to find out that this graph, which can be easily produced with standard plotting software and which is depicted in Fig. 1.2, is a constant.

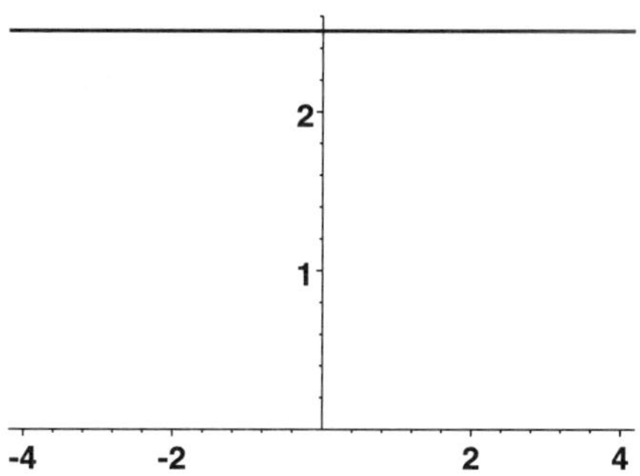

FIGURE 1.2. Graph of $f(x)$

In fact, this superficial impression proves to be wrong. If the scale of the y-axis is changed as in Fig. 1.3, then we see that $f(x)$ is not constant; it oscillates between 2.50662826 and 2.50662829.

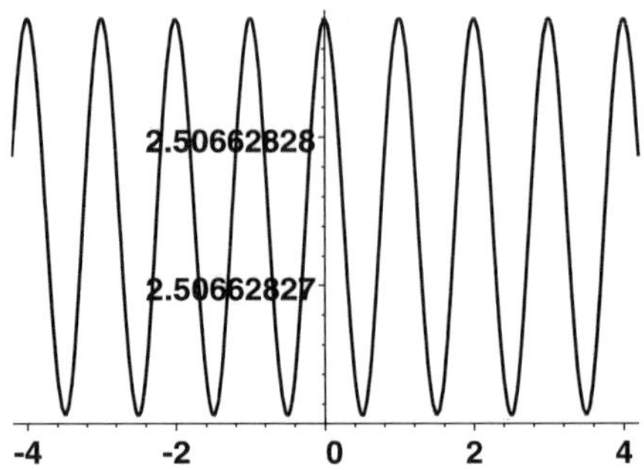

FIGURE 1.3. Zoomed graph of $f(x)$

One obtains the same picture if this procedure is repeated for the sum

$$f_{\mathcal{D}}(x) = \sum_{m=-\infty}^{\infty} \mathrm{e}^{-(x-m)^2/\mathcal{D}}$$

with different values of the parameter $\mathcal{D} > 0$. Figs. 1.4 and 1.5 show the graph of $f_{\mathcal{D}}$ for the parameters $\mathcal{D} = 0.5$ and $\mathcal{D} = 4$, respectively. The plot of the function

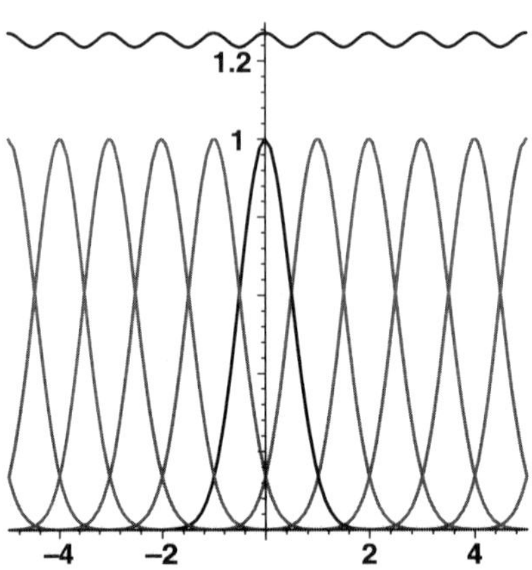

FIGURE 1.4. $f_{1/2}(x)$ and individual terms

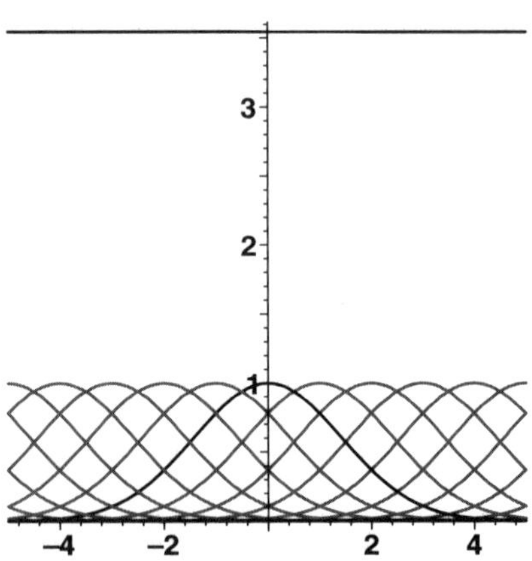

FIGURE 1.5. $f_4(x)$ and individual terms

$f_{1/2}(x)$ shows the oscillating behavior, whereas f_4 looks like a constant. In fact f_4 is also oscillating between $3.54490770181103205 \pm 1.43 \cdot 10^{-15}$, which is very hard to depict. One can conjecture, that the oscillating function $f_{\mathcal{D}}$ tends to a constant if $\mathcal{D}$ increases.

To rigorously explain peculiarities of the graphs, let us consider the Fourier series of the function

$$\theta(x,\mathcal{D}) = \frac{1}{\sqrt{\pi\mathcal{D}}} \sum_{m=-\infty}^{\infty} \mathrm{e}^{-(x-m)^2/\mathcal{D}}, \ \mathcal{D} > 0 . \tag{1.1}$$

Its coefficients can be computed as follows:

$$\begin{aligned} \frac{1}{\sqrt{\pi\mathcal{D}}} \int_0^1 \sum_{m=-\infty}^{\infty} \mathrm{e}^{-(x-m)^2/\mathcal{D}} \, \mathrm{e}^{-2\pi i\nu x} \, dx &= \frac{1}{\sqrt{\pi\mathcal{D}}} \sum_{m=-\infty}^{\infty} \int_m^{m+1} \mathrm{e}^{-x^2/\mathcal{D}} \, \mathrm{e}^{-2\pi i\nu x} \, dx \\ = \frac{1}{\sqrt{\pi\mathcal{D}}} \int_{-\infty}^{\infty} \mathrm{e}^{-x^2/\mathcal{D}} \, \mathrm{e}^{-2\pi i\nu x} \, dx &= \frac{\mathrm{e}^{-\pi^2\mathcal{D}\nu^2}}{\sqrt{\pi\mathcal{D}}} \int_{-\infty}^{\infty} \mathrm{e}^{-(x/\sqrt{\mathcal{D}}+i\pi\sqrt{\mathcal{D}}\nu)^2} \, dx \\ = \frac{\mathrm{e}^{-\pi^2\mathcal{D}\nu^2}}{\sqrt{\pi\mathcal{D}}} \int_{-\infty}^{\infty} \mathrm{e}^{-x^2/\mathcal{D}} \, dx &= \mathrm{e}^{-\pi^2\mathcal{D}\nu^2} . \end{aligned}$$

The order of summation and integration can be changed here because of the absolute convergence of the infinite sum. Hence we obtain the Fourier series

$$\theta(x,\mathcal{D}) = \sum_{\nu=-\infty}^{\infty} \mathrm{e}^{-\pi^2\mathcal{D}\nu^2} \, \mathrm{e}^{2\pi i\nu x} . \tag{1.2}$$

This representation of the function θ is a special case of the so-called Poisson summation formula

$$\sum_{m=-\infty}^{\infty} u(x+m) = \sum_{\nu=-\infty}^{\infty} \mathcal{F}u(\nu) \, \mathrm{e}^{2\pi i\nu x} , \tag{1.3}$$

where $\mathcal{F}u$ denotes the Fourier transform of the function u. The definition of the Fourier transform will be given in Section 2.1, where we also discuss some properties of this important formula.

From (1.2), we have

$$\theta(x,\mathcal{D}) = 1 + 2\sum_{\nu=1}^{\infty} \mathrm{e}^{-\pi^2\mathcal{D}\nu^2} \cos 2\pi\nu x ,$$

i.e., our function $\theta(x,\mathcal{D})$ differs from 1 by the infinite series

$$2\sum_{\nu=1}^{\infty} \mathrm{e}^{-\pi^2\mathcal{D}\nu^2} \cos 2\pi\nu x . \tag{1.4}$$

The coefficients $\mathrm{e}^{-\pi^2\mathcal{D}\nu^2}$, $\nu = 1, 2, \ldots$, can be very small depending on $\mathcal{D}$, as seen from the relation $\mathrm{e}^{-\pi^2} = 0.000051723\ldots$. In particular, if $\mathcal{D} \geq 1$, then for any x the modulus of (1.4) is less than $1.04 \cdot 10^{-4\mathcal{D}}$. Note that in the cases $\mathcal{D} = 2$ and $\mathcal{D} = 4$ the difference is comparable to the single and, respectively, double precision in the arithmetics of most modern computers, i.e., in these cases where the function $\theta(x,\mathcal{D})$ is numerically the constant function 1. Moreover, the difference $|\theta(x,\mathcal{D}) - 1|$ can be made less than any prescribed positive tolerance ε by choosing $\mathcal{D}$ large enough. For this it suffices to take

$$\mathcal{D} > \pi^{-2}(\log|\varepsilon| - \log 2) .$$

REMARK 1.1. The function θ is closely connected with Jacobi's Theta function ϑ_3, which is defined as (see [**1**, 16.27])

$$\vartheta_3(z|\tau) := \sum_{n=-\infty}^{\infty} \mathrm{e}^{i\pi\tau n^2}\,\mathrm{e}^{2inz}, \tag{1.5}$$

by the relation $\theta(x,\mathcal{D}) = \vartheta_3(\pi x|i\pi\mathcal{D})$.

1.1.2. Simple approximation formula. We have seen that for "large" $\mathcal{D}$ the integer shifts

$$\left\{\frac{1}{\sqrt{\pi\mathcal{D}}}\,\mathrm{e}^{-(x-m)^2/\mathcal{D}}, m \in \mathbb{Z}\right\} \tag{1.6}$$

form an approximate partition of unity, i.e., the sum of these functions is approximatively equal to the constant function 1. In addition, the functions in the family (1.6) decay very rapidly if $|x-m| \to \infty$. Hence, in the sum (1.1), one has to take into account only a small number of terms, if one wants to compute the value at a given point x. This leads to the idea of introducing an approximation formula using the usual scaling and translation operations with a "small" parameter h for the family of functions $\mathrm{e}^{-x^2/\mathcal{D}}$

$$\mathcal{M}_{h,\mathcal{D}}u(x) = \frac{1}{\sqrt{\pi\mathcal{D}}}\sum_{m=-\infty}^{\infty} u(mh)\,\mathrm{e}^{-(x-mh)^2/\mathcal{D}h^2}. \tag{1.7}$$

Formulas of this type are known as *quasi-interpolants* and we are interested in their behavior as $h \to 0$.

Let us suppose that the function u is twice continuously differentiable with bounded derivatives. The Taylor expansion of u at the point mh has the form

$$u(mh) = u(x) + u'(x)(mh-x) + u''(x_m)\frac{(mh-x)^2}{2}$$

for some x_m between x and mh. Putting this into (1.7), we derive

$$\begin{aligned}\mathcal{M}_{h,\mathcal{D}}u(x) = {} & \frac{u(x)}{\sqrt{\pi\mathcal{D}}}\sum_{m=-\infty}^{\infty} \mathrm{e}^{-(x-mh)^2/\mathcal{D}h^2} \\ & + \frac{u'(x)}{\sqrt{\pi\mathcal{D}}}\sum_{m=-\infty}^{\infty} (mh-x)\,\mathrm{e}^{-(x-mh)^2/\mathcal{D}h^2} \\ & + \frac{1}{2\sqrt{\pi\mathcal{D}}}\sum_{m=-\infty}^{\infty} u''(x_m)(mh-x)^2\,\mathrm{e}^{-(x-mh)^2/\mathcal{D}h^2}.\end{aligned} \tag{1.8}$$

The sum of the first term on the right-hand side is the function $\theta(x/h,\mathcal{D})$, whereas the sum in the second term can be expressed, for example, by the derivative

$$\theta'\big(\frac{x}{h},\mathcal{D}\big) = \frac{2}{\sqrt{\pi\mathcal{D}}\mathcal{D}h}\sum_{m=-\infty}^{\infty} (mh-x)\,\mathrm{e}^{-(x-mh)^2/\mathcal{D}h^2},$$

which provides the relation

$$\frac{1}{\sqrt{\pi\mathcal{D}}}\sum_{m=-\infty}^{\infty} (mh-x)\,\mathrm{e}^{-(x-mh)^2/\mathcal{D}h^2} = -2\pi\mathcal{D}h\sum_{\nu=1}^{\infty} \nu\,\mathrm{e}^{-\pi^2\mathcal{D}\nu^2}\sin 2\pi\nu\frac{x}{h}.$$

Therefore, by using (1.2), we can write the quasi-interpolant in the form

$$\mathcal{M}_{h,\mathcal{D}}u(x) = u(x) + C_{\mathcal{D},h}(x) + R_h(x) \tag{1.9}$$

with the function

$$C_{\mathcal{D},h}(x) = 2u(x)\sum_{\nu=1}^{\infty} \mathrm{e}^{-\pi^2\mathcal{D}\nu^2}\cos 2\pi\nu\frac{x}{h} - 2u'(x)\pi\mathcal{D}h\sum_{\nu=1}^{\infty} \nu\,\mathrm{e}^{-\pi^2\mathcal{D}\nu^2}\sin 2\pi\nu\frac{x}{h} \tag{1.10}$$

and the remainder term

$$R_h(x) = \frac{1}{2\sqrt{\pi\mathcal{D}}}\sum_{m=-\infty}^{\infty} u''(x_m)(mh-x)^2\,\mathrm{e}^{-(x-mh)^2/\mathcal{D}h^2},$$

which obviously satisfies

$$|R_h(x)| \le \max_{t\in\mathbb{R}}|u''(t)|\,\frac{1}{2\sqrt{\pi\mathcal{D}}}\sum_{m=-\infty}^{\infty}(mh-x)^2\,\mathrm{e}^{-(x-mh)^2/\mathcal{D}h^2}.$$

The Fourier series of the last sum can be calculated similarly to the case $\theta(x,\mathcal{D})$, and it holds that

$$\frac{1}{\sqrt{\pi\mathcal{D}}}\sum_{m=-\infty}^{\infty}(mh-x)^2\,\mathrm{e}^{-(x-mh)^2/\mathcal{D}h^2} = \frac{\mathcal{D}h^2}{2}\sum_{\nu=-\infty}^{\infty}(1-2\pi^2\mathcal{D}\nu^2)\,\mathrm{e}^{-\pi^2\mathcal{D}\nu^2}\,\mathrm{e}^{2\pi i\nu x/h},$$

which leads to the estimate

$$|R_h(x)| \le \max_{t\in\mathbb{R}}|u''(t)|\frac{\mathcal{D}h^2}{4}\Big(1+2\sum_{\nu=1}^{\infty}\Big|(1-2\pi^2\mathcal{D}\nu^2)\cos 2\pi\nu\frac{x}{h}\Big|\,\mathrm{e}^{-\pi^2\mathcal{D}\nu^2}\Big).$$

Hence, the difference between u and the quasi-interpolant $\mathcal{M}_{h,\mathcal{D}}u$ can be estimated for any $x \in \mathbb{R}$ by

$$\begin{aligned}\big|\mathcal{M}_{h,\mathcal{D}}u(x) - u(x)\big| \le \frac{\mathcal{D}h^2}{4}\Big(1+\sum_{\nu=1}^{\infty}|4\pi^2\mathcal{D}\nu^2-2|\,\mathrm{e}^{-\pi^2\mathcal{D}\nu^2}\Big)\max_{t\in\mathbb{R}}|u''(t)| \\ + \big|C_{\mathcal{D},h}(x)\big|.\end{aligned} \tag{1.11}$$

This inequality is valid for all values of the positive parameters $\mathcal{D}$ and h. Here we find the special feature of approximate approximations, mentioned in the Preface. The approximation error consists of a term of the order $\mathcal{O}(\mathcal{D}h^2)$ and the term $\big|C_{\mathcal{D},h}(x)\big|$, which is called the *saturation error*, because it does not converge to zero as $h \to 0$. However, we obtain from (1.10) that

$$\big|C_{\mathcal{D},h}(x)\big| \le 2|u(x)|\sum_{\nu=1}^{\infty}\mathrm{e}^{-\pi^2\mathcal{D}\nu^2} + 2\pi\mathcal{D}h|u'(x)|\sum_{\nu=1}^{\infty}\nu\,\mathrm{e}^{-\pi^2\mathcal{D}\nu^2}.$$

Therefore, owing to the rapid decay of $\mathrm{e}^{-\pi^2\mathcal{D}\nu^2}$, $\nu = 1, 2, \ldots$, for any $\varepsilon > 0$ one can fix $\mathcal{D} > 0$ such that the saturation error satisfies

$$|C_{\mathcal{D},h}(x)| < \varepsilon\,\big(|u(x)| + h|u'(x)|\big).$$

Since the first term of the right-hand side of (1.11) with a fixed $\mathcal{D}$ converges to zero, we see that $\mathcal{M}_{h,\mathcal{D}}u$ approximates u with the order $\mathcal{O}(h^2)$ as long as the saturation bound $\varepsilon\big(|u(x)| + h|u'(x)|\big)$ is reached. Hence, choosing the parameter $\mathcal{D}$ such that

ε is less than the precision of the computing system, formula $\mathcal{M}_{h,\mathcal{D}}u$ behaves in numerical computations as a usual second-order approximation.

Let us emphasize the structure of $C_{\mathcal{D},h}$, which is the sum of $u(x)$ and $hu'(x)$ multiplied by oscillating functions with period h. For sufficiently large $\mathcal{D}$ the main term of $C_{\mathcal{D}}$ is given by

$$2u(x)\,\mathrm{e}^{-\pi^2\mathcal{D}}\cos 2\pi\frac{x}{h}\,.$$

This is a fast oscillating simple harmonics modulated by the slowly varying value of the approximated function.

In the following we show that formulas of type (1.7), where the Gaussian e^{-x^2} is replaced by more general basis functions, can provide similar or even better approximation properties. We give some one- and multi-dimensional examples of those approximation formulas and define approximate quasi-interpolation on uniform grids in the next section.

1.2. Further examples

1.2.1. Errors of approximate quasi-interpolation. We illustrate here the approximation properties of the quasi-interpolant $\mathcal{M}_{h,\mathcal{D}}u$ defined in (1.7) for the function $u(x) = \sin x$ using different values of the parameters $\mathcal{D}$ and h. Figs. 1.6 and 1.7 show the particular form of the terms

$$\frac{1}{\sqrt{\pi\mathcal{D}}}\sin(mh)\,\mathrm{e}^{-(x-mh)^2/\mathcal{D}h^2} \tag{1.12}$$

and its sum

$$(\mathcal{M}_{h,\mathcal{D}}\sin)(x) = \frac{1}{\sqrt{\pi\mathcal{D}}}\sum_{m=-\infty}^{\infty}\sin(mh)\,\mathrm{e}^{-(x-mh)^2/\mathcal{D}h^2} \tag{1.13}$$

for $h = 0.4$ and two different values $\mathcal{D} = 1$ and $\mathcal{D} = 2$. Visually the sums are good approximations of $\sin x$ for this rather large step h.

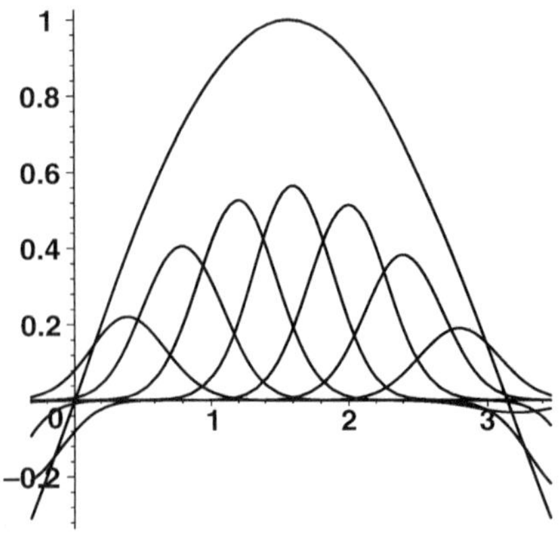

FIGURE 1.6. $\mathcal{D} = 1$, $h = 0.4$

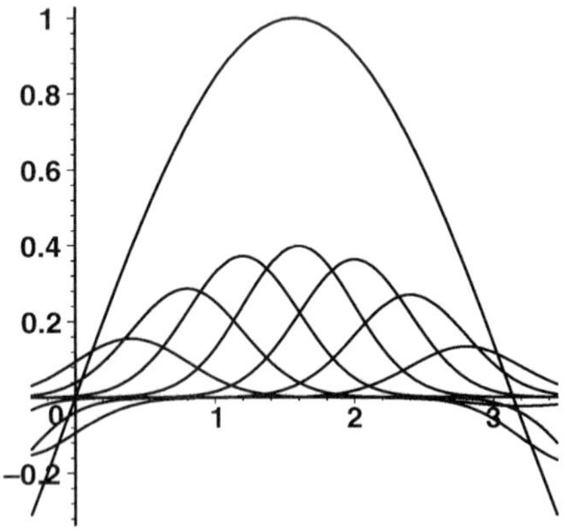

FIGURE 1.7. $\mathcal{D} = 2$, $h = 0.4$

Although the functions $\mathrm{e}^{-(x-mh)^2/\mathcal{D}h^2}$ are supported by the whole real axis, one needs only a few terms in the sum (1.7) to compute the value of $\mathcal{M}_{h,\mathcal{D}}u$ at a given point x within a given accuracy. For a fixed tolerance $\delta > 0$ one has to sum only over the integers m for which

$$|m - x/h| \leq \sqrt{-\mathcal{D} \log \delta}\,.$$

Hence the number of terms, necessary to compute $\mathcal{M}_{h,\mathcal{D}}u(x)$ for fixed h, increases proportionally to $\sqrt{\mathcal{D}}$.

On the other hand, if $\mathcal{D}$ is fixed, then this number does not depend on h. For example, if $\delta = 10^{-6}$, then one has to sum up 7 and 11 terms in (1.7) if $\mathcal{D} = 1$ and $\mathcal{D} = 2$, respectively.

The differences between $\sin x$ and the quasi-interpolants (1.13) are plotted in Figs. 1.8 and 1.9 for the values $\mathcal{D} = 1, 2$ and $h = 0.4, 0.2$, respectively. The graphs confirm the second-order convergence from estimate (1.11). However, the case of

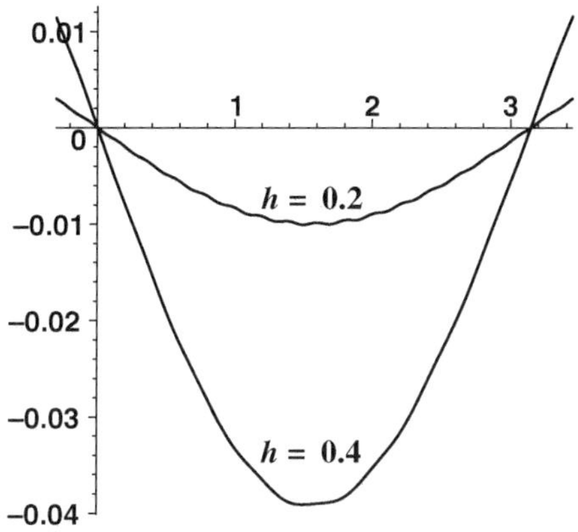

FIGURE 1.8. $(\mathcal{M}_{h,1} - I) \sin x$

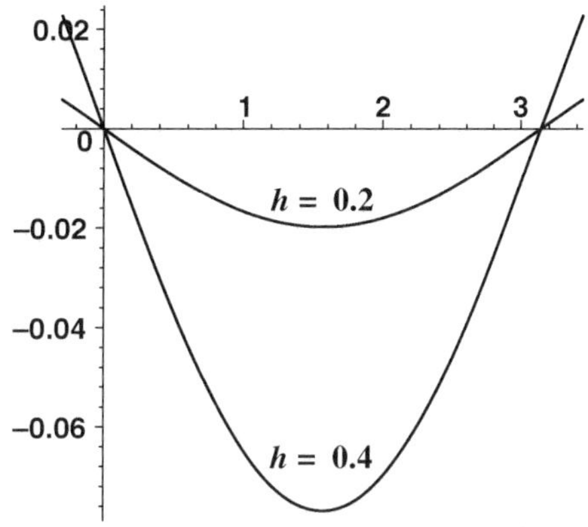

FIGURE 1.9. $(\mathcal{M}_{h,2} - I) \sin x$

smaller step h already gives different pictures. Figs. 1.10–1.15 depict the quasi-interpolation error of $\sin x$ with smaller h and for $\mathcal{D} = 1$ and 2.

The plotted errors confirm the second-order convergence, but the error for $\mathcal{D} = 1$ oscillates very fast, with frequency depending of h. In Figs. 1.10 and 1.12 the saturation error is already visible.

It can be seen from Fig. 1.14, that for $\mathcal{D} = 1$ the quasi-interpolation error has reached its saturation bound, since it does not decrease if h becomes smaller. On the other hand, Fig. 1.15 shows that the approximation with $\mathcal{M}_{h,2}$ for the same values of h is of the second order, that the saturation is not reached, yet.

The behavior of the quasi-interpolants $\mathcal{M}_{h,\mathcal{D}}$, predicted by the estimate (1.11), is confirmed also in Table 1.1, where the quasi-interpolation error in the maximum norm for different h and $\mathcal{D}$ and the convergence rate calculated as

$$\log_2 \frac{\|u - \mathcal{M}_{2h,\mathcal{D}}u\|_{L^\infty}}{\|u - \mathcal{M}_{h,\mathcal{D}}u\|_{L^\infty}} \tag{1.14}$$

are given.

Recall that the main term of the saturation error is $1.04 \cdot 10^{-4\mathcal{D}}\,|u(x)|$. If $\mathcal{D} = 1$, then we have the second-order approximation only for relative large h. In the cases $\mathcal{D} = 2$ and $\mathcal{D} = 4$ the saturation error is still negligible compared to the first term of estimate (1.11).

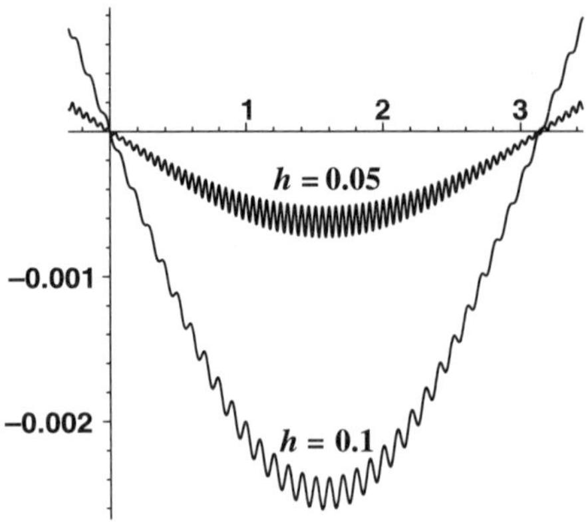

FIGURE 1.10. $(\mathcal{M}_{h,1} - I)\sin x$

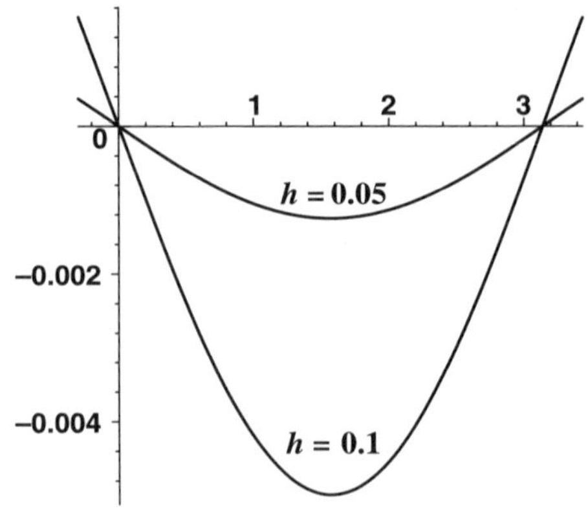

FIGURE 1.11. $(\mathcal{M}_{h,2} - I)\sin x$

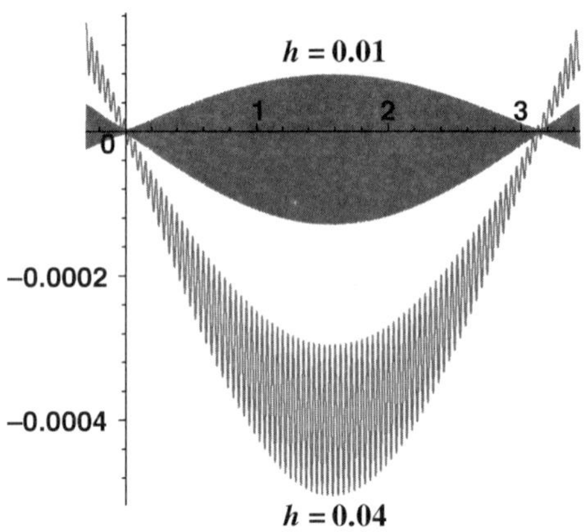

FIGURE 1.12. $(\mathcal{M}_{h,1} - I)\sin x$

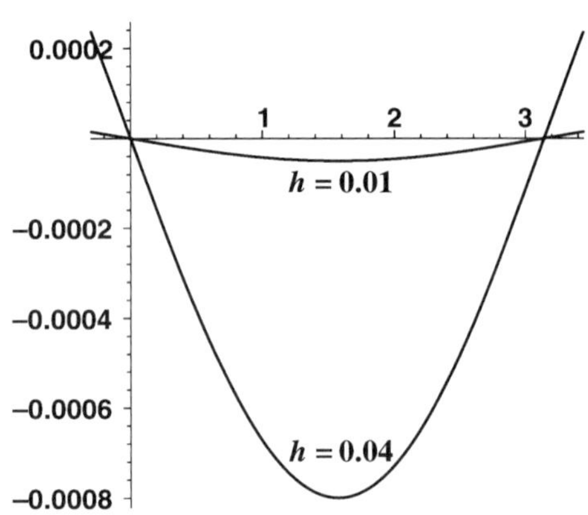

FIGURE 1.13. $(\mathcal{M}_{h,2} - I)\sin x$

1.2.2. A simple application of the approximation formula (1.7)**.** Consider the initial value problem for the heat equation

$$u_t(x,t) - u_{xx}(x,t) = 0\,, \quad t > 0\,, \quad u(x,0) = \varphi(x), \quad x \in \mathbb{R}\,. \tag{1.15}$$

Its solution is given by the *Poisson integral*

$$u(x,t) = \mathcal{P}_t\varphi(x) := \frac{1}{\sqrt{4\pi t}} \int\limits_{\mathbb{R}} \mathrm{e}^{-(x-y)^2/4t}\, \varphi(y)\, dy\,.$$

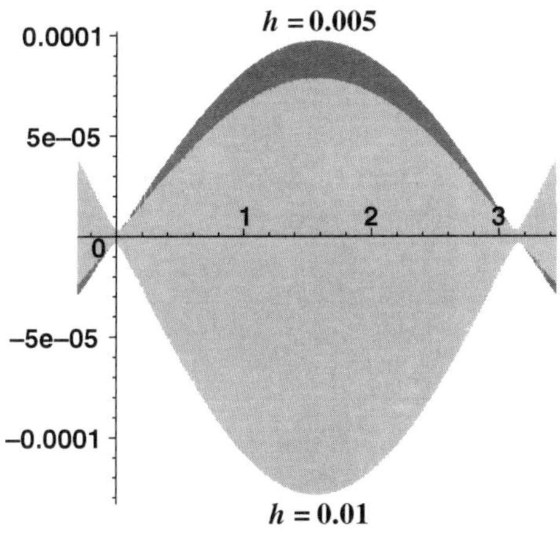

FIGURE 1.14. $(\mathcal{M}_{h,1} - I)\sin x$

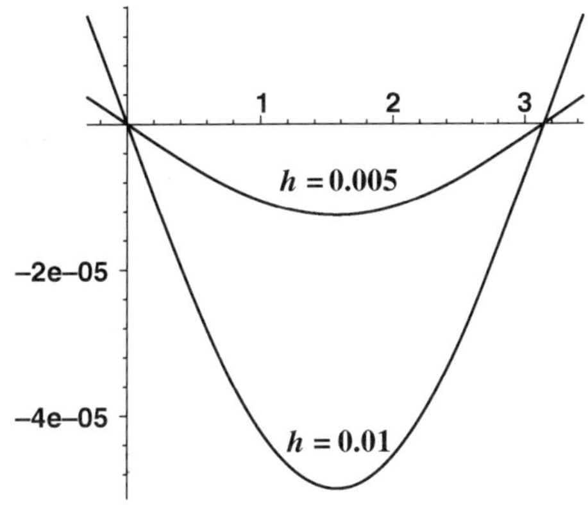

FIGURE 1.15. $(\mathcal{M}_{h,2} - I)\sin x$

h	$\mathcal{D}=1$	rate	$\mathcal{D}=2$	rate	$\mathcal{D}=4$	rate
0.4	$3.91 \cdot 10^{-2}$		$7.69 \cdot 10^{-2}$		$1.48 \cdot 10^{-1}$	
0.2	$1.00 \cdot 10^{-2}$	1.96	$1.98 \cdot 10^{-2}$	1.96	$3.92 \cdot 10^{-2}$	1.91
0.1	$2.60 \cdot 10^{-3}$	1.95	$4.99 \cdot 10^{-3}$	1.99	$9.95 \cdot 10^{-3}$	1.98
0.05	$7.29 \cdot 10^{-4}$	1.84	$1.25 \cdot 10^{-3}$	2.00	$2.50 \cdot 10^{-3}$	1.99
0.025	$2.60 \cdot 10^{-4}$	1.49	$3.12 \cdot 10^{-4}$	2.00	$6.25 \cdot 10^{-4}$	2.00
0.0125	$1.42 \cdot 10^{-4}$	0.87	$7.81 \cdot 10^{-5}$	2.00	$1.56 \cdot 10^{-4}$	2.00
0.00625	$1.11 \cdot 10^{-4}$	0.36	$1.95 \cdot 10^{-5}$	2.00	$3.91 \cdot 10^{-5}$	2.00

TABLE 1.1. Approximation error for the function $u(x) = \sin x$ using the quasi-interpolant (1.7)

This integral cannot be taken in a closed form, in general, but this is possible for some functions φ, for example, for the Gaussian function. In particular,

$$\frac{1}{\sqrt{4\pi t}} \int\limits_{\mathbb{R}} \mathrm{e}^{-(x-y)^2/4t}\ \mathrm{e}^{-y^2/\mathcal{D}h^2}\, dy = \frac{\sqrt{\mathcal{D}}h}{\sqrt{\mathcal{D}h^2+4t}}\ \mathrm{e}^{-x^2/(\mathcal{D}h^2+4t)}\,. \tag{1.16}$$

Hence, if we replace the initial value φ by the quasi-interpolant $\mathcal{M}_{h,\mathcal{D}}\varphi$ defined by (1.7), then we obtain the exact solution

$$\mathcal{P}_t(\mathcal{M}_{h,\mathcal{D}}\varphi)(x) = \frac{h}{\sqrt{\pi(\mathcal{D}h^2+4t)}} \sum_{m=-\infty}^{\infty} \varphi(hm)\ \mathrm{e}^{-(x-hm)^2/(\mathcal{D}h^2+4t)} \tag{1.17}$$

of the heat equation (1.15) with the modified initial condition $u(x,0) = \mathcal{M}_{h,\mathcal{D}}\varphi(x)$. Since

$$\big|\mathcal{P}_t\varphi(x) - \mathcal{P}_t(\mathcal{M}_{h,\mathcal{D}}\varphi)(x)\big| \le \sup_{y\in\mathbb{R}} \big|\varphi(y) - \mathcal{M}_{h,\mathcal{D}}\varphi(y)\big| \frac{1}{\sqrt{4\pi t}} \int_{\mathbb{R}} \mathrm{e}^{-(x-y)^2/4t}\, dy$$
$$= \sup_{y\in\mathbb{R}} \big|\varphi(y) - \mathcal{M}_{h,\mathcal{D}}\varphi(y)\big|\,,$$

the estimate (1.11) shows that the function $u_h(x,t) = \mathcal{P}_t(\mathcal{M}_{h,\mathcal{D}}\varphi)(x)$ approximates the solution $u(x,t)$ of the original problem (1.15) with the error

$$\begin{aligned} |u(x,t) - u_h(x,t)| \le{}& \frac{\mathcal{D}h^2}{4}\big(1 + 4\mathcal{D}\pi^2\,\mathrm{e}^{-\pi^2\mathcal{D}}\big) \max_{y\in\mathbb{R}} |\varphi''(y)| \\ &+ 2 \max_{y\in\mathbb{R}} \big(|\varphi(y)| + |\varphi'(y)|\pi\mathcal{D}h\big)\,\mathrm{e}^{-\pi^2\mathcal{D}} + \mathcal{O}(\mathrm{e}^{-2\pi^2\mathcal{D}})\,. \end{aligned} \tag{1.18}$$

This very simple example is in many ways typical in applying approximate approximations to the solution of partial differential equations; one replaces some function in the original problem by an approximant such that the solution of the equation can be performed very efficiently, either analytically or by some other numerical method.

Let us mention that (1.18) is only a rough error estimate for the approximate solution of the heat equation. This can be seen from Table 1.2 which contains numerical results for the heat equation (1.15) with the initial value $\phi(x) = \mathrm{e}^{-x^2}$. It provides the maximum error

$$\max_x |u(x,t) - u_h(x,t)|\,, \quad t = 10\,,$$

for different values of $\mathcal{D}$ and h.

h	$\mathcal{D}=1$	rate	$\mathcal{D}=2$	rate	$\mathcal{D}=4$	rate
0.4	$3.04\cdot10^{-4}$		$6.06\cdot10^{-4}$		$1.20\cdot10^{-3}$	
0.2	$7.61\cdot10^{-5}$	2.00	$1.52\cdot10^{-4}$	2.00	$3.04\cdot10^{-4}$	1.99
0.1	$1.90\cdot10^{-5}$	2.00	$3.81\cdot10^{-5}$	2.00	$7.61\cdot10^{-5}$	2.00
0.05	$4.76\cdot10^{-6}$	2.00	$9.52\cdot10^{-6}$	2.00	$1.90\cdot10^{-5}$	2.00
0.025	$1.19\cdot10^{-6}$	2.00	$2.38\cdot10^{-6}$	2.00	$4.76\cdot10^{-6}$	2.00
0.0125	$2.98\cdot10^{-7}$	2.00	$5.95\cdot10^{-7}$	2.00	$1.19\cdot10^{-6}$	2.00
0.00625	$7.44\cdot10^{-8}$	2.00	$1.49\cdot10^{-7}$	2.00	$2.98\cdot10^{-7}$	2.00

TABLE 1.2. Numerical error for the initial value problem (1.15) with $\phi(x) = \mathrm{e}^{-x^2}$ and $t = 10$ using the approximate solution (1.17)

In contrast to the quasi-interpolation results, given in Table 1.1, a saturation error cannot be seen. We will show in Subsection 6.2.1 that due to the properties of the Poisson integral and the structure of the saturation error the approximate solution $u_h(x,t)$ converges to $u(x,t)$.

1.2.3. Other basis functions. The simplicity of formulas of the form

$$\mathcal{Q}_h u(x) := \sum_{m=-\infty}^{\infty} u(mh)\,\eta\Big(\frac{x}{h} - m\Big) \tag{1.19}$$

makes them very attractive for approximation processes. Suppose, for example, that η is a *Lagrangian function*, which means that η is subject to

$$\eta(0) = 1 \quad \text{and} \quad \eta(m) = 0 \quad \text{for all } m \in \mathbb{Z}\backslash 0 .$$

Then the sum (1.19) satisfies $\mathcal{Q}_h u(mh) = u(mh)$, $m \in \mathbb{Z}$, i.e., $\mathcal{Q}_h u$ interpolates u. As two representative examples, we mention here the piecewise linear hat function and the sinc function

$$\operatorname{sinc} x = \frac{\sin \pi x}{\pi x} .$$

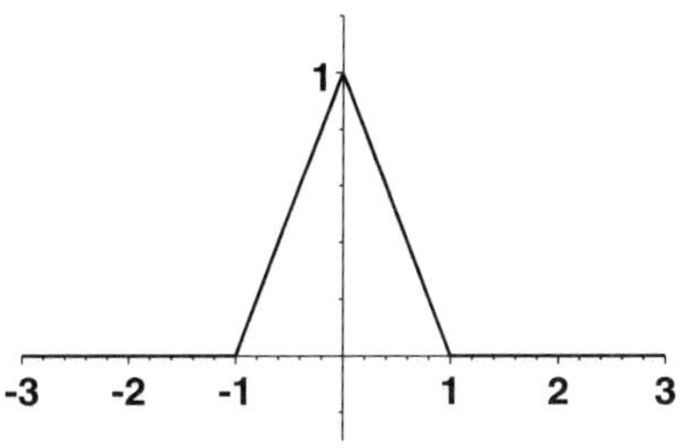

FIGURE 1.16. Hat function

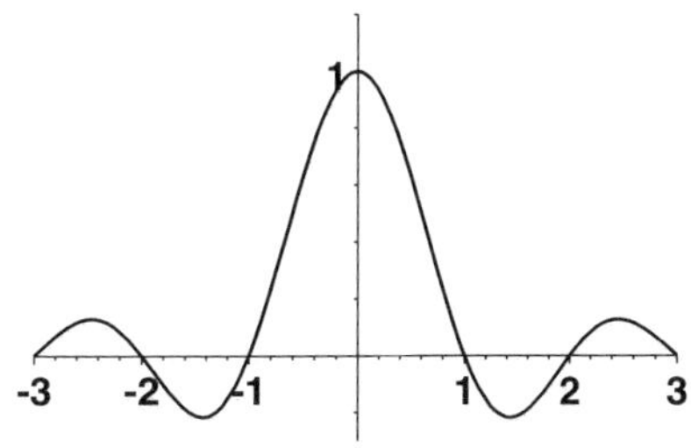

FIGURE 1.17. sinc function

If η is the hat function as shown in Fig. 1.16, then the resulting sum is the polygonal line connecting the points $(hm, u(hm))$. This piecewise linear function approximates u with the order $\mathcal{O}(h^2)$. To find the approximant at a given point x, one has to sum up only two terms of (1.19). But it is visually not very nice to approximate a smooth curve by some piecewise linear function.

On the other hand, the sinc function (depicted in Fig. 1.17) generates an interpolant which is smooth and even provides an exponential order of convergence (see [**91**]). However, since the generating function decreases very slowly, it is practically impossible to compute the approximant (1.19) if the function u is not compactly supported.

Let us mention that in Chapter 7 we introduce another Lagrangian function

$$\Psi_{\mathcal{D}}(x) = \frac{\sin \pi x}{\pi \mathcal{D} \sinh \dfrac{x}{\mathcal{D}}}$$

depending on the parameter $\mathcal{D} > 0$. This function is a small perturbation of the Lagrangian function from the family of shifted Gaussians (1.6). The corresponding interpolant approximates smooth functions with exponential order, but similar to the approximation formula (1.7) only up to a saturation error of the order $\mathcal{O}(\mathrm{e}^{-\pi^2 \mathcal{D}})$. Therefore $\Psi_{\mathcal{D}}$ can be considered as approximate sinc function, providing similar approximation properties but decaying exponentially for $|x| \to \infty$.

There exists, of course, a variety of other basis functions η for interpolation formulas (1.19). However, the Lagrangian functions for those bases have, in general, large supports. For example, the Lagrangian function for the class of smooth cubic splines, which are cubic polynomials on the intervals $(m, m+1)$, $m \in \mathbb{Z}$, and two-times continuously differentiable, is supported on the whole real line.

It turns out that good approximations can also be obtained by replacing the Lagrangian function in (1.19) by some simpler function of the same class. In the case of smooth cubic splines one can choose η as the cubic B-spline

$$b(x) = \frac{1}{12}\left(|x+2|^3 - 4|x+1|^3 + 6|x|^3 - 4|x-1|^3 + |x-2|^3\right) \tag{1.20}$$

depicted in Fig. 1.18, which gives a C^2-approximant of the order $\mathcal{O}(h^2)$.

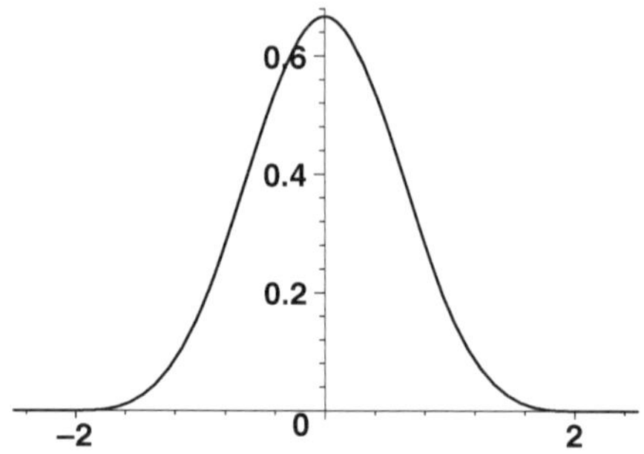

FIGURE 1.18. Cubic B-spline

But clearly the resulting approximant does not interpolate; therefore approximation formulas (1.19) with non-Lagrangian functions η are called *quasi-interpolants*.

Thus, $\mathcal{M}_{h,\mathcal{D}}$ in (1.7) represents a quasi-interpolant with

$$\eta(x) = \frac{e^{-x^2/\mathcal{D}}}{\sqrt{\pi\mathcal{D}}}.$$

We have seen in (1.11) that for a fixed $\mathcal{D}$ the sum $\mathcal{M}_{h,\mathcal{D}}u$ is a smooth approximation to u of order $\mathcal{O}(h^2)$ until the saturation error is reached, which can be neglected in numerical computations if $\mathcal{D}$ is sufficiently large.

It is important that for a quite general class of basis functions the quasi-interpolants have similar properties as in the case of the Gaussian. Take, for example, the function

$$\operatorname{sech} x = \frac{1}{\cosh x}.$$

Putting the Taylor expansion of u into

$$\mathcal{M}_h u(x) = \frac{1}{\pi\sqrt{\mathcal{D}}} \sum_{m=-\infty}^{\infty} u(mh) \operatorname{sech} \frac{x-mh}{\sqrt{\mathcal{D}}h}, \tag{1.21}$$

we obtain as in (1.8)

$$\mathcal{M}_h u(x) = \frac{u(x)}{\pi\sqrt{\mathcal{D}}} \sum_{m=-\infty}^{\infty} \operatorname{sech} \frac{x-mh}{\sqrt{\mathcal{D}}h} + \frac{u'(x)}{\pi\sqrt{\mathcal{D}}} \sum_{m=-\infty}^{\infty} (mh-x) \operatorname{sech} \frac{x-mh}{\sqrt{\mathcal{D}}h}$$
$$+ \frac{1}{2\pi\sqrt{\mathcal{D}}} \sum_{m=-\infty}^{\infty} u''(x_m)(mh-x)^2 \operatorname{sech} \frac{x-mh}{\sqrt{\mathcal{D}}h}.$$

The infinite sums in the first and second term on the right-hand side can be transformed by using Poisson's summation formula (1.3) and the Fourier transform of $\operatorname{sech} x$,

$$(\mathcal{F}\operatorname{sech})(\lambda) = \pi \operatorname{sech} \pi^2 \lambda .$$

Then we obtain the relations

$$I_0 := \frac{1}{\pi\sqrt{\mathcal{D}}} \sum_{m=-\infty}^{\infty} \operatorname{sech} \frac{x-m}{\sqrt{\mathcal{D}}} = 1 + 2\sum_{\nu=1}^{\infty} \operatorname{sech}(\pi^2\sqrt{\mathcal{D}}\nu) \cos 2\pi\nu x ,$$

$$\begin{aligned} I_1 &:= \frac{1}{\pi\sqrt{\mathcal{D}}} \sum_{m=-\infty}^{\infty} \frac{x-m}{\sqrt{\mathcal{D}}} \operatorname{sech} \frac{x-m}{\sqrt{\mathcal{D}}} \\ &= \pi \sum_{\nu=1}^{\infty} \operatorname{sech}(\pi^2\sqrt{\mathcal{D}}\nu) \tanh(\pi^2\sqrt{\mathcal{D}}\nu) \sin 2\pi\nu x , \end{aligned}$$

which shows that

$$|I_0 - 1| < 2\varepsilon(\mathcal{D}) \quad \text{and} \quad |I_1| < \pi\varepsilon(\mathcal{D}) ,$$

where we use the notation

$$\varepsilon(\mathcal{D}) := \sum_{\nu=1}^{\infty} \operatorname{sech}(\pi^2\sqrt{\mathcal{D}}\nu) .$$

Moreover,

$$\frac{1}{2\pi\sqrt{\mathcal{D}}} \left| \sum_{m=-\infty}^{\infty} u''(x_m) \frac{(mh-x)^2}{\mathcal{D}h^2} \operatorname{sech} \frac{x-mh}{\sqrt{\mathcal{D}}h} \right| \le \frac{5}{4} \sup_{t\in\mathbb{R}} |u''(t)| ,$$

so that

$$|u(x) - \mathcal{M}_h u(x)| \le \frac{5}{4} \mathcal{D}h^2 \max_{\mathbb{R}} |u''| + \varepsilon(\mathcal{D})\big(2\ |u(x)| + \pi\sqrt{\mathcal{D}}h\, |u'(x)|\big) . \tag{1.22}$$

As in the example with the Gaussian function the quasi-interpolant (1.21) does not converge to $u(x)$, but the number $\varepsilon(\mathcal{D})$ is an upper bound for the saturation error and can be made arbitrarily small by choosing $\mathcal{D}$ large enough. For example, if $\mathcal{D} = 4$, then $\varepsilon(\mathcal{D}) = 0.000000005351$.

Again, the inequality (1.22) shows that the quasi-interpolant $\mathcal{M}_h u$ approximates any C^2-function u like a second-order approximant above the tolerance

$$\varepsilon(\mathcal{D})\big(2\,|u(x)| + \pi\sqrt{\mathcal{D}}h\,|u'(x)|\big) ,$$

and that any prescribed accuracy can be reached if $\mathcal{D}$ is chosen sufficiently large. In Table 1.3 we give the L^∞-error of the quasi-interpolation of $\sin x$ with formula (1.22) for different h and $\mathcal{D}$ and the convergence rate obtained using (1.14).

1.2.4. Examples of higher-order quasi-interpolants. There exist approximants with approximation orders larger than 2 up to some prescribed accuracy which have the same simple form as second-order approximate quasi-interpolants. Consider, for example, the quasi-interpolant

$$u_h(x) := \mathcal{D}^{-1/2} \sum_{m=-\infty}^{\infty} u(mh)\, \eta\Big(\frac{x-mh}{\sqrt{\mathcal{D}}h}\Big) \tag{1.23}$$

h	$\mathcal{D}=1$	rate	$\mathcal{D}=2$	rate	$\mathcal{D}=4$	rate
0.4	$1.69\cdot10^{-1}$		$2.96\cdot10^{-1}$		$4.73\cdot10^{-1}$	
0.2	$4.75\cdot10^{-2}$	1.83	$9.12\cdot10^{-2}$	1.70	$1.69\cdot10^{-1}$	1.48
0.1	$1.24\cdot10^{-2}$	1.93	$2.42\cdot10^{-2}$	1.92	$4.74\cdot10^{-2}$	1.84
0.05	$3.32\cdot10^{-3}$	1.91	$6.14\cdot10^{-3}$	1.98	$1.22\cdot10^{-2}$	1.96
0.025	$1.01\cdot10^{-3}$	1.71	$1.54\cdot10^{-3}$	1.99	$3.08\cdot10^{-3}$	1.99
0.0125	$4.37\cdot10^{-4}$	1.22	$3.89\cdot10^{-4}$	1.99	$7.71\cdot10^{-4}$	2.00
0.00625	$2.18\cdot10^{-4}$	1.01	$9.98\cdot10^{-5}$	1.96	$1.93\cdot10^{-4}$	2.00

TABLE 1.3. Error of approximating $u(x)=\sin x$ with formula (1.21)

with one of the two generating functions

$$\eta_1(x) = (3/2 - x^2)\,\frac{\mathrm{e}^{-x^2}}{\sqrt{\pi}} \quad \text{or} \quad \eta_2(x) = \sqrt{\frac{\mathrm{e}}{\pi}}\,\mathrm{e}^{-x^2}\cos\sqrt{2}x\,, \tag{1.24}$$

shown in Figs. 1.19 and 1.20.

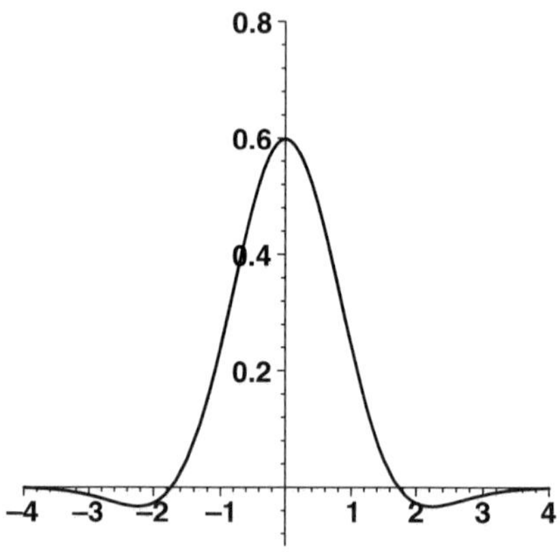

FIGURE 1.19. $\eta_1(x/\sqrt{2})$

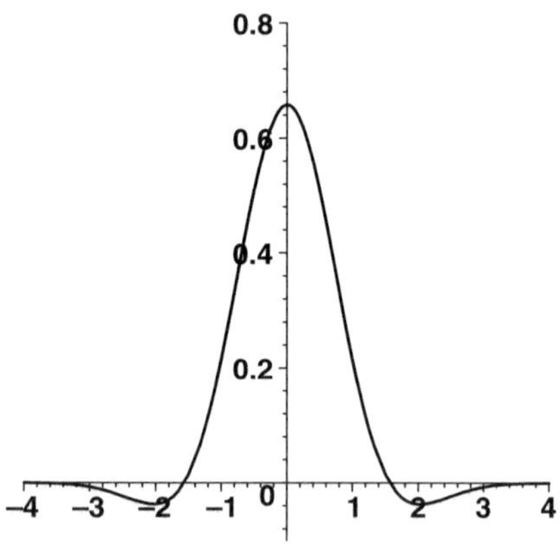

FIGURE 1.20. $\eta_2(x/\sqrt{2})$

Figs. 1.21 – 1.26 repeat the error plots of Subsection 1.2.1 for the approximation of the function $\sin x$ with

$$\mathcal{N}_{h,\mathcal{D}}(\mathbf{x}) := \left(\frac{\mathrm{e}}{\pi\mathcal{D}}\right)^{1/2} \sum_{m=-\infty}^{\infty} u(mh)\,\cos\left(\sqrt{2/\mathcal{D}}(x/h-m)\right)\,\mathrm{e}^{-(x-mh)^2/\mathcal{D}h^2}\,, \tag{1.25}$$

where now the values $\mathcal{D}=1.5$ and $\mathcal{D}=2.5$ are used.

The absolute errors given in Figs. 1.21 and 1.22 are much smaller than those plotted in Figs. 1.8 and 1.9. Moreover, the graphs indicate approximation with the order 4.

The visible oscillations of the errors in Fig. 1.21 (the case $\mathcal{D}=1.5$ for quite large steps h) are caused by the relatively large saturation error. The error plots in Figs. 1.23 and 1.25 show clearly that $\mathcal{N}_{0.1,1.5}$ has reached the saturation and that

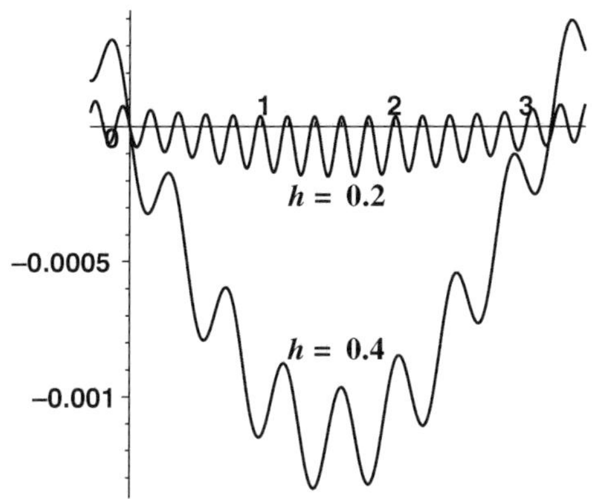

FIGURE 1.21. $(\mathcal{N}_{h,1.5} - I) \sin x$

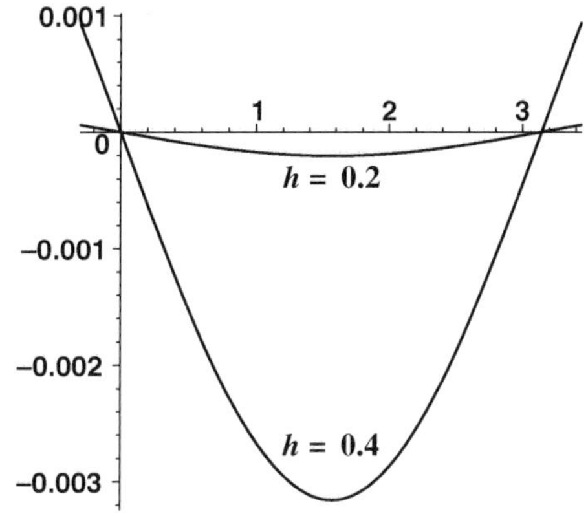

FIGURE 1.22. $(\mathcal{N}_{h,2.5} - I) \sin x$

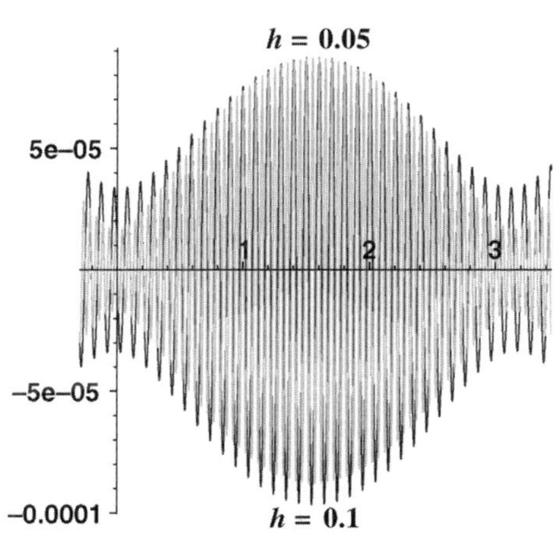

FIGURE 1.23. $(\mathcal{N}_{h,1.5} - I) \sin x$

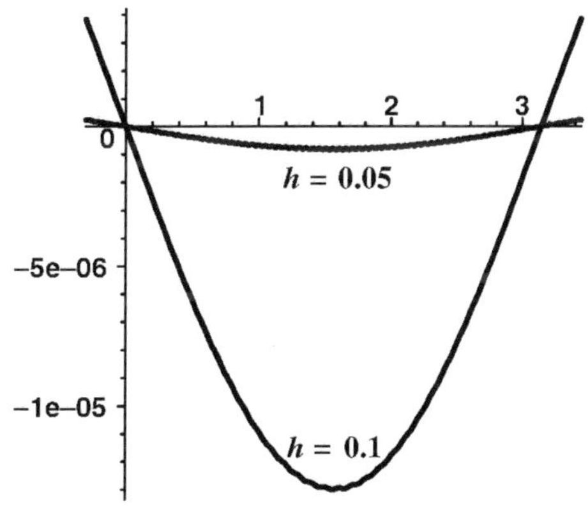

FIGURE 1.24. $(\mathcal{N}_{h,2.5} - I) \sin x$

any h smaller than 0.1 does not give more accurate results for the quasi-interpolant $\mathcal{N}_{h,1.5}$.

The situation is much better for $\mathcal{N}_{h,2.5}$, as indicated in Figs. 1.24 and 1.26. The approximation is much more accurate for small h; the approximation error is dominated by the saturation only if $h \leq 0.01$.

The approximation errors of the function $\sin x$ with the basis functions (1.24) are given in the Tables 1.4 and 1.5 which confirm an approximation with the order 4 up to some saturation error.

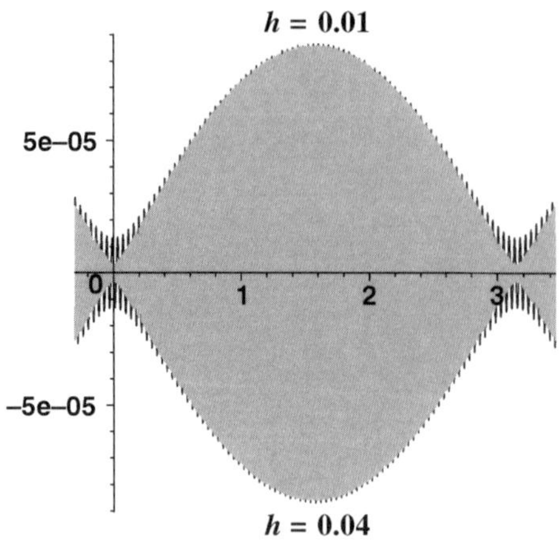

FIGURE 1.25. $(\mathcal{N}_{h,1.5} - I)\sin x$

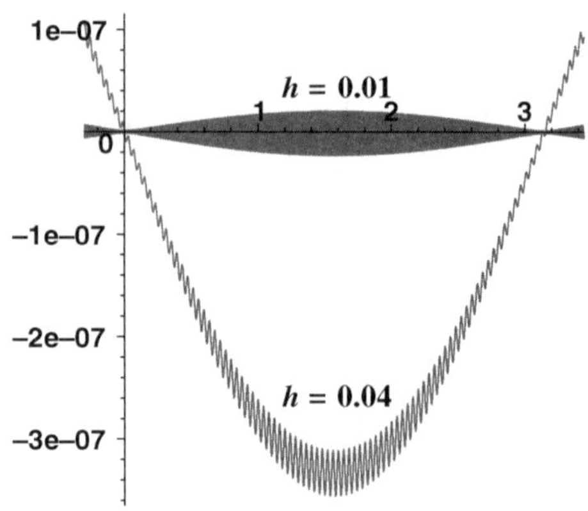

FIGURE 1.26. $(\mathcal{N}_{h,2.5} - I)\sin x$

h	$\mathcal{D}=2$	ord	$\mathcal{D}=3$	ord	$\mathcal{D}=4$	ord
0.4	$3.03 \cdot 10^{-3}$		$6.65 \cdot 10^{-3}$		$1.15 \cdot 10^{-2}$	
0.2	$1.97 \cdot 10^{-4}$	3.85	$4.41 \cdot 10^{-4}$	3.76	$7.79 \cdot 10^{-4}$	3.70
0.1	$1.26 \cdot 10^{-5}$	3.92	$2.80 \cdot 10^{-5}$	3.94	$4.97 \cdot 10^{-5}$	3.92
0.05	$8.96 \cdot 10^{-7}$	3.51	$1.75 \cdot 10^{-6}$	3.99	$3.12 \cdot 10^{-6}$	3.98
0.025	$1.60 \cdot 10^{-7}$	1.39	$1.09 \cdot 10^{-7}$	4.00	$1.95 \cdot 10^{-7}$	4.00
0.0125	$9.72 \cdot 10^{-8}$	0.41	$7.63 \cdot 10^{-9}$	3.98	$1.24 \cdot 10^{-8}$	4.03

TABLE 1.4. Error of approximating $u(x) = \sin x$ with $\eta(x) = \dfrac{3/2 - x^2}{\pi^{1/2}}\, \mathrm{e}^{-x^2}$

h	$\mathcal{D}=2$	ord	$\mathcal{D}=3$	ord	$\mathcal{D}=4$	ord
0.4	$2.04 \cdot 10^{-3}$		$4.50 \cdot 10^{-3}$		$7.84 \cdot 10^{-3}$	
0.2	$1.34 \cdot 10^{-4}$	3.81	$2.95 \cdot 10^{-4}$	3.81	$5.22 \cdot 10^{-4}$	3.75
0.1	$9.94 \cdot 10^{-6}$	3.37	$1.87 \cdot 10^{-5}$	3.95	$3.32 \cdot 10^{-5}$	3.94
0.05	$2.00 \cdot 10^{-6}$	1.24	$1.17 \cdot 10^{-6}$	3.99	$2.08 \cdot 10^{-6}$	3.98
0.025	$1.48 \cdot 10^{-6}$	0.34	$7.43 \cdot 10^{-8}$	3.97	$1.30 \cdot 10^{-7}$	4.00
0.0125	$1.33 \cdot 10^{-6}$	0.28	$5.38 \cdot 10^{-9}$	3.65	$8.42 \cdot 10^{-9}$	4.07

TABLE 1.5. Error of approximating $u(x) = \sin x$ with $\eta(x) = \dfrac{\mathrm{e}^{1/2-x^2}}{\pi^{1/2}} \cos \sqrt{2}x$

We see in Section 3.3, Example 3.2, that

$$\eta_{10}(x) = \pi^{-1/2}\, \mathrm{e}^{-x^2} \left(\frac{315}{128} - \frac{105}{16}\, x^2 + \frac{63}{16}\, x^4 - \frac{3}{4}\, x^6 + \frac{1}{24}\, x^8 \right) \tag{1.26}$$

generates a quasi-interpolant which approximates smooth functions with the order $N = 10$ up to some small saturation. This theoretical result is confirmed in Table 1.6.

h	$\mathcal{D} = 3$	ord	$\mathcal{D} = 5$	ord	$\mathcal{D} = 6$	
0.8	$1.41 \cdot 10^{-4}$		$1.41 \cdot 10^{-3}$		$3.08 \cdot 10^{-3}$	
0.7	$4.17 \cdot 10^{-5}$	9.11	$4.33 \cdot 10^{-4}$	8.85	$9.74 \cdot 10^{-4}$	8.62
0.6	$1.00 \cdot 10^{-5}$	9.24	$1.06 \cdot 10^{-4}$	9.13	$2.45 \cdot 10^{-4}$	8.96
0.5	$1.87 \cdot 10^{-6}$	9.22	$1.92 \cdot 10^{-5}$	9.38	$4.53 \cdot 10^{-5}$	9.25
0.4	$3.03 \cdot 10^{-7}$	8.16	$2.26 \cdot 10^{-6}$	9.58	$5.44 \cdot 10^{-6}$	9.50
0.3	$6.68 \cdot 10^{-8}$	5.26	$1.37 \cdot 10^{-7}$	9.75	$3.34 \cdot 10^{-7}$	9.70
0.2	$2.67 \cdot 10^{-8}$	2.26	$2.57 \cdot 10^{-9}$	9.80	$6.20 \cdot 10^{-9}$	9.83
0.1	$1.39 \cdot 10^{-8}$	0.94	$7.52 \cdot 10^{-11}$	5.10	$2.07 \cdot 10^{-11}$	8.29
0.05	$1.12 \cdot 10^{-8}$	0.31	$7.12 \cdot 10^{-11}$	0.07	$8.52 \cdot 10^{-12}$	1.34

TABLE 1.6. Error of approximating $u(x) = \sin x$ with the basis function (1.26)

1.2.5. Examples of multi-dimensional quasi-interpolants. One important feature of approximate quasi-interpolation is the simplicity of its multi-dimensional generalization. In the next chapters, we shall see that sufficiently smooth and rapidly decaying functions with non-vanishing mean value can be taken as generating functions for quasi-interpolants on uniform grids in $\mathbb{R}^n$. So we have access to a large class of appropriate functions, which generate high-order approximants with simple analytic representations. This is, for example, in contrast to the case of spline functions, where n-dimensional generalizations have quite complicated analytic expressions.

One possibility is, for example, to use the radial counterparts of one-dimensional generating functions. So the n-dimensional analogue of (1.7) is the formula

$$\mathcal{M}_{h,\mathcal{D}}u(\mathbf{x}) = \frac{1}{(\pi\mathcal{D})^{n/2}} \sum_{\mathbf{m}\in\mathbb{Z}^n} u(h\mathbf{m})\, \mathrm{e}^{-|\mathbf{x}-h\mathbf{m}|^2/\mathcal{D}h^2}, \tag{1.27}$$

whereas (1.21) can be extended to the approximation formula

$$\frac{1}{c_n\mathcal{D}^{n/2}} \sum_{\mathbf{m}\in\mathbb{Z}^n} u(h\mathbf{m}) \operatorname{sech} \frac{|\mathbf{x}-h\mathbf{m}|}{\sqrt{\mathcal{D}}h}, \qquad c_n = \int_{\mathbb{R}^n} \operatorname{sech}|\mathbf{x}|\, d\mathbf{x}.$$

Here and in what follows we make the notational convention that finite-dimensional vectors are denoted by bold face symbols, i.e., $\mathbf{x} = (x_1, \dots, x_n) \in \mathbb{R}^n$, $\mathbf{m} = (m_1, \dots, m_n)$, $m_j \in \mathbb{Z}$. The scalar product of two vectors $\mathbf{x} = (x_1, \dots, x_n)$ and $\mathbf{y} = (y_1, \dots, y_n)$ in the Euclidean space $\mathbb{R}^n$ is denoted by

$$\langle \mathbf{x}, \mathbf{y} \rangle = \sum_{j=1}^{n} x_j y_j .$$

For the Euclidean norm of $\mathbf{x} \in \mathbb{R}^n$, we use the notation

$$|\mathbf{x}| = |\mathbf{x}|_2 = \sqrt{\langle \mathbf{x}, \mathbf{x} \rangle}\,.$$

We shall see in the next chapter that both formulas approximate with order $\mathcal{O}(h^2)$ up to some saturation bound.

A fourth-order approximation up to some small saturation is given by the formula

$$\frac{1}{(\pi\mathcal{D})^{n/2}}\sum_{\mathbf{m}\in\mathbb{Z}^n} u(\mathbf{m}h)\Big(\frac{n+2}{2}-\frac{|\mathbf{x}-\mathbf{m}h|^2}{\mathcal{D}h^2}\Big)\,\mathrm{e}^{-|\mathbf{x}-h\mathbf{m}|^2/\mathcal{D}h^2}$$

whereas the sixth-order can be obtained with the generating function

$$\eta(\mathbf{x})=\frac{\mathrm{e}^{-|\mathbf{x}|^2}}{2\pi^{n/2}}\Big(\frac{(n+4)(n+2)}{4}-(n+4)|\mathbf{x}|^2+|\mathbf{x}|^4\Big).$$

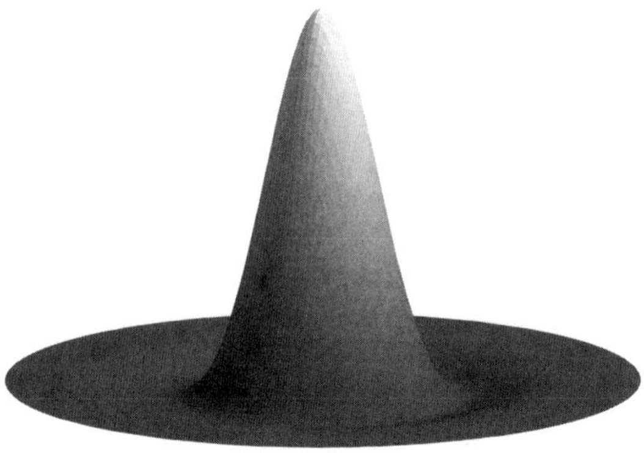

FIGURE 1.27. Sixth-order generating function in $\mathbb{R}^2$

Furthermore, in some applications, we use generating functions of the form

$$\eta(\mathbf{x})=\phi(\langle A\mathbf{x},\mathbf{x}\rangle)\,,$$

where A is an $n\times n$ matrix.

CHAPTER 2

Error estimates for quasi-interpolation

This chapter is devoted to the theoretical foundation of approximate quasi-interpolation. In Section 2.1, we introduce notation and mention basic results, which will be used throughout the book. Error estimations of general approximate quasi-interpolants in uniform and integral norms are obtained in Sections 2.2 – 2.4. We formulate conditions on the generating functions, which ensure high-order approximations up to a prescribed precision.

2.1. Auxiliary results

Here we introduce notation and mention some classical facts, which can be found in textbooks on functional analysis and which will be used in the following.

2.1.1. Function spaces. The space $L_p(\mathbb{R}^n)$, where $1 \le p < \infty$, consists of measurable complex-valued functions $u(\mathbf{x})$, $\mathbf{x} \in \mathbb{R}^n$, for which

$$\|u\|_{L_p} = \Big(\int\limits_{\mathbb{R}^n} |u(\mathbf{x})|^p \, d\mathbf{x} \Big)^{1/p} < \infty .$$

As is common in approximation theory, we suppose that $L_\infty(\mathbb{R}^n)$ is the space of bounded continuous functions equipped with the norm

$$\|u\|_{L_\infty} = \sup_{\mathbf{x} \in \mathbb{R}^n} |u(\mathbf{x})| .$$

Henceforth, we reserve the symbol L_p to denote the space $L_p(\mathbb{R}^n)$. For the L_p-spaces of functions defined on a measurable set $\Omega \subset \mathbb{R}^n$ we will use the notation $L_p(\Omega)$.

By $\ell_p(\mathbb{Z}^n)$, $1 \le p \le \infty$, we denote the space of sequences $\{a_{\mathbf{m}}\}_{\mathbf{m} \in \mathbb{Z}^n}$ with

$$\|\{a_{\mathbf{m}}\}\|_{\ell_p} = \Big(\sum_{\mathbf{m} \subset \mathbb{Z}^n} |a_{\mathbf{m}}|^p \Big)^{1/p} < \infty .$$

Throughout the book we use multi-index notation: A multi-index $\boldsymbol{\alpha}$ is a vector with non-negative integer components, $\boldsymbol{\alpha} = (\alpha_1, \ldots, \alpha_n) \in \mathbb{Z}^n_{\ge 0}$, its length is

$$[\boldsymbol{\alpha}] := \alpha_1 + \cdots + \alpha_n .$$

We write $\boldsymbol{\alpha} \le \boldsymbol{\beta}$ for two multi-indices if $\alpha_j \le \beta_j$ for all $j = 1, \ldots, n$. Furthermore, we denote

$$\boldsymbol{\alpha}! = \alpha_1! \ldots \alpha_n! , \quad \mathbf{x}^{\boldsymbol{\alpha}} = x_1^{\alpha_1} \cdots x_n^{\alpha_n}$$

and

$$\partial^{\boldsymbol{\alpha}} u(\mathbf{x}) = \frac{\partial^{[\boldsymbol{\alpha}]}}{\partial x_1^{\alpha_1} \cdots \partial x_n^{\alpha_n}} u(\mathbf{x}) .$$

In the following, $\nabla_k u$ denotes the vector of partial derivatives $\{\partial^{\boldsymbol{\alpha}} u\}_{[\boldsymbol{\alpha}]=k}$. Using multi-indices, many multi-variate formulas have the form of the one-dimensional analogues. For example,

$$(\mathbf{x}+\mathbf{y})^{\boldsymbol{\alpha}} = \sum_{\boldsymbol{\beta}\le\boldsymbol{\alpha}} \frac{\boldsymbol{\alpha}!}{(\boldsymbol{\alpha}-\boldsymbol{\beta})!\,\boldsymbol{\beta}!}\,\mathbf{x}^{\boldsymbol{\alpha}-\boldsymbol{\beta}}\,\mathbf{y}^{\boldsymbol{\beta}}\,.$$

As usual, the Sobolev space W_p^N, $N \in \mathbb{N}$, consists of L_p-functions whose generalized derivatives up to the order N also belong to L_p. The norm in W_p^N is defined by

$$\|u\|_{W_p^N(\Omega)} = \sum_{[\boldsymbol{\alpha}]\le N} \|\partial^{\boldsymbol{\alpha}} u\|_{L_p(\Omega)}.$$

If the derivatives of order N of a function u are integrable, then its *Taylor expansion* has the form

$$u(\mathbf{y}) = \sum_{[\boldsymbol{\alpha}]=0}^{N-1} \frac{(\mathbf{y}-\mathbf{x})^{\boldsymbol{\alpha}}}{\boldsymbol{\alpha}!}\,\partial^{\boldsymbol{\alpha}} u(\mathbf{x}) + \sum_{[\boldsymbol{\alpha}]=N} \frac{(\mathbf{y}-\mathbf{x})^{\boldsymbol{\alpha}}}{\boldsymbol{\alpha}!}\,U_{\boldsymbol{\alpha}}(\mathbf{x},\mathbf{y})\,, \tag{2.1}$$

where the remainder terms are given by

$$U_{\boldsymbol{\alpha}}(\mathbf{x},\mathbf{y}) = N \int_0^1 s^{N-1} \partial^{\boldsymbol{\alpha}} u(s\mathbf{x}+(1-s)\mathbf{y})\, ds\,. \tag{2.2}$$

If the derivatives of order N of u are continuous, then by the mean value theorem, the remainder can also be given as

$$U_{\boldsymbol{\alpha}}(\mathbf{x},\mathbf{y}) = \partial^{\boldsymbol{\alpha}} u(s\mathbf{x}+(1-s)\mathbf{y}) \tag{2.3}$$

for some $s \in [0,1]$.

By $\mathcal{S}(\mathbb{R}^n)$ we denote the *Schwartz space* of smooth and rapidly decaying functions, i.e.,

$$u \in \mathcal{S}(\mathbb{R}^n) \quad \text{if and only if} \quad \sup_{[\boldsymbol{\alpha}]\le N}\ \sup_{\mathbf{x}\in\mathbb{R}^n} (1+|\mathbf{x}|)^N \big|\partial^{\boldsymbol{\alpha}} u(\mathbf{x})\big| < \infty$$

for $N = 0, 1, \ldots$.

2.1.2. Fourier transform. The *Fourier transform* of an absolutely integrable function $u \in L_1(\mathbb{R}^n)$ is defined as

$$\mathcal{F}u(\boldsymbol{\lambda}) = \widehat{u}(\boldsymbol{\lambda}) = \int_{\mathbb{R}^n} u(\mathbf{x})\ \mathrm{e}^{-2\pi i\langle \mathbf{x},\boldsymbol{\lambda}\rangle}\ d\mathbf{x}\,. \tag{2.4}$$

Here

$$\langle \mathbf{x},\mathbf{y}\rangle = \sum_{j=1}^{n} x_j y_j$$

denotes the usual scalar product of the vectors $\mathbf{x} = (x_1,\ldots,x_n)$, $\mathbf{y} = (y_1,\ldots,y_n) \in \mathbb{R}^n$. The Euclidean norm of $\mathbf{x} \in \mathbb{R}^n$ is denoted by

$$|\mathbf{x}| = |\mathbf{x}|_2 := \sqrt{\langle \mathbf{x},\mathbf{x}\rangle}\,.$$

For $u \in L_1(\mathbb{R}^n)$ the Fourier transform $\mathcal{F}u$ is continuous and $\mathcal{F}u(\boldsymbol{\lambda}) \to 0$ as $|\boldsymbol{\lambda}| \to \infty$. The *inverse Fourier transform* of $u \in L_1(\mathbb{R}^n)$ is defined as

$$\mathcal{F}^{-1}u(\boldsymbol{\lambda}) := \int_{\mathbb{R}^n} u(\mathbf{x})\ \mathrm{e}^{2\pi i\langle \mathbf{x},\boldsymbol{\lambda}\rangle}\ d\mathbf{x}\,.$$

If $u \in \mathcal{S}(\mathbb{R}^n)$, then also $\mathcal{F}u \in \mathcal{S}(\mathbb{R}^n)$ and the inverse Fourier transform gives

$$u(\mathbf{x}) = \int_{\mathbb{R}^n} \mathcal{F}u(\boldsymbol{\lambda})\, \mathrm{e}^{2\pi i \langle \mathbf{x}, \boldsymbol{\lambda} \rangle}\, d\boldsymbol{\lambda}. \tag{2.5}$$

The Fourier transforms $\mathcal{F}$ and $\mathcal{F}^{-1}$ can be extended to operators on the space of tempered distributions $\mathcal{S}'(\mathbb{R}^n)$ dual to $\mathcal{S}(\mathbb{R}^n)$, by

$$(u, \mathcal{F}^{\pm 1}\varphi) = (\mathcal{F}^{\pm 1}u, \varphi), \quad \text{for all} \quad u \in \mathcal{S}(\mathbb{R}^n)\,, \varphi \in \mathcal{S}'(\mathbb{R}^n)$$

and to isometries on the space $L_2(\mathbb{R}^n)$.

2.1.3. Convolutions. For two functions $f(\mathbf{x})$ and $g(\mathbf{x})$, $\mathbf{x} \in \mathbb{R}^n$, such that the integral in (2.6) is convergent, the *continuous convolution* $f * g$ is defined as the function

$$f * g(\mathbf{x}) = \int_{\mathbb{R}^n} f(\mathbf{y})\, g(\mathbf{x} - \mathbf{y})\, d\mathbf{y}. \tag{2.6}$$

For two vectors $u = (u_{\mathbf{m}})$ and $v = (v_{\mathbf{m}})$, $\mathbf{m} \in \mathbb{Z}^n$, such that the summation in (2.7) is convergent, the *discrete convolution* $u * v$ is defined as the vector

$$\forall \mathbf{k} \in \mathbb{Z}^n, \quad (u * v)_{\mathbf{k}} = \sum_{\mathbf{m} \in \mathbb{Z}^n} u_{\mathbf{m}}\, v_{\mathbf{k}-\mathbf{m}}. \tag{2.7}$$

For a vector $u = (u_{\mathbf{m}})$ and a function g such that the summation in (2.8) is convergent, we denote by $u *_h g$ the *hybrid convolution* with step h of u and g as the function

$$u *_h g(\mathbf{x}) = \sum_{\mathbf{m} \in \mathbb{Z}^n} u_{\mathbf{m}}\, g(\mathbf{x}/h - \mathbf{m}). \tag{2.8}$$

The *semi-discrete convolution with step* h of the functions f and g is defined by the equality

$$f *_h g(\mathbf{x}) = \sum_{\mathbf{m} \in \mathbb{Z}^n} f(h\mathbf{m})\, g(\mathbf{x}/h - \mathbf{m}). \tag{2.9}$$

It is the hybrid convolution with the vector given by the values of the function f, $u_{\mathbf{m}} = f(h\mathbf{m})$.

The Fourier transform of a continuous convolution is the product of the Fourier transform of its parts:

$$\mathcal{F}(f * g) = \mathcal{F}f\, \mathcal{F}g. \tag{2.10}$$

Continuous and discrete convolutions can be estimated by *Young's inequality*, which has the following form for continuous convolutions:

$$\|f * g\|_q \le \|f\|_r \|g\|_p\,, \tag{2.11}$$

where $1 \le p, q, r \le \infty$ and $q^{-1} = p^{-1} + r^{-1} - 1$ (see e.g., [**90**, A.2]).

2.1.4. Radial functions. We will often deal with functions of $\mathbf{x} \in \mathbb{R}^n$ which depend only on the Euclidean norm $|\mathbf{x}|$, $f(\mathbf{x}) = f_0(r)$, $r = |\mathbf{x}|$. The Fourier transform $\mathcal{F}f$ is also a radial function, $\mathcal{F}f(\boldsymbol{\lambda}) = F_0(t)$ for $t = |\boldsymbol{\lambda}|$ with

$$F_0(t) = \frac{2\pi}{t^{n/2-1}} \int_0^\infty f_0(r)\, J_{n/2-1}(2\pi r t)\, r^{n/2}\, dr\,, \tag{2.12}$$

where J_ν are the *Bessel functions of the first kind* (see [**90**, Thm. IV.3.3]). Formula (2.12) follows from the integral representation

$$J_\nu(z) = \frac{z^\nu}{2^\nu \Gamma(\nu + \frac{1}{2})\sqrt{\pi}} \int_0^\pi \mathrm{e}^{iz\cos\theta} \sin^{2\nu}\theta\, d\theta \tag{2.13}$$

for $z \in \mathbb{C}$ and $\operatorname{Re}\nu > 0$ ([**96**, 3.3(6)]), which implies in particular

$$\int_{S_{n-1}} \mathrm{e}^{2\pi i z \langle \boldsymbol{\omega}, \boldsymbol{\omega}' \rangle}\, d\sigma_{\boldsymbol{\omega}'} = \frac{2\pi}{z^{n/2-1}} J_{n/2-1}(2\pi z)\,, \tag{2.14}$$

where S_{n-1} denotes the unit sphere in $\mathbb{R}^n$ and $\boldsymbol{\omega} \in S_{n-1}$.

To simplify notation, we will denote a radial function $f(\mathbf{x})$ by the same symbol $f(|\mathbf{x}|)$ if considered as univariate function depending on $|\mathbf{x}|$. The convolution of two radial functions

$$Q * f(\mathbf{x}) = \int_{\mathbb{R}^n} Q(|\mathbf{x} - \mathbf{y}|)\, f(|\mathbf{y}|)\, d\mathbf{y} = \int_{\mathbb{R}^n} \mathcal{F}Q(\boldsymbol{\lambda})\, \mathcal{F}f(\boldsymbol{\lambda})\, \mathrm{e}^{2\pi i \langle \mathbf{x}, \boldsymbol{\lambda} \rangle}\, d\boldsymbol{\lambda}$$

is also radial and can be expressed by the one-dimensional integral

$$Q * f(\mathbf{x}) = \frac{2\pi}{|\mathbf{x}|^{n/2-1}} \int_0^\infty \mathcal{F}Q(r)\, \mathcal{F}f(r)\, J_{n/2-1}(2\pi r |\mathbf{x}|)\, r^{n/2}\, dr\,. \tag{2.15}$$

Since for integer $k \geq 0$

$$J_{k+1/2}(z) = (-1)^k \sqrt{\frac{2}{\pi}}\, z^{k+1/2} \Big(\frac{1}{z} \frac{d}{dz} \Big)^k \frac{\sin z}{z} \tag{2.16}$$

(cf. [**1**, 10.1.25]), the convolution of two radial functions is expressed as a one-dimensional integral of elementary functions when the space dimension is an odd number. In many cases this integral can be taken analytically. But also the case of even space dimensions can be handled successfully as well, since the integral in question is a Bessel transform and a variety of tables of the Bessel transform are available (cf. for example [**7**]).

2.1.5. Multi-dimensional Poisson summation formula. Throughout the book we will make use of *Poisson's summation formula* (see [**90**, Thm. VII.2.4]):

$$\sum_{\mathbf{m} \in \mathbb{Z}^n} u(\mathbf{x} + \mathbf{m}) = \sum_{\mathbf{m} \in \mathbb{Z}^n} \mathcal{F}u(\mathbf{m})\, \mathrm{e}^{2\pi i \langle \mathbf{m}, \mathbf{x} \rangle}\,. \tag{2.17}$$

We formulate two sufficient conditions on u ensuring this equality for all $\mathbf{x} \in \mathbb{R}^n$.

(1) The function u and its Fourier transform $\mathcal{F}u$ satisfy

$$|u(\mathbf{x})| \leq A(1 + |\mathbf{x}|)^{-n-\delta} \quad \text{and} \quad |\mathcal{F}u(\boldsymbol{\lambda})| \leq A(1 + |\boldsymbol{\lambda}|)^{-n-\delta} \tag{2.18}$$

for some $\delta > 0$, which implies continuity for both functions.

(2) The continuous function u satisfies for some $\delta > 0$

$$|u(\mathbf{x})| \le A(1+|\mathbf{x}|)^{-n-\delta} \quad \text{and} \quad \{\mathcal{F}u(\mathbf{m})\} \in \ell_1(\mathbb{Z}^n)\,. \tag{2.19}$$

In particular, (2.17) implies

$$\sum_{\mathbf{m}\in\mathbb{Z}^n} u(\mathbf{m}) = \sum_{\mathbf{m}\in\mathbb{Z}^n} \mathcal{F}u(\mathbf{m}) \tag{2.20}$$

and

$$\sum_{\mathbf{m}\in\mathbb{Z}^n} u(\mathbf{x}+\mathbf{m})\, \mathrm{e}^{2\pi i\langle \mathbf{x}+\mathbf{m},\boldsymbol{\lambda}\rangle} = \sum_{\mathbf{m}\in\mathbb{Z}^n} \mathcal{F}u(\mathbf{m}-\boldsymbol{\lambda})\, \mathrm{e}^{2\pi i\langle \mathbf{m},\mathbf{x}\rangle}\,. \tag{2.21}$$

If $A = \|a_{jk}\|_{j,k=1}^n : \mathbb{R}^n \to \mathbb{R}^n$ is a non-singular real matrix, then

$$\sum_{\mathbf{m}\in\mathbb{Z}^n} u(\mathbf{x}+A^{-1}\mathbf{m}) = \det A \sum_{\mathbf{m}\in\mathbb{Z}^n} \mathcal{F}u(A^t\mathbf{m})\, \mathrm{e}^{2\pi i\langle A^t\mathbf{m},\mathbf{x}\rangle}\,, \tag{2.22}$$

where A^t denotes the transposed matrix

$$A^t = \|b_{jk}\|_{j,k=1}^n \quad \text{with} \quad b_{jk} = a_{kj}\,.$$

2.2. Some properties of quasi-interpolants

In this section we consider some properties of general quasi-interpolation formulas of the form

$$\mathcal{M}_{h,\mathcal{D}}u(\mathbf{x}) = \mathcal{D}^{-n/2} \sum_{\mathbf{m}\in\mathbb{Z}^n} u(h\mathbf{m})\, \eta\Big(\frac{\mathbf{x}-h\mathbf{m}}{\sqrt{\mathcal{D}}\,h}\Big), \quad \mathbf{x}\in\mathbb{R}^n\,. \tag{2.23}$$

Here and in the following, it is always assumed that the generating function η is continuous. Using the notation (2.9), the quasi-interpolant $\mathcal{M}_{h,\mathcal{D}}u$ represents the semi-discrete convolution

$$\mathcal{M}_{h,\mathcal{D}}u = u *_h \eta_{\mathcal{D}}$$

with the function $\eta_{\mathcal{D}} = \eta(\cdot/\mathcal{D})$ and step h.

2.2.1. Young's inequality for semi-discrete convolutions.

Let us define the norm

$$\|u\|_{p,h} := \begin{cases} \Big(h^n \sum_{\mathbf{m}\in\mathbb{Z}^n} |u(h\mathbf{m})|^p\Big)^{1/p}, & 1 \le p < \infty\,, \\ \sup_{\mathbf{m}\in\mathbb{Z}^n} |u(h\mathbf{m})|\,, & p=\infty\,. \end{cases} \tag{2.24}$$

LEMMA 2.1.

$$\|\mathcal{M}_{h,\mathcal{D}}u\|_{L_q} \le (\sqrt{\mathcal{D}}h)^{n(1/r-1)} \|\eta\|_{L_r}^{1/q} \Big\|\mathcal{D}^{-n/2} \sum_{\mathbf{m}\in\mathbb{Z}^n} \Big|\eta\Big(\frac{\cdot-\mathbf{m}}{\sqrt{\mathcal{D}}}\Big)\Big|^r \Big\|_{L_\infty}^{1/r-1/q} \|u\|_{p,h}$$

for $1 \le p,q,r \le \infty$ *with* $q^{-1} = p^{-1} + r^{-1} - 1$.

PROOF. Note that $p=\infty$ implies $q=\infty$ and $r=1$, so that the assertion follows immediately. Similarly, if $q=\infty$, then $1/p+1/r=1$, and Hölder's inequality yields the result.

Let $p, q, r < \infty$. Then the numbers λ, μ, ν given by $\lambda = q$, $1/r - 1/\lambda = 1/\mu$, $1/p - 1/\lambda = 1/\nu$, are positive and satisfy $\lambda^{-1} + \mu^{-1} + \nu^{-1} = 1$. We write

$$\Big|\eta\Big(\frac{\mathbf{x} - h\mathbf{m}}{\sqrt{\mathcal{D}}\,h}\Big)\Big||u(h\mathbf{m})| = \Big|\eta\Big(\frac{\mathbf{x} - h\mathbf{m}}{\sqrt{\mathcal{D}}\,h}\Big)\Big|^{r/\lambda}|u(h\mathbf{m})|^{p/\lambda} \cdot \Big|\eta\Big(\frac{\mathbf{x} - h\mathbf{m}}{\sqrt{\mathcal{D}}\,h}\Big)\Big|^{r(1/r-1/\lambda)}|u(h\mathbf{m})|^{p(1/p-1/\lambda)}$$

and we apply Hölder's inequality with the exponents λ, μ, ν. Then

$$|\mathcal{M}_{h,\mathcal{D}}u(\mathbf{x})| \le \mathcal{D}^{-n/2}\Big(\sum_{\mathbf{m}\in\mathbb{Z}^n}\Big|\eta\Big(\frac{\mathbf{x} - h\mathbf{m}}{\sqrt{\mathcal{D}}\,h}\Big)\Big|^{r}|u(h\mathbf{m})|^{p}\Big)^{1/\lambda} \times \Big(\sum_{\mathbf{m}\in\mathbb{Z}^n}\Big|\eta\Big(\frac{\mathbf{x} - h\mathbf{m}}{\sqrt{\mathcal{D}}\,h}\Big)\Big|^{r\mu(1/r-1/\lambda)}\Big)^{1/\mu}\Big(\sum_{\mathbf{m}\in\mathbb{Z}^n}|u(h\mathbf{m})|^{p\nu(1/p-1/\lambda)}\Big)^{1/\nu},$$

and therefore

$$\|\mathcal{M}_{h,\mathcal{D}}u\|_{L_q}^q \le \mathcal{D}^{-nq/2}\sup_{\mathbf{x}}\Big(\sum_{\mathbf{m}\in\mathbb{Z}^n}\Big|\eta\Big(\frac{\mathbf{x} - \mathbf{m}}{\sqrt{\mathcal{D}}}\Big)\Big|^{r}\Big)^{q(1/r-1/\lambda)} \times \Big(\sum_{\mathbf{m}\in\mathbb{Z}^n}|u(h\mathbf{m})|^{p}\Big)^{q(1/p-1/\lambda)}\int_{\mathbb{R}^n}\sum_{\mathbf{m}\in\mathbb{Z}^n}\Big|\eta\Big(\frac{\mathbf{x} - h\mathbf{m}}{\sqrt{\mathcal{D}}\,h}\Big)\Big|^{r}|u(h\mathbf{m})|^{p}\,d\mathbf{x}.$$

The last integral equals

$$\sum_{\mathbf{m}\in\mathbb{Z}^n}|u(h\mathbf{m})|^{p}\int_{\mathbb{R}^n}\Big|\eta\Big(\frac{\mathbf{x} - h\mathbf{m}}{\sqrt{\mathcal{D}}\,h}\Big)\Big|^{r}d\mathbf{x} = (\sqrt{\mathcal{D}}\,h)^n\sum_{\mathbf{m}\in\mathbb{Z}^n}|u(h\mathbf{m})|^{p}\int_{\mathbb{R}^n}|\eta(\mathbf{x})|^{r}\,d\mathbf{x},$$

so that

$$\|\mathcal{M}_{h,\mathcal{D}}u\|_{L_q}^q \le \mathcal{D}^{-nq/2}(\sqrt{\mathcal{D}}\,h)^n\|\eta\|_{L_r}\Big(\sum_{\mathbf{m}\in\mathbb{Z}^n}|u(h\mathbf{m})|^{p}\Big)^{q/p} \times \sup_{\mathbf{x}}\Big(\sum_{\mathbf{m}\in\mathbb{Z}^n}\Big|\eta\Big(\frac{\mathbf{x} - \mathbf{m}}{\sqrt{\mathcal{D}}}\Big)\Big|^{r}\Big)^{q/r-1},$$

which gives the desired inequality. □

COROLLARY 2.2. *Let $1 \le p \le \infty$ and let the function u be such that $\|u\|_{p,h} \le C$ with a constant C independent of $h > 0$. Then the quasi-interpolants $\mathcal{M}_{h,\mathcal{D}}u$ are bounded in L_p uniformly in h if the continuous generating function $\eta \in L_1(\mathbb{R}^n)$ and if it satisfies additionally*

$$\Big\|\mathcal{D}^{-n/2}\sum_{\mathbf{m}\in\mathbb{Z}^n}\Big|\eta\Big(\frac{\cdot - \mathbf{m}}{\sqrt{\mathcal{D}}}\Big)\Big|\Big\|_{L_\infty} < \infty. \tag{2.25}$$

Then

$$\|\mathcal{M}_{h,\mathcal{D}}u\|_{L_p} \le C_p\|u\|_{p,h} \quad \text{with} \quad C_p = \|\eta\|_{L_1}^{1/p}\Big\|\mathcal{D}^{-n/2}\sum_{\mathbf{m}\in\mathbb{Z}^n}\Big|\eta\Big(\frac{\cdot - \mathbf{m}}{\sqrt{\mathcal{D}}}\Big)\Big|\Big\|_{L_\infty}^{1-1/p}.$$

REMARK 2.3. From the definition of the Riemann integral it follows immediately that the norms $\|u\|_{p,h}$ are uniformly bounded with respect to $h > 0$ if the function $|u(\mathbf{x})|^p$, $1 \le p < \infty$, is Riemann integrable. If $p = \infty$, then

$$\|u\|_{\infty,h} = \sup_{\mathbf{m}\in\mathbb{Z}^n}|u(h\mathbf{m})|$$

is obviously bounded for a bounded function u.

The relation (2.25) is satisfied under the following *decay condition* on the generating function η, which is assumed throughout the book: There exist constants $A > 0$ and $K > n$ such that

$$|\eta(\mathbf{x})| \le A\,(1+|\mathbf{x}|)^{-K} \text{ for all } \mathbf{x} \in \mathbb{R}^n\,. \tag{2.26}$$

The decay of η ensures that for given $\mathcal{D} > 0$ and any $\delta > 0$, there exists a finite subset $Z_\delta \subset \mathbb{Z}^n$ such that

$$\mathcal{D}^{-n/2} \sup_{\mathbf{x}\in Q} \sum_{\mathbf{m}\in\mathbb{Z}^n\setminus Z_\delta} \Big|\eta\Big(\frac{\mathbf{x}-\mathbf{m}}{\sqrt{\mathcal{D}}}\Big)\Big| < \delta\,,$$

where Q denotes the cube $\big[-\frac{1}{2},\frac{1}{2}\big]^n$. Then for bounded functions u, the value of the quasi-interpolant $\mathcal{M}_{h,\mathcal{D}}u(\mathbf{x})$ is, within the precision $\delta\|u\|_{L_\infty}$, determined by the finite sum

$$\mathcal{D}^{-n/2} \sum_{\mathbf{m}\in\mathbf{k}+Z_\delta} u(h\mathbf{m})\,\eta\Big(\frac{\mathbf{x}-h\mathbf{m}}{\sqrt{\mathcal{D}}\,h}\Big)\,,$$

where the integer vector $\mathbf{k} \in \mathbb{Z}^n$ is chosen such that $\mathbf{x}/h - \mathbf{k} \in Q$.

2.2.2. An auxiliary function. Let us introduce the function

$$g_{\mathbf{x},h}(\mathbf{y}) := \mathcal{D}^{-n/2}u(\mathbf{y})\,\eta\Big(\frac{\mathbf{x}-\mathbf{y}}{\sqrt{\mathcal{D}}h}\Big) \tag{2.27}$$

for given $\mathbf{x} \in \mathbb{R}^n$, h and $\mathcal{D}$.

LEMMA 2.4. *Suppose that η is subjected to (2.26) and the function u satisfies*

$$|u(\mathbf{x})| \le B(1+|\mathbf{x}|)^L\,,\ \mathbf{x}\in\mathbb{R}^n\,, \tag{2.28}$$

for some positive $L < K - n$. Then for fixed parameters $\mathcal{D}$ and h and any $\mathbf{x} \in \mathbb{R}^n$, the series

$$\sum_{\mathbf{m}\in\mathbb{Z}^n} g_{\mathbf{x},h}(h\mathbf{m}+\mathbf{z}) = \mathcal{D}^{-n/2}\sum_{\mathbf{m}\in\mathbb{Z}^n} u(h\mathbf{m}+\mathbf{z})\eta\Big(\frac{\mathbf{x}-h\mathbf{m}-\mathbf{z}}{\sqrt{\mathcal{D}}h}\Big) \tag{2.29}$$

converges absolutely for all $\mathbf{z}$ to a bounded h-periodic function $G_{\mathbf{x}}(\mathbf{z})$ with the Fourier series

$$\sum_{\boldsymbol{\nu}\in\mathbb{Z}^n} \mathrm{e}^{\frac{2\pi i}{h}\langle\mathbf{z},\boldsymbol{\nu}\rangle}(\sqrt{\mathcal{D}}h)^{-n}\int_{\mathbb{R}^n} u(\mathbf{y})\eta\Big(\frac{\mathbf{x}-\mathbf{y}}{\sqrt{\mathcal{D}}h}\Big)\,\mathrm{e}^{-\frac{2\pi i}{h}\langle\mathbf{y},\boldsymbol{\nu}\rangle}\,d\mathbf{y}\,. \tag{2.30}$$

PROOF. Note first that in view of (2.26) and (2.28)

$$\begin{aligned}\sum_{\mathbf{m}\in\mathbb{Z}^n}|g_{\mathbf{x},h}(h\mathbf{m}+\mathbf{z})| &= \mathcal{D}^{-n/2}\sum_{\mathbf{m}\in\mathbb{Z}^n}\Big|u(h\mathbf{m}+\mathbf{z})\eta\Big(\frac{\mathbf{x}-h\mathbf{m}-\mathbf{z}}{\sqrt{\mathcal{D}}h}\Big)\Big|\\
&\le AB\mathcal{D}^{-n/2}\sum_{\mathbf{m}\in\mathbb{Z}^n}\big(1+|h\mathbf{m}+\mathbf{z}|\big)^L\Big(1+\Big|\frac{\mathbf{x}-h\mathbf{m}-\mathbf{z}}{\sqrt{\mathcal{D}}h}\Big|\Big)^{-K}\\
&\le AB\mathcal{D}^{-n/2}\sum_{\mathbf{m}\in\mathbb{Z}^n}\big(|\mathbf{x}-h\mathbf{m}-\mathbf{z}|+1+|\mathbf{x}|\big)^L\Big(1+\Big|\frac{\mathbf{x}-h\mathbf{m}-\mathbf{z}}{\sqrt{\mathcal{D}}h}\Big|\Big)^{-K}.\end{aligned}$$

Since

$$\big(|\mathbf{x}-h\mathbf{m}-\mathbf{z}|\;+1+|\mathbf{x}|\big)^L=(\sqrt{\mathcal{D}}h)^L\Big(1+\Big|\frac{\mathbf{x}-h\mathbf{m}-\mathbf{z}}{\sqrt{\mathcal{D}}h}\Big|+\frac{1+|\mathbf{x}|}{\sqrt{\mathcal{D}}h}-1\Big)$$
$$=\sum_{j=0}^{L}\frac{L!\,(\sqrt{\mathcal{D}}h)^j}{j!\,(L-j)!}\,\big(1+|\mathbf{x}|-\sqrt{\mathcal{D}}h\big)^{L-j}\Big(1+\Big|\frac{\mathbf{x}-h\mathbf{m}-\mathbf{z}}{\sqrt{\mathcal{D}}h}\Big|\Big)^j,$$

we obtain

$$\sum_{\mathbf{m}\in\mathbb{Z}^n}|g_{\mathbf{x},h}(h\mathbf{m}+\mathbf{z})|$$
$$\le AB\mathcal{D}^{-n/2}\sum_{j=0}^{L}\frac{L!\,(\sqrt{\mathcal{D}}h)^j}{j!\,(L-j)!}\,\big(1+|\mathbf{x}|-\sqrt{\mathcal{D}}h\big)^{L-j}\Big(1+\Big|\frac{\mathbf{x}-h\mathbf{m}-\mathbf{z}}{\sqrt{\mathcal{D}}h}\Big|\Big)^{j-K}.$$

Let us define the function

$$\phi_\mu(s):=\sup_{\mathbb{R}^n}s^{-n}\sum_{\mathbf{m}\in\mathbb{Z}^n}\big(1+|s^{-1}\mathbf{m}-\mathbf{x}|\big)^{-n-\mu}\,,\qquad \mu>0\,,\tag{2.31}$$

which depends continuously on $s\in(0,\infty)$. Note that

$$\phi_\mu(s)\to\int_{\mathbb{R}^n}(1+|\mathbf{x}|)^{-n-\mu}\,d\mathbf{x}=\frac{2\pi^{n/2}}{\Gamma(n/2)}\,\frac{\Gamma(n)\Gamma(\mu)}{\Gamma(n+\mu)}\qquad\text{as}\quad s\to\infty\,,$$

which shows that $\phi_\mu(s)$ is bounded if $s\ge s_0>0$.

With the help of ϕ_μ, we estimate

$$\sum_{\mathbf{m}\in\mathbb{Z}^n}|g_{\mathbf{x},h}(h\mathbf{m}+\mathbf{z})|\le AB\sum_{j=0}^{L}\frac{L!\,(\sqrt{\mathcal{D}}h)^j}{j!\,(L-j)!}\,(1-\sqrt{\mathcal{D}}h+|\mathbf{x}|)^{L-j}\,\phi_{K-j-n}(\sqrt{\mathcal{D}})\,,$$

which shows that the sum on the left-hand side is bounded by a constant not depending on $\mathbf{z}$ for fixed $\mathcal{D}$, h, and $\mathbf{x}$. Therefore the sum (2.29) converges absolutely to the bounded h-periodic function $G_{\mathbf{x}}(\mathbf{z})$, which has the Fourier coefficients

$$\gamma_{\boldsymbol{\nu}}(\mathbf{x})=h^{-n}\int_{hQ}\sum_{\mathbf{m}\in\mathbb{Z}^n}g_{x,h}(h\mathbf{m}+\mathbf{z})\,\mathrm{e}^{-\frac{2\pi i}{h}\langle\mathbf{z},\boldsymbol{\nu}\rangle}\,d\mathbf{z}\,,$$

where Q is the cube $\big[-\frac12,\frac12\big]^n$. Due to the absolute convergence, we can change the integration and summation and obtain

$$\gamma_{\boldsymbol{\nu}}(\mathbf{x})=(\sqrt{\mathcal{D}}h)^{-n}\sum_{\mathbf{m}\in\mathbb{Z}^n}\int_{h(\mathbf{m}+Q)}u(\mathbf{y})\eta\Big(\frac{\mathbf{x}-\mathbf{y}}{\sqrt{\mathcal{D}}h}\Big)\,\mathrm{e}^{-\frac{2\pi i}{h}\langle\mathbf{y}-h\mathbf{m},\boldsymbol{\nu}\rangle}\,d\mathbf{y}$$
$$=(\sqrt{\mathcal{D}}h)^{-n}\int_{\mathbb{R}^n}u(\mathbf{y})\eta\Big(\frac{\mathbf{x}-\mathbf{y}}{\sqrt{\mathcal{D}}h}\Big)\,\mathrm{e}^{-\frac{2\pi i}{h}\langle\mathbf{y},\boldsymbol{\nu}\rangle}\,d\mathbf{y}\,.\qquad\square$$

Since the generating function η is supposed to be continuous, the continuity of u implies that $G_{\mathbf{x}}(\mathbf{z})$ is continuous. If, moreover, the Fourier series (2.30) converges absolutely, then we have the equality

$$G_{\mathbf{x}}(\mathbf{z})=(\sqrt{\mathcal{D}}h)^{-n}\sum_{\boldsymbol{\nu}\in\mathbb{Z}^n}\mathrm{e}^{\frac{2\pi i}{h}\langle\mathbf{z},\boldsymbol{\nu}\rangle}\int_{\mathbb{R}^n}u(\mathbf{y})\eta\Big(\frac{\mathbf{x}-\mathbf{y}}{\sqrt{\mathcal{D}}h}\Big)\,\mathrm{e}^{-\frac{2\pi i}{h}\langle\mathbf{y},\boldsymbol{\nu}\rangle}\,d\mathbf{y}\,.$$

For the general case, the following summation formula holds.

LEMMA 2.5. *Let $\varphi(\mathbf{x})$ be a continuous function with*

$$|\varphi(\mathbf{x})| \le c\,(1+|\mathbf{x}|)^{-n-\delta}\,, \int_{\mathbb{R}^n} \varphi(\mathbf{x}) d\mathbf{x} = 1\,, \; |\mathcal{F}\varphi(\boldsymbol{\lambda})| \le c\,(1+|\boldsymbol{\lambda}|)^{-n-\delta}\,, \; \delta > 0\,.$$

Then the equality

$$G_{\mathbf{x}}(\mathbf{z}) = (\sqrt{\mathcal{D}}h)^{-n} \lim_{\varepsilon\to 0} \sum_{\boldsymbol{\nu}\in\mathbb{Z}^n} \mathcal{F}\varphi(\varepsilon\boldsymbol{\nu})\, \mathrm{e}^{\frac{2\pi i}{h}\langle \mathbf{z},\boldsymbol{\nu}\rangle} \int_{\mathbb{R}^n} u(\mathbf{y})\eta\Big(\frac{\mathbf{x}-\mathbf{y}}{\sqrt{\mathcal{D}}h}\Big)\, \mathrm{e}^{-\frac{2\pi i}{h}\langle \mathbf{y},\boldsymbol{\nu}\rangle}\, d\mathbf{y}$$

holds for any z belonging to the Lebesgue set of the function $G_{\mathbf{x}}(\mathbf{z})$.

PROOF. The assertion follows from [**90**, Thm. VII.2.11], where the convergence of general summation formulas for Fourier series of L_1-functions is proved. □

2.2.3. Representations of the quasi-interpolant $\mathcal{M}_{h,\mathcal{D}}$. If we set

$$u(\mathbf{y}) = \Big(\frac{\mathbf{x}-\mathbf{y}}{\sqrt{\mathcal{D}}h}\Big)^{\boldsymbol{\alpha}}, \quad 0 \le [\boldsymbol{\alpha}] \le L,$$

in the definition (2.27) of the function $g_{\mathbf{x},h}$, then the Fourier coefficients in (2.30) can be expressed by the Fourier transform of the generating function η

$$\begin{aligned}(\sqrt{\mathcal{D}}h)^{-n} \int_{\mathbb{R}^n} \Big(\frac{\mathbf{x}-\mathbf{y}}{\sqrt{\mathcal{D}}h}\Big)^{\boldsymbol{\alpha}} \eta\Big(\frac{\mathbf{x}-\mathbf{y}}{\sqrt{\mathcal{D}}h}\Big)\, \mathrm{e}^{-\frac{2\pi i}{h}\langle \mathbf{y},\boldsymbol{\nu}\rangle}\, d\mathbf{y} \\ = \mathrm{e}^{-\frac{2\pi i}{h}\langle \mathbf{x},\boldsymbol{\nu}\rangle} \int_{\mathbb{R}^n} \mathbf{y}^{\boldsymbol{\alpha}}\, \eta(\mathbf{y})\, \mathrm{e}^{2\pi i\langle \mathbf{y},\sqrt{\mathcal{D}}\boldsymbol{\nu}\rangle}\, d\mathbf{y} \\ = \Big(\frac{i}{2\pi}\Big)^{[\boldsymbol{\alpha}]} \mathrm{e}^{-\frac{2\pi i}{h}\langle \mathbf{x},\boldsymbol{\nu}\rangle}\, \partial^{\boldsymbol{\alpha}}\mathcal{F}\eta(-\sqrt{\mathcal{D}}\boldsymbol{\nu})\,,\end{aligned}$$

where we use the relation

$$\int_{\mathbb{R}^n} \mathbf{y}^{\boldsymbol{\alpha}}\, \eta(\mathbf{y})\, \mathrm{e}^{-2\pi i\langle \mathbf{y},\boldsymbol{\lambda}\rangle}\, d\mathbf{y} = \Big(\frac{i}{2\pi}\Big)^{[\boldsymbol{\alpha}]} \partial^{\boldsymbol{\alpha}}\mathcal{F}\eta(\boldsymbol{\lambda})\,.$$

The choice of $u(\mathbf{y})$ gives rise to some infinite sums which will occur at different places. We introduce the notation

$$\sigma_{\boldsymbol{\alpha}}(\mathbf{x},\eta,\mathcal{D}) := \mathcal{D}^{-n/2} \sum_{\mathbf{m}\in\mathbb{Z}^n} \Big(\frac{\mathbf{x}-\mathbf{m}}{\sqrt{\mathcal{D}}}\Big)^{\boldsymbol{\alpha}} \eta\Big(\frac{\mathbf{x}-\mathbf{m}}{\sqrt{\mathcal{D}}}\Big), \tag{2.32}$$

$$\rho_{\boldsymbol{\alpha}}(\mathbf{x},\eta,\mathcal{D}) := \mathcal{D}^{-n/2} \sum_{\mathbf{m}\in\mathbb{Z}^n} \Big|\Big(\frac{\mathbf{x}-\mathbf{m}}{\sqrt{\mathcal{D}}}\Big)^{\boldsymbol{\alpha}} \eta\Big(\frac{\mathbf{x}-\mathbf{m}}{\sqrt{\mathcal{D}}}\Big)\Big|, \tag{2.33}$$

for positive $\mathcal{D}$ and multi-index $\boldsymbol{\alpha} = (\alpha_1, ..., \alpha_n) \in \mathbb{Z}^n_{\ge 0}$. In view of the decay condition (2.26) with $K > L+n$, the sums converge absolutely if $[\boldsymbol{\alpha}] \le L$. Hence, both functions $\sigma_{\boldsymbol{\alpha}}$ and $\rho_{\boldsymbol{\alpha}}$ are continuous and periodic in each space direction with period 1.

LEMMA 2.6. *For any $\mathcal{D}_0 > 0$ and $\boldsymbol{\alpha}$, $0 \le [\boldsymbol{\alpha}] \le L$, there exist constants $c_{\boldsymbol{\alpha}}$ such that for all $\mathcal{D} \ge \mathcal{D}_0$*

$$\|\sigma_{\boldsymbol{\alpha}}(\cdot\,,\eta,\mathcal{D})\|_{L^\infty} \le \|\rho_{\boldsymbol{\alpha}}(\cdot\,,\eta,\mathcal{D})\|_{L^\infty} \le c_{\boldsymbol{\alpha}}\,.$$

PROOF. The functions $\sigma_{\boldsymbol\alpha}(\mathbf{x},\eta,\mathcal{D})$ and $\rho_{\boldsymbol\alpha}(\mathbf{x},\eta,\mathcal{D})$ are majorized by

$$\mathcal{D}^{-n/2}\sum_{\mathbf{m}\in\mathbb{Z}^n}\Big(1+\frac{|\mathbf{x}-\mathbf{m}|}{\sqrt{\mathcal{D}}}\Big)^{[\boldsymbol\alpha]-K}\le\phi_{K-n-[\boldsymbol\alpha]}(\sqrt{\mathcal{D}}),\tag{2.34}$$

where the function ϕ_μ is defined by (2.31). There we have shown that $\phi_\mu(s)$, $\mu>0$, is bounded if $s\ge s_0>0$. □

Suppose now that the function $u\in W^L_\infty(\mathbb{R}^n)$. Taking the Taylor expansion (2.1) for each node $h\mathbf{m}$, i.e.,

$$u(h\mathbf{m})=\sum_{[\boldsymbol\alpha]=0}^{L-1}\frac{(h\mathbf{m}-\mathbf{x})^{\boldsymbol\alpha}}{\boldsymbol\alpha!}\partial^{\boldsymbol\alpha}u(\mathbf{x})+\sum_{[\boldsymbol\alpha]=L}\frac{(h\mathbf{m}-\mathbf{x})^{\boldsymbol\alpha}}{\boldsymbol\alpha!}U_{\boldsymbol\alpha}(\mathbf{x},h\mathbf{m}),$$

we write the quasi-interpolant $\mathcal{M}_{h,\mathcal{D}}u$ in the form

$$\begin{aligned}\mathcal{M}_{h,\mathcal{D}}u(\mathbf{x})=&\,\mathcal{D}^{-n/2}\sum_{\mathbf{m}\in\mathbb{Z}^n}\sum_{[\boldsymbol\alpha]=0}^{L-1}\frac{(h\mathbf{m}-\mathbf{x})^{\boldsymbol\alpha}}{\boldsymbol\alpha!}\eta\Big(\frac{\mathbf{x}-h\mathbf{m}}{\sqrt{\mathcal{D}}h}\Big)\partial^{\boldsymbol\alpha}u(\mathbf{x})\\&+\mathcal{D}^{-n/2}\sum_{\mathbf{m}\in\mathbb{Z}^n}\sum_{[\boldsymbol\alpha]=L}\frac{(h\mathbf{m}-\mathbf{x})^{\boldsymbol\alpha}}{\boldsymbol\alpha!}\eta\Big(\frac{\mathbf{x}-h\mathbf{m}}{\sqrt{\mathcal{D}}h}\Big)U_{\boldsymbol\alpha}(\mathbf{x},h\mathbf{m}).\end{aligned}$$

Because of the absolute convergence of the sums (2.32) and (2.33) and the boundedness of

$$\begin{aligned}|U_{\boldsymbol\alpha}(\mathbf{x},\mathbf{y})|&=L\Big|\int_0^1 s^{L-1}\partial^{\boldsymbol\alpha}u(s\mathbf{x}+(1-s)\mathbf{y})\,ds\Big|\\&\le L\,\|\partial^{\boldsymbol\alpha}u\|_{L_\infty}\int_0^1 s^{L-1}ds=\|\partial^{\boldsymbol\alpha}u\|_{L_\infty},\end{aligned}\tag{2.35}$$

the order of summation can be changed. Hence we obtain, by using the definition (2.32) of $\sigma_{\boldsymbol\alpha}$,

$$\begin{aligned}\mathcal{M}_{h,\mathcal{D}}u(\mathbf{x})=&\sum_{[\boldsymbol\alpha]=0}^{L-1}(-\sqrt{\mathcal{D}}h)^{[\boldsymbol\alpha]}\,\sigma_{\boldsymbol\alpha}\Big(\frac{\mathbf{x}}{h},\eta,\mathcal{D}\Big)\frac{\partial^{\boldsymbol\alpha}u(\mathbf{x})}{\boldsymbol\alpha!}\\&+(-\sqrt{\mathcal{D}}h)^L\sum_{[\boldsymbol\alpha]=L}\mathcal{D}^{-n/2}\sum_{\mathbf{m}\in\mathbb{Z}^n}\Big(\frac{\mathbf{x}-h\mathbf{m}}{\sqrt{\mathcal{D}}h}\Big)^{\boldsymbol\alpha}\eta\Big(\frac{\mathbf{x}-h\mathbf{m}}{\sqrt{\mathcal{D}}h}\Big)\frac{U_{\boldsymbol\alpha}(\mathbf{x},h\mathbf{m})}{\boldsymbol\alpha!}.\end{aligned}\tag{2.36}$$

Now, (2.35) together with (2.33) implies

$$\begin{aligned}\Big|\sum_{[\boldsymbol\alpha]=L}\mathcal{D}^{-n/2}\sum_{\mathbf{m}\in\mathbb{Z}^n}\Big(\frac{\mathbf{x}-h\mathbf{m}}{\sqrt{\mathcal{D}}h}\Big)^{\boldsymbol\alpha}\eta\Big(\frac{\mathbf{x}-h\mathbf{m}}{\sqrt{\mathcal{D}}h}\Big)\frac{U_{\boldsymbol\alpha}(\mathbf{x},h\mathbf{m})}{\boldsymbol\alpha!}\Big|&\\\le\sum_{[\boldsymbol\alpha]=L}\frac{\|\partial^{\boldsymbol\alpha}u\|_{L_\infty}}{\boldsymbol\alpha!}\rho_{\boldsymbol\alpha}\Big(\frac{\mathbf{x}}{h},\eta,\mathcal{D}\Big),&\end{aligned}\tag{2.37}$$

and so we have proved the following.

LEMMA 2.7. *Suppose that η is subjected to (2.26) and let $u \in W_\infty^L(\mathbb{R}^n)$ for some positive $L < K - n$. Then*

$$\Big|\mathcal{M}_{h,\mathcal{D}}u(\mathbf{x}) - \sum_{[\boldsymbol{\alpha}]=0}^{L-1} (-\sqrt{\mathcal{D}}h)^{[\boldsymbol{\alpha}]}\sigma_{\boldsymbol{\alpha}}\Big(\frac{\mathbf{x}}{h}, \eta, \mathcal{D}\Big) \frac{\partial^{\boldsymbol{\alpha}} u(\mathbf{x})}{\boldsymbol{\alpha}!}\Big|$$
$$\leq (\sqrt{\mathcal{D}}h)^L \sum_{[\boldsymbol{\alpha}]=L} \frac{\|\partial^{\boldsymbol{\alpha}} u\|_{L_\infty}}{\boldsymbol{\alpha}!} \rho_{\boldsymbol{\alpha}}\Big(\frac{\mathbf{x}}{h}, \eta, \mathcal{D}\Big).$$

2.2.4. Relations to continuous convolutions. Let us consider the continuous counterpart of the semi-discrete convolution $\mathcal{M}_{h,\mathcal{D}}u$, the continuous convolution operator

$$\mathcal{C}_\delta u(\mathbf{x}) := \delta^{-n} \int_{\mathbb{R}^n} \eta\Big(\frac{\mathbf{x}-\mathbf{y}}{\delta}\Big)\, u(\mathbf{y})\, d\mathbf{y}\,, \quad \delta > 0\,. \tag{2.38}$$

The close connection of the operators $\mathcal{M}_{h,\mathcal{D}}$ and $\mathcal{C}_{\sqrt{\mathcal{D}}h}$ can be seen if we write

$$\mathcal{C}_{\sqrt{\mathcal{D}}h} u(\mathbf{x}) = (\sqrt{\mathcal{D}}h)^{-n} \sum_{\mathbf{m}\in\mathbb{Z}^n} \int_{h(\mathbf{m}+Q)} u(\mathbf{y})\, \eta\Big(\frac{\mathbf{x}-\mathbf{y}}{\sqrt{\mathcal{D}}h}\Big) d\mathbf{y}\,,$$

where Q denotes the cube $\big[-\frac{1}{2}, \frac{1}{2}\big]^n$. Comparing this with the definition (2.23) of $\mathcal{M}_{h,\mathcal{D}}u$, we see that the quasi-interpolant $\mathcal{M}_{h,\mathcal{D}}u(\mathbf{x})$ is, for a fixed $\mathbf{x}$, the simple midpoint cubature formula of the integral $\mathcal{C}_{\sqrt{\mathcal{D}}h}u(\mathbf{x})$, where the integrals

$$\int_{h(\mathbf{m}+Q)} u(\mathbf{y})\, \eta\Big(\frac{\mathbf{x}-\mathbf{y}}{\sqrt{\mathcal{D}}h}\Big) d\mathbf{y}$$

for all $\mathbf{m} \in \mathbb{Z}^n$ are replaced by

$$h^n\, u(h\mathbf{m})\, \eta\Big(\frac{\mathbf{x}-h\mathbf{m}}{\sqrt{\mathcal{D}}h}\Big).$$

However, standard estimates for cubature methods do not give useful bounds for the difference between the semi-discrete and continuous convolutions. For example, the midpoint rule is of second order (cf., e.g., [**23**]). Hence, we obtain

$$|\mathcal{M}_{\sqrt{\mathcal{D}}h}u(\mathbf{x}) - \mathcal{C}_{\sqrt{\mathcal{D}}h}u(\mathbf{x})| \leq ch^2 \sup_{\mathbf{y}\in R^n} \sum_{[\boldsymbol{\alpha}]=2} \Big|\partial_{\mathbf{y}}^{\boldsymbol{\alpha}}\Big(u(\mathbf{y})\eta\Big(\frac{\mathbf{x}-\mathbf{y}}{\sqrt{\mathcal{D}}h}\Big)\Big)\Big|$$

and the second derivatives yield the factor $(\mathcal{D}h^2)^{-1}$.

The results of the previous subsection help us to obtain better estimations of this difference. From Lemma 2.5, we derive

COROLLARY 2.8. *Let φ be a function as in Lemma 2.5. If u is subjected to the growth condition (2.28) and η satisfies the decay condition (2.26) with $K > L + n$, then*

$$\mathcal{M}_{h,\mathcal{D}}\, u(\mathbf{x}) - \mathcal{C}_{\sqrt{\mathcal{D}}h}u(\mathbf{x})$$
$$= (\sqrt{\mathcal{D}}h)^{-n} \lim_{\varepsilon\to 0} \sum_{\boldsymbol{\nu}\in\mathbb{Z}^n\setminus\{\mathbf{0}\}} \mathcal{F}\varphi(\varepsilon\boldsymbol{\nu}) \int_{\mathbb{R}^n} u(\mathbf{y})\eta\Big(\frac{\mathbf{x}-\mathbf{y}}{\sqrt{\mathcal{D}}h}\Big)\, \mathrm{e}^{-\frac{2\pi i}{h}\langle \mathbf{y}, \boldsymbol{\nu}\rangle}\, d\mathbf{y}\,.$$

PROOF. From the definition of the function $G_{\mathbf{x}}$ in Lemma 2.4, we see that $\mathcal{M}_{h,\mathcal{D}}u(\mathbf{x}) = G_{\mathbf{x}}(\mathbf{0})$ and therefore by Lemma 2.5

$$\mathcal{M}_{h,\mathcal{D}}u(\mathbf{x}) = (\sqrt{\mathcal{D}}h)^{-n} \lim_{\varepsilon\to 0} \sum_{\boldsymbol{\nu}\in\mathbb{Z}^n} \mathcal{F}\varphi(\varepsilon\boldsymbol{\nu}) \int_{\mathbb{R}^n} u(\mathbf{y})\eta\Big(\frac{\mathbf{x}-\mathbf{y}}{\sqrt{\mathcal{D}}h}\Big)\, \mathrm{e}^{-\frac{2\pi i}{h}\langle \mathbf{y},\boldsymbol{\nu}\rangle}\, d\mathbf{y}\,. \qquad \square$$

In the next theorem, we derive a series expansion into powers of $\sqrt{\mathcal{D}}h$ of the difference between the semi-discrete and continuous convolutions for sufficiently smooth functions u. To this end, we introduce the functions

$$\varepsilon_{\boldsymbol{\alpha}}(\mathbf{x},\eta,\mathcal{D}) := \sigma_{\boldsymbol{\alpha}}(\mathbf{x},\eta,\mathcal{D}) - \int_{\mathbb{R}^n} \mathbf{y}^{\boldsymbol{\alpha}}\eta(\mathbf{y})\,d\mathbf{y}\,. \tag{2.39}$$

THEOREM 2.9. *Suppose that η satisfies (2.26) with $K > L+n$, $L\in\mathbb{N}$. Then for any function $u \in W^L_\infty(\mathbb{R}^n)$*

$$|\mathcal{M}_{h,\mathcal{D}}u(\mathbf{x}) - \mathcal{C}_{\sqrt{\mathcal{D}}h}u(\mathbf{x})| \le \sum_{[\boldsymbol{\alpha}]=0}^{L-1} (\sqrt{\mathcal{D}}h)^{[\boldsymbol{\alpha}]}\, \|\varepsilon_{\boldsymbol{\alpha}}(\cdot,\eta,\mathcal{D})\|_{L_\infty} \frac{|\partial^{\boldsymbol{\alpha}}u(\mathbf{x})|}{\boldsymbol{\alpha}!}$$
$$+(\sqrt{\mathcal{D}}h)^L \sum_{[\boldsymbol{\alpha}]=L} \Big(\|\rho_{\boldsymbol{\alpha}}(\cdot,\eta,\mathcal{D})\|_{L_\infty} + \int_{\mathbb{R}^n} |\mathbf{y}^{\boldsymbol{\alpha}}\eta(\mathbf{y})|\,d\mathbf{y}\Big) \frac{\|\partial^{\boldsymbol{\alpha}}u\|_{L_\infty}}{\boldsymbol{\alpha}!}\,.$$

PROOF. From (2.36), we know that for $u \in W^L_\infty(\mathbb{R}^n)$

$$\mathcal{M}_{h,\mathcal{D}}u(\mathbf{x}) = \sum_{[\boldsymbol{\alpha}]=0}^{L-1} \frac{\partial^{\boldsymbol{\alpha}}u(\mathbf{x})}{\boldsymbol{\alpha}!}(-\sqrt{\mathcal{D}}h)^{[\boldsymbol{\alpha}]}\sigma_{\boldsymbol{\alpha}}\Big(\frac{\mathbf{x}}{h},\eta,\mathcal{D}\Big)$$
$$+ (-\sqrt{\mathcal{D}}h)^L \sum_{[\boldsymbol{\alpha}]=L} \mathcal{D}^{-n/2} \sum_{\mathbf{m}\in\mathbb{Z}^n} \Big(\frac{\mathbf{x}-h\mathbf{m}}{\sqrt{\mathcal{D}}h}\Big)^{\boldsymbol{\alpha}} \eta\Big(\frac{\mathbf{x}-h\mathbf{m}}{\sqrt{\mathcal{D}}h}\Big) \frac{U_{\boldsymbol{\alpha}}(\mathbf{x},h\mathbf{m})}{\boldsymbol{\alpha}!}\,,$$

and similarly, the Taylor expansion (2.1) leads to the representation

$$\begin{aligned}\mathcal{C}_\delta u(\mathbf{x}) &= \sum_{[\boldsymbol{\alpha}]=0}^{L-1} \frac{\partial^{\boldsymbol{\alpha}}u(\mathbf{x})}{\boldsymbol{\alpha}!}(-\delta)^{[\boldsymbol{\alpha}]} \int_{\mathbb{R}^n} \mathbf{y}^{\boldsymbol{\alpha}}\eta(\mathbf{y})\,d\mathbf{y}\\ &+ (-\delta)^L \sum_{[\boldsymbol{\alpha}]=L} \frac{1}{\boldsymbol{\alpha}!} \int_{\mathbb{R}^n} \mathbf{y}^{\boldsymbol{\alpha}}\eta(\mathbf{y})\, U_{\boldsymbol{\alpha}}(\mathbf{x},\mathbf{x}-\delta\mathbf{y})\,d\mathbf{y}\,.\end{aligned} \tag{2.40}$$

Setting $\delta = \sqrt{\mathcal{D}}h$ in (2.40), we obtain

$$\mathcal{M}_{h,\mathcal{D}}u(\mathbf{x}) - \mathcal{C}_{\sqrt{\mathcal{D}}h}u(\mathbf{x}) = \sum_{[\boldsymbol{\alpha}]=0}^{L-1} \frac{\partial^{\boldsymbol{\alpha}}u(\mathbf{x})}{\boldsymbol{\alpha}!}(-\sqrt{\mathcal{D}}h)^{[\boldsymbol{\alpha}]} \Big(\sigma_{\boldsymbol{\alpha}}\Big(\frac{\mathbf{x}}{h},\eta,\mathcal{D}\Big) - \int_{\mathbb{R}^n} \mathbf{y}^{\boldsymbol{\alpha}}\eta(\mathbf{y})\,d\mathbf{y}\Big)$$
$$+ (-\sqrt{\mathcal{D}}h)^L \sum_{[\boldsymbol{\alpha}]=L} \frac{1}{\boldsymbol{\alpha}!}\mathcal{D}^{-n/2} \sum_{\mathbf{m}\in\mathbb{Z}^n} \Big(\frac{\mathbf{x}-h\mathbf{m}}{\sqrt{\mathcal{D}}h}\Big)^{\boldsymbol{\alpha}} \eta\Big(\frac{\mathbf{x}-h\mathbf{m}}{\sqrt{\mathcal{D}}h}\Big) U_{\boldsymbol{\alpha}}(\mathbf{x},h\mathbf{m})$$
$$- (-\sqrt{\mathcal{D}}h)^L \sum_{[\boldsymbol{\alpha}]=L} \frac{1}{\boldsymbol{\alpha}!} \int_{\mathbb{R}^n} \mathbf{y}^{\boldsymbol{\alpha}}\eta(\mathbf{y})\, U_{\boldsymbol{\alpha}}(\mathbf{x},\mathbf{x}-\sqrt{\mathcal{D}}h\mathbf{y})\,d\mathbf{y}\,.$$

Now the assertion follows from (2.35). $\square$

We see that the quasi-interpolants $\mathcal{M}_{h,\mathcal{D}}u$, for a fixed $\mathcal{D}$ and any $u \in W^L_\infty(\mathbb{R}^n)$, approximate the continuous convolutions $\mathcal{C}_{\sqrt{\mathcal{D}}h}u$ with the order $\mathcal{O}(h^L)$ if and only if $\varepsilon_{\boldsymbol{\alpha}}(\mathbf{x},\eta,\mathcal{D}) = 0$ for all $[\boldsymbol{\alpha}] < L$. On the other hand, a good approximation can be obtained for all $\mathbf{x}$, if the norms $\|\varepsilon_{\boldsymbol{\alpha}}(\cdot,\eta,\mathcal{D})\|_{L_\infty}$ are small for all $[\boldsymbol{\alpha}] < L$.

We see from Corollary 2.8 that for all $\mathbf{x}$ and any function φ mentioned in Lemma 2.5

$$\varepsilon_{\boldsymbol{\alpha}}(\mathbf{x},\eta,\mathcal{D}) = \Big(\frac{i}{2\pi}\Big)^{[\boldsymbol{\alpha}]} \lim_{\varepsilon\to 0} \sum_{\boldsymbol{\nu}\in\mathbb{Z}^n\setminus\{\mathbf{0}\}} \mathcal{F}\varphi(\varepsilon\boldsymbol{\nu})\,\partial^{\boldsymbol{\alpha}}\mathcal{F}\eta(\sqrt{\mathcal{D}}\boldsymbol{\nu})\,\mathrm{e}^{2\pi i\langle\mathbf{x},\boldsymbol{\nu}\rangle}\,. \tag{2.41}$$

Thus we derive

LEMMA 2.10. *Suppose that the continuous function η satisfies the decay condition (2.26) and let $0 \le [\boldsymbol{\alpha}] \le K-n$. If $\{\partial^{\boldsymbol{\alpha}}\mathcal{F}\eta(\sqrt{\mathcal{D}}\,\cdot)\} \in \ell_1(\mathbb{Z}^n)$, then*

$$\begin{aligned}\|\varepsilon_{\boldsymbol{\alpha}}(\cdot,\eta,\mathcal{D})\|_{L_\infty} &= \Big\|\sigma_{\boldsymbol{\alpha}}(\cdot,\eta,\mathcal{D}) - \int_{\mathbb{R}^n} \mathbf{y}^{\boldsymbol{\alpha}}\eta(\mathbf{y})\,d\mathbf{y}\Big\|_{L_\infty}\\ &\le (2\pi)^{-[\boldsymbol{\alpha}]} \sum_{\boldsymbol{\nu}\in\mathbb{Z}^n\setminus\{\mathbf{0}\}} |\partial^{\boldsymbol{\alpha}}\mathcal{F}\eta(\sqrt{\mathcal{D}}\boldsymbol{\nu})|\,.\end{aligned}$$

The integral

$$\int_{\mathbb{R}^n} \mathbf{x}^{\boldsymbol{\alpha}}\,\eta(\mathbf{x})\,d\mathbf{x}\,, \quad \boldsymbol{\alpha}\in\mathbb{Z}^n_{\ge 0} \text{ a multi-index}\,, \tag{2.42}$$

is called the $\boldsymbol{\alpha}$-th *moment* of η.

2.2.5. Poisson's summation formula for $\sigma_{\boldsymbol{\alpha}}$. Formula (2.41) represents a weak form of Poisson's summation formula (2.17) applied to the function $\sigma_{\boldsymbol{\alpha}}$:

$$\mathcal{D}^{-n/2}\sum_{\mathbf{m}\in\mathbb{Z}^n}\Big(\frac{\mathbf{x}-\mathbf{m}}{\sqrt{\mathcal{D}}}\Big)^{\boldsymbol{\alpha}}\eta\Big(\frac{\mathbf{x}-\mathbf{m}}{\sqrt{\mathcal{D}}}\Big) = \Big(\frac{i}{2\pi}\Big)^{[\boldsymbol{\alpha}]}\sum_{\boldsymbol{\nu}\in\mathbb{Z}^n}\partial^{\boldsymbol{\alpha}}\mathcal{F}\eta(\sqrt{\mathcal{D}}\boldsymbol{\nu})\,\mathrm{e}^{2\pi i\langle\mathbf{x},\boldsymbol{\nu}\rangle}\,. \tag{2.43}$$

The next lemma gives a sufficient condition on η ensuring (2.43).

LEMMA 2.11. *Let $\mu = [n/2]+1$. Suppose that the derivatives $\partial^{\boldsymbol{\beta}}\eta$, $[\boldsymbol{\beta}] \le \mu$, exist and satisfy the decay condition*

$$|\partial^{\boldsymbol{\beta}}\eta(\mathbf{x})| \le A\,(1+|\mathbf{x}|)^{-K} \textit{ for all } \mathbf{x}\in\mathbb{R}^n$$

with constants $A>0$ and $K>n$. Then for all $\boldsymbol{\alpha}$, $[\boldsymbol{\alpha}] < K-n$, and all $\mathcal{D}>0$, the sequence of Fourier coefficients $\{\partial^{\boldsymbol{\alpha}}\mathcal{F}\eta(\sqrt{\mathcal{D}}\boldsymbol{\nu})\} \in \ell_1(\mathbb{Z}^n)$. Moreover, for any $\varepsilon>0$ there exists $\mathcal{D}_0>0$ such that for all $\mathcal{D}>\mathcal{D}_0$

$$\sum_{\boldsymbol{\nu}\in\mathbb{Z}^n\setminus\{\mathbf{0}\}} |\partial^{\boldsymbol{\alpha}}\mathcal{F}\eta(\sqrt{\mathcal{D}}\boldsymbol{\nu})| < \varepsilon\,. \tag{2.44}$$

PROOF. Fix a multi-index $\boldsymbol{\alpha}$ and consider the partial derivative $\partial^{\boldsymbol{\beta}}\sigma_{\boldsymbol{\alpha}}$ with $[\boldsymbol{\beta}] = \mu$, i.e.,

$$\partial^{\boldsymbol{\beta}}_{\mathbf{x}}\sigma_{\boldsymbol{\alpha}}(\mathbf{x},\eta,\mathcal{D}) = \mathcal{D}^{-\mu/2}\partial^{\boldsymbol{\beta}}_{\mathbf{y}}\mathcal{D}^{-n/2}\sum_{\mathbf{m}\in\mathbb{Z}^n}\Big(\mathbf{y}-\frac{\mathbf{m}}{\sqrt{\mathcal{D}}}\Big)^{\boldsymbol{\alpha}}\eta\Big(\mathbf{y}-\frac{\mathbf{m}}{\sqrt{\mathcal{D}}}\Big)\Big|_{\mathbf{y}=\mathbf{x}/\sqrt{\mathcal{D}}}\,.$$

Since the decay of η and its derivatives ensures the absolute convergence of all series, we can write

$$\partial^{\boldsymbol{\beta}}_{\mathbf{x}} \sigma_{\boldsymbol{\alpha}}(\mathbf{x}, \eta, \mathcal{D})$$
$$= \mathcal{D}^{-\mu/2} \mathcal{D}^{-n/2} \sum_{\mathbf{m}\in\mathbb{Z}^n} \sum_{\boldsymbol{\gamma}\le\boldsymbol{\beta}} \frac{\boldsymbol{\beta}!}{(\boldsymbol{\beta}-\boldsymbol{\gamma})!\,\boldsymbol{\beta}!}\, \partial^{\boldsymbol{\gamma}}_{\mathbf{y}} \Big(\mathbf{y} - \frac{\mathbf{m}}{\sqrt{\mathcal{D}}}\Big)^{\boldsymbol{\alpha}} \partial^{\boldsymbol{\beta}-\boldsymbol{\gamma}}_{\mathbf{y}} \eta\Big(\mathbf{y} - \frac{\mathbf{m}}{\sqrt{\mathcal{D}}}\Big)\Big|_{\mathbf{y}=\mathbf{x}/\sqrt{\mathcal{D}}} .$$

It is clear that $\partial^{\boldsymbol{\gamma}}_{\mathbf{y}} \mathbf{y}^{\boldsymbol{\alpha}} = 0$ if $\boldsymbol{\gamma} \not\le \boldsymbol{\alpha}$. We know from Lemma 2.6 that the functions

$$\mathcal{D}^{-n/2} \sum_{\mathbf{m}\in\mathbb{Z}^n} \Big(\frac{\mathbf{x}-\mathbf{m}}{\sqrt{\mathcal{D}}}\Big)^{\boldsymbol{\alpha}-\boldsymbol{\gamma}} \partial^{\boldsymbol{\beta}-\boldsymbol{\gamma}} \eta\Big(\frac{\mathbf{x}-\mathbf{m}}{\sqrt{\mathcal{D}}}\Big)$$

are uniformly bounded for all $\mathcal{D} \ge \mathcal{D}_0 > 0$. Consequently

$$\|\partial^{\boldsymbol{\beta}} \sigma_{\boldsymbol{\alpha}}(\,\cdot\,, \eta, \mathcal{D})\|_{L^\infty} \le c_{\boldsymbol{\beta}}\, \mathcal{D}^{-\mu/2} \qquad \text{for all } [\boldsymbol{\beta}] = \mu$$

with some constants $c_{\boldsymbol{\beta}}$ not depending on $\mathcal{D}$. On the other hand, the Fourier coefficients of $\partial^{\boldsymbol{\beta}} \sigma_{\boldsymbol{\alpha}}$ are $(2\pi i \boldsymbol{\nu})^{\boldsymbol{\beta}} \mathcal{F}\eta(\sqrt{\mathcal{D}}\boldsymbol{\nu})$; hence Parseval's equality gives

$$\sum_{[\boldsymbol{\beta}]=\mu} \sum_{\boldsymbol{\nu}\in\mathbb{Z}^n} (2\pi\boldsymbol{\nu})^{2\boldsymbol{\beta}} \, |\partial^{\boldsymbol{\alpha}} \mathcal{F}\eta(\sqrt{\mathcal{D}}\boldsymbol{\nu})|^2 = \sum_{[\boldsymbol{\beta}]=\mu} \|\partial^{\boldsymbol{\beta}} \sigma_{\boldsymbol{\alpha}}(\,\cdot\,, \eta, \mathcal{D})\|^2_{L^2([0,1]^n)} \le c^2\, \mathcal{D}^{-\mu} ,$$

where the constant $c > 0$ depends only on η, n, and μ. Now, we note that

$$\sum_{[\boldsymbol{\beta}]=\mu} (2\pi\boldsymbol{\nu})^{2\boldsymbol{\beta}} \ge c_1^2\, |\boldsymbol{\nu}|^{2\mu}$$

with some constant c_1, depending only on n and μ, which leads to

$$\sum_{\boldsymbol{\nu}\in\mathbb{Z}^n\setminus\{\mathbf{0}\}} |\partial^{\boldsymbol{\alpha}} \mathcal{F}\eta(\sqrt{\mathcal{D}}\boldsymbol{\nu})| \le \sum_{\boldsymbol{\nu}\in\mathbb{Z}^n\setminus\{\mathbf{0}\}} |\partial^{\boldsymbol{\alpha}} \mathcal{F}\eta(\sqrt{\mathcal{D}}\boldsymbol{\nu})| \Big(\sum_{[\boldsymbol{\beta}]=\mu} (2\pi\boldsymbol{\nu})^{2\boldsymbol{\beta}}\Big)^{1/2} \frac{|\boldsymbol{\nu}|^{-\mu}}{c_1}$$
$$\le \frac{1}{c_1} \Big(\sum_{\boldsymbol{\nu}\in\mathbb{Z}^n\setminus\{\mathbf{0}\}} |\partial^{\boldsymbol{\alpha}} \mathcal{F}\eta(\sqrt{\mathcal{D}}\boldsymbol{\nu})|^2 \sum_{[\boldsymbol{\beta}]=\mu} (2\pi\boldsymbol{\nu})^{2\boldsymbol{\beta}}\Big)^{1/2} \Big(\sum_{\boldsymbol{\nu}\in\mathbb{Z}^n\setminus\{\mathbf{0}\}} |\boldsymbol{\nu}|^{-2\mu}\Big)^{1/2}$$
$$\le \frac{c^{1/2}\, \mathcal{D}^{-\mu/2}}{c_1} \Big(\sum_{\boldsymbol{\nu}\in\mathbb{Z}^n\setminus\{\mathbf{0}\}} |\boldsymbol{\nu}|^{-2\mu}\Big)^{1/2} . \qquad \square$$

Since we are interested in generating functions subject to the inequality (2.44), we introduce the following extension of the *decay condition*:

CONDITION 2.12. Let $\mu = [n/2] + 1$ be the smallest integer greater than $n/2$. Suppose that $\partial^{\boldsymbol{\beta}}\eta$, $0 \le [\boldsymbol{\beta}] \le \mu$, are continuous and satisfy the decay condition: There exist constants $A > 0$ and $K > n$ such that

$$|\partial^{\boldsymbol{\beta}}\eta(\mathbf{x})| \le A\,(1+|\mathbf{x}|)^{-K} \text{ for all } \mathbf{x} \in \mathbb{R}^n . \tag{2.45}$$

COROLLARY 2.13. *Assume Condition 2.12. Then for any multi-index* $\boldsymbol{\alpha}$ *with* $[\boldsymbol{\alpha}] < K - n$

$$\varepsilon_{\boldsymbol{\alpha}}(\mathbf{x}, \eta, \mathcal{D}) = \Big(\frac{i}{2\pi}\Big)^{[\boldsymbol{\alpha}]} \sum_{\boldsymbol{\nu}\in\mathbb{Z}^n\setminus\{\mathbf{0}\}} \partial^{\boldsymbol{\alpha}} \mathcal{F}\eta(\sqrt{\mathcal{D}}\boldsymbol{\nu})\, \mathrm{e}^{2\pi i \langle \mathbf{x}, \boldsymbol{\nu}\rangle} , \tag{2.46}$$

and for any $\varepsilon > 0$ *there exists* $\mathcal{D}_0 > 0$ *such that for all* $\mathcal{D} > \mathcal{D}_0$

$$\|\varepsilon_{\boldsymbol{\alpha}}(\cdot, \eta, \mathcal{D})\|_{L_\infty} < \varepsilon .$$

COROLLARY 2.14. *If the generating function η fulfills Condition 2.12, then for any $\varepsilon > 0$, there exists $\mathcal{D} > 0$ such that*

$$|\mathcal{M}_{h,\mathcal{D}}u(\mathbf{x}) - \mathcal{C}_{\sqrt{\mathcal{D}}h}u(\mathbf{x})| \le \varepsilon \sum_{[\boldsymbol{\alpha}]=0}^{L-1} (\sqrt{\mathcal{D}}h)^{[\boldsymbol{\alpha}]} \frac{|\partial^{\boldsymbol{\alpha}} u(\mathbf{x})|}{\boldsymbol{\alpha}!}$$
$$+ (\sqrt{\mathcal{D}}h)^L \sum_{[\boldsymbol{\alpha}]=L} \Big(\|\rho_{\boldsymbol{\alpha}}(\cdot\,,\eta,\mathcal{D})\|_{L_\infty} + \int_{\mathbb{R}^n} |\mathbf{y}^{\boldsymbol{\alpha}}\eta(\mathbf{y})|\, d\mathbf{y}\Big) \frac{\|\partial^{\boldsymbol{\alpha}} u\|_{L_\infty}}{\boldsymbol{\alpha}!}$$

for all $u \in W_\infty^L(\mathbb{R}^n)$.

2.3. Pointwise error estimates for quasi-interpolation

2.3.1. Using the moments of η. Let us introduce another important condition for the generating function η.

CONDITION 2.15. We say that η satisfies the *moment condition of order* $N \in \mathbb{N}$ if

$$\int_{\mathbb{R}^n} \eta(\mathbf{x})\, d\mathbf{x} = 1\,, \quad \int_{\mathbb{R}^n} \mathbf{x}^{\boldsymbol{\alpha}}\eta(\mathbf{x})\, d\mathbf{x} = 0, \quad \forall \boldsymbol{\alpha}\,,\ 1 \le [\boldsymbol{\alpha}] < N\,. \tag{2.47}$$

Let us remark that the moment Condition 2.15 is equivalent to

$$\partial^{\boldsymbol{\alpha}} \mathcal{F}\eta(\mathbf{0}) = \delta_{0[\boldsymbol{\alpha}]}, \quad \forall \boldsymbol{\alpha}\,,\ 0 \le [\boldsymbol{\alpha}] < N\,, \tag{2.48}$$

where δ_{jk} denotes the Kronecker sign

$$\delta_{jk} = \begin{cases} 1, & j = k\,, \\ 0, & j \ne k\,. \end{cases}$$

Thus a function η satisfies the moment Condition 2.15 if the function $\mathcal{F}\eta(\boldsymbol{\lambda}) - 1$ has a zero of order N at the origin.

LEMMA 2.16. *If η satisfies the decay condition (2.45) with $K > N + n$ and the moment Condition 2.15 of order N, then for any $u \in W_\infty^N(\mathbb{R}^n)$*

$$\sup_{\mathbb{R}^n} \big|\mathcal{C}_\delta u(\mathbf{x}) - u(\mathbf{x})\big| \le \delta^N \sum_{[\boldsymbol{\alpha}]=N} \frac{\|\partial^{\boldsymbol{\alpha}} u\|_{L_\infty}}{\boldsymbol{\alpha}!} \int_{\mathbb{R}^n} |\mathbf{y}^{\boldsymbol{\alpha}}\eta(\mathbf{y})|\, d\mathbf{y}\,.$$

PROOF. The representation (2.40) with L replaced by N gives

$$\mathcal{C}_\delta u(\mathbf{x}) = \sum_{[\boldsymbol{\alpha}]=0}^{N-1} \frac{\partial^{\boldsymbol{\alpha}} u(\mathbf{x})}{\boldsymbol{\alpha}!} (-\delta)^{[\boldsymbol{\alpha}]} \int_{\mathbb{R}^n} \mathbf{y}^{\boldsymbol{\alpha}}\eta(\mathbf{y})\, d\mathbf{y}$$
$$+ (-\delta)^N \sum_{[\boldsymbol{\alpha}]=N} \frac{1}{\boldsymbol{\alpha}!} \int_{\mathbb{R}^n} \mathbf{y}^{\boldsymbol{\alpha}}\eta(\mathbf{y})\, U_{\boldsymbol{\alpha}}(\mathbf{x}, \mathbf{x} - \delta\mathbf{y})\, d\mathbf{y}\,,$$

and the first term on the right-hand side is equal to $u(\mathbf{x})$ in view of (2.47). □

Lemma 2.16 together with Corollary 2.14 implies the main feature of the approximate quasi-interpolation:

THEOREM 2.17. *Suppose that η satisfies the decay and moment Conditions 2.12 and 2.15 with $K > N+n$. Then for any $\varepsilon > 0$ there exists $\mathcal{D} > 0$ such that for any $u \in W_\infty^L(\mathbb{R}^n)$ the approximation error of the quasi-interpolation can be estimated pointwise by*

$$|u(\mathbf{x}) - \mathcal{M}_{h,\mathcal{D}}u(\mathbf{x})| \le c_\eta \, (\sqrt{\mathcal{D}}h)^M \, \|\nabla_M u\|_{L_\infty} + \varepsilon \sum_{k=0}^{M-1} (\sqrt{\mathcal{D}}h)^k |\nabla_k u(\mathbf{x})| \,, \tag{2.49}$$

where $M = \min(L, N)$ and the constant c_η does not depend on u, h, and $\mathcal{D}$.

Let us comment on the estimate (2.49). Representation (2.36) and relation (2.46) show that the quasi-interpolant $\mathcal{M}_{h,\mathcal{D}}u$ for $u \in W_\infty^L(\mathbb{R}^n)$ can be expanded under the conditions of Theorem 2.17 in the form

$$\begin{aligned}\mathcal{M}_{h,\mathcal{D}}u(\mathbf{x}) &= u(\mathbf{x}) + \sum_{[\boldsymbol{\alpha}]=0}^{M-1} (-\sqrt{\mathcal{D}}h)^{[\boldsymbol{\alpha}]} \, \frac{\partial^{\boldsymbol{\alpha}} u(\mathbf{x})}{\boldsymbol{\alpha}!} \, \varepsilon_{\boldsymbol{\alpha}}\Big(\frac{\mathbf{x}}{h}, \eta, \mathcal{D}\Big) \\ &+ \sum_{[\boldsymbol{\alpha}]=N}^{L-1} (-\sqrt{\mathcal{D}}h)^{[\boldsymbol{\alpha}]} \, \frac{\partial^{\boldsymbol{\alpha}} u(\mathbf{x})}{\boldsymbol{\alpha}!} \, \sigma_{\boldsymbol{\alpha}}\Big(\frac{\mathbf{x}}{h}, \eta, \mathcal{D}\Big) + R_{L,h}(\mathbf{x}) \,,\end{aligned} \tag{2.50}$$

where $M = \min(N, L)$ and the remainder $R_{L,h}$ is of the form

$$R_{L,h}(\mathbf{x}) = (-\sqrt{\mathcal{D}}h)^L \sum_{[\boldsymbol{\alpha}]=L} \mathcal{D}^{-n/2} \sum_{\mathbf{m}\in\mathbb{Z}^n} \Big(\frac{\mathbf{x}-h\mathbf{m}}{\sqrt{\mathcal{D}}h}\Big)^{\boldsymbol{\alpha}} \eta\Big(\frac{\mathbf{x}-h\mathbf{m}}{\sqrt{\mathcal{D}}h}\Big) \, \frac{U_{\boldsymbol{\alpha}}(\mathbf{x}, h\mathbf{m})}{\boldsymbol{\alpha}!} \tag{2.51}$$

(compare with (2.37)). If $L < N$, then the second sum in the expansion (2.50) is absent, of course. Since

$$\varepsilon_{\boldsymbol{\alpha}}\Big(\frac{\mathbf{x}}{h}, \eta, \mathcal{D}\Big) = \Big(\frac{i}{2\pi}\Big)^{[\boldsymbol{\alpha}]} \sum_{\boldsymbol{\nu}\in\mathbb{Z}^n\setminus\{\mathbf{0}\}} \partial^{\boldsymbol{\alpha}} \mathcal{F}\eta(\sqrt{\mathcal{D}}\boldsymbol{\nu}) \, \mathrm{e}^{\frac{2\pi i}{h}\langle \mathbf{x}, \boldsymbol{\nu}\rangle} \,,$$

the quasi-interpolant $\mathcal{M}_{h,\mathcal{D}}u$ differs from u by the sum of $M-1$ fast oscillating functions and a remainder of order $\mathcal{O}((\sqrt{\mathcal{D}}h)^M)$.

In view of Corollary 2.13, the maximum norms of the oscillating functions

$$\sum_{\boldsymbol{\nu}\in\mathbb{Z}^n\setminus\mathbf{0}} \partial^{\boldsymbol{\alpha}} \mathcal{F}\eta(\sqrt{\mathcal{D}}\boldsymbol{\nu}) \, \mathrm{e}^{\frac{2\pi i}{h}\langle \mathbf{x}, \boldsymbol{\nu}\rangle} \,, \quad 0 \le [\boldsymbol{\alpha}] < N \,,$$

can be made arbitrarily small if $\mathcal{D}$ is large enough. This implies that for fixed $\mathcal{D} > 0$ and $h \to 0$ the sum $\mathcal{M}_{h,\mathcal{D}}u$ does not converge to the function u. However, it approximates u with the order $\mathcal{O}((\sqrt{\mathcal{D}}h)^M)$ as long as the difference $\mathcal{M}_{h,\mathcal{D}}u(\mathbf{x}) - u(\mathbf{x})$ attains the *saturation error*, which has the representation

$$\sum_{j=0}^{M-1} \Big(\frac{i\sqrt{\mathcal{D}}h}{2\pi}\Big)^j \sum_{[\boldsymbol{\alpha}]=j} \frac{\partial^{\boldsymbol{\alpha}} u(\mathbf{x})}{\boldsymbol{\alpha}!} \sum_{\boldsymbol{\nu}\in\mathbb{Z}^n\setminus\mathbf{0}} \partial^{\boldsymbol{\alpha}} \mathcal{F}\eta(\sqrt{\mathcal{D}}\boldsymbol{\nu}) \, \mathrm{e}^{\frac{2\pi i}{h}\langle \mathbf{x}, \boldsymbol{\nu}\rangle} \,. \tag{2.52}$$

Since this error can be made smaller than any prescribed accuracy, for example, the accuracy of the computing system, the absence of convergence is not important in practical applications.

REMARK 2.18. Due to the terms $(\sqrt{\mathcal{D}}h)^k$ in the estimate of (2.49) the step h must be chosen such that $\sqrt{\mathcal{D}}h < 1$. Hence, for the practical application of approximate quasi-interpolation it is necessary, especially for the multi-dimensional case, to use generating functions subject to

$$\Big| \sum_{\boldsymbol{\nu}\in\mathbb{Z}^n\setminus\mathbf{0}} \partial^{\boldsymbol{\alpha}} \mathcal{F}\eta(\sqrt{\mathcal{D}}\boldsymbol{\nu})\, \mathrm{e}^{\frac{2\pi i}{h}\langle \mathbf{x},\boldsymbol{\nu}\rangle} \Big| \ll 1$$

even for relatively small $\mathcal{D}$, at least if $[\boldsymbol{\alpha}] = 0$.

2.3.2. Truncation of summation. Since, in general, the support of the generating function η of the quasi-interpolant $\mathcal{M}_{h,\mathcal{D}}u$ is the whole space, one has to truncate the summation in (2.23). Here we consider the proper truncation of the quasi-interpolant (2.23) such that the error estimates remain valid.

To be more specific, denoting the closed ball of radius κ centered at $\mathbf{x}$ by $B(\mathbf{x},\kappa) = \{\mathbf{y}\in\mathbb{R}^n : |\mathbf{y}-\mathbf{x}| \le \kappa\}$, we consider the truncated quasi-interpolant

$$\mathcal{M}^{(\kappa)}_{h,\mathcal{D}}u(\mathbf{x}) = \mathcal{D}^{-n/2} \sum_{h\mathbf{m}\in B(\mathbf{x},\kappa)} u(h\mathbf{m})\eta\Big(\frac{\mathbf{x}-h\mathbf{m}}{\sqrt{\mathcal{D}}h}\Big). \tag{2.53}$$

The difference to $\mathcal{M}_{h,\mathcal{D}}u$ can be estimated as follows:

$$\begin{aligned} |\mathcal{M}^{(\kappa)}_{h,\mathcal{D}}u(\mathbf{x}) - \mathcal{M}_{h,\mathcal{D}}u(\mathbf{x})| &\le \sup_{h\mathbf{m}\notin B(\mathbf{x},\kappa)} |u(h\mathbf{m})|\, \mathcal{D}^{-n/2} \sum_{h\mathbf{m}\notin B(\mathbf{x},\kappa)} \Big|\eta\Big(\frac{\mathbf{x}-h\mathbf{m}}{\sqrt{\mathcal{D}}h}\Big)\Big| \\ &\le g_{\mathcal{D}}(\kappa/h,\eta)\, \|u\|_{L_\infty}, \end{aligned} \tag{2.54}$$

where the non-negative function $g_{\mathcal{D}}(t,\eta)$ is defined by

$$g_{\mathcal{D}}(t,\eta) = \sup_{\mathbf{x}\in\mathbb{R}^n} \mathcal{D}^{-n/2} \sum_{|\mathbf{x}-\mathbf{m}|>t} \Big|\eta\Big(\frac{\mathbf{x}-\mathbf{m}}{\sqrt{\mathcal{D}}}\Big)\Big|. \tag{2.55}$$

For functions η satisfying the decay condition (2.26) we have

$$\begin{aligned} g_{\mathcal{D}}(t,\eta) &\le A \sup_{\mathbf{x}\in\mathbb{R}^n} \mathcal{D}^{-n/2} \sum_{|\mathbf{x}-\mathbf{m}|>t} \Big(1 + \frac{|\mathbf{x}-\mathbf{m}|}{\sqrt{\mathcal{D}}}\Big)^{-K} \\ &\le A\mathcal{D}^{(K-n)/2} \sup_{\mathbf{x}\in\mathbb{R}^n} \sum_{|\mathbf{x}-\mathbf{m}|>t} |\mathbf{x}-\mathbf{m}|^{-K} \le B\Big(\frac{t}{\sqrt{\mathcal{D}}}\Big)^{n-K} \end{aligned} \tag{2.56}$$

with a constant B depending on η and the space dimension n. Hence by (2.54), the truncation error is bounded by

$$|\mathcal{M}^{(\kappa)}_{h,\mathcal{D}}u(\mathbf{x}) - \mathcal{M}_{h,\mathcal{D}}u(\mathbf{x})| \le B\Big(\frac{\sqrt{\mathcal{D}}h}{\kappa}\Big)^{K-n} \|u\|_{L_\infty}. \tag{2.57}$$

Thus Theorem 2.17 leads immediately to the next corollary.

COROLLARY 2.19. *Suppose that η satisfies the decay and moment Conditions 2.12 and 2.15 with $K > N + n$. For any $\varepsilon > 0$, there exists $\mathcal{D} > 0$ such that for fixed $\kappa > 0$ and any $u \in W^L_\infty(\mathbb{R}^n)$*

$$\begin{aligned} |(I - \mathcal{M}^{(\kappa)}_{h,\mathcal{D}})u(\mathbf{x})| \le\; & c_\eta(\sqrt{\mathcal{D}}h)^M \big(\|\nabla_M u\|_{L_\infty} + \|u\|_{L_\infty}\big) \\ & + \varepsilon \sum_{k=0}^{M-1} (\sqrt{\mathcal{D}}h)^k |\nabla_k u(\mathbf{x})|, \end{aligned} \tag{2.58}$$

where $M = \min(L,N)$ and c_η depends on η and κ.

Another consequence of (2.54) can be obtained for the case that the parameter κ is proportional to h. Let $\kappa = ht$, $t > 0$. Then

$$\mathcal{M}^{(ht)}_{h,\mathcal{D}}u(\mathbf{x}) = \mathcal{D}^{-n/2} \sum_{|\mathbf{x}-h\mathbf{m}|<ht} u(h\mathbf{m})\eta\Big(\frac{\mathbf{x}-h\mathbf{m}}{\sqrt{\mathcal{D}}h}\Big), \tag{2.59}$$

i.e., the summation is restricted to a small number of terms not depending on h. We have

$$|\mathcal{M}^{(ht)}_{h,\mathcal{D}}u(\mathbf{x}) - \mathcal{M}_{h,\mathcal{D}}u(\mathbf{x})| \le g_{\mathcal{D}}(t,\eta)\,\|u\|_{L_\infty}\,,$$

and by choosing t sufficiently large, the error caused by the truncation of the summation is comparable with the saturation error of $\mathcal{M}_{h,\mathcal{D}}u$. For example, if

$$g_{\mathcal{D}}(t,\eta) \le \|\varepsilon_{\mathbf{0}}(\,\cdot\,,\eta,\mathcal{D})\|_{L_\infty}\,,$$

then

$$|(I - \mathcal{M}^{(ht)}_{h,\mathcal{D}})u(\mathbf{x})| \le |(I - \mathcal{M}_{h,\mathcal{D}})u(\mathbf{x})| + \|\varepsilon_{\mathbf{0}}(\,\cdot\,,\eta,\mathcal{D})\|_{L_\infty}\|u\|_{L_\infty}\,,$$

and we obtain

COROLLARY 2.20. *Suppose that η satisfies the decay and moment Conditions 2.12 and 2.15. For any $\varepsilon > 0$, there exist positive $\mathcal{D}$ and t such that for any $u \in W^L_\infty(\mathbb{R}^n)$*

$$\begin{aligned} |(I - \mathcal{M}^{(ht)}_{h,\mathcal{D}})u(\mathbf{x})| &\le c_\eta(\sqrt{\mathcal{D}}h)^M\,\|\nabla_M u\|_{L_\infty} \\ &+ \varepsilon\big(\|u\|_{L_\infty} + \sum_{k=1}^{M-1}(\sqrt{\mathcal{D}}h)^k|\nabla_k u(\mathbf{x})|\big)\,, \end{aligned} \tag{2.60}$$

where $M = \min(L,N)$ and c_η depends only on η.

REMARK 2.21. It follows from Corollary 2.20 that the computation of $\mathcal{M}_{h,\mathcal{D}}u(\mathbf{x})$ requires taking only the terms in (2.23) for which $|\mathbf{x}/h - \mathbf{m}| \le t$. Thus the number of summands is proportional to t^n and does not depend on the mesh size h.

The function $g_{\mathcal{D}}$ can be used to characterize the behavior of $\mathcal{M}_{h,\mathcal{D}}u$ outside the support of the function u.

LEMMA 2.22. *For any bounded function u vanishing outside some domain Ω and for $\mathbf{x} \in \mathbb{R}^n \setminus \overline{\Omega}$ we have*

$$|\mathcal{M}_{h,\mathcal{D}}u(\mathbf{x})| \le g_{\mathcal{D}}(h^{-1}\operatorname{dist}(\mathbf{x},\Omega),\eta)\,\sup_{\Omega}|u|\,.$$

2.3.3. Local estimate of the quasi-interpolation.

We have seen that $\mathcal{M}_{h,\mathcal{D}}$ and its properly truncated version $\mathcal{M}^{(t)}_{h,\mathcal{D}}$ provide similar approximation properties, i.e., the error estimate depends on the supremum of $\nabla_M u$ on the whole space. However, since $\mathcal{M}^{(t)}_{h,\mathcal{D}}u(\mathbf{x})$ depends only on the values $u(h\mathbf{m})$ with $|h\mathbf{m} - \mathbf{x}| \le ht$, this local procedure should enjoy a local error estimate.

Those estimates can be established if the truncation radius is slightly enlarged. For the generating function η satisfying the usual decay and moment conditions of order N, we introduce another monotone function:

$$r_{\mathcal{D}}(t,\eta) := \sup_{\mathbf{x}\in\mathbb{R}^n} \mathcal{D}^{-n/2} \sum_{|\mathbf{x}-\mathbf{m}|>t} |\mathbf{x}-\mathbf{m}|^{N-1}\Big|\eta\Big(\frac{\mathbf{x}-\mathbf{m}}{\sqrt{\mathcal{D}}}\Big)\Big|\,. \tag{2.61}$$

Obviously, $r_{\mathcal{D}}(t,\eta) = \mathcal{O}(t^{N+n-1-K})$ as $t \to \infty$.

Suppose that $u \in W_\infty^N(B(\mathbf{x}, ht))$ and use the Taylor expansion (2.1) of u in $B(\mathbf{x}, ht)$ to represent the truncated sum

(2.62)

$$\mathcal{M}_{h,\mathcal{D}}^{(t)} u(\mathbf{x}) = \sum_{[\boldsymbol{\alpha}]=0}^{N-1} (-\sqrt{\mathcal{D}}h)^{[\boldsymbol{\alpha}]} \frac{\partial^{\boldsymbol{\alpha}} u(\mathbf{x})}{\boldsymbol{\alpha}!} \mathcal{D}^{-n/2} \sum_{h\mathbf{m}\in B(\mathbf{x},ht)} \Big(\frac{\mathbf{x}-h\mathbf{m}}{\sqrt{\mathcal{D}}h}\Big)^{\boldsymbol{\alpha}} \eta\Big(\frac{\mathbf{x}-h\mathbf{m}}{\sqrt{\mathcal{D}}h}\Big)$$

$$+ (-\sqrt{\mathcal{D}}h)^N \sum_{[\boldsymbol{\alpha}]=N} \frac{\mathcal{D}^{-n/2}}{\boldsymbol{\alpha}!} \sum_{h\mathbf{m}\in B(\mathbf{x},ht)} \Big(\frac{\mathbf{x}-h\mathbf{m}}{\sqrt{\mathcal{D}}h}\Big)^{\boldsymbol{\alpha}} \eta\Big(\frac{\mathbf{x}-h\mathbf{m}}{\sqrt{\mathcal{D}}h}\Big) U_{\boldsymbol{\alpha}}(\mathbf{x}, h\mathbf{m}).$$

By (2.39) and (2.47) we can estimate

$$(\sqrt{\mathcal{D}}h)^{[\boldsymbol{\alpha}]} \Big| \mathcal{D}^{-n/2} \sum_{h\mathbf{m}\in B(\mathbf{x},ht)} \Big(\frac{\mathbf{x}-h\mathbf{m}}{\sqrt{\mathcal{D}}h}\Big)^{\boldsymbol{\alpha}} \eta\Big(\frac{\mathbf{x}-h\mathbf{m}}{\sqrt{\mathcal{D}}h}\Big) - \delta_{0[\boldsymbol{\alpha}]} \Big|$$

$$\leq (\sqrt{\mathcal{D}}h)^{[\boldsymbol{\alpha}]} |\varepsilon_{\boldsymbol{\alpha}}(\mathbf{x}, \eta, \mathcal{D})| + h^{[\boldsymbol{\alpha}]} \mathcal{D}^{-n/2} \sum_{|\mathbf{x}-\mathbf{m}|>t} \Big|\frac{\mathbf{x}-h\mathbf{m}}{h}\Big|^{[\boldsymbol{\alpha}]} \Big|\eta\Big(\frac{\mathbf{x}-h\mathbf{m}}{\sqrt{\mathcal{D}}h}\Big)\Big|$$

$$\leq (\sqrt{\mathcal{D}}h)^{[\boldsymbol{\alpha}]} \|\varepsilon_{\boldsymbol{\alpha}}(\,\cdot\,, \eta, \mathcal{D})\|_{L_\infty} + h^{[\boldsymbol{\alpha}]} t^{[\boldsymbol{\alpha}]+1-N} r_{\mathcal{D}}(t, \eta).$$

Now we define the truncation radius $\kappa > 0$ satisfying

$$\kappa^{[\boldsymbol{\alpha}]+1-N}\, r_{\mathcal{D}}(\kappa, \eta) \leq \mathcal{D}^{[\boldsymbol{\alpha}]/2} \|\varepsilon_{\boldsymbol{\alpha}}(\,\cdot\,, \eta, \mathcal{D})\|_{L_\infty}. \tag{2.63}$$

The existence of a bounded κ follows from $t^{[\boldsymbol{\alpha}]+1-N} r_{\mathcal{D}}(t, \eta) = \mathcal{O}(t^{[\boldsymbol{\alpha}]+n-K}) \to 0$ for all $[\boldsymbol{\alpha}] < N$. Since, obviously,

$$g_{\mathcal{D}}(\kappa, \eta) \leq \|\varepsilon_{\mathbf{0}}(\,\cdot\,, \eta, \mathcal{D})\|_{L_\infty}, \tag{2.64}$$

where $g_{\mathcal{D}}$ is defined in (2.55), the parameter κ can be used to control the error estimates for quasi-interpolation in domains. We will discuss this problem in Chapter 9.

To proceed with the estimation, we note that (2.63) implies

$$(\sqrt{\mathcal{D}}h)^{[\boldsymbol{\alpha}]} \Big| \mathcal{D}^{-n/2} \sum_{h\mathbf{m}\in B(\mathbf{x},h\kappa)} \Big(\frac{\mathbf{x}-h\mathbf{m}}{\sqrt{\mathcal{D}}h}\Big)^{\boldsymbol{\alpha}} \eta\Big(\frac{\mathbf{x}-h\mathbf{m}}{\sqrt{\mathcal{D}}h}\Big) - \delta_{0[\boldsymbol{\alpha}]} \Big|$$

$$\leq 2(\sqrt{\mathcal{D}}h)^{[\boldsymbol{\alpha}]} \|\varepsilon_{\boldsymbol{\alpha}}(\,\cdot\,, \eta, \mathcal{D})\|_{L_\infty},$$

which gives the following.

LEMMA 2.23. *Suppose that η satisfies the decay and moment Conditions* 2.12 *and* 2.15 *with $K > N + n$ and the conditions of Lemma* 2.11, *and let $\kappa > 0$ be such that* (2.63) *holds. If u is continuously differentiable in $B(\mathbf{x}, h\kappa)$ up to order N, then*

$$|(I - \mathcal{M}_{h,\mathcal{D}}^{(\kappa)}) u(\mathbf{x})| \leq (\sqrt{\mathcal{D}}h)^N \sum_{[\boldsymbol{\alpha}]=N} \frac{\|\rho_{\boldsymbol{\alpha}}(\cdot\,, \eta, \mathcal{D})\|_{L_\infty}}{\boldsymbol{\alpha}!} \|\partial^{\boldsymbol{\alpha}} u\|_{L_\infty(B(\mathbf{x},h\kappa))}$$
$$+ 2 \sum_{[\boldsymbol{\alpha}]=0}^{N-1} (\sqrt{\mathcal{D}}h)^{[\boldsymbol{\alpha}]} \frac{|\partial^{\boldsymbol{\alpha}} u(\mathbf{x})|}{\boldsymbol{\alpha}!} \|\varepsilon_{\boldsymbol{\alpha}}(\cdot\,, \eta, \mathcal{D})\|_{L_\infty}. \tag{2.65}$$

Because of

$$|(\mathcal{M}_{h,\mathcal{D}} - \mathcal{M}_{h,\mathcal{D}}^{(\kappa)}) u(\mathbf{x})| \leq \|\varepsilon_{\mathbf{0}}(\cdot\,, \eta, \mathcal{D})\|_{L_\infty} \|u\|_{L_\infty},$$

which is a consequence of (2.64), the following local estimate can be established.

COROLLARY 2.24. *Under the conditions of Lemma 2.23 for any $\varepsilon > 0$ there exists positive $\mathcal{D}$ and κ such that for any $u \in W_\infty^L(B(\mathbf{x}, h\kappa))$*

$$\begin{aligned}|(I - \mathcal{M}_{h,\mathcal{D}})u(\mathbf{x})| \le & c_\eta(\sqrt{\mathcal{D}}h)^M \, \|\nabla_M u\|_{L_\infty(B(\mathbf{x},h\kappa))} \\ & + \varepsilon\Big(\|u\|_{L_\infty} + \sum_{k=1}^{M-1} (\sqrt{\mathcal{D}}h)^k |\nabla_k u(\mathbf{x})|\Big),\end{aligned}$$

with $M = \min(L, N)$ and the constant c_η depending only on η. Moreover, if $u \in W_\infty^L(\Omega)$ in some domain $\Omega \subset \mathbb{R}^n$, then

$$\begin{aligned}|(I - \mathcal{M}_{h,\mathcal{D}})u(\mathbf{x})| \le & c_\eta(\sqrt{\mathcal{D}}h)^M \, \|\nabla_M u\|_{L_\infty(\Omega)} \\ & + \varepsilon\Big(\|u\|_{L_\infty(\Omega)} + \sum_{k=1}^{M-1} (\sqrt{\mathcal{D}}h)^k |\nabla_k u(\mathbf{x})|\Big)\end{aligned} \tag{2.66}$$

for all $\mathbf{x} \in \Omega_{\kappa h}$, where we use the notation

$$\Omega_\tau = \{\mathbf{x} : B(\mathbf{x}, \tau) \subset \Omega\}. \tag{2.67}$$

2.3.4. Hölder continuous functions. The proposed quasi-interpolation formula approximates non-smooth functions as well. The Figs. 2.1 and 2.2 depict the difference between the function $u = |x|^{1/2}$ and its quasi-interpolant $\mathcal{M}_{h,2}u$ with the Gaussian basis function (cf. (1.7)). We see that the error behaves near $x = 0$ like $h^{1/2}$.

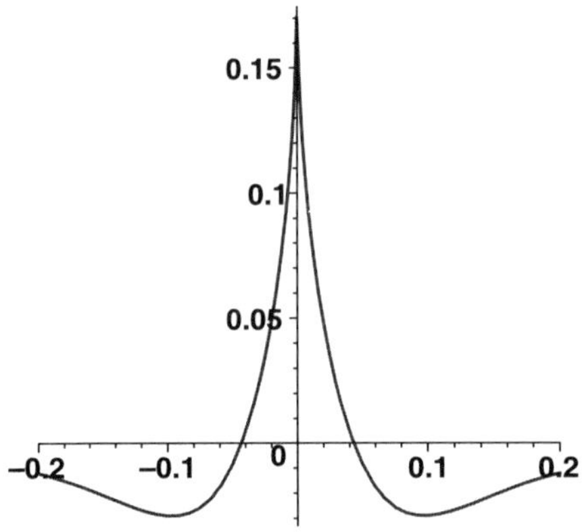

FIGURE 2.1. Error for $h = 0.08$

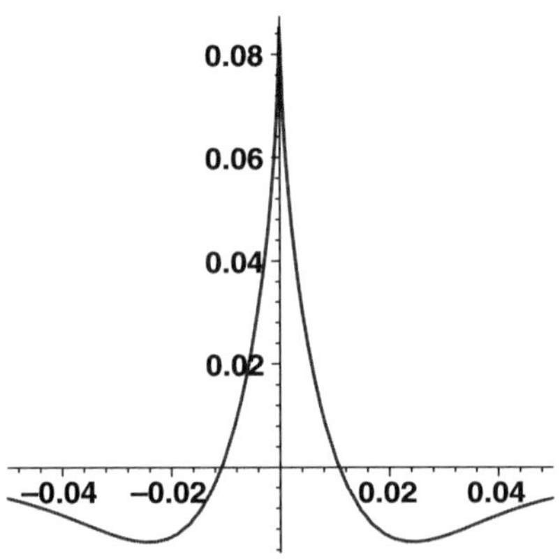

FIGURE 2.2. Error for $h = 0.02$

Such convergence behavior can be established also in the general case. Let, for example, $u \in W_\infty^{L+\gamma}(\mathbb{R}^n)$, i.e.,

$$\frac{\|\partial^{\boldsymbol{\alpha}} u(\cdot + \mathbf{x}) - \partial^{\boldsymbol{\alpha}} u(\cdot)\|_{L_\infty}}{|\mathbf{x}|^\gamma} \le c_{\boldsymbol{\alpha}} \quad \text{for all } \mathbf{x} \in \mathbb{R}^n, \quad [\boldsymbol{\alpha}] = L,$$

with some constants $c_{\boldsymbol{\alpha}}$ and $L < N$.

THEOREM 2.25. *If η is as in Theorem 2.17, then the estimate*

$$|u(\mathbf{x}) - \mathcal{M}_{h,\mathcal{D}}u(\mathbf{x})| \le c_{L,\gamma}\,(\sqrt{\mathcal{D}}h)^{L+\gamma} \sum_{[\boldsymbol{\alpha}]=L} \frac{c_{\boldsymbol{\alpha}}\,\rho_{\boldsymbol{\alpha},\gamma}}{\alpha!} + \sum_{[\boldsymbol{\alpha}]=0}^{L} \frac{|\partial^{\boldsymbol{\alpha}} u(\mathbf{x})|}{\boldsymbol{\alpha}!}(\sqrt{\mathcal{D}}h)^{[\boldsymbol{\alpha}]}\,\varepsilon_{\boldsymbol{\alpha}}(\eta,\mathcal{D})$$

holds, where

$$c_{L,\gamma} = L\frac{\Gamma(L)\Gamma(\gamma+1)}{\Gamma(L+\gamma+1)}$$

and

$$\rho_{\boldsymbol{\alpha},\gamma} = \Big\|\mathcal{D}^{-n/2}\sum_{\mathbf{m}\in\mathbb{Z}^n}\Big|\frac{\cdot-\mathbf{m}}{\sqrt{\mathcal{D}}}\Big|^{\gamma}\Big|\Big(\frac{\cdot-\mathbf{m}}{\sqrt{\mathcal{D}}}\Big)^{\boldsymbol{\alpha}}\eta\Big(\frac{\cdot-\mathbf{m}}{\sqrt{\mathcal{D}}}\Big)\Big|\Big\|_{L_\infty}.$$

PROOF. From

$$U_{\boldsymbol{\alpha}}(\mathbf{x},h\mathbf{m}) - \partial^{\boldsymbol{\alpha}}u(\mathbf{x}) = L\int_0^1 s^{L-1}\big(\partial^{\boldsymbol{\alpha}}u(s\mathbf{x}+(1-s)h\mathbf{m}) - \partial^{\boldsymbol{\alpha}}u(\mathbf{x})\big)ds\,,$$

one obtains the estimate

$$\begin{aligned}
&\big|U_{\boldsymbol{\alpha}}(\mathbf{x},h\mathbf{m}) - \partial^{\boldsymbol{\alpha}}u(\mathbf{x})\big|\\
&\quad\le |\mathbf{x}-h\mathbf{m}|^{\gamma}L\int_0^1 s^{L-1}(1-s)^{\gamma}\frac{|\partial^{\boldsymbol{\alpha}}u(s\mathbf{x}+(1-s)h\mathbf{m}) - \partial^{\boldsymbol{\alpha}}u(\mathbf{x})|}{|s\mathbf{x}+(1-s)h\mathbf{m}-\mathbf{x}|}\,ds\\
&\quad\le |\mathbf{x}-h\mathbf{m}|^{\gamma}c_{\boldsymbol{\alpha}}L\int_0^1 s^{L-1}(1-s)^{\gamma}\,ds = c_{\boldsymbol{\alpha}}\,L\frac{\Gamma(L)\Gamma(\gamma+1)}{\Gamma(L+\gamma+1)}\,|\mathbf{x}-h\mathbf{m}|^{\gamma}.
\end{aligned}$$

Hence using (2.51), we derive

$$\begin{aligned}
&\Big|R_{L,h}(\mathbf{x}) - (-\sqrt{\mathcal{D}}h)^L\sum_{[\boldsymbol{\alpha}]=L}\frac{\partial^{\boldsymbol{\alpha}}u(\mathbf{x})}{\boldsymbol{\alpha}!}\sigma_{\boldsymbol{\alpha}}\Big(\frac{\mathbf{x}}{h},\eta,\mathcal{D}\Big)\Big|\\
&\le c_{L,\gamma}\,(\sqrt{\mathcal{D}}h)^{L+\gamma}\frac{c_{\boldsymbol{\alpha}}}{\alpha!}\mathcal{D}^{-n/2}\sum_{\mathbf{m}\in\mathbb{Z}^n}\Big|\frac{\mathbf{x}-h\mathbf{m}}{\sqrt{\mathcal{D}}h}\Big|^{\gamma}\Big|\Big(\frac{\mathbf{x}-h\mathbf{m}}{\sqrt{\mathcal{D}}h}\Big)^{\boldsymbol{\alpha}}\eta\Big(\frac{\mathbf{x}-h\mathbf{m}}{\sqrt{\mathcal{D}}h}\Big)\Big|. \qquad\square
\end{aligned}$$

REMARK 2.26. If the bounded function u satisfies

$$|u(\mathbf{x}+\mathbf{y}) - u(\mathbf{x})| \le C\,|\mathbf{y}|^{\gamma}$$

for some $0<\gamma\le 1$ and all $\mathbf{x}\in\mathbb{R}^n$, then

$$|u(\mathbf{x}) - \mathcal{M}_{h,\mathcal{D}}u(\mathbf{x})| \le c_{\gamma}\,(\sqrt{\mathcal{D}}h)^{\gamma}C + \varepsilon_{\mathbf{0}}(\eta,\mathcal{D})|u(\mathbf{x})|\,.$$

The following sufficient assumption on η, which ensures that $\varepsilon_{\mathbf{0}}(\eta,\mathcal{D})<\varepsilon$ for prescribed $\varepsilon>0$ and sufficiently large $\mathcal{D}$, can be easily derived from Corollary 2.13.

The generating function η has to satisfy

$$|\partial^{\boldsymbol{\beta}}\eta(\mathbf{x})|(1+|\mathbf{x}|)^K < \infty\,,\quad \mathbf{x}\in\mathbb{R}^n\,,$$

for some $K>n$ and all $0\le[\boldsymbol{\beta}]\le\mu$, where μ is the smallest integer greater than $n/2$, and $\mathcal{F}\eta(\mathbf{0})=1$.

2.3.5. Approximation of derivatives. If the derivative $\partial^{\boldsymbol\beta}\eta$ exists and satisfies the decay condition (2.45), then

$$\partial^{\boldsymbol\beta}\mathcal{M}_{h,\mathcal{D}}u(\mathbf{x}) = \mathcal{D}^{-n/2}(\sqrt{\mathcal{D}}h)^{-[\boldsymbol\beta]}\sum_{\mathbf{m}\in\mathbb{Z}^n} u(h\mathbf{m})\,\partial^{\boldsymbol\beta}\eta\Big(\frac{\mathbf{x}-h\mathbf{m}}{\sqrt{\mathcal{D}}\,h}\Big).$$

We note that the continuous convolution satisfies

$$\begin{aligned}\partial^{\boldsymbol\beta}\mathcal{C}_{\sqrt{\mathcal{D}}h}u(\mathbf{x}) &= (\sqrt{\mathcal{D}}h)^{-n-[\boldsymbol\beta]}\int_{\mathbb{R}^n} u(\mathbf{y})\,\partial^{\boldsymbol\beta}\eta\Big(\frac{\mathbf{x}-\mathbf{y}}{\sqrt{\mathcal{D}}h}\Big)d\mathbf{y}\\ &= (\sqrt{\mathcal{D}}h)^{-n}\int_{\mathbb{R}^n}\partial^{\boldsymbol\beta}u(\mathbf{y})\,\eta\Big(\frac{\mathbf{x}-\mathbf{y}}{\sqrt{\mathcal{D}}h}\Big)\,d\mathbf{y} = \mathcal{C}_{\sqrt{\mathcal{D}}h}\partial^{\boldsymbol\beta}u(\mathbf{x}).\end{aligned} \tag{2.68}$$

Proceeding as in the proof of Theorem 2.9, we obtain

$$\begin{aligned}&(\sqrt{\mathcal{D}}h)^{[\boldsymbol\beta]}\,\big|\partial^{\boldsymbol\beta}\mathcal{M}_{h,\mathcal{D}}u(\mathbf{x}) - \partial^{\boldsymbol\beta}\mathcal{C}_{\sqrt{\mathcal{D}}h}u(\mathbf{x})\big|\\ &\quad\le (\sqrt{\mathcal{D}}h)^{L}\sum_{[\boldsymbol\alpha]=L}\Big(\big\|\rho_{\boldsymbol\alpha}(\cdot\,,\partial^{\boldsymbol\beta}\eta,\mathcal{D})\big\|_{L_\infty} + \int_{\mathbb{R}^n}\big|\mathbf{y}^{\boldsymbol\alpha}\,\partial^{\boldsymbol\beta}\eta(\mathbf{y})\big|\,d\mathbf{y}\Big)\,\frac{\|\partial^{\boldsymbol\alpha}u\|_{L_\infty}}{\boldsymbol\alpha!}\\ &\qquad + \sum_{[\boldsymbol\alpha]=0}^{L-1}(-\sqrt{\mathcal{D}}h)^{[\boldsymbol\alpha]}\Big|\sigma_{\boldsymbol\alpha}\Big(\frac{\mathbf{x}}{h},\partial^{\boldsymbol\beta}\eta,\mathcal{D}\Big) - \int_{\mathbb{R}^n}\mathbf{y}^{\boldsymbol\alpha}\,\partial^{\boldsymbol\beta}\eta(\mathbf{y})\,d\mathbf{y}\Big|\,\frac{|\partial^{\boldsymbol\alpha}u(\mathbf{x})|}{\boldsymbol\alpha!}.\end{aligned}$$

If η satisfies the moment Condition 2.15 of order N, then in view of (2.68) and Lemma 2.16,

$$\big|\partial^{\boldsymbol\beta}\mathcal{C}_{\sqrt{\mathcal{D}}h}u(\mathbf{x}) - \partial^{\boldsymbol\beta}u(\mathbf{x})\big| \le \sum_{[\boldsymbol\alpha]=N}\frac{\|\partial^{\boldsymbol\alpha+\boldsymbol\beta}u\|_{L_\infty}}{\boldsymbol\alpha!}\int_{\mathbb{R}^n}\big|\mathbf{y}^{\boldsymbol\alpha}\eta(\mathbf{y})\big|\,d\mathbf{y},$$

and we obtain the estimate

$$\begin{aligned}&(\sqrt{\mathcal{D}}h)^{[\boldsymbol\beta]}\big|\partial^{\boldsymbol\beta}\mathcal{M}_{h,\mathcal{D}}u(\mathbf{x}) - \partial^{\boldsymbol\beta}u(\mathbf{x})\big|\\ &\quad\le (\sqrt{\mathcal{D}}h)^{N+[\boldsymbol\beta]}\sum_{[\boldsymbol\alpha]=N}\frac{\|\partial^{\boldsymbol\alpha+\boldsymbol\beta}u\|_{L_\infty}}{\boldsymbol\alpha!}\int_{\mathbb{R}^n}\big|\mathbf{y}^{\boldsymbol\alpha}\eta(\mathbf{y})\big|\,d\mathbf{y}\\ &\qquad + (\sqrt{\mathcal{D}}h)^{L}\sum_{[\boldsymbol\alpha]=L}\Big(\big\|\rho_{\boldsymbol\alpha}(\cdot\,,\eta,\mathcal{D})\big\|_{L_\infty} + \int_{\mathbb{R}^n}\big|\mathbf{y}^{\boldsymbol\alpha}\,\partial^{\boldsymbol\beta}\eta(\mathbf{y})\big|\,d\mathbf{y}\Big)\,\frac{\|\partial^{\boldsymbol\alpha}u\|_{L_\infty}}{\boldsymbol\alpha!}\\ &\qquad + \sum_{[\boldsymbol\alpha]=0}^{L-1}(\sqrt{\mathcal{D}}h)^{[\boldsymbol\alpha]}\Big|\sigma_{\boldsymbol\alpha}\Big(\frac{\mathbf{x}}{h},\partial^{\boldsymbol\beta}\eta,\mathcal{D}\Big) - \int_{\mathbb{R}^n}\mathbf{y}^{\boldsymbol\alpha}\,\partial^{\boldsymbol\beta}\eta(\mathbf{y})\,d\mathbf{y}\Big|\,\frac{|\partial^{\boldsymbol\alpha}u(\mathbf{x})|}{\boldsymbol\alpha!}.\end{aligned}$$

Hence we have proved

THEOREM 2.27. *Suppose that η satisfies* (2.45), (2.47), *and suppose that the partial derivative $\partial^{\boldsymbol\beta}\eta$ satisfies the conditions of Lemma* 2.11. *Then for any $\varepsilon > 0$,* *there exists $\mathcal{D} > 0$ such that for any $u \in W^L_\infty(\mathbb{R}^n)$, $L \geq N + [\boldsymbol{\beta}]$,*

$$\begin{aligned}
&\big|\partial^{\boldsymbol{\beta}} \mathcal{M}_{h,\mathcal{D}}u(\mathbf{x}) - \partial^{\boldsymbol{\beta}}u(\mathbf{x})\big| \leq (\sqrt{\mathcal{D}}h)^N \sum_{[\boldsymbol{\alpha}]=N} \frac{\|\partial^{\boldsymbol{\alpha}+\boldsymbol{\beta}}u\|_{L_\infty}}{\boldsymbol{\alpha}!} \int_{\mathbb{R}^n} \big|\mathbf{y}^{\boldsymbol{\alpha}}\eta(\mathbf{y})\big|\, d\mathbf{y} \\
&+ (\sqrt{\mathcal{D}}h)^{L-[\boldsymbol{\beta}]} \sum_{[\boldsymbol{\alpha}]=L} \Big(\big\|\rho_{\boldsymbol{\alpha}}(\cdot\,, \partial^{\boldsymbol{\beta}}\eta, \mathcal{D})\big\|_{L_\infty} + \int_{\mathbb{R}^n} \big|\mathbf{y}^{\boldsymbol{\alpha}}\, \partial^{\boldsymbol{\beta}}\eta(\mathbf{y})\big|\, d\mathbf{y}\Big) \frac{\|\partial^{\boldsymbol{\alpha}}u\|_{L_\infty}}{\boldsymbol{\alpha}!} \\
&+ \varepsilon \sum_{[\boldsymbol{\alpha}]=0}^{L-1} (\sqrt{\mathcal{D}}h)^{[\boldsymbol{\alpha}]-[\boldsymbol{\beta}]} \frac{|\partial^{\boldsymbol{\alpha}}u(\mathbf{x})|}{\boldsymbol{\alpha}!}\,.
\end{aligned}$$

2.4. L_p-estimates of the quasi-interpolation error

To justify applications of approximate approximations in numerical methods, it is important to study approximation properties of approximate quasi-interpolants also in integral norms. In this section we concentrate on estimates in the norm of the space $L_p = L_p(\mathbb{R}^n)$, $1 \leq p < \infty$.

2.4.1. Formulation of the result. We use the representation of the approximate quasi-interpolant

$$\begin{aligned}
\mathcal{M}_{h,\mathcal{D}}u(\mathbf{x}) \;-\; u(\mathbf{x}) &= \sum_{[\boldsymbol{\alpha}]=0}^{M-1} (-\sqrt{\mathcal{D}}h)^{[\boldsymbol{\alpha}]} \frac{\partial^{\boldsymbol{\alpha}}u(\mathbf{x})}{\boldsymbol{\alpha}!}\, \varepsilon_{\boldsymbol{\alpha}}\Big(\frac{\mathbf{x}}{h}, \eta, \mathcal{D}\Big) \\
&+ \sum_{[\boldsymbol{\alpha}]=N}^{L-1} (-\sqrt{\mathcal{D}}h)^{[\boldsymbol{\alpha}]} \frac{\partial^{\boldsymbol{\alpha}}u(\mathbf{x})}{\boldsymbol{\alpha}!}\, \sigma_{\boldsymbol{\alpha}}\Big(\frac{\mathbf{x}}{h}, \eta, \mathcal{D}\Big) + R_{L,h}(\mathbf{x})\,,
\end{aligned} \tag{2.69}$$

which is valid for any sufficiently smooth function u (see (2.50)). It is supposed here that η is subject to the decay Condition 2.12 with $K > \max(L, N) + n$ and the moment Condition 2.15 of order N. The integer M in (2.69) is given as $M = \min(N, L)$ and the remainder $R_{L,h}$ is defined by (2.51).

Under the assumption that $\partial^{\boldsymbol{\alpha}}u \in L_p(\mathbb{R}^n)$, $0 \leq [\boldsymbol{\alpha}] < L$, we obtain, by using Lemma 2.6,

$$\begin{aligned}
&\|u - \mathcal{M}_{h,\mathcal{D}}u\|_{L_p} \leq \sum_{[\boldsymbol{\alpha}]=0}^{N-1} \frac{(\sqrt{\mathcal{D}}h)^{[\boldsymbol{\alpha}]}}{\boldsymbol{\alpha}!} \|\partial^{\boldsymbol{\alpha}}u\|_{L_p}\, \|\varepsilon_{\boldsymbol{\alpha}}(\cdot\,, \eta, \mathcal{D})\|_{L_\infty} \\
&+ \sum_{[\boldsymbol{\alpha}]=N}^{L-1} \frac{(\sqrt{\mathcal{D}}h)^{[\boldsymbol{\alpha}]}}{\boldsymbol{\alpha}!} \|\partial^{\boldsymbol{\alpha}}u\|_{L_p}\, \|\sigma_{\boldsymbol{\alpha}}(\cdot\,, \eta, \mathcal{D})\|_{L_\infty} \\
&+ (\sqrt{\mathcal{D}}h)^L \Big\| \sum_{[\boldsymbol{\alpha}]=L} \frac{\mathcal{D}^{-n/2}}{\boldsymbol{\alpha}!} \sum_{\mathbf{m}\in\mathbb{Z}^n} \Big(\frac{\mathbf{x} - h\mathbf{m}}{\sqrt{\mathcal{D}}h}\Big)^{\boldsymbol{\alpha}} \eta\Big(\frac{\mathbf{x} - h\mathbf{m}}{\sqrt{\mathcal{D}}h}\Big)\, U_{\boldsymbol{\alpha}}(\mathbf{x}, h\mathbf{m})\Big\|_{L_p}.
\end{aligned}$$

The first two sums on the right-hand side can be estimated by using Corollary 2.13 and Lemma 2.6. In particular,

$$\begin{aligned}
&\sum_{[\boldsymbol{\alpha}]=k} \frac{(\sqrt{\mathcal{D}}h)^{[\boldsymbol{\alpha}]}}{\boldsymbol{\alpha}!} \|\partial^{\boldsymbol{\alpha}}u\|_{L_p} \|\sigma_{\boldsymbol{\alpha}}(\cdot\,, \eta, \mathcal{D})\|_{L_\infty} \\
&\leq \frac{(\sqrt{\mathcal{D}}h)^k}{(2\pi)^k} \sum_{[\boldsymbol{\alpha}]=k} \|\partial^{\boldsymbol{\alpha}}u\|_{L_p} \sum_{\boldsymbol{\nu}\in\mathbb{Z}^n\setminus\{\mathbf{0}\}} \big|\partial^{\boldsymbol{\alpha}}\mathcal{F}\eta(\sqrt{\mathcal{D}}\boldsymbol{\nu})\big| \leq (\sqrt{\mathcal{D}}h)^k \frac{\varepsilon_k(\mathcal{D})}{(2\pi)^k} \|\nabla_k u\|_{L_p}
\end{aligned}$$

with the numbers

$$\varepsilon_k(\mathcal{D}) = \max_{[\boldsymbol{\alpha}]=k} \sum_{\boldsymbol{\nu}\in\mathbb{Z}^n\setminus\{\mathbf{0}\}} \left|\partial^{\boldsymbol{\alpha}}\mathcal{F}\eta(\sqrt{\mathcal{D}}\boldsymbol{\nu})\right| . \tag{2.70}$$

We show in Lemma 2.29 below that

$$\Big\| \mathcal{D}^{-n/2} \sum_{\mathbf{m}\in\mathbb{Z}^n} \Big(\frac{\mathbf{x}-h\mathbf{m}}{\sqrt{\mathcal{D}}h}\Big)^{\boldsymbol{\alpha}} \eta\Big(\frac{\mathbf{x}-h\mathbf{m}}{\sqrt{\mathcal{D}}h}\Big) U_{\boldsymbol{\alpha}}(\mathbf{x},h\mathbf{m})\Big\|_{L_p} \leq c_\eta \|\partial^{\boldsymbol{\alpha}} u\|_{L_p} \tag{2.71}$$

if $\partial^{\boldsymbol{\alpha}} u \in L_p(\mathbb{R}^n)$ and $[\boldsymbol{\alpha}] = L > n/p$, where the constant c_η does not depend on u, h, and $\mathcal{D}$. Note that the condition $L > n/p$ guarantees that $u \in W_p^L(\mathbb{R}^n)$ can be identified with a continuous function.

THEOREM 2.28. *Suppose that η satisfies the decay and moment Conditions* 2.12 *and* 2.15 *of order N. Then for any continuous function $u \in W_p^L$, $1 \leq p \leq \infty$ and $n/p < L < K$, the quasi-interpolant* (2.23) *satisfies*

$$\|u - \mathcal{M}_{h,\mathcal{D}} u\|_{L_p} \leq \; c_\eta\, (\sqrt{\mathcal{D}}h)^M \; \|\nabla_M u\|_{L_p} + \sum_{k=0}^{M-1} (\sqrt{\mathcal{D}}h)^k \,\frac{\varepsilon_k(\mathcal{D})}{(2\pi)^k} \|\nabla_k u\|_{L_p}$$

where $M = \min(L, N)$, the constant c_η does not depend on u, h, and $\mathcal{D}$, and the numbers ε_k are defined by (2.70). *Moreover, for any $\varepsilon > 0$, there exists $\mathcal{D}_0 > 0$ such that for all $\mathcal{D} > \mathcal{D}_0$*

$$\|u - \mathcal{M}_{h,\mathcal{D}} u\|_{L_p} \leq \; c_\eta\, (\sqrt{\mathcal{D}}h)^M \; \|\nabla_M u\|_{L_p} + \varepsilon \sum_{k=0}^{M-1} (\sqrt{\mathcal{D}}h)^k \|\nabla_k u\|_{L_p} .$$

2.4.2. Estimation of the remainder term. To prove (2.71), we introduce the functions

$$S_{\boldsymbol{\alpha},h}(\mathbf{x}) := \mathcal{D}^{-n/2} \sum_{\mathbf{m}\in\mathbb{Z}^n} \Big(\frac{\mathbf{x}-h\mathbf{m}}{\sqrt{\mathcal{D}}h}\Big)^{\boldsymbol{\alpha}} \eta\Big(\frac{\mathbf{x}-h\mathbf{m}}{\sqrt{\mathcal{D}}h}\Big) U_{\boldsymbol{\alpha}}(\mathbf{x},h\mathbf{m}) . \tag{2.72}$$

LEMMA 2.29. *Suppose that η satisfies the decay condition* (2.45) *and that the function u is such that $\partial^{\boldsymbol{\alpha}} u \in L_p$ with $[\boldsymbol{\alpha}] = L$, where $1 \leq p < \infty$ and $n/p < L < K - n$. Then the L_p-norm of $S_{\boldsymbol{\alpha},h}$ admits the estimate*

$$\|S_{\boldsymbol{\alpha},h}\|_{L_p} \leq c_\eta \|\partial^{\boldsymbol{\alpha}} u\|_{L_p}$$

with some constant c_η not depending on u, h, and $\mathcal{D}$.

PROOF. We use the functions ϕ_μ defined by (2.31). Since

$$\phi_\mu(s) \;\rightarrow\; \frac{2\pi^{n/2}}{\Gamma(n/2)} \frac{\Gamma(n)\Gamma(\mu)}{\Gamma(n+\mu)} \quad \text{as} \quad s \to \infty ,$$

there exist constants $\Phi_\mu(t)$ for any $t \in (0,\infty)$ such that

$$\phi_\mu(s) \leq \Phi_\mu(t) \quad \text{if } t \leq s < \infty .$$

On the other hand, $s^n \phi_\mu(s)$ is an increasing function of s; hence

$$\phi_\mu(s) \leq (t/s)^n \,\Phi_\mu(t) , \quad \text{for } 0 < s \leq t ,$$

which, together with the decay condition (2.45), implies

$$\Big\| (\sqrt{\mathcal{D}}s)^{-n} \sum_{\mathbf{m}\in\mathbb{Z}^n} \Big| \Big(\frac{\cdot - \mathbf{m}}{\sqrt{\mathcal{D}}s}\Big)^{\boldsymbol{\alpha}} \eta\Big(\frac{\cdot - \mathbf{m}}{\sqrt{\mathcal{D}}s}\Big) \Big| \Big\|_{L_\infty} \leq A\, s^{-n}\, \Phi_{K-L}(\sqrt{\mathcal{D}}) \tag{2.73}$$

for any $s \in (0,1]$ and $[\boldsymbol{\alpha}] = L$.

First let $1 < p < \infty$. Applying Hölder's inequality and (2.73) with $s = 1$, we obtain

$$\|S_{\boldsymbol{\alpha},h}\|_{L_p}^p \le \int\limits_{\mathbb{R}^n}\Big(\sum_{\mathbf{m}\in\mathbb{Z}^n}\Big|\mathcal{D}^{-n/2}\Big(\frac{\mathbf{x}-h\mathbf{m}}{\sqrt{\mathcal{D}}h}\Big)^{\boldsymbol{\alpha}}\eta\Big(\frac{\mathbf{x}-h\mathbf{m}}{\sqrt{\mathcal{D}}h}\Big)\Big|^{1/p+1/p'}|U_{\boldsymbol{\alpha}}(\mathbf{x},h\mathbf{m})|\Big)^p d\mathbf{x}$$

$$\le \big(A\Phi_{K-L}(\sqrt{\mathcal{D}})\big)^{p/p'}\int\limits_{\mathbb{R}^n}\mathcal{D}^{-n/2}\sum_{\mathbf{m}\in\mathbb{Z}^n}\Big|\Big(\frac{\mathbf{x}-h\mathbf{m}}{\sqrt{\mathcal{D}}h}\Big)^{\boldsymbol{\alpha}}\eta\Big(\frac{\mathbf{x}-h\mathbf{m}}{\sqrt{\mathcal{D}}h}\Big)\Big|\,\big|U_{\boldsymbol{\alpha}}(\mathbf{x},h\mathbf{m})\big|^p d\mathbf{x}\,,$$

with $p' = (1-p)/p$. Choose $\theta \in (0, L - n/p)$ and apply Hölder's inequality once more to get

$$\big|U_{\boldsymbol{\alpha}}(\mathbf{x},h\mathbf{m})\big|^p = L^p\,\Big|\int\limits_0^1 s^{L-1}\partial^{\boldsymbol{\alpha}}u(s\mathbf{x}+(1-s)h\mathbf{m})ds\Big|^p$$

$$\le L^p\int\limits_0^1 s^{(L-1-1/p'-\theta)p}|\partial^{\boldsymbol{\alpha}}u(s\mathbf{x}+(1-s)h\mathbf{m})|^p ds\,\Big\{\int\limits_0^1 s^{-1+\theta p'}ds\Big\}^{p/p'}$$

$$= \frac{L^p}{(\theta p')^{p/p'}}\int\limits_0^1 s^{(L-\theta)p-1}|\partial^{\boldsymbol{\alpha}}u(s\mathbf{x}+(1-s)h\mathbf{m})|^p ds\,.$$

Hence it remains to estimate

$$\int\limits_{\mathbb{R}^n}\mathcal{D}^{-n/2}\sum_{\mathbf{m}\in\mathbb{Z}^n}\Big|\Big(\frac{\mathbf{x}-h\mathbf{m}}{\sqrt{\mathcal{D}}h}\Big)^{\boldsymbol{\alpha}}\eta\Big(\frac{\mathbf{x}-h\mathbf{m}}{\sqrt{\mathcal{D}}h}\Big)\Big|\int\limits_0^1 s^{(L-\theta)p-1}|\partial^{\boldsymbol{\alpha}}u(s\mathbf{x}+(1-s)h\mathbf{m})|^p ds\,d\mathbf{x}$$

$$= \int\limits_0^1 s^{(L-\theta)p-1}\int\limits_{\mathbb{R}^n}|\partial^{\boldsymbol{\alpha}}u(\mathbf{t})|^p\,(\sqrt{\mathcal{D}}s)^{-n}\sum_{\mathbf{m}\in\mathbb{Z}^n}\Big|\Big(\frac{\mathbf{t}-h\mathbf{m}}{\sqrt{\mathcal{D}}hs}\Big)^{\boldsymbol{\alpha}}\eta\Big(\frac{\mathbf{t}-h\mathbf{m}}{\sqrt{\mathcal{D}}hs}\Big)\Big|d\mathbf{t}\,ds\,,$$

where we changed the order of integrations and summation in the last equality several times, which is justified since the integrands are non-negative, and substituted $\mathbf{t} = s\mathbf{x} + (1-s)h\mathbf{m}$. The application of (2.73) results in

$$\int\limits_0^1 s^{(L-\theta)p-1}\int\limits_{\mathbb{R}^n}|\partial^{\boldsymbol{\alpha}}u(\mathbf{t})|^p\,(\sqrt{\mathcal{D}}s)^{-n}\sum_{\mathbf{m}\in\mathbb{Z}^n}\Big|\Big(\frac{\mathbf{t}-h\mathbf{m}}{\sqrt{\mathcal{D}}hs}\Big)^{\boldsymbol{\alpha}}\eta\Big(\frac{\mathbf{t}-h\mathbf{m}}{\sqrt{\mathcal{D}}hs}\Big)\Big|d\mathbf{t}\,ds$$

$$\le A\,\Phi_{K-L}(\sqrt{\mathcal{D}})\int\limits_0^1 s^{(L-\theta)p-1-n}ds\int\limits_{\mathbb{R}^n}|\partial^{\boldsymbol{\alpha}}u(\mathbf{t})|^p d\mathbf{t} = \frac{A\,\Phi_{K-L}(\sqrt{\mathcal{D}})}{(L-\theta)p-n}\,\|\partial^{\boldsymbol{\alpha}}u\|_{L_p}^p\,,$$

which leads to

$$\|S_{\boldsymbol{\alpha},h}\|_{L_p}^p \le \big(A\Phi_{K-L}(\sqrt{\mathcal{D}})\big)^{p/p'}\,\frac{L^p}{(\theta p')^{p/p'}}\,\frac{A\,\Phi_{K-L}(\sqrt{\mathcal{D}})}{(L-\theta)p-n}\,\|\partial^{\boldsymbol{\alpha}}u\|_{L_p}^p$$

for arbitrary $\theta \in (0, L - n/p)$. Because of

$$\min_{0<\theta<L-n/p}\frac{1}{(\theta p')^{1/p'}\,((L-\theta)p-n)^{1/p}} = \frac{p}{Lp-n}\,,$$

we obtain

$$\|S_{\boldsymbol{\alpha},h}\|_{L_p} \le A\,\Phi_{K-L}(\sqrt{\mathcal{D}})\frac{Lp}{Lp-n}\,\|\partial^{\boldsymbol{\alpha}}u\|_{L_p}\,.$$

If $p = \infty$, then using (2.73) with $s = 1$ and the inequality $|U_{\boldsymbol{\alpha}}(\mathbf{x}, \mathbf{y})| \le \|\partial^{\boldsymbol{\alpha}} u\|_{L_\infty}$, one obtains

$$\|S_{\boldsymbol{\alpha},h}\|_{L_\infty} \le A\Phi_{K-L}(\sqrt{\mathcal{D}})\|\partial^{\boldsymbol{\alpha}} u\|_{L_\infty} .$$

If $p = 1$, then clearly

$$\|S_{\boldsymbol{\alpha},h}\|_{L_1} \le L \int_0^1 s^{L-1} \int_{\mathbb{R}^n} |\partial^{\boldsymbol{\alpha}} u(\mathbf{t})|(\sqrt{\mathcal{D}}s)^{-n} \sum_{\mathbf{m}\in\mathbb{Z}^n} \Big|\Big(\frac{\mathbf{t}-h\mathbf{m}}{\sqrt{\mathcal{D}}hs}\Big)^{\boldsymbol{\alpha}} \eta\Big(\frac{\mathbf{t}-h\mathbf{m}}{\sqrt{\mathcal{D}}hs}\Big)\Big| \, d\mathbf{t}\, ds$$

$$\le A\, L\, \Phi_{K-L}(\sqrt{\mathcal{D}}) \int_0^1 s^{L-1-n} ds \int_{\mathbb{R}^n} |\partial^{\boldsymbol{\alpha}} u(\mathbf{t})| d\mathbf{t} \le \frac{A\, L\, \Phi_{K-L}(\sqrt{\mathcal{D}})}{L-n} \|\partial^{\boldsymbol{\alpha}} u\|_{L_1} .$$

Thus the assertion is proved for $1 \le p \le \infty$ and we see that the constant c_η is bounded by

$$c_\eta \le A\, \Phi_{K-L}(\sqrt{\mathcal{D}}) \frac{Lp}{Lp-n} . \qquad \square$$

2.4.3. Remark on the best approximation. Using properties of the quasi-interpolants, we can draw some preliminary conclusions concerning approximation properties of the sets

$$\operatorname{span}\Big\{\eta\Big(\frac{\mathbf{x}-h\mathbf{m}}{h\sqrt{\mathcal{D}}}\Big), \mathbf{m} \in \mathbb{Z}^n\Big\}$$

if $h \to 0$.

A space S of complex-valued functions defined on $\mathbb{R}^n$ is called *shift-invariant* if, for each $f \in S$, the space S contains the shifts $f(\cdot + \mathbf{m})$, $\mathbf{m} \in \mathbb{Z}^n$. In other words, S contains all the integer translates of f if it contains f. A particularly simple example is provided by the space of all finite linear combinations of shifts of a single function φ. The closure of this set in some Banach space X denoted by

$$S(\varphi) = \operatorname{closure}_X\big(\operatorname{span}\{\varphi(\cdot - \mathbf{m}), \mathbf{m} \in \mathbb{Z}^n\}\big)$$

is called the *principal shift-invariant subspace* generated by φ. The space $S(\varphi)$ can be dilated by the parameter $h > 0$ to obtain

$$S^h(\varphi) := \{f(\cdot/h) : f \in S(\varphi)\},$$

and the family $(S^h(\varphi))_h$ is called a *ladder* of principal shift-invariant spaces. Thus the quasi-interpolants $\mathcal{M}_{h,\mathcal{D}}u$ are elements from the ladder $(S^h(\eta(\cdot/\mathcal{D})))_h$, and Theorem 2.28 implies, for example, that for any $\varepsilon > 0$, there exists $\mathcal{D}_0 > 0$ such that for all $\mathcal{D} > \mathcal{D}_0$

$$\operatorname{dist}(u, S^h(\eta(\cdot/\mathcal{D}); L_p)) \le c((\sqrt{\mathcal{D}}h)^N + \varepsilon)\|u\|_{W_p^N} \tag{2.74}$$

for any $u \in W_p^N(\mathbb{R}^n)$, $N > n/p$, if η satisfies the decay and moment Conditions 2.12 and 2.15. Here the distance of the subspace $S^h(\varphi)$ to f is defined as

$$\operatorname{dist}(f, S^h(\varphi); X) = \inf_{f_h \in S_h(\varphi)} \|f - f_h\|_X . \tag{2.75}$$

In the next chapter, we shall see that (2.74) is valid if the moment Condition 2.15 is replaced by the requirement $\mathcal{F}\eta(\mathbf{0}) \ne 0$.

Let us note that this requirement together with the boundedness

$$|\partial^{\boldsymbol{\beta}}\eta(\mathbf{x})|(1 + |\mathbf{x}|)^K < \infty, \quad \mathbf{x} \in \mathbb{R}^n,$$

for some $K > n$ and all $0 \le [\beta] \le \mu$, where μ is the smallest integer greater than $n/2$, ensures that for any $\varepsilon > 0$ there exists $\mathcal{D} > 0$ such that

$$\lim_{h\to 0} \operatorname{dist}(u, S^h(\eta(\cdot/\mathcal{D}); L_p)) < \varepsilon \|u\|_{L_p} . \tag{2.76}$$

This follows from Remark 2.26 and the density of compactly supported smooth functions in L_p.

2.4.4. Local L_p-estimates. Here, we estimate the quasi-interpolation error in L_p for functions, which are given on a bounded domain $\Omega \subset \mathbb{R}^n$. Let $u \in W_p^L(\Omega)$. We know from (2.62) and the proof of Lemma 2.23 that the truncated quasi-interpolant $\mathcal{M}_{h,D}^{(\kappa)} u$ allows the representation

$$\mathcal{M}_{h,D}^{(\kappa)} u(\mathbf{x}) = u(\mathbf{x}) + \sum_{[\boldsymbol{\alpha}]=0}^{M-1} (\sqrt{\mathcal{D}}h)^{[\boldsymbol{\alpha}]} f_{\boldsymbol{\alpha}}(\mathbf{x}) \frac{\partial^{\boldsymbol{\alpha}} u(\mathbf{x})}{\boldsymbol{\alpha}!}$$
$$+ (-\sqrt{\mathcal{D}}h)^M \sum_{[\boldsymbol{\alpha}]=M} \frac{\mathcal{D}^{-n/2}}{\boldsymbol{\alpha}!} \sum_{h\mathbf{m}\in B(\mathbf{x},ht)} \Big(\frac{\mathbf{x}-h\mathbf{m}}{\sqrt{\mathcal{D}}h}\Big)^{\boldsymbol{\alpha}} \eta\Big(\frac{\mathbf{x}-h\mathbf{m}}{\sqrt{\mathcal{D}}h}\Big) U_{\boldsymbol{\alpha}}(\mathbf{x}, h\mathbf{m}),$$

with sufficiently small functions satisfying $f_{\boldsymbol{\alpha}}(\mathbf{x}) \le 2\|\varepsilon_{\boldsymbol{\alpha}}(\cdot\,,\eta,\mathcal{D})\|_{L_\infty}$ if the truncation parameter κ satisfies (2.63) and $\mathbf{x} \in \Omega_{\kappa h}$ (see (2.67)). Here $M = \min(L, N)$. Proceeding as in the proof of Theorem 2.28 and Lemma 2.29, one can deduce

$$\|(I - \mathcal{M}_{h,D}^{(\kappa)})u\|_{L_p(\Omega_{\kappa h})} \le c\,(\sqrt{\mathcal{D}}h)^M \, \|\nabla_M u\|_{L_p(\Omega)}$$
$$+ 2 \sum_{[\boldsymbol{\alpha}]=0}^{M-1} (\sqrt{\mathcal{D}}h)^{[\boldsymbol{\alpha}]} \frac{\|\partial^{\boldsymbol{\alpha}} u(\mathbf{x})\|_{L_p(\Omega)}}{\boldsymbol{\alpha}!} \|\varepsilon_{\boldsymbol{\alpha}}(\cdot\,,\eta,\mathcal{D})\|_{L_\infty}.$$

The difference $(\mathcal{M}_{h,D} - \mathcal{M}_{h,D}^{(\kappa)})u$ can be estimated in view of Lemma 2.1 and (2.63) by

$$\|(\mathcal{M}_{h,\mathcal{D}} - \mathcal{M}_{h,\mathcal{D}}^{(\kappa)})u\|_{L_p(\Omega_{\kappa h})} \le \Big(\int_{|\mathbf{x}|>\kappa\sqrt{\mathcal{D}}} |\eta|\, d\mathbf{x}\Big)^{1/p} (\kappa^{1-N}\, r_{\mathcal{D}}(\kappa,\eta))^{1-1/p} \|u\|_{p,h}$$
$$\le \|\varepsilon_{\mathbf{0}}(\cdot\,,\eta,\mathcal{D})\|_{L_\infty} \|u\|_{p,h}\,.$$

Hence we derive

LEMMA 2.30. *Suppose that η satisfies the conditions of Theorem 2.28. Let Ω be a domain in $\mathbb{R}^n$ and let $u \in W_p^L(\Omega)$ with $L > n/p$, $1 \le p \le \infty$. Then for any $\varepsilon > 0$ there exist positive $\mathcal{D}$ and κ such that*

$$\|(I - \mathcal{M}_{h,D})u\|_{L_p(\Omega_{\kappa h})} \le c\,(\sqrt{\mathcal{D}}h)^M \, \|\nabla_M u\|_{L_p(\Omega)} + \varepsilon \sum_{k=0}^{M-1} (\sqrt{\mathcal{D}}h)^k \|\nabla_k u\|_{L_p(\Omega)}\,,$$

where $\Omega_{\kappa h}$ is the subdomain defined in (2.67) *and the constant c does not depend on u, h, and $\mathcal{D}$.*

2.5. Notes

The error estimates for the quasi-interpolants $\mathcal{M}_{h,\mathcal{D}}u$ obtained in this chapter have been announced and proved, partially in a different form, in [**62**], [**64**], [**66**], [**67**]. The proof of Lemma 2.11 is taken from [**71**].

It can be seen from (2.50) and Theorem 2.17 that the quasi-interpolant $\mathcal{M}_{h,\mathcal{D}}u$ converges to u with the order $\mathcal{O}((\sqrt{\mathcal{D}}h)^N)$ if and only if η satisfies, besides the moment and decay Conditions 2.15 and 2.12, the conditions

$$\partial^{\boldsymbol{\alpha}} \mathcal{F}\eta(\sqrt{\mathcal{D}}\boldsymbol{\nu}) = 0\,, \quad \forall \boldsymbol{\nu} \in \mathbb{Z}^n \backslash \{\mathbf{0}\}\,,\ 0 \le [\boldsymbol{\alpha}] < N\,. \tag{2.77}$$

They are known as the *Strang-Fix conditions*, which play an important role in the theory of approximations from shift-invariant spaces. The determination of the approximation powers of principal shift invariant spaces, i.e., the rate of decay for sufficiently smooth functions, has received considerable attention in the literature. We mention some results, which are relevant for the fast decaying generating functions η considered here.

In [**92**] Strang and Fix analyzed the L_2-approximation properties of the (non-closed) subspace $S_2(\varphi) = S_*(\varphi) \cap L_2(\mathbb{R}^n)$, where $S_*(\varphi)$ denotes the set of all linear combinations of the integer shifts of a compactly supported function φ. They proved that the condition

$$\mathcal{F}\varphi(\mathbf{0}) \neq 0\,, \quad \partial^{\boldsymbol{\alpha}} \mathcal{F}\varphi(\boldsymbol{\nu}) = 0\,, \quad \forall \boldsymbol{\nu} \in \mathbb{Z}^n \backslash \{\mathbf{0}\}\,,\ 0 \le [\boldsymbol{\alpha}] < N\,, \tag{2.78}$$

is necessary and sufficient for the *controlled approximation of order* N. This means that for a given $f \in W_2^N(\mathbb{R}^n)$ there exist approximants

$$f_h = \sum_{\mathbf{m} \in \mathbb{Z}^n} c_{\mathbf{m}}^h \varphi(\cdot - \mathbf{m}) \in S_2^h(\varphi)$$

such that

$$\|f - f_h\|_{L_2} \le c h^N \|f\|_{W_2^N}$$

and the coefficients of f_h satisfy

$$\|\{c_{\mathbf{m}}^h\}\|_{\ell_2(\mathbb{Z}^n)} \le c h^{-n/2} \|f\|_{L_2}\,, \tag{2.79}$$

with constants not independent of f and h. They showed also that (2.78) is equivalent to the polynomial reproducing property, i.e., for any algebraic polynomial $g \in \Pi_{N-1}$ of degree $< N$

$$\sum_{\mathbf{m} \in \mathbb{Z}^n} g(\mathbf{m}) \varphi(\cdot - \mathbf{m}) \in \Pi_{N-1}\,. \tag{2.80}$$

Note that the Strang-Fix conditions (2.78) have been considered previously for $n = 1$ by Schoenberg in [**86**]. He showed that all polynomials of degree $< N$ can be written as $\sum_{m \in \mathbb{Z}} c_m \varphi(\cdot - m)$, if the piecewise continuous function φ with the exponential decay at infinity satisfies (2.78).

There is a rich literature where Strang-Fix conditions and the connections to the approximation power of subspaces have been clarified and extended in various directions. The extensions include shift-invariant spaces which are generated by a collection of generating functions, generating functions with non-compact support, and approximation orders in other than L_2-norms. The description of these results is beyond the scope of this book. A good overview concerning the historical development of L_2- and L_∞-approximation orders is given, for example, in [**11**], which develops a general approach to this topic. It contains also a rather large bibliography and additional information. Approximation orders in L_p have been investigated, in particular, in [**12**], [**38**], where it is shown that the Strang-Fix conditions are equivalent to certain "controlled" approximation orders of principal shift-invariant spaces, too.

In [**11**] de Boor, DeVore, and Ron proved the following result, which shows that one cannot expect the convergence by the ladder $(S^h(\eta(\cdot/\mathcal{D})))_h$ if $\mathcal{D}$ is fixed: *Assume that $\mathcal{F}\varphi$ is bounded on some neighborhood of the origin. If $S^h(\varphi)_h$ provides the approximation order N in $L_2(\mathbb{R}^n)$, then $\mathcal{F}\varphi$ has a zero of order N at every $\boldsymbol{\nu} \in \mathbb{Z}^n \setminus \{\mathbf{0}\}$.*

The concept of approximate quasi-interpolation replaces the Strang-Fix conditions (2.77) by the weaker conditions

$$\sum_{\boldsymbol{\nu}\in\mathbb{Z}^n\setminus\{\mathbf{0}\}} |\partial^{\boldsymbol{\alpha}}\mathcal{F}\eta(\sqrt{\mathcal{D}}\boldsymbol{\nu})| \to 0\,, \quad 0 \le [\boldsymbol{\alpha}] < N\,, \quad \text{if } \mathcal{D}\to\infty\,. \tag{2.81}$$

Depending on η, one can fix the parameter $\mathcal{D}$ such that the saturation term satisfies

$$\Big|\sum_{[\boldsymbol{\alpha}]=0}^{N-1} \frac{\partial^{\boldsymbol{\alpha}}u(\mathbf{x})}{\boldsymbol{\alpha}!}\Big(\frac{-i\sqrt{\mathcal{D}}h}{2\pi}\Big)^{[\boldsymbol{\alpha}]} \sum_{\boldsymbol{\nu}\in\mathbb{Z}^n\setminus\mathbf{0}} \partial^{\boldsymbol{\alpha}}\mathcal{F}\eta(\sqrt{\mathcal{D}}\boldsymbol{\nu})\, \mathrm{e}^{\frac{2\pi i}{h}\langle\mathbf{x},\boldsymbol{\nu}\rangle}\Big|$$
$$< \varepsilon \sum_{k=0}^{N-1} (\sqrt{\mathcal{D}}h)^k |\nabla_k u(\mathbf{x})|$$

for some prescribed $\varepsilon > 0$. Therefore, if ε is sufficiently small, the quasi-interpolant $\mathcal{M}_{h,\mathcal{D}}u$ behaves in numerical computations like a converging approximation process.

This approach allows one to enlarge the class of generating functions for the practical application of quasi-interpolants of the form (2.23) to numerical methods for the solution of partial differential and multi-dimensional integral equations. In fact, the action of important multi-dimensional integral or even pseudodifferential operators on many functions with fast decaying Fourier transform can be given analytically or by other efficient methods, and these functions satisfy (2.81). In the next chapter, we discuss several methods to obtain new basis functions with simple analytic structure, which satisfy the moment condition (2.47) with large N. This provides high-order approximate approximations with the property that many analytic operations can be performed very efficiently.

A straightforward consequence of the estimates obtained in this chapter is the following observation: If (2.77) is violated, then in order to derive convergence, the parameter $\mathcal{D}$ should dependent on h, $\mathcal{D} = \mathcal{D}(h)$, so that

$$\sum_{\boldsymbol{\nu}\in\mathbb{Z}^n\setminus\{\mathbf{0}\}} |\partial^{\boldsymbol{\alpha}}\mathcal{F}\eta(\sqrt{\mathcal{D}(h)}\,\boldsymbol{\nu})| \to 0 \quad \text{as} \quad h\to 0\,, \tag{2.82}$$

at least for $\boldsymbol{\alpha} = \mathbf{0}$. Then the conclusion of Theorem 2.17, for example, can be reformulated as follows: The quasi-interpolant $\mathcal{M}_{h,\mathcal{D}(h)}u$ satisfies the estimate

$$\|\mathcal{M}_{h,\mathcal{D}(h)}u - u\| \le c\big(h^M\mathcal{D}(h)^{M/2}\|\nabla_M u\| + \sum_{k=0}^{M-1} h^k\mathcal{D}(h)^{k/2}\delta_k(h)\,\|\nabla_k u\|\big)\,, \tag{2.83}$$

where

$$\delta_k(h) := \max_{[\boldsymbol{\alpha}]=k} \|\varepsilon_{\boldsymbol{\alpha}}(\cdot,\eta,\mathcal{D}(h))\| \to 0 \quad \text{as} \quad h\to 0\,.$$

Hence, $\mathcal{M}_{h,\mathcal{D}(h)}u$ converges to u if the function $\mathcal{D}(h)$ is chosen to satisfy

$$\mathcal{D}(h)\to\infty \quad \text{and} \quad h^2\mathcal{D}(h)\to 0 \quad \text{as} \quad h\to 0\,.$$

Then the convergence rate is determined by the minimum of $h^M \mathcal{D}(h)^{M/2}$ and $h^k \mathcal{D}(h)^{k/2} \delta_k(h)$, $k = 0, \dots, M-1$. It is clear that one has to check these conditions for any generating function under consideration individually and that the search for an optimal function $\mathcal{D}(h)$ can be rather complicated.

One can find some sporadic results in the literature, where this approach to the quasi-interpolation was studied for special generating functions. Stenger, in his book [**91**, Section 5.8], studies one-dimensional quasi-interpolants with special basis functions satisfying the moment condition for $N = 2$ and with the functional dependence $\mathcal{D} = h^{2(\beta-1)}$, $0<\beta<1$. He proved the convergence of the approximations for continuous functions. In [**8**] Beatson and Light analyze multi-dimensional quasi-interpolants with variable $\mathcal{D}$, which use tensor products of linear combinations of univariate Gaussians or exponentials $\mathrm{e}^{-|\cdot|}$ as basis functions. It is shown that the quasi-interpolant with linear combinations of the Gaussian as generating function converges to $u \in W_\infty^N(\mathbb{R}^n)$ with the order $\mathcal{O}(h^N |\log h|^N)$ if $\mathcal{D}(h) = N\,|\log h|/\pi^2$.

By dealing with fixed parameters $\mathcal{D}$ in the quasi-interpolation formulas, we avoid the problem of finding the optimal dependence of h and can treat different generating functions simultaneously. Although approximate quasi-interpolation does not converge, we obtain approximations with certain order, which can be rather large, up to a prescribed accuracy for appropriately chosen $\mathcal{D}$.

Moreover, it is advantageous in various numerical applications of the method to choose $\mathcal{D}$ not depending on h. First of all, the number of terms in the quasi-interpolation formula which are necessary to compute the approximate value at a fixed point within a given tolerance does not depend on h. Hence, if one fixes $\mathcal{D}$, one can use this number of summands for any step size. But more importantly, the fixed $\mathcal{D}$ is very useful in the cubature of integral and pseudodifferential operators, which is one of the main applications of approximate quasi-interpolation and will be discussed in the following at different places. Here a fixed parameter $\mathcal{D}$ allows one to reuse already calculated terms of the cubature formula with a certain step size h for other discretization levels.

Let us note that in the abstract framework of approximation properties of shift invariant spaces there is the notion of *non-stationary* ladders $(S^h(\varphi_h))_h$, where the h-entry of the ladder is the h-dilate of an h-dependent principal shift-invariant space $S(\varphi_h)$. It generalizes the notion of *stationary* ladders $(S^h(\varphi))_h$ described above, which are obtained by dilating the same space $S(\varphi)$. Obviously, the situation corresponding to the quasi-interpolation operator $\mathcal{M}_{h,\mathcal{D}(h)}$ satisfying (2.83) fits into the setting of the non-stationary ladder $(S^h(\eta(\cdot/\mathcal{D}(h))))_h$.

CHAPTER 3

Various basis functions — examples and constructions

3.1. Introduction

We have seen that approximation rates of order N can be guaranteed up to some saturation error for the quasi-interpolant (2.23) if η is a sufficiently smooth and rapidly decaying function which satisfies the moment Condition 2.15. For example, any function $\eta \in \mathcal{S}(\mathbb{R}^n)$ which is symmetric and satisfies $\mathcal{F}\eta(\mathbf{0}) \neq 0$ can be used as a generating function for approximate quasi-interpolation operators of the second order. It is the aim of this chapter to develop different methods to construct generating functions satisfying the moment condition with large N from simpler ones.

First we give examples of functions depending on one or several variables which generate approximate quasi-interpolants of the second or fourth order. These functions provide small terms $\|\varepsilon_{\boldsymbol{\alpha}}(\cdot\,,\eta,\mathcal{D})\|_{L_\infty}$ even for relatively small $\mathcal{D}$, which makes them suitable for practical applications as mentioned in Remark 2.18.

3.2. Examples

3.2.1. One-dimensional examples. The table below lists some basis functions, defined on $\mathbb{R}$, and their Fourier transforms. The last column of the table contains a lower bound $\mathcal{D}_{min}$ which guarantees that the main term of the saturation error satisfies

$$\sum_{\nu\in\mathbb{Z}\setminus\{0\}} |\mathcal{F}\eta(\sqrt{\mathcal{D}}\nu)| < 10^{-8}$$

for any $\mathcal{D} \geq \mathcal{D}_{min}$.

$\eta(x)$	$\mathcal{F}\eta(\lambda)$	$\mathcal{D}_{min}$
$\pi^{-1/2}\,\mathrm{e}^{-x^2}$	$\mathrm{e}^{-\pi^2\lambda^2}$	2.0
$\dfrac{1}{\pi\cosh x}$	$\dfrac{1}{\cosh \pi^2\lambda}$	4.1
$\dfrac{2x}{\pi^2\,\sinh x}$	$\dfrac{1}{\cosh^2\,\pi^2\lambda}$	1.1
$\dfrac{2}{\pi(1+x^2)^2}$	$(1+2\pi\|\lambda\|)\,\mathrm{e}^{-2\pi\|\lambda\|}$	12.0
$\sqrt{\dfrac{\mathrm{e}}{\pi}}\,\mathrm{e}^{-x^2}\cos\sqrt{2}x$	$\mathrm{e}^{-\pi^2\lambda^2}\cosh\sqrt{2}\pi\lambda$	2.6

The first four functions satisfy the moment condition (2.47) with order $N = 2$, and the last function with $N = 4$.

3.2.2. Examples of multi-variate basis functions. For an arbitrary n, one can take tensor products of functions mentioned in the one-dimensional case. Some further examples which satisfy the moment condition with $N = 2$ are contained in the following table.

$\eta(\mathbf{x})$	$\mathcal{F}\eta(\boldsymbol{\lambda})$	dim
$\pi^{-n/2}\,\mathrm{e}^{-\lvert\mathbf{x}\rvert^2}$	$\mathrm{e}^{-\pi^2\lvert\boldsymbol{\lambda}\rvert^2}$	any
$\operatorname{sech}(\lvert\mathbf{x}\rvert)$	$\dfrac{\pi^2\tanh(\pi^2\lvert\boldsymbol{\lambda}\rvert)}{\lvert\boldsymbol{\lambda}\rvert\cosh(\pi^2\lvert\boldsymbol{\lambda}\rvert)}$	$n = 3$
$\dfrac{4}{3\pi^{n+1/2}}\,\dfrac{\Gamma(\frac{n+5}{2})}{(1+\lvert\mathbf{x}\rvert^2)^{(n+5)/2}}$	$(1+2\pi\lvert\boldsymbol{\lambda}\rvert+\frac{4}{3}\pi^2\lvert\boldsymbol{\lambda}\rvert^2)\,\mathrm{e}^{-2\pi\lvert\boldsymbol{\lambda}\rvert}$	any
$\dfrac{\Gamma(k+1+\frac{n}{2})}{\pi^{n/2}\Gamma(k+1)}\,(1-\lvert\mathbf{x}\rvert^2)^k\,\chi(\mathbf{x})$	$\Gamma(k+1+\frac{n}{2})\,\dfrac{J_{k+n/2}(2\pi\lvert\boldsymbol{\lambda}\rvert)}{(\pi\lvert\boldsymbol{\lambda}\rvert)^{k+n/2}}$	any
$(-1)^k\dfrac{\pi^{(n+1)/2}}{\Gamma(k+\frac{n+1}{2})}\,\dfrac{\partial^k}{\partial\tau^k}\,\dfrac{\mathrm{e}^{-2\pi\sqrt{\tau}\lvert\mathbf{x}\rvert}}{\sqrt{\tau}}\Bigg\rvert_{\tau=1}$	$(1+\lvert\boldsymbol{\lambda}\rvert^2)^{-k-(n+1)/2}$	any

Here χ denotes the characteristic function of the unit ball $B(\mathbf{0}, 1)$ and J_ν denotes the Bessel function of the first kind.

In the following we describe some methods to construct functions satisfying the moment condition (2.47) of high order N. A general analytic formula, to be obtained in Section 3.3, leads to interesting function systems especially in the case of radial η. In Section 3.4 we consider other analytic methods, which employ the parameter $\mathcal{D}$. Section 3.5 deals with linear combinations of translates of a given simple basis function. We focus on the construction of multi-dimensional basis functions with minimal support and on the calculation of the terms which determine the saturation error.

Finally, in Section 3.7 we consider methods of diminishing the value of the parameter $\mathcal{D}$ that can be reduced without increasing the saturation error.

3.3. Basis functions for higher-order approximations

3.3.1. A general formula. We obtain a general analytic formula for the construction of generating functions which satisfy the moment condition for an arbitrarily given N. We use the notation

$$\partial^{\alpha}(\mathcal{F}\eta)^{-1}(\mathbf{0}) := \partial^{\alpha}\frac{1}{\mathcal{F}\eta(\boldsymbol{\lambda})}\bigg\rvert_{\boldsymbol{\lambda}=\mathbf{0}} . \tag{3.1}$$

THEOREM 3.1. *Suppose that η satisfies the decay condition (2.26) with $K \geq N + n$, that it is continuously differentiable up to order $N - 1$, and that*

$$\int_{\mathbb{R}^n} \eta(\mathbf{x})\,d\mathbf{x} \neq 0\,, \quad \int_{\mathbb{R}^n} \lvert\mathbf{x}\rvert^{N-1}\lvert\partial^{\alpha}\eta(\mathbf{x})\rvert\,d\mathbf{x} < \infty,\ 0 \leq [\boldsymbol{\alpha}] \leq N-1\,.$$

Then the function

$$\eta_N(\mathbf{x}) = \sum_{[\boldsymbol{\alpha}]=0}^{N-1} \frac{\partial^{\boldsymbol{\alpha}}(\mathcal{F}\eta)^{-1}(\mathbf{0})}{\boldsymbol{\alpha}!\,(2\pi i)^{[\boldsymbol{\alpha}]}}\,\partial^{\boldsymbol{\alpha}}\eta(\mathbf{x}) \tag{3.2}$$

satisfies the moment Condition 2.15 *of order* N.

PROOF. Because of (2.26) the Fourier transform $\mathcal{F}\eta$ is continuously differentiable up to order N and moreover $\mathcal{F}\eta(\mathbf{0}) \neq 0$. We denote the N-th order Taylor polynomial of $\dfrac{1}{\mathcal{F}\eta(\boldsymbol{\lambda})}$ by

$$P_N(\boldsymbol{\lambda}) = \sum_{[\boldsymbol{\alpha}]=0}^{N-1} \partial^{\boldsymbol{\alpha}}(\mathcal{F}\eta)^{-1}(\mathbf{0})\,\frac{\boldsymbol{\lambda}^{\boldsymbol{\alpha}}}{\boldsymbol{\alpha}!}\,.$$

Since

$$\partial^{\boldsymbol{\alpha}} P_N(\mathbf{0}) = \partial^{\boldsymbol{\alpha}}(\mathcal{F}\eta)^{-1}(\mathbf{0})\,, \quad 0 \le [\boldsymbol{\alpha}] < N\,,$$

we obtain the relations

$$\partial^{\boldsymbol{\beta}}\big(\mathcal{F}\eta(\boldsymbol{\lambda})P_N(\boldsymbol{\lambda})\big)\big|_{\boldsymbol{\lambda}=\mathbf{0}} = \partial^{\boldsymbol{\beta}}\Big(\mathcal{F}\eta(\boldsymbol{\lambda})\frac{1}{\mathcal{F}\eta(\boldsymbol{\lambda})}\Big)\Big|_{\boldsymbol{\lambda}=\mathbf{0}} = 0$$

for all $0 \le [\boldsymbol{\beta}] < N$. Hence the function $P_N(\boldsymbol{\lambda})\mathcal{F}\eta(\boldsymbol{\lambda}) - 1$ has a zero of order N at the origin, i.e., it satisfies the moment Condition 2.15. By

$$P_N(\boldsymbol{\lambda})\mathcal{F}\eta(\boldsymbol{\lambda}) = \mathcal{F}\Big(\sum_{[\boldsymbol{\alpha}]=0}^{N-1} \partial^{\boldsymbol{\alpha}}(\mathcal{F}\eta)^{-1}(\mathbf{0})\frac{1}{\boldsymbol{\alpha}!}\Big(\frac{1}{2\pi i}\Big)^{[\boldsymbol{\alpha}]}\,\partial^{\boldsymbol{\alpha}}\eta\Big)(\boldsymbol{\lambda})\,,$$

the assertion is shown. □

Note that the Fourier transform of η_N is given by

$$\mathcal{F}\eta_N(\boldsymbol{\lambda}) = \mathcal{F}\eta(\boldsymbol{\lambda}) \sum_{[\boldsymbol{\alpha}]=0}^{N-1} \frac{\partial^{\boldsymbol{\alpha}}(\mathcal{F}\eta)^{-1}(\mathbf{0})}{\boldsymbol{\alpha}!}\,\boldsymbol{\lambda}^{\boldsymbol{\alpha}}\,.$$

Hence, if η is such that η_N satisfies the decay Condition 2.12, then the saturation error of the quasi-interpolant (2.23) with the generating function η_N is controlled by the values $\partial^{\boldsymbol{\alpha}}\mathcal{F}\eta(\sqrt{\mathcal{D}}\boldsymbol{\nu})$ and $\partial^{\boldsymbol{\alpha}}(\mathcal{F}\eta)^{-1}(\mathbf{0})$.

3.3.2. Symmetric basis functions. If the basis function η has some additional symmetries, then the general formula (3.2) becomes simpler. Let η be symmetric with respect to the coordinate planes $x_i = 0$, i.e.,

$$\eta(x_1, \ldots, x_i, \ldots, x_n) = \eta(x_1, \ldots, -x_i, \ldots, x_n)\,,\ i = 1, \ldots, n\,. \tag{3.3}$$

Then automatically

$$\int_{\mathbb{R}^n} \mathbf{x}^{\boldsymbol{\alpha}}\eta(\mathbf{x})\,d\mathbf{x} = 0$$

for any multi-index $\boldsymbol{\alpha} = (\alpha_1, \ldots, \alpha_n)$ containing at least one odd α_i. Hence all indices α_j in formula (3.2) must be even integers and this formula simplifies to

$$\eta_{2M}(\mathbf{x}) = \sum_{[\boldsymbol{\alpha}]=0}^{M-1} \frac{\partial^{2\boldsymbol{\alpha}}(\mathcal{F}\eta)^{-1}(\mathbf{0})}{(2\boldsymbol{\alpha})!\,(-4\pi^2)^{[\boldsymbol{\alpha}]}}\,\partial^{2\boldsymbol{\alpha}}\eta(\mathbf{x})\,, \quad N = 2M\,. \tag{3.4}$$

EXAMPLE 3.2. Consider the one-dimensional Gaussian function $\eta(x) = \mathrm{e}^{-x^2}$. Here formula (3.4) leads to

$$\eta_{2M}(x) = \frac{(-1)^{M-1}}{\sqrt{\pi}\,(M-1)!\,2^{2M-1}} \frac{H_{2M-1}(x)\,\mathrm{e}^{-x^2}}{x} \tag{3.5}$$

where H_k denotes the *Hermite polynomial*

$$H_k(\tau) = (-1)^k\,\mathrm{e}^{\tau^2}\left(\frac{d}{d\tau}\right)^k \mathrm{e}^{-\tau^2}\,. \tag{3.6}$$

Indeed, $\dfrac{1}{\mathcal{F}\eta(\lambda)} = \pi^{-1/2}\,\mathrm{e}^{\pi^2\lambda^2}$ and from

$$\left(\frac{d}{d\lambda}\right)^{2j} \mathrm{e}^{\pi^2\lambda^2}\bigg|_{\lambda=0} = (-1)^j\pi^{2j}\left(\frac{d}{d\lambda}\right)^{2j}\mathrm{e}^{-\lambda^2}\bigg|_{\lambda=0} = (-1)^j\pi^{2j}\,H_{2j}(0)\,, \tag{3.7}$$

we obtain

$$\frac{\partial^{2\alpha}(\mathcal{F}\eta)^{-1}(0)}{\pi^{2\alpha}} = \frac{(-1)^\alpha}{\sqrt{\pi}}\,H_{2\alpha}(0)\,,$$

which implies

$$\eta_{2M}(x) = \frac{\mathrm{e}^{-x^2}}{\sqrt{\pi}}\sum_{j=0}^{M-1}\frac{H_{2j}(0)}{(2j)!\,4^j}\,H_{2j}(x)\,.$$

From [**1**, 22.4.8]

$$H_{2j}(0) = (-1)^j\frac{(2j)!}{j!}\,, \qquad H_{2j+1}(0) = 0\,, \tag{3.8}$$

so that

$$\eta_{2M}(x) = \frac{\mathrm{e}^{-x^2}}{\sqrt{\pi}}\sum_{j=0}^{M-1}\frac{(-1)^j}{j!\,4^j}\,H_{2j}(x)\,. \tag{3.9}$$

The recurrence relation [**1**, 22.7.13]

$$H_{k+1}(x) = 2xH_k(x) - 2kH_{k-1}(x)$$

implies the summation formula

$$\frac{(-1)^k}{k!\,2^{2k+1}}\,H_{2k+1}(x) = x\sum_{j=0}^{k}\frac{(-1)^j}{j!\,2^{2j}}\,H_{2j}(x)$$

and hence formula (3.5) follows.

Figs. 3.1 and 3.2 show two of the one-dimensional generating functions which provide quasi-interpolants of order 6 and 12, respectively.

3.3.3. Radial basis functions. Suppose that the function η is radial, i.e., $\eta(\mathbf{x}) = \psi(r)$ with $r = |\mathbf{x}|$. Here we give a formula for the radial function $\eta_{2M}(\mathbf{x}) = \psi_{2M}(r)$ satisfying the moment Condition 2.15.

The moments of the radial function η_{2M} are

$$\int\limits_{\mathbb{R}^n} \mathbf{x}^{2\alpha}\eta_{2M}(\mathbf{x})\,d\mathbf{x} = \int\limits_0^\infty r^{2[\alpha]+n-1}\,\psi_{2M}(r)\,dr \int\limits_{S_{n-1}} \boldsymbol{\omega}^{2\alpha}\,d\boldsymbol{\omega}$$

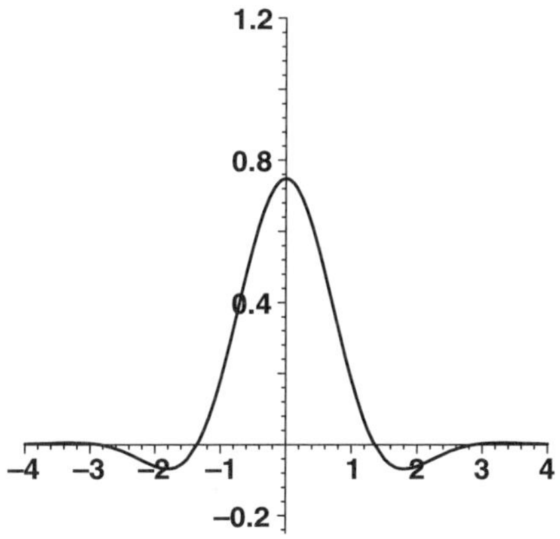

FIGURE 3.1. Graph of $\eta_6(x)$

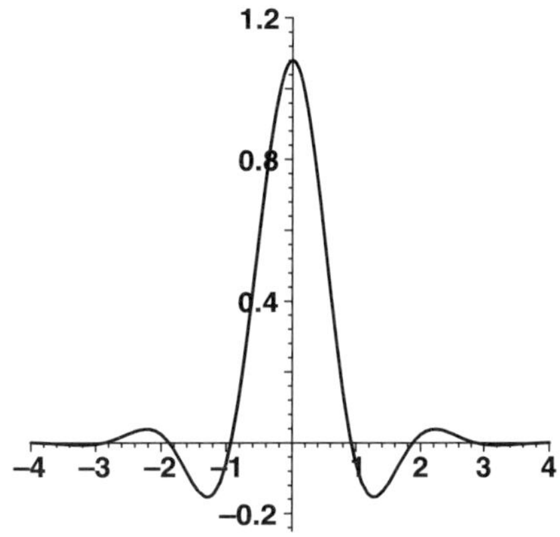

FIGURE 3.2. Graph of $\eta_{12}(x)$

where S_{n-1} denotes the unit sphere in $\mathbb{R}^n$ and $\boldsymbol{\omega} = \mathbf{x}/|\mathbf{x}| \in S_{n-1}$. Therefore it is sufficient that ψ_{2M} fulfills

$$\int_0^\infty r^{2k+n-1}\,\psi_{2M}(r)\,dr = \frac{\delta_{0k}}{\omega_n}\,, \quad k = 0, \ldots, M-1\,. \tag{3.10}$$

Here

$$\omega_n = |S_{n-1}| = \frac{2\pi^{n/2}}{\Gamma(n/2)} \tag{3.11}$$

is the surface area of S_{n-1}. Furthermore,

$$\omega_n \int_0^\infty r^{2k+n-1}\,\psi_{2M}(r)\,dr = \int_{\mathbb{R}^n} |\mathbf{x}|^{2k}\eta_{2M}(\mathbf{x})\,d\mathbf{x} = (-2\pi)^{-2k}\Delta^k\mathcal{F}\eta_{2M}(\mathbf{0})$$

with the Laplace operator

$$\Delta = \sum_{j=1}^n \frac{\partial^2}{\partial x_j^2}\,.$$

Hence the moment condition (3.10) for the radial function η_{2M} is equivalent to

$$\mathcal{F}\eta_{2M}(\mathbf{0}) = 1\,,\ \Delta^k\mathcal{F}\eta_{2M}(\mathbf{0}) = 0\,, \quad k = 1, \ldots, M-1\,.$$

Since the polynomial

$$P_{2M}(\boldsymbol{\lambda}) = \sum_{j=0}^{M-1} \frac{\Gamma\big(\frac{n}{2}\big)\ \Delta^j(\mathcal{F}\eta)^{-1}(\mathbf{0})}{j!\,2^{2j}\,\Gamma\big(j+\frac{n}{2}\big)}\ |\boldsymbol{\lambda}|^{2j}$$

satisfies

$$\Delta^k P_{2M}(\mathbf{0}) = \Delta^k(\mathcal{F}\eta)^{-1}(\mathbf{0}) := \Delta^k \frac{1}{\mathcal{F}\eta(\boldsymbol{\xi})}\bigg|_{\boldsymbol{\xi}=\mathbf{0}}\,, \quad k = 0, \ldots, M-1\,,$$

we derive the following.

THEOREM 3.3. *Suppose that η satisfies the conditions of Theorem 3.1 and is a radial function. Then*

$$\eta_{2M}(\mathbf{x}) = \Gamma\Big(\frac{n}{2}\Big) \sum_{j=0}^{M-1} \frac{(-1)^j \, \Delta^j (\mathcal{F}\eta)^{-1}(\mathbf{0})}{j!\,(4\pi)^{2j}\,\Gamma\big(j+\frac{n}{2}\big)} \, \Delta^j \eta(\mathbf{x}) \tag{3.12}$$

is subject to the moment condition (2.47) with $N = 2M$ and it has the Fourier transform

$$\mathcal{F}\eta_{2M}(\boldsymbol{\lambda}) = \mathcal{F}\eta(\boldsymbol{\lambda}) \sum_{j=0}^{M-1} \frac{\Gamma\big(\frac{n}{2}\big)\, \Delta^j (\mathcal{F}\eta)^{-1}(\mathbf{0})}{j!\, 4^j\, \Gamma\big(j+\frac{n}{2}\big)} \, |\boldsymbol{\lambda}|^{2j} \, .$$

REMARK 3.4. An interesting feature of (3.12) is its additive structure. The approximation order of a given quasi-interpolant

$$\mathcal{D}^{-n/2} \sum_{\mathbf{m}\in\mathbb{Z}^n} u(h\mathbf{m})\, \eta_{2M}\Big(\frac{\mathbf{x}-h\mathbf{m}}{\sqrt{\mathcal{D}}\,h}\Big)$$

can be increased by 2 if a new sum of the form

$$\mathcal{D}^{-n/2} \sum_{\mathbf{m}\in\mathbb{Z}^n} u(h\mathbf{m})\, \Phi\Big(\frac{\mathbf{x}-h\mathbf{m}}{\sqrt{\mathcal{D}}\,h}\Big)$$

with the function

$$\Phi(\mathbf{x}) = \Gamma\Big(\frac{n}{2}\Big) \frac{(-1)^M \, \Delta^M (\mathcal{F}\eta)^{-1}(\mathbf{0})}{M!\,(4\pi)^{2M}\,\Gamma\big(M+\frac{n}{2}\big)} \, \Delta^M \eta(\mathbf{x})$$

is added to the quasi-interpolant.

3.3.4. Example. We apply formula (3.12) to the n-dimensional Gaussian $\eta(\mathbf{x}) = \mathrm{e}^{-|\mathbf{x}|^2}$. From

$$\begin{aligned}\Delta^j (\mathcal{F}\eta)^{-1}(\mathbf{0}) = \pi^{-n/2}\, \Delta^j\, \mathrm{e}^{\pi^2|\boldsymbol{\xi}|^2}\Big|_{\boldsymbol{\xi}=\mathbf{0}} &= (-1)^j \pi^{-n/2}\, \Delta^j\, \mathrm{e}^{-\pi^2|\boldsymbol{\xi}|^2}\Big|_{\boldsymbol{\xi}=\mathbf{0}} \\ &= (-1)^j \pi^{-n} \Delta^j \mathcal{F}\eta(\mathbf{0})\,,\end{aligned}$$

we see that

$$\Delta^j (\mathcal{F}\eta)^{-1}(\mathbf{0}) = \frac{(2\pi)^{2j}}{\pi^n} \int_{\mathbb{R}^n} |\mathbf{x}|^{2j}\, \mathrm{e}^{-|\mathbf{x}|^2}\, d\mathbf{x} = \frac{(2\pi)^{2j}\omega_n}{\pi^n} \int_0^\infty r^{2j+n-1}\, \mathrm{e}^{-r^2}\, dr\,.$$

Using

$$\int_0^\infty r^{k-1}\, \mathrm{e}^{-r^2}\, dr = \frac{\Gamma(k/2)}{2}$$

and making use of (3.11), we obtain

$$\Delta^j (\mathcal{F}\eta)^{-1}(\mathbf{0}) = \frac{(4\pi^2)^j\, \pi^{n/2}\Gamma\big(j+\frac{n}{2}\big)}{\Gamma(n/2)}\,. \tag{3.13}$$

Then formula (3.12) becomes

$$
\begin{aligned}
\eta_{2M}(\mathbf{x}) &= \Gamma\Big(\frac{n}{2}\Big) \sum_{j=0}^{M-1} \frac{(-1)^j \Delta^j (\mathcal{F}\eta)^{-1}(\mathbf{0})}{j!\,(4\pi)^{2j}\,\Gamma\big(j+\frac{n}{2}\big)}\, \Delta^j \,\mathrm{e}^{-|\mathbf{x}|^2} \\
&= \pi^{-n/2} \sum_{j=0}^{M-1} \frac{(-1)^j}{j!\,4^j}\, \Delta^j \,\mathrm{e}^{-|\mathbf{x}|^2} .
\end{aligned} \tag{3.14}
$$

Because

$$
\mathcal{F}(\Delta^j \,\mathrm{e}^{-|\cdot|^2})(\boldsymbol{\lambda}) = \pi^{n/2}(-1)^j (2\pi)^{2j} |\boldsymbol{\lambda}|^{2j}\,\mathrm{e}^{-\pi^2|\boldsymbol{\lambda}|^2}
$$

and from (2.12), we obtain

$$
\begin{aligned}
\Delta^j \,\mathrm{e}^{-|\mathbf{x}|^2} &= \frac{\pi^{n/2}(-1)^j (2\pi)^{2j+1}}{|\mathbf{x}|^{\,n/2-1}} \int_0^\infty r^{\,2j}\,\mathrm{e}^{-\pi^2 r^2}\, J_{n/2-1}(2\pi r|\mathbf{x}|)\, r^{\,n/2} dr \\
&= (-1)^j\, j!\, 4^j\,\mathrm{e}^{-|\mathbf{x}|^2}\, L_j^{(n/2-1)}(|\mathbf{x}|^2) .
\end{aligned} \tag{3.15}
$$

The last integral can be found in [**7**, 8.6.13], and $L_j^{(\gamma)}$ are the *generalized Laguerre polynomials*, which are defined by

$$
L_k^{(\gamma)}(y) = \frac{\mathrm{e}^{\,y}\, y^{-\gamma}}{k!} \Big(\frac{d}{dy}\Big)^k (\mathrm{e}^{-y}\, y^{k+\gamma}) , \quad \gamma > -1 , \tag{3.16}
$$

and are orthogonal in the space $L_2(0,\infty)$ with weight $y^\gamma\,\mathrm{e}^{-y}$ (see, e.g., [**93**]).

Thus we can rewrite (3.14) as

$$
\eta_{2M}(\mathbf{x}) = \pi^{-n/2}\,\mathrm{e}^{-|\mathbf{x}|^2} \sum_{j=0}^{M-1} L_j^{(n/2-1)}(|\mathbf{x}|^2) ,
$$

which transforms, by using the recurrence relation [**93**, (5.1.14)]

$$
L_j^{(\gamma-1)}(y) = L_j^{(\gamma)}(y) - L_{j-1}^{(\gamma)}(y) \tag{3.17}
$$

to the function

$$
\eta_{2M}(\mathbf{x}) = \pi^{-n/2}\, L_{M-1}^{(n/2)}(|\mathbf{x}|^2)\,\mathrm{e}^{-|\mathbf{x}|^2} .
$$

Note that η_{2M} is the unique function of the form $p(|\mathbf{x}|^2)\,\mathrm{e}^{-|\mathbf{x}|^2}$ with a polynomial p of degree $M-1$, which satisfies the moment Condition 2.15 for $2M$. Indeed, by (3.10), p is subject to

$$
\int_0^\infty y^{k-1+n/2}\, p(y)\,\mathrm{e}^{-y}\, dy = \delta_{0k} \frac{\Gamma(n/2)}{\pi^{n/2}} , \quad k = 0, \dots, M-1 ,
$$

i.e., the polynomial p of degree $M-1$ is orthogonal to all polynomials of degree less than $M-1$ with respect to the weight $y^{n/2}\,\mathrm{e}^{-y}$. Hence $p(y) = cL_{M-1}^{(n/2)}(y)$, and the last integral with $k=0$ and (3.16) lead to $c = \pi^{-n/2}$.

THEOREM 3.5. *An n-dimensional approximate quasi-interpolant of the form (2.23) with approximation order $\mathcal{O}(h^{2M})$ can be constructed with the generating* *function*

$$
\begin{aligned}
\eta_{2M}(\mathbf{x}) &= \pi^{-n/2}\, L_{M-1}^{(n/2)}(|\mathbf{x}|^2)\, \mathrm{e}^{-|\mathbf{x}|^2} = \pi^{-n/2} \sum_{j=0}^{M-1} L_j^{(n/2-1)}(|\mathbf{x}|^2)\, \mathrm{e}^{-|\mathbf{x}|^2} \\
&= \pi^{-n/2} \sum_{j=0}^{M-1} \frac{(-1)^j}{j!\, 4^j}\, \Delta^j\, \mathrm{e}^{-|\mathbf{x}|^2}\,,
\end{aligned}
\tag{3.18}
$$

which has the Fourier transform

$$
\mathcal{F}\eta_{2M}(\boldsymbol{\lambda}) = \mathrm{e}^{-\pi^2|\boldsymbol{\lambda}|^2} \sum_{j=0}^{M-1} \frac{(\pi^2|\boldsymbol{\lambda}|^2)^j}{j!}\,.
\tag{3.19}
$$

The functions $\{\eta_{2M}\}$, $M = 1, 2, \ldots$, *form the unique system* $\{p(|\mathbf{x}|^2)\,\mathrm{e}^{-|\mathbf{x}|^2}\}$ *with univariate polynomials* p *of degree* $M-1$ *satisfying the moment Condition* 2.15 *for* $N = 2M$.

REMARK 3.6. The property mentioned in Remark 3.4 takes the form

$$
\eta_{2M+2}(\mathbf{x}) = \eta_{2M}(\mathbf{x}) + \pi^{-n/2}\, L_M^{(n/2-1)}(|\mathbf{x}|^2)\, \mathrm{e}^{-|\mathbf{x}|^2}\,.
$$

Therefore the quasi-interpolants

$$
\mathcal{M}_{h,\mathcal{D}}^{(2M)} u(\mathbf{x}) := \mathcal{D}^{-n/2} \sum_{\mathbf{m}\in\mathbb{Z}^n} u(h\mathbf{m})\, \eta_{2M}\Big(\frac{\mathbf{x}-h\mathbf{m}}{\sqrt{\mathcal{D}}\,h}\Big)\,, \qquad M = 1, 2, \ldots,
\tag{3.20}
$$

allow the representations

$$
\begin{aligned}
\mathcal{M}_{h,\mathcal{D}}^{(2M+2)} u(\mathbf{x}) &= \mathcal{M}_{h,\mathcal{D}}^{(2M)} u(\mathbf{x}) \\
&\quad + \frac{1}{(\pi\mathcal{D})^{n/2}} \sum_{\mathbf{m}\in\mathbb{Z}^n} u(h\mathbf{m})\, L_M^{(n/2-1)}\Big(\frac{|\mathbf{x}-h\mathbf{m}|^2}{\mathcal{D}h^2}\Big)\, \mathrm{e}^{-|\mathbf{x}-h\mathbf{m}|^2/\mathcal{D}h^2} \\
&= \mathcal{M}_{h,\mathcal{D}}^{(2M)} u(\mathbf{x}) + \frac{(-1)^M}{M!}\Big(\frac{\mathcal{D}h^2}{4}\Big)^M \Delta^M \mathcal{M}_{h,\mathcal{D}}^{(2)} u(\mathbf{x})\,.
\end{aligned}
\tag{3.21}
$$

REMARK 3.7. Note that formula (3.5) corresponding to $n = 1$ is in accordance with the relation

$$
L_k^{(1/2)}(\tau^2) = \frac{(-1)^k}{k!\, 2^{2k+1}}\, \frac{H_{2k+1}(\tau)}{\tau}\,;
\tag{3.22}
$$

see [**1**, 22.5.39].

3.3.5. Anisotropic Gaussian function. An important generalization of the Gaussian is given by the exponential function

$$
\mathrm{e}^{-\langle A\mathbf{x},\mathbf{x}\rangle}\,, \quad \mathbf{x}\in\mathbb{R}^n\,,
$$

where A is a non-singular $n\times n$ matrix of the form $A = A_R + iA_I$ with two real symmetric matrices A_R and A_I satisfying $\operatorname{Re} A = A_R \geq 0$. Note that the inverse matrix A^{-1} belongs to the same set of matrices. The Fourier transform of this *anisotropic Gaussian function* is given by

$$
\mathcal{F}(\mathrm{e}^{-\langle A\cdot,\cdot\rangle})(\boldsymbol{\lambda}) = \frac{\pi^{n/2}}{(\det A)^{1/2}}\, \mathrm{e}^{-\pi^2\langle A^{-1}\boldsymbol{\lambda},\boldsymbol{\lambda}\rangle}\,,
\tag{3.23}
$$

where the square root of $\det A$ is defined as the value of the analytic branch of $(\det B)^{-1/2}$ on the convex set of symmetric matrices $\{B\}$ with $\operatorname{Re} B > 0$ which satisfies $(\det B)^{-1/2} > 0$ for real B (cf. [**34**], where further explanations are given).

Then for any symmetric matrix A satisfying $\operatorname{Re} A > 0$ the function

$$\eta^A(\mathbf{x}) := \frac{\mathrm{e}^{-\langle A^{-1}\mathbf{x},\mathbf{x}\rangle}}{\pi^{n/2}(\det A)^{1/2}} \quad \text{with} \quad \mathcal{F}\eta^A(\boldsymbol{\lambda}) = \mathrm{e}^{-\pi^2\langle A\boldsymbol{\lambda},\boldsymbol{\lambda}\rangle} \tag{3.24}$$

generates an approximate quasi-interpolant with approximation order 2. As in the above construction, the function η^A_{2M} with the Fourier transform

$$\mathcal{F}\eta^A_{2M}(\boldsymbol{\lambda}) := \mathrm{e}^{-\pi^2\langle A\boldsymbol{\lambda},\boldsymbol{\lambda}\rangle} \sum_{j=0}^{M-1} \frac{(\pi^2\langle A\boldsymbol{\lambda},\boldsymbol{\lambda}\rangle)^j}{j!}$$

satisfies the moment Condition 2.15 of order $2M$,

$$\eta^A_{2M}(\mathbf{0}) = 1\,, \ \partial^{\boldsymbol{\alpha}}\mathcal{F}\eta^A_{2M}(\mathbf{0}) = 0\,, \quad \forall\, \boldsymbol{\alpha} \text{ with } 0 < [\boldsymbol{\alpha}] < 2M\,.$$

Using the notation

$$\langle A\nabla,\nabla\rangle = \sum_{j,k=1}^{n} a_{jk}\frac{\partial^2}{\partial x_j\partial x_k}$$

and the properties of the Fourier transform, the function η^A_{2M} can be given as

$$\eta^A_{2M}(\mathbf{x}) = \frac{1}{\pi^{n/2}(\det A)^{1/2}} \sum_{j=0}^{M-1} \frac{(-1)^j}{j!\,4^j}\,\langle A\nabla,\nabla\rangle^j\,\mathrm{e}^{-\langle A^{-1}\mathbf{x},\mathbf{x}\rangle}\,.$$

Similarly to the proof of Theorem 3.5, one establishes

$$\begin{aligned}\eta^A_{2M}(\mathbf{x}) &= \frac{1}{\pi^{n/2}(\det A)^{1/2}} \sum_{j=0}^{M-1} L_j^{(n/2-1)}(\langle A^{-1}\mathbf{x},\mathbf{x}\rangle)\ \mathrm{e}^{-\langle A^{-1}\mathbf{x},\mathbf{x}\rangle}\\ &= \frac{1}{\pi^{n/2}(\det A)^{1/2}}\, L_{M-1}^{(n/2)}(\langle A^{-1}\mathbf{x},\mathbf{x}\rangle)\ \mathrm{e}^{-\langle A^{-1}\mathbf{x},\mathbf{x}\rangle}\,.\end{aligned} \tag{3.25}$$

If $\operatorname{Re} A > 0$, then this function generates, due to Theorem 2.17, an approximate quasi-interpolant of order $\mathcal{O}(h^{2M})$.

3.4. Some other methods

In this section, we consider some other analytic methods which provide generating functions η_N such that $\mathcal{F}\eta - 1$ has a zero of higher order N. In particular, we show that $\mathcal{D}$ may serve not only as a scaling parameter to control the saturation error. This parameter is also useful to obtain other approximating functions adopted to special situations. In the following we suppose that the basis function η is symmetric with respect to coordinate planes, i.e. satisfies (3.26), and has unit 0-th moment

$$\int_{\mathbb{R}^n} \eta(\mathbf{x})\,d\mathbf{x} = 1\,. \tag{3.26}$$

3.4.1. Using a collection of different $\mathcal{D}$. We consider the function

$$\tilde{\eta}_{I\!D}(\mathbf{x}) = \sum_{j=1}^{M} a_j \mathcal{D}_j^{-n/2} \eta\Big(\frac{\mathbf{x}}{\sqrt{\mathcal{D}_j}}\Big)$$

for a given M-tuple $I\!D = (\mathcal{D}_1, \dots, \mathcal{D}_M)$, $\mathcal{D}_j > 0$, $\mathcal{D}_j \neq \mathcal{D}_l$, $j \neq l$. The $\boldsymbol{\alpha}$-th moment of $\tilde{\eta}_{I\!D}$ evaluates to

$$\int_{\mathbb{R}^n} \mathbf{x}^{\boldsymbol{\alpha}} \tilde{\eta}_{I\!D}(\mathbf{x})\, d\mathbf{x} = \sum_{j=1}^{M} a_j \mathcal{D}_j^{-n/2} \int_{\mathbb{R}^n} \mathbf{x}^{\boldsymbol{\alpha}} \eta\Big(\frac{\mathbf{x}}{\sqrt{\mathcal{D}_j}}\Big)\, d\mathbf{x} = \int_{\mathbb{R}^n} \mathbf{x}^{\boldsymbol{\alpha}} \eta(\mathbf{x})\, d\mathbf{x} \sum_{j=1}^{M} a_j \mathcal{D}_j^{[\boldsymbol{\alpha}]/2} .$$

Because of the symmetry (3.26),

$$\int_{\mathbb{R}^n} \mathbf{x}^{\boldsymbol{\alpha}} \eta(\mathbf{x})\, d\mathbf{x} = 0$$

for odd values of $[\boldsymbol{\alpha}]$, the function $\tilde{\eta}_{I\!D}$ satisfies the moment Condition 2.15 of order $N = 2M$ if the coefficients a_j satisfy the linear system

$$\sum_{j=1}^{M} a_j \mathcal{D}_j^{\,k} = \delta_{0k} , \quad k = 0, \dots, M-1 .$$

The unique solution of this system is

$$a_j = \prod_{\substack{k=1\\k\neq j}}^{M} \frac{\mathcal{D}_k}{\mathcal{D}_k - \mathcal{D}_j} .$$

THEOREM 3.8. *Suppose that η satisfies* (3.26), (3.3), *and the decay condition* (2.26) *with $K \geq 2M + n$. Then for any M-tuple of pairwise different positive numbers $\mathcal{D}_j$ the generating function*

$$\tilde{\eta}_{I\!D}(\mathbf{x}) := \sum_{j=1}^{M} \mathcal{D}_j^{-n/2} \eta\Big(\frac{\mathbf{x}}{\sqrt{\mathcal{D}_j}}\Big) \prod_{\substack{k=1\\k\neq j}}^{M} \frac{\mathcal{D}_k}{\mathcal{D}_k - \mathcal{D}_j} \tag{3.27}$$

satisfies the moment condition (2.47) *with $N = 2M$.*

REMARK 3.9. It can be easily seen from formula (3.27) that the generating function $\eta_{I\!D}$ is subjected to the extended decay Condition 2.12 together with η. Since the Fourier transform of $\eta_{I\!D}$ equals

$$\mathcal{F}\tilde{\eta}_{I\!D}(\boldsymbol{\lambda}) = \sum_{j=1}^{M} \mathcal{F}\eta(\sqrt{\mathcal{D}_j}\boldsymbol{\lambda}) \prod_{\substack{k=1\\k\neq j}}^{M} \frac{\mathcal{D}_k}{\mathcal{D}_k - \mathcal{D}_j} ,$$

we conclude from Theorem 2.17 that the approximate quasi-interpolant

$$\mathcal{D}^{-n/2} \sum_{\mathbf{m}\in\mathbb{Z}^n} u(h\mathbf{m})\, \tilde{\eta}_{I\!D}\Big(\frac{\mathbf{x} - h\mathbf{m}}{\sqrt{\mathcal{D}}\, h}\Big)$$

has the approximation order $2M$ and the saturation error can be controlled by the values of

$$\sum_{\boldsymbol{\nu}\in\mathbb{Z}^n\setminus\{\mathbf{0}\}} \Big| \sum_{j=1}^{M} \partial^{\boldsymbol{\alpha}} \mathcal{F}\eta(\sqrt{\mathcal{D}\mathcal{D}_j}\boldsymbol{\nu}) \prod_{\substack{k=1\\k\neq j}}^{M} \frac{\mathcal{D}_k}{\mathcal{D}_k - \mathcal{D}_j} \Big| , \quad 0 \leq [\boldsymbol{\alpha}] < 2M .$$

Note that the coefficients in (3.27) do not depend on η. Hence, Theorem 3.8 provides a very simple method to enlarge the approximation orders which can be applied without further effort to any sufficiently smooth and decaying function η.

A practical realization of this construction is the following: Suppose that for given values $\{u(h\mathbf{m}),\ \mathbf{m} \in \mathbb{Z}^n\}$ and different $\mathcal{D}_j$, we have approximate quasi-interpolants

$$\mathcal{M}_{j,h}u(\mathbf{x}) = \mathcal{D}_j^{-n/2} \sum_{\mathbf{m}\in\mathbb{Z}^n} u(h\mathbf{m})\, \eta\Big(\frac{\mathbf{x}-h\mathbf{m}}{\sqrt{\mathcal{D}_j}\,h}\Big)\,,\ j=1,\dots,M\,,$$

of the order $\mathcal{O}(h^2)$. Then by Remark 3.9, the linear combination

$$\sum_{j=1}^{M} \prod_{\substack{k=1\\k\neq j}}^{M} \frac{\mathcal{D}_l}{\mathcal{D}_k-\mathcal{D}_j} \mathcal{M}_{j,h}u(\mathbf{x})$$

provides an approximate approximation to u of the order $\mathcal{O}(h^{2M})$.

3.4.2. Derivatives with respect to $\mathcal{D}$. Another construction can be derived from formula (3.27) letting $\mathcal{D}_j \to \tau > 0\ ,\ j=1,\dots,M$. The expression

$$\sum_{j=1}^{M} \prod_{\substack{k=1\\k\neq j}}^{M} \frac{x_k}{x_k-x_j} f(x_j)$$

is the value at the point $x=0$ of the polynomial interpolating the function f at x_j, $j=1,\dots,M$. If all $x_j \to y$, then this polynomial tends to the polynomial interpolating the derivatives $f^{(k)}(y)$, $k=0,\dots,M-1$, i.e.,

$$\sum_{j=1}^{M} \prod_{\substack{k=1\\k\neq j}}^{M} \frac{x_k}{x_k-x_j} f(x_j) \longrightarrow \sum_{k=0}^{M-1} \frac{(-y)^k}{k!} f^{(k)}(y) \quad \text{as} \quad x_j \to y\,,\ j=1,\dots,M\,.$$

Thus, if $\mathcal{D}_j \to \tau > 0$, $j=1,\dots,M$, then

$$\sum_{j=1}^{M} \prod_{\substack{k=1\\k\neq j}}^{M} \frac{\mathcal{D}_k}{\mathcal{D}_k-\mathcal{D}_j} \mathcal{D}_j^{-n/2} \eta\Big(\frac{\mathbf{x}}{\sqrt{\mathcal{D}_j}}\Big) \longrightarrow \sum_{j=0}^{M-1} \frac{(-\tau)^j}{j!} \frac{\partial^j}{\partial\tau^j}\Big(\tau^{-n/2}\eta\Big(\frac{\mathbf{x}}{\sqrt{\tau}}\Big)\Big)\,.$$

To show that the moment condition holds, we note that

$$\sum_{j=0}^{M-1} \frac{(-\tau)^j}{j!} \frac{\partial^j}{\partial\tau^j}\Big(\tau^{-n/2}\eta\Big(\frac{\mathbf{x}}{\sqrt{\tau}}\Big)\Big) = \frac{(-1)^{M-1}\tau^M}{(M-1)!}\Big(\frac{\partial}{\partial\tau}\Big)^{M-1}\Big(\tau^{-1-n/2}\eta\Big(\frac{\mathbf{x}}{\sqrt{\tau}}\Big)\Big)\,.$$

Hence for $\tau > 0$ and any multi-index $\boldsymbol{\alpha}$, $[\boldsymbol{\alpha}] < 2M$, we obtain

$$\begin{aligned}\frac{(-1)^{M-1}\tau^M}{(M-1)!}\Big(\frac{d}{d\tau}\Big)^{M-1}&\Big(\tau^{-1-n/2}\int_{\mathbb{R}^n}\eta\Big(\frac{\mathbf{x}}{\sqrt{\tau}}\Big)\mathbf{x}^{\boldsymbol{\alpha}}\,d\mathbf{x}\Big)\\ &= \frac{(-1)^{M-1}}{(M-1)!}\int_{\mathbb{R}^n}\eta(\mathbf{x})\,\mathbf{x}^{\boldsymbol{\alpha}}\,d\mathbf{x}\,\tau^M\Big(\frac{d}{d\tau}\Big)^{M-1}\tau^{[\boldsymbol{\alpha}]/2-1} = \delta_{[\boldsymbol{\alpha}]0}\,,\end{aligned}$$

where we use the symmetry (3.3) if $[\boldsymbol{\alpha}]$ is an odd number.

THEOREM 3.10. *Suppose that η is continuously differentiable up to order $M-1$ and that it satisfies (3.26), (3.3), and (2.45) with $K \ge 2M+n$. Then the generating function*

$$\tilde{\eta}_M(\mathbf{x}) := \frac{(-1)^{M-1}}{(M-1)!}\Big(\frac{\partial}{\partial\tau}\Big)^{M-1}\Big(\tau^{-1-n/2}\eta\Big(\frac{\mathbf{x}}{\sqrt{\tau}}\Big)\Big)\Big|_{\tau=1} \tag{3.28}$$

satisfies the moment condition (2.47) with $N = 2M$. The Fourier transform of $\tilde{\eta}_M(\mathbf{x})$ is given by

$$\mathcal{F}\tilde{\eta}_M(\boldsymbol{\lambda}) = \frac{(-1)^{M-1}}{(M-1)!}\Big(\frac{\partial}{\partial\tau}\Big)^{M-1}\Big(\tau^{-1}\mathcal{F}\eta(\sqrt{\tau}\boldsymbol{\lambda})\Big)\Big|_{\tau=1}.$$

REMARK 3.11. Note that formula (3.28) applied to $e^{-|\mathbf{x}|^2}$ gives the function system described in Theorem 3.5 by the uniqueness of this system, but, in general, (3.4) and (3.28) lead to different systems.

3.5. Linear combinations of translates

We consider quasi-interpolation formulas

$$\sum_{\mathbf{m}\in\mathbb{Z}^n} u(h\mathbf{m})\,\eta_{\mathcal{D}}\Big(\frac{\mathbf{x}}{h}-\mathbf{m}\Big), \tag{3.29}$$

where for given $\mathcal{D}$ the generating function $\eta_{\mathcal{D}}$ is a linear combination of shifts

$$\eta_{\mathcal{D}}(\mathbf{x}) = \mathcal{D}^{-n/2}\sum_{\mathbf{k}\in\Lambda}\gamma_{\mathbf{k}}\,\eta\Big(\frac{\mathbf{x}-\mathbf{k}}{\sqrt{\mathcal{D}}}\Big) \tag{3.30}$$

for a finite subset $\Lambda \subset \mathbb{Z}^n$. Then one can rewrite

$$\sum_{\mathbf{m}\in\mathbb{Z}^n} u(h\mathbf{m})\,\eta_{\mathcal{D}}\Big(\frac{\mathbf{x}}{h}-\mathbf{m}\Big) = \mathcal{D}^{-n/2}\sum_{\mathbf{m}\in\mathbb{Z}^n} F^h_{\mathbf{m}}(u)\,\eta\Big(\frac{\mathbf{x}-h\mathbf{m}}{\sqrt{\mathcal{D}}h}\Big) \tag{3.31}$$

with the functionals

$$F^h_{\mathbf{m}}(u) = \sum_{\mathbf{k}\in\Lambda}\gamma_{\mathbf{k}}\,u(h(\mathbf{m}-\mathbf{k}))\,.$$

Hence, the quasi-interpolant (3.29) belongs to the principal shift-invariant subspace spanned by $\eta(\cdot/\sqrt{\mathcal{D}})$. If η satisfies the decay Condition 2.12 with sufficiently large K and if $\Lambda \subset \mathbb{Z}^n$ and the coefficients $\gamma_{\mathbf{k}}$ are chosen such that $\eta_{\mathcal{D}}$ satisfies the moment Condition 2.15 for $N < K-n$, then one can expect an approximation rate $\mathcal{O}(h^N)$ plus some saturation error.

3.5.1. Construction of the generating function (3.30).

THEOREM 3.12. *Suppose that η satisfies (2.45) and let $\mathcal{F}\eta(\mathbf{0}) \neq 0$. For any positive $N < K-n$, there exist a finite set $\Lambda_N \subset \mathbb{Z}^n$ with the cardinality $|\Lambda_N| = \dfrac{(N+n-1)!}{(N-1)!\,n!}$ and coefficients $\gamma_{\mathbf{k}}$, $\mathbf{k}\in\Lambda_N$, such that the function*

$$\eta_{\mathcal{D}}(\mathbf{x}) = \mathcal{D}^{-n/2}\sum_{\mathbf{k}\in\Lambda_N}\gamma_{\mathbf{k}}\,\eta\Big(\frac{\mathbf{x}-\mathbf{k}}{\sqrt{\mathcal{D}}}\Big) \tag{3.32}$$

satisfies the moment Condition 2.15 of order N. Moreover, for all $[\boldsymbol{\alpha}] < N$ and $\boldsymbol{\nu}\in\mathbb{Z}^n$

$$\partial^{\boldsymbol{\alpha}}\mathcal{F}\eta_{\mathcal{D}}(\boldsymbol{\nu}) = \mathcal{D}^{[\boldsymbol{\alpha}]/2}\sum_{\boldsymbol{\beta}\le\boldsymbol{\alpha}}\frac{\boldsymbol{\alpha}!}{\boldsymbol{\beta}!\,(\boldsymbol{\alpha}-\boldsymbol{\beta})!}\,\partial^{\boldsymbol{\beta}}\mathcal{F}\eta(\sqrt{\mathcal{D}}\boldsymbol{\nu})\,\partial^{\boldsymbol{\alpha}-\boldsymbol{\beta}}(\mathcal{F}\eta)^{-1}(\mathbf{0})\,.$$

PROOF. The Fourier transform of (3.32) equals

$$\mathcal{F}\eta_{\mathcal{D}}(\boldsymbol{\lambda}) = \mathcal{F}\eta(\sqrt{\mathcal{D}}\boldsymbol{\lambda}) \sum_{\mathbf{k}\in\Lambda} \gamma_{\mathbf{k}}\, \mathrm{e}^{-2\pi i \langle \mathbf{k}\,,\,\lambda\rangle} \,.$$

We want to find a trigonometric polynomial

$$P_\Lambda(\boldsymbol{\lambda}) = \sum_{\mathbf{k}\in\Lambda} \gamma_{\mathbf{k}}\, \mathrm{e}^{-2\pi i \langle \mathbf{k}\,,\,\boldsymbol{\lambda}\rangle}$$

such that $\eta_{\mathcal{D}} = \mathcal{F}^{-1}(P_\Lambda\, \mathcal{F}\eta(\sqrt{\mathcal{D}}\,\cdot\,))$ satisfies the moment Condition 2.15 of order N, or equivalently, such that

$$\partial^{\boldsymbol{\alpha}}_{\boldsymbol{\lambda}} \big(P_\Lambda(\boldsymbol{\lambda})\mathcal{F}\eta(\sqrt{\mathcal{D}}\boldsymbol{\lambda}) - 1\big)\Big|_{\boldsymbol{\lambda}=\mathbf{0}} = 0\,, \quad \forall\, \boldsymbol{\alpha}\,,\ 0 \le [\boldsymbol{\alpha}] < N\,.$$

It can be shown similarly to the proof of Theorem 3.1 that this requirement can be achieved with any trigonometric polynomial satisfying

$$\partial^{\boldsymbol{\alpha}} P_\Lambda(\mathbf{0}) = \mathcal{D}^{[\boldsymbol{\alpha}]/2}\, \partial^{\boldsymbol{\alpha}} (\mathcal{F}\eta)^{-1}(\mathbf{0})\,, \quad \forall\, \boldsymbol{\alpha}\,,\ 0 \le [\boldsymbol{\alpha}] < N\,. \tag{3.33}$$

The notation $\partial^{\boldsymbol{\alpha}} (\mathcal{F}\eta)^{-1}(\mathbf{0})$ is explained in (3.1).

This means that the coefficients $\gamma_{\mathbf{k}}$ have to satisfy the system of linear algebraic equations

$$\sum_{\mathbf{k}\in\Lambda} \gamma_{\mathbf{k}} \mathbf{k}^{\boldsymbol{\alpha}} = \Big(-\frac{\sqrt{\mathcal{D}}}{2\pi i}\Big)^{[\boldsymbol{\alpha}]} \partial^{\boldsymbol{\alpha}} (\mathcal{F}\eta)^{-1}(0)\,, \quad 0 \le [\boldsymbol{\alpha}] < N\,. \tag{3.34}$$

It is known (cf., e.g., [**88**, XIV.1]) that for any N there exist various sets $\Lambda \subset \mathbb{Z}^n$ with $|\Lambda| = \dfrac{(N+n-1)!}{(N-1)!\,n!}$ such that the linear system (3.34) is uniquely solvable. Taking one of these sets Λ, the first assertion is established.

Now, we note that

$$\partial^{\boldsymbol{\alpha}} \mathcal{F}\eta_{\mathcal{D}}(\boldsymbol{\lambda}) = \sum_{\boldsymbol{\beta}\le\boldsymbol{\alpha}} \frac{\boldsymbol{\alpha}!}{\boldsymbol{\beta}!\,(\boldsymbol{\alpha}-\boldsymbol{\beta})!}\, \mathcal{D}^{[\boldsymbol{\beta}]/2}\, \partial^{\boldsymbol{\beta}} \mathcal{F}\eta(\sqrt{\mathcal{D}}\boldsymbol{\lambda})\, \partial^{\boldsymbol{\alpha}-\boldsymbol{\beta}} P_\Lambda(\boldsymbol{\lambda})\,;$$

hence if $\boldsymbol{\lambda} = \boldsymbol{\nu} \in \mathbb{Z}^n$, the periodicity of $P_\Lambda(\lambda)$ implies

$$\partial^{\boldsymbol{\alpha}} \mathcal{F}\eta_{\mathcal{D}}(\boldsymbol{\nu}) = \sum_{\boldsymbol{\beta}\le\boldsymbol{\alpha}} \frac{\boldsymbol{\alpha}!}{\boldsymbol{\beta}!\,(\boldsymbol{\alpha}-\boldsymbol{\beta})!}\, \mathcal{D}^{[\boldsymbol{\beta}]/2}\, \partial^{\boldsymbol{\beta}} \mathcal{F}\eta(\sqrt{\mathcal{D}}\boldsymbol{\nu})\, \partial^{\boldsymbol{\alpha}-\boldsymbol{\beta}} P_\Lambda(\mathbf{0})\,.$$

Taking into account (3.33), we get

$$\partial^{\boldsymbol{\alpha}} \mathcal{F}\eta_{\mathcal{D}}(\boldsymbol{\nu}) = \sum_{\boldsymbol{\beta}\le\boldsymbol{\alpha}} \frac{\boldsymbol{\alpha}!}{\boldsymbol{\beta}!\,(\boldsymbol{\alpha}-\boldsymbol{\beta})!}\, \mathcal{D}^{[\boldsymbol{\beta}]/2}\, \partial^{\boldsymbol{\beta}} \mathcal{F}\eta(\sqrt{\mathcal{D}}\boldsymbol{\nu})\, \mathcal{D}^{[\boldsymbol{\alpha}-\boldsymbol{\beta}]/2} \partial^{\boldsymbol{\alpha}-\boldsymbol{\beta}} (\mathcal{F}\eta)^{-1}(\mathbf{0})\,,$$

which establishes the second assertion. □

Let us mention that the quasi-interpolant

$$\mathcal{M}_h^{\mathcal{D}} u(\mathbf{x}) := \sum_{\mathbf{m}\in\mathbb{Z}^n} u(h\mathbf{m})\eta_{\mathcal{D}}\Big(\frac{\mathbf{x}}{h} - \mathbf{m}\Big) \tag{3.35}$$

differs from the quasi-interpolants considered up to now: The matter is that the dependence of the generating function $\eta_{\mathcal{D}}$ on the parameter $\mathcal{D}$ is more complicated than for quasi-interpolants of the form (2.23). The quasi-interpolant $\mathcal{M}_h^{\mathcal{D}}$ is not generated by a scaled function

$$\mathcal{D}^{-n/2}\, \eta(\cdot/\sqrt{\mathcal{D}}),$$

but by a linear combination of translates of this scaled function with coefficients depending on $\mathcal{D}$, but not on h. However, the more complicated dependence is compensated by the fact that the function η has to satisfy the moment Condition 2.15 with large N in the case (2.23), whereas here only the 0-th moment η must be different from zero.

Nevertheless, the approximation properties of $\mathcal{M}_h^{\mathcal{D}}$ are similar to those of $\mathcal{M}_{h,\mathcal{D}}$. Since we have no scaling parameter $\mathcal{D}$ in (3.35), the expansion (2.50) applied to $\mathcal{M}_h^{\mathcal{D}} u$ leads to the representation

$$\mathcal{M}_h^{\mathcal{D}} u(\mathbf{x}) = u(\mathbf{x}) + \sum_{[\boldsymbol{\alpha}]=0}^{N-1} \frac{\partial^{\boldsymbol{\alpha}} u(\mathbf{x})}{\boldsymbol{\alpha}!} \Big(\frac{h}{2\pi i}\Big)^{[\boldsymbol{\alpha}]} \sum_{\boldsymbol{\nu}\in\mathbb{Z}^n\setminus\{\mathbf{0}\}} \partial^{\boldsymbol{\alpha}} \mathcal{F}\eta_{\mathcal{D}}(\boldsymbol{\nu})\, \mathrm{e}^{\frac{2\pi i}{h}\langle \mathbf{x},\boldsymbol{\nu}\rangle} + R_{N,h}(\mathbf{x})$$

with the remainder term

$$R_{N,h}(\mathbf{x}) = (-h)^N \sum_{[\boldsymbol{\alpha}]=N} \frac{1}{\boldsymbol{\alpha}!} \sum_{\mathbf{m}\in\mathbb{Z}^n} \Big(\frac{\mathbf{x}}{h} - \mathbf{m}\Big)^{\boldsymbol{\alpha}} \eta_{\mathcal{D}}\Big(\frac{\mathbf{x}}{h} - \mathbf{m}\Big)\, U_{\boldsymbol{\alpha}}(\mathbf{x}, h\mathbf{m})\,.$$

Thus one obtains the estimate

$$|u(\mathbf{x}) - \mathcal{M}_h^{\mathcal{D}} u(\mathbf{x})| \le c_{\mathcal{D}}\, h^N \,\|\nabla_N u\|_{L_\infty(\mathbb{R}^n)} + \sum_{k=0}^{N-1} \varepsilon_k h^k |\nabla_k u(\mathbf{x})|)\,, \tag{3.36}$$

with the constants

$$c_{\mathcal{D}} = \max_{[\boldsymbol{\alpha}]=N} \frac{1}{\boldsymbol{\alpha}!} \sum_{\mathbf{m}\in\mathbb{Z}^n} \big|(\mathbf{x}-\mathbf{m})^{\boldsymbol{\alpha}}\, \eta_{\mathcal{D}}(\mathbf{x}-\mathbf{m})\big|\,,$$

and

$$\begin{aligned} \varepsilon_k &= (2\pi)^{-k} \Big| \max_{[\boldsymbol{\alpha}]=k} \frac{1}{\boldsymbol{\alpha}!} \sum_{\boldsymbol{\nu}\in\mathbb{Z}^n} \partial^{\boldsymbol{\alpha}} \mathcal{F}\eta_{\mathcal{D}}(\boldsymbol{\nu})\, \mathrm{e}^{\frac{2\pi i}{h}\langle \mathbf{x},\boldsymbol{\nu}\rangle} \Big| \\ &\le \Big(\frac{\sqrt{\mathcal{D}}}{2\pi}\Big)^k \max_{[\boldsymbol{\alpha}]=k} \sum_{\boldsymbol{\beta}\le\boldsymbol{\alpha}} \frac{|\partial^{\boldsymbol{\alpha}-\boldsymbol{\beta}}(\mathcal{F}\eta)^{-1}(\mathbf{0})|}{\boldsymbol{\beta}!\,(\boldsymbol{\alpha}-\boldsymbol{\beta})!} \sum_{\boldsymbol{\nu}\in\mathbb{Z}^n} |\partial^{\boldsymbol{\beta}} \mathcal{F}\eta(\sqrt{\mathcal{D}}\boldsymbol{\nu})|\,. \end{aligned}$$

THEOREM 3.13. *Let $u \in W_\infty^N(\mathbb{R}^n)$ for some $N \in \mathbb{N}$ and suppose that η satisfies the extended decay Condition 2.12 and its zeroth moment is different from 0. Then for any $\varepsilon > 0$ there exist $\mathcal{D} > 0$, a discrete set $\Lambda_N \in \mathbb{Z}^n$, and coefficients $\gamma_{\mathbf{k}}$, $\mathbf{k} \in \Lambda_N$, such that the quasi-interpolant (3.35) generated by*

$$\eta_{\mathcal{D}}(\mathbf{x}) = \mathcal{D}^{-n/2} \sum_{\mathbf{k}\in\Lambda_N} \gamma_{\mathbf{k}}\, \eta\Big(\frac{\mathbf{x}-\mathbf{k}}{\sqrt{\mathcal{D}}}\Big)$$

satisfies the estimate

$$\big|u(\mathbf{x}) - \mathcal{M}_h^{\mathcal{D}} u(\mathbf{x})\big| \le c_{\mathcal{D}}\, h^M \,\|\nabla_N u\|_{L_\infty} + \varepsilon \sum_{k=0}^{N-1} h^k |\nabla_k u(\mathbf{x})| \tag{3.37}$$

with a constant $c_{\mathcal{D}}$ independent of u and h .

REMARK 3.14. A useful feature of the quasi-interpolant (3.31) is a consequence of fixing the parameter $\mathcal{D}$. We have already mentioned that the coefficients $\gamma_{\mathbf{k}}$, which solve the system (3.34), depend on $\mathcal{D}$ but not on h. Hence, if these coefficients are determined, then they can be used for different h to obtain high-order approximations, which is impossible in the case of $\mathcal{D}$ varying together with h.

From the proof of Theorem 2.28 it is clear that the assertion of Theorem 3.13 can be generalized to estimates in the L_p-norm, $1 \le p < \infty$. Then, as a direct consequence, we get an estimate of the best approximation announced in Subsection 2.4.3.

LEMMA 3.15. *Suppose that $\eta \in \mathcal{S}(\mathbb{R}^n)$ with $\mathcal{F}\eta(\mathbf{0}) \neq 0$. Then for any N and any $\varepsilon > 0$ there exist $\mathcal{D} > 0$ and a constant c_N such that for all $u \in W_p^N(\mathbb{R}^n)$, $1 \le p \le \infty$,*

$$\operatorname{dist}(u,\, S^h(\eta(\cdot/\mathcal{D}); L_\infty)) \le c_N\, (\sqrt{\mathcal{D}}h)^N\, \|\nabla_N u\|_{L_p} + \varepsilon \sum_{k=0}^{N-1} (\sqrt{\mathcal{D}}h)^k \|\nabla_k u\|_{L_p}\,.$$

3.5.2. The case of symmetric η. The construction of $\eta_\mathcal{D}$ in the case of symmetric η and $N = 2M$ can be performed using the trigonometric polynomial

$$P_M(\boldsymbol{\lambda}) = \sum_{[\boldsymbol{\beta}]<M} a_{\boldsymbol{\beta}} \prod_{j=1}^{n} \cos 2\pi\beta_j\lambda_j\,, \qquad \boldsymbol{\beta} = (\beta_1,\ldots,\beta_n) \in \mathbb{Z}^n_{\ge 0}\,. \tag{3.38}$$

Then $\eta_\mathcal{D} = \mathcal{F}^{-1}(P_\Lambda\, \mathcal{F}\eta(\sqrt{\mathcal{D}}\cdot))$ is symmetric in the sense of (3.3). Therefore, (3.34) leads to the linear system

$$\sum_{[\boldsymbol{\beta}]<M} a_{\boldsymbol{\beta}} \boldsymbol{\beta}^{2\boldsymbol{\alpha}} = \Big(-\frac{\mathcal{D}}{4\pi^2}\Big)^{[\boldsymbol{\alpha}]} \partial^{2\boldsymbol{\alpha}} (\mathcal{F}\eta)^{-1}(\mathbf{0})\,, \quad 0 < [\boldsymbol{\alpha}] < M\,. \tag{3.39}$$

EXAMPLE 3.16. Consider the basis function for approximation order $2M$ derived from $\eta(x) = \operatorname{sech} x$, $x \in \mathbb{R}$. Here (3.39) is the linear system

$$\sum_{|\beta|<M} a_\beta \beta^{2\alpha} = \frac{1}{\pi}\Big(-\frac{\mathcal{D}}{4\pi^2}\Big)^{\alpha} \Big(\frac{d}{d\lambda}\Big)^{2\alpha} \cosh \pi^2\lambda\Big|_{\lambda=0}\,, \quad 0 \le \alpha < M\,, \tag{3.40}$$

with the Vandermonde matrix $\big(\beta^{2\alpha}\big)_{\alpha,\beta=0}^{M-1}$. Thus, there exists a unique solution of (3.40) for any M. In the case $M = 4$, for example, the function

$$\eta_\mathcal{D}(x) = \sum_{|\beta|<4} a_\beta \operatorname{sech} \frac{x-\beta}{\mathcal{D}}$$

with the coefficients

$$a_0 = \frac{1}{\pi} + \frac{49\pi\mathcal{D}}{144} + \frac{7\pi^3\mathcal{D}^2}{288} + \frac{\pi^5\mathcal{D}^3}{2304}, \qquad a_{\pm 1} = -\frac{3\pi\mathcal{D}}{8} - \frac{13\pi^3\mathcal{D}^2}{384} - \frac{\pi^5\mathcal{D}^3}{1536},$$

$$a_{\pm 2} = \frac{3\pi\mathcal{D}}{80} + \frac{\pi^3\mathcal{D}^2}{96} + \frac{\pi^5\mathcal{D}^3}{3840}, \qquad a_{\pm 3} = -\frac{\pi\mathcal{D}}{360} - \frac{\pi^3\mathcal{D}^2}{1152} - \frac{\pi^5\mathcal{D}^3}{23040}$$

generates a quasi-interpolant of order $\mathcal{O}(h^8)$.

A set of multi-indices $J \subset \mathbb{Z}^n_{\ge 0}$ is said to be normal if $\boldsymbol{\alpha} \in J$ and $\boldsymbol{\beta} \le \boldsymbol{\alpha}$ imply $\boldsymbol{\beta} \in J$. The linear space of n-variate polynomials associated with J is denoted by

$$\mathcal{P}_J = \operatorname{span}\{\mathbf{x}^{\boldsymbol{\alpha}} : \boldsymbol{\alpha} \in J\}.$$

Furthermore, for $1 \le j \le n$ let there be given sequences $T_j = \{t_{j,l}\}_{l=0}^{\infty}$ of different real numbers and consider the lattice

$$\mathcal{T}_J = \{\mathbf{t}_{\boldsymbol{\alpha}} = (t_{1,\alpha_1},\ldots,t_{n,\alpha_n}) : \boldsymbol{\alpha} \in J\}\,,$$

which is determined by the sequences T_j and the index set J.

In the multi-variate case the matrix $(\boldsymbol{\beta}^{2\boldsymbol{\alpha}})_{[\boldsymbol{\alpha}],[\boldsymbol{\beta}]=0}^{M-1}$, $\boldsymbol{\alpha},\boldsymbol{\beta} \in \mathbb{Z}^n_{\geq 0}$, is non-singular owing to the following general result of Hakopian [**32**]:

THEOREM 3.17 ([**32**]). *Let $J \subset \mathbb{Z}^n_{\geq 0}$ be a normal set. Then each polynomial $p \in \mathcal{P}_J$ is uniquely determined by its values on the lattice $\mathcal{T}_J$.*

Note that

$$\prod_{j=1}^{n} \cos 2\pi\beta_j\lambda_j = 2^{-\kappa(\boldsymbol{\beta})} \sum_{\xi(\mathbf{k})=\boldsymbol{\beta}} \mathrm{e}^{2\pi i\langle \mathbf{k},\boldsymbol{\lambda}\rangle} ,$$

where $\xi(\mathbf{k})$ denotes the vector $(|k_1|,\ldots,|k_n|) \in \mathbb{Z}^n_{\geq}$ and $\kappa(\boldsymbol{\beta})$ stands for the number of non-zero components of the multi-index $\boldsymbol{\beta}$. Thus, the unique solution $\{a_{\boldsymbol{\beta}}\}$ of (3.39) provides the generating function

$$\eta_{\mathcal{D}}(\mathbf{x}) = \sum_{[\boldsymbol{\beta}]<M} 2^{-\kappa(\boldsymbol{\beta})} a_{\boldsymbol{\beta}} \sum_{\xi(\mathbf{k})=\boldsymbol{\beta}} \eta\Big(\frac{\mathbf{x}-\mathbf{k}}{\sqrt{\mathcal{D}}}\Big), \tag{3.41}$$

which satisfies the moment condition (2.47) with $N = 2M$. This construction leads to the approximation formula

$$\mathcal{D}^{-n/2} \sum_{\mathbf{m}\in\mathbb{Z}^n} \eta\Big(\frac{\mathbf{x}-h\mathbf{m}}{\sqrt{\mathcal{D}}h}\Big)\Big(\sum_{|\xi(\mathbf{k})|<M} c_{\mathbf{k}}\, u(h(\mathbf{m}-\mathbf{k}))\Big)$$

where $|\xi(\mathbf{k})| = |k_1| + \cdots + |k_n|$ and $c_{\mathbf{k}} = 2^{-\kappa(\xi(\mathbf{k}))} a_{\xi(\mathbf{k})}$. This formula has approximation order $\mathcal{O}(h^{2M})$ plus saturation term and is based on the values of u at the mesh points $h(\mathbf{m}-\mathbf{k})$ with the minimal distance to $h\mathbf{m}$.

REMARK 3.18. For any radial function $\eta(\mathbf{x})$, the solution $\{a_{\boldsymbol{\beta}}\}$ of the linear system (3.39) is independent of any permutation $\sigma(\boldsymbol{\beta})$ of the components of the multi-index $\boldsymbol{\beta}$. Then formula (3.41) can be written in the form

$$\eta_{\mathcal{D}}(\mathbf{x}) = \sum_{\substack{[\boldsymbol{\beta}]<M \\ \beta_1\geq\ldots\geq\beta_n}} 2^{-\kappa(\boldsymbol{\beta})} a_{\boldsymbol{\beta}} \sum_{\sigma(\xi(\mathbf{k}))=\boldsymbol{\beta}} \eta\Big(\frac{\mathbf{x}-\mathbf{k}}{\sqrt{\mathcal{D}}}\Big) .$$

3.6. Matrix-valued basis functions

An interesting consequence of formula (3.18) is the identity

$$\Delta^{-1}\eta_{2M}(\mathbf{x}) = \pi^{-n/2}\Big(\Delta^{-1}\,\mathrm{e}^{-|\mathbf{x}|^2} - \sum_{j=0}^{M-2} \frac{(-1)^j\,\Delta^j\,\mathrm{e}^{-|\mathbf{x}|^2}}{(j+1)!\,4^{j+1}}\Big), \tag{3.42}$$

which will be used in Section 4.3 to obtain cubature formulas for the inverse of the Laplacian.

This formula can be generalized to other partial differential operators. Let, for example, L be a positive definite $m \times m$ matrix of homogeneous partial differential operators of second order with constant coefficients. Then, for any vector function $\mathbf{f}$ with elements from $\mathcal{S}(\mathbb{R}^n)$, the Fourier transform of $L\mathbf{f}$ has the form

$$\mathcal{F}L\mathbf{f}(\boldsymbol{\lambda}) = 4\pi^2\,|\boldsymbol{\lambda}|^2 A(\boldsymbol{\omega})\,\mathcal{F}\mathbf{f}(\boldsymbol{\lambda}), \quad \boldsymbol{\lambda} \in \mathbb{R}^n\,, \; \boldsymbol{\omega} \in S_{n-1}\,,$$

with a positive definite matrix $A(\boldsymbol{\omega}) \geq c > 0$. We can introduce the exponential matrix function $\mu(\boldsymbol{\lambda}) = \mathrm{e}^{-\pi^2|\boldsymbol{\lambda}|^2 A(\boldsymbol{\omega})}$ and similarly to (3.19), the modified function

$$\mu_{2M}(\boldsymbol{\lambda}) = \mu(\boldsymbol{\lambda}) \sum_{j=0}^{M-1} \frac{(\pi^2|\boldsymbol{\lambda}|^2 A(\boldsymbol{\omega}))^j}{j!},$$

which satisfies

$$\partial^{\boldsymbol{\alpha}}(\mu_{2M}(\boldsymbol{\lambda}) - I)\Big|_{\boldsymbol{\lambda}=0} = 0, \quad 0 \leq [\boldsymbol{\alpha}] < 2M.$$

Hence the inverse Fourier transform of μ_{2M} satisfies the decay and moment Conditions 2.12 and 2.15, which allows us to construct high order approximate quasi-interpolants. This generating matrix function is given by

$$\check{\mu}_{2M}(\mathbf{x}) := \mathcal{F}^{-1}\mu_{2M}(\mathbf{x}) = \sum_{j=0}^{M-1} \frac{1}{j!\,4^j} L^j\,(\mathcal{F}^{-1}\mu)(\mathbf{x}) \tag{3.43}$$

and can be easily derived if $\check{\mu} := \mathcal{F}^{-1}\mu$ is analytically known. Moreover, and this is more important, it is then possible to derive high-order approximations for the inverse of L by

$$L^{-1}\mu_{2M}(\mathbf{x}) = L^{-1}\check{\mu}(\mathbf{x}) - \sum_{j=0}^{M-2} \frac{L^j\,\check{\mu}(\mathbf{x})}{(j+1)!\,4^{j+1}}, \tag{3.44}$$

which is the analogue of (3.42).

3.7. Diminishing the parameter $\mathcal{D}$

In this section we consider the problem of how the parameter $\mathcal{D}$ can be diminished without loss of accuracy of the quasi-interpolant (2.23).

Since $\mathcal{D}$ determines the number of summands in (2.23) necessary to evaluate the approximate value of $u(\mathbf{x})$ within a given tolerance, it is of practical interest, especially for the multi-dimensional case, that this number be as small as possible. On the other hand, the saturation terms are determined by the values of $\partial^{\boldsymbol{\alpha}}\mathcal{F}\eta(\sqrt{\mathcal{D}}\boldsymbol{\nu})$, $\boldsymbol{\nu} \in \mathbb{Z}^n\backslash\{\mathbf{0}\}$, $[\boldsymbol{\alpha}] < N$, such that smaller values of $\mathcal{D}$ enlarge this error, in general. This concerns especially the main part of this error,

$$u(\mathbf{x}) \sum_{\boldsymbol{\nu}\in\mathbb{Z}^n\backslash\{\mathbf{0}\}} \mathcal{F}\eta(\sqrt{\mathcal{D}}\boldsymbol{\nu})\,\mathrm{e}^{-\frac{2\pi i}{h}\langle\mathbf{x},\boldsymbol{\nu}\rangle},$$

which does not decrease together with h.

3.7.1. Perturbation of η.

We start with the equality

$$\begin{aligned}&\mathcal{D}^{-n/2} \sum_{\mathbf{m}\in\mathbb{Z}^n} u(h\mathbf{m})\eta\Big(\frac{\mathbf{x}-h\mathbf{m}}{\sqrt{\mathcal{D}}h}\Big)\\ &\quad= u(\mathbf{x})\,g(\mathbf{x}/h) + \mathcal{D}^{-n/2} \sum_{\mathbf{m}\in\mathbb{Z}^n} (u(h\mathbf{m}) - u(\mathbf{x}))\,\eta\Big(\frac{\mathbf{x}-h\mathbf{m}}{\sqrt{\mathcal{D}}h}\Big),\end{aligned} \tag{3.45}$$

where η satisfies the usual conditions and $g(\mathbf{x})$ denotes the sum

$$g(\mathbf{x}) := \mathcal{D}^{-n/2} \sum_{\mathbf{m}\in\mathbb{Z}^n} \eta\Big(\frac{\mathbf{x}-\mathbf{m}}{\sqrt{\mathcal{D}}}\Big) = \sum_{\boldsymbol{\nu}\in\mathbb{Z}^n} \mathcal{F}\eta(\sqrt{\mathcal{D}}\boldsymbol{\nu})\,\mathrm{e}^{-2\pi i\langle\mathbf{x},\boldsymbol{\nu}\rangle}.$$

From (2.36) and Theorem 2.17 it follows easily that the second term in the right-hand side of (3.45) is bounded by

(3.46)

$$\varepsilon_{u,h} := \sum_{[\boldsymbol{\alpha}]=1}^{N-1} (\sqrt{\mathcal{D}}h)^{[\boldsymbol{\alpha}]} \|\varepsilon_{\boldsymbol{\alpha}}(\cdot,\mathcal{D})\|_{L_\infty} \frac{|\partial^{\boldsymbol{\alpha}} u(\mathbf{x})|}{\boldsymbol{\alpha}!} + c_\eta\, (\sqrt{\mathcal{D}}h)^N \sum_{[\boldsymbol{\alpha}]=N} \frac{\|\partial^{\boldsymbol{\alpha}} u\|_{L_\infty}}{\boldsymbol{\alpha}!}$$

converging to zero together with h.

Since we require from η that its integer shifts span an approximate partition of unity, it is natural to assume that $|g(\mathbf{x})| \geq c > 0$. Hence

$$\Big|\frac{\mathcal{D}^{-n/2}}{g(\mathbf{x}/h)} \sum_{\mathbf{m}\in\mathbb{Z}^n} u(h\mathbf{m})\, \eta\Big(\frac{\mathbf{x}-h\mathbf{m}}{\sqrt{\mathcal{D}}h}\Big) - u(\mathbf{x})\Big| \leq c^{-1}\, \varepsilon_{u,h}\,. \tag{3.47}$$

However, the function $g(\mathbf{x})$ is one-periodic and therefore

$$\eta\Big(\frac{\mathbf{x}-h\mathbf{m}}{\sqrt{\mathcal{D}}h}\Big)\Big/ g\Big(\frac{\mathbf{x}}{h}-\mathbf{m}\Big) = \eta\Big(\frac{\mathbf{x}-h\mathbf{m}}{\sqrt{\mathcal{D}}h}\Big)\Big/ g\Big(\frac{\mathbf{x}}{h}\Big)\,.$$

If we define

$$\tilde{\eta}(\mathbf{x}) := \frac{\eta(\mathbf{x})}{g(\sqrt{\mathcal{D}}\mathbf{x})}\,,$$

we derive a new quasi-interpolant

$$\widetilde{\mathcal{M}}_h u(\mathbf{x}) = \mathcal{D}^{-n/2} \sum_{\mathbf{m}\in\mathbb{Z}^n} u(h\mathbf{m})\, \tilde{\eta}\Big(\frac{\mathbf{x}-h\mathbf{m}}{\sqrt{\mathcal{D}}h}\Big) \tag{3.48}$$

which satisfies the error estimate

$$\big|u(\mathbf{x}) - \widetilde{\mathcal{M}}_h u(\mathbf{x})\big| \leq c^{-1}\varepsilon_{u,h}$$

by (3.47), i.e., even the saturation error of (3.48) tends to zero as $h \to 0$.

In general, it is impossible to get an effective expression for g^{-1}, but a good approximation can be found, for example

$$g_1(\mathbf{x}) = 1 - \sum_{|\boldsymbol{\nu}|=1} \mathcal{F}\eta(\sqrt{\mathcal{D}}\boldsymbol{\nu})\, \mathrm{e}^{-2\pi i \langle \mathbf{x},\boldsymbol{\nu}\rangle}\,.$$

LEMMA 3.19. *Suppose that η satisfies* (2.45), (2.47), *and the conditions of Lemma* 2.11. *Then*

$$\mathcal{D}^{-n/2} \sum_{\mathbf{m}\in\mathbb{Z}^n} u(h\mathbf{m})\, \hat{\eta}\Big(\frac{\mathbf{x}-h\mathbf{m}}{\sqrt{\mathcal{D}}h}\Big),$$

with $\hat{\eta}(\mathbf{x}) = \eta(\mathbf{x})g_1(\sqrt{\mathcal{D}}\mathbf{x})$, *approximates u with the order $\mathcal{O}((\sqrt{\mathcal{D}}h)^N)$ plus some saturation error and the main part of this saturation can be estimated by*

$$\varepsilon_{\mathbf{0}}(\hat{\eta},\mathcal{D}) \leq n \sum_{|\boldsymbol{\nu}|=1} |\mathcal{F}\eta(\sqrt{\mathcal{D}}\boldsymbol{\nu})|^2 + c_1 \sum_{|\boldsymbol{\nu}|\geq 2} |\mathcal{F}\eta(\sqrt{\mathcal{D}}\boldsymbol{\nu})|\,,$$

where $c_1 = \max|g_1(\mathbf{x})|$.

PROOF. Since

$$\eta\Big(\frac{\mathbf{x}-h\mathbf{m}}{\sqrt{\mathcal{D}}h}\Big)\, g_1\Big(\frac{\mathbf{x}-h\mathbf{m}}{h}\Big) = \eta\Big(\frac{\mathbf{x}-h\mathbf{m}}{\sqrt{\mathcal{D}}h}\Big)\, g_1\Big(\frac{\mathbf{x}}{h}\Big)\,,$$

we obtain after multiplying (3.45) by $g_1(\mathbf{x}/h)$

$$\mathcal{D}^{-n/2} \sum_{\mathbf{m}\in\mathbb{Z}^n} u(h\mathbf{m})\,\eta\Big(\frac{\mathbf{x}-h\mathbf{m}}{\sqrt{\mathcal{D}}h}\Big)\, g_1\Big(\frac{\mathbf{x}-h\mathbf{m}}{h}\Big)$$
$$= u(\mathbf{x})\, g_1\Big(\frac{\mathbf{x}}{h}\Big)\, g\Big(\frac{\mathbf{x}}{h}\Big) + g_1\Big(\frac{\mathbf{x}}{h}\Big)\, \mathcal{D}^{-n/2} \sum_{\mathbf{m}\in\mathbb{Z}^n} (u(h\mathbf{m}) - u(\mathbf{x}))\,\eta\Big(\frac{\mathbf{x}-h\mathbf{m}}{\sqrt{\mathcal{D}}h}\Big)\,.$$

Now,

$$g_1(\mathbf{x})g(\mathbf{x}) = 1 - \Big(\sum_{|\boldsymbol{\nu}|=1} \mathcal{F}\eta(\sqrt{\mathcal{D}}\boldsymbol{\nu})\, \mathrm{e}^{-2\pi i\langle \mathbf{x},\boldsymbol{\nu}\rangle}\Big)^2 + g_1(\mathbf{x})\Big(\sum_{|\boldsymbol{\nu}|\geq 2} \mathcal{F}\eta(\sqrt{\mathcal{D}}\boldsymbol{\nu})\, \mathrm{e}^{-2\pi i\langle \mathbf{x},\boldsymbol{\nu}\rangle}\Big),$$

implying the inequality

$$\Big|u(\mathbf{x}) - \mathcal{D}^{-n/2} \sum_{\mathbf{m}\in\mathbb{Z}^n} u(h\mathbf{m})\,\hat{\eta}\Big(\frac{\mathbf{x}-h\mathbf{m}}{\sqrt{\mathcal{D}}h}\Big)\Big|$$
$$\leq |u(\mathbf{x})|\Big(n \sum_{|\boldsymbol{\nu}|=1} |\mathcal{F}\eta(\sqrt{\mathcal{D}}\boldsymbol{\nu})|^2 + c_1 \sum_{|\boldsymbol{\nu}|\geq 2} |\mathcal{F}\eta(\sqrt{\mathcal{D}}\boldsymbol{\nu})|\Big) + c_1 \varepsilon_{u,h}\,. \qquad \square$$

3.7.2. Combinations of translates. The following approach is in the spirit of (3.38). Suppose that η satisfies the usual moment and decay conditions (2.47) and (2.45), suppose it is symmetric as in (3.3), and let $K \geq 2M$. We seek a trigonometric polynomial

$$T_M(\boldsymbol{\lambda}) = \sum_{|\boldsymbol{\beta}|<M} a_{\boldsymbol{\beta}} \prod_{j=1}^{n} \cos \frac{\pi(\beta_j + \frac{1}{2})\lambda_j}{\sqrt{\mathcal{D}}}$$

such that

$$\partial^{2\boldsymbol{\alpha}} T_M(\mathbf{0}) = \partial^{2\boldsymbol{\alpha}} (\mathcal{F}\eta)^{-1}(\mathbf{0})\,, \quad \forall\, \boldsymbol{\alpha},\ [\boldsymbol{\alpha}] < M\,, \tag{3.49}$$

and we define the generating function $\hat{\eta}_{2M}(\mathbf{x}) := \mathcal{F}^{-1}(T_M\, \mathcal{F}\eta)(\mathbf{x})$. Clearly, this function satisfies (3.3), the moment condition

$$\mathcal{F}\hat{\eta}_{2M}(\mathbf{0}) = 1\,,\ \partial^{\boldsymbol{\alpha}} \mathcal{F}\hat{\eta}_{2M}(\mathbf{0}) = 0\,, \quad 1 \leq [\boldsymbol{\alpha}] \leq 2M\,,$$

and additionally, based on the special structure of T_M,

$$\mathcal{F}\hat{\eta}_{2M}(\sqrt{\mathcal{D}}\boldsymbol{\nu}) = 0$$

if at least one component ν_j of $\boldsymbol{\nu}\in\mathbb{Z}^n$ is an odd number.

Thus, $\hat{\eta}_{2M}$ generates an approximate approximation of order $\mathcal{O}(h^{2M})$ and the main part $\varepsilon_{\mathbf{0}}(\hat{\eta}_{2M}, \mathcal{D})$ of the saturation error is bounded by

$$\varepsilon_{\mathbf{0}}(\hat{\eta}_{2M}, \mathcal{D}) \leq \sum_{\boldsymbol{\nu}\in\mathbb{Z}^n\setminus\{\mathbf{0}\}} |\mathcal{F}\hat{\eta}_{2M}(2\sqrt{\mathcal{D}}\boldsymbol{\nu})|\,.$$

The equations (3.49) lead to the system

$$\sum_{|\boldsymbol{\beta}|<M} a_{\boldsymbol{\beta}} (\boldsymbol{\beta} + \frac{\mathbf{e}}{2})^{2\boldsymbol{\alpha}} = \Big(-\frac{\mathcal{D}}{\pi^2}\Big)^{[\boldsymbol{\alpha}]} \partial^{2\boldsymbol{\alpha}} (\mathcal{F}\eta)^{-1}(\mathbf{0})\,, \quad [\boldsymbol{\alpha}] < M\,,$$

where $\mathbf{e} = (1, \ldots, 1)$. This system is similar in form to (3.39) and it is uniquely solvable for all n and M.

We apply this method to the example $\pi^{-3/2}\,\mathrm{e}^{-|\mathbf{x}|^2}$, $n = 3$, $M = 2$, and we derive the generating function

$$\hat{\eta}_4(\mathbf{x}) = \pi^{-3/2}\,\mathrm{e}^{-|\mathbf{x}|^2}\Big(\mathrm{e}^{3/16\mathcal{D}}\,(11/8+3\mathcal{D})\prod_{j=1}^{3}\cosh\frac{x_j}{2\sqrt{\mathcal{D}}} - \mathrm{e}^{11/16\mathcal{D}}\,(1/8+\mathcal{D})\sum_{|\boldsymbol{\beta}|=1}\prod_{j=1}^{3}\cosh\frac{(2\beta_j+1)x_j}{2\sqrt{\mathcal{D}}}\Big)$$

providing the order $\mathcal{O}(h^4)$. Here even the choice $\mathcal{D} = 1$ gives

$$\varepsilon_{\mathbf{0}}(\hat{\eta}_4,\mathcal{D}) < 4.2943\cdot 10^{-17},\quad \varepsilon_{\boldsymbol{\alpha}}(\hat{\eta}_4,\mathcal{D}) < 3.8793\cdot 10^{-5}\,,\ [\boldsymbol{\alpha}] = 1\,.$$

3.8. Notes

This chapter is based on the articles [**66**] and [**70**], where some other methods of constructing generating functions for high-order quasi-interpolation have also been considered.

Note that the analytic methods discussed in Sections 3.3 and 3.4 lead to quasi-interpolants which do not belong to the shift-invariant space generated by the starting basis function η. This is in contrast to the method described in Section 3.5, which is well known for basis functions satisfying the Strang-Fix conditions (cf. [**92**], [**85**]).

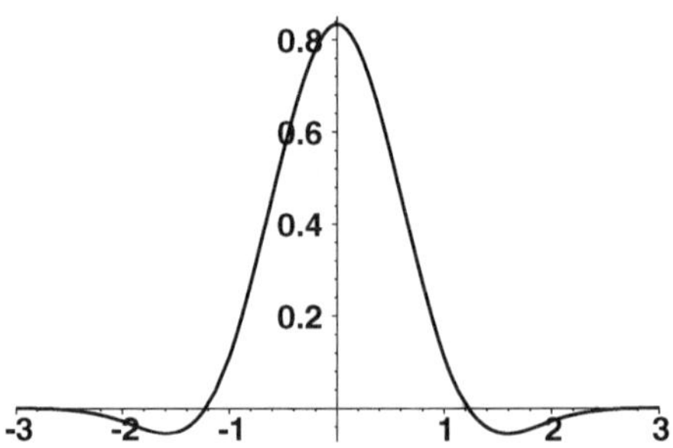

FIGURE 3.3. Graph of $c(x)$

Probably the simplest example in this respect is shown in Fig. 3.3. This function is the linear combination of cubic B-splines (cf. (1.20))

$$c(x) = \frac{1}{6}(8b(x) - b(x-1) - b(x+1))\,,$$

which satisfies the moment Condition 2.15 with $N = 4$ and generates therefore a quasi-interpolant of approximation order $\mathcal{O}(h^4)$.

CHAPTER 4

Approximation of integral operators

4.1. Introduction

In the next three chapters, we discuss the probably most important applications of approximate approximations, which lead to new classes of cubature formulas for integral operators of mathematical physics and to closed formulas for approximate solutions of initial boundary value problems for classical partial differential equations.

Many integral operators of mathematical physics are convolutions with singular kernel functions, for example with fundamental solutions of partial differential operators. Because of the singularity of the integrand, the numerical computation of those integrals by standard methods is an involved and time-consuming task.

Here the use of quasi-interpolants (2.23) with adapted basis functions can be very advantageous. Consider, for example, the convolution operator

$$\mathcal{K}u(\mathbf{x}) := \int_{\mathbb{R}^n} k(\mathbf{x}-\mathbf{y})\, u(\mathbf{y})\, d\mathbf{y}\,. \tag{4.1}$$

A cubature of this multi-dimensional integral can be obtained if the density u is approximated by a quasi-interpolant

$$u_h(\mathbf{x}) := \mathcal{D}^{-n/2} \sum_{\mathbf{m}\in\mathbb{Z}^n} u(h\mathbf{m})\, \eta\Big(\frac{\mathbf{x}-h\mathbf{m}}{\sqrt{\mathcal{D}}h}\Big) \tag{4.2}$$

and the generating function η is chosen such that $\mathcal{K}\eta(\mathbf{x})$ can be computed efficiently. Then the sum

$$\mathcal{K}u_h(\mathbf{x}) = h^n \sum_{\mathbf{m}\in\mathbb{Z}^n} u(h\mathbf{m}) \int_{\mathbb{R}^n} k\Big(\sqrt{\mathcal{D}}h\Big(\frac{\mathbf{x}-h\mathbf{m}}{\sqrt{\mathcal{D}}h} - \mathbf{y}\Big)\Big)\, \eta(\mathbf{y})\, d\mathbf{y} \tag{4.3}$$

provides a simple cubature formula for the convolution (4.1). If the integral $\mathcal{K}\eta(\mathbf{x})$ is expressed analytically, then (4.3) becomes a semi-analytic cubature formula. This allows one to apply other analytic operations to the cubature formula (4.3).

Another advantage of (4.3) is that it retains the convolutional structure of the integral operator. Very often, one has to compute a potential on a given uniform grid as one part of a more involved algorithm. This leads to the computation of a discrete convolution

$$\mathcal{K}u_h(h\mathbf{k}) = h^n \sum_{\mathbf{m}\in\mathbb{Z}^n} a_{\mathbf{k}-\mathbf{m}} u(h\mathbf{m})\,, \quad \mathbf{k}\in\mathbb{Z}^n\,, \tag{4.4}$$

with the coefficients

$$a_{\mathbf{k}} = \int_{\mathbb{R}^n} k\Big(\sqrt{\mathcal{D}}h\Big(\frac{\mathbf{k}}{\sqrt{\mathcal{D}}} - \mathbf{y}\Big)\Big)\, \eta(\mathbf{y})\, d\mathbf{y}\,, \tag{4.5}$$

which can be done efficiently using the Fast Fourier Transform.

If additionally the kernel function k is positive homogeneous of degree β, then

$$\mathcal{K}u_h(h\mathbf{k}) = h^n(\sqrt{\mathcal{D}}h)^\beta \sum_{\mathbf{m}\in\mathbb{Z}^n} b_{\mathbf{k}-\mathbf{m}} u(h\mathbf{m}) \tag{4.6}$$

where the coefficients

$$b_{\mathbf{k}} = \int_{\mathbb{R}^n} k\Big(\frac{\mathbf{k}}{\sqrt{\mathcal{D}}} - \mathbf{y}\Big)\, \eta(\mathbf{y})\, d\mathbf{y} \tag{4.7}$$

do not depend on the step size h. Thus, if the parameter $\mathcal{D}$ is fixed, then one can precompute the required values of $b_{\mathbf{k}}$ and use them for the cubature of (4.1) with different h.

These ideas can be best demonstrated on convolutions with a radial kernel $k(\mathbf{x}) = k(|\mathbf{x}|)$. In Section 2.1 we noted that the Fourier transform $\mathcal{F}k(\boldsymbol{\lambda})$, which can be obtained by (2.12), is also radial. It is convenient to choose η as a radial function also, with $\mathcal{F}\eta(\boldsymbol{\lambda}) = \mathcal{F}\eta(|\boldsymbol{\lambda}|)$. Then by (2.15), the convolution integral in the sum (4.3) reduces to an integral over the half-line

$$\int_{\mathbb{R}^n} k\Big(\sqrt{\mathcal{D}}h\Big(\frac{\mathbf{x}-h\mathbf{m}}{\sqrt{\mathcal{D}}h} - \mathbf{y}\Big)\Big)\, \eta(\mathbf{y})\, d\mathbf{y}$$

$$= \frac{2\pi}{|\mathbf{x}-h\mathbf{m}|^{n/2-1}} \int_0^\infty \mathcal{F}k(r)\mathcal{F}\eta(\sqrt{\mathcal{D}}hr)\, J_{n/2-1}(2\pi r|\mathbf{x}-h\mathbf{m}|) r^{n/2} dr\,.$$

It remains to find a proper radial function η such that this integral can be taken analytically or at least by some efficient quadrature method. So the evaluation of the multi-dimensional integral operator with possibly singular kernel function is reduced to the quadrature of a one-dimensional integral or even to the evaluation of explicit analytic expressions.

In this chapter, we give a detailed study of the already-mentioned approach for two important integral operators.

In the next section, we consider the Hilbert transform, i.e., a one-dimensional singular integral operator. It is shown that the derived semi-analytic cubature formula approximates with high order up to some saturation bound.

Sections 4.3 – 4.5 are devoted to harmonic potentials. First, we derive analytic expressions of their action on the Gaussian function in any space dimension, which are extended later to basis functions of higher-order approximate quasi-interpolation. This results in semi-analytic cubature formulas for harmonic potentials and related pseudodifferential operators.

In Section 4.4, dealing with approximation properties of these formulas, we state the interesting effect that the cubature converges as the mesh size tends to zero, whereas the density is approximated with some saturation error. This is caused by the special structure of the saturation error and its convergence in weak norms.

Finally, in Section 4.5 we describe a direct application of the high-order cubature of harmonic potentials to the solution of boundary value problems for Poisson's equation, using boundary integral equation methods.

4.2. Hilbert transform

Here we consider the approximation of the singular integral

$$\mathcal{H}u(x) := \frac{1}{\pi} \int_{-\infty}^{\infty} \frac{u(y)}{y-x}\, dy\,, \tag{4.8}$$

which can be treated by our approach quite simply. The Hilbert transform is a bounded operator in $L_p(\mathbb{R})$, $1 < p < \infty$. Therefore we derive from Theorem 2.28 the estimate

$$\|\mathcal{H}(u - \mathcal{M}_{h,\mathcal{D}}u)\|_{L_p} \le \|\mathcal{H}\|_{L_p} \Big(c_\eta\, (\sqrt{\mathcal{D}}h)^M \, \|\nabla_M u\|_{L_p} + \varepsilon \sum_{k=0}^{M-1} (\sqrt{\mathcal{D}}h)^k \|\nabla_k u\|_{L_p} \Big)\,,$$

provided the generating function of the quasi-interpolant $\mathcal{M}_{h,\mathcal{D}}$ satisfies the decay and moment Conditions 2.12 and 2.15 of order N, $u \in W_p^L(\mathbb{R})$ and $M = \min(L, N)$. Since

$$\mathcal{H}(\mathcal{M}_{h,\mathcal{D}}u)(x) = \frac{1}{\sqrt{\mathcal{D}}} \sum_{m\in\mathbb{Z}} u(hm) \mathcal{H}\eta\Big(\frac{x-hm}{\sqrt{\mathcal{D}}h}\Big)\,,$$

we obtain a semi-analytic quadrature formula for the Hilbert transform if $\mathcal{H}\eta$ has an explicit analytic expression.

This is the case if η is the Gaussian function, for example. Since the Fourier transform of $(\pi x)^{-1}$ is $i \operatorname{sgn} \lambda$, we get from (2.10)

$$\begin{aligned} \frac{1}{\pi} \int_{-\infty}^{\infty} \frac{\mathrm{e}^{-y^2}}{y-x}\, dy &= i\sqrt{\pi} \int_{-\infty}^{\infty} \operatorname{sgn}\lambda\, \mathrm{e}^{-\pi^2\lambda^2}\, \mathrm{e}^{2\pi i x\lambda}\, d\lambda = -2\sqrt{\pi} \int_0^{\infty} \mathrm{e}^{-\pi^2\lambda^2} \sin 2\pi x\lambda\, d\lambda \\ &= i\, \mathrm{e}^{-x^2} \operatorname{erf}(ix) \end{aligned}$$

(see [**7**, 2.4(18)]), where erf is the *error function*

$$\operatorname{erf}(\tau) = \frac{2}{\sqrt{\pi}} \int_0^{\tau} \mathrm{e}^{-t^2}\, dt\,. \tag{4.9}$$

The *imaginary error function* is defined by

$$\operatorname{erfi}(\tau) = -i \operatorname{erf}(i\tau) = \frac{2}{\sqrt{\pi}} \int_0^{\tau} \mathrm{e}^{t^2}\, dt\,;$$

therefore we can write

$$\frac{1}{\pi} \int_{-\infty}^{\infty} \frac{\mathrm{e}^{-y^2}}{y-x}\, dy = -\,\mathrm{e}^{-x^2} \operatorname{erfi}(x)\,. \tag{4.10}$$

We see that the quasi-interpolant with Gaussians provides the quadrature formula of second order (modulo saturation terms)

$$\mathcal{H}^{(2)}_{h,\mathcal{D}}u(x) = -\frac{1}{\sqrt{\pi\mathcal{D}}} \sum_{m\in\mathbb{Z}} u(mh)\, \mathrm{e}^{-(x-mh)^2/\mathcal{D}h^2} \operatorname{erfi}\Big(\frac{x-mh}{\sqrt{\mathcal{D}}h}\Big)\,.$$

Next, we determine

$$\mathcal{H}\eta_{2M}(x) = \frac{1}{\pi^{3/2}} \sum_{j=0}^{M-1} \frac{(-1)^j}{j!4^j} \int_{-\infty}^{\infty} \frac{H_{2j}(y)\,\mathrm{e}^{-y^2}}{y-x}\,dy$$
$$= \frac{1}{\pi^{3/2}} \sum_{j=0}^{M-1} \frac{(-1)^j}{j!4^j} \Big(\frac{d}{dx}\Big)^{2j} \int_{-\infty}^{\infty} \frac{\mathrm{e}^{-y^2}}{y-x}\,dy\,,$$

where the formulas (3.6) and (3.9) for the Hermite polynomials and for η_{2M} are used. From (4.10), we obtain

$$\frac{1}{\pi^{3/2}}\Big(\frac{d}{dx}\Big)^{2j} \int_{-\infty}^{\infty} \frac{\mathrm{e}^{-y^2}}{y-x}\,dy = -\frac{2}{\pi}\Big(\frac{d}{dx}\Big)^{2j} \mathrm{e}^{-x^2} \int_0^x \mathrm{e}^{t^2}\,dt$$
$$= -\frac{2}{\pi} \sum_{k=0}^{2j} \frac{(2j)!}{k!(2j-k)!} \Big(\frac{d}{dx}\Big)^{k} \mathrm{e}^{-x^2} \Big(\frac{d}{dx}\Big)^{2j-k} \int_0^x \mathrm{e}^{t^2}\,dt$$
$$= -\frac{2}{\pi} H_{2j}(x)\,\mathrm{e}^{-x^2} \int_0^x \mathrm{e}^{t^2}\,dt$$
$$-\frac{2}{\pi} \sum_{k=0}^{2j-1} \frac{(2j)!}{k!(2j-k)!} (-1)^k H_k(x)\,\mathrm{e}^{-x^2} \Big(\frac{d}{dx}\Big)^{2j-k-1} \mathrm{e}^{x^2}$$
$$= -\frac{H_{2j}(x)\,\mathrm{e}^{-x^2}}{\sqrt{\pi}} \operatorname{erfi}(x) + \frac{2}{\pi} \sum_{k=0}^{2j-1} \frac{(2j)!\,i^{2j-k-1}}{k!(2j-k)!} H_k(x)\,H_{2j-k-1}(ix)\,,$$

which leads to the following representation for the Hilbert transform of η_{2M}:

$$\mathcal{H}\eta_{2M}(x) = -\eta_{2M}(x)\operatorname{erfi}(x) + \frac{2}{\pi} \sum_{j=0}^{M-1} \frac{(-1)^j}{j!4^j} \sum_{k=0}^{2j-1} \frac{(2j)!\,i^{2j-k-1}}{k!(2j-k)!} H_k(x)\,H_{2j-k-1}(ix).$$

The function

$$p_M(x) = \frac{2}{\pi} \sum_{j=0}^{M-1} \frac{(-1)^j}{j!4^j} \sum_{k=0}^{2j-1} \frac{(2j)!\,i^{2j-k-1}}{k!(2j-k)!} H_k(x)\,H_{2j-k-1}(ix)$$

is a polynomial of degree $2M-3$, and it can be determined by using that $\mathcal{H}\eta_{2M}(x) \to 0$ as $|x| \to \infty$. Equality (4.10) provides the asymptotic expansion

$$\mathrm{e}^{-x^2}\operatorname{erfi}(x) = \frac{1}{\pi x} \int_{-\infty}^{\infty} \frac{\mathrm{e}^{-y^2}}{1-y/x}\,dy = \frac{1}{\pi x} \sum_{k=0}^{\infty} x^{-2k} \int_{-\infty}^{\infty} y^{2k}\,\mathrm{e}^{-y^2}\,dy$$
$$= \frac{1}{\sqrt{\pi}} \sum_{k=0}^{\infty} \frac{(-1)^k}{4^k\,x^{2k+1}} H_{2k}(0)$$

for $|x| \to \infty$. Hence the polynomial p_M satisfies

$$p_M(x) - \eta_{2M}(x)\,\mathrm{e}^{x^2}\,\frac{1}{\sqrt{\pi}} \sum_{k=0}^{\infty} \frac{(-1)^k}{4^k\,x^{2k+1}} H_{2k}(0) \to 0$$

as $|x| \to \infty$. Now we note that by (3.5)

$$\frac{\eta_{2M}(x)\,\mathrm{e}^{x^2}}{\sqrt{\pi}} = \frac{(-1)^{M-1}}{\pi\,(M-1)!\,2^{2M-1}}\,\frac{H_{2M-1}(x)}{x},$$

which together with the relation

$$H_{2k}(0) = (-1)^k\,\frac{(2k)!}{k!}$$

(cf. (3.8)) leads to

$$\eta_{2M}(x)\,\mathrm{e}^{x^2}\,\frac{1}{\sqrt{\pi}}\sum_{k=0}^{\infty}\frac{(-1)^k}{4^k\,x^{2k+1}}H_{2k}(0) = \frac{(-1)^{M-1}H_{2M-1}(x)}{\pi(M-1)!\;2^{2M-1}}\sum_{k=0}^{\infty}\frac{(2k)!\,x^{-(2k+2)}}{4^k\,k!}.$$

After substituting here the explicit expression

$$H_{2M-1}(x) = \sum_{m=0}^{M-1}\frac{(-1)^m}{m!(2M-2m-1)!}(2x)^{2M-2m-1}$$

and some transformations, we get the final form of the polynomial

$$p_M(x) = \frac{(2M-1)!}{\pi(M-1)!\,2^{2M-1}}\sum_{k=0}^{M-2}(2x)^{2k+1}\sum_{m=k}^{M-2}\frac{(-1)^{m+1}\,(2m-2k)!}{(2m+3)!\,(M-2-m)!\,(m-k)!}.$$

Hence we obtain

THEOREM 4.1. *The semi-analytic formula*

$$\boxed{\mathcal{H}_{h,\mathcal{D}}^{(2M)}u(x) = \frac{1}{\sqrt{\mathcal{D}}}\sum_{m\in\mathbb{Z}}u(mh)\left(p_M\Big(\frac{x-mh}{\sqrt{\mathcal{D}}h}\Big) - \eta_{2M}\Big(\frac{x-mh}{\sqrt{\mathcal{D}}h}\Big)\mathrm{erfi}\Big(\frac{x-mh}{\sqrt{\mathcal{D}}h}\Big)\right)}$$

approximates the Hilbert transform of any $u \in W_p^L(\mathbb{R})$, $1 < p < \infty$, $L \geq 1$, with

$$\|\mathcal{H}u - \mathcal{H}_{h,\mathcal{D}}^{(2M)}u\|_{L_p} \leq c(\sqrt{\mathcal{D}}h)^K\|\nabla_K u\|_{L_p} + \varepsilon\sum_{k=0}^{K-1}(\sqrt{\mathcal{D}}h)^k\|\nabla_k u\|_{L_p},$$

where $K = \min(L, 2M+2)$. The number $\varepsilon > 0$ can be made arbitrarily small by choosing $\mathcal{D}$ sufficiently large and the constant c depends only on p, L, and M.

For example, a sixth-order approximation for the Hilbert transform is obtained by the simple formula

$$\frac{1}{\sqrt{\pi\mathcal{D}}}\sum_{m\in\mathbb{Z}}u(mh)\Big(\frac{9\tau_m}{16\sqrt{\pi}} - \frac{\tau_m^3}{8\sqrt{\pi}} - \big(\frac{\tau_m^4}{2} - \frac{5\tau_m^2}{2} + \frac{15}{8}\big)\,\mathrm{e}^{-\tau_m^2}\,\mathrm{erfi}(\tau_m)\Big)$$

with $\tau_m = (x-mh)/\sqrt{\mathcal{D}}h$.

4.3. Harmonic potentials

Here, as a multivariate example, we consider the inverse of the Laplace operator, the harmonic potential

$$\mathcal{L}_n u(\mathbf{x}) := \frac{\Gamma(\frac{n}{2}-1)}{4\pi^{n/2}} \int\limits_{\mathbb{R}^n} \frac{u(\mathbf{y})}{|\mathbf{x}-\mathbf{y}|^{n-2}}\, d\mathbf{y}\,, \quad n \geq 3\,, \tag{4.11}$$

which provides the unique solution of the Poisson equation

$$-\Delta f = u \quad \text{in } \mathbb{R}^n\,, \quad |f(\mathbf{x})| \leq C|\mathbf{x}|^{n-2} \text{ as } |\mathbf{x}| \to \infty\,,$$

for integrable right-hand sides u. In the case $n = 2$, the harmonic potential is given by the integral

$$\mathcal{L}_2 u(\mathbf{x}) = \frac{1}{2\pi} \int\limits_{\mathbb{R}^2} \log \frac{1}{|\mathbf{x}-\mathbf{y}|}\, u(\mathbf{y})\, d\mathbf{y}\,. \tag{4.12}$$

The Fourier transform of the kernel is $(4\pi^2|\boldsymbol{\lambda}|^2)^{-1}$ and we have from (2.15) that

$$\begin{aligned} \mathcal{L}_n \eta(\mathbf{x}) &= \frac{2\pi}{|\mathbf{x}|^{n/2-1}} \int\limits_0^\infty \frac{\mathcal{F}\eta(r)}{4\pi^2 r^2} J_{n/2-1}(2\pi r|\mathbf{x}|)\, r^{n/2} dr \\ &= \frac{1}{2\pi|\mathbf{x}|^{n/2-1}} \int\limits_0^\infty \mathcal{F}\eta(r)\, J_{n/2-1}(2\pi r|\mathbf{x}|)\, r^{(n-4)/2} dr \end{aligned} \tag{4.13}$$

for any radial function η. In the following we approximate the density $u(\mathbf{x})$ of the harmonic potential with the quasi-interpolant $\mathcal{M}_{h,\mathcal{D}}^{(2M)} u$ defined by (3.20), which is based on the functions η_{2M}.

4.3.1. Action on Gaussians. First, we determine the harmonic potential of the Gaussian function $\mathrm{e}^{-|\mathbf{x}|^2}$. From (4.13), we have

$$\mathcal{L}_n\big(\mathrm{e}^{-|\cdot|^2}\big)(\mathbf{x}) = \frac{\pi^{n/2-1}}{2|\mathbf{x}|^{n/2-1}} \int\limits_0^\infty \mathrm{e}^{-\pi^2 r^2} J_{n/2-1}(2\pi r|\mathbf{x}|)\, r^{(n-4)/2} dr\,. \tag{4.14}$$

This integral exists for any $n \geq 3$ and can be expressed by the *lower incomplete Gamma function*

$$\gamma(a,x) := \int\limits_0^x \tau^{a-1}\, \mathrm{e}^{-\tau}\, d\tau \tag{4.15}$$

as follows:

$$\boxed{\mathcal{L}_n\big(\mathrm{e}^{-|\cdot|^2}\big)(\mathbf{x}) = \frac{1}{4|\mathbf{x}|^{n-2}} \int\limits_0^{|\mathbf{x}|^2} \tau^{n/2-2}\, \mathrm{e}^{-\tau}\, d\tau = \frac{1}{4|\mathbf{x}|^{n-2}}\, \gamma\Big(\frac{n}{2}-1, |\mathbf{x}|^2\Big)} \tag{4.16}$$

(see for example [**7**, 8.6.11]). Let us mention that in the case $k \in \mathbb{N}$

$$\gamma(k,x) = (k-1)!\Big(1 - \mathrm{e}^{-x} \sum_{j=0}^{k-1} \frac{x^j}{j!}\Big)\,,$$

i.e., the harmonic potential of the Gaussian is expressed by elementary functions. In the case of odd space dimension n, we see from

$$\gamma\Big(\frac{1}{2}, x\Big) = \sqrt{\pi}\,\mathrm{erf}(x)$$

with the error function erf given in (4.9), and from the recurrence relation

$$\gamma(a+1,x) = a\,\gamma(a,x) - \mathrm{e}^{-x}\,x^a\,, \tag{4.17}$$

that the harmonic potential of the Gaussian is expressed using the error function erf. In particular,

$$\mathcal{L}_4\big(\mathrm{e}^{-|\cdot|^2}\big)(\mathbf{x}) = \frac{1}{4\pi^2}\int\limits_{\mathbb{R}^4} \frac{\mathrm{e}^{-|\mathbf{y}|^2}}{|\mathbf{x}-\mathbf{y}|^2}\,d\mathbf{y} = \frac{1}{4|\mathbf{x}|^2}\,(1-\mathrm{e}^{-|\mathbf{x}|^2}) \tag{4.18}$$

and

$$\mathcal{L}_3\big(\mathrm{e}^{-|\cdot|^2}\big)(\mathbf{x}) = \frac{1}{4\pi}\int\limits_{\mathbb{R}^3} \frac{\mathrm{e}^{-|\mathbf{y}|^2}}{|\mathbf{x}-\mathbf{y}|}\,d\mathbf{y} = \frac{1}{2|\mathbf{x}|}\int\limits_0^{|\mathbf{x}|} \mathrm{e}^{-\tau^2}\,d\tau = \frac{\sqrt{\pi}}{4|\mathbf{x}|}\,\mathrm{erf}(|\mathbf{x}|)\,. \tag{4.19}$$

If $n=2$, then we use that the radial function $f(|\mathbf{x}|) = \mathcal{L}_2\big(\mathrm{e}^{-|\cdot|^2}\big)(\mathbf{x})$ is the solution of the differential equation

$$\frac{1}{r}\frac{d}{dr}\Big(r\frac{d}{dr}\Big)f(r) = -\,\mathrm{e}^{-r^2}\,, \quad 0<r<\infty\,,$$

satisfying

$$f(0) = -\int\limits_0^\infty r\,\log r\;\mathrm{e}^{-r^2}\,dr = \frac{\gamma}{4}\,,$$

with the Euler constant $\gamma = 0.577215\ldots$, and

$$f(r) = O\Big(\log\frac{1}{r}\Big) \quad \text{as } r\to\infty\,.$$

From [**1**, 5.1.39]

$$f(r) = \frac{1}{4}\int\limits_0^{r^2} \frac{\mathrm{e}^{-t}-1}{t}\,dt + \frac{\gamma}{4} = -\,\frac{E_1(r^2)+\log r^2}{4}\,,$$

where

$$E_1(x) = \int\limits_x^\infty \frac{\mathrm{e}^{-t}}{t}\,dt \tag{4.20}$$

is the exponential integral, and therefore

$$\mathcal{L}_2\big(\mathrm{e}^{-|\cdot|^2}\big)(\mathbf{x}) = -\frac{1}{4}\,\mathrm{E}_1(|\mathbf{x}|^2) - \frac{1}{2}\log|\mathbf{x}|\,. \tag{4.21}$$

Thus, we obtain the semi-analytic approximation of the harmonic potential for any dimension $n \geq 2$

$$(4.22) \qquad \frac{\mathcal{D}h^2}{(\pi\mathcal{D})^{n/2}} \sum_{\mathbf{m}\in\mathbb{Z}^n} u(h\mathbf{m})\, \mathcal{L}_n\big(\mathrm{e}^{-|\cdot|^2}\big)(\mathbf{r}_\mathbf{m}) \quad \text{with} \quad \mathbf{r}_\mathbf{m} = \frac{\mathbf{x}-h\mathbf{m}}{\sqrt{\mathcal{D}}h},$$

using the quasi-interpolant based on the Gaussian function.

4.3.2. Action on higher-order basis functions. Now we consider the harmonic potential of the radial function

$$\eta_{2M}(\mathbf{x}) = \pi^{-n/2} L_{M-1}^{(n/2)}(|\mathbf{x}|^2)\, \mathrm{e}^{-|\mathbf{x}|^2}.$$

We know from (3.18) that

$$L_{M-1}^{(n/2)}(|\mathbf{x}|^2)\, \mathrm{e}^{-|\mathbf{x}|^2} = \sum_{j=0}^{M-1} \frac{(-1)^j}{j!\,4^j} \Delta^j\, \mathrm{e}^{-|\mathbf{x}|^2},$$

which, because of $\mathcal{L}_n\Delta = -I$, gives

$$(4.23) \qquad \begin{aligned} \mathcal{L}_n\big(L_{M-1}^{(n/2)}(|\cdot|^2)\, \mathrm{e}^{-|\cdot|^2}\big)(\mathbf{x}) &= \mathcal{L}_n\big(\mathrm{e}^{-|\cdot|^2}\big)(\mathbf{x}) - \sum_{j=1}^{M-1} \frac{(-1)^j}{j!\,4^j} \Delta^{j-1}\, \mathrm{e}^{-|\mathbf{x}|^2} \\ &= \mathcal{L}_n\big(\mathrm{e}^{-|\cdot|^2}\big)(\mathbf{x}) + \sum_{j=1}^{M-1} \frac{1}{4j}\frac{(-1)^{j-1}}{(j-1)!\,4^{j-1}} \Delta^{j-1}\, \mathrm{e}^{-|\mathbf{x}|^2}. \end{aligned}$$

Furthermore, applying (3.15), we get

$$(4.24) \qquad \frac{(-1)^j}{j!\,4^j} \Delta^j\, \mathrm{e}^{-|\mathbf{x}|^2} = L_j^{(n/2-1)}(|\mathbf{x}|^2)\, \mathrm{e}^{-|\mathbf{x}|^2},$$

so that (4.23) can be written in the form

$$\mathcal{L}_n\big(L_{M-1}^{(n/2)}(|\cdot|^2)\, \mathrm{e}^{-|\cdot|^2}\big)(\mathbf{x}) = \mathcal{L}_n\big(\mathrm{e}^{-|\cdot|^2}\big)(\mathbf{x}) + \mathrm{e}^{-|\mathbf{x}|^2} \sum_{j=0}^{M-2} \frac{L_j^{(n/2-1)}(|\mathbf{x}|^2)}{4(j+1)}.$$

Hence, the harmonic potential of the generating function η_{2M} is given by the finite sum

$$(4.25) \qquad \mathcal{L}_n\eta_{2M}(\mathbf{x}) = \frac{1}{\pi^{n/2}} \mathcal{L}_n\big(\mathrm{e}^{-|\cdot|^2}\big)(\mathbf{x}) + \frac{\mathrm{e}^{-|\mathbf{x}|^2}}{\pi^{n/2}} \sum_{j=0}^{M-2} \frac{L_j^{(n/2-1)}(|\mathbf{x}|^2)}{4(j+1)}.$$

In the classical cases $n = 2$ and $n = 3$, one obtains the following formulas using classical orthogonal polynomials: For the Newton potential, i.e., for $n = 3$, relations (3.22) and (4.19) give

$$(4.26) \qquad \mathcal{L}_3\eta_{2M}(\mathbf{x}) = \frac{1}{4\pi^{3/2}|\mathbf{x}|}\Big(\sqrt{\pi}\,\mathrm{erf}\,(|\mathbf{x}|) + \mathrm{e}^{-|\mathbf{x}|^2} \sum_{j=0}^{M-2} \frac{(-1)^j H_{2j+1}(|\mathbf{x}|)}{(j+1)!\,2^{2j+1}}\Big)$$

with the Hermite polynomials H_j. If $n = 2$, then (4.21) leads to

$$(4.27) \qquad \mathcal{L}_2\eta_{2M}(\mathbf{x}) = \frac{1}{4\pi}\Big(2\log\frac{1}{|\mathbf{x}|} - \mathrm{E}_1(|\mathbf{x}|^2) + \mathrm{e}^{-|\mathbf{x}|^2} \sum_{j=0}^{M-2} \frac{L_j(|\mathbf{x}|^2)}{j+1}\Big)$$

with the Laguerre polynomials $L_j = L_j^{(0)}$.

A very simple formula can be obtained in the case $n = 4$ if a recurrence relation of generalized Laguerre polynomials is used. We provide two of those relations, which will be used in the following:

$$L_j^{(\gamma-1)}(y) = L_j^{(\gamma)}(y) - L_{j-1}^{(\gamma)}(y)\,, \tag{4.28}$$

$$L_j^{(\gamma+1)}(y) = \frac{1}{y}\Big((j+\gamma+1)\,L_j^{(\gamma)}(y) - (j+1)\,L_{j+1}^{(\gamma)}(y)\Big)\,; \tag{4.29}$$

see, for example, [**1**, 22.7.30,31]. From (4.23), we have

$$\mathcal{L}_4\,\eta_{2M}(\mathbf{x}) = \frac{1}{\pi^2}\mathcal{L}_4\big(\mathrm{e}^{-|\cdot|^2}\big)(\mathbf{x}) + \frac{\mathrm{e}^{-|\mathbf{x}|^2}}{\pi^2}\sum_{j=0}^{M-2}\frac{L_j^{(1)}(|\mathbf{x}|^2)}{4(j+1)}\,.$$

Since by (4.29)

$$L_j^{(1)}(|\mathbf{x}|^2) = \frac{1}{|\mathbf{x}|^2}\Big((j+1)\,L_j^{(0)}(|\mathbf{x}|^2) - (j+1)\,L_{j+1}^{(0)}(|\mathbf{x}|^2)\Big)\,,$$

we derive

$$\sum_{j=0}^{M-2}\frac{L_j^{(1)}(|\mathbf{x}|^2)}{4(j+1)} = \frac{1}{4|\mathbf{x}|^2}\big(1 - L_{M-1}^{(0)}(|\mathbf{x}|^2)\big)\,,$$

which together with (4.18) provides

$$\begin{aligned}\mathcal{L}_4\,\eta_{2M}(\mathbf{x}) &= \frac{1}{4\pi^2|\mathbf{x}|^2}\big(1-\mathrm{e}^{-|\mathbf{x}|^2}\big) + \frac{\mathrm{e}^{-|\mathbf{x}|^2}}{4\pi^2|\mathbf{x}|^2}\big(1 - L_{M-1}(|\mathbf{x}|^2)\big)\\ &= \frac{1-\mathrm{e}^{-|\mathbf{x}|^2}\,L_{M-1}(|\mathbf{x}|^2)}{4\pi^2|\mathbf{x}|^2}\,.\end{aligned} \tag{4.30}$$

4.3.3. Semi-analytic cubature formulas. Summarizing the results of the preceding subsections, we conclude that the approximation of the density by the quasi-interpolation operator $\mathcal{M}_{h,\mathcal{D}}^{(2M)}$ leads via (4.3) to the cubature formula for the harmonic potential

$$\mathcal{L}_{n,h}^{M}u(\mathbf{x}) := \frac{\mathcal{D}h^2}{4(\pi\mathcal{D})^{n/2}}\sum_{\mathbf{m}\in\mathbb{Z}^n}u(h\mathbf{m})\Big(\mathcal{L}_n\big(\mathrm{e}^{-|\cdot|^2}\big)(\mathbf{r_m}) + \mathrm{e}^{-|\mathbf{r_m}|^2}\sum_{j=0}^{M-2}\frac{L_j^{(n/2-1)}(|\mathbf{r_m}|^2)}{j+1}\Big), \tag{4.31}$$

where we set

$$\mathbf{r_m} = \frac{\mathbf{x}-h\mathbf{m}}{\sqrt{\mathcal{D}}h} \tag{4.32}$$

and the corresponding analytic expression (4.16) or (4.21) has to be used for $\mathcal{L}_n\big(\mathrm{e}^{-|\cdot|^2}\big)(\mathbf{r_m})$.

In particular, we obtain the cubature formulas for the practically important two- and three-dimensional cases

(4.33)
$$\mathcal{L}_{2,h}^{M} u(\mathbf{x}) = \frac{h^2}{4\pi} \sum_{\mathbf{m}\in\mathbb{Z}^2} u(h\mathbf{m}) \left(2\log\frac{1}{|\mathbf{r}_\mathbf{m}|} - \mathrm{E}_1(|\mathbf{r}_\mathbf{m}|^2) + \mathrm{e}^{-|\mathbf{r}_\mathbf{m}|^2} \sum_{j=0}^{M-2} \frac{L_j(|\mathbf{r}_\mathbf{m}|^2)}{j+1} \right)$$

(4.34)
$$\mathcal{L}_{3,h}^{M} u(\mathbf{x}) = \frac{h^2}{4\pi\sqrt{\pi\mathcal{D}}} \sum_{\mathbf{m}\in\mathbb{Z}^3} u(h\mathbf{m}) \left(\frac{\sqrt{\pi}\,\mathrm{erf}\,(|\mathbf{r}_\mathbf{m}|)}{|\mathbf{r}_\mathbf{m}|} + \mathrm{e}^{-|\mathbf{r}_\mathbf{m}|^2} \sum_{j=0}^{M-2} \frac{(-1)^j\, H_{2j+1}(|\mathbf{r}_\mathbf{m}|)}{(j+1)!\, 2^{2j+1}\, |\mathbf{r}_\mathbf{m}|} \right)$$

The cubature errors of these formulas will be analyzed in Section 4.4.

4.3.4. Numerical example. We provide some results of numerical tests to approximate the Newton potential. We applied (4.34) with different M and fixed $\mathcal{D} \geq 2$ to sufficiently smooth densities of the explicitly known Newton potential. Fig. 4.1 depicts the maximum cubature error for $M = 1, \dots, 5$, different mesh sizes h and the densities $(\Delta((1-|\mathrm{x}|^2)_+^8)$ and $(1-|\mathrm{x}|^2)_+^4)$ measured at the unit ball B_1.

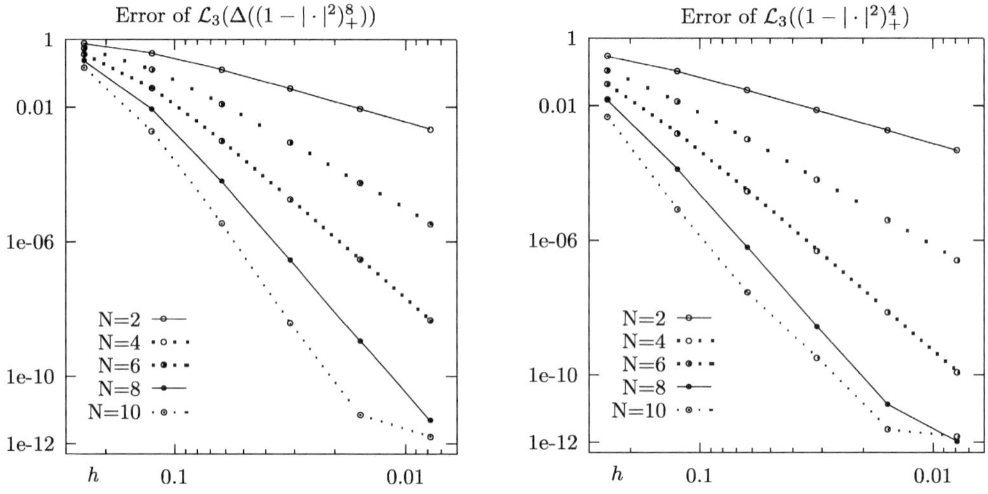

FIGURE 4.1. Cubature error of $\mathcal{L}_{3,h}^{M}$, $N = 2M$, using $\mathcal{D} = 3$.

Note that the slope of the error curves coincides with the approximation order of the corresponding quasi-interpolation of the density. Only for very small h and already very small relative errors, the slope decreases. This can be seen in more detail in Table 4.1, which contains the corresponding values of the approximation rate.

Let us discuss some features of the cubature formulas (4.31). The first term $\mathcal{L}_n(\mathrm{e}^{-|\cdot|^2})(\mathbf{r}_\mathbf{m})$, which provides the approximation order 2, decays slowly. Therefore the approximation of the harmonic potential requires the summation over all nodes $h\mathbf{m}$ within the support of the density u. The remaining terms in (4.31), which increase the approximation order, are only local corrections due to the factor

M	1	2	3	4	5	1	2	3	4	5
h^{-1}	$\mathcal{L}_3(\Delta((1-\|\cdot\|^2)_+^8))$					$\mathcal{L}_3((1-\|\cdot\|^2)_+^4)$				
8	0.98	2.12	3.38	4.78	6.34	1.53	3.09	4.82	6.83	9.23
16	1.62	3.35	5.18	7.09	9.10	1.86	3.75	5.72	7.76	8.28
32	1.89	3.83	5.79	7.78	9.81	1.96	3.94	5.93	7.82	6.34
64	1.97	3.96	5.95	7.95	9.03	1.99	3.98	5.98	7.64	7.02
128	1.99	3.99	5.99	7.85	2.12	2.00	4.00	6.01	3.73	0.72

TABLE 4.1. Approximation rate of the cubature $\mathcal{L}_{3,h}^M$ for the densities $\Delta((1-|\mathbf{x}|^2)_+^8)$ and $(1-|\mathbf{x}|^2)_+^4$ using $\mathcal{D}=3$.

$e^{-|\mathbf{r_m}|^2}$. Hence, in order to obtain higher approximation orders at a given point $\mathbf{x}$, one has to add only a small number of new terms associated with the grid points $h\mathbf{m}$ in a neighborhood of $\mathbf{x}$. Hence, these corrections depend only on the density in a neighborhood of this point.

Another feature is the fast computation of (4.31) for uniform grids. Since the kernel of the harmonic potential is homogeneous of degree -2, the computation on the grid $h\mathbf{k}$, $\mathbf{k}\in\mathbb{Z}^n$, can be performed in accordance with (4.6) using the coefficients

$$b_{\mathbf{k}} = \mathcal{L}_n\big(e^{-|\cdot|^2}\big)(\mathbf{r_k}) + e^{-|\mathbf{r_k}|^2}\sum_{j=0}^{M-2}\frac{L_j^{(n/2-1)}(|\mathbf{r_k}|^2)}{j+1} \quad \text{with} \quad \mathbf{r_k}=\mathbf{k}/\sqrt{\mathcal{D}},$$

which do not depend on the grid size h.

4.3.5. Gradient of the harmonic potential. Let us give a simple application which utilizes the semi-analytic nature of the cubature formula (4.31). The gradient $\nabla(\mathcal{L}_n u)$ of the harmonic potential can be approximated by the gradient of the cubature formula $\mathcal{L}_{n,h}^M u$ given in (4.31), which is a semi-analytic formula, too. The estimate of the approximation error will be derived in Theorem 4.11. Here we derive a simple analytic expression for $\nabla(\mathcal{L}_{n,h}^M u)$.

First we consider $\nabla(\mathcal{L}_n\eta_{2M})$. From (4.25) we get

$$\frac{\partial}{\partial x_k}\mathcal{L}_n\eta_{2M}(\mathbf{x}) = \frac{1}{\pi^{n/2}}\frac{\partial}{\partial x_k}\Big(\mathcal{L}_n\big(e^{-|\cdot|^2}\big)(\mathbf{x}) + e^{-|\mathbf{x}|^2}\sum_{j=0}^{M-2}\frac{L_j^{(n/2-1)}(|\mathbf{x}|^2)}{4(j+1)}\Big). \tag{4.35}$$

If $n\geq 3$, then the first term evaluates in view of (4.16) to

$$\begin{aligned}\frac{\partial}{\partial x_k}\mathcal{L}_n\big(e^{-|\cdot|^2}\big)(\mathbf{x}) &= \frac{\partial}{\partial x_k}\frac{1}{4|\mathbf{x}|^{n-2}}\gamma\Big(\frac{n}{2}-1,|\mathbf{x}|^2\Big)\\ &= \frac{(2-n)x_k}{4|\mathbf{x}|^n}\gamma\Big(\frac{n}{2}-1,|\mathbf{x}|^2\Big) + \frac{x_k}{2|\mathbf{x}|^{n-2}}|\mathbf{x}|^{n-4}e^{-|\mathbf{x}|^2}\\ &= -\frac{x_k}{2|\mathbf{x}|^n}\gamma\Big(\frac{n}{2},|\mathbf{x}|^2\Big),\end{aligned} \tag{4.36}$$

and the last equation is a consequence of (4.17). One easily obtains from (4.21) that

$$\frac{\partial}{\partial x_k}\mathcal{L}_2\big(e^{-|\cdot|^2}\big)(\mathbf{x}) = -\frac{x_k}{2|\mathbf{x}|^2}(1-e^{-|\mathbf{x}|^2}) = -\frac{x_k}{2|\mathbf{x}|^2}\gamma(1,|\mathbf{x}|^2);$$

hence, for all $n \geq 2$

$$\frac{\partial}{\partial x_k}\mathcal{L}_n\big(\mathrm{e}^{-|\cdot|^2}\big)(\mathbf{x}) = -\frac{x_k}{2|\mathbf{x}|^n}\,\gamma\Big(\frac{n}{2},|\mathbf{x}|^2\Big) = -2x_k\,\mathcal{L}_{n+2}\big(\mathrm{e}^{-|\cdot|^2}\big)(|\mathbf{x}|). \tag{4.37}$$

Here by $\mathcal{L}_{n+2}\big(\mathrm{e}^{-|\cdot|^2}\big)(|\mathbf{x}|)$ we denote the radial function on $\mathbb{R}^n$, which coincides with the value of the harmonic potential of the Gaussian in $\mathbb{R}^{n+2}$.

In particular,

$$\frac{\partial}{\partial x_k}\mathcal{L}_3\big(\mathrm{e}^{-|\cdot|^2}\big)(\mathbf{x}) = -\frac{x_k}{2|\mathbf{x}|^2}\Big(\frac{\sqrt{\pi}\,\mathrm{erf}(|\mathbf{x}|)}{2|\mathbf{x}|} - \mathrm{e}^{-|\mathbf{x}|^2}\Big),$$

$$\frac{\partial}{\partial x_k}\mathcal{L}_4\big(\mathrm{e}^{-|\cdot|^2}\big)(\mathbf{x}) = -\frac{x_k}{2|\mathbf{x}|^4}\big(1-(|\mathbf{x}|^2+1)\,\mathrm{e}^{-|\mathbf{x}|^2}\big).$$

To compute the derivative of the sum in (4.35), we use the generating function of the Laguerre polynomials

$$\sum_{j=0}^{\infty} L_j^{(\gamma)}(y)\,z^j = (1-z)^{-\gamma-1}\,\mathrm{e}^{yz/(z-1)}$$

(see [**1**, 22.19.5]). Hence, differentiating the sum

$$(1-z)^{-\gamma-1}\,\mathrm{e}^{y/(z-1)} = \sum_{j=0}^{\infty} L_j^{(\gamma)}(y)\,\mathrm{e}^{-y}\,z^j,$$

we obtain the formula

$$\Big(\frac{d}{dy}\Big)^m\Big(L_j^{(\gamma)}(y)\,\mathrm{e}^{-y}\Big) = (-1)^m L_j^{(\gamma+m)}(y)\,\mathrm{e}^{-y}. \tag{4.38}$$

Therefore the terms of the sum in (4.35) can be transformed to

$$\frac{\partial}{\partial x_k}\frac{L_j^{(n/2-1)}(|\mathbf{x}|^2)\,\mathrm{e}^{-|\mathbf{x}|^2}}{4(j+1)} = -\,x_k\,\frac{L_j^{(n/2)}(|\mathbf{x}|^2)\,\mathrm{e}^{-|\mathbf{x}|^2}}{2(j+1)} \tag{4.39}$$

which gives

$$\begin{aligned}\frac{\partial}{\partial x_k}\mathcal{L}_n\eta_{2M}(\mathbf{x}) &= \frac{-2x_k}{\pi^{n/2}}\Big(\mathcal{L}_{n+2}\big(\mathrm{e}^{-|\cdot|^2}\big)(|\mathbf{x}|) + \sum_{j=0}^{M-2}\frac{L_j^{(n/2)}(|\mathbf{x}|^2)\,\mathrm{e}^{-|\mathbf{x}|^2}}{4(j+1)}\Big)\\ &= -2\pi x_k\,\mathcal{L}_{n+2}\eta_{2M}(|\mathbf{x}|),\end{aligned} \tag{4.40}$$

where $\mathcal{L}_{n+2}\eta_{2M}(|\mathbf{x}|)$ is the radial function on $\mathbb{R}^n$ which is equal to the harmonic potential of the radial function η_{2M} on $\mathbb{R}^{n+2}$.

There exists another formula for $\nabla\mathcal{L}_n\eta_{2M}$ for $n \geq 3$, which may be useful in computations and will be derived next. We can rewrite (4.36) as

$$\frac{\partial}{\partial x_k}\mathcal{L}_n\big(\mathrm{e}^{-|\cdot|^2}\big)(\mathbf{x}) = \frac{x_k}{2|\mathbf{x}|^2}\Big((4-2n)\mathcal{L}_n\big(\mathrm{e}^{-|\cdot|^2}\big)(|\mathbf{x}|) + \mathrm{e}^{-|\mathbf{x}|^2}\Big).$$

Using (4.38) and (4.29), the terms in the sum in (4.35) are transformed to

$$\begin{aligned}\frac{\partial}{\partial x_k}\frac{L_j^{(n/2-1)}(|\mathbf{x}|^2)\,\mathrm{e}^{-|\mathbf{x}|^2}}{4(j+1)} &= -\,x_k\,\frac{L_j^{(n/2)}(|\mathbf{x}|^2)\,\mathrm{e}^{-|\mathbf{x}|^2}}{2(j+1)}\\ &= -\frac{x_k\,\mathrm{e}^{-|\mathbf{x}|^2}}{2|\mathbf{x}|^2}\Big(L_j^{(n/2-1)}(|\mathbf{x}|^2) - L_{j+1}^{(n/2-1)}(|\mathbf{x}|^2) + \frac{n-2}{2(j+1)}L_j^{(n/2-1)}(|\mathbf{x}|^2)\Big),\end{aligned}$$

which, by noting that $L_0^{(\gamma)}(y) = 1$, implies

$$\frac{\partial}{\partial x_k} \sum_{j=0}^{M-2} \frac{L_j^{(n/2-1)}(|\mathbf{x}|^2)\,\mathrm{e}^{-|\mathbf{x}|^2}}{4(j+1)} = (2-n)\frac{x_k}{|\mathbf{x}|^2} \sum_{j=0}^{M-2} \frac{L_j^{(n/2-1)}(|\mathbf{x}|^2)\,\mathrm{e}^{-|\mathbf{x}|^2}}{4(j+1)} + \frac{x_k\,\mathrm{e}^{-|\mathbf{x}|^2}}{2|\mathbf{x}|^2}\Big(L_{M-1}^{(n/2-1)}(|\mathbf{x}|^2) - 1\Big).$$

Therefore, if $n \geq 3$, then (4.35) can be transformed to

$$\begin{aligned} \frac{\partial}{\partial x_k}\mathcal{L}_n\eta_{2M}(\mathbf{x}) &= \pi^{-n/2}\frac{x_k}{2|\mathbf{x}|^2}\Big((4-2n)\mathcal{L}_n\big(\mathrm{e}^{-|\cdot|^2}\big)(\mathbf{x}) + \mathrm{e}^{-|\mathbf{x}|^2}\Big) \\ &\quad + (2-n)\frac{x_k\,\mathrm{e}^{-|\mathbf{x}|^2}}{|\mathbf{x}|^2}\sum_{j=0}^{M-2}\frac{L_j^{(n/2-1)}(|\mathbf{x}|^2)}{4(j+1)} + \frac{x_k\,\mathrm{e}^{-|\mathbf{x}|^2}}{2|\mathbf{x}|^2}\Big(L_{M-1}^{(n/2-1)}(|\mathbf{x}|^2) - 1\Big) \\ &= \frac{x_k}{|\mathbf{x}|^2}\Big((2-n)\mathcal{L}_n\eta_{2M}(\mathbf{x}) + \frac{L_{M-1}^{(n/2-1)}(|\mathbf{x}|^2)\,\mathrm{e}^{-|\mathbf{x}|^2}}{2\pi^{n/2}}\Big). \end{aligned} \tag{4.41}$$

Compared with formula (4.40), this formula has the advantage that it allows one to compute both the harmonic potential and its gradient with minimal additional effort.

If $n = 2$, then we obtain the partial derivatives of the harmonic potential as

$$\begin{aligned} \frac{\partial}{\partial x_k}\mathcal{L}_2\eta_{2M}(\mathbf{x}) &= -\frac{x_k}{2\pi|\mathbf{x}|^2}(1 - \mathrm{e}^{-|\mathbf{x}|^2}) + \frac{x_k\,\mathrm{e}^{-|\mathbf{x}|^2}}{2\pi|\mathbf{x}|^2}(L_{M-1}^{(0)}(|\mathbf{x}|^2) - 1) \\ &= \frac{x_k}{2\pi|\mathbf{x}|^2}(L_{M-1}^{(0)}(|\mathbf{x}|^2)\,\mathrm{e}^{-|\mathbf{x}|^2} - 1). \end{aligned} \tag{4.42}$$

Summarizing, we arrive at the following formulas for the gradient of $\mathcal{L}_{n,h}^M u$:

$$\begin{aligned} &\nabla(\mathcal{L}_{n,h}^M u)(\mathbf{x}) \\ &= \frac{h}{\mathcal{D}^{(n-1)/2}}\sum_{\mathbf{m}\in\mathbb{Z}^n}\frac{u(h\mathbf{m})\mathbf{r}_\mathbf{m}}{|\mathbf{r}_\mathbf{m}|^2}\Big((2-n)\mathcal{L}_n\eta_{2M}(\mathbf{r}_\mathbf{m}) + \frac{L_{M-1}^{(n/2-1)}(|\mathbf{r}_\mathbf{m}|^2)\,\mathrm{e}^{-|\mathbf{r}_\mathbf{m}|^2}}{2\pi^{n/2}}\Big) \\ &= \frac{h}{2\pi^{n/2}\mathcal{D}^{(n-1)/2}}\sum_{\mathbf{m}\in\mathbb{Z}^n} u(h\mathbf{m})\mathbf{r}_\mathbf{m}\Big(\frac{\gamma(n/2,|\mathbf{r}_\mathbf{m}|^2)}{|\mathbf{r}_\mathbf{m}|^n} + \mathrm{e}^{-|\mathbf{r}_\mathbf{m}|^2}\sum_{j=0}^{M-2}\frac{L_j^{(n/2)}(|\mathbf{r}_\mathbf{m}|^2)}{j+1}\Big) \end{aligned} \tag{4.43}$$

with $\mathbf{r}_\mathbf{m}$ from (4.32), which is valid if $n \geq 3$. For $n = 2$, we obtain

$$\nabla(\mathcal{L}_{2,h}^M u)(\mathbf{x}) = \frac{h}{2\pi\mathcal{D}}\sum_{\mathbf{m}\in\mathbb{Z}^2}\frac{u(h\mathbf{m})\mathbf{r}_\mathbf{m}}{|\mathbf{r}_\mathbf{m}|^2}\,(L_{M-1}^{(0)}(|\mathbf{r}_\mathbf{m}|^2)\,\mathrm{e}^{-|\mathbf{r}_\mathbf{m}|^2} - 1). \tag{4.44}$$

4.4. Approximation properties

4.4.1. Rough error estimate. Let us estimate the error of the cubature formula (4.31) for the harmonic potential in the case $n \geq 3$. By construction, the cubature formula is the action of the integral onto the quasi-interpolant

$$\mathcal{L}_{n,h}^{M}u(\mathbf{x}) = \mathcal{L}_n(\mathcal{M}_{h,\mathcal{D}}^{(2M)}u)(\mathbf{x})$$

and therefore the cubature error equals

$$\mathcal{L}_{n,h}^{M}u(\mathbf{x}) - \mathcal{L}_n u(\mathbf{x}) = \mathcal{L}_n(\mathcal{M}_{h,\mathcal{D}}^{(2M)} - I)u(\mathbf{x}) .$$

It is known from Sobolev's Theorem that for $n \geq 3$, $1 < p < n/2$, and $q = np/(n-2p)$ the operator $\mathcal{L}_n$ as the inverse of the Laplacian is a bounded mapping from $L_p(\mathbb{R}^n)$ into $L_q(\mathbb{R}^n)$ (cf. [**89**, Theorem V.1]). Hence

$$\|\mathcal{L}_{n,h}^{M}u - \mathcal{L}_n u\|_{L_q} \leq A_{p,q}\|u - \mathcal{M}_{h,\mathcal{D}}^{(2M)}u\|_{L_p} ,$$

where $A_{p,q}$ denotes the norm of $\mathcal{L}_n : L_p(\mathbb{R}^n) \to L_q(\mathbb{R}^n)$. Then Theorem 2.28 gives immediately

LEMMA 4.2. *Let $1 < p < n/2$, $q = np/(n-2p)$ and let $u \in W_p^L$ with $L > n/p$ and choose $2M \geq L$. Then for any $\varepsilon > 0$, there exists $\mathcal{D} > 0$ such that*

$$\|\mathcal{L}_{n,h}^{M}u - \mathcal{L}_n u\|_{L_q} \leq A_{p,q}\Big(c_\eta\, (\sqrt{\mathcal{D}}h)^L\, \|\nabla_L u\|_{L_p} + \varepsilon \sum_{k=0}^{L-1} (\sqrt{\mathcal{D}}h)^k \|\nabla_k u\|_{L_p}\Big) .$$

However, the numerical tests do not show that there exists a saturation error, which does not converge to zero. We shall see in the next section that this is because $\mathcal{L}_n$ is a smoothing operator.

4.4.2. Quasi-interpolation error in Sobolev spaces of negative order. We study error estimates in dual spaces of Sobolev spaces of functions with derivatives in L_p. For $r \in \mathbb{N}$, we denote the space of linear functionals on $W_{p'}^r$, $p' = p/(p-1)$, $1 < p < \infty$, by $W_p^{-r} = W_p^{-r}(\mathbb{R}^n)$.

LEMMA 4.3. *Let $u \in W_p^{2r}$, $1 < p < \infty$, $r \in \mathbb{N}$, and $\{a_{\boldsymbol{\nu}}\} \in l_1(\mathbb{Z}^n)$. Then there exists a constant c depending only on n, r, and p such that*

$$\Big\| u \sum_{\boldsymbol{\nu}\in\mathbb{Z}^n\setminus\{\mathbf{0}\}} a_{\boldsymbol{\nu}}\, \mathrm{e}^{\frac{2\pi i}{h}\langle\,\cdot\,,\boldsymbol{\nu}\rangle} \Big\|_{W_p^{-2r}} \leq c\, h^{2r} \sum_{\boldsymbol{\nu}\in\mathbb{Z}^n\setminus\{\mathbf{0}\}} \frac{|a_{\boldsymbol{\nu}}|}{(4\pi^2|\boldsymbol{\nu}|^2)^r}\, \|u\|_{W_p^{2r}} . \tag{4.45}$$

PROOF. For $|\boldsymbol{\nu}| \neq 0$, the norm of the multiplier

$$\mathrm{e}^{\frac{2\pi i}{h}\langle\cdot,\boldsymbol{\nu}\rangle} : W_p^{2r} \to W_p^{-2r}$$

can be estimated by

$$\begin{aligned}
\Big\| \mathrm{e}^{\frac{2\pi i}{h}\langle\,\cdot\,,\boldsymbol{\nu}\rangle}\, u \Big\|_{W_p^{-2r}} &= \sup_{\|v\|_{W_{p'}^{2r}}=1} \Big| \int_{\mathbb{R}^n} \mathrm{e}^{\frac{2\pi i}{h}\langle\mathbf{x},\boldsymbol{\nu}\rangle}\, u(\mathbf{x})\,\overline{v(\mathbf{x})}\, d\mathbf{x} \Big| \\
&= \sup_{\|v\|_{W_{p'}^{2r}}=1} \Big| \frac{h^{2r}}{(4\pi^2|\boldsymbol{\nu}|^2)^r} \int_{\mathbb{R}^n} u(\mathbf{x})\,\overline{v(\mathbf{x})}\,(-\Delta)^r\, \mathrm{e}^{\frac{2\pi i}{h}\langle\mathbf{x},\boldsymbol{\nu}\rangle}\, d\mathbf{x} \Big| \\
&\leq \frac{h^{2r}}{(4\pi^2|\boldsymbol{\nu}|^2)^r} \sup_{\|v\|_{W_{p'}^{2r}}=1} \int_{\mathbb{R}^n} \big|\Delta^r (u(\mathbf{x})\,\overline{v(\mathbf{x})})\big|\, d\mathbf{x} .
\end{aligned}$$

Since

$$\Delta^r(u\,\overline{v}) = \sum_{[\boldsymbol{\alpha}]=r} \frac{r!}{\boldsymbol{\alpha}!}\, \partial^{2\boldsymbol{\alpha}}(u\,\overline{v}) = \sum_{[\boldsymbol{\alpha}]=r} \frac{r!}{\boldsymbol{\alpha}!} \sum_{\boldsymbol{\beta}\leq 2\boldsymbol{\alpha}} \frac{(2\boldsymbol{\alpha})!}{\boldsymbol{\beta}!\,(2\boldsymbol{\alpha}-\boldsymbol{\beta})!}\, \partial^{\boldsymbol{\beta}} u\, \partial^{2\boldsymbol{\alpha}-\boldsymbol{\beta}}\overline{v} ,$$

we see that there exists a constant depending only on n, r, and p such that

$$\int_{\mathbb{R}^n} \left|\Delta^r (u(\mathbf{x})\,\overline{v(\mathbf{x})})\right| d\mathbf{x} \le c\,\|u\|_{W_p^{2r}}\,\|v\|_{W_{p'}^{2r}}\,,$$

which implies

$$\left\| \mathrm{e}^{\frac{2\pi i}{h}\langle\,\cdot\,,\boldsymbol{\nu}\rangle}\, u \right\|_{W_p^{-2r}} \le \frac{c\,h^{2r}}{(4\pi^2|\boldsymbol{\nu}|^2)^r}\,\|u\|_{W_p^{2r}} \tag{4.46}$$

for any $\boldsymbol{\nu} \in \mathbb{Z}^n$ with $|\boldsymbol{\nu}| \ne 0$. □

REMARK 4.4. The constant c in (4.45) for the special case $r = 1$ is bounded by

$$c \le \max\,(2, n^{1/\tilde{p}})\,, \quad \tilde{p} = \min\,(p, p')\,.$$

By interpolation arguments, Lemma 4.3 can be generalized for norms in function spaces of arbitrary negative order. The *Bessel potential space* $H_p^s = H_p^s(\mathbb{R}^n)$ is defined as the closure of compactly supported smooth functions with respect to the norm

$$\|u\|_{H_p^s} = \|\mathcal{F}^{-1}(1+4\pi^2|\cdot|^2)^{s/2}\mathcal{F}u\|_{L_p} = \|(I-\Delta)^{s/2}u\|_{L_p}\,. \tag{4.47}$$

COROLLARY 4.5. *Let $u \in H_p^s$, $1 < p < \infty$, $s > 0$, and $\{a_{\boldsymbol{\nu}}\} \in l_1(\mathbb{Z}^n)$. Then there exists a constant c depending only on n, s, and p such that*

$$\left\| u \sum_{\boldsymbol{\nu}\in\mathbb{Z}^n\setminus\{\mathbf{0}\}} a_{\boldsymbol{\nu}}\, \mathrm{e}^{\frac{2\pi i}{h}\langle\,\cdot\,,\boldsymbol{\nu}\rangle} \right\|_{H_p^{-s}} \le c\,h^s \sum_{\boldsymbol{\nu}\in\mathbb{Z}^n\setminus\{\mathbf{0}\}} \frac{|a_{\boldsymbol{\nu}}|}{(2\pi|\boldsymbol{\nu}|)^s}\,\|u\|_{H_p^s}\,.$$

PROOF. The spaces H_p^s are interpolation spaces that coincide, for $r \in \mathbb{Z}$, with the Sobolev spaces, $H_p^r = W_p^r$, $1 < p < \infty$, and are obtained from Sobolev spaces by means of the complex interpolation method; cf., e.g., [**94**]. Hence, the multiplication operator

$$A_{h,\boldsymbol{\nu}}u(\mathbf{x}) = \mathrm{e}^{2\pi i\langle \mathbf{x},\boldsymbol{\nu}\rangle/h}\,u(\mathbf{x})$$

satisfies

$$\|A_{h,\boldsymbol{\nu}}\|_{H_p^{2r\theta}\to H_p^{-2r\theta}} \le \|A_{h,\boldsymbol{\nu}}\|^{\theta}_{H_p^{2r}\to H_p^{-2r}} \left\|A_{h,\boldsymbol{\nu}}\right\|^{1-\theta}_{L_p\to L_p}$$

for $0 < \theta < 1$. Since $A_{h,\boldsymbol{\nu}}$ is an isometric operator in L_p, the assertion follows from (4.46). □

This leads to the following error estimate for the quasi-interpolation process.

THEOREM 4.6. *Suppose that η satisfies the decay and moments Conditions 2.12 and 2.15 of order N. Then for any $u \in H_p^L$, $1 < p < \infty$, $L \ge N$ with $n/p < L \le K - n$ and positive $s \le L$, there exist constants c_η and $c_{s,p}$ not depending on u and h such that the quasi-interpolant $\mathcal{M}_{h,\mathcal{D}}u$ defined by (2.23) satisfies*

$$\|u - \mathcal{M}_{h,\mathcal{D}}u\|_{H_p^{-s}} \le c_\eta\,(\sqrt{\mathcal{D}}h)^N\,\|u\|_{H_p^L}$$

$$+\, c_{s,p}\,h^s \sum_{[\boldsymbol{\alpha}]=0}^{\min(N-1,[L-s])} \left(\frac{\sqrt{\mathcal{D}}h}{2\pi}\right)^{[\boldsymbol{\alpha}]} \frac{\|\partial^{\boldsymbol{\alpha}}u\|_{H_p^s}}{\boldsymbol{\alpha}!} \sum_{\boldsymbol{\nu}\in\mathbb{Z}^n\setminus\{\mathbf{0}\}} \frac{|\partial^{\boldsymbol{\alpha}}\mathcal{F}\eta(\sqrt{\mathcal{D}}\boldsymbol{\nu})|}{(2\pi|\boldsymbol{\nu}|)^s}\,.$$

REMARK 4.7. The previous estimate is of the usual form $\mathcal{O}((\sqrt{\mathcal{D}}h)^N)$ plus a small saturation error which is controlled by $\mathcal{D}$. However, if $s < 0$, then one has even the convergence of the quasi-interpolant $\mathcal{M}_{h,\mathcal{D}}u$ in H_p^s as $h \to 0$, since the saturation error is multiplied by h^{-s}. The convergence of the quasi-interpolants can be justified also for other spaces of negative order. Note that in the case $s > 0$ a similar estimate is valid, which implies that the saturation term increases with the factor h^{-s}. This is in accordance with the assertion of Theorem 2.27 on the approximation of derivatives.

PROOF OF THEOREM 4.6. Since $\|\cdot\|_{H_p^s} \le \|\cdot\|_{H_p^t}$ for $s < t$, from the representation (2.69) we obtain the estimate

$$\|u - \mathcal{M}_{h,\mathcal{D}}u\|_{H_p^{-s}} \le \|R_{L,h}\|_{L_p} + \sum_{[\boldsymbol{\alpha}]=N}^{L-1} \frac{(\sqrt{\mathcal{D}}h)^{[\boldsymbol{\alpha}]}}{\boldsymbol{\alpha}!} \|\sigma_{\boldsymbol{\alpha}}(h^{-1}\cdot, \eta, \mathcal{D})\, \partial^{\boldsymbol{\alpha}} u\,\|_{H_p^{-s}}$$
$$+ \sum_{[\boldsymbol{\alpha}]=0}^{N-1} \frac{(\sqrt{\mathcal{D}}h)^{[\boldsymbol{\alpha}]}}{\boldsymbol{\alpha}!} \|\varepsilon_{\boldsymbol{\alpha}}(h^{-1}\cdot, \eta, \mathcal{D})\, \partial^{\boldsymbol{\alpha}} u\,\|_{H_p^{-s}} .$$

Obviously, the first two terms on the right-hand side of this inequality are bounded by $c\,(\sqrt{\mathcal{D}}h)^N \|u\|_{H_p^L}$ with some constant c not depending on u, h, and $\mathcal{D}$. To estimate the last term, we note that by Corollary 2.13

$$\varepsilon_{\boldsymbol{\alpha}}\Big(\frac{\mathbf{x}}{h}, \eta, \mathcal{D}\Big) = \Big(\frac{i}{2\pi}\Big)^{[\boldsymbol{\alpha}]} \sum_{\boldsymbol{\nu}\in\mathbb{Z}^n\setminus\{\mathbf{0}\}} \partial^{\boldsymbol{\alpha}} \mathcal{F}\eta(\sqrt{\mathcal{D}}\boldsymbol{\nu})\, \mathrm{e}^{-\frac{2\pi i}{h}\langle \mathbf{x},\boldsymbol{\nu}\rangle} .$$

In the case $[\boldsymbol{\alpha}] + s \le L$ we use Corollary 4.5 to derive

$$\|\varepsilon_{\boldsymbol{\alpha}}(h^{-1}\cdot, \eta, \mathcal{D})\, \partial^{\boldsymbol{\alpha}} u\|_{H_p^{-s}} \le c_{s,p}\, h^s\, \|\partial^{\boldsymbol{\alpha}} u\|_{H_p^s} (2\pi)^{-[\boldsymbol{\alpha}]} \sum_{\boldsymbol{\nu}\in\mathbb{Z}^n\setminus\{\mathbf{0}\}} \frac{|\partial^{\boldsymbol{\alpha}}\mathcal{F}\eta(\sqrt{\mathcal{D}}\boldsymbol{\nu})|}{(2\pi|\boldsymbol{\nu}|)^s} .$$

If $[\boldsymbol{\alpha}] + s > L$, then Corollary 4.5 with $u \in H_p^{L-[\boldsymbol{\alpha}]}$ implies

$$\|\varepsilon_{\boldsymbol{\alpha}}(h^{-1}\cdot, \eta, \mathcal{D})\, \partial^{\boldsymbol{\alpha}} u\|_{H_p^{-s}}$$
$$\le c_{L-[\boldsymbol{\alpha}],p}\, h^{L-[\boldsymbol{\alpha}]}\, \|u\|_{H_p^L} (2\pi)^{-[\boldsymbol{\alpha}]} \sum_{\boldsymbol{\nu}\in\mathbb{Z}^n\setminus\{\mathbf{0}\}} \frac{|\partial^{\boldsymbol{\alpha}}\mathcal{F}\eta(\sqrt{\mathcal{D}}\boldsymbol{\nu})|}{(2\pi|\boldsymbol{\nu}|)^{L-[\boldsymbol{\alpha}]}} ,$$

which shows that

$$\sum_{[\boldsymbol{\alpha}]>L-s}^{N-1} \Big(\frac{\sqrt{\mathcal{D}}h}{2\pi}\Big)^{[\boldsymbol{\alpha}]} \frac{\|\varepsilon_{\boldsymbol{\alpha}}(h^{-1}\cdot, \eta, \mathcal{D})\, \partial^{\boldsymbol{\alpha}} u\|_{H_p^{-s}}}{\boldsymbol{\alpha}!} \le c\,(\sqrt{\mathcal{D}}h)^N \|u\|_{H_p^L} . \qquad \square$$

Note that (4.47) yields in the case $s = 2$

$$\|u\|_{H_p^{-2}} \le B_{p'} \|u\|_{W_p^{-2}} \quad \text{with the constant} \quad B_{p'} = \sup_{v\in W_{p'}^2} \frac{\|v\|_{W_{p'}^2}}{\|(I-\Delta)v\|_{L_{p'}}} .$$

Therefore, by using Remark 4.4, we can formulate the following.

PROPOSITION 4.8. *Suppose that η satisfies the decay and moments Conditions 2.12 and 2.15 of order N. Then for any $\varepsilon > 0$ there exists $\mathcal{D} > 0$ such that for all $u \in W_p^L$, $L \geq N \geq 2$ with $n/p < L < K - n$*

$$\|u - \mathcal{M}_{h,\mathcal{D}} u\|_{H_p^{-2}} \leq c\,(\sqrt{\mathcal{D}}h)^N \,\|u\|_{W_p^L} + \varepsilon\, c_p\, h^2 \sum_{[\boldsymbol{\alpha}]=0}^{N-3} \frac{(\sqrt{\mathcal{D}}h)^{[\boldsymbol{\alpha}]}}{\boldsymbol{\alpha}!} \,\|\partial^{\boldsymbol{\alpha}} u\|_{W_p^2}$$

with constants $c_p = B_{p'} \cdot \max(2, n^{1/\tilde{p}})$ and c not depending on u, h, and $\mathcal{D}$.

REMARK 4.9. The weak convergence of the quasi-interpolants in L_p leads to an interesting consequence concerning the density of the ladder of principal shift-invariant subspaces

$$S^h(\eta(\cdot/\mathcal{D})) = \Big\{\eta\Big(\frac{\mathbf{x} - h\mathbf{m}}{\sqrt{\mathcal{D}}h}\Big) : \mathbf{m} \in \mathbb{Z}^n, h > 0\Big\}$$

in L_p-spaces, $1 < p < \infty$. By the Banach-Saks Theorem, the sequence of the arithmetic means of a series, weakly convergent in L_p, converges strongly. Therefore, it follows from Theorem 4.6 that for any $u \in W_p^L$, $L > n/p$, and any sequence $h_j \to 0$

$$\Big\|u - \lim_{N\to\infty} \frac{1}{N} \sum_{j=1}^{N} \mathcal{M}_{h_j,\mathcal{D}} u\Big\|_{L_p} \to 0\,.$$

Hence, if we introduce the linear subspace

$$S^{(H)}(\eta(\cdot/\mathcal{D})) = \bigcup_{h_j} S^{h_j}(\eta(\cdot/\mathcal{D}))\,,$$

where $H = \{h_j\}$ with $h_j \to 0$, we conclude from the results of Subsection 2.4.3 that for any $u \in L_p$, $p < 1 < \infty$,

$$\mathrm{dist}(u, S^{(H)}(\eta(\cdot/\mathcal{D})); L_p) = 0\,,$$

if $\mathcal{F}\eta(\mathbf{0}) \neq 0$ and for some $K > n$ and $0 \leq [\boldsymbol{\beta}] \leq \mu$

$$|\partial^{\boldsymbol{\beta}} \eta(\mathbf{x})|(1 + |\mathbf{x}|)^K < \infty\,, \quad \mathbf{x} \in \mathbb{R}^n\,,$$

where μ is the smallest integer greater than $n/2$.

4.4.3. Cubature error for the harmonic potential.

THEOREM 4.10. *Let $1 < p < n/2$, $q = np/(n - 2p)$. Then for any function $u \in W_p^L$, where $L \geq 2M$ and $L > n/p$, there exists a constant c_η not depending on u, h, and $\mathcal{D}$ such that the cubature formula (4.31) for the harmonic potential satisfies*

$$\|\mathcal{L}_n u - \mathcal{L}_{n,h}^M u\|_{L_q} \leq c_\eta\,(\sqrt{\mathcal{D}}h)^{2M} \|u\|_{W_p^L} + h^2 \sum_{k=0}^{2M-3} (\sqrt{\mathcal{D}}h)^k \,\frac{\varepsilon_k(\mathcal{D})}{(2\pi)^{k+2}} \sum_{l=0}^{2} (A_{p,q}\, c_p\, |u|_{W_p^{k+l}} + c_q\, |u|_{W_q^{k+l}})$$

with the norm $A_{p,q}$ of $\mathcal{L}_n : L_p(\mathbb{R}^n) \to L_q(\mathbb{R}^n)$, the constant c_p of Proposition 4.8, and the numbers $\varepsilon_k(\mathcal{D})$ defined by (2.70).

PROOF. The assertion follows immediately from Theorem 2.28, Proposition 4.8, and the mapping properties of the operator $\mathcal{L}_n$. Since

$$\mathcal{L}_n u - \mathcal{L}_{n,h}^M u = \mathcal{L}_n(u - \mathcal{M}_{h,\mathcal{D}}^{(2M)} u) ,$$

we obtain

$$\begin{aligned}\|\mathcal{L}_n u - \mathcal{L}_{n,h}^M u\|_{L_q} &= \|(-\Delta)^{-1}(I-\Delta)(I-\Delta)^{-1}(u - \mathcal{M}_{h,\mathcal{D}}^{(2M)} u)\|_{L_q} \\ &\leq \|(-\Delta)^{-1}(I-\Delta)^{-1}(u - \mathcal{M}_{h,\mathcal{D}}^{(2M)} u)\|_{L_q} + \|(I-\Delta)^{-1}(u - \mathcal{M}_{h,\mathcal{D}}^{(2M)} u)\|_{L_q} \\ &\leq A_{p,q}\,\|u - \mathcal{M}_{h,\mathcal{D}}^{(2M)} u\|_{H_p^{-2}} + \|u - \mathcal{M}_{h,\mathcal{D}}^{(2M)} u\|_{H_q^{-2}} .\end{aligned}$$

Using the continuous embedding $W_p^L \subset W_q^{L-2}$, we have only to apply the estimate of Proposition 4.8. □

Let us estimate the approximation error of the gradient $\nabla(\mathcal{L}_n u)$ by the formulas (4.43) and (4.44) for $\nabla(\mathcal{L}_{n,h}^M u)$, respectively.

THEOREM 4.11. *Let $1 < p < n$, $q = np/(n-p)$. Then for any function $u \in W_p^L$, where $L \geq N+1$ with $n/p < L$, there exist constants c_η and c_p independent on u, h, and $\mathcal{D}$ such that*

$$\begin{aligned}\|\nabla(\mathcal{L}_n u) - \nabla(\mathcal{L}_{n,h}^M u)\|_{L_q} &\leq c_\eta\,(\sqrt{\mathcal{D}}h)^{2M}\,\|u\|_{H_p^L} \\ &+ c_p\,h \sum_{[\boldsymbol{\alpha}]=0}^{2M-2} \Big(\frac{\sqrt{\mathcal{D}}h}{2\pi}\Big)^{[\boldsymbol{\alpha}]} \frac{\|\partial^{\boldsymbol{\alpha}} u\|_{H_p^1}}{\boldsymbol{\alpha}!} \sum_{\boldsymbol{\nu}\in\mathbb{Z}^n\setminus\{\mathbf{0}\}} \frac{|\partial^{\boldsymbol{\alpha}}\mathcal{F}\eta_{2M}(\sqrt{\mathcal{D}}\boldsymbol{\nu})|}{2\pi|\boldsymbol{\nu}|} .\end{aligned}$$

PROOF. The norm $\|\nabla u\|_{L_q}$ is equivalent to the norm $\|(-\Delta)^{1/2} u\|_{L_q}$. Acting as in the proof of the previous theorem, we get

$$\|(-\Delta)^{1/2}(\mathcal{L}_n u - \mathcal{L}_{n,h}^M u)\|_{L_q} \leq B_{p,q}\,\|u - \mathcal{M}_{h,\mathcal{D}}^{(2M)} u\|_{H_p^{-1}} + \|u - \mathcal{M}_{h,\mathcal{D}}^{(2M)} u\|_{H_q^{-1}} ,$$

where $B_{p,q}$ denotes the norm of the bounded mapping $(-\Delta)^{-1/2} : L_p \to L_q$ (**[89]**). Hence by Theorem 4.6, the assertion follows immediately. □

Theorems 4.10 and 4.11 show that if the parameter $\mathcal{D}$ is chosen so that the values $\varepsilon_k(\mathcal{D})$ are sufficiently small, then both the cubature $\mathcal{L}_{n,h}^M u$ and its gradient $\nabla(\mathcal{L}_{n,h}^M u)$ approximate with the order h^{2M} up to the prescribed accuracy. Moreover, by the smoothing properties of the harmonic potential and its gradient, the corresponding small saturation errors converge with the rate h^2 and h, respectively, as h tends to zero. This property holds, in general, for other pseudodifferential operators of negative order, whereas for singular integral operators, the corresponding cubatures approximate with the order N for $h \geq h_0$, but they do not converge.

4.5. Application to boundary integral equation methods

Let us consider a direct application of results of the previous section to the numerical solution of boundary value problems with boundary integral equation methods (BIM).

Since the 1980s these methods have become more and more popular for solving boundary value problems for partial differential equations with constant coefficients which occur in mechanics, electromagnetics, and other fields of mathematical physics. We will discuss certain BIM in more detail in Section 12.2. Here we consider only a particular problem that arises when the differential equation has an inhomogeneous right-hand side.

4.5.1. Transformation of inhomogeneous problems. Let L be a partial differential operator with the fundamental solution $\mathcal{E}$. Consider the equation

$$Lu = f \quad \text{in } \Omega\,,$$

complemented by some boundary conditions. This boundary value problem can be formulated as an integral equation over the boundary of the domain by different methods, using for example Green's formulas or a surface potential ansatz. However, non-homogeneous terms f are included in the formulation by means of domain integrals, thus making the technique lose the attraction of its 'boundary-only' character.

The simplest way to avoid domain integrals in the BIM is to represent the solution u as the sum

$$u(\mathbf{x}) = u_0(\mathbf{x}) + U(\mathbf{x}),$$

where U is a particular solution

$$LU = f \quad \text{in } \Omega$$

and u_0 satisfies the homogeneous equation

$$Lu_0 = 0 \quad \text{in } \Omega$$

with boundary conditions adjusted such that the total solution u satisfies the boundary conditions of the original problem. The remainder u_0 is obtained by solving the corresponding boundary integral equations, involving the new boundary data for u_0. In order to find these data accurately, one must be able to compute a particular solution (and, very often, its derivatives) with high precision.

Since the construction of a closed-form particular solution is possible only for some special inhomogeneities, it must be approximated. The standard way to do so is the cubature of the volume potential

$$\mathcal{K}\tilde{f}(\mathbf{x}) = \int\limits_{\mathbb{R}^n} \tilde{f}(\mathbf{y})\, \mathcal{E}(\mathbf{x},\mathbf{y})\, d\mathbf{y}$$

where $\tilde{f}$ is a compactly supported function with $\tilde{f}(\mathbf{x}) = f(\mathbf{x})$, $\mathbf{x} \in \Omega$.

Here, we discuss the application of Theorems 4.10 and 4.11 for the particular case $L = -\Delta$ and mixed boundary conditions are imposed:

$$\begin{gathered} -\Delta u = f, \\ u|_{\partial\Omega_D} = g_0\,, \quad \partial_n u|_{\partial\Omega_N} = g_1\,. \end{gathered} \tag{4.48}$$

The domain $\Omega \subset \mathbb{R}^n$ is bounded by a Lipschitz boundary $\partial\Omega$, which is split into two parts, where Dirichlet or Neumann boundary conditions are imposed, $\overline{\partial\Omega_D \cup \partial\Omega_N} = \partial\Omega$. The normal derivative $\partial_n u$ is taken with respect to the exterior normal $\mathbf{n}$ to $\partial\Omega$.

Let us suppose that the right-hand side f of the Poisson equation (4.48) can be continued outside Ω to a compactly supported function $\tilde{f}$ of preserved regularity, say $\tilde{f} \in W_2^L(\widetilde{\Omega})$ and $\operatorname{supp} \tilde{f} \subset \widetilde{\Omega}$. Then the harmonic potential $U = \mathcal{L}_n \tilde{f}$ is a particular solution of (4.48) and we denote its approximant by $U_h = \mathcal{L}_{n,h}^M \tilde{f}$ with the cubature formula given in (4.34).

Since $\widetilde{\Omega}$ is bounded, U_h is a finite sum, and, by Theorems 4.10 and 4.11, the values of $U|_\Gamma$ and $\partial_n U|_\Gamma$ can be approximated by $U_h(\mathbf{x})$ and $\mathbf{n} \cdot \nabla U_h(\mathbf{x})$, $\mathbf{x} \in \Gamma$, with high accuracy using appropriate chosen values of the parameters $\mathcal{D}$, h, and M. Thus, the remainder u_0, subject to

$$\Delta u_0 = 0 \quad \text{in } \Omega\,,$$
$$u_0|_{\Gamma_D} = g_0 - U|_{\Gamma_D}\,, \quad \partial_n u_0|_{\Gamma_N} = g_1 - \partial_n U|_{\Gamma_N}\,,$$

can be approximated by the solution of the modified problem

$$\Delta u_h = 0 \quad \text{in } \Omega,$$
$$h_h|_{\Gamma_D} = g_0 - U_h|_{\Gamma_D}\,, \quad \partial_n u_h|_{\Gamma_N} = g_1 - \partial_n U_h|_{\Gamma_N}\,.$$

In order to estimate the difference between u_0 and u_h, we use the Sobolev space $W_2^{1/2}(\Gamma)$, which is defined as the set of restrictions of functions $u \in W_2^1(\mathbb{R}^n)$ onto $\Gamma = \partial\Omega$. We consider the spaces

$$W_2^{1/2}(\Gamma_D) = \{u|_{\Gamma_D}, u \in W_2^{1/2}(\Gamma)\}\,,$$
$$W_2^{-1/2}(\Gamma_N) = \big(\mathring{W}_2^{1/2}(\Gamma_N)\big)'\,,$$

the second space being the dual of

$$\mathring{W}_2^{1/2}(\Gamma_N) = \{u \in W_2^{1/2}(\Gamma) : \operatorname{supp} u \subset \overline{\Gamma_N}\}$$

with respect to the L_2-duality.

Assuming $\Gamma_D \neq \emptyset$, $f \in L_2(\Omega)$, the boundary data $g_0 \in W_2^{1/2}(\Gamma_D)$ and $g_1 \in W_2^{-1/2}(\Gamma_N)$, there exists a unique weak solution $u \in W_2^1(\Omega)$ of the problem (4.48) and we have the estimate

$$\|u\|_{W_2^1(\Omega)} \le c\,(\|f\|_{L_2(\Omega)} + \|g_0\|_{W_2^{1/2}(\Gamma_D)} + \|g_1\|_{W_2^{-1/2}(\Gamma_N)})\,, \tag{4.49}$$

with a constant c independent of the data. This classical result can be found in [**30**].

So, we find that the difference between u_0 and u_h is bounded by

$$\|u_0 - u_h\|_{W_2^1(\Omega)} \le c\,(\|U - U_h\|_{W_2^{1/2}(\Gamma_D)} + \|\partial_n U - \partial_n U_h\|_{W_2^{-1/2}(\Gamma_N)})\,.$$

4.5.2. Error estimates.

THEOREM 4.12. *If the right-hand side of the Poisson equation* (4.48) *satisfies* $f \in W_2^L(\Omega)$, $L \ge 2M$, *then*

$$\|u_0 - u_h\|_{W_2^1(\Omega)} \le c\Big((\sqrt{\mathcal{D}}h)^{2M}\, \|f\|_{W_2^L(\Omega)}$$
$$+ h \sum_{[\boldsymbol{\alpha}]=0}^{2M-2} \Big(\frac{\sqrt{\mathcal{D}}h}{2\pi}\Big)^{[\boldsymbol{\alpha}]} \frac{\|\partial^{\boldsymbol{\alpha}} f\|_{W_2^1(\Omega)}}{\boldsymbol{\alpha}!} \sum_{\boldsymbol{\nu}\in\mathbb{Z}^n\setminus\{\mathbf{0}\}} \frac{|\partial^{\boldsymbol{\alpha}}\mathcal{F}\eta_{2M}(\sqrt{\mathcal{D}}\boldsymbol{\nu})|}{2\pi|\boldsymbol{\nu}|}\Big)\,,$$

where

$$\mathcal{F}\eta_{2M}(\boldsymbol{\lambda}) = \mathrm{e}^{-\pi^2|\boldsymbol{\lambda}|^2} \sum_{k=0}^{M-1} \frac{|\boldsymbol{\lambda}|^{2k}}{k!}\,.$$

PROOF. By a trace theorem (cf., e.g., [**30**]), we obtain

$$\|U - U_h\|_{W_2^{1/2}(\Gamma_D)} + \|\partial_n U - \partial_n U_h\|_{W_2^{-1/2}(\Gamma_N)}$$
$$\le c\|U - U_h\|_{W_2^1(\Omega)} = c\|\mathcal{L}_n\tilde{u} - \mathcal{L}_{n,h}^M\tilde{u}\|_{W_2^1(\Omega)}\,,$$

and the last norm is estimated in the following lemma. □

LEMMA 4.13. *Let Ω, $\mathrm{G} \subset \mathbb{R}^n$ be bounded domains, and let $f \in \mathring{W}_2^L(\Omega)$, $L > n/2$ and $L \geq 2M$. Then the cubature error of the harmonic potential can be estimated in the norm of the Sobolev space $W_2^1(\mathrm{G})$ by*

$$\|\mathcal{L}_n f - \mathcal{L}_{n,h}^M f\|_{W_2^1(\mathrm{G})} \leq c\Big((\sqrt{\mathcal{D}}h)^{2M} \, \|f\|_{W_2^L(\Omega)} + h \sum_{[\boldsymbol{\alpha}]=0}^{2M-2} \Big(\frac{\sqrt{\mathcal{D}}h}{2\pi}\Big)^{[\boldsymbol{\alpha}]} \frac{\|\partial^{\boldsymbol{\alpha}} f\|_{W_2^1(\Omega)}}{\boldsymbol{\alpha}!} \sum_{\boldsymbol{\nu}\in\mathbb{Z}^n\setminus\{\mathbf{0}\}} \frac{|\partial^{\boldsymbol{\alpha}} \mathcal{F}\eta_{2M}(\sqrt{\mathcal{D}}\boldsymbol{\nu})|}{2\pi|\boldsymbol{\nu}|}\Big) . \tag{4.50}$$

Note that $\mathcal{L}_n f - \mathcal{L}_{n,h}^M f = \mathcal{L}_n(f - \mathcal{M}_{h,\mathcal{D}}^{(2M)} f)$. Owing to Theorem 4.6 and $W_2^{-1}(\mathbb{R}^n) = H_2^{-1}(\mathbb{R}^n)$, the norm

$$\|f - \mathcal{M}_{h,\mathcal{D}}^{(2M)} f\|_{W_2^{-1}(\mathbb{R}^n)}$$

can be estimated by (4.50). Unfortunately, the harmonic potential $\mathcal{L}_n$ does not map $W_2^{-1}(\mathbb{R}^n)$ boundedly into $W_2^1(\mathbb{R}^n)$. Therefore we consider sufficiently smooth functions f with $\operatorname{supp} f \subseteq \overline{\Omega}$.

PROOF OF LEMMA 4.13. Since the support of the quasi-interpolant $\mathcal{M}_{h,\mathcal{D}}^{(2M)} f$ is the whole $\mathbb{R}^n$, we have to split the estimation of $\mathcal{L}_n f - \mathcal{L}_{n,h}^M f$. Let us define the r-neighborhood Ω_r^+, $r > 0$, of Ω by

$$\Omega_r^+ = \{\mathbf{x} : \operatorname{dist}(\mathbf{x}, \Omega) < r\} , \tag{4.51}$$

where $\operatorname{dist}(\mathbf{x}, \Omega)$ denotes the distance between $\mathbf{x}$ and the closure $\overline{\Omega} \supseteq \operatorname{supp} f$. Furthermore, we introduce a smooth cut-off function φ with $\varphi(\mathbf{x}) = 1$ for $\mathbf{x} \in \Omega_r^+$ and $\operatorname{supp} \varphi \subseteq \overline{\Omega_{2r}^+}$, and we write

$$\mathcal{L}_n(f - \mathcal{M}_{h,\mathcal{D}}^{(2M)} f) = \mathcal{L}_n((\varphi - 1)\mathcal{M}_{h,\mathcal{D}}^{(2M)} f) + \mathcal{L}_n(\varphi(f - \mathcal{M}_{h,\mathcal{D}}^{(2M)} f)) . \tag{4.52}$$

Note that $\mathcal{M}_{h,\mathcal{D}}^{(2M)} f$ decays rapidly outside the support of f. There exists a constant c_M such that for all $\mathbf{x} \in \mathbb{R}^n$

$$\mathrm{e}^{-|\mathbf{x}|^2} \, |L_{M-1}^{(n/2)}(|\mathbf{x}|^2)| \leq c_M \, \mathrm{e}^{-|\mathbf{x}|^2/2} ,$$

which implies

$$\begin{aligned} |\mathcal{M}_{h,\mathcal{D}}^{(2M)} f(\mathbf{x})| &= \frac{1}{(\pi\mathcal{D})^{n/2}} \Big| \sum_{h\mathbf{m}\in\Omega} f(h\mathbf{m}) \, \mathrm{e}^{-|\mathbf{x}-h\mathbf{m}|^2/(\mathcal{D}h^2)} \, L_{M-1}^{(n/2)}\Big(\frac{|\mathbf{x}-h\mathbf{m}|^2}{\mathcal{D}h^2}\Big) \Big| \\ &\leq \frac{c_M}{(\pi\mathcal{D})^{n/2}} \sum_{h\mathbf{m}\in\Omega} |f(h\mathbf{m})| \, \mathrm{e}^{-|\mathbf{x}-h\mathbf{m}|^2/(2\mathcal{D}h^2)} \\ &\leq c\, h^{-1} \, \mathrm{e}^{-\operatorname{dist}(\mathbf{x},\Omega)^2/(2\mathcal{D}h^2)} \max |f| \end{aligned}$$

for $\mathbf{x} \notin \overline{\Omega}$. The constant c depends only on $\mathcal{D}$, M, and the measure of Ω. Since

$$\mathcal{L}_n((\varphi - 1)\mathcal{M}_{h,\mathcal{D}}^{(2M)} f)(\mathbf{y}) = \frac{\Gamma(\frac{n}{2} - 1)}{4\pi^{n/2}} \int\limits_{\operatorname{dist}(\mathbf{x},\Omega)\geq r} \frac{(\varphi(\mathbf{x}) - 1)\mathcal{M}_{h,\mathcal{D}}^{(2M)} f(\mathbf{x})}{|\mathbf{y} - \mathbf{x}|^{n-2}} \, d\mathbf{x} ,$$

we obtain for $\mathbf{y} \in \Omega$

$$|\mathcal{L}_n((1-\varphi)\mathcal{M}^{(2M)}_{h,\mathcal{D}} f)(\mathbf{y})| \le c\, h^{-1} \max|f| \int\limits_{\mathrm{dist}(\mathbf{x},\Omega)\ge r} \frac{\mathrm{e}^{-\mathrm{dist}(\mathbf{x},\Omega)^2/(2\mathcal{D}h^2)}}{|\mathbf{y}-\mathbf{x}|^{n-2}}\, d\mathbf{x}$$

$$= c\, \max|f|\, \frac{\mathrm{e}^{-r^2/(2\mathcal{D}h^2)}}{r^{n-2}} \int\limits_{\mathrm{dist}(\mathbf{x},h^{-1}\Omega)\ge r/h} \mathrm{e}^{-(\mathrm{dist}(\mathbf{x},h^{-1}\Omega)^2-r^2/h^2)/(2\mathcal{D})}\, d\mathbf{x}\,,$$

$$|\nabla\mathcal{L}_n((1-\varphi)\mathcal{M}^{(2M)}_{h,\mathcal{D}} f)(\mathbf{y})|$$
$$\le c\, \max|f|\, \frac{\mathrm{e}^{-r^2/(2\mathcal{D}h^2)}}{r^{n-1}} \int\limits_{\mathrm{dist}(\mathbf{x},h^{-1}\Omega)\ge r/h} \mathrm{e}^{-(\mathrm{dist}(\mathbf{x},h^{-1}\Omega)^2-r^2/h^2)/(2\mathcal{D})}\, d\mathbf{x}\,,$$

which, in view of

$$\int\limits_{\mathrm{dist}(\mathbf{x},h^{-1}\Omega)\ge r/h} \mathrm{e}^{-(\mathrm{dist}(\mathbf{x},h^{-1}\Omega)^2-r^2/h^2)/(2\mathcal{D})}\, d\mathbf{x} \le \int\limits_{|\mathbf{t}|\ge r/h} \mathrm{e}^{-(|\mathbf{t}|^2-r^2/h^2)/(2\mathcal{D})}\, d\mathbf{t}$$
$$\le (2\pi\mathcal{D})^{n/2}\,,$$

leads to the estimate

$$\|\mathcal{L}_n((1-\varphi)\mathcal{M}^{(2M)}_{h,\mathcal{D}} f)\|_{W_2^1(\mathrm{G})} \le c\, \max|f(\mathbf{x})|\, \frac{\mathrm{e}^{-r^2/(2\mathcal{D}h^2)}}{r^{n-1}} \tag{4.53}$$

with a constant c independent of $h>0$ and $r>0$. The estimation of the second term of (4.52) is based on the inequality

$$\|\mathcal{L}_n v\|_{W_2^1(\mathrm{G})} \le c\, \|v\|_{(W_2^1(\Omega))'} \tag{4.54}$$

with a constant $c = c(\mathrm{G},\Omega)$, which holds for all v with compact support $\mathrm{supp}\, v \subseteq \overline{\Omega}$. This follows by interpolation from the well-known boundedness of the harmonic potential $\mathcal{L}_n : L_2(\Omega) \to W_2^2(\mathrm{G})$ for any bounded domains Ω, $\mathrm{G} \subset \mathbb{R}^n$; cf., e.g., [**75**, Satz 12.5.3]. Since $\mathcal{L}_n$ has a symmetric kernel, we obtain by duality that $\mathcal{L}_n : (W_2^2(\Omega))' \to L_2(\mathrm{G})$ is bounded, too.

Hence we obtain by (4.54)

$$\|\mathcal{L}_n(\varphi(f-\mathcal{M}^{(2M)}_{h,\mathcal{D}} f))\|_{W_2^1(\mathrm{G})} \le c\, \|\varphi(f-\mathcal{M}^{(2M)}_{h,\mathcal{D}} f)\|_{(W_2^1(\Omega^+_{2r}))'}$$
$$\le c \sup_{w\in W_2^1(\Omega^+_{2r})} \frac{(\varphi w, f-\mathcal{M}^{(2M)}_{h,\mathcal{D}} f)_{L_2}}{\|w\|_{W_2^1(\Omega^+_{2r})}}$$
$$\le c\, \|f-\mathcal{M}^{(2M)}_{h,\mathcal{D}} f\|_{W_2^{-1}(\mathbb{R}^n)} \sup_{w\in W_2^1(\Omega^+_{2r})} \frac{\|\varphi w\|_{W_2^1(\mathbb{R}^n)}}{\|w\|_{W_2^1(\Omega^+_{2r})}}\,,$$

which together with (4.53) implies for every $r>0$

$$\|\mathcal{L}_n(f-\mathcal{M}^{(2M)}_{h,\mathcal{D}} f)\|_{W_2^1(\mathrm{G})}$$
$$\le c\Big(\|f-\mathcal{M}^{(2M)}_{h,\mathcal{D}} f\|_{W_2^{-1}(\mathbb{R}^n)} + \|f\|_{W_2^L(\Omega)}\, r^{1-n}\, \mathrm{e}^{-r^2/2\mathcal{D}h^2}\Big). \qquad \square$$

4.6. Notes

The main formulas and some of the results of Sections 4.2, 4.3, and 4.4 have been obtained in [**67**].

It is obvious that the error estimates for the cubature of the harmonic potentials given in Section 4.4 can be generalized to potentials of symmetric and positive definite differential operators with constant coefficients. The fact that the dilated shifts $\eta(\mathbf{x}/h_j - \mathbf{m})$, $\mathbf{m} \in \mathbb{Z}^n$, $h_j \to 0$, are dense in the space L_p, mentioned in Remark 4.9, was recently extended by Bui and Laugesen [**17**] under rather weak conditions on η for $p \in [1, \infty)$.

The problem of avoiding domain integrals in boundary integral formulations has attracted much attention within the boundary element community and different approaches have been developed to solve this problem; cf. [**77**]. The method proposed in Section 4.5 differs from these approaches in using quasi-interpolation instead of interpolating the densities, which provides high-order accuracy without solving linear systems.

CHAPTER 5

Cubature of diffraction, elastic, and hydrodynamic potentials

In this chapter we develop semi-analytic cubature formulas for other volume potentials which are used to solve classical problems of mathematical physics. We derive analytic formulas for these potentials applied to the generating functions η_{2M} defined in (3.18). Analogously to the preceding chapter, these formulas can be used to construct high-order cubature formulas for the corresponding potentials.

Diffraction potentials are studied in Section 5.1. We obtain semi-analytic cubature formulas for the one- and three-dimensional cases. Formulas for the potential of a more general advection-diffusion operator applied to the Gaussian are given in Section 5.2. In the remaining sections we derive semi-analytic cubature formulas for two- and three-dimensional elastic and hydrodynamic potentials.

5.1. Diffraction potentials

In this section, we consider the volume potential

$$\mathcal{S}_n u(\mathbf{x}) = \int\limits_{\mathbb{R}^n} \mathcal{E}_k(\mathbf{x}-\mathbf{y})\, u(\mathbf{y})\, d\mathbf{y}\,, \tag{5.1}$$

with the fundamental solution of the Helmholtz equation

$$\mathcal{E}_k(\mathbf{x}) = \frac{i}{4}\Big(\frac{k}{2\pi|\mathbf{x}|}\Big)^{n/2-1} H^{(1)}_{n/2-1}(k|\mathbf{x}|)\,, \quad \operatorname{Im} k \geq 0\,. \tag{5.2}$$

Here $H^{(1)}_n = J_n + iY_n$ is the n-th order Hankel function of the first kind.

The function $f = \mathcal{S}u$ is the solution of the Helmholtz equation

$$\Delta f + k^2 f = -u\,, \tag{5.3}$$

satisfying Sommerfeld's radiation condition

$$\lim_{|\mathbf{x}|\to\infty} |\mathbf{x}|^{(n-1)/2}\Big(\Big\langle \frac{\mathbf{x}}{|\mathbf{x}|}, \nabla f(\mathbf{x})\Big\rangle - ik\, f(\mathbf{x})\Big) = 0\,. \tag{5.4}$$

The diffraction potential appears frequently in problems of acoustics, electromagnetics, and optics. Besides the singularity of the kernel function $\mathcal{E}_k$ the approximation of this integral operator is challenging because of the fast oscillations of $\mathcal{E}_k$ for high wave numbers k. The application of approximate approximations to this problem provides semi-analytic cubature formulas, which reduce these problems to the efficient computation of certain special functions.

5.1.1. Higher-order formulas. Since the kernel $\mathcal{E}_k$ is a radial function, the diffraction potential of radial generating functions η can be obtained by one-dimensional quadrature. Moreover, from $\mathcal{S}_n = -(\Delta + k^2)^{-1}$, we have

$$\mathcal{S}_n(-\Delta)^j = (-\Delta)^j \mathcal{S} = k^{2j}\mathcal{S} + \sum_{p=0}^{j-1} k^{2(j-p-1)}(-\Delta)^p .$$

Hence, similarly to Section 4.3, it is easy to determine the diffraction potentials of generating functions constructed via the general formula (3.12). Consider, for example, the basis function $\eta_{2M}(\mathbf{x}) = \pi^{-n/2}\, L_{M-1}^{(n/2)}(|\mathbf{x}|^2)\, \mathrm{e}^{-|\mathbf{x}|^2}$. Then from formulas (3.12) and (3.13), we obtain

$$\begin{aligned}\mathcal{S}_n\eta_{2M} &= \Gamma\big(\frac{n}{2}\big) \sum_{j=0}^{M-1} \frac{\Delta^j(\mathcal{F}\eta)^{-1}(\mathbf{0})}{j!\,(4\pi)^{2j}\,\Gamma(j+\frac{n}{2})}\, \mathcal{S}_n(-\Delta)^j(\mathrm{e}^{-|\cdot|^2}) \\ &= \pi^{-n/2}\sum_{j=0}^{M-1}\frac{1}{4^j j!}\Big(k^{2j}\mathcal{S}_n + \sum_{p=0}^{j-1} k^{2(j-p-1)}(-\Delta)^p\Big)(\mathrm{e}^{-|\cdot|^2}) .\end{aligned}$$

By using (4.24), we obtain the representation

$$(5.5)\qquad \begin{aligned}\mathcal{S}_n\eta_{2M}(\mathbf{x}) &= \frac{\mathcal{S}_n(\mathrm{e}^{-|\cdot|^2})(\mathbf{x})}{\pi^{n/2}} \sum_{j=0}^{M-1}\frac{k^{2j}}{4^j j!} \\ &\quad + \frac{\mathrm{e}^{-|\mathbf{x}|^2}}{\pi^{n/2}k^2}\sum_{j=0}^{M-1}\sum_{m=0}^{j-1}\frac{k^{2(j-m)}m!}{2^{2(j-m)}j!}\, L_m^{(n/2-1)}(|\mathbf{x}|^2) ,\end{aligned}$$

which is the basis for approximate cubature formulas of order $\mathcal{O}(h^{2M})$ for the diffraction potential.

5.1.2. Action of the one-dimensional diffraction potential on the Gaussian. In order to find an explicit form of (5.5), it remains to compute $\mathcal{S}_n\big(\mathrm{e}^{-|\cdot|^2}\big)$. At least in the case of odd space dimensions, where the fundamental solution $\mathcal{E}_k$ can be obtained from

$$(5.6)\qquad \begin{aligned} &H^{(1)}_{-1/2} = \sqrt{\frac{2}{\pi}}\frac{\mathrm{e}^{iz}}{\sqrt{z}}, \quad H^{(1)}_{1/2} = -i\sqrt{\frac{2}{\pi}}\frac{\mathrm{e}^{iz}}{\sqrt{z}}, \\ &H^{(1)}_{j+1/2} = -i(-1)^j\sqrt{\frac{2}{\pi}}\, z^{j+1/2}\Big(\frac{1}{z}\frac{d}{dz}\Big)^n \frac{\mathrm{e}^{iz}}{z}, \qquad j = 1,2,\dots ,\end{aligned}$$

one can find analytic formulas of these potentials rather straightforwardly. In particular, if $n = 1$, then

$$\mathcal{E}_k(x) = \frac{\mathrm{e}^{ik|x|}}{2ik} ,$$

and the diffraction potentials can be expressed using the *scaled complementary error function*

$$(5.7)\qquad \mathrm{w}(z) = \mathrm{e}^{-z^2}\,\mathrm{erfc}(-iz) = \mathrm{e}^{-z^2}\Big(1 + \frac{2i}{\sqrt{\pi}}\int_0^z \mathrm{e}^{t^2}\,dt\Big)$$

(see [**1**, 7.1.3]), where

$$\operatorname{erfc}(\tau) = 1 - \operatorname{erf}(\tau) \tag{5.8}$$

is the *complementary error function.* The function $\mathrm{w}(z)$ is also known as the *Faddeeva function* and will appear throughout the book at different places. It can also be written in the form

$$\mathrm{w}(z) = \frac{2}{\sqrt{\pi}} \int_0^\infty \mathrm{e}^{-t^2} \, \mathrm{e}^{2izt} \, dt \tag{5.9}$$

and it has the property

$$\mathrm{w}(-\bar{z}) = \overline{\mathrm{w}(z)} \quad \text{and} \quad \mathrm{w}(-z) = 2\,\mathrm{e}^{-z^2} - \mathrm{w}(z) \quad \text{for any } z \in \mathbb{C} \tag{5.10}$$

(cf. [**1**, 7.1.11/12]).

If $\operatorname{Im} z > 0$, then the integral representation

$$\mathrm{w}(z) = \frac{i}{\pi} \int_{-\infty}^{\infty} \frac{\mathrm{e}^{-t^2}}{z - t} \, dt \tag{5.11}$$

holds [**1**, 7.1.4]. Then the real and imaginary parts of the Faddeeva function

$$\mathrm{w}(x + iy) = K(x, y) + iL(x, y), \quad y > 0,$$

coincide with the *Voigt functions*

$$K(x, y) = \frac{y}{\pi} \int_{-\infty}^{\infty} \frac{\mathrm{e}^{-t^2}}{(x - t)^2 + y^2} \, dt, \quad L(x, y) = \frac{1}{\pi} \int_{-\infty}^{\infty} \frac{(x - t)\,\mathrm{e}^{-t^2}}{(x - t)^2 + y^2} \, dt, \tag{5.12}$$

which appear in several fields of physics.

It follows immediately from (5.9) that for $n = 1$ the diffraction potential of the Gaussian is equal to

$$\frac{1}{2ik} \int_{-\infty}^{\infty} \mathrm{e}^{ik|x-y|} \, \mathrm{e}^{-y^2} \, dy = \frac{\sqrt{\pi}\,\mathrm{e}^{-x^2}}{4ik} \left(\mathrm{w}\left(\frac{k}{2} + ix\right) + \mathrm{w}\left(\frac{k}{2} - ix\right) \right). \tag{5.13}$$

Thus, from (5.5), we obtain the diffraction potential of the higher-order generating function

$$\begin{aligned} \frac{1}{2ik} \int_{-\infty}^{\infty} \mathrm{e}^{ik|x-y|} \, \eta_{2M}(y) \, dy = & \frac{\mathrm{e}^{-x^2}}{4ik} \left(\mathrm{w}\left(\frac{k}{2} + ix\right) + \mathrm{w}\left(\frac{k}{2} - ix\right) \right) \sum_{j=0}^{M-1} \frac{k^{2j}}{4^j j!} \\ & + \frac{\mathrm{e}^{-x^2}}{\sqrt{\pi}} \sum_{m=0}^{M-1} \frac{(-1)^m H_{2m}(x)}{k^{2(m+1)}} \sum_{j=m+1}^{M-1} \frac{k^{2j}}{4^j j!}, \end{aligned} \tag{5.14}$$

where the relation

$$H_{2m}(\tau) = (-1)^m \, m! \, 4^m L_m^{(-1/2)}(\tau^2)$$

between the Hermite polynomials H_j and the generalized Laguerre polynomials (cf. [**1**, 22.5.38]) has been used.

5.1.3. General form of the convolution with Gaussian. The fundamental solution of the Helmholtz operator has the Fourier transform $(4\pi^2|\boldsymbol{\lambda}|^2 - k^2)^{-1}$. Therefore, if the formula for convolutions of radial functions (2.15) is applied, then the diffraction potential is expressed as a one-dimensional singular integral, if k is real. Here we introduce another method to determine convolutions of general radial functions with the Gaussian, which is applied to the three-dimensional diffraction potential in the following.

Consider the convolution with a radial kernel $Q(\mathbf{x}) = Q(|\mathbf{x}|)$, i.e.,

$$\left(Q * \mathrm{e}^{-|\cdot|^2}\right)(\mathbf{x}) = \int_{\mathbb{R}^n} Q(|\mathbf{x}-\mathbf{y}|)\, \mathrm{e}^{-|\mathbf{y}|^2}\, d\mathbf{y} = \int_{\mathbb{R}^n} Q(|\mathbf{y}|)\, \mathrm{e}^{-|\mathbf{x}-\mathbf{y}|^2}\, d\mathbf{y}\,.$$

By introducing spherical coordinates in $\mathbb{R}^n$, we write

$$\left(Q * \mathrm{e}^{-|\cdot|^2}\right)(\mathbf{x}) = \mathrm{e}^{-|\mathbf{x}|^2} \int_0^\infty Q(r)\, \mathrm{e}^{-r^2}\, r^{n-1}\, dr \int_{S^{n-1}} \mathrm{e}^{2|\mathbf{x}|r\langle\boldsymbol{\omega},\boldsymbol{\omega}'\rangle}\, d\sigma_{\boldsymbol{\omega}'}\,,$$

where $\mathbf{x} = |\mathbf{x}|\,\boldsymbol{\omega}$, $\mathbf{y} = r\,\boldsymbol{\omega}'$, and S^{n-1} is the unit sphere in $\mathbb{R}^n$. From (2.14), one derives

$$\int_{S^{n-1}} \mathrm{e}^{2|\mathbf{x}|r\langle\boldsymbol{\omega},\boldsymbol{\omega}'\rangle}\, d\sigma_{\boldsymbol{\omega}'} = \frac{2\pi^{n/2} J_{n/2-1}(2i|\mathbf{x}|r)}{(i|\mathbf{x}|r)^{n/2-1}} = \frac{2\pi^{n/2} I_{n/2-1}(2|\mathbf{x}|r)}{(|\mathbf{x}|r)^{n/2-1}}$$

with the modified Bessel function of the first kind $I_\nu(t) = i^{-\nu} J_\nu(it)$. Hence, the convolution of the Gaussian with a radial function can be obtained from the formula

$$\left(Q * \mathrm{e}^{-|\cdot|^2}\right)(\mathbf{x}) = \frac{2\pi^{n/2}\, \mathrm{e}^{-|\mathbf{x}|^2}}{|\mathbf{x}|^{n/2-1}} \int_0^\infty Q(r)\, \mathrm{e}^{-r^2}\, I_{n/2-1}(2|\mathbf{x}|r)\, r^{n/2}\, dr\,. \tag{5.15}$$

Note that from (2.16)

$$I_{j+1/2}(z) = \sqrt{\frac{2}{\pi}}\, z^{j+1/2} \left(\frac{1}{z}\frac{d}{dz}\right)^j \frac{\sinh z}{z}\,. \tag{5.16}$$

5.1.4. Analytic expressions of the diffraction potential. Applying (5.2) and (5.15), we derive the representation

$$\begin{aligned}
\mathcal{S}_n\left(\mathrm{e}^{-|\cdot|^2}\right) &= \int_{\mathbb{R}^n} \frac{i}{4}\left(\frac{k}{2\pi|\mathbf{x}-\mathbf{y}|}\right)^{n/2-1} H^{(1)}_{n/2-1}(k|\mathbf{x}-\mathbf{y}|)\, \mathrm{e}^{-|\mathbf{y}|^2}\, d\mathbf{y} \\
&= \frac{i}{4}\left(\frac{k}{2\pi}\right)^{n/2-1} \frac{2\pi^{n/2}\, \mathrm{e}^{-|\mathbf{x}|^2}}{|\mathbf{x}|^{n/2-1}} \int_0^\infty \frac{H^{(1)}_{n/2-1}(kr)}{r^{n/2-1}}\, \mathrm{e}^{-r^2}\, I_{n/2-1}(2|\mathbf{x}|r)\, r^{n/2}\, dr \\
&= \frac{\pi i\, \mathrm{e}^{-|\mathbf{x}|^2}}{2}\left(\frac{k}{2|\mathbf{x}|}\right)^{n/2-1} \int_0^\infty H^{(1)}_{n/2-1}(kr)\, I_{n/2-1}(2|\mathbf{x}|r)\, \mathrm{e}^{-r^2}\, r\, dr\,.
\end{aligned}$$

In the special case $n = 3$, we obtain from (5.6)

$$\mathcal{E}_k(\mathbf{x}) = \frac{\mathrm{e}^{ik|\mathbf{x}|}}{4\pi|\mathbf{x}|}\,;$$

hence (5.15) leads to

$$\frac{1}{4\pi}\int\limits_{\mathbb{R}^3}\frac{e^{ik|\mathbf{x}-\mathbf{y}|}}{|\mathbf{x}-\mathbf{y}|}\,e^{-|\mathbf{y}|^2}\,d\mathbf{y}=\frac{1}{4\pi}\frac{2\pi^{3/2}\,e^{-|\mathbf{x}|^2}}{|\mathbf{x}|^{1/2}}\int\limits_0^\infty\frac{e^{ikr}}{r}\,e^{-r^2}\,r^{3/2}\,I_{1/2}(2|\mathbf{x}|r)\,dr\,.$$

By (5.16) and (2.16),

$$I_{1/2}(z)=\sqrt{\frac{2}{\pi}}\,\frac{\sinh z}{\sqrt{z}}\,,$$

which together with (5.15) leads to

$$\frac{1}{4\pi}\int\limits_{\mathbb{R}^3}\frac{e^{ik|\mathbf{x}-\mathbf{y}|}}{|\mathbf{x}-\mathbf{y}|}\,e^{-|\mathbf{y}|^2}\,d\mathbf{y}=\frac{e^{-|\mathbf{x}|^2}}{4|\mathbf{x}|}\int\limits_0^\infty e^{-r^2}\left(e^{r(2|\mathbf{x}|+ik)}-e^{-r(2|\mathbf{x}|-ik)}\right)dr\,. \tag{5.17}$$

Hence, by (5.9) we can write the right-hand side as

$$\frac{1}{4\pi}\int\limits_{\mathbb{R}^3}\frac{e^{ik|\mathbf{x}-\mathbf{y}|}}{|\mathbf{x}-\mathbf{y}|}\,e^{-|\mathbf{y}|^2}\,d\mathbf{y}=\frac{\sqrt{\pi}}{2}\frac{e^{-|\mathbf{x}|^2}}{4|\mathbf{x}|}\left(\mathrm{w}\left(\frac{k}{2}-i|\mathbf{x}|\right)-\mathrm{w}\left(\frac{k}{2}+i|\mathbf{x}|\right)\right). \tag{5.18}$$

If $|\mathbf{x}|=0$, then

$$\frac{1}{4\pi}\int\limits_{\mathbb{R}^3}\frac{e^{ik|\mathbf{y}|}}{|\mathbf{y}|}\,e^{-|\mathbf{y}|^2}\,d\mathbf{y}=\frac{1}{2}+\frac{ik\sqrt{\pi}}{4}\,\mathrm{w}\left(\frac{k}{2}\right).$$

Combining (5.18) with (5.5) and (3.22) gives the analytic formula for $n=3$:

$$\begin{aligned}\mathcal{S}_3\eta_{2M}(\mathbf{x})=&\frac{e^{-|\mathbf{x}|^2}}{8\pi|\mathbf{x}|}\left(\mathrm{w}\left(\frac{k}{2}-i|\mathbf{x}|\right)-\mathrm{w}\left(\frac{k}{2}+i|\mathbf{x}|\right)\right)\sum_{j=0}^{M-1}\frac{k^{2j}}{4^j j!}\\&+\frac{e^{-|\mathbf{x}|^2}}{2\pi^{3/2}}\sum_{p=0}^{M-2}\frac{(-1)^p H_{2p+1}(|\mathbf{x}|)}{k^{2(p+1)}|\mathbf{x}|}\sum_{j=p+1}^{M-1}\frac{k^{2j}}{4^j j!}\,.\end{aligned} \tag{5.19}$$

Note that numerical computations with $\mathrm{w}(z)$ can lead to overflow problems if $\operatorname{Im} z<0$, which can be seen from the representation (5.9). This does not concern the case $\operatorname{Im} z\geq 0$, where reliable and efficient implementations for computing $\mathrm{w}(z)$ with double precision are available. Fortunately, by using (5.10), the formulas (5.14) and (5.19) can be modified so that the arguments of the Faddeeva function have non-negative imaginary part. In particular, in the case $k\in\mathbb{R}$, the diffraction potentials $\mathcal{S}_1$ and $\mathcal{S}_3$ of the Gaussian can be expressed by the Voigt functions (5.12):

$$\begin{aligned}&\frac{1}{2ik}\int\limits_{-\infty}^{\infty}e^{ik|x-y|}\,e^{-y^2}\,dy=\frac{\sqrt{\pi}}{2k}\Big(\frac{e^{ik|x|}\,e^{-k^2/4}}{i}+e^{-x^2}\,L\Big(\frac{k}{2}+i|x|\Big)\Big)\,,\\&\frac{1}{4\pi}\int\limits_{\mathbb{R}^3}\frac{e^{ik|\mathbf{x}-\mathbf{y}|}}{|\mathbf{x}-\mathbf{y}|}\,e^{-|\mathbf{y}|^2}\,d\mathbf{y}=\frac{\sqrt{\pi}}{4}\Big(\frac{e^{ik|\mathbf{x}|}\,e^{-k^2/4}}{|\mathbf{x}|}-\frac{e^{-|\mathbf{x}|^2}}{|\mathbf{x}|}\,K\Big(\frac{k}{2}+i|\mathbf{x}|\Big)\Big).\end{aligned} \tag{5.20}$$

5.2. Potentials of advection-diffusion operators

Here, we consider the volume potential of the differential operator with constant coefficients in $\mathbb{R}^n$

$$\mathcal{A}_n = -\Delta + 2\mathbf{b}\cdot\nabla + c\,, \tag{5.21}$$

where $\mathbf{b} \in \mathbb{C}^n$ and $c \in \mathbb{C}$. The fundamental solution $\kappa_\lambda(\mathbf{x})$ of the operator $\mathcal{A}_n$ depends on the value of $\vartheta = c + |\mathbf{b}|^2$. We use the notation

$$\langle \mathbf{y}, \mathbf{z}\rangle = \sum_{j=1}^{n} y_j z_j \quad \text{and} \quad |\mathbf{z}|^2 = \langle \mathbf{z}, \mathbf{z}\rangle$$

also for complex-valued vectors $\mathbf{y}, \mathbf{z} \in \mathbb{C}^n$.

If $\vartheta \neq 0$, then

$$\kappa_\lambda(\mathbf{x}) := \frac{\mathrm{e}^{\langle \mathbf{b},\mathbf{x}\rangle}}{(2\pi)^{n/2}} \Big(\frac{|\mathbf{x}|}{\lambda}\Big)^{1-n/2} K_{n/2-1}(\lambda|\mathbf{x}|)\,, \quad \mathbf{x} \neq \mathbf{0}\,,$$

where $\lambda \in \mathbb{C} \setminus (-\infty, 0]$ with $\lambda^2 = \vartheta$ and K_ν is the modified Bessel function of the second kind, also known as the Macdonald function [**1**, 9.6].
If $\vartheta = 0$, then

$$\kappa_0(\mathbf{x}) := \begin{cases} \dfrac{\mathrm{e}^{\langle \mathbf{b},\mathbf{x}\rangle}}{2\pi} \log \dfrac{1}{|\mathbf{x}|}\,, & n = 2\,, \\[2ex] \dfrac{1}{(n-2)\omega_n} \dfrac{\mathrm{e}^{\langle \mathbf{b},\mathbf{x}\rangle}}{|\mathbf{x}|^{n-2}}\,, & n \geq 3\,, \end{cases}$$

where ω_n is the surface area of the unit sphere in $\mathbb{R}^n$.

LEMMA 5.1. *Let $\vartheta := c + |\mathbf{b}|^2 \neq 0$. Then the solution of the equation*

$$-\Delta u + 2\mathbf{b}\cdot\nabla u + cu = \mathrm{e}^{-|\mathbf{x}|^2}, \ \mathbf{x} \in \mathbb{R}^n\,, \tag{5.22}$$

is given as the one-dimensional integral

$$u(\mathbf{x}) = \frac{\mathrm{e}^{-|\mathbf{x}|^2}\,\lambda^{n/2-1}}{|2\mathbf{x}+\mathbf{b}|^{n/2-1}} \int_0^\infty K_{n/2-1}(\lambda r)\, I_{n/2-1}(|2\mathbf{x}+\mathbf{b}|r)\, r\, \mathrm{e}^{-r^2}\, dr \tag{5.23}$$

where $\lambda^2 = c + |\mathbf{b}|^2$. In particular, for $n = 3$,

$$\boxed{u(\mathbf{x}) = \frac{\sqrt{\pi}}{4} \frac{\mathrm{e}^{-|\mathbf{x}|^2}}{|2\mathbf{x}+\mathbf{b}|} \Big(\mathrm{w}\Big(\frac{i}{2}(\lambda - |2\mathbf{x}+\mathbf{b}|)\Big) - \mathrm{w}\Big(\frac{i}{2}(\lambda + |2\mathbf{x}+\mathbf{b}|)\Big)\Big).} \tag{5.24}$$

PROOF. We have to simplify the integral

$$\begin{aligned} u(\mathbf{x}) &= \int_{\mathbb{R}^n} \kappa_\lambda(\mathbf{x}-\mathbf{y})\,\mathrm{e}^{-|\mathbf{y}|^2}\, d\mathbf{y} \\ &= \frac{\mathrm{e}^{\langle \mathbf{b},\mathbf{x}\rangle}}{(2\pi)^{n/2}} \int_{\mathbb{R}^n} \Big(\frac{|\mathbf{x}-\mathbf{y}|}{\lambda}\Big)^{1-n/2} K_{n/2-1}(\lambda|\mathbf{x}-\mathbf{y}|)\,\mathrm{e}^{-\langle \mathbf{b},\mathbf{y}\rangle - |\mathbf{y}|^2}\, d\mathbf{y} \\ &= \frac{\mathrm{e}^{\langle \mathbf{b},\mathbf{x}\rangle}\,\mathrm{e}^{|\mathbf{b}|^2/4}}{(2\pi)^{n/2}} \int_{\mathbb{R}^n} \Big(\frac{|\mathbf{x}+\mathbf{b}/2-\mathbf{y}|}{\lambda}\Big)^{1-n/2} K_{n/2-1}(\lambda|\mathbf{x}+\mathbf{b}/2-\mathbf{y}|)\,\mathrm{e}^{-|\mathbf{y}|^2}\, d\mathbf{y}\,. \end{aligned}$$

By using (5.15), we derive

$$u(\mathbf{x}-\mathbf{b}/2) = \frac{e^{\langle \mathbf{b},(\mathbf{x}-\mathbf{b}/2)\rangle}\, e^{|\mathbf{b}|^2/4}}{(2\pi)^{n/2}} \int\limits_{\mathbb{R}^n} \Big(\frac{|\mathbf{x}-\mathbf{y}|}{\lambda}\Big)^{1-n/2} K_{n/2-1}(\lambda|\mathbf{x}-\mathbf{y}|)\, e^{-|\mathbf{y}|^2}\, d\mathbf{y}$$

$$= \frac{e^{\langle \mathbf{b},(\mathbf{x}-\mathbf{b}/2)\rangle}\, e^{|\mathbf{b}|^2/4}}{(2\pi)^{n/2}} \frac{2\pi^{n/2}\, e^{-|\mathbf{x}|^2}}{|\mathbf{x}|^{n/2-1}}$$

$$\times \int\limits_0^\infty \Big(\frac{r}{\lambda}\Big)^{1-n/2} K_{n/2-1}(\lambda r)\, e^{-r^2}\, I_{n/2-1}(2|\mathbf{x}|r)\, r^{n/2}\, dr$$

$$= \frac{e^{-|\mathbf{x}-\mathbf{b}/2|^2}\, \lambda^{n/2-1}}{(2|\mathbf{x}|)^{n/2-1}} \int\limits_0^\infty K_{n/2-1}(\lambda r)\, I_{n/2-1}(2|\mathbf{x}|r)\, r\, e^{-r^2}\, dr\,,$$

which gives formula (5.23).

Noting that

$$K_{1/2}(\lambda r) = \sqrt{\frac{\pi}{2}}\, \frac{e^{-\lambda r}}{\sqrt{\lambda r}}\,, \qquad I_{1/2}(|2\mathbf{x}+\mathbf{b}|r) = \sqrt{\frac{2}{\pi}}\, \frac{\sinh(|2\mathbf{x}+\mathbf{b}|r)}{\sqrt{|2\mathbf{x}+\mathbf{b}|r}}\,,$$

we see that for $n = 3$, the integral (5.23) takes the form

$$\frac{e^{-|\mathbf{x}|^2}}{|2\mathbf{x}+\mathbf{b}|} \int\limits_0^\infty e^{-r^2}\, e^{-\lambda r}\, \sinh(|2\mathbf{x}+\mathbf{b}|r)\, dr$$

$$= \frac{e^{-|\mathbf{x}|^2}}{|4\mathbf{x}+2\mathbf{b}|} \int\limits_0^\infty e^{-r^2} \big(e^{r(|2\mathbf{x}+\mathbf{b}|-\lambda)} - e^{-r(|2\mathbf{x}+\mathbf{b}|+\lambda)}\big)\, dr\,,$$

which gives formula (5.24) in view of (5.9). □

5.3. Elastic and hydrodynamic potentials

In the following, we consider volume potentials which arise in the solution of two- and three-dimensional problems in elasticity and hydrodynamics.

Linear isotropic and homogeneous elastic problems are governed by the Lamé system

$$\mu\, \Delta\mathbf{f} + (\lambda+\mu)\, \mathrm{grad}\,\mathrm{div}\, \mathbf{f} = -\mathbf{u}\,, \tag{5.25}$$

where $\mathbf{f}$ is the displacement vector, $\mathbf{u}$ is the volume force, λ and μ are the Lamé constants. The equations are considered in the whole space, either $\mathbb{R}^2$ or $\mathbb{R}^3$. Correspondingly, the vector functions $\mathbf{f}$ and $\mathbf{u}$ have two or three components.

The hydrodynamic potentials correspond to the linearized classical Navier-Stokes equations

$$\nu\Delta\mathbf{f} - \mathrm{grad}\, p = \mathbf{u}\,, \quad \mathrm{div}\, \mathbf{f} = 0\,, \tag{5.26}$$

where $\mathbf{f}$ is the velocity vector, p denotes the pressure, ν is the constant viscosity coefficient.

5.4. Two-dimensional potentials

5.4.1. Fundamental solutions. A solution of the two-dimensional Lamé system (5.25) is given by the volume potential

$$\mathbf{f}(\mathbf{x}) = \int\limits_{\mathbb{R}^2} \mathbf{\Gamma}(\mathbf{x}-\mathbf{y})\, \mathbf{u}(\mathbf{y})\, d\mathbf{y}\,,$$

where $\mathbf{\Gamma} = \|\Gamma_{kl}\|_{2\times 2}$ is the Boussinesq fundamental matrix with the components

$$\Gamma_{kl}(\mathbf{x}) = \frac{\lambda+\mu}{4\pi\mu(\lambda+2\mu)}\Big(\delta_{kl}\frac{\lambda+3\mu}{\lambda+\mu}\log\frac{1}{|\mathbf{x}|} + \frac{x_k x_l}{|\mathbf{x}|^2}\Big).$$

A solution of the Stokes system (5.26) is given by the hydrodynamic potentials

$$f_k(\mathbf{x}) = \int\limits_{\mathbb{R}^2}\sum_{l=1}^{2}\Psi_{kl}(\mathbf{x}-\mathbf{y})\, u_l(\mathbf{y})\, d\mathbf{y}\,, \quad p(\mathbf{x}) = \int\limits_{\mathbb{R}^3}\langle\Theta(\mathbf{x}-\mathbf{y}), \mathbf{u}(\mathbf{y})\rangle d\mathbf{y}$$

with the fundamental matrix

$$\Psi_{kl}(\mathbf{x}) = \frac{1}{4\pi\nu}\Big(\delta_{kl}\log\frac{1}{|\mathbf{x}|} + \frac{x_k x_l}{|\mathbf{x}|^2}\Big)\,, \quad \Theta(\mathbf{x}) = \frac{\mathbf{x}}{2\pi|\mathbf{x}|^2}\,. \tag{5.27}$$

The aim of this section is to derive analytic formulas for these two-dimensional potentials applied to the generating functions η_{2M} defined in (3.18). Note that

$$\begin{aligned} \frac{x_k x_l}{|\mathbf{x}|^2} &= \delta_{kl}\log\frac{1}{|\mathbf{x}|} - \frac{1}{2}\frac{\partial^2}{\partial x_k \partial x_l}\Big(|\mathbf{x}|^2\Big(\log\frac{1}{|\mathbf{x}|}+\frac{1}{2}\Big)\Big)\,,\\ \frac{\mathbf{x}}{|\mathbf{x}|^2} &= -\nabla\log\frac{1}{|\mathbf{x}|}\,. \end{aligned} \tag{5.28}$$

5.4.2. An auxiliary formula. Since explicit representations of the harmonic potentials

$$\mathcal{L}_2\eta_{2M}(\mathbf{x}) = \frac{1}{2\pi}\int\limits_{\mathbb{R}^2}\log\frac{1}{|\mathbf{x}-\mathbf{y}|}\,\eta_{2M}(\mathbf{y})\, d\mathbf{y}$$

are already known from (4.27), it suffices to determine the integrals

$$I_{kl}\eta_{2M}(\mathbf{x}) = \frac{\partial^2}{\partial x_k\partial x_l}\frac{1}{4\pi}\int\limits_{\mathbb{R}^2}|\mathbf{x}-\mathbf{y}|^2\Big(\log\frac{1}{|\mathbf{x}-\mathbf{y}|}+\frac{1}{2}\Big)\,\eta_{2M}(\mathbf{y})\, d\mathbf{y}\,.$$

Then, obviously,

$$\int\limits_{\mathbb{R}^2}\Gamma_{kl}(\mathbf{x}-\mathbf{y})\eta_{2M}(\mathbf{y})\, d\mathbf{y} = \frac{\delta_{kl}}{\mu}\mathcal{L}_2\eta_{2M}(\mathbf{x}) - \frac{\lambda+\mu}{2\mu(\lambda+2\mu)}I_{kl}\eta_{2M}(\mathbf{x}) \tag{5.29}$$

and

$$\begin{aligned} &\int\limits_{\mathbb{R}^2}\Psi_{kl}(\mathbf{x}-\mathbf{y})\eta_{2M}(\mathbf{y})\, d\mathbf{y} = \frac{\delta_{kl}}{\nu}\mathcal{L}_2\eta_{2M}(\mathbf{x}) - \frac{1}{2\nu}I_{kl}\eta_{2M}(\mathbf{x})\,,\\ &\frac{x_k}{2\pi}\int\limits_{\mathbb{R}^2}\frac{\eta_{2M}(\mathbf{y})}{|\mathbf{x}-\mathbf{y}|^2}\, d\mathbf{y} = -\frac{\partial}{\partial x_k}\mathcal{L}_2\eta_{2M}(\mathbf{x})\,. \end{aligned} \tag{5.30}$$

Explicit expressions of $I_{kl}\eta_{2M}$ are based on the expansion

$$\eta_{2M}(\mathbf{x}) = L^{(1)}_{M-1}(|\mathbf{x}|^2)\, \mathrm{e}^{-|\mathbf{x}|^2} = \frac{1}{\pi}\sum_{j=0}^{M-1} \frac{(-1)^j}{j!\,4^j}\, \Delta^j\, \mathrm{e}^{-|\mathbf{x}|^2},$$

which is derived in Theorem 3.5. We have to consider the integrals

$$\begin{aligned} I_{kl}\eta_{2M}(\mathbf{x}) &= \frac{\partial^2}{\partial x_k \partial x_l}\frac{1}{4\pi^2}\int_{\mathbb{R}^2} |\mathbf{x}-\mathbf{y}|^2\Big(\log\frac{1}{|\mathbf{x}-\mathbf{y}|}+\frac{1}{2}\Big)\, \mathrm{e}^{-|\mathbf{y}|^2}\, d\mathbf{y} \\ &+ \sum_{j=1}^{M-1}\frac{(-1)^j}{j!\,4^j}\frac{\partial^2}{\partial x_k \partial x_l}\frac{1}{4\pi^2}\int_{\mathbb{R}^2} |\mathbf{x}-\mathbf{y}|^2\Big(\log\frac{1}{|\mathbf{x}-\mathbf{y}|}+\frac{1}{2}\Big)\, \Delta^j\, \mathrm{e}^{-|\mathbf{y}|^2}\, d\mathbf{y}. \end{aligned} \tag{5.31}$$

5.4.3. Potentials of the Gaussian. Using integration by parts, it is easy to check that

$$\begin{aligned} &\frac{\partial}{\partial x_l}\int_{\mathbb{R}^2} |\mathbf{x}-\mathbf{y}|^2\Big(\log\frac{1}{|\mathbf{x}-\mathbf{y}|}+\frac{1}{2}\Big)\, \mathrm{e}^{-|\mathbf{y}|^2}\, d\mathbf{y} \\ &\qquad = (2x_l + \frac{\partial}{\partial x_l})\int_{\mathbb{R}^2} \log\frac{1}{|\mathbf{x}-\mathbf{y}|}\, \mathrm{e}^{-|\mathbf{y}|^2}\, d\mathbf{y}. \end{aligned} \tag{5.32}$$

Therefore, we derive from (4.21) that

$$\begin{aligned} I_{kl}(\mathrm{e}^{-|\cdot|^2})(\mathbf{x}) &= \frac{\partial^2}{\partial x_k \partial x_l}\frac{1}{4\pi}\int_{\mathbb{R}^2} |\mathbf{x}-\mathbf{y}|^2\Big(\log\frac{1}{|\mathbf{x}-\mathbf{y}|}+\frac{1}{2}\Big)\, \mathrm{e}^{-|\mathbf{y}|^2}\, d\mathbf{y} \\ &= \delta_{kl}\,\mathcal{L}_2(\mathrm{e}^{-|\cdot|^2})(\mathbf{x}) + (x_l\frac{\partial}{\partial x_k} + \frac{1}{2}\frac{\partial^2}{\partial x_k \partial x_l})\mathcal{L}_2(\mathrm{e}^{-|\cdot|^2})(\mathbf{x}) \\ &= \delta_{kl}\Big(\mathcal{L}_2(\mathrm{e}^{-|\cdot|^2})(\mathbf{x}) + \frac{\mathrm{e}^{-|\mathbf{x}|^2}-1}{4|\mathbf{x}|^2}\Big) + \frac{x_k x_l}{2|\mathbf{x}|^2}\Big(\frac{1-\mathrm{e}^{-|\mathbf{x}|^2}}{|\mathbf{x}|^2}-1\Big), \end{aligned} \tag{5.33}$$

which leads to the formulas

$$\begin{aligned} &\int_{\mathbb{R}^2} \Gamma_{kl}(\mathbf{x}-\mathbf{y})\, \mathrm{e}^{-|\mathbf{y}|^2}\, d\mathbf{y} = \frac{\delta_{kl}}{\mu}\mathcal{L}_2(\mathrm{e}^{-|\cdot|^2})(\mathbf{x}) - \frac{\lambda+\mu}{2\mu(\lambda+2\mu)}\, I_{kl}(\mathrm{e}^{-|\cdot|^2})(\mathbf{x}) \\ &= -\frac{\lambda+3\mu}{8\mu(\lambda+2\mu)}\,\delta_{kl}\,(\,\mathrm{E}_1(|\mathbf{x}|^2) + 2\log|\mathbf{x}|) \\ &\quad + \delta_{kl}\frac{\lambda+\mu}{8\mu(\lambda+2\mu)}\frac{1-\mathrm{e}^{-|\mathbf{x}|^2}}{|\mathbf{x}|^2} + \frac{\lambda+\mu}{4\mu(\lambda+2\mu)}\frac{x_k x_l}{|\mathbf{x}|^2}\Big(\frac{\mathrm{e}^{-|\mathbf{x}|^2}-1}{|\mathbf{x}|^2}+1\Big) \end{aligned}$$

and

$$\int_{\mathbb{R}^2} \Psi_{kl}(\mathbf{x}-\mathbf{y})\, \mathrm{e}^{-|\mathbf{y}|^2}\, d\mathbf{y} = \frac{\delta_{kl}}{\nu}\mathcal{L}_2(\mathrm{e}^{-|\cdot|^2})(\mathbf{x}) - \frac{1}{2\nu}\, I_{kl}(\mathrm{e}^{-|\cdot|^2})(\mathbf{x})$$

$$= -\frac{\delta_{kl}}{2\nu}\Big(\mathrm{E}_1(|\mathbf{x}|^2) + 2\log|\mathbf{x}| + \frac{\mathrm{e}^{-|\mathbf{x}|^2}-1}{|\mathbf{x}|^2}\Big) + \frac{x_k x_l}{4\nu|\mathbf{x}|^2}\Big(\frac{\mathrm{e}^{-|\mathbf{x}|^2}-1}{|\mathbf{x}|^2}+1\Big),$$

$$\frac{x_k}{2\pi}\int_{\mathbb{R}^2} \frac{\mathrm{e}^{-|\mathbf{y}|^2}}{|\mathbf{x}-\mathbf{y}|^2}\, d\mathbf{y} = -\frac{\partial}{\partial x_k}\mathcal{L}_2(\mathrm{e}^{-|\cdot|^2})(\mathbf{x}) = \frac{x_k}{2|\mathbf{x}|^2}(1-\mathrm{e}^{-|\mathbf{x}|^2}).$$

Thus, the elastic potential of $\eta_2(\mathbf{x}) = \pi^{-1}\mathrm{e}^{-|\mathbf{x}|^2}$ is given by

$$\int_{\mathbb{R}^2} \Gamma_{kl}(\mathbf{x}-\mathbf{y})\,\eta_2(\mathbf{y})\, d\mathbf{y} = \frac{\lambda+\mu}{4\pi\mu(\lambda+2\mu)}\frac{x_k x_l}{|\mathbf{x}|^2}\Big(\frac{\mathrm{e}^{-|\mathbf{x}|^2}-1}{|\mathbf{x}|^2}+1\Big)$$
$$+\,\delta_{kl}\Big(\frac{\lambda+3\mu}{2\mu(\lambda+2\mu)}\mathcal{L}_2\eta_2(\mathbf{x}) + \frac{\lambda+\mu}{8\pi\mu(\lambda+2\mu)}\frac{1-\mathrm{e}^{-|\mathbf{x}|^2}}{|\mathbf{x}|^2}\Big), \tag{5.34}$$

whereas the hydrodynamic potential can be derived from

$$\int_{\mathbb{R}^2} \Psi_{kl}(\mathbf{x}-\mathbf{y})\,\eta_2(\mathbf{y})\, d\mathbf{y} = \frac{\delta_{kl}}{2\nu}\Big(\mathcal{L}_2\eta_2(\mathbf{x}) + \frac{1-\mathrm{e}^{-|\mathbf{x}|^2}}{2\pi|\mathbf{x}|^2}\Big)$$
$$+\,\frac{x_k x_l}{4\pi\nu|\mathbf{x}|^2}\Big(\frac{\mathrm{e}^{-|\mathbf{x}|^2}-1}{|\mathbf{x}|^2}+1\Big), \tag{5.35}$$
$$\frac{x_k}{2\pi}\int_{\mathbb{R}^2}\frac{\eta_2(\mathbf{y})}{|\mathbf{x}-\mathbf{y}|^2}\, d\mathbf{y} = \frac{x_k}{2\pi|\mathbf{x}|^2}(1-\mathrm{e}^{-|\mathbf{x}|^2}).$$

5.4.4. Potentials of higher-order generating functions. We consider the integrals $I_{kl}\eta_{2M}$, defined by (5.31), for $M \geq 2$. We write

$$I_{kl}\eta_{2M}(\mathbf{x}) = I_{kl}\eta_4(\mathbf{x})$$
$$+ \sum_{j=2}^{M-1} \frac{(-1)^j}{j!\,4^j}\frac{\partial^2}{\partial x_k \partial x_l}\frac{1}{4\pi^2}\int_{\mathbb{R}^2} |\mathbf{x}-\mathbf{y}|^2\Big(\log\frac{1}{|\mathbf{x}-\mathbf{y}|}+\frac{1}{2}\Big)\Delta^j\,\mathrm{e}^{-|\mathbf{y}|^2}\, d\mathbf{y}.$$

Using the relation

$$\Delta\,|\mathbf{x}-\mathbf{y}|^2\Big(\log\frac{1}{|\mathbf{x}-\mathbf{y}|}+\frac{1}{2}\Big) = 4\log\frac{1}{|\mathbf{x}-\mathbf{y}|} - 2$$

and Green's second formula, we have for $j \geq 2$

$$\frac{1}{4\pi^2}\int_{\mathbb{R}^2} |\mathbf{x}-\mathbf{y}|^2\Big(\log\frac{1}{|\mathbf{x}-\mathbf{y}|}+\frac{1}{2}\Big)\Delta^j\,\mathrm{e}^{-|\mathbf{y}|^2}\, d\mathbf{y} = \frac{1}{\pi^2}\int_{\mathbb{R}^2} \log\frac{1}{|\mathbf{x}-\mathbf{y}|}\,\Delta^{j-1}\,\mathrm{e}^{-|\mathbf{y}|^2}\, d\mathbf{y}.$$

Thus (5.31) can be written in the form

$$
\begin{aligned}
I_{kl}\eta_{2M}(\mathbf{x}) &= I_{kl}\eta_4(\mathbf{x}) \\
&\quad + \frac{\partial^2}{\partial x_k \partial x_l} \sum_{j=2}^{M-1} \frac{(-1)^j}{j!\,4^{j-1}} \frac{1}{4\pi^2} \int\limits_{\mathbb{R}^2} \log \frac{1}{|\mathbf{x}-\mathbf{y}|} \, \Delta^{j-1} \, \mathrm{e}^{-|\mathbf{y}|^2} \, d\mathbf{y} \,.
\end{aligned} \tag{5.36}
$$

Because

$$
\frac{1}{4\pi^2} \int\limits_{\mathbb{R}^2} |\mathbf{x}-\mathbf{y}|^2 \Big(\log \frac{1}{|\mathbf{x}-\mathbf{y}|} + \frac{1}{2} \Big) \Delta \, \mathrm{e}^{-|\mathbf{y}|^2} \, d\mathbf{y} = \frac{1}{\pi^2} \int\limits_{\mathbb{R}^2} \Big(\log \frac{1}{|\mathbf{x}-\mathbf{y}|} - \frac{1}{2} \Big) \, \mathrm{e}^{-|\mathbf{y}|^2} \, d\mathbf{y},
$$

we obtain

$$
I_{kl}\eta_4(\mathbf{x}) = \frac{\partial^2}{\partial x_k \partial x_l} \frac{1}{4\pi^2} \int\limits_{\mathbb{R}^2} \bigg(|\mathbf{x}-\mathbf{y}|^2 \Big(\log \frac{1}{|\mathbf{x}-\mathbf{y}|} + \frac{1}{2} \Big) - \log \frac{1}{|\mathbf{x}-\mathbf{y}|} \bigg) \, \mathrm{e}^{-|\mathbf{y}|^2} \, d\mathbf{y} \,,
$$

which can be simplified, in view of (5.32), to

$$
\begin{aligned}
I_{kl}\eta_4(\mathbf{x}) &= \frac{\partial}{\partial x_k} \frac{x_l}{2\pi^2} \int\limits_{\mathbb{R}^2} \log \frac{1}{|\mathbf{x}-\mathbf{y}|} \, \mathrm{e}^{-|\mathbf{y}|^2} \, d\mathbf{y} \\
&= \Big(\delta_{kl} + x_l \frac{\partial}{\partial x_k} \Big) \frac{1}{2\pi^2} \int\limits_{\mathbb{R}^2} \log \frac{1}{|\mathbf{x}-\mathbf{y}|} \, \mathrm{e}^{-|\mathbf{y}|^2} \, d\mathbf{y} \,.
\end{aligned}
$$

From (4.21), we have

$$
x_l \frac{\partial}{\partial x_k} \mathcal{L}_2 \big(\mathrm{e}^{-|\cdot|^2} \big)(\mathbf{x}) = \frac{x_k x_l}{2|\mathbf{x}|^2} (\mathrm{e}^{-|\mathbf{x}|^2} - 1) \,, \tag{5.37}
$$

so that the integral $I_{kl}\eta_4$ has the simple form

$$
I_{kl}\,\eta_4(\mathbf{x}) = \frac{1}{\pi} \Big(\delta_{kl} \mathcal{L}_2 \big(\mathrm{e}^{-|\cdot|^2} \big)(\mathbf{x}) + \frac{x_k x_l}{2|\mathbf{x}|^2} (\mathrm{e}^{-|\mathbf{x}|^2} - 1) \Big). \tag{5.38}
$$

To treat the second term in (5.36), which is present only if $M > 2$, we recall that for $j \geq 2$

$$
\frac{1}{2\pi} \int\limits_{\mathbb{R}^2} \log \frac{1}{|\mathbf{x}-\mathbf{y}|} \, \Delta^{j-1} \, \mathrm{e}^{-|\mathbf{y}|^2} \, d\mathbf{y} = -\Delta^{j-2} \, \mathrm{e}^{-|\mathbf{x}|^2} \,,
$$

so that we obtain, by using (4.24),

$$
I_{kl}\eta_{2M}(\mathbf{x}) = I_{kl}\,\eta_4(\mathbf{x}) - \frac{1}{\pi} \frac{\partial^2}{\partial x_k \partial x_l} \sum_{j=0}^{M-3} \frac{L_j^{(0)}(|\mathbf{x}|^2) \, \mathrm{e}^{-|\mathbf{x}|^2}}{8(j+1)(j+2)} \,. \tag{5.39}
$$

Sums of Laguerre polynomials. The expression (5.39) for $I_{kl}\eta_{2M}$, $M > 2$, can be simplified further by using the properties of the Laguerre polynomials, which are given in Subsection 4.3.3.

First, we rewrite the sum

$$
S_M := \frac{\partial^2}{\partial x_k \partial x_l} \sum_{j=0}^{M-3} \frac{L_j^{(0)}(|\mathbf{x}|^2) \, \mathrm{e}^{-|\mathbf{x}|^2}}{8(j+1)(j+2)} \,.
$$

It follows from formula (4.38) that

$$
\frac{\partial^2}{\partial x_k \partial x_l} L_j^{(\gamma)}(|\mathbf{x}|^2) \, \mathrm{e}^{-|\mathbf{x}|^2} = \Big(-2\delta_{kl} L_j^{(\gamma+1)}(|\mathbf{x}|^2) + 4 x_k x_l L_j^{(\gamma+2)}(|\mathbf{x}|^2) \Big) \, \mathrm{e}^{-|\mathbf{x}|^2}, \tag{5.40}
$$

which implies

$$S_M = x_k x_l \sum_{j=0}^{M-3} \frac{L_j^{(2)}(|\mathbf{x}|^2)\,\mathrm{e}^{-|\mathbf{x}|^2}}{2(j+1)(j+2)} - \delta_{kl} \sum_{j=0}^{M-3} \frac{L_j^{(1)}(|\mathbf{x}|^2)\,\mathrm{e}^{-|\mathbf{x}|^2}}{4(j+1)(j+2)}.$$

We simplify the first sum by using (4.29), which gives

$$L_j^{(2)}(|\mathbf{x}|^2) = \frac{1}{|\mathbf{x}|^2}\big((j+2)L_j^{(1)}(|\mathbf{x}|^2) - (j+1)L_{j+1}^{(1)}(|\mathbf{x}|^2)\big),$$

and therefore

$$\begin{aligned}
x_k x_l \sum_{j=0}^{M-3} \frac{L_j^{(2)}(|\mathbf{x}|^2)}{2(j+1)(j+2)} &= \frac{x_k x_l}{|\mathbf{x}|^2} \sum_{j=0}^{M-3} \frac{(j+2)L_j^{(1)}(|\mathbf{x}|^2) - (j+1)L_{j+1}^{(1)}(|\mathbf{x}|^2)}{2(j+1)(j+2)} \\
&= \frac{x_k x_l}{2|\mathbf{x}|^2} \sum_{j=0}^{M-3} \Big(\frac{L_j^{(1)}(|\mathbf{x}|^2)}{j+1} - \frac{L_{j+1}^{(1)}(|\mathbf{x}|^2)}{j+2}\Big) = \frac{x_k x_l}{2|\mathbf{x}|^2}\Big(1 - \frac{L_{M-2}^{(1)}(|\mathbf{x}|^2)}{M-1}\Big).
\end{aligned}$$

To simplify the second sum, we use (4.28) and the fact that $L_0^{(\gamma)} = 1$. Then we can write

$$\begin{aligned}
\sum_{j=0}^{M-3} \frac{L_j^{(1)}(|\mathbf{x}|^2)}{4(j+1)(j+2)} &= \frac{1}{4} \sum_{j=0}^{M-3} \Big(\frac{1}{j+1} - \frac{1}{j+2}\Big) L_j^{(1)}(|\mathbf{x}|^2) \\
&= \frac{1}{4} L_0^{(1)}(|\mathbf{x}|^2) + \sum_{j=1}^{M-3} \frac{\big(L_j^{(1)}(|\mathbf{x}|^2) - L_{j-1}^{(1)}(|\mathbf{x}|^2)\big)}{4(j+1)} - \frac{L_{M-3}^{(1)}(|\mathbf{x}|^2)}{4(M-1)} \qquad (5.41) \\
&= \sum_{j=0}^{M-3} \frac{L_j^{(0)}(|\mathbf{x}|^2)}{4(j+1)} - \frac{L_{M-3}^{(1)}(|\mathbf{x}|^2)}{4(M-1)}.
\end{aligned}$$

To derive a more compact form of the potentials, we note that because of (4.28)

$$\sum_{j=0}^{M-3} \frac{L_j^{(0)}(|\mathbf{x}|^2)}{4(j+1)} - \frac{L_{M-3}^{(1)}(|\mathbf{x}|^2)}{4(M-1)} = \sum_{j=0}^{M-2} \frac{L_j^{(0)}(|\mathbf{x}|^2)}{4(j+1)} - \frac{L_{M-2}^{(1)}(|\mathbf{x}|^2)}{4(M-1)}.$$

Hence, we obtain

$$\begin{aligned}
S_M = &\frac{x_k x_l}{2|\mathbf{x}|^2}\Big(\mathrm{e}^{-|\mathbf{x}|^2} - \frac{L_{M-2}^{(1)}(|\mathbf{x}|^2)\,\mathrm{e}^{-|\mathbf{x}|^2}}{M-1}\Big) \\
&- \delta_{kl}\Big(\sum_{j=0}^{M-2} \frac{L_j^{(0)}(|\mathbf{x}|^2)\,\mathrm{e}^{-|\mathbf{x}|^2}}{4(j+1)} - \frac{L_{M-2}^{(1)}(|\mathbf{x}|^2)\,\mathrm{e}^{-|\mathbf{x}|^2}}{4(M-1)}\Big).
\end{aligned}$$

Final form of $I_{kl}\eta_{2M}$. We have transformed (5.39) to

$$\begin{aligned}
I_{kl}\eta_{2M}(\mathbf{x}) &= \frac{1}{\pi}\Big(\delta_{kl} + x_l \frac{\partial}{\partial x_k}\Big)\mathcal{L}_2(\mathrm{e}^{-|\cdot|^2})(\mathbf{x}) - \frac{1}{\pi} S_M \\
&= \frac{\delta_{kl}}{\pi}\Big(\mathcal{L}_2(\mathrm{e}^{-|\cdot|^2})(\mathbf{x}) + \sum_{j=0}^{M-2} \frac{L_j^{(0)}(|\mathbf{x}|^2)\,\mathrm{e}^{-|\mathbf{x}|^2}}{4(j+1)} - \frac{L_{M-2}^{(1)}(|\mathbf{x}|^2)\,\mathrm{e}^{-|\mathbf{x}|^2}}{4(M-1)}\Big) \\
&\quad + \frac{1}{\pi}\Big(x_l \frac{\partial}{\partial x_k} \mathcal{L}_2(\mathrm{e}^{-|\cdot|^2})(\mathbf{x}) - \frac{x_k x_l}{2|\mathbf{x}|^2}\Big(\mathrm{e}^{-|\mathbf{x}|^2} - \frac{L_{M-2}^{(1)}(|\mathbf{x}|^2)\,\mathrm{e}^{-|\mathbf{x}|^2}}{M-1}\Big)\Big).
\end{aligned}$$

Now we use the results of Section 4.3 for the two-dimensional harmonic potential $\mathcal{L}_2(\mathrm{e}^{-|\cdot|^2})$. From (5.37), we have

$$x_l \frac{\partial}{\partial x_k} \mathcal{L}_2(\mathrm{e}^{-|\cdot|^2})(\mathbf{x}) - \frac{x_k x_l}{2|\mathbf{x}|^2}\Big(\mathrm{e}^{-|\mathbf{x}|^2} - \frac{L_{M-2}^{(1)}(|\mathbf{x}|^2)\,\mathrm{e}^{-|\mathbf{x}|^2}}{M-1}\Big) = \frac{x_k x_l}{2|\mathbf{x}|^2}\Big(\frac{L_{M-2}^{(1)}(|\mathbf{x}|^2)\,\mathrm{e}^{-|\mathbf{x}|^2}}{M-1} - 1\Big)$$

and by (4.25)

$$\mathcal{L}_2(\mathrm{e}^{-|\cdot|^2})(\mathbf{x}) + \sum_{j=0}^{M-2} \frac{L_j^{(0)}(|\mathbf{x}|^2)\,\mathrm{e}^{-|\mathbf{x}|^2}}{4(j+1)} = \mathcal{L}_2(L_{M-1}^{(1)}(|\cdot|^2)\,\mathrm{e}^{-|\cdot|^2})(\mathbf{x}).$$

So, $I_{kl}\eta_{2M}$ can be written in the form

$$\begin{aligned} I_{kl}\eta_{2M}(\mathbf{x}) &= \frac{\delta_{kl}}{\pi}\Big(\mathcal{L}_2(L_{M-1}^{(1)}(|\cdot|^2)\,\mathrm{e}^{-|\cdot|^2})(\mathbf{x}) - \frac{L_{M-2}^{(1)}(|\mathbf{x}|^2)\,\mathrm{e}^{-|\mathbf{x}|^2}}{4(M-1)}\Big) \\ &\quad + \frac{x_k x_l}{2\pi|\mathbf{x}|^2}\Big(\frac{L_{M-2}^{(1)}(|\mathbf{x}|^2)\,\mathrm{e}^{-|\mathbf{x}|^2}}{M-1} - 1\Big) \\ &= \delta_{kl}\Big(\mathcal{L}_2\eta_{2M}(\mathbf{x}) - \frac{\eta_{2M-2}(\mathbf{x})}{4(M-1)}\Big) + \frac{x_k x_l}{2|\mathbf{x}|^2}\Big(\frac{\eta_{2M-2}(\mathbf{x})}{M-1} - \frac{1}{\pi}\Big). \end{aligned} \tag{5.42}$$

By (5.38), this formula is valid also if $2M = 4$.

5.4.5. Final form of the elastic and hydrodynamic potentials. By using (5.42), we obtain the explicit expression of the elastic potential from the representations (5.29) and (5.30), i.e.,

$$\begin{aligned} \int_{\mathbb{R}^2} \Gamma_{kl}(\mathbf{x}-\mathbf{y})\,\eta_{2M}(\mathbf{y})\,d\mathbf{y} &= \frac{\lambda+3\mu}{2\mu(\lambda+2\mu)}\,\delta_{kl}\,\mathcal{L}_2\eta_{2M}(\mathbf{x}) \\ &\quad + \frac{\lambda+\mu}{4\mu(\lambda+2\mu)}\Big(\Big(\frac{\delta_{kl}}{2} - \frac{x_k x_l}{|\mathbf{x}|^2}\Big)\frac{\eta_{2M-2}(\mathbf{x})}{M-1} + \frac{x_k x_l}{\pi\,|\mathbf{x}|^2}\Big). \end{aligned} \tag{5.43}$$

Using (4.42), the hydrodynamic potentials can be written as

$$\begin{aligned} \int_{\mathbb{R}^2} \Psi_{kl}(\mathbf{x}-\mathbf{y})\,\eta_{2M}(\mathbf{y})\,d\mathbf{y} &= -\frac{\delta_{kl}}{2\nu}\,\mathcal{L}_2\eta_{2M}(\mathbf{x}) \\ &\quad + \frac{1}{4\nu}\Big(\Big(\frac{\delta_{kl}}{2} - \frac{x_k x_l}{|\mathbf{x}|^2}\Big)\frac{\eta_{2M-2}(\mathbf{x})}{M-1} + \frac{x_k x_l}{\pi\,|\mathbf{x}|^2}\Big), \\ \frac{x_k}{2\pi}\int_{\mathbb{R}^2} \frac{\eta_{2M}(\mathbf{y})}{|\mathbf{x}-\mathbf{y}|^2}\,d\mathbf{y} &= \frac{x_k}{2\pi|\mathbf{x}|^2}(1 - L_{M-1}^{(0)}(|\mathbf{x}|^2)\,\mathrm{e}^{-|\mathbf{x}|^2}). \end{aligned} \tag{5.44}$$

REMARK 5.2. We note that by simple differentiation of the expressions (5.34) and (5.43), it is possible to obtain effective approximations of the stress tensor

$$\sigma_k = \lambda(\frac{\partial u_1}{\partial x_1} + \frac{\partial u_2}{\partial x_2}) + 2\mu \frac{\partial u_k}{\partial x_k}, \quad \tau_{1,2} = \mu(\frac{\partial u_1}{\partial x_2} + \frac{\partial u_2}{\partial x_1}) .$$

REMARK 5.3. Note that if $M \geq 2$, then we obtain

$$I_{kl}\eta_{2M+2}(\mathbf{x}) - I_{kl}\eta_{2M}(\mathbf{x}) = \frac{\mathrm{e}^{-|\mathbf{x}|^2}}{4\pi M(M-1)} \Big(\delta_{kl} L_{M-2}^{(1)}(|\mathbf{x}|^2) - 2x_k x_l L_{M-2}^{(2)}(|\mathbf{x}|^2)\Big)$$

from (5.39) and (5.40). This shows that as in the case of harmonic potentials, the elastic and hydrodynamic potentials of η_{2M+2} can be obtained from those of η_{2M} by adding some rapidly decaying term. Therefore, to improve the approximation accuracy to the potentials at a given point, only the values of the density in a small neighborhood are required.

This is not true if $M = 1$, since (5.33) and (5.38) imply

$$I_{kl}\eta_4(\mathbf{x}) - I_{kl}\eta_2(\mathbf{x}) = -\frac{\delta_{kl}}{4\pi} \frac{\mathrm{e}^{-|\mathbf{x}|^2} - 1}{|\mathbf{x}|^2} + \frac{x_k x_l}{2|\mathbf{x}|^4} \Big((|\mathbf{x}|^2 + 1)\,\mathrm{e}^{-|\mathbf{x}|^2} - 1\Big).$$

5.4.6. Using matrix-valued basis functions. Following the idea suggested in Section 3.6, we introduce matrix-valued basis functions adapted to the Lamé system. In the two-dimensional case the symbol matrix of the operator

$$L := \frac{1}{\mu} E = n - (\Delta + \sigma \operatorname{grad} \operatorname{div}) , \quad \sigma = \frac{\lambda + \mu}{\mu}, \tag{5.45}$$

has the form

$$4\pi^2 \begin{pmatrix} |\boldsymbol{\xi}|^2 + \sigma\xi_1^2 & \sigma\xi_1\xi_2 \\ \sigma\xi_1\xi_2 & \mu|\boldsymbol{\xi}|^2 + \sigma\xi_2^2 \end{pmatrix} = 4\pi^2 |\boldsymbol{\xi}|^2 A(\sigma, \gamma) ,$$

with the matrix function

$$A(\sigma, \gamma) := \begin{pmatrix} 1 + \sigma\cos^2\gamma & \sigma\cos\gamma\sin\gamma \\ \sigma\cos\gamma\sin\gamma & 1 + \sigma\sin^2\gamma \end{pmatrix}$$

and $\cos\gamma = \xi_1/|\boldsymbol{\xi}|$. The matrix $A(\sigma, \gamma)$ has the remarkable property

$$A(\sigma_1, \gamma)A(\sigma_2, \gamma) = A((1+\sigma_1)(1+\sigma_2) - 1, \gamma) ,$$

so that

$$A^k(\sigma, \gamma) = A((1+\sigma)^k - 1, \gamma) , \ k \in \mathbb{Z} .$$

Hence, it is easy to find the exponential matrix

$$\mathrm{e}^{-\pi^2|\boldsymbol{\xi}|^2 A(\sigma_1,\gamma)} = \sum_{k=0}^{\infty} \frac{(-1)^k \pi^{2k} |\boldsymbol{\xi}|^{2k}}{k!} A^k(\sigma, \gamma) = \sum_{k=0}^{\infty} \frac{(-1)^k \pi^{2k} |\boldsymbol{\xi}|^{2k}}{k!} A((1-\sigma)^k, \gamma)$$

$$= \sum_{k=0}^{\infty} \frac{(-1)^k \pi^{2k} |\boldsymbol{\xi}|^{2k}}{k!} \begin{pmatrix} 1 + \big((1+\sigma)^k - 1\big)\cos^2\gamma & \big((1+\sigma)^k - 1\big)\cos\gamma\sin\gamma \\ \big((1+\sigma)^k - 1\big)\cos\gamma\sin\gamma & 1 + \big((1+\sigma)^k - 1\big)\sin^2\gamma \end{pmatrix}$$

$$= \mathrm{e}^{-\pi^2|\boldsymbol{\xi}|^{2k}} \begin{pmatrix} 1 + \Big(\mathrm{e}^{-\pi^2\sigma|\boldsymbol{\xi}|^2} - 1\Big)\cos^2\gamma & \Big(\mathrm{e}^{-\pi^2\sigma|\boldsymbol{\xi}|^2} - 1\Big)\cos\gamma\sin\gamma \\ \Big(\mathrm{e}^{-\pi^2\sigma|\boldsymbol{\xi}|^2} - 1\Big)\cos\gamma\sin\gamma & 1 + \Big(\mathrm{e}^{-\pi^2\sigma|\boldsymbol{\xi}|^2} - 1\Big)\sin^2\gamma \end{pmatrix}.$$

Therefore we choose the inverse Fourier transform of the matrix

$$\mathcal{F}\boldsymbol{\eta}(\boldsymbol{\xi}) := \mathrm{e}^{-\pi^2|\boldsymbol{\xi}|^2} \begin{pmatrix} 1 + \dfrac{\xi_1^2}{|\boldsymbol{\xi}|^2}\Big(\mathrm{e}^{-\pi^2\sigma|\boldsymbol{\xi}|^2} - 1\Big) & \dfrac{\xi_1\xi_2}{|\boldsymbol{\xi}|^2}\Big(\mathrm{e}^{-\pi^2\sigma|\boldsymbol{\xi}|^2} - 1\Big) \\ \dfrac{\xi_1\xi_2}{|\boldsymbol{\xi}|^2}\Big(\mathrm{e}^{-\pi^2\sigma|\boldsymbol{\xi}|^2} - 1\Big) & 1 + \dfrac{\xi_2^2}{|\boldsymbol{\xi}|^2}\Big(\mathrm{e}^{-\pi^2\sigma|\boldsymbol{\xi}|^2} - 1\Big) \end{pmatrix}$$

as generating matrix function $\boldsymbol{\eta}$ for the cubature of the inverse of the operator L. The elements of this matrix $\boldsymbol{\eta}$ can be given analytically as

$$\begin{aligned} \eta_{11} &= \frac{1}{\pi}\Big(\mathrm{e}^{-|\mathbf{x}|^2} + \frac{x_1^2}{|\mathbf{x}|^2}\Big(\mathrm{e}^{-|\mathbf{x}|^2} - \frac{\mathrm{e}^{-|\mathbf{x}|^2/(1+\sigma)}}{1+\sigma}\Big) \\ &\qquad + \Big(\frac{x_1^2}{|\mathbf{x}|^2} - \frac{1}{2}\Big)\frac{\mathrm{e}^{-|\mathbf{x}|^2} - \mathrm{e}^{-|\mathbf{x}|^2/(1+\sigma)}}{|\mathbf{x}|^2}\Big), \\ \eta_{12} &= \frac{x_1x_2}{\pi|\mathbf{x}|^2}\Big(\mathrm{e}^{-|\mathbf{x}|^2} - \frac{\mathrm{e}^{-|\mathbf{x}|^2/(1+\sigma)}}{1+\sigma} + \frac{\mathrm{e}^{-|\mathbf{x}|^2} - \mathrm{e}^{-|\mathbf{x}|^2/(1+\sigma)}}{|\mathbf{x}|^2}\Big), \\ \eta_{22} &= \frac{1}{\pi}\Big(\mathrm{e}^{-|\mathbf{x}|^2} + \frac{x_2^2}{|\mathbf{x}|^2}\Big(\mathrm{e}^{-|\mathbf{x}|^2} - \frac{\mathrm{e}^{-|\mathbf{x}|^2/(1+\sigma)}}{1+\sigma}\Big) \\ &\qquad + \Big(\frac{x_2^2}{|\mathbf{x}|^2} - \frac{1}{2}\Big)\frac{\mathrm{e}^{-|\mathbf{x}|^2} - \mathrm{e}^{-|\mathbf{x}|^2/(1+\sigma)}}{|\mathbf{x}|^2}\Big). \end{aligned}$$

Now, in accordance with (3.43), the matrix function

$$\boldsymbol{\eta}_{2M}(\mathbf{x}) = \sum_{j=0}^{M-1} \frac{1}{j!\,4^j} L^j\,\boldsymbol{\eta}(\mathbf{x})$$

generates an approximate cubature formula of order $\mathcal{O}((\sqrt{\mathcal{D}}h)^{2M})$ for $E^{-1}\mathbf{u}$ by the sum

$$\frac{h^2}{\mu}\sum_{\mathbf{m}\in\mathbb{Z}^2} \widetilde{\boldsymbol{\eta}}_{2M}\Big(\frac{\mathbf{x}-h\mathbf{m}}{\sqrt{\mathcal{D}}h}\Big)\,\mathbf{u}(h\mathbf{m})$$

with

$$\widetilde{\boldsymbol{\eta}}_{2M}(\mathbf{x}) := L^{-1}\boldsymbol{\eta}_{2M}(\mathbf{x}) = L^{-1}\boldsymbol{\eta}(\mathbf{x}) - \sum_{j=0}^{M-2} \frac{L^j\,\boldsymbol{\eta}(\mathbf{x})}{(j+1)!\,4^{j+1}}.$$

It remains to determine the elements of the matrix $L^{-1}\boldsymbol{\eta}$. Since the symbolic matrix of L^{-1} has the form

$$\frac{A^{-1}(\sigma,\gamma)}{4\pi^2|\boldsymbol{\xi}|^2} = \frac{1}{4\pi^2|\boldsymbol{\xi}|^2(1+\sigma)}\begin{pmatrix} 1+\sigma\dfrac{\xi_2^2}{|\boldsymbol{\xi}|^2} & -\sigma\dfrac{\xi_1\xi_2}{|\boldsymbol{\xi}|^2} \\ -\sigma\dfrac{\xi_1\xi_2}{|\boldsymbol{\xi}|^2} & 1+\sigma\dfrac{\xi_1^2}{|\boldsymbol{\xi}|^2} \end{pmatrix}, \tag{5.46}$$

it follows that $L^{-1}\boldsymbol{\eta}$ is given by the Fourier transform of

$$\frac{A^{-1}(\sigma,\gamma)\mathcal{F}\boldsymbol{\eta}(\boldsymbol{\xi})}{4\pi^2|\boldsymbol{\xi}|^2} = \frac{\mathrm{e}^{-\pi^2|\boldsymbol{\xi}|^2}}{4\pi^2|\boldsymbol{\xi}|^2}\begin{pmatrix} \dfrac{\xi_2^2}{|\boldsymbol{\xi}|^2} + \dfrac{\xi_1^2}{|\boldsymbol{\xi}|^2}\dfrac{\mathrm{e}^{-\pi^2\sigma|\boldsymbol{\xi}|^2}}{1+\sigma} & \dfrac{\xi_1\xi_2}{|\boldsymbol{\xi}|^2}\Big(\dfrac{\mathrm{e}^{-\pi^2\sigma|\boldsymbol{\xi}|^2}}{1+\sigma} - 1\Big) \\ \dfrac{\xi_1\xi_2}{|\boldsymbol{\xi}|^2}\Big(\dfrac{\mathrm{e}^{-\pi^2\sigma|\boldsymbol{\xi}|^2}}{1+\sigma} - 1\Big) & \dfrac{\xi_1^2}{|\boldsymbol{\xi}|^2} + \dfrac{\xi_2^2}{|\boldsymbol{\xi}|^2}\dfrac{\mathrm{e}^{-\pi^2\sigma|\boldsymbol{\xi}|^2}}{1+\sigma} \end{pmatrix}.$$

We note that the relation

$$\mathcal{F}\Big(\frac{\partial^2 f}{\partial x_k \partial x_l}\Big)(\boldsymbol{\xi}) = \frac{\xi_k \xi_l}{4\pi^2 |\boldsymbol{\xi}|^4}\, \mathrm{e}^{-\pi^2 |\boldsymbol{\xi}|^2}$$

is valid for the solution f of the biharmonic equation

$$\Delta^2 f = -\frac{\mathrm{e}^{-|\mathbf{x}|^2}}{\pi}.$$

Therefore, we obtain

$$\begin{aligned}\Phi_{kl}(\mathbf{x}) := \frac{\partial^2 f(\mathbf{x})}{\partial x_k \partial x_l} &= \frac{1}{8\pi^2} \frac{\partial^2}{\partial x_k \partial x_l} \int\limits_{\mathbb{R}^2} |\mathbf{x}-\mathbf{y}|^2 \Big(\log \frac{1}{|\mathbf{x}-\mathbf{y}|} + \frac{1}{2}\Big)\, \mathrm{e}^{-|\mathbf{y}|^2}\, d\mathbf{y} \\ &= \frac{\delta_{kl}}{2\pi} \mathcal{L}_2\Big(\mathrm{e}^{-|\cdot|^2}\Big)(\mathbf{x}) + \frac{\delta_{kl}}{8\pi} \frac{\mathrm{e}^{-|\mathbf{x}|^2} - 1}{|\mathbf{x}|^2} + \frac{1}{4\pi} \frac{x_k x_l}{|\mathbf{x}|^2} \Big(\frac{1-\mathrm{e}^{-|\mathbf{x}|^2}}{|\mathbf{x}|^2} - 1\Big),\end{aligned}$$

which implies the following expressions for the elements of the matrix $L^{-1}\boldsymbol{\eta}$:

$$\begin{aligned}(L^{-1}\boldsymbol{\eta})_{11}(\mathbf{x}) &= \Phi_{22}(\mathbf{x}) + \frac{1}{1+\sigma} \Phi_{11}\Big(\frac{\mathbf{x}}{1+\sigma}\Big) \\ &= -\frac{1}{8\pi} \Big(E_1(|\mathbf{x}|^2) + \frac{1}{1+\sigma} E_1\Big(\frac{|\mathbf{x}|^2}{1+\sigma}\Big) + \frac{2(2+\sigma)}{1+\sigma} \log|\mathbf{x}| - \frac{\log(1+\sigma)}{1+\sigma}\Big) \\ &+ \frac{1}{4\pi} \Big(\Big(\frac{x_1^2}{|\mathbf{x}|^2} - \frac{1}{2}\Big) \frac{\mathrm{e}^{-|\mathbf{x}|^2} - \mathrm{e}^{-|\mathbf{x}|^2/(1+\sigma)}}{|\mathbf{x}|^2} + \frac{\sigma}{1+\sigma} \frac{x_1^2}{|\mathbf{x}|^2} - 1\Big),\end{aligned}$$

$$\begin{aligned}(L^{-1}\boldsymbol{\eta})_{12}(\mathbf{x}) = (L^{-1}\boldsymbol{\eta})_{21}(\mathbf{x}) &= \frac{1}{1+\sigma} \Phi_{21}\Big(\frac{\mathbf{x}}{1+\sigma}\Big) - \Phi_{12}(\mathbf{x}) \\ &= \frac{1}{4\pi} \frac{x_1 x_2}{|\mathbf{x}|^2} \Big(\frac{\mathrm{e}^{-|\mathbf{x}|^2} - \mathrm{e}^{-|\mathbf{x}|^2/(1+\sigma)}}{|\mathbf{x}|^2} + \frac{\sigma}{1+\sigma}\Big),\end{aligned}$$

$$\begin{aligned}(L^{-1}\boldsymbol{\eta})_{22}(\mathbf{x}) &= \Phi_{11}(\mathbf{x}) + \frac{1}{1+\sigma} \Phi_{22}\Big(\frac{\mathbf{x}}{1+\sigma}\Big) \\ &= -\frac{1}{8\pi} \Big(E_1(|\mathbf{x}|^2) + \frac{1}{1+\sigma} E_1\Big(\frac{|\mathbf{x}|^2}{1+\sigma}\Big) + \frac{2(2+\sigma)}{1+\sigma} \log|\mathbf{x}| - \frac{\log(1+\sigma)}{1+\sigma}\Big) \\ &+ \frac{1}{4\pi} \Big(\Big(\frac{x_2^2}{|\mathbf{x}|^2} - \frac{1}{2}\Big) \frac{\mathrm{e}^{-|\mathbf{x}|^2} - \mathrm{e}^{-|\mathbf{x}|^2/(1+\sigma)}}{|\mathbf{x}|^2} + \frac{\sigma}{1+\sigma} \frac{x_2^2}{|\mathbf{x}|^2} - 1\Big).\end{aligned}$$

5.5. Three-dimensional potentials

Let us turn to the elastic and hydrodynamic volume potentials which provide solutions of the equations (5.25) and (5.26) in $\mathbb{R}^3$. The solution of the three-dimensional Lamé system is given by the volume potentials

$$u_k(\mathbf{x}) = \int\limits_{\mathbb{R}^3} \Gamma_{kl}(\mathbf{x}-\mathbf{y})\, f_l(\mathbf{y})\, d\mathbf{y}\,,$$

where $\|\Gamma_{jk}\|_{3\times 3}$ is the Kelvin-Somigliana fundamental matrix with

$$\Gamma_{kl}(\mathbf{x}) = \frac{\lambda+\mu}{8\pi\mu(\lambda+2\mu)} \Big(\frac{\lambda+3\mu}{\lambda+\mu} \frac{\delta_{kl}}{|\mathbf{x}|} + \frac{x_k\, x_l}{|\mathbf{x}|^3}\Big). \tag{5.47}$$

The solution of the Stokes problem in $\mathbb{R}^3$ can be expressed by the hydrodynamic potentials

$$u_k(\mathbf{x}) = \int_{\mathbb{R}^3} \sum_{l=1}^{3} \Psi_{kl}(\mathbf{x}-\mathbf{y})\, f_l(\mathbf{y})\, d\mathbf{y}\,, \quad p(\mathbf{x}) = \int_{\mathbb{R}^3} \langle \Theta(\mathbf{x}-\mathbf{y}), \mathbf{f}(\mathbf{y})\rangle d\mathbf{y} \tag{5.48}$$

with the fundamental solution

$$\Psi_{kl}(\mathbf{x}) = \frac{1}{8\pi\nu}\Big(\frac{\delta_{kl}}{|\mathbf{x}|} + \frac{x_k x_l}{|\mathbf{x}|^3}\Big), \quad \Theta(\mathbf{x}) = \frac{\mathbf{x}}{4\pi|\mathbf{x}|^3}\,. \tag{5.49}$$

In the following, we determine the values of the integral operators with the kernels (5.47) and (5.49) acting on the basis functions $\eta_{2M}(\mathbf{x})$. We use that

$$\frac{x_k x_l}{|\mathbf{x}|^3} = \frac{\delta_{kl}}{|\mathbf{x}|} - \frac{\partial^2}{\partial x_k \partial x_l}|\mathbf{x}|\,, \quad \frac{\mathbf{x}}{|\mathbf{x}|^3} = -\nabla \frac{1}{|\mathbf{x}|}\,. \tag{5.50}$$

Hence, the potentials can be obtained from the harmonic potentials (4.26) and the integrals

$$I_{kl}\eta_{2M}(\mathbf{x}) = \frac{\partial^2}{\partial x_k \partial x_l} \frac{1}{4\pi} \int_{\mathbb{R}^3} |\mathbf{x}-\mathbf{y}|\, \eta_{2M}(\mathbf{y})\, d\mathbf{y}\,. \tag{5.51}$$

Then similarly to the two-dimensional case

$$\int_{\mathbb{R}^3} \Gamma_{kl}(\mathbf{x}-\mathbf{y})\eta_{2M}(\mathbf{y})\, d\mathbf{y} = \frac{\delta_{kl}}{\mu}\mathcal{L}_3\eta_{2M}(\mathbf{x}) - \frac{\lambda+\mu}{2\mu(\lambda+2\mu)} I_{kl}\eta_{2M}(\mathbf{x}) \tag{5.52}$$

and

$$\begin{aligned} &\int_{\mathbb{R}^3} \Psi_{kl}(\mathbf{x}-\mathbf{y})\eta_{2M}(\mathbf{y})\, d\mathbf{y} = \frac{\delta_{kl}}{\nu}\mathcal{L}_3\eta_{2M}(\mathbf{x}) - \frac{1}{2\nu} I_{kl}\eta_{2M}(\mathbf{x})\,, \\ &\frac{x_k}{4\pi}\int_{\mathbb{R}^3} \frac{\eta_{2M}(\mathbf{y})}{|\mathbf{x}-\mathbf{y}|^3}\, d\mathbf{y} = -\frac{\partial}{\partial x_k}\mathcal{L}_3\eta_{2M}(\mathbf{x})\,. \end{aligned} \tag{5.53}$$

5.5.1. Potentials of the Gaussian.

We apply the expansion (3.18) to write

$$\begin{aligned} I_{kl}(L^{(1)}_{M-1}(|\cdot|^2)\,\mathrm{e}^{-|\cdot|^2})(\mathbf{x}) &= \frac{\partial^2}{\partial x_k \partial x_l}\frac{1}{4\pi}\int_{\mathbb{R}^3} |\mathbf{x}-\mathbf{y}|\,\mathrm{e}^{-|\mathbf{y}|^2}\, d\mathbf{y} \\ &+ \sum_{j=1}^{M-1} \frac{(-1)^j}{j!\,4^j}\frac{\partial^2}{\partial x_k \partial x_l}\frac{1}{4\pi}\int_{\mathbb{R}^2} |\mathbf{x}-\mathbf{y}|\,\Delta^j\,\mathrm{e}^{-|\mathbf{y}|^2}\, d\mathbf{y}\,. \end{aligned} \tag{5.54}$$

The relation

$$\frac{\partial}{\partial x_l}\int_{\mathbb{R}^3} |\mathbf{x}-\mathbf{y}|\,\mathrm{e}^{-|\mathbf{y}|^2}\, d\mathbf{y} = (x_l + \frac{1}{2}\frac{\partial}{\partial x_l})\int_{\mathbb{R}^3} \frac{\mathrm{e}^{-|\mathbf{y}|^2}}{|\mathbf{x}-\mathbf{y}|} d\mathbf{y} \tag{5.55}$$

implies, in view of (4.19), that

$$\begin{aligned} I_{kl}(e^{-|\cdot|^2})(\mathbf{x}) &= \delta_{kl}\mathcal{L}_3(e^{-|\cdot|^2})(\mathbf{x}) + \Big(x_l + \frac{1}{2}\frac{\partial}{\partial x_l}\Big)\frac{\partial}{\partial x_k}\mathcal{L}_3(e^{-|\cdot|^2})(\mathbf{x}) \\ &= \frac{1}{4|\mathbf{x}|^2}\Big(\delta_{kl} - \frac{3x_k x_l}{|\mathbf{x}|^2}\Big)\Big(e^{-|\mathbf{x}|^2} - \frac{\sqrt{\pi}\,\mathrm{erf}(|\mathbf{x}|)}{2|\mathbf{x}|}\Big) \\ &\quad + \frac{\sqrt{\pi}\,\mathrm{erf}(|\mathbf{x}|)}{4|\mathbf{x}|}\Big(\delta_{kl} - \frac{x_k x_l}{|\mathbf{x}|^2}\Big). \end{aligned}$$

Thus, the three-dimensional elastic potential of the generating function η_2 equals

$$\begin{aligned} \int_{\mathbb{R}^3} \Gamma_{kl}(\mathbf{x}-\mathbf{y})\eta_2(\mathbf{y})\,d\mathbf{y} &= \frac{\mathrm{erf}(|\mathbf{x}|)}{8\pi\mu(\lambda+2\mu)|\mathbf{x}|}\Big((\lambda+3\mu)\delta_{kl} + (\lambda+\mu)\frac{x_k x_l}{|\mathbf{x}|^2}\Big) \\ &+ \frac{\lambda+\mu}{8\pi^{3/2}\mu(\lambda+2\mu)|\mathbf{x}|^2}\Big(\frac{3x_k x_l}{|\mathbf{x}|^2} - \delta_{kl}\Big)\Big(e^{-|\mathbf{x}|^2} - \frac{\sqrt{\pi}\,\mathrm{erf}(|\mathbf{x}|)}{2|\mathbf{x}|}\Big), \end{aligned} \tag{5.56}$$

whereas the hydrodynamic potential can be derived from

$$\begin{aligned} \int_{\mathbb{R}^3} \Psi_{kl}(\mathbf{x}-\mathbf{y})\eta_2(\mathbf{y})\,d\mathbf{y} &= \frac{\mathrm{erf}(|\mathbf{x}|)}{8\pi\nu|\mathbf{x}|}\Big(\frac{x_j x_k}{|\mathbf{x}|^2} + \delta_{jk}\Big) \\ &+ \frac{1}{8\pi^{3/2}\nu|\mathbf{x}|^2}\Big(\frac{3x_k x_l}{|\mathbf{x}|^2} - \delta_{kl}\Big)\Big(e^{-|\mathbf{x}|^2} - \frac{\sqrt{\pi}\,\mathrm{erf}(|\mathbf{x}|)}{2|\mathbf{x}|}\Big), \\ \frac{1}{4\pi}\int_{\mathbb{R}^3} \frac{x_k - y_k}{|\mathbf{x}-\mathbf{y}|^3}\eta_2(\mathbf{y})\,d\mathbf{y} &= \frac{x_k}{2\pi^{3/2}|\mathbf{x}|^2}\Big(\frac{\sqrt{\pi}\,\mathrm{erf}(|\mathbf{x}|)}{2|\mathbf{x}|} - e^{-|\mathbf{x}|^2}\Big). \end{aligned} \tag{5.57}$$

5.5.2. Elastic potential of higher-order generating functions. We note that

$$\frac{1}{8\pi}\int_{\mathbb{R}^3} |\mathbf{x}-\mathbf{y}|\,\Delta^j e^{-|\mathbf{y}|^2}\,d\mathbf{y} = \frac{\Delta^{j-1}}{4\pi}\int_{\mathbb{R}^3} \frac{e^{-|\mathbf{y}|^2}}{|\mathbf{x}-\mathbf{y}|}\,d\mathbf{y}\,,$$

which for $j \ge 2$ is equal to $-\Delta^{j-2} e^{-|\mathbf{x}|^2}$. Thus, for $M \ge 2$

$$\begin{aligned} I_{kl}(L^{(1)}_{M-1}(|\cdot|^2)\,e^{-|\cdot|^2})(\mathbf{x}) &= \frac{\partial^2}{\partial x_k \partial x_l}\frac{1}{4\pi}\Big(\int_{\mathbb{R}^3} |\mathbf{x}-\mathbf{y}|\,e^{-|\mathbf{y}|^2}\,d\mathbf{y} - \frac{1}{2}\int_{\mathbb{R}^3}\frac{e^{-|\mathbf{y}|^2}}{|\mathbf{x}-\mathbf{y}|}d\mathbf{y}\Big) \\ &\quad - 2\frac{\partial^2}{\partial x_k \partial x_l}\sum_{j=2}^{M-1}\frac{(-1)^j}{j!\,4^j}\Delta^{j-2} e^{-|\mathbf{x}|^2}\,. \end{aligned}$$

Using (5.55), the first term, which is equal to $I_{kl}(L^{(1)}_1(|\cdot|^2)\,e^{-|\cdot|^2})$, simplifies to

$$\frac{\partial^2}{\partial x_k \partial x_l}\frac{1}{4\pi}\int_{\mathbb{R}^3}\Big(|\mathbf{x}-\mathbf{y}| - \frac{1}{2|\mathbf{x}-\mathbf{y}|}\Big)e^{-|\mathbf{y}|^2}\,d\mathbf{y} = \Big(\delta_{kl} + x_l\frac{\partial}{\partial x_k}\Big)\frac{1}{4\pi}\int_{\mathbb{R}^3}\frac{e^{-|\mathbf{y}|^2}}{|\mathbf{x}-\mathbf{y}|}d\mathbf{y},$$

and because

$$x_l\frac{\partial}{\partial x_k}\mathcal{L}_3(e^{-|\cdot|^2})(\mathbf{x}) = \frac{x_k x_l}{2|\mathbf{x}|^2}\Big(e^{-|\mathbf{x}|^2} - \frac{\sqrt{\pi}\,\mathrm{erf}(|\mathbf{x}|)}{2|\mathbf{x}|}\Big), \tag{5.58}$$

the integral $I_{kl}\eta_4$ has the simple form

$$I_{kl}\eta_4(\mathbf{x}) = \pi^{-3/2}\Big(\delta_{kl} + x_l\frac{\partial}{\partial x_k}\Big)\mathcal{L}_3(\,\mathrm{e}^{-|\cdot|^2}\,)$$
$$= \frac{x_k x_l\,\mathrm{e}^{-|\mathbf{x}|^2}}{2\pi^{3/2}|\mathbf{x}|^2} + \frac{\operatorname{erf}(|\mathbf{x}|)}{4\pi|\mathbf{x}|}\Big(\delta_{kl} - \frac{x_k x_l}{|\mathbf{x}|^2}\Big).$$

Moreover, we can apply (3.15) to obtain

$$I_{kl}\eta_{2M}(\mathbf{x}) = I_{kl}\eta_4(\mathbf{x}) - \frac{\partial^2}{\partial x_k \partial x_l}\sum_{j=0}^{M-3}\frac{L_j^{(1/2)}(|\mathbf{x}|^2)\,\mathrm{e}^{-|\mathbf{x}|^2}}{8(j+1)(j+2)}. \tag{5.59}$$

Sums of Laguerre polynomials. One can simplify the second term in (5.59) similarly to the two-dimensional case:

$$S_M = \frac{\partial^2}{\partial x_k \partial x_l}\sum_{j=0}^{M-3}\frac{L_j^{(1/2)}(|\mathbf{x}|^2)\,\mathrm{e}^{-|\mathbf{x}|^2}}{8(j+1)(j+2)}$$
$$= x_k x_l \sum_{j=0}^{M-3}\frac{L_j^{(5/2)}(|\mathbf{x}|^2)\,\mathrm{e}^{-|\mathbf{x}|^2}}{2(j+1)(j+2)} - \delta_{kl}\sum_{j=0}^{M-3}\frac{L_j^{(3/2)}(|\mathbf{x}|^2)\,\mathrm{e}^{-|\mathbf{x}|^2}}{4(j+1)(j+2)},$$

where the expression (5.40) for the second derivatives is taken into account. The first sum can be transformed by using

$$L_j^{(5/2)}(|\mathbf{x}|^2) = \frac{1}{|\mathbf{x}|^2}\Big((j+\frac{5}{2})L_j^{(3/2)}(|\mathbf{x}|^2) - (j+1)L_{j+1}^{(3/2)}(|\mathbf{x}|^2)\Big),$$

which follows from (4.29). This leads to

$$x_k x_l \sum_{j=0}^{M-3}\frac{L_j^{(5/2)}(|\mathbf{x}|^2)}{2(j+1)(j+2)}$$
$$= \frac{x_k x_l}{|\mathbf{x}|^2}\Big(\sum_{j=0}^{M-3}\Big(\frac{L_j^{(3/2)}(|\mathbf{x}|^2)}{2(j+1)} - \frac{L_{j+1}^{(3/2)}(|\mathbf{x}|^2)}{2(j+2)}\Big) + \sum_{j=0}^{M-3}\frac{L_j^{(3/2)}(|\mathbf{x}|^2)}{4(j+1)(j+2)}\Big)$$
$$= \frac{x_k x_l}{|\mathbf{x}|^2}\Big(\frac{1}{2} - \frac{L_{M-2}^{(3/2)}(|\mathbf{x}|^2)}{2(M-1)} + \sum_{j=0}^{M-3}\frac{L_j^{(3/2)}(|\mathbf{x}|^2)}{4(j+1)(j+2)}\Big).$$

Similarly to (5.41), we obtain

$$\sum_{j=0}^{M-3}\frac{L_j^{(3/2)}(|\mathbf{x}|^2)}{4(j+1)(j+2)} = \sum_{j=0}^{M-2}\frac{L_j^{(1/2)}(|\mathbf{x}|^2)}{4(j+1)} - \frac{L_{M-2}^{(3/2)}(|\mathbf{x}|^2)}{4(M-1)}, \tag{5.60}$$

resulting in

$$x_k x_l \sum_{j=0}^{M-3}\frac{L_j^{(5/2)}(|\mathbf{x}|^2)}{2(j+1)(j+2)} = \frac{x_k x_l}{|\mathbf{x}|^2}\Big(\frac{1}{2} + \sum_{j=0}^{M-2}\frac{L_j^{(1/2)}(|\mathbf{x}|^2)}{4(j+1)} - \frac{3L_{M-2}^{(3/2)}(|\mathbf{x}|^2)}{4(M-1)}\Big).$$

Hence we arrive at the representation

$$S_M = \frac{x_k x_l}{2|\mathbf{x}|^2}\Big(\mathrm{e}^{-|\mathbf{x}|^2} + \sum_{j=0}^{M-2} \frac{L_j^{(1/2)}(|\mathbf{x}|^2)\,\mathrm{e}^{-|\mathbf{x}|^2}}{2(j+1)} - \frac{3L_{M-2}^{(3/2)}(|\mathbf{x}|^2)\,\mathrm{e}^{-|\mathbf{x}|^2}}{2(M-1)}\Big)$$
$$- \delta_{kl}\Big(\sum_{j=0}^{M-2} \frac{L_j^{(1/2)}(|\mathbf{x}|^2)\,\mathrm{e}^{-|\mathbf{x}|^2}}{4(j+1)} - \frac{L_{M-2}^{(3/2)}(|\mathbf{x}|^2)\,\mathrm{e}^{-|\mathbf{x}|^2}}{4(M-1)}\Big).$$

Final form of $I_{kl}\eta_{2M}$. In view of (5.59) and (5.58), we can write the integral $I_{kl}\eta_{2M}$ in the form

$$I_{kl}\eta_{2M}(\mathbf{x}) = \frac{1}{\pi^{3/2}}\Big(\Big(\delta_{kl} + x_l \frac{\partial}{\partial x_k}\Big)\mathcal{L}_3(\mathrm{e}^{-|\cdot|^2})(\mathbf{x}) - S_M\Big)$$
$$= \frac{1}{\pi^{3/2}}\Big(\delta_{kl}\mathcal{L}_3(\mathrm{e}^{-|\cdot|^2})(\mathbf{x}) + \frac{x_k x_l}{2|\mathbf{x}|^2}\Big(\mathrm{e}^{-|\mathbf{x}|^2} - 2\mathcal{L}_3(\mathrm{e}^{-|\cdot|^2})(\mathbf{x})\Big) - S_M\Big).$$

Collecting the terms, we see that $I_{kl}\eta_{2M}$ is the sum of

$$\frac{\delta_{kl}}{\pi^{3/2}}\Big(\mathcal{L}_3(\mathrm{e}^{-|\cdot|^2})(\mathbf{x}) + \sum_{j=0}^{M-2} \frac{L_j^{(1/2)}(|\mathbf{x}|^2)\,\mathrm{e}^{-|\mathbf{x}|^2}}{4(j+1)} - \frac{L_{M-2}^{(3/2)}(|\mathbf{x}|^2)\,\mathrm{e}^{-|\mathbf{x}|^2}}{4(M-1)}\Big)$$
$$= \delta_{kl}\Big(\mathcal{L}_3\eta_{2M}(\mathbf{x}) - \frac{\eta_{2M-2}(\mathbf{x})}{4(M-1)}\Big)$$

and

$$\frac{x_k x_l}{\pi^{3/2}|\mathbf{x}|^2}\Big(- \mathcal{L}_3(\mathrm{e}^{-|\cdot|^2})(\mathbf{x}) - \sum_{j=0}^{M-2} \frac{L_j^{(1/2)}(|\mathbf{x}|^2)\,\mathrm{e}^{-|\mathbf{x}|^2}}{4(j+1)} + \frac{3L_{M-2}^{(3/2)}(|\mathbf{x}|^2)\,\mathrm{e}^{-|\mathbf{x}|^2}}{4(M-1)}\Big)$$
$$= \frac{x_k x_l}{|\mathbf{x}|^2}\Big(- \mathcal{L}_3\eta_{2M}(\mathbf{x}) + \frac{3\eta_{2M-2}(\mathbf{x})}{4(M-1)}\Big),$$

so the final form of $I_{kl}\eta_{2M}$ is

$$I_{kl}\eta_{2M}(\mathbf{x}) = \Big(\delta_{kl} - \frac{x_k x_l}{|\mathbf{x}|^2}\Big)\mathcal{L}_3\eta_{2M}(\mathbf{x}) + \Big(\frac{3x_k x_l}{|\mathbf{x}|^2} - \delta_{kl}\Big)\frac{\eta_{2M-2}(\mathbf{x})}{4(M-1)}. \tag{5.61}$$

5.5.3. Final form of the elastic and hydrodynamic potentials. An explicit representation of the integral

$$\frac{x_k}{4\pi}\int_{\mathbb{R}^3} \frac{\eta_{2M}(\mathbf{y})}{|\mathbf{x}-\mathbf{y}|^3}\,d\mathbf{y} = -\frac{\partial}{\partial x_k}\mathcal{L}_3\eta_{2M}(\mathbf{x})$$

can be derived from (4.41), which shows that

$$\frac{x_k}{4\pi}\int_{\mathbb{R}^3} \frac{\eta_{2M}(\mathbf{y})}{|\mathbf{x}-\mathbf{y}|^3}\,d\mathbf{y} = \frac{x_k}{|\mathbf{x}|^2}\Big(\mathcal{L}_3\eta_{2M}(\mathbf{x}) - \frac{L_{M-1}^{(1/2)}(|\mathbf{x}|^2)\,\mathrm{e}^{-|\mathbf{x}|^2}}{2\pi^{3/2}}\Big).$$

Hence, together with (5.52) and (5.61), we obtain the elastic potential of η_{2M}

$$\int\limits_{\mathbb{R}^3} \Gamma_{kl}(\mathbf{x}-\mathbf{y})\,\eta_{2M}(\mathbf{y})\,d\mathbf{y} = \frac{\lambda+\mu}{8\mu(\lambda+2\mu)}\Big(\delta_{kl} - 3\frac{x_k x_l}{|\mathbf{x}|^2}\Big)\frac{\eta_{2M-2}(\mathbf{x})}{M-1} + \Big(\frac{\lambda+3\mu}{2\mu(\lambda+2\mu)}\,\delta_{kl} + \frac{\lambda+\mu}{2\mu(\lambda+2\mu)}\,\frac{x_k x_l}{|\mathbf{x}|^2}\Big)\mathcal{L}_3\eta_{2M}(\mathbf{x})\,, \tag{5.62}$$

and the 3-dimensional hydrodynamic potentials in the form

$$\begin{aligned} \int\limits_{\mathbb{R}^3} \Psi_{kl}(\mathbf{x}-\mathbf{y})\,\eta_{2M}(\mathbf{y})\,d\mathbf{y} &= \frac{1}{2\nu}\Big(\delta_{kl} + \frac{x_k x_l}{|\mathbf{x}|^2}\Big)\mathcal{L}_3\eta_{2M}(\mathbf{x}) \\ &\quad + \frac{1}{8\nu}\Big(\delta_{kl} - \frac{3x_k x_l}{|\mathbf{x}|^2}\Big)\frac{\eta_{2M-2}(\mathbf{x})}{M-1}\,, \\ \frac{x_k}{4\pi}\int\limits_{\mathbb{R}^3} \frac{\eta_{2M}(\mathbf{y})}{|\mathbf{x}-\mathbf{y}|^3}\,d\mathbf{y} &= \frac{x_k}{|\mathbf{x}|^2}\Big(\mathcal{L}_3\eta_{2M}(\mathbf{x}) - \frac{L_{M-1}^{(1/2)}(|\mathbf{x}|^2)\,\mathrm{e}^{-|\mathbf{x}|^2}}{2\pi^{3/2}}\Big). \end{aligned} \tag{5.63}$$

REMARK 5.4. Analogously to the two-dimensional case, the elastic and hydrodynamic potentials of η_{2M+2} can be obtained from those of η_{2M} by adding some rapidly decaying term if $M \geq 2$. In fact, by (5.59) and (5.40),

$$I_{kl}\eta_{2M+2}(\mathbf{x}) - I_{kl}\eta_{2M}(\mathbf{x}) = \frac{\mathrm{e}^{-|\mathbf{x}|^2}}{4\pi^{3/2}M(M-1)}\Big(\delta_{kl}L_{M-2}^{(3/2)}(|\mathbf{x}|^2) - 2x_k x_l L_{M-2}^{(5/2)}(|\mathbf{x}|^2)\Big)\,.$$

5.6. Notes

Various applications of potential methods to scattering, elastic, and hydrodynamic problems were developed by many authors. For details, we refer to the classical monographs [**51**], [**52**]. The main formulas for diffraction and elastic potentials applied to η_{2M} have been obtained in [**67**].

Approximation properties of the cubature (5.19) for diffraction potentials are studied in Section 12.1. Since the mapping properties of the elastic and hydrodynamic potentials are similar to those of the harmonic potential, the estimation of the cubature error for elastic and hydrodynamic potentials can be carried out analogously to Theorems 4.10 and 4.11.

CHAPTER 6

Some other cubature problems

In this chapter, we consider applications of approximate quasi-interpolation to various problems, which can be solved by using integral representations.

In Section 6.1, some integral and pseudodifferential operators, occurring frequently in applications, are treated. We obtain explicit formulas for the action of these operators onto the basis functions η_{2M}.

In Section 6.2, we use integral representations of solutions to initial value problems to derive approximate solutions via approximate approximations. As model problems, the heat, wave, and plate equations are considered and semi-analytic approximate solutions of these problems are obtained.

Based on integral representations of solutions to parabolic and hyperbolic problems, in Section 6.3 we derive efficient formulas for the potentials of elliptic equations applied to general Gaussian functions. These formulas lead to cubature formulas for higher-order generating functions derived from anisotropic Gaussians, considered in Section 6.4, and from approximate wavelets, considered in Section 8.7.

Finally, in Section 6.5, we derive formulas for harmonic and diffraction potentials of products of Gaussians with the characteristic function of a half-space, which are applied to the cubature of potentials over the half-space.

6.1. Cubature of some pseudodifferential operators

6.1.1. Square root of the Laplacian. In fracture mechanics, electrostatics, and hydrodynamics, one encounters the pseudodifferential operator of order 1

$$(-\Delta)^{1/2}u(\mathbf{x}) = \frac{-\Delta}{2\pi}\int\limits_{\mathbb{R}^2} \frac{u(\mathbf{y})\,d\mathbf{y}}{|\mathbf{x}-\mathbf{y}|}, \quad \mathbf{x}\in\mathbb{R}^2, \tag{6.1}$$

where Δ is the two-dimensional Laplacian. Using (5.15), one can transform the integral

$$\frac{1}{2\pi}\int\limits_{\mathbb{R}^2} \frac{e^{-|\mathbf{y}|^2}}{|\mathbf{x}-\mathbf{y}|}\,d\mathbf{y} = e^{-|\mathbf{x}|^2}\int\limits_0^\infty e^{-r^2} I_0(2|\mathbf{x}|r)\,dr = \frac{\sqrt{\pi}}{2}\,e^{-|\mathbf{x}|^2/2} I_0\Big(\frac{|\mathbf{x}|^2}{2}\Big)$$

(the last relation is taken from [**82**, 2.15.5.2]). Then

$$\frac{-\Delta}{2\pi}\int\limits_{\mathbb{R}^2} \frac{e^{-|\mathbf{y}|^2}}{|\mathbf{x}-\mathbf{y}|}\,d\mathbf{y} = \sqrt{\pi}\;e^{-|\mathbf{x}|^2/2}\Big((1-|\mathbf{x}|^2)I_0\Big(\frac{|\mathbf{x}|^2}{2}\Big) + |\mathbf{x}|^2 I_1\Big(\frac{|\mathbf{x}|^2}{2}\Big)\Big) \tag{6.2}$$

with the modified Bessel functions I_0 and I_1. Thus, a second-order approximation of (6.1) is given by the formula

$$(-\Delta)^{1/2}u(\mathbf{x}) \approx \frac{(\pi\mathcal{D})^{-1/2}}{\mathcal{D}h}\sum_{\mathbf{m}\in\mathbb{Z}^2} u(h\mathbf{m})\,e^{-r_{\mathbf{m}}/2}\Big((1-r_{\mathbf{m}})I_0\Big(\frac{r_{\mathbf{m}}}{2}\Big) + r_{\mathbf{m}} I_1\Big(\frac{r_{\mathbf{m}}}{2}\Big)\Big)$$

with

$$r_{\mathbf{m}} = \frac{|\mathbf{x} - h\mathbf{m}|^2}{\mathcal{D}h^2} .$$

In view of

$$\eta_{2M}(\mathbf{x}) = \frac{1}{\pi} \sum_{j=0}^{M-1} \frac{(-1)^j}{j!\, 4^j} \Delta^j \, \mathrm{e}^{-|\mathbf{x}|^2} ,$$

from (6.2) one obtains

$$(-\Delta)^{1/2} \eta_{2M}(\mathbf{x})$$
$$= \frac{1}{\sqrt{\pi}} \sum_{j=0}^{M-1} \frac{(-1)^j}{j!\, 4^j} \Delta^j \Big(\mathrm{e}^{-|\mathbf{x}|^2/2} \Big(I_0\Big(\frac{|\mathbf{x}|^2}{2}\Big)(1 - |\mathbf{x}|^2) + |\mathbf{x}|^2 I_1\Big(\frac{|\mathbf{x}|^2}{2}\Big)\Big)\Big),$$

which can be written in the form

$$(-\Delta)^{1/2}\eta_{2M}(\mathbf{x}) = \mathrm{e}^{-|\mathbf{x}|^2/2} \Big(P_M(|\mathbf{x}|^2) I_0\Big(\frac{|\mathbf{x}|^2}{2}\Big) + Q_M(|\mathbf{x}|^2) I_1\Big(\frac{|\mathbf{x}|^2}{2}\Big)\Big)$$

with polynomials P_M and Q_M of degree $M + 1$. Thus, an approximate approximation of the square root of the two-dimensional Laplacian (6.1) with the order $\mathcal{O}(h^{2M})$ modulo saturation error is given by the formula

$$\frac{1}{\pi^{1/2}\mathcal{D}^{3/2}h} \sum_{\mathbf{m}\in\mathbb{Z}^2} u(h\mathbf{m})\, \mathrm{e}^{-r_{\mathbf{m}}/2} \Big(P_M(r_{\mathbf{m}}) I_0\Big(\frac{r_{\mathbf{m}}}{2}\Big) + Q_M(r_{\mathbf{m}}) I_1\Big(\frac{r_{\mathbf{m}}}{2}\Big)\Big) .$$

6.1.2. Higher-dimensional singular integrals. Some problems from mechanics, electromagnetics, and hydrodynamics can be solved by using second derivatives of the harmonic potential

$$\frac{\partial^2}{\partial x_i \partial x_k} \mathcal{L}_n u(\mathbf{x}) = -\frac{\partial^2}{\partial x_i \partial x_k} \Delta^{-1} u(\mathbf{x}) , \quad \mathbf{x} \in \mathbb{R}^n , \tag{6.3}$$

which represent pseudodifferential operators of order 0. In Section 13.2 we apply approximation formulas for these operators to the numerical solution of non-stationary Navier-Stokes equations.

Using the results of Subsection 4.3.5 concerning the gradient of the harmonic potential, it is simple to obtain high-order approximation formulas for (6.3). From (4.36), we have

$$\frac{\partial}{\partial x_k} \frac{1}{4|\mathbf{x}|^{n-2}} \gamma\Big(\frac{n}{2} - 1, |\mathbf{x}|^2\Big) = -\frac{x_k}{2|\mathbf{x}|^n} \gamma\Big(\frac{n}{2}, |\mathbf{x}|^2\Big).$$

Hence by (4.37)

$$\begin{aligned} \frac{\partial^2}{\partial x_i \partial x_k} \mathcal{L}_n(\mathrm{e}^{-|\cdot|^2})(\mathbf{x}) &= -\frac{\partial}{\partial x_i} \frac{x_k}{2|\mathbf{x}|^n} \gamma\Big(\frac{n}{2}, |\mathbf{x}|^2\Big) \\ &= -\frac{\delta_{ik}}{2|\mathbf{x}|^n} \gamma\Big(\frac{n}{2}, |\mathbf{x}|^2\Big) + \frac{x_i x_k}{|\mathbf{x}|^{n+2}} \gamma\Big(\frac{n}{2} + 1, |\mathbf{x}|^2\Big), \end{aligned}$$

which is valid for $n \geq 2$. From (4.17), we get

$$\gamma\Big(\frac{n}{2} + 1, |\mathbf{x}|^2\Big) = \frac{n}{2} \gamma\Big(\frac{n}{2}, |\mathbf{x}|^2\Big) - |\mathbf{x}|^n \, \mathrm{e}^{-|\mathbf{x}|^2}$$

leading to

$$\boxed{(6.4)\qquad \frac{\partial^2}{\partial x_i \partial x_k}\mathcal{L}_n(\mathrm{e}^{-|\cdot|^2})(\mathbf{x}) = \frac{n x_i x_k - \delta_{ik}|\mathbf{x}|^2}{2|\mathbf{x}|^{n+2}}\,\gamma\Big(\frac{n}{2}, |\mathbf{x}|^2\Big) - \frac{x_i x_k}{|\mathbf{x}|^2}\,\mathrm{e}^{-|\mathbf{x}|^2}\,.}$$

Similarly we obtain from (4.39)

$$\begin{aligned}\frac{\partial^2}{\partial x_i \partial x_k}\frac{L_j^{(n/2-1)}(|\mathbf{x}|^2)\,\mathrm{e}^{-|\mathbf{x}|^2}}{4(j+1)} &= -\frac{\partial}{\partial x_i} x_k \frac{L_j^{(n/2)}(|\mathbf{x}|^2)\,\mathrm{e}^{-|\mathbf{x}|^2}}{2(j+1)} \\ &= x_i x_k \frac{L_j^{(n/2+1)}(|\mathbf{x}|^2)\,\mathrm{e}^{-|\mathbf{x}|^2}}{j+1} - \delta_{ik}\frac{L_j^{(n/2)}(|\mathbf{x}|^2)\,\mathrm{e}^{-|\mathbf{x}|^2}}{2(j+1)}\,.\end{aligned}$$

Since by (4.38)

$$\frac{L_j^{(n/2+1)}(|\mathbf{x}|^2)}{j+1} = \frac{L_j^{(n/2)}(|\mathbf{x}|^2) - L_{j+1}^{(n/2)}(|\mathbf{x}|^2)}{|\mathbf{x}|^2} + \frac{n\,L_j^{(n/2)}(|\mathbf{x}|^2)}{2|\mathbf{x}|^2(j+1)}\,,$$

the second derivative transforms to

$$\begin{aligned}&\frac{\partial^2}{\partial x_i \partial x_k}\frac{L_j^{(n/2-1)}(|\mathbf{x}|^2)\,\mathrm{e}^{-|\mathbf{x}|^2}}{4(j+1)} \\ &= \frac{\mathrm{e}^{-|\mathbf{x}|^2}}{|\mathbf{x}|^2}\Big(x_i x_k (L_j^{(n/2)}(|\mathbf{x}|^2) - L_{j+1}^{(n/2)}(|\mathbf{x}|^2)) + (n x_i x_k - \delta_{ik}|\mathbf{x}|^2)\frac{L_j^{(n/2)}(|\mathbf{x}|^2)}{2(j+1)}\Big).\end{aligned}$$

Thus,

$$\begin{aligned}\frac{\partial^2}{\partial x_i \partial x_k}\sum_{j=0}^{M-2}\frac{L_j^{(n/2-1)}(|\mathbf{x}|^2)\,\mathrm{e}^{-|\mathbf{x}|^2}}{4(j+1)} &= \Big(n\frac{x_i x_k}{|\mathbf{x}|^2} - \delta_{ik}\Big)\sum_{j=0}^{M-2}\frac{L_j^{(n/2)}(|\mathbf{x}|^2)\,\mathrm{e}^{-|\mathbf{x}|^2}}{2(j+1)} \\ &\quad + \frac{x_i x_k\,\mathrm{e}^{-|\mathbf{x}|^2}}{|\mathbf{x}|^2}(1 - L_{M-1}^{(n/2)}(|\mathbf{x}|^2)),\end{aligned}$$

and therefore, in view of (4.25) and (6.4),

$$\begin{aligned}\frac{\partial^2}{\partial x_i \partial x_k}\mathcal{L}_n \eta_{2M}(\mathbf{x}) &= -\frac{x_i x_k}{\pi^{n/2}|\mathbf{x}|^2} L_{M-1}^{(n/2)}(|\mathbf{x}|^2)\,\mathrm{e}^{-|\mathbf{x}|^2} \\ &\quad + \Big(n\frac{x_i x_k}{|\mathbf{x}|^2} - \delta_{ik}\Big)\Big(\frac{1}{2\pi^{n/2}|\mathbf{x}|^n}\gamma\Big(\frac{n}{2}, |\mathbf{x}|^2\Big) + \sum_{j=0}^{M-2}\frac{L_j^{(n/2)}(|\mathbf{x}|^2)\,\mathrm{e}^{-|\mathbf{x}|^2}}{2\pi^{n/2}(j+1)}\Big) \\ &= 2\pi\Big(n\frac{x_i x_k}{|\mathbf{x}|^2} - \delta_{ik}\Big)\mathcal{L}_{n+2}\eta_{2M}(|\mathbf{x}|) - \frac{x_i x_k}{\pi^{n/2}|\mathbf{x}|^2} L_{M-1}^{(n/2)}(|\mathbf{x}|^2)\,\mathrm{e}^{-|\mathbf{x}|^2}\,.\end{aligned}$$

Hence the action of the operator (6.3) on the quasi-interpolant $\mathcal{M}_{h,\mathcal{D}}^{(2M)}u$ can be given as

(6.5)
$$-\frac{\partial^2}{\partial x_i \partial x_k}\,\Delta^{-1}\mathcal{M}_{h,\mathcal{D}}^{(2M)}u(\mathbf{x})$$
$$=\frac{1}{(\pi\mathcal{D})^{n/2}}\sum_{\mathbf{m}\in\mathbb{Z}^n}u(h\mathbf{m})\Big\{-\frac{(x_i-hm_i)(x_k-hm_k)}{|\mathbf{x}-h\mathbf{m}|^2}L_{M-1}^{(n/2)}(|\mathbf{r_m}|^2)\,\mathrm{e}^{-|\mathbf{r_m}|^2}$$
$$+\Big(n\frac{(x_i-hm_i)(x_k-hm_k)}{|\mathbf{x}-h\mathbf{m}|^2}-\delta_{ik}\Big)\Big(\frac{\gamma(n/2,|\mathbf{r_m}|^2)}{2|\mathbf{r_m}|^n}+\mathrm{e}^{-|\mathbf{r_m}|^2}\sum_{j=0}^{M-2}\frac{L_j^{(n/2)}(|\mathbf{r_m}|^2)}{2(j+1)}\Big)\Big\},$$

where $\mathbf{r_m}=|\mathbf{x}-h\mathbf{m}|/(\sqrt{\mathcal{D}}h)$. Since (6.3) is a pseudodifferential operator of order 0, we obtain the estimate

$$\Big\|\frac{\partial^2}{\partial x_i \partial x_k}\,\Delta^{-1}(u-\mathcal{M}_{h,\mathcal{D}}^{(2M)}u)\Big\|=\mathcal{O}(h^{2M})+\varepsilon\,,$$

with small ε if $\mathcal{D}$ is large enough.

As an example, we provide the cubature formula of the two-dimensional singular integral

(6.6)
$$S_2u(\mathbf{x}):=\frac{2}{\pi}\int\limits_{\mathbb{R}^2}\frac{(x_1-y_1)(x_2-y_2)}{|\mathbf{x}-\mathbf{y}|^4}\,u(\mathbf{y})\,d\mathbf{y}\,,\quad \mathbf{x}\in\mathbb{R}^2\,.$$

Noting that

$$S_2u(\mathbf{x})=\frac{\partial^2}{\partial x_1\partial x_2}\mathcal{L}_2u(\mathbf{x})\,,$$

we obtain from (6.5) that

$$S_2u(\mathbf{x})\approx\frac{1}{\pi\mathcal{D}}\sum_{\mathbf{m}\in\mathbb{Z}^2}u(h\mathbf{m})\Big\{-\frac{(x_1-hm_1)(x_2-hm_2)}{|\mathbf{x}-h\mathbf{m}|^2}L_{M-1}^{(1)}(|\mathbf{r_m}|^2)\,\mathrm{e}^{-|\mathbf{r_m}|^2}$$
$$+\frac{(x_1-hm_1)(x_2-hm_2)}{|\mathbf{x}-h\mathbf{m}|^2}\Big(\frac{\gamma(1,|\mathbf{r_m}|^2)}{|\mathbf{r_m}|^2}+\mathrm{e}^{-|\mathbf{r_m}|^2}\sum_{j=0}^{M-2}\frac{L_j^{(1)}(|\mathbf{r_m}|^2)}{j+1}\Big)\Big\}.$$

Note that by (4.30)

$$\frac{\gamma(1,|\mathbf{r_m}|^2)}{|\mathbf{r_m}|^2}+\mathrm{e}^{-|\mathbf{r_m}|^2}\sum_{j=0}^{M-2}\frac{L_j^{(1)}(|\mathbf{r_m}|^2)}{j+1}=\frac{1-\mathrm{e}^{-|\mathbf{r_m}|^2}\,L_{M-1}(|\mathbf{r_m}|^2)}{|\mathbf{r_m}|^2}\,.$$

Thus, the singular integral (6.6) can be approximated with the order $\mathcal{O}(h^{2M})$ modulo saturation error by

$$S_2u(\mathbf{x})\approx\frac{1}{\pi\mathcal{D}}\sum_{\mathbf{m}\in\mathbb{Z}^2}u(h\mathbf{m})\frac{(x_1-hm_1)(x_2-hm_2)}{|\mathbf{x}-h\mathbf{m}|^2}$$
$$\times\frac{1-\mathrm{e}^{-|\mathbf{r_m}|^2}\,(L_{M-1}(|\mathbf{r_m}|^2)+|\mathbf{r_m}|^2L_{M-1}^{(1)}(|\mathbf{r_m}|^2))}{|\mathbf{r_m}|^2}\,.$$

6.1.3. Biharmonic potential. Consider the bi-Laplace equation

$$\Delta^2 f(\mathbf{x}) = -u(\mathbf{x})\,, \quad \mathbf{x} \in \mathbb{R}^3\,,$$

which has the solution

$$f(\mathbf{x}) = \frac{1}{8\pi} \int\limits_{\mathbb{R}^3} |\mathbf{x} - \mathbf{y}|\, u(\mathbf{y})\, d\mathbf{y}\,.$$

Here, we determine the biharmonic potential of the generating function η_{2M}:

$$\mathcal{H}_3 \eta_{2M}(\mathbf{x}) := \frac{1}{8\pi} \int\limits_{\mathbb{R}^3} |\mathbf{x} - \mathbf{y}|\, \eta_{2M}(\mathbf{y})\, d\mathbf{y}\,. \tag{6.7}$$

Recall that by (3.18)

$$\pi^{3/2} \eta_{2M}(\mathbf{x}) = L_{M-1}^{(3/2)}(|\mathbf{x}|^2)\, \mathrm{e}^{-|\mathbf{x}|^2} = \sum_{j=0}^{M-1} \frac{(-1)^j}{j!\, 4^j} \Delta^j\, \mathrm{e}^{-|\mathbf{x}|^2}\,;$$

hence, because $\mathcal{H}_3 \Delta^2 = -I$, we obtain

$$\begin{aligned}
\mathcal{H}_3 \eta_{2M}(\mathbf{x}) &= \pi^{-3/2} \Big(\mathcal{H}_3 (\, \mathrm{e}^{-|\cdot|^2}\,)(\mathbf{x}) - \frac{1}{4} \mathcal{H}_3 (\Delta\, \mathrm{e}^{-|\cdot|^2}\,)(\mathbf{x}) - \sum_{j=2}^{M-1} \frac{(-1)^j}{j!\, 4^j} \Delta^{j-2}\, \mathrm{e}^{-|\mathbf{x}|^2} \Big) \\
&= \pi^{-3/2} \Big(\mathcal{H}_3 (\, \mathrm{e}^{-|\cdot|^2}\,)(\mathbf{x}) - \frac{1}{4} \mathcal{L}_3 (\, \mathrm{e}^{-|\cdot|^2}\,)(\mathbf{x}) - \sum_{j=2}^{M-1} \frac{(-1)^j}{j!\, 4^j} \Delta^{j-2}\, \mathrm{e}^{-|\mathbf{x}|^2} \Big),
\end{aligned}$$

where we use the relation $\mathcal{H}_3 \Delta = -\mathcal{H}_3 \Delta^2 \mathcal{L}_3 = \mathcal{L}_3$. To determine the action of the biharmonic potential on the Gaussian, we use the general formula (5.15) with $Q(r) = r/8\pi$, which gives, in view of (5.16),

$$\begin{aligned}
\frac{1}{8\pi} \int\limits_{\mathbb{R}^3} |\mathbf{x} - \mathbf{y}|\, \mathrm{e}^{-|\mathbf{y}|^2}\, d\mathbf{y} &= \frac{\pi^{1/2}\, \mathrm{e}^{-|\mathbf{x}|^2}}{4 |\mathbf{x}|^{1/2}} \int\limits_0^\infty \mathrm{e}^{-r^2}\, I_{1/2}(2|\mathbf{x}| r)\, r^{5/2}\, dr \\
&= \frac{\mathrm{e}^{-|\mathbf{x}|^2}}{8|\mathbf{x}|} \int\limits_0^\infty r^2\, \mathrm{e}^{-r^2} \left(\mathrm{e}^{2|\mathbf{x}| r} - \mathrm{e}^{-2|\mathbf{x}| r} \right) dr \\
&= \frac{1}{8|\mathbf{x}|} \int\limits_0^\infty r^2 \left(\mathrm{e}^{-(r-|\mathbf{x}|)^2} - \mathrm{e}^{-(r+|\mathbf{x}|)^2} \right) dr \\
&= \frac{\mathrm{e}^{-|\mathbf{x}|^2}}{8} + \frac{\sqrt{\pi}\, \mathrm{erf}(|\mathbf{x}|)}{16|\mathbf{x}|} (2|\mathbf{x}|^2 + 1)\,.
\end{aligned} \tag{6.8}$$

Thus, we conclude from (4.19) that

$$\begin{aligned}
\mathcal{H}_3 \eta_4(\mathbf{x}) &= \pi^{-3/2} \Big(\mathcal{H}_3 (\, \mathrm{e}^{-|\cdot|^2}\,)(\mathbf{x}) - \frac{1}{4} \mathcal{L}_3 (\, \mathrm{e}^{-|\cdot|^2}\,)(\mathbf{x}) \Big) \\
&= \frac{1}{8\pi^{3/2}} \big(\mathrm{e}^{-|\mathbf{x}|^2} + \sqrt{\pi} |\mathbf{x}|\, \mathrm{erf}(|\mathbf{x}|) \big).
\end{aligned} \tag{6.9}$$

Note that the expression for $\mathcal{H}_3 \eta_4$ is even simpler than that for $\mathcal{H}_3 \eta_2$. The expansion (6.8) also indicates that improving the approximation order pointwise by using the values of the density in a small neighborhood is only possible for η_{2M} with $M \geq 2$.

This was already mentioned in Remark 5.4 for the case of elastic and hydrodynamic potentials.

Next, we consider

$$\sum_{j=2}^{M-1} \frac{(-1)^j}{j!\,4^j} \Delta^{j-2} \mathrm{e}^{-|\mathbf{x}|^2} = \frac{1}{16} \sum_{j=0}^{M-3} \frac{(-1)^j}{(j+2)!\,4^j} \Delta^j \mathrm{e}^{-|\mathbf{x}|^2}$$
$$= \frac{\mathrm{e}^{-|\mathbf{x}|^2}}{16} \sum_{j=0}^{M-3} \frac{L_j^{(1/2)}(|\mathbf{x}|^2)}{(j+1)(j+2)},$$

where the last equality is ensured by (4.24). This shows that, as in the case of harmonic potentials for $M \geq 3$, the biharmonic potential of η_{2M+2} can be obtained from that of η_{2M} by adding some rapidly decaying term. Therefore, to improve the approximation accuracy to the biharmonic potential at a given point, only the values of the density in a small neighborhood are required.

The sum (6.9) can be simplified further owing to

$$\sum_{j=0}^{M-3} \frac{L_j^{(1/2)}(|\mathbf{x}|^2)}{(j+1)(j+2)} = \sum_{j=0}^{M-3} \Big(\frac{1}{j+1} - \frac{1}{j+2}\Big) L_j^{(1/2)}(|\mathbf{x}|^2)$$
$$= L_0^{(1/2)}(|\mathbf{x}|^2) + \sum_{j=1}^{M-3} \frac{1}{j+1} \big(L_j^{(1/2)}(|\mathbf{x}|^2) - L_{j-1}^{(1/2)}(|\mathbf{x}|^2)\big) - \frac{1}{M-1} L_{M-3}^{(1/2)}(|\mathbf{x}|^2)$$
$$= \sum_{j=0}^{M-3} \frac{L_j^{(-1/2)}(|\mathbf{x}|^2)}{j+1} - \frac{L_{M-3}^{(1/2)}(|\mathbf{x}|^2)}{M-1},$$

where we use (4.28) and the fact that $L_0^{(\gamma)} = 1$. Hence, we obtain

$$\mathcal{H}_3\eta_{2M}(\mathbf{x}) = \frac{\mathrm{e}^{-|\mathbf{x}|^2} + \sqrt{\pi}|\mathbf{x}|\operatorname{erf}(|\mathbf{x}|)}{8\pi^{3/2}} + \frac{L_{M-3}^{(1/2)}(|\mathbf{x}|^2)\,\mathrm{e}^{-|\mathbf{x}|^2}}{16\pi^{3/2}(M-1)} - \sum_{j=0}^{M-3} \frac{L_j^{(-1/2)}(|\mathbf{x}|^2)\,\mathrm{e}^{-|\mathbf{x}|^2}}{16\pi^{3/2}(j+1)}. \tag{6.10}$$

6.2. Approximate solution of non-stationary problems

The method of approximate approximations can be employed to solve initial value problems for classical partial differential equations efficiently. As examples, we consider the heat, wave, and plate equations.

6.2.1. The Cauchy problem for the heat equation. We introduce the n-dimensional analogue of the example given in Subsection 1.2.2

$$\frac{\partial u}{\partial t} - a\Delta_{\mathbf{x}} u = 0\,, \quad t > 0\,, \quad u(\mathbf{x},0) = \varphi(\mathbf{x}), \quad \mathbf{x} \in \mathbb{R}^n\,. \tag{6.11}$$

This Cauchy problem can be solved by the Poisson integral

$$u(\mathbf{x},t) = \mathcal{P}_t\varphi(\mathbf{x}) = \frac{1}{(4\pi a t)^{n/2}} \int_{\mathbb{R}^n} \mathrm{e}^{-|\mathbf{x}-\mathbf{y}|^2/(4at)}\, \varphi(\mathbf{y})\, d\mathbf{y}\,; \tag{6.12}$$

cf., e.g., [**75**, Kapitel 25]. Note that

$$(\mathcal{P}_t \, \mathrm{e}^{-|\cdot|^2/\mathcal{D}})(\mathbf{x}) = \Big(\frac{\mathcal{D}}{\mathcal{D}+4at}\Big)^{n/2} \mathrm{e}^{-|\mathbf{x}|^2/(\mathcal{D}+4at)} \, . \tag{6.13}$$

If the initial value $\varphi(\mathbf{x})$ is approximated by the quasi-interpolant (3.20), i.e.,

$$\mathcal{M}^{(2M)}_{\mathcal{D},h}\varphi(\mathbf{x}) = \mathcal{D}^{-n/2} \sum_{\mathbf{m}\in\mathbb{Z}^n} \varphi(h\mathbf{m}) \, \eta_{2M}\Big(\frac{|\mathbf{x}-h\mathbf{m}|}{\sqrt{\mathcal{D}}h}\Big) ,$$

then the function

$$u_h(\mathbf{x},t) := \mathcal{P}_t(\mathcal{M}^{(2M)}_{\mathcal{D},h}\varphi)(\mathbf{x})$$

is a high-order approximate solution of the initial value problem for the heat equation (6.11). Since

$$\frac{1}{(4\pi at)^{n/2}} \int_{\mathbb{R}^n} \mathrm{e}^{-|\mathbf{x}|^2/(4at)} \, d\mathbf{x} = 1 \, , \; t > 0 \, ,$$

we obtain from (2.11) that

$$\|u(\cdot,t)-u_h(\cdot,t)\|_{L_p(\mathbb{R}^n)} = \|\mathcal{P}_t(\varphi-\mathcal{M}^{(2M)}_{\mathcal{D},h}\varphi)\|_{L_p(\mathbb{R}^n)} \le \|\varphi-\mathcal{M}^{(2M)}_{\mathcal{D},h}\varphi\|_{L_p(\mathbb{R}^n)} \tag{6.14}$$

for any $t > 0$ and $1 \le p \le \infty$. Using the relation

$$\eta_{2M}\Big(\frac{|\mathbf{x}|}{\sqrt{\mathcal{D}}h}\Big) = \pi^{-n/2} \sum_{j=0}^{M-1} \frac{(-1)^j}{j!} \Big(\frac{\mathcal{D}h^2}{4}\Big)^j \Delta^j \, \mathrm{e}^{-|\mathbf{x}|^2/\mathcal{D}h^2} \, ,$$

we derive the analytic expression

$$\begin{aligned}
&\mathcal{P}_t\eta_{2M}\Big(\frac{|\cdot|}{\sqrt{\mathcal{D}}h}\Big)(\mathbf{x}) \\
&\quad = \frac{1}{\pi^n(4at)^{n/2}} \sum_{j=0}^{M-1} \frac{(-1)^j}{j!} \Big(\frac{\mathcal{D}h^2}{4}\Big)^j \Delta^j \int_{\mathbb{R}^n} \mathrm{e}^{-|\mathbf{x}-\mathbf{y}|^2/(4at)} \, \mathrm{e}^{-|\mathbf{y}|^2/(\mathcal{D}h^2)} \, d\mathbf{y} \\
&\quad = \frac{(\mathcal{D}h^2)^{n/2}}{\pi^{n/2}(\mathcal{D}h^2+4at)^{n/2}} \sum_{j=0}^{M-1} \frac{(-1)^j}{j!} \Big(\frac{\mathcal{D}h^2}{4}\Big)^j \Delta^j \, \mathrm{e}^{-|\mathbf{x}|^2/(\mathcal{D}h^2+4at)} \, .
\end{aligned}$$

We see from (3.15) that

$$\Delta^j \, \mathrm{e}^{-|\mathbf{x}|^2/(\mathcal{D}h^2+4at)} = \frac{(-1)^j \, j! \, 4^j}{(\mathcal{D}h^2+4at)^j} \, \mathrm{e}^{-|\mathbf{x}|^2/(\mathcal{D}h^2+4at)} \, L_j^{(n/2-1)}\Big(\frac{|\mathbf{x}|^2}{\mathcal{D}h^2+4at}\Big) ;$$

hence

$$\begin{aligned}
&\mathcal{P}_t \, \eta_{2M}\Big(\frac{|\cdot|}{\sqrt{\mathcal{D}}h}\Big)(\mathbf{x}) \\
&\quad = \frac{(\mathcal{D}h^2)^{n/2} \, \mathrm{e}^{-|\mathbf{x}|^2/(\mathcal{D}h^2+4at)}}{\pi^{n/2}(\mathcal{D}h^2+4at)^{n/2}} \sum_{j=0}^{M-1} \Big(\frac{\mathcal{D}h^2}{\mathcal{D}h^2+4at}\Big)^j L_j^{(n/2-1)}\Big(\frac{|\mathbf{x}|^2}{\mathcal{D}h^2+4at}\Big).
\end{aligned} \tag{6.15}$$

Thus, the approximate solution of the initial value problem for the heat equation (6.11) can be given by the sum

$$
(6.16)\qquad \boxed{\begin{aligned} u_h(\mathbf{x},t) &= \frac{h^n}{\pi^{n/2}(\mathcal{D}h^2+4at)^{n/2}} \\ &\quad \times \sum_{\mathbf{m}\in\mathbb{Z}^n} \varphi(h\mathbf{m})\, \mathrm{e}^{-z_{\mathbf{m}}(t)} \sum_{j=0}^{M-1} \Big(\frac{\mathcal{D}h^2}{\mathcal{D}h^2+4at}\Big)^j L_j^{(n/2-1)}(z_{\mathbf{m}}(t))\,, \end{aligned}}
$$

where the notation

$$
(6.17)\qquad z_{\mathbf{m}}(t) = \frac{|\mathbf{x}-h\mathbf{m}|^2}{\mathcal{D}h^2+4at}
$$

is used.

Let us note that the inequality (6.14) gives only a rough error estimate for the approximate solution (6.16) of the heat equation. It states that if $\varphi \in W_p^{2M}(\mathbb{R}^n)$, then for any fixed $t>0$ the formula (6.16) approximates the solution $u(\mathbf{x},t)$ of (6.11) with the order $\mathcal{O}(h^{2M})$ up to a saturation error, which does not converge to zero. However, since the Poisson integral is a smoothing integral operator, one can expect that the saturation error also tends to zero as $h\to 0$.

This can be easily seen from the representation (2.69) which implies

$$
\begin{aligned}
&\mathcal{P}_t(\mathcal{M}_{h,\mathcal{D}}^{(2M)}\varphi)(\mathbf{x}) - \mathcal{P}_t\varphi(\mathbf{x}) \\
&= (4\pi a t)^{-n/2} \sum_{[\boldsymbol{\alpha}]=0}^{2M-1} \frac{(-\sqrt{\mathcal{D}}h)^{[\boldsymbol{\alpha}]}}{\boldsymbol{\alpha}!} \int_{\mathbb{R}^n} \mathrm{e}^{-|\mathbf{x}-\mathbf{y}|^2/4at}\, \partial^{\boldsymbol{\alpha}}\varphi(\mathbf{y})\, \varepsilon_{\boldsymbol{\alpha}}\Big(\frac{\mathbf{y}}{h},\eta_{2M},\mathcal{D}\Big)\, d\mathbf{y} \\
&\quad + \mathcal{P}_t(R_{2M,h})(\mathbf{x}),
\end{aligned}
$$

where the last term $\mathcal{P}_t(R_{2M,h})$ is of order $\mathcal{O}(h^{2M})$. The first terms, which constitute the saturation error, can be written by (2.46) in the form

$$
\begin{aligned}
&\frac{1}{(4\pi a t)^{n/2}} \int_{\mathbb{R}^n} \mathrm{e}^{-|\mathbf{x}-\mathbf{y}|^2/4at}\, \partial^{\boldsymbol{\alpha}}\varphi(\mathbf{y})\, \varepsilon_{\boldsymbol{\alpha}}\Big(\frac{\mathbf{y}}{h},\eta_{2M},\mathcal{D}\Big)\, d\mathbf{y} \\
&= \Big(\frac{i}{2\pi}\Big)^{[\boldsymbol{\alpha}]} \sum_{\boldsymbol{\nu}\in\mathbb{Z}^n\setminus\{\mathbf{0}\}} \frac{\partial^{\boldsymbol{\alpha}}\mathcal{F}\eta_{2M}(\sqrt{\mathcal{D}}\boldsymbol{\nu})}{(4\pi a t)^{n/2}} \int_{\mathbb{R}^n} \mathrm{e}^{-|\mathbf{x}-\mathbf{y}|^2/4at}\, \mathrm{e}^{\frac{2\pi i}{h}\langle \mathbf{y},\boldsymbol{\nu}\rangle}\, \partial^{\boldsymbol{\alpha}}\varphi(\mathbf{y})\, d\mathbf{y} \\
&= \Big(\frac{i}{2\pi}\Big)^{[\boldsymbol{\alpha}]} \sum_{\boldsymbol{\nu}\in\mathbb{Z}^n\setminus\{\mathbf{0}\}} \partial^{\boldsymbol{\alpha}}\mathcal{F}\eta_{2M}(\sqrt{\mathcal{D}}\boldsymbol{\nu})\, \mathrm{e}^{\frac{2\pi i}{h}\langle \mathbf{x},\boldsymbol{\nu}\rangle} \\
&\qquad \times \frac{1}{\pi^{n/2}} \int_{\mathbb{R}^n} \mathrm{e}^{-|\mathbf{y}|^2}\, \mathrm{e}^{\frac{2\pi i\sqrt{4at}}{h}\langle \mathbf{y},\boldsymbol{\nu}\rangle}\, \partial^{\boldsymbol{\alpha}}\varphi(\mathbf{x}-\sqrt{4at}\,\mathbf{y})\, d\mathbf{y}\,.
\end{aligned}
$$

Now we note that for any $\mathbf{x}\in\mathbb{R}^n$ the function

$$
f_{\boldsymbol{\alpha}}(\mathbf{y}) := \frac{\mathrm{e}^{-|\mathbf{y}|^2}}{\pi^{n/2}}\, \partial^{\boldsymbol{\alpha}}\varphi(\mathbf{x}-\sqrt{4at}\,\mathbf{y})
$$

decays rapidly as $|\mathbf{y}|\to\infty$ and its derivatives up to the order $2M-[\boldsymbol{\alpha}]$ belong to $L_1(\mathbb{R}^n)$. Moreover, the L_1-norm of these derivatives are bounded uniformly in $\mathbf{x}$. Hence the Riemann-Lebesgue Theorem gives

$$
\Big|\int_{\mathbb{R}^n} \mathrm{e}^{-|\mathbf{y}|^2}\, \mathrm{e}^{2\pi i\langle \mathbf{y},\boldsymbol{\lambda}\rangle}\, \partial^{\boldsymbol{\alpha}}\varphi(\mathbf{x}-\sqrt{4at}\,\mathbf{y})\, d\mathbf{y}\Big| = |\mathcal{F}f_{\boldsymbol{\alpha}}(\boldsymbol{\lambda})| \le c_{\boldsymbol{\alpha}}|\boldsymbol{\lambda}|^{[\boldsymbol{\alpha}]-2M}
$$

with a constant $c_{\boldsymbol{\alpha}}$ depending on φ and t. Thus, the terms of the saturation error can be estimated as follows:

$$\begin{aligned}
&\Big|\frac{1}{(4\pi at)^{n/2}}\int_{\mathbb{R}^n} \mathrm{e}^{-|\mathbf{x}-\mathbf{y}|^2/4at}\,\partial^{\boldsymbol{\alpha}}\varphi(\mathbf{y})\,\varepsilon_{\boldsymbol{\alpha}}\Big(\frac{\mathbf{y}}{h},\eta_{2M},\mathcal{D}\Big)\,d\mathbf{y}\Big| \\
&= (2\pi)^{-[\boldsymbol{\alpha}]}\Big|\sum_{\boldsymbol{\nu}\in\mathbb{Z}^n\setminus\{\mathbf{0}\}} \partial^{\boldsymbol{\alpha}}\mathcal{F}\eta_{2M}(\sqrt{\mathcal{D}}\boldsymbol{\nu})\,\mathrm{e}^{\frac{2\pi i}{h}\langle\mathbf{x},\boldsymbol{\nu}\rangle}\,\mathcal{F}f_{\boldsymbol{\alpha}}\Big(\frac{2\pi i\sqrt{4at}\boldsymbol{\nu}}{h}\Big)\Big| \\
&\le \frac{c_{\boldsymbol{\alpha}}h^{2M-[\boldsymbol{\alpha}]}}{(2\pi)^{2M}(4at)^{M-[\boldsymbol{\alpha}]/2}}\sum_{\boldsymbol{\nu}\in\mathbb{Z}^n\setminus\{\mathbf{0}\}} \big|\partial^{\boldsymbol{\alpha}}\mathcal{F}\eta_{2M}(\sqrt{\mathcal{D}}\boldsymbol{\nu})\big|\,|\boldsymbol{\nu}|^{[\boldsymbol{\alpha}]-2M}.
\end{aligned}$$

Hence, we have proved

THEOREM 6.1. *If the initial values of the parabolic problem (6.11) satisfy $\varphi \in W_p^{2M}(\mathbb{R}^n)$, then the approximate solution (6.16) converges for any fixed $t > 0$ with the order $\mathcal{O}(h^{2M})$ to the solution of the problem.*

In Table 6.1 we provide the error estimates and approximation rates for the approximate solution of the heat equation (1.15) using formula (6.16) with $M = 3$.

h	$\mathcal{D}=1$	rate	$\mathcal{D}=2$	rate	$\mathcal{D}=4$	rate
0.8	$8.69\cdot10^{-8}$		$1.28\cdot10^{-6}$		$1.00\cdot10^{-5}$	
0.4	$2.87\cdot10^{-9}$	4.92	$2.27\cdot10^{-8}$	5.81	$1.78\cdot10^{-7}$	5.82
0.2	$4.52\cdot10^{-11}$	5.99	$3.61\cdot10^{-10}$	5.98	$2.87\cdot10^{-9}$	5.96
0.1	$7.08\cdot10^{-13}$	6.00	$5.66\cdot10^{-12}$	5.99	$4.52\cdot10^{-11}$	5.99
0.05	$1.14\cdot10^{-14}$	5.96	$8.90\cdot10^{-14}$	5.99	$7.08\cdot10^{-13}$	6.00

TABLE 6.1. Numerical error for the initial value problem (1.15) with $\phi(x) = \mathrm{e}^{-x^2}$ and $t = 10$ using the approximate solution (6.16) with $M = 3$

6.2.2. The Cauchy problem for the wave equation. The next example concerns the initial value problem for the wave equation

$$\begin{aligned}
&u_{tt}(\mathbf{x},t) - \Delta_{\mathbf{x}}u(\mathbf{x},t) = 0\,, \quad t>0\,, \quad \mathbf{x}\in\mathbb{R}^n\,, \\
&u(\mathbf{x},0) = g_1(\mathbf{x})\,, \quad u_t(\mathbf{x},0) = g_2(\mathbf{x})\,.
\end{aligned} \tag{6.18}$$

If u satisfies (6.18), then the Fourier transform $\hat{u}(\boldsymbol{\lambda},t) = \mathcal{F}_{\mathbf{x}\to\boldsymbol{\lambda}}u(\cdot,t)$ is a solution of the Cauchy problem

$$\begin{aligned}
&\hat{u}_{tt}(\boldsymbol{\lambda},t) + 4\pi^2|\boldsymbol{\lambda}|^2\hat{u}(\boldsymbol{\lambda},t) = 0\,, \quad t>0\,, \quad \boldsymbol{\lambda}\in\mathbb{R}^n\,, \\
&\hat{u}(\boldsymbol{\lambda},0) = \hat{g}_1(\boldsymbol{\lambda})\,, \quad \hat{u}_t(\boldsymbol{\lambda},0) = \hat{g}_2(\boldsymbol{\lambda})\,.
\end{aligned}$$

Hence

$$\hat{u}(\boldsymbol{\lambda},t) = \hat{g}_1(\boldsymbol{\lambda})\cos 2\pi|\boldsymbol{\lambda}|t + \hat{g}_2(\boldsymbol{\lambda})\frac{\sin 2\pi|\boldsymbol{\lambda}|t}{2\pi|\boldsymbol{\lambda}|}\,. \tag{6.19}$$

To find an approximate solution of (6.18), the initial values g_1 and g_2 are replaced by high-order quasi-interpolants. To determine the solution of the wave equation with generating functions for those approximants, we apply the results of Subsection 3.4.2. In Theorem 3.10, we have shown that the function

$$\eta_M(\mathbf{x}) = \frac{(-1)^{M-1}}{(M-1)!}\Big(\frac{\partial}{\partial\tau}\Big)^{M-1}\Big(\tau^{-1-n/2}\eta\Big(\frac{\mathbf{x}}{\sqrt{\tau}}\Big)\Big)\Big|_{\tau=1}$$

generates an approximate quasi-interpolant of order $\mathcal{O}(h^{2M})$ for a suitable η.

Again, we start with the Gaussian function. Let $g_j(\mathbf{x}) = \tau^{-1-n/2}\,\mathrm{e}^{-|\mathbf{x}|^2/\tau}$. Using formula (2.12), the inverse Fourier transform $\mathcal{F}^{-1}_{\boldsymbol{\lambda}\to\mathbf{x}}$ of

$$\tau^{-1}\,\mathrm{e}^{-\pi^2\tau|\boldsymbol{\lambda}|^2}\cos 2\pi|\boldsymbol{\lambda}|t \quad\text{and}\quad \tau^{-1}\,\mathrm{e}^{-\pi^2\tau|\boldsymbol{\lambda}|^2}\frac{\sin 2\pi|\boldsymbol{\lambda}|t}{2\pi|\boldsymbol{\lambda}|}$$

can be determined from the integrals

$$\frac{2\pi}{\tau|\mathbf{x}|^{\,n/2-1}}\int\limits_0^\infty J_{n/2-1}(2\pi r|\mathbf{x}|)\,\mathrm{e}^{-\pi^2\tau r^2}\cos(2\pi r t)\,r^{\,n/2}dr\,,$$

$$\frac{1}{\tau|\mathbf{x}|^{\,n/2-1}}\int\limits_0^\infty J_{n/2-1}(2\pi r|\mathbf{x}|)\,\mathrm{e}^{-\pi^2\tau r^2}\sin(2\pi r t)\,r^{\,n/2-1}dr\,,$$

respectively. In view of (2.16), at least for odd n, these integrals can be taken analytically. In particular, if $n = 3$, then

$$J_{1/2}(z) = \sqrt{\frac{2}{\pi z}}\,\sin z$$

and [**7**, 1.4.11, 2.4.19] imply that

$$\begin{aligned}
&\frac{2\pi}{\tau|\mathbf{x}|^{\,1/2}}\int\limits_0^\infty \sqrt{\frac{2}{2\pi^2 r|\mathbf{x}|}}\,\sin(2\pi r|\mathbf{x}|)\,\mathrm{e}^{-\pi^2\tau r^2}\cos(2\pi r t)\,r^{3/2}dr\\
&\quad= \frac{1}{\tau|\mathbf{x}|}\int\limits_0^\infty r\,\mathrm{e}^{-\pi^2\tau r^2}\big(\sin 2\pi r(t+|\mathbf{x}|) - \sin 2\pi r(t-|\mathbf{x}|)\big)\,dr\\
&\quad= \frac{1}{2\pi^{3/2}\tau^{5/2}|\mathbf{x}|}\Big((t+|\mathbf{x}|)\,\mathrm{e}^{-(t+|\mathbf{x}|)^2/\tau} - (t-|\mathbf{x}|)\,\mathrm{e}^{-(t-|\mathbf{x}|)^2/\tau}\Big)
\end{aligned}$$

as well as

$$\begin{aligned}
&\frac{1}{\tau|\mathbf{x}|^{\,1/2}}\int\limits_0^\infty \sqrt{\frac{2}{2\pi^2 r|\mathbf{x}|}}\,\sin(2\pi r|\mathbf{x}|)\,\mathrm{e}^{-\pi^2\tau r^2}\sin(2\pi r t)\,r^{1/2}dr\\
&\quad= \frac{1}{2\pi\tau|\mathbf{x}|}\int\limits_0^\infty \mathrm{e}^{-\pi^2\tau r^2}\big(\cos 2\pi r(t-|\mathbf{x}|) - \cos 2\pi r(t+|\mathbf{x}|)\big)\,dr\\
&\quad= \frac{1}{4(\pi\tau)^{3/2}|\mathbf{x}|}\Big(\mathrm{e}^{-(t-|\mathbf{x}|)^2/\tau} - \mathrm{e}^{-(t+|\mathbf{x}|)^2/\tau}\Big)\,.
\end{aligned}$$

Thus, for $n = 3$ we obtain

$$\mathcal{F}^{-1}_{\boldsymbol{\lambda}\to\mathbf{x}}(\mathcal{F}\eta_M\,\cos(2\pi|\cdot|t)) = \frac{1}{2\pi^{3/2}|\mathbf{x}|}\Big(\mathcal{A}_M(t+|\mathbf{x}|) - \mathcal{A}_M(t-|\mathbf{x}|)\Big)$$

with

$$\mathcal{A}_M(\xi) = \frac{(-1)^{M-1}}{(M-1)!} \Big(\frac{\partial}{\partial \tau}\Big)^{M-1} \Big(\frac{\xi \, \mathrm{e}^{-\xi^2/\tau}}{\tau^{5/2}}\Big)\Big|_{\tau=1} \tag{6.20}$$

and

$$\mathcal{F}^{-1}_{\boldsymbol{\lambda}\to\mathbf{x}}\Big(\frac{\mathcal{F}\eta_M \sin(2\pi|\cdot|t)}{2\pi|\cdot|}\Big) = \frac{1}{4\pi^{3/2}|\mathbf{x}|}\Big(\mathcal{B}_M(t-|\mathbf{x}|) - \mathcal{B}_M(t+|\mathbf{x}|)\Big)$$

with

$$\mathcal{B}_M(\xi) = \frac{(-1)^{M-1}}{(M-1)!} \Big(\frac{\partial}{\partial \tau}\Big)^{M-1} \Big(\frac{\mathrm{e}^{-\xi^2/\tau}}{\tau^{3/2}}\Big)\Big|_{\tau=1} . \tag{6.21}$$

In particular,

$$\mathcal{A}_2(\xi) = \frac{\mathrm{e}^{-\xi^2}}{2}(\xi^3 - 5\xi)\,, \qquad \mathcal{A}_3(\xi) = \frac{\mathrm{e}^{-\xi^2}}{8}(28\xi^3 - 4\xi^5 - 35\xi)\,,$$

$$\mathcal{B}_2(\xi) = \frac{\mathrm{e}^{-\xi^2}}{2}(\xi^2 - 3)\,, \qquad \mathcal{B}_3(\xi) = \frac{\mathrm{e}^{-\xi^2}}{8}(20\xi^2 - 4\xi^4 - 15)\,.$$

Thus, an approximate solution to the wave equation (6.18) in $\mathbb{R}^3$ with the order $\mathcal{O}((\sqrt{\mathcal{D}}h)^{2M})$ modulo a saturation error can be obtained from the summation formula

$$u_h(\mathbf{x},t) = \frac{1}{(\pi\mathcal{D})^{3/2}} \sum_{\mathbf{m}\in\mathbb{Z}^3} \Big(\frac{\sqrt{\mathcal{D}}h}{d_{\mathbf{m}}}\Big(\mathcal{A}_M\Big(\frac{t+d_{\mathbf{m}}}{\sqrt{\mathcal{D}}h}\Big) - \mathcal{A}_M\Big(\frac{t-d_{\mathbf{m}}}{\sqrt{\mathcal{D}}h}\Big)\Big) g_1(h\mathbf{m}) - \frac{\mathcal{D}h^2}{d_{\mathbf{m}}}\Big(\mathcal{B}_M\Big(\frac{t+d_{\mathbf{m}}}{\sqrt{\mathcal{D}}h}\Big) - \mathcal{B}_M\Big(\frac{t-d_{\mathbf{m}}}{\sqrt{\mathcal{D}}h}\Big)\Big) g_2(h\mathbf{m})\Big),$$

where $d_{\mathbf{m}} = |\mathbf{x} - h\mathbf{m}|$ and the functions $\mathcal{A}_M$ and $\mathcal{B}_M$ are defined by (6.20) and (6.21).

6.2.3. Vibrations of a plate. Similarly, one can study the initial value problem for free vibrations of a thin elastic plate which is modeled by

$$\begin{aligned} &u_{tt}(\mathbf{x},t) + \Delta^2_{\mathbf{x}} u(\mathbf{x},t) = 0\,, \quad t>0\,, \quad \mathbf{x}\in\mathbb{R}^2\,, \\ &u(\mathbf{x},0) = g_1(\mathbf{x})\,, \quad u_t(\mathbf{x},0) = g_2(\mathbf{x})\,. \end{aligned} \tag{6.22}$$

The two-dimensional Fourier transform $\hat{u}(\boldsymbol{\lambda},t)$ of the solution to (6.22) satisfies

$$\begin{aligned} &\hat{u}_{tt}(\boldsymbol{\lambda},t) + 16\pi^4|\boldsymbol{\lambda}|^4 \hat{u}(\boldsymbol{\lambda},t) = 0\,, \quad t>0\,, \quad \boldsymbol{\lambda}\in\mathbb{R}^2\,, \\ &\hat{u}(\boldsymbol{\lambda},0) = \hat{g}_1(\boldsymbol{\lambda})\,, \quad \hat{u}_t(\boldsymbol{\lambda},0) = \hat{g}_2(\boldsymbol{\lambda})\,. \end{aligned}$$

It is therefore of the form

$$\hat{u}(\boldsymbol{\lambda},t) = \hat{g}_1(\boldsymbol{\lambda}) \cos 4\pi^2|\boldsymbol{\lambda}|^2 t + \hat{g}_2(\boldsymbol{\lambda}) \frac{\sin 4\pi^2|\boldsymbol{\lambda}|^2 t}{4\pi^2|\boldsymbol{\lambda}|^2}\,.$$

As in the previous example, it remains to determine the inverse Fourier transform of

$$\tau^{-1}\,\mathrm{e}^{-\pi^2\tau|\boldsymbol{\lambda}|^2} \cos 4\pi^2|\boldsymbol{\lambda}|^2 t \quad \text{and} \quad \frac{\mathrm{e}^{-\pi^2\tau|\boldsymbol{\lambda}|^2} \sin 4\pi^2|\boldsymbol{\lambda}|^2 t}{4\pi^2\tau|\boldsymbol{\lambda}|^2}\,, \quad \boldsymbol{\lambda}\in\mathbb{R}^2\,.$$

Obviously,

$$f_1(\mathbf{x},t,\tau) := \tau^{-1}\mathcal{F}^{-1}_{\boldsymbol{\lambda}\to\mathbf{x}}\Big(\mathrm{e}^{-\pi^2\tau|\cdot|^2}\cos 4\pi^2|\cdot|^2 t\Big) = \mathrm{Re}\,\frac{\mathrm{e}^{-|\mathbf{x}|^2/(\tau-4it)}}{\pi\tau(\tau-4it)}, \tag{6.23}$$

whereas the smooth function

$$f_2(\mathbf{x},t,\tau) := \mathcal{F}^{-1}_{\boldsymbol{\lambda}\to\mathbf{x}}\Big(\frac{\mathrm{e}^{-\pi^2\tau|\boldsymbol{\lambda}|^2}\sin 4\pi^2|\boldsymbol{\lambda}|^2 t}{4\pi^2\tau|\boldsymbol{\lambda}|^2}\Big)$$

is the harmonic potential of

$$\tau^{-1}\mathcal{F}^{-1}_{\boldsymbol{\lambda}\to\mathbf{x}}\Big(\mathrm{e}^{-\pi^2\tau|\boldsymbol{\lambda}|^2}\sin 4\pi^2|\boldsymbol{\lambda}|^2 t\Big)$$

and therefore satisfies the Poisson equation

$$\Delta f_2(\mathbf{x},t,\tau) = -\,\mathrm{Im}\,\frac{\mathrm{e}^{-|\mathbf{x}|^2/(\tau-4it)}}{\pi\tau(\tau-4it)} \quad \text{with} \quad f_2(\mathbf{x},t,\tau)\to 0 \quad \text{as} \quad |\mathbf{x}|\to\infty.$$

Since by (4.21),

$$\Delta\Big(E_1\Big(\frac{|\mathbf{x}|^2}{\tau\pm 4it}\Big) + \log\frac{|\mathbf{x}|^2}{\tau\pm 4it}\Big) = 4\,\frac{\mathrm{e}^{-|\mathbf{x}|^2/(\tau\pm 4it)}}{\tau\pm 4it}$$

and

$$\mathrm{Im}\,\Big(E_1\Big(\frac{|\mathbf{x}|^2}{\tau-4it}\Big) + \log\frac{|\mathbf{x}|^2}{\tau-4it}\Big) = \mathrm{Im}\,E_1\Big(\frac{|\mathbf{x}|^2}{\tau-4it}\Big) + \frac{1}{2}\arctan\frac{8\tau t}{\tau^2+16t^2},$$

it follows that

$$f_2(\mathbf{x},t,\tau) = -\frac{1}{4\pi\tau}\,\mathrm{Im}\,E_1\Big(\frac{|\mathbf{x}|^2}{\tau-4it}\Big). \tag{6.24}$$

We introduce the functions

$$\mathcal{A}_M(\xi,t) = \frac{(-1)^{M-1}}{(M-1)!}\Big(\frac{\partial}{\partial\tau}\Big)^{M-1} f_1(\xi,t,\tau)\Big|_{\tau=1} \tag{6.25}$$

and

$$\mathcal{B}_M(\xi,t) = \frac{(-1)^{M-1}}{(M-1)!}\Big(\frac{\partial}{\partial\tau}\Big)^{M-1} f_2(\xi,t,\tau)\Big|_{\tau=1} \tag{6.26}$$

which are the solutions of (6.22) with high-order generating functions as initial values.

Thus, an approximate solution to (6.22) of order $\mathcal{O}((\sqrt{\mathcal{D}}h)^{2M})$ is given by

$$\begin{aligned} u_h(\mathbf{x},t) = {} & \frac{1}{\pi\mathcal{D}}\sum_{\mathbf{m}\in\mathbb{Z}^2}\mathcal{A}_M\Big(\frac{|\mathbf{x}-h\mathbf{m}|}{\sqrt{\mathcal{D}}h},\frac{t}{\mathcal{D}h^2}\Big)g_1(h\mathbf{m}) \\ & + \frac{h^2}{\pi}\sum_{\mathbf{m}\in\mathbb{Z}^2}\mathcal{B}_M\Big(\frac{|\mathbf{x}-h\mathbf{m}|}{\sqrt{\mathcal{D}}h},\frac{t}{\mathcal{D}h^2}\Big)g_2(h\mathbf{m}). \end{aligned}$$

6.3. Potentials of anisotropic Gaussians

We introduce a method to determine potentials of anisotropic Gaussians defined by formula (3.24). This method is based on the solution of initial value problems for linear parabolic equations.

6.3.1. Second-order problems. First, we solve the differential equation

$$-\sum_{j,k=1}^{n} b_{jk}\partial_{x_j}\partial_{x_k} u(\mathbf{x}) = \mathrm{e}^{-\langle A^{-1}\mathbf{x},\mathbf{x}\rangle}, \quad \mathbf{x}\in\mathbb{R}^n,\ n\geq 3, \tag{6.27}$$

where the constant matrices A and $B := \|b_{jk}\|_{j,k=1}^{n}$ are supposed to be non-singular and complex symmetric satisfying $\operatorname{Re} A > 0$, $\operatorname{Re} B \geq 0$.

As mentioned in (3.23), the Fourier transform of the right-hand side of (6.27) has the form

$$\mathcal{F}\big(\mathrm{e}^{-\langle A^{-1}\cdot,\cdot\rangle}\big)(\boldsymbol{\lambda}) = \pi^{n/2}\sqrt{\det A}\ \mathrm{e}^{-\pi^2\langle A\boldsymbol{\lambda},\boldsymbol{\lambda}\rangle}.$$

The solution of (6.27) can be represented as a one-dimensional integral.

THEOREM 6.2. *Suppose that the complex* $n\times n$ *matrices* A *and* B *are non-singular and symmetric and that they satisfy* $\operatorname{Re} A > 0$, $\operatorname{Re} B \geq 0$. *Then for* $n\geq 3$, *the function*

$$u(\mathbf{x}) = \frac{1}{4}\int_0^\infty \frac{\mathrm{e}^{-\langle (A+tB)^{-1}\mathbf{x},\mathbf{x}\rangle}}{\sqrt{\det(A+tB)}}\,dt \tag{6.28}$$

is a bounded solution of the equation

$$-\langle B\nabla,\nabla\rangle u(\mathbf{x}) = \frac{\mathrm{e}^{-\langle A^{-1}\mathbf{x},\mathbf{x}\rangle}}{\sqrt{\det A}}, \quad \mathbf{x}\in\mathbb{R}^n. \tag{6.29}$$

PROOF. Consider the Cauchy problem for the parabolic equation in $\mathbb{R}^n$:

$$\frac{\partial v(\mathbf{x},t)}{\partial t} - \langle B\nabla_{\mathbf{x}},\nabla_{\mathbf{x}}\rangle\, v(\mathbf{x},t) = 0, \quad t>0, \quad v(\mathbf{x},0) = \frac{\mathrm{e}^{-\langle A^{-1}\mathbf{x},\mathbf{x}\rangle}}{\pi^{n/2}\sqrt{\det A}}. \tag{6.30}$$

Applying the Fourier transformation to (6.30), we conclude from (3.23) that $\hat{v}(\boldsymbol{\lambda},t) = \mathcal{F}_{\mathbf{x}\to\boldsymbol{\lambda}} v(\cdot,t)$ satisfies the differential equation

$$\hat{v}_t + 4\pi^2\langle B\boldsymbol{\lambda},\boldsymbol{\lambda}\rangle\hat{v} = 0, \quad \hat{v}(\boldsymbol{\lambda},0) = \mathrm{e}^{-\pi^2\langle A\boldsymbol{\lambda},\boldsymbol{\lambda}\rangle}, \quad \boldsymbol{\lambda}\in\mathbb{R}^n,$$

which gives $\hat{v}(\boldsymbol{\lambda},t) = \mathrm{e}^{-\pi^2\langle (A+4tB)\boldsymbol{\lambda},\boldsymbol{\lambda}\rangle}$. Since $\operatorname{Re}(A+4tB) > 0$, it follows from (3.23) that

$$v(\mathbf{x},t) = \frac{\mathrm{e}^{-\langle (A+4tB)^{-1}\mathbf{x},\mathbf{x}\rangle}}{\pi^{n/2}\sqrt{\det(A+4tB)}}. \tag{6.31}$$

Integrating (6.30) in t, we arrive at

$$\int_0^T v_t(\mathbf{x},t)\,dt = \int_0^T \langle B\nabla_{\mathbf{x}},\nabla_{\mathbf{x}}\rangle\, v(\mathbf{x},t)\,dt = \langle B\nabla_{\mathbf{x}},\nabla_{\mathbf{x}}\rangle \int_0^T v(\mathbf{x},t)\,dt = v(\mathbf{x},T) - v(\mathbf{x},0).$$

The asymptotics $|v(\mathbf{x},t)| = \mathcal{O}(t^{-n/2})$ as $t\to\infty$, which is uniform in $\mathbf{x}$, implies that for $n\geq 3$ and $T\to\infty$

$$-\langle B\nabla,\nabla\rangle \int_0^\infty \frac{\mathrm{e}^{-\langle (A+4tB)^{-1}\mathbf{x},\mathbf{x}\rangle}}{\pi^{n/2}\sqrt{\det(A+4tB)}}\,dt = \frac{\mathrm{e}^{-\langle A^{-1}\mathbf{x},\mathbf{x}\rangle}}{\pi^{n/2}\sqrt{\det A}},$$

which establishes the assertion. □

COROLLARY 6.3. *Let the matrices A and B be as in Theorem 6.2 and let $\boldsymbol{\alpha}$ be some multi-index. Then a bounded solution of the equation*

$$-\langle B\nabla, \nabla\rangle u(\mathbf{x}) = \partial^{\boldsymbol{\alpha}} \mathrm{e}^{-\langle A^{-1}\mathbf{x},\mathbf{x}\rangle}, \quad \mathbf{x} \in \mathbb{R}^n, \tag{6.32}$$

is given by the integral

$$u(\mathbf{x}) = \frac{1}{4}\int_0^\infty \partial_{\mathbf{x}}^{\boldsymbol{\alpha}} \frac{\mathrm{e}^{-\langle (A+tB)^{-1}\mathbf{x},\mathbf{x}\rangle}}{\sqrt{\det(I+tA^{-1}B)}}\, dt\,. \tag{6.33}$$

The assertions of Theorem 6.2 can be extended to other second-order elliptic equations.

THEOREM 6.4. *Under the assumptions of Theorem 6.2 on the matrices A and B a solution of the elliptic equation*

$$-\langle B\nabla, \nabla\rangle u(\mathbf{x}) + au(\mathbf{x}) = \frac{\mathrm{e}^{-\langle A^{-1}\mathbf{x},\mathbf{x}\rangle}}{\sqrt{\det A}}, \quad \mathbf{x} \in \mathbb{R}^n,\ n \geq 1, \tag{6.34}$$

with constant $a \in \mathbb{C}$, $\operatorname{Re} a > 0$, is given by the integral

$$u(\mathbf{x}) = \frac{1}{4}\int_0^\infty \frac{\mathrm{e}^{-\langle (A+tB)^{-1}\mathbf{x},\mathbf{x}\rangle}}{\sqrt{\det(A+tB)}}\, \mathrm{e}^{-at/4}\, dt\,. \tag{6.35}$$

If $\operatorname{Re} a = 0$, *then the solution formula* (6.35) *holds for $n \geq 3$.*

PROOF. Let v be a solution of (6.30). Then the function

$$w(\mathbf{x},t) := v(\mathbf{x},t)\, \mathrm{e}^{-at} = \frac{\mathrm{e}^{-\langle (A+4tB)^{-1}\mathbf{x},\mathbf{x}\rangle}\, \mathrm{e}^{-at}}{\pi^{n/2}\sqrt{\det(A+4tB)}}$$

obviously satisfies

$$\frac{\partial w}{\partial t} - \langle B\nabla_{\mathbf{x}}, \nabla_{\mathbf{x}}\rangle w + a\, w = 0, \quad t > 0, \quad w(\mathbf{x},0) = \frac{\mathrm{e}^{-\langle A^{-1}\mathbf{x},\mathbf{x}\rangle}}{\pi^{n/2}\sqrt{\det A}}, \quad \mathbf{x} \in \mathbb{R}^n, \tag{6.36}$$

and hence,

$$(\langle B\nabla_{\mathbf{x}}, \nabla_{\mathbf{x}}\rangle - a)\int_0^T w(\mathbf{x},t)\, dt = w(\mathbf{x},T) - w(\mathbf{x},0)\,.$$

Note that in the case $\operatorname{Re} a > 0$, the limits as $T \to \infty$ exist for any space dimension $n \geq 1$. □

6.3.2. Some special cases.

1. If $A = B = I$, then (6.28) is another way to determine the explicit expression (4.16) of the harmonic potential of the Gaussian

$$\mathcal{L}_n\big(\mathrm{e}^{-|\cdot|^2}\big)(\mathbf{x}) = \frac{1}{4}\int_0^\infty \frac{\mathrm{e}^{-|\mathbf{x}|^2/(1+t)}}{(1+t)^{n/2}}\, dt = \frac{1}{4|\mathbf{x}|^{n-2}}\int_0^{|\mathbf{x}|^2} t^{n/2-2}\, \mathrm{e}^{-t}\, dt\,.$$

2. The case $B = I$ corresponds to the harmonic potential of $\mathrm{e}^{-\langle A^{-1}\mathbf{x},\mathbf{x}\rangle}$, which can be obtained from the one-dimensional integral

$$\mathcal{L}_n\big(\mathrm{e}^{-\langle A^{-1}\cdot,\cdot\rangle}\big)(\mathbf{x}) = \frac{1}{4}\int_0^\infty \frac{\mathrm{e}^{-\langle (A+tI)^{-1}\mathbf{x},\mathbf{x}\rangle}}{\sqrt{\det(I+tA^{-1})}}\, dt\,. \tag{6.37}$$

In particular, in the case of a diagonal matrix $A = \mathrm{diag}(a_j^{-1})_{j=1}^n$, $a_j > 0$, we obtain the harmonic potential of the orthotropic Gaussians

$$\psi(\mathbf{x}) = \prod_{j=1}^{n} \mathrm{e}^{-a_j x_j^2}$$

as the one-dimensional integral

$$\mathcal{L}_n \psi(\mathbf{x}) = \frac{1}{4} \int_0^\infty \prod_{j=1}^{n} \frac{\mathrm{e}^{-a_j x_j^2/(1+a_j t)}}{(1+a_j t)^{1/2}}\, dt \,. \tag{6.38}$$

3. If $B = I$ and $a = 1$, then the equation (6.34) takes the form

$$-\Delta u + u = \frac{\mathrm{e}^{-\langle A^{-1}\mathbf{x},\mathbf{x}\rangle}}{\sqrt{\det A}}\,, \tag{6.39}$$

and by Theorem 6.4, the L_2-solution has the form

$$u(\mathbf{x}) = \frac{1}{4} \int_0^\infty \frac{\mathrm{e}^{-t/4}\, \mathrm{e}^{-\langle (A+tI)^{-1}\mathbf{x},\mathbf{x}\rangle}}{\sqrt{\det(A+tI)}}\, dt \,.$$

4. If $B = iI$ and $a = \varepsilon - ik^2$ with $\varepsilon > 0$, then (6.34) is the Helmholtz equation

$$\Delta u + (k^2 + i\varepsilon)u = \frac{i\,\mathrm{e}^{-\langle A^{-1}\mathbf{x},\mathbf{x}\rangle}}{\sqrt{\det A}}\,,$$

and by (6.35), the L_2-solution has the form

$$u_\varepsilon(\mathbf{x}) = \frac{1}{4} \int_0^\infty \frac{\mathrm{e}^{-(\varepsilon - ik^2)t/4}\, \mathrm{e}^{-\langle (A+itI)^{-1}\mathbf{x},\mathbf{x}\rangle}}{\sqrt{\det(A+itI)}}\, dt \,. \tag{6.40}$$

If $n \geq 3$, then

$$\lim_{\varepsilon \to 0} u_\varepsilon(\mathbf{x}) = \frac{1}{4} \int_0^\infty \frac{\mathrm{e}^{ik^2 t/4}\, \mathrm{e}^{-\langle (A+itI)^{-1}\mathbf{x},\mathbf{x}\rangle}}{\sqrt{\det(A+itI)}}\, dt := u(\mathbf{x})\,, \tag{6.41}$$

which satisfies, by the limiting absorption principle (see e.g. [**25**]), the equation

$$\Delta u + k^2 u = \frac{i\,\mathrm{e}^{-\langle A^{-1}\mathbf{x},\mathbf{x}\rangle}}{\sqrt{\det A}} \tag{6.42}$$

and Sommerfeld's radiation condition (5.4). Since the solution of

$$\Delta u + k^2 u = -\varphi(\mathbf{x})$$

is given by the diffraction potential

$$\mathcal{S}_n \varphi(\mathbf{x}) := \int_{\mathbb{R}^n} \mathcal{E}_k(\mathbf{x} - \mathbf{y})\, \varphi(\mathbf{y})\, d\mathbf{y}\,,$$

we obtain the one-dimensional integral representation of the diffraction potential of anisotropic Gaussians

$$\mathcal{S}_n\big(\mathrm{e}^{-\langle A^{-1}\cdot,\cdot\rangle}\big)(\mathbf{x}) = \frac{i}{4}\int_0^{\infty} \frac{\mathrm{e}^{ik^2t/4}\,\mathrm{e}^{-\langle (A+itI)^{-1}\mathbf{x},\mathbf{x}\rangle}}{\sqrt{\det(I+itA^{-1})}}\,dt$$
$$= \frac{1}{4}\int_0^{i\infty} \frac{\mathrm{e}^{k^2z/4}\,\mathrm{e}^{-\langle (A+zI)^{-1}\mathbf{x},\mathbf{x}\rangle}}{\sqrt{\det(I+zA^{-1})}}\,dz\,.$$

In particular, the diffraction potential of the orthotropic Gaussian can be obtained from the integral

$$(6.43)\qquad \int_{\mathbb{R}^n} \mathcal{E}_k(\mathbf{x}-\mathbf{y})\,\mathrm{e}^{-(y_1^2/a_1+\ldots+y_n^2/a_n)}\,d\mathbf{y} = \frac{1}{4}\int_0^{i\infty} \mathrm{e}^{k^2z/4}\prod_{j=1}^{n}\frac{\mathrm{e}^{-x_j^2/(a_j+az)}}{(1+z/a_j)^{1/2}}\,dz\,.$$

In the special case of the isotropic Gaussian, we get the formula

$$(6.44)\qquad \mathcal{S}_n\big(\mathrm{e}^{-|\cdot|^2}\big)(\mathbf{x}) = \frac{1}{4}\int_0^{i\infty}\frac{\mathrm{e}^{-|\mathbf{x}|^2/(1+t)}}{(1+t)^{n/2}}\,\mathrm{e}^{k^2t/4}\,dt\,.$$

Let $n=3$. Then

$$\int_0^{i\infty}\frac{\mathrm{e}^{-|\mathbf{x}|^2/(1+t)}}{(1+t)^{3/2}}\,\mathrm{e}^{k^2(1+t)/4}\,dt = 2\int_\Gamma \frac{\mathrm{e}^{-|\mathbf{x}|^2/\tau^2+k^2\tau^2/4}}{\tau^2}\,d\tau\,,$$

where $\Gamma=\{\tau=\sqrt{1+it}, t\in(0,\infty)\}$, and hence $|\arg\tau|<\pi/4$. Then the integral formula

$$\int \mathrm{e}^{-a^2/t^2+b^2t^2}\,\frac{dt}{t^2} = \frac{\sqrt{\pi}}{4a}\,\mathrm{e}^{-a^2/t^2+b^2t^2}\left(\mathrm{w}\left(bt+\frac{ia}{t}\right)-\mathrm{w}\left(bt-\frac{ia}{t}\right)\right)+C\,,$$

which follows from [**1**, 7.4.34], provides the analytic expression (5.19) of the three-dimensional diffraction potential of the Gaussian.

Furthermore, (6.44) allows one to obtain explicit expressions of diffraction potential of isotropic Gaussians for all odd space dimensions n. Since

$$\int_0^{i\infty}\frac{\mathrm{e}^{-a^2/(1+t)}}{(1+t)^{j+3/2}}\,\mathrm{e}^{k^2t/4}\,dt = (-1)^j\left(\frac{1}{2a}\frac{d}{da}\right)^j\int_0^{i\infty}\frac{\mathrm{e}^{-a^2/(1+t)}}{(1+t)^{3/2}}\,\mathrm{e}^{k^2t/4}\,dt\,,$$

we obtain

$$\mathcal{S}_{2j+3}\big(\mathrm{e}^{-|\cdot|^2}\big)(\mathbf{x}) = \frac{(-1)^j\sqrt{\pi}}{8}\left(\frac{1}{2a}\frac{d}{da}\right)^j\left(\frac{\mathrm{e}^{-a^2}}{a}\left(\mathrm{w}\left(\frac{k}{2}-ia\right)-\mathrm{w}\left(\frac{k}{2}+ia\right)\right)\right)\bigg|_{a=|\mathbf{x}|}\,,$$

which, for example, evaluates for $n=5$ to

$$\mathcal{S}_5\big(\mathrm{e}^{-|\cdot|^2}\big)(\mathbf{x})$$
$$= -\frac{\mathrm{e}^{-|\mathbf{x}|^2}}{4|\mathbf{x}|} + \frac{\sqrt{\pi}\,\mathrm{e}^{-|\mathbf{x}|^2}}{16|\mathbf{x}|^3}\Big((1-ik|\mathbf{x}|)\,\mathrm{w}\left(\frac{k}{2}-i|\mathbf{x}|\right)-(1+ik|\mathbf{x}|)\,\mathrm{w}\left(\frac{k}{2}+i|\mathbf{x}|\right)\Big).$$

6.3.3. Elastic and hydrodynamic potentials of anisotropic Gaussians. In this part, we obtain one-dimensional integral representations of the three-dimensional elastic and hydrodynamic potentials, if the density is an anisotropic Gaussian function.

We know from the considerations in Section 5.5 that it suffices to find the integrals

$$I_{kl}(\mathbf{x}) := \frac{\partial^2}{\partial x_k \partial x_l} \frac{1}{8\pi} \int\limits_{\mathbb{R}^3} |\mathbf{x}-\mathbf{y}| \, \mathrm{e}^{-\langle A^{-1}\mathbf{y},\mathbf{y}\rangle} \, d\mathbf{y}$$
$$= \frac{1}{8\pi} \int\limits_{\mathbb{R}^3} |\mathbf{x}-\mathbf{y}| \frac{\partial^2}{\partial y_k \partial y_l} \, \mathrm{e}^{-\langle A^{-1}\mathbf{y},\mathbf{y}\rangle} \, d\mathbf{y}\,.$$

Since $|\mathbf{x}|/8\pi$ is the fundamental solution of the biharmonic equation, the function I_{kl} is a solution of the bi-Laplace equation

$$\Delta^2 w(\mathbf{x}) = -\frac{\partial^2}{\partial x_k \partial x_l} \, \mathrm{e}^{-\langle A^{-1}\mathbf{x},\mathbf{x}\rangle}\,, \quad \mathbf{x} \in \mathbb{R}^3\,.$$

THEOREM 6.5. *Let $n \geq 3$ and let the $n \times n$ matrix A satisfy the assumptions of Theorem 6.2. The unique solution w_{jk} of the bi-Laplace equation*

$$-\Delta^2 w\,(\mathbf{x}) = \frac{\partial^2}{\partial x_k \partial x_l} \frac{\mathrm{e}^{-\langle A^{-1}\mathbf{x},\mathbf{x}\rangle}}{\sqrt{\det A}}\,, \quad \mathbf{x} \in \mathbb{R}^n\,, \tag{6.45}$$

satisfying $w\,(\mathbf{x}) \to 0$ as $|\mathbf{x}| \to \infty$, is given by the one-dimensional integral

$$w_{kl}(\mathbf{x}) = -\frac{1}{16} \int\limits_0^\infty t \frac{\partial^2}{\partial x_k \partial x_l} \frac{\mathrm{e}^{-\langle (A+tI)^{-1}\mathbf{x},\mathbf{x}\rangle}}{\sqrt{\det(A+tI)}} \, dt\,. \tag{6.46}$$

PROOF. Similarly to the proof of Theorem 6.2, we find the solution of (6.45) by solving the Cauchy problem

$$\begin{aligned} &v_{tt}(\mathbf{x},t) + \Delta_{\mathbf{x}}^2 v(\mathbf{x},t) = 0\,, \quad t>0\,, \quad \mathbf{x} \in \mathbb{R}^n\,, \\ &v(\mathbf{x},0) = \frac{\partial^2}{\partial x_k \partial x_l} \frac{\mathrm{e}^{-\langle A^{-1}\mathbf{x},\mathbf{x}\rangle}}{\pi^{n/2}\sqrt{\det A}}\,, \quad v_t(\mathbf{x},0) = 0\,. \end{aligned} \tag{6.47}$$

The Fourier transformed problem

$$\hat{v}_{tt}(\boldsymbol{\lambda},t) + 16\pi^4 |\boldsymbol{\lambda}|^4 \hat{v}(\boldsymbol{\lambda},t) = 0\,, \quad t>0\,, \quad \boldsymbol{\lambda} \in \mathbb{R}^n\,,$$
$$\hat{v}(\boldsymbol{\lambda},0) = -4\pi^2 \lambda_k \lambda_l \, \mathrm{e}^{-\pi^2\langle A\boldsymbol{\lambda},\boldsymbol{\lambda}\rangle}\,, \quad \hat{v}_t(\boldsymbol{\lambda},0) = 0\,,$$

has the solution

$$\hat{v}_{kl}(\boldsymbol{\lambda},t) = -4\pi^2 \lambda_k \lambda_l \, \mathrm{e}^{-\pi^2\langle A\boldsymbol{\lambda},\boldsymbol{\lambda}\rangle} \cos 4\pi^2 |\boldsymbol{\lambda}|^2 t$$
$$= -2\pi^2 \lambda_k \lambda_l \Big(\mathrm{e}^{-\pi^2\langle (A+4itI)\boldsymbol{\lambda},\boldsymbol{\lambda}\rangle} + \mathrm{e}^{-\pi^2\langle (A-4itI)\boldsymbol{\lambda},\boldsymbol{\lambda}\rangle} \Big)\,.$$

Hence, from (3.23) the solution of (6.47) is

$$v_{kl}(\mathbf{x},t) = \frac{\partial^2}{\partial x_k \partial x_l} \frac{1}{2\pi^{n/2}} \left(\frac{\mathrm{e}^{-\langle (A+4itI)^{-1}\mathbf{x},\mathbf{x}\rangle}}{\sqrt{\det(A+4itI)}} + \frac{\mathrm{e}^{-\langle (A-4itI)^{-1}\mathbf{x},\mathbf{x}\rangle}}{\sqrt{\det(A-4itI)}} \right).$$

Multiplying the differential equation in (6.47) by t and integrating, we arrive at

$$-\int_0^T t\,\Delta_{\mathbf{x}}^2 v_{kl}(\mathbf{x},t)\,dt = \int_0^T t\,\frac{\partial^2 v_{kl}(\mathbf{x},t)}{\partial^2 t}\,dt = T\,\frac{\partial v_{kl}(\mathbf{x},T)}{\partial t} - v_{kl}(\mathbf{x},T) + v_{kl}(\mathbf{x},0)\,.$$

Noting that for $t \to \infty$

$$v_{kl}(\mathbf{x},t) = \begin{cases} \mathcal{O}(t^{-n/2-1})\,, & j = k\,, \\ \mathcal{O}(t^{-n/2-2})\,, & j \neq k\,, \end{cases} \quad \text{and} \quad \frac{\partial^j v_{jk}(\mathbf{x},t)}{\partial t^j} = \mathcal{O}(t^{-j} v_{kl}(\mathbf{x},t))\,,\ j \in \mathbb{N}\,,$$

uniformly in $\mathbf{x}$, and letting $T \to \infty$, we obtain

$$\Delta^2 \int_0^\infty t\,v_{jk}(\mathbf{x},t)\,dt = -\frac{\partial^2}{\partial x_k \partial x_l}\frac{\mathrm{e}^{-\langle A^{-1}\mathbf{x},\mathbf{x}\rangle}}{\pi^{n/2}\sqrt{\det A}}\,.$$

Now, we note that

$$\begin{aligned} w_{kl}(\mathbf{x}) &= \int_0^\infty t\,v_{kl}(\mathbf{x},t)\,dt \\ &= \frac{1}{2}\int_0^\infty t\,\frac{\partial^2}{\partial x_k \partial x_l}\Big(\frac{\mathrm{e}^{-\langle (A+4itI)^{-1}\mathbf{x},\mathbf{x}\rangle}}{\sqrt{\det(A+4itI)}} + \frac{\mathrm{e}^{-\langle (A-4itI)^{-1}\mathbf{x},\mathbf{x}\rangle}}{\sqrt{\det(A-4itI)}}\Big)\,dt \\ &= -\frac{1}{32}\Big(\int_0^{i\infty} z\,\frac{\partial^2}{\partial x_k \partial x_l}\frac{\mathrm{e}^{-\langle (A+zI)^{-1}\mathbf{x},\mathbf{x}\rangle}}{\sqrt{\det(A+zI)}}\,dz + \int_0^{-i\infty} z\,\frac{\partial^2}{\partial x_k \partial x_l}\frac{\mathrm{e}^{-\langle (A+zI)^{-1}\mathbf{x},\mathbf{x}\rangle}}{\sqrt{\det(A+zI)}}\,dz\Big). \end{aligned}$$

The function

$$g(z) := z\,\frac{\partial^2}{\partial x_k \partial x_l}\frac{\mathrm{e}^{-\langle (A+zI)^{-1}\mathbf{x},\mathbf{x}\rangle}}{\sqrt{\det(A+zI)}}$$

is holomorphic in the half-plane $\{\operatorname{Re} z > -\lambda_{\min}\}$, where $\lambda_{\min} > 0$ is the minimal eigenvalue of the symmetric matrix $\operatorname{Re} A$. Furthermore, $|g(z)| \le cR^{-n/2}$ for $|z| = R \to \infty$; hence if $n \ge 3$, then

$$\int_{\substack{|z|=R \\ \operatorname{Re} z \ge 0}} g(z)\,dz \to 0 \quad \text{as} \quad R \to \infty\,.$$

Therefore, by Cauchy's integral theorem

$$\int_0^{\pm i\infty} g(z)\,dz = \int_0^\infty g(z)\,dz\,,$$

which implies

$$w_{kl}(\mathbf{x}) = -\frac{1}{16}\int_0^\infty z\,\frac{\partial^2}{\partial x_k \partial x_l}\frac{\mathrm{e}^{-\langle (A+zI)^{-1}\mathbf{x},\mathbf{x}\rangle}}{\sqrt{\det(A+zI)}}\,dz\,. \qquad \square$$

The elastic potential of the anisotropic Gaussian can be obtained, in accordance with (5.52), from

$$\begin{aligned}\int_{\mathbb{R}^3} \Gamma_{kl}(\mathbf{x}-\mathbf{y})\,\mathrm{e}^{-\langle A^{-1}\mathbf{y},\mathbf{y}\rangle}\,d\mathbf{y} &= \frac{\delta_{kl}}{\mu}\mathcal{L}_3\big(\mathrm{e}^{-\langle A^{-1}\cdot,\cdot\rangle}\big)(\mathbf{x}) - \frac{\lambda+\mu}{\mu(\lambda+2\mu)}I_{kl}(\mathbf{x})\\ &= \frac{1}{4\mu}\int_0^\infty \Big(\delta_{kl} + \frac{t(\lambda+\mu)}{4(\lambda+2\mu)}\frac{\partial^2}{\partial x_k \partial x_l}\Big)\frac{\mathrm{e}^{-\langle (A+tI)^{-1}\mathbf{x},\mathbf{x}\rangle}}{\sqrt{\det(I+tA^{-1})}}\,dt\,,\end{aligned} \tag{6.48}$$

whereas the hydrodynamic potentials are derived from

$$\begin{aligned}\int_{\mathbb{R}^3} \Psi_{kl}(\mathbf{x}-\mathbf{y})\,\mathrm{e}^{-\langle A^{-1}\mathbf{y},\mathbf{y}\rangle}\,d\mathbf{y} &= \frac{\delta_{kl}}{\nu}\mathcal{L}_3\big(\mathrm{e}^{-\langle A^{-1}\cdot,\cdot\rangle}\big)(\mathbf{x}) - \frac{1}{\nu}I_{kl}(\mathbf{x})\\ &= \frac{1}{4\nu}\int_0^\infty \Big(\delta_{kl} + \frac{t}{4}\frac{\partial^2}{\partial x_k \partial x_l}\Big)\frac{\mathrm{e}^{-\langle (A+tI)^{-1}\mathbf{x},\mathbf{x}\rangle}}{\sqrt{\det(I+tA^{-1})}}\,dt\,,\\ \frac{x_k}{4\pi}\int_{\mathbb{R}^3}\frac{\mathrm{e}^{-\langle A^{-1}\mathbf{y},\mathbf{y}\rangle}}{|\mathbf{x}-\mathbf{y}|^3}\,d\mathbf{y} &= -\frac{1}{4}\int_0^\infty \frac{\partial}{\partial x_k}\frac{\mathrm{e}^{-\langle (A+tI)^{-1}\mathbf{x},\mathbf{x}\rangle}}{\sqrt{\det(I+tA^{-1})}}\,dt\,.\end{aligned} \tag{6.49}$$

6.3.4. Potentials of orthotropic Gaussians. We specify the formulas (6.48) and (6.49) for the case of the orthotropic Gaussian

$$\psi(\mathbf{x}) = \prod_{j=1}^{3}\mathrm{e}^{-a_j x_j^2}\,.$$

Since $A = \mathrm{diag}(a_j^{-1})_{j=1}^n$, it holds that

$$\frac{\partial^2}{\partial x_k \partial x_l}\frac{\mathrm{e}^{-\langle (A+tI)^{-1}\mathbf{x},\mathbf{x}\rangle}}{\sqrt{\det(I+tA^{-1})}} = \Big(\frac{4x_k x_l a_k a_l}{(1+a_k t)(1+a_l t)} - \frac{2\delta_{kl}a_k}{1+a_k t}\Big)\prod_{j=1}^{3}\frac{\mathrm{e}^{-a_j x_j^2/(1+a_j t)}}{(1+a_j t)^{1/2}}\,.$$

Hence we derive from (6.48) together with (6.38)

$$\begin{aligned}\int_{\mathbb{R}^3}\Gamma_{kl}(\mathbf{x}-\mathbf{y})\psi(\mathbf{y})\,d\mathbf{y} = {}& \frac{\delta_{kl}}{4\mu}\int_0^\infty\Big(1+\frac{(\lambda+3\mu)a_k t}{2(\lambda+2\mu)}\Big)\frac{1}{1+a_k t}\prod_{j=1}^{3}\frac{\mathrm{e}^{-a_j x_j^2/(1+a_j t)}}{(1+a_j t)^{1/2}}\,dt\\ &+\frac{(\lambda+\mu)x_k x_l}{4\mu(\lambda+2\mu)}\int_0^\infty\frac{t a_k a_l}{(1+a_k t)(1+a_l t)}\prod_{j=1}^{3}\frac{\mathrm{e}^{-a_j x_j^2/(1+a_j t)}}{(1+a_j t)^{1/2}}\,dt\end{aligned}$$

and the hydrodynamic potentials

$$\int\limits_{\mathbb{R}^3} \Psi_{kl}(\mathbf{x}-\mathbf{y})\psi(\mathbf{y})\,d\mathbf{y}$$

$$= \frac{1}{4\nu}\int\limits_0^\infty \Big(\frac{\delta_{kl}(2+a_kt)}{2(1+a_kt)} + \frac{x_kx_la_ka_lt}{(1+a_kt)(1+a_lt)}\Big)\prod_{j=1}^3 \frac{\mathrm{e}^{-a_jx_j^2/(1+a_jt)}}{(1+a_jt)^{1/2}}\,dt\,,$$

$$\frac{x_k}{4\pi}\int\limits_{\mathbb{R}^3}\frac{\psi(\mathbf{y})}{|\mathbf{x}-\mathbf{y}|^3}\,d\mathbf{y} = \frac{a_kx_k}{2}\int\limits_0^\infty \frac{1}{1+a_kt}\prod_{j=1}^3 \frac{\mathrm{e}^{-a_jx_j^2/(1+a_jt)}}{(1+a_jt)^{1/2}}\,dt\,.$$

6.4. Potentials of higher-order generating functions for orthotropic Gaussians

In this section, we determine potentials of the higher order generating function

$$\eta_{2M}^A(\mathbf{x}) = \frac{1}{\pi^{n/2}(\det A)^{1/2}}\sum_{j=0}^{M-1}\frac{(-1)^j}{j!\,4^j}\,\langle A\nabla,\nabla\rangle^j\,\mathrm{e}^{-\langle A^{-1}\mathbf{x},\mathbf{x}\rangle}$$

(see (3.25)) in the case of a diagonal matrix $A = \operatorname{diag}(a_1,\dots,a_n)$, $a_j > 0$. This function can be written in the form

$$\eta_{2M}^A(\mathbf{x}) = \frac{1}{\pi^{n/2}\sqrt{a_1\dots a_n}}\sum_{j=0}^{M-1}\frac{(-1)^j}{j!\,4^j}\,\psi_j(\mathbf{x})$$

with

$$\psi_j(\mathbf{x}) := \Big(\sum_{k=1}^n a_k\frac{\partial^2}{\partial x_k^2}\Big)^j\prod_{\ell=1}^n \mathrm{e}^{-x_\ell^2/a_\ell}\,. \tag{6.50}$$

Recall that the corresponding quasi-interpolants generate approximate cubature formulas of order $\mathcal{O}(h^{2M})$.

Our aim is to obtain one-dimensional integral representations of the harmonic and diffraction potentials acting on η_{2M}^A. We start with the harmonic potential

$$\mathcal{L}_n\eta_{2M}^A(\mathbf{x}) = \frac{1}{\pi^{n/2}\sqrt{a_1\dots a_n}}\sum_{j=0}^{M-1}\frac{(-1)^j}{j!\,4^j}\,\mathcal{L}_n\psi_j(\mathbf{x})\,. \tag{6.51}$$

From formula (6.38), we have

$$\mathcal{L}_n\psi_0(\mathbf{x}) = \frac{1}{4}\int\limits_0^\infty\prod_{\ell=1}^n\frac{\mathrm{e}^{-x_\ell^2/(a_\ell+t)}}{\sqrt{1+t/a_\ell}}\,dt\,,$$

and, in view of (6.50), Corollary 6.3 provides

$$\mathcal{L}_n\psi_j(\mathbf{x}) = \frac{1}{4}\int\limits_0^\infty\Big(\sum_{k=1}^n a_k\frac{\partial^2}{\partial x_k^2}\Big)^j\prod_{\ell=1}^n\frac{\mathrm{e}^{-x_\ell^2/(a_\ell+t)}}{\sqrt{1+t/a_\ell}}\,dt\,.$$

Since

$$\Big(\sum_{k=1}^n a_k\frac{\partial^2}{\partial x_k^2}\Big)^j = \sum_{[\beta]=j}\frac{j!}{\beta!}a_k^{\beta_k}\frac{\partial^{2\beta_k}}{\partial x_k^{2\beta_k}}$$

with the multi-index $\boldsymbol{\beta} = (\beta_1, \dots, \beta_n) \in Z^n_{\geq 0}$ and

$$a_k^{\beta_k} \frac{\partial^{2\beta_k}}{\partial x_k^{2\beta_k}} \frac{\mathrm{e}^{-x_k^2/(a_k+t)}}{\sqrt{1+t/a_k}} = \frac{a_k^{\beta_k}}{(a_k+t)^{\beta_k}} \frac{\mathrm{e}^{-x_k^2/(a_k+t)}}{\sqrt{1+t/a_k}} H_{2\beta_k}\Big(\frac{x_k}{\sqrt{a_k+t}}\Big),$$

the action of the differential operator results in

$$\begin{aligned} &\Big(\sum_{k=1}^n a_k \frac{\partial^2}{\partial x_k^2}\Big)^j \prod_{\ell=1}^n \frac{\mathrm{e}^{-x_\ell^2/(a_\ell+t)}}{\sqrt{1+t/a_\ell}} \\ &\qquad = \sum_{[\boldsymbol{\beta}]=j} \frac{j!}{\boldsymbol{\beta}!} \prod_{k=1}^n \frac{\mathrm{e}^{-x_k^2/(a_k+t)}}{(1+t/a_k)^{\beta_k+1/2}} H_{2\beta_k}\Big(\frac{x_k}{\sqrt{a_k+t}}\Big) \end{aligned} \tag{6.52}$$

which leads to the equality

$$\begin{aligned} &\sum_{j=0}^{M-1} \frac{(-1)^j}{j!\,4^j} \mathcal{L}_n \psi_j(\mathbf{x}) \\ &\quad = \sum_{j=0}^{M-1} \frac{(-1)^j}{4^{j+1}} \sum_{[\boldsymbol{\beta}]=j} \frac{1}{\boldsymbol{\beta}!} \int_0^\infty \prod_{k=1}^n \frac{\mathrm{e}^{-x_k^2/(a_k+t)}}{(1+t/a_k)^{\beta_k+1/2}} H_{2\beta_k}\Big(\frac{x_k}{\sqrt{a_k+t}}\Big)\, dt\,. \end{aligned}$$

Hence, from (6.51), we obtain the representation of the harmonic potential of higher-order generating functions as a one-dimensional integral

$$\boxed{\mathcal{L}_n \eta^A_{2M}(\mathbf{x}) = \sum_{j=0}^{M-1} \frac{(-1)^j}{\pi^{n/2}\, 4^{j+1}} \sum_{[\boldsymbol{\beta}]=j} \int_0^\infty \prod_{k=1}^n \frac{a_k^{\beta_k}\, \mathrm{e}^{-x_k^2/(a_k+t)}}{\beta_k!\,(a_k+t)^{\beta_k+1/2}} H_{2\beta_k}\Big(\frac{x_k}{\sqrt{a_k+t}}\Big)\, dt\,.}$$

Consider the diffraction potential. By formula (6.43), we have

$$\mathcal{S}_n \psi_0(\mathbf{x}) = \int_{\mathbb{R}^n} \mathcal{E}_k(\mathbf{x}-\mathbf{y})\, \psi_0(\mathbf{y})\, d\mathbf{y} = \frac{1}{4} \int_0^{i\infty} \mathrm{e}^{k^2 z/4} \prod_{\ell=1}^n \frac{\mathrm{e}^{-x_\ell^2/(a_\ell+z)}}{(1+z/a_\ell)^{1/2}}\, dz\,,$$

and therefore

$$\mathcal{S}_n \psi_j(\mathbf{x}) = \frac{1}{4} \int_0^{i\infty} \mathrm{e}^{k^2 z/4} \Big(\sum_{k=1}^n a_k \frac{\partial^2}{\partial x_k^2}\Big)^j \prod_{\ell=1}^n \frac{\mathrm{e}^{-x_\ell^2/(a_\ell+z)}}{(1+z/a_\ell)^{1/2}}\, dz\,.$$

Now, (6.52) leads to the equality

$$\begin{aligned} &\sum_{j=0}^{M-1} \frac{(-1)^j}{j!\,4^j} \mathcal{S}_n \psi_j(\mathbf{x}) \\ &\quad = \sum_{j=0}^{M-1} \frac{(-1)^j}{4^{j+1}} \sum_{[\boldsymbol{\beta}]=j} \frac{1}{\boldsymbol{\beta}!} \int_0^{i\infty} \mathrm{e}^{k^2 z/4} \prod_{k=1}^n \frac{\mathrm{e}^{-x_k^2/(a_k+z)}}{(1+z/a_k)^{\beta_k+1/2}} H_{2\beta_k}\Big(\frac{x_k}{\sqrt{a_k+z}}\Big)\, dz\,. \end{aligned}$$

Hence, the diffraction potential of higher-order generating functions can be obtained from the one-dimensional integral

$$\mathcal{S}_n \eta_{2M}^A(\mathbf{x}) = \sum_{j=0}^{M-1} \frac{(-1)^j}{\pi^{n/2}\, 4^{j+1}} \sum_{[\beta]=j} \int_0^{i\infty} \mathrm{e}^{k^2 z/4} \prod_{k=1}^{n} \frac{a_k^{\beta_k}\, \mathrm{e}^{-x_k^2/(a_k+z)}}{\beta_k!\,(a_k+z)^{\beta_k+1/2}} H_{2\beta_k}\Big(\frac{x_k}{\sqrt{a_k+z}}\Big)\, dz\,.$$

Note that for the functions corresponding to isotropic Gaussians, i.e., $A = I$, we derive

$$\mathcal{S}_n \eta_{2M}(\mathbf{x}) = \sum_{j=0}^{M-1} \frac{(-1)^j}{\pi^{n/2}\, 4^{j+1}} \sum_{[\beta]=j} \int_0^{i\infty} \mathrm{e}^{k^2 z/4} \frac{\mathrm{e}^{-|\mathbf{x}|^2/(1+z)}}{\beta!\,(1+z)^{[\beta]+n/2}} H_{2\beta}\Big(\frac{\mathbf{x}}{\sqrt{1+z}}\Big)\, dz\,.$$

6.5. Potentials of truncated Gaussians

6.5.1. Integral operators over bounded domains. In this section, we derive one-dimensional integral representations of the harmonic and diffraction potential of the Gaussian kernel restricted to a half-space.

The approximation formulas for integral operators which have been considered previously provide certain approximation orders up to a small saturation term, if the density is compactly supported and if it is sufficiently smooth on the whole space. Suppose that one wants to compute the integral

$$\mathcal{K}u(\mathbf{x}) = \int_{\Omega} k(\mathbf{x}-\mathbf{y})\, u(\mathbf{y})\, d\mathbf{y} \tag{6.53}$$

over a bounded domain Ω, where the density $u \neq 0$ at the boundary $\partial\Omega$. The direct application of the method described in Chapter 4, which is based on replacing the density by a quasi-interpolant $\mathcal{M}_{h,\mathcal{D}}u$ and on the known values of $\mathcal{K}\eta$, does not give good approximations of (6.53), in general. One reason is that $\mathcal{M}_{h,\mathcal{D}}u$ approximates u only in a subdomain $\{\mathbf{x} \in \Omega : \operatorname{dist}(\mathbf{x}, \partial\Omega) \geq \kappa h\}$ with some $\kappa > 1$ (see, for example, Corollary 2.20). This difficulty can be tackled by the continuation of u with preserved smoothness to a larger domain, say $\widetilde{\Omega} = \{\mathbf{x} \in \mathbb{R}^n : \operatorname{dist}(\mathbf{x}, \Omega) \leq \kappa h\}$. Then, obviously, the quasi-interpolant of the continuation $\tilde{u}$ of u to $\widetilde{\Omega}$ approximates u in Ω and the integral

$$\int_{\Omega} g(\mathbf{x}-\mathbf{y})\, \mathcal{M}_{h,\mathcal{D}}\tilde{u}(\mathbf{y})\, d\mathbf{y} = \mathcal{D}^{-n/2} \sum_{h\mathbf{m}\in\widetilde{\Omega}} \tilde{u}(h\mathbf{m}) \int_{\Omega} k(\mathbf{x}-\mathbf{y})\eta\Big(\frac{\mathbf{y}-h\mathbf{m}}{\sqrt{\mathcal{D}}h}\Big)\, d\mathbf{y}$$

would give a cubature of (6.53), if the integrals

$$\int_{\Omega} k(\mathbf{x}-\mathbf{y})\eta\Big(\frac{\mathbf{y}-h\mathbf{m}}{\sqrt{\mathcal{D}}h}\Big)\, d\mathbf{y} \tag{6.54}$$

could be computed efficiently. Owing to the strong decay of η, these integrals can be replaced without loss of accuracy by the known values

$$\int_{\mathbb{R}^n} k(\mathbf{x}-\mathbf{y})\eta\Big(\frac{\mathbf{y}-h\mathbf{m}}{\sqrt{\mathcal{D}}h}\Big)\, d\mathbf{y}\,,$$

if the centers $h\mathbf{m} \in \{\mathbf{x} \in \Omega : \operatorname{dist}(\mathbf{x}, \partial\Omega) \geq \kappa h\}$. However, if the points are located in a strip near the boundary $h\mathbf{m} \in \widetilde{\Omega} \setminus \{\mathbf{x} \in \Omega : \operatorname{dist}(\mathbf{x}, \partial\Omega) \geq \kappa h\}$, then it is necessary to compute (6.54) directly. This requires, in general, the use of other n-dimensional cubature formulas and degrades the advantages of our cubature approach. In Chapter 9, we propose an alternative method for the cubature of integral operators over bounded domains, which avoids the computation of domain integrals of the form (6.54) and works for rather general domains and kernels k.

However, in some special cases potentials of the form (6.54) can be reduced to one-dimensional integrals and therefore can be computed efficiently.

6.5.2. Integral operators over the half-space. Suppose that the bounded domain $\Omega \subset \mathbb{R}^n_+ = \{\mathbf{x} = (\mathbf{x}', x_n) \in \mathbb{R}^n, x_n \geq 0\}$ with $\partial\Omega \cap \{\mathbf{x} \in \mathbb{R}^n, x_n = 0\} \neq \emptyset$, and that $u \in C^2(\overline{\Omega})$ can be extended smoothly by zero onto $\mathbb{R}^n_+$, but $u(\mathbf{x}', 0) \neq 0$.

To derive a second-order quasi-interpolant to u in Ω, we extend this function onto $\mathbb{R}^n_-$, using the formula

$$\tilde{u}(\mathbf{x}', x_n) = \sum_{j=1}^{3} c_j u(\mathbf{x}', -j x_n)\,, \ x_n < 0\,,$$

where the vector $\{c_j\}$ is a solution of the algebraic system

$$\sum_{j=1}^{3} j^{k-1} c_j = (-1)^{k-1}\,, \ k = 1, 2, 3\,.$$

Then $\tilde{u}$ is a C^2 extension of u for $x_n \leq 0$ (see [**33**]) and, in view of Corollary 2.20, for given $\varepsilon > 0$ there exist $\mathcal{D}$ and κ such that for any $\mathbf{x} \in \Omega$

$$|u(\mathbf{x}) - \mathcal{M}^{(h\kappa)}_{h,\mathcal{D}} \tilde{u}(\mathbf{x})| \leq c_\eta (\sqrt{\mathcal{D}} h)^2 \, \|\nabla_N u\|_{L_\infty(\Omega)} + \varepsilon \big(\|u\|_{L_\infty(\Omega)} + (\sqrt{\mathcal{D}} h) |\nabla u(\mathbf{x})| \big)$$

with the quasi-interpolant

$$\mathcal{M}^{(h\kappa)}_{h,\mathcal{D}} \tilde{u}(\mathbf{x}) = \mathcal{D}^{-n/2} \sum_{h\mathbf{m} \in B(\mathbf{x}, h\kappa)} \tilde{u}(h\mathbf{m}) \, \mathrm{e}^{-|\mathbf{x} - h\mathbf{m}|^2 / \mathcal{D} h^2}\,.$$

Hence, a cubature formula of the integral

$$\mathcal{K}u(\mathbf{x}) = \int_\Omega k(\mathbf{x} - \mathbf{y}) \, u(\mathbf{y}) \, d\mathbf{y}$$

is given by the sum

$$\begin{aligned} K_h u(\mathbf{x}) =& \mathcal{D}^{-n/2} \sum_{\substack{h\mathbf{m} \in \Omega \\ m_n > \kappa}} u(h\mathbf{m}) \int_{\mathbb{R}^n} k(\mathbf{x} - \mathbf{y}) \, \mathrm{e}^{-|\mathbf{y} - h\mathbf{m}|^2 / \mathcal{D} h^2} \, d\mathbf{y} \\ &+ \mathcal{D}^{-n/2} \sum_{\substack{\tilde{u}(h\mathbf{m}) \neq 0 \\ |m_n| \leq \kappa}} \tilde{u}(h\mathbf{m}) \int_{\mathbb{R}^n_+} k(\mathbf{x} - \mathbf{y}) \, \mathrm{e}^{-|\mathbf{y} - h\mathbf{m}|^2 / \mathcal{D} h^2} \, d\mathbf{y}\,, \end{aligned} \tag{6.55}$$

which is of order $\mathcal{O}(\mathcal{D} h^2)$ modulo small saturation terms. Since the integrals in the first sum are supposed to be known, it remains to give explicit expressions for the integrals over the half-space, appearing in the second sum of (6.55).

6.5.3. Harmonic potentials. Let us compute the harmonic potential of truncated Gaussians

$$v_a(\mathbf{x}) = \frac{\Gamma(\frac{n}{2}-1)}{4\pi^{n/2}} \int\limits_{\mathbb{R}^n \cap \{y_n > a\}} \frac{e^{-|\mathbf{y}|^2}}{|\mathbf{x}-\mathbf{y}|^{n-2}} \, d\mathbf{y} \,, \quad n \geq 3 \,.$$

We have

$$\Delta v_a = \begin{cases} -e^{-|\mathbf{x}|^2}, & x_n > a \,, \\ 0 \,, & x_n < a \,. \end{cases}$$

Taking the Fourier transform with respect to the variables $\mathbf{x}' = (x_1, \dots, x_{n-1})$, one obtains the differential equation for the function $\hat{v}_a(\boldsymbol{\lambda}', x_n) = (\mathcal{F}_{\mathbf{x}' \to \boldsymbol{\lambda}'} v_a)(\boldsymbol{\lambda}', x_n)$:

$$\frac{\partial^2 \hat{v}_a}{\partial x_n^2} - 4\pi^2 |\boldsymbol{\lambda}'|^2 \, \hat{v}_a = \begin{cases} -\pi^{(n-1)/2} e^{-x_n^2} e^{-\pi^2 |\boldsymbol{\lambda}'|^2}, & x_n > a \,, \\ 0 \,, & x_n < a \,. \end{cases}$$

Its solution is given by the integral

$$\hat{v}_a(\boldsymbol{\lambda}', x_n) = \frac{\pi^{(n-1)/2} e^{-\pi^2 |\boldsymbol{\lambda}'|^2}}{4\pi |\boldsymbol{\lambda}'|} \int\limits_a^\infty e^{-2\pi |\boldsymbol{\lambda}'| \, |x_n - t|} e^{-t^2} \, dt \,.$$

If $x_n < a$, then

$$\begin{aligned} \hat{v}_a(\boldsymbol{\lambda}', x_n) &= \frac{\pi^{(n-1)/2} e^{-\pi^2 |\boldsymbol{\lambda}'|^2}}{4\pi |\boldsymbol{\lambda}'|} \int\limits_a^\infty e^{-2\pi |\boldsymbol{\lambda}'| \, (t - x_n)} e^{-t^2} \, dt \\ &= \frac{\pi^{(n-1)/2} e^{2\pi x_n |\boldsymbol{\lambda}'|}}{4\pi |\boldsymbol{\lambda}'|} \int\limits_a^\infty e^{-(t + \pi |\boldsymbol{\lambda}'|)^2} \, dt \\ &= \frac{\pi^{n/2} e^{2\pi x_n |\boldsymbol{\lambda}'|}}{8\pi |\boldsymbol{\lambda}'|} \operatorname{erfc}\left(a + \pi |\boldsymbol{\lambda}'|\right). \end{aligned}$$

If $x_n > a$, then

$$\begin{aligned} \hat{v}_a(\boldsymbol{\lambda}', x_n) &= \frac{\pi^{(n-1)/2} e^{-\pi^2 |\boldsymbol{\lambda}'|^2}}{4\pi |\boldsymbol{\lambda}'|} \Bigg(\int\limits_a^{x_n} e^{-2\pi |\boldsymbol{\lambda}'| \, (x_n - t)} e^{-t^2} \, dt \\ &\qquad\qquad + \int\limits_{x_n}^\infty e^{-2\pi |\boldsymbol{\lambda}'| \, (t - x_n)} e^{-t^2} \, dt \Bigg) \\ &= \frac{\pi^{(n-1)/2}}{4\pi |\boldsymbol{\lambda}'|} \Bigg(e^{-2\pi x_n |\boldsymbol{\lambda}'|} \int\limits_a^{x_n} e^{-(t - \pi |\boldsymbol{\lambda}'|)^2} \, dt + e^{2\pi x_n |\boldsymbol{\lambda}'|} \int\limits_{x_n}^\infty e^{-(t + \pi |\boldsymbol{\lambda}'|)^2} \, dt \Bigg) \\ &= \frac{\pi^{n/2}}{8\pi |\boldsymbol{\lambda}'|} \Big(e^{2\pi x_n |\boldsymbol{\lambda}'|} \operatorname{erfc}(x_n + \pi |\boldsymbol{\lambda}'|) \\ &\qquad + e^{-2\pi x_n |\boldsymbol{\lambda}'|} \left(\operatorname{erfc}(a - \pi |\boldsymbol{\lambda}'|) - \operatorname{erfc}(x_n - \pi |\boldsymbol{\lambda}'|) \right) \Big). \end{aligned}$$

Let us define the function

$$\begin{aligned} g_a(t, x_n) := {} & e^{2 x_n t} \operatorname{erfc}\left(\max(a, x_n) + t\right) \\ & + e^{-2 x_n t} \Big(\operatorname{erfc}\left(a - t\right) - \operatorname{erfc}\left(\max(a, x_n) - t\right) \Big). \end{aligned} \tag{6.56}$$

Then

$$\hat{v}_a(\boldsymbol{\lambda}', x_n) = \frac{\pi^{n/2}}{8\pi|\boldsymbol{\lambda}'|}\, g_a(\pi|\boldsymbol{\lambda}'|, x_n) \tag{6.57}$$

and, consequently, via the inverse Fourier transform (2.12),

$$v_a(\mathbf{x}) = \frac{\pi^{n/2}}{4|\mathbf{x}'|^{(n-3)/2}} \int_0^\infty g_a(\pi t, x_n)\; J_{(n-3)/2}(2\pi|\mathbf{x}'|t)\, t^{(n-3)/2}\, dt\,.$$

Therefore, the harmonic potential of the truncated Gaussian is expressed by the one-dimensional integral

$$\frac{\Gamma(\frac{n}{2}-1)}{4\pi^{n/2}} \int\limits_{\mathbb{R}^n \cap \{y_n > a\}} \frac{\mathrm{e}^{-|\mathbf{y}|^2}}{|\mathbf{x}-\mathbf{y}|^{n-2}}\, d\mathbf{y} = \frac{\sqrt{\pi}}{4|\mathbf{x}'|^{(n-3)/2}} \int_0^\infty g_a(t, x_n)\; J_{(n-3)/2}(2|\mathbf{x}'|t)\, t^{(n-3)/2}\, dt\,,$$

which gives, in the case of the harmonic potential, for the integrals of the second sum in the formula (6.55)

$$\frac{\Gamma(\frac{n}{2}-1)}{4\pi^{n/2}} \int\limits_{\mathbb{R}^n_+} \frac{\mathrm{e}^{-|\mathbf{y}-h\mathbf{m}|^2/\mathcal{D}h^2}}{|\mathbf{x}-\mathbf{y}|^{n-2}}\, d\mathbf{y}$$
$$= \frac{(\sqrt{\mathcal{D}}h)^{(n+5)/2}\sqrt{\pi}}{4|\mathbf{x}'-h\mathbf{m}'|^{(n-3)/2}} \int_0^\infty g_{hm_n}\Big(t, \frac{x_n - hm_n}{\sqrt{\mathcal{D}}h}\Big)\, J_{(n-3)/2}\Big(\frac{2|\mathbf{x}'-h\mathbf{m}'|t}{\sqrt{\mathcal{D}}h}\Big)\, t^{(n-3)/2}\, dt\,.$$

6.5.4. Diffraction potentials. Let us apply the same method to determine the diffraction potential of truncated Gaussians. Consider the equation

$$\Delta u - cu = \begin{cases} -\,\mathrm{e}^{-|\mathbf{x}|^2}, & x_n > a\,,\\ \;0\,, & x_n < a\,, \end{cases} \tag{6.58}$$

where the constant c has a non-zero imaginary part, $\operatorname{Im} c \neq 0$. There exists a unique L_2-solution $w_a(\mathbf{x})$ of (6.58), whose Fourier transform

$$\widehat{w}_a(\boldsymbol{\lambda}', x_n) = (\mathcal{F}_{\mathbf{x}'\to\boldsymbol{\lambda}'}\, w_a)(\boldsymbol{\lambda}', x_n)$$

is a solution of the differential equation

$$\frac{\partial^2 \widehat{w}_a}{\partial x_n^2} - (4\pi^2|\boldsymbol{\lambda}'|^2 + c)\widehat{w}_a = \begin{cases} -\pi^{(n-1)/2}\,\mathrm{e}^{-x_n^2}\,\mathrm{e}^{-\pi^2|\boldsymbol{\lambda}'|^2}\,, & x_n > a\,,\\ 0\,, & x_n < a\,. \end{cases}$$

Hence,

$$\widehat{w}_a(\boldsymbol{\lambda}', x_n) = \frac{\pi^{(n-1)/2}\,\mathrm{e}^{-\pi^2|\boldsymbol{\lambda}'|^2}}{2(4\pi^2|\boldsymbol{\lambda}'|^2 + c)^{1/2}} \int_a^\infty \mathrm{e}^{-(4\pi^2|\boldsymbol{\lambda}'|^2+c)^{1/2}\,|x_n - t|}\;\mathrm{e}^{-t^2}\, dt\,, \tag{6.59}$$

where the branch of the square root is chosen so that $(4\pi^2|\boldsymbol{\lambda}'|^2 + c)^{1/2} > 0$ for $c > -4\pi^2|\boldsymbol{\lambda}'|^2$ and its branch-cut is $(-\infty, 0)$. Since $\operatorname{Im} c \neq 0$, it follows that

$$\operatorname{Re}(4\pi^2|\boldsymbol{\lambda}'|^2 + c)^{1/2} > 0\,,$$

which confirms the validity of formula (6.59). Noting that

$$\mathrm{e}^{-\pi^2|\boldsymbol{\lambda}'|^2}\int\limits_b^d \mathrm{e}^{\pm(4\pi^2|\boldsymbol{\lambda}'|^2+c)^{1/2}\,t}\,\mathrm{e}^{-t^2}\,dt = \mathrm{e}^{c/4}\int\limits_b^d \mathrm{e}^{-(t\mp(4\pi^2|\boldsymbol{\lambda}'|^2+c)^{1/2}/2)^2}\,dt$$

$$= \frac{\sqrt{\pi}\,\mathrm{e}^{c/4}}{2}\Big(\operatorname{erfc}\Big(b\mp\frac{(4\pi^2|\boldsymbol{\lambda}'|^2+c)^{1/2}}{2}\Big)-\operatorname{erfc}\Big(d\mp\frac{(4\pi^2|\boldsymbol{\lambda}'|^2+c)^{1/2}}{2}\Big)\Big),$$

we obtain similarly to (6.57)

$$\widehat{w}_a(\boldsymbol{\lambda}',x_n) = \frac{\pi^{n/2}\,\mathrm{e}^{c/4}}{4(4\pi^2|\boldsymbol{\lambda}'|^2+c)^{1/2}}\,g_a\Big(\frac{(4\pi^2|\boldsymbol{\lambda}'|^2+c)^{1/2}}{2},x_n\Big)$$

with the function g_a defined by (6.56). Hence, the solution of (6.58) can be obtained via inverse Fourier transform and is given by the integral

$$w_a(\mathbf{x}) = \frac{2\pi\,\mathrm{e}^{c/4}}{|\mathbf{x}'|^{(n-3)/2}}\int\limits_0^\infty \frac{\pi^{n/2}\,t^{(n-1)/2}}{4(4\pi^2t^2+c)^{1/2}}\,g_a\Big(\frac{(4\pi^2t^2+c)^{1/2}}{2},x_n\Big)\,J_{(n-3)/2}(2\pi|\mathbf{x}'|t)\,dt$$

$$= \frac{\sqrt{\pi}\,\mathrm{e}^{c/4}}{4|\mathbf{x}'|^{(n-3)/2}}\int\limits_0^\infty g_a((t^2+c/4)^{1/2},x_n)\,J_{(n-3)/2}(2|\mathbf{x}'|t)\,\frac{t^{(n-1)/2}}{(t^2+c/4)^{1/2}}\,dt\,.$$

Making the substitution $z=(t^2+c/4)^{1/2}$, we derive

(6.60)

$$w_a(\mathbf{x}) = \frac{\sqrt{\pi}\,\mathrm{e}^{c/4}}{4\,|2\mathbf{x}'|^{(n-3)/2}}\int\limits_{\Gamma_c} g_a(z,x_n)\,J_{(n-3)/2}(|\mathbf{x}'|(4z^2-c)^{1/2})\,(4z^2-c)^{(n-3)/4}\,dz\,,$$

with the integration contour

$$\Gamma_c = \{(t^2+c/4)^{1/2},\ t\in(0,\infty)\}\,.$$

Using the series representation of the Bessel functions

$$J_\nu(t) = \Big(\frac{t}{2}\Big)^\nu\sum_{k=0}^\infty\frac{(-1)^k t^{2k}}{2^{2k}k!\Gamma(\nu+k+1)}\,,$$

we see that

$$J_{(n-3)/2}(|\mathbf{x}'|(4z^2-c)^{1/2})\,(4z^2-c)^{(n-3)/4}$$

$$= \Big(\frac{|\mathbf{x}'|(4z^2-c)}{2}\Big)^{(n-3)/2}\sum_{k=0}^\infty\frac{|\mathbf{x}'|^{2k}(c-4z^2)^k}{2^{2k}k!\Gamma(k+(n-1)/2)}\,.$$

Thus, for fixed a and x_n, the integrand of (6.60) is a holomorphic function of $z\in\mathbb{C}\setminus(-\infty,0]$, and the integration contour Γ_c of (6.60) can be deformed to an arbitrary contour going from $\sqrt{c}/2$ to $+\infty$ and not crossing the negative real axis $\operatorname{Re} z<0$.

As a simple application of the formula (6.60), we compute the diffraction potential of truncated Gaussians

(6.61)
$$v_a(\mathbf{x}) = \int\limits_{\mathbb{R}^n\cap\{y_n>a\}} \mathcal{E}_k(\mathbf{x}-\mathbf{y})\,\mathrm{e}^{-|\mathbf{y}|^2}\,d\mathbf{y}$$

with the wave number $k > 0$. Since the solutions w_a of (6.58) with $c = -(k^2 + i\varepsilon)$, $\varepsilon > 0$ and $\varepsilon \to 0$, converge by the limiting absorption principle to the integral (6.61), we obtain

$$\boxed{v_a(\mathbf{x}) = \frac{\sqrt{\pi}\,e^{-k^2/4}}{4\,|2\mathbf{x}'|^{(n-3)/2}} \int\limits_{\Gamma_k} g_a(z, x_n)\, J_{(n-3)/2}(|\mathbf{x}'|(4z^2 + k^2)^{1/2})\,(4z^2 + k^2)^{(n-3)/4}\, dz}$$

where the integration contour Γ_k connects $-ik/2$ and $+\infty$ and lies in the closed half-plane $\operatorname{Re} z \geq 0$.

6.6. Notes

In Subsections 6.1.1 and 6.1.2, we extend formulas from [**64**] to higher approximation orders and arbitrary space dimension. The formula (6.4) for $n = 2$ and $n = 3$ has also been given by Kanaun and Kochekseraii in [**44**] and has been used for the solution of singular volume integral equations arising in thermo- and electrostatic problems, where the unknown function is a linear combination of Gaussian kernels. In this paper the integrals of so-called *edge Gaussian functions* are calculated, which are the truncated Gaussians considered in Section 6.5.

The approximation method for solving the Cauchy problem for the heat and wave equations considered in Subsections 6.2.1 and 6.2.2 was proposed in [**64**]. Here, we extend the method with the formulas for high-order approximation rates in the n-dimensional case.

The results of Section 6.3 appeared first in [**72**].

CHAPTER 7

Approximation by Gaussians

Using approximate quasi-interpolation in Chapters 1 and 2, we obtained approximation formulas based on semi-discrete convolutions with Gaussians. Let us emphasize that the concept of approximate quasi-interpolation is applicable to any function in the Schwartz space with non-vanishing moment of order zero. However, the Gaussian function is distinguished by the property that many integral operators acting on this function provide simple analytical expressions. In this chapter we consider some other remarkable approximation properties of semi-discrete convolutions with the Gaussian.

In Section 7.1, we show that there exist semi-discrete convolutions which approximate analytic functions very accurately even for large step sizes. We give, in particular, estimates for the approximation of polynomials and exponential functions.

The results of Subsection 3.5.2 are applied in Section 7.2 to obtain explicit error estimates for the high-order approximate quasi-interpolation using linear combinations of translates of the Gaussian as basis function.

In Section 7.3, we consider the interpolation with linear combinations of Gaussians on a uniform grid. Based on the explicit formulas for the interpolants, we deduce the new class of Lagrangian functions

$$\Psi_{\mathcal{D}}(x) = \frac{\sin \pi x}{\pi \mathcal{D} \sinh \dfrac{x}{\mathcal{D}}} \,.$$

It is shown that the interpolants spanned by $\Psi_{\mathcal{D}}$ have approximation properties, similar to those of the sinc function. In particular, we show the exponential approximation order of this interpolation up to some saturation error.

7.1. Approximation of entire functions

7.1.1. Approximants on coarse grids. Suppose that $u(\mathbf{x})$ is the restriction to $\mathbb{R}^n$ of an entire function $u(\mathbf{z})$, $\mathbf{z} = (z_1, \ldots, z_n) \in \mathbb{C}^n$, which satisfies the estimate

$$|u(\mathbf{x} + i\pi\mathcal{D}h\mathbf{y})\, \mathrm{e}^{-\pi^2\mathcal{D}|\mathbf{y}|^2}\,| \le A(1 + |\mathbf{y}|)^{-n-\delta}\,, \quad \delta > 0, \tag{7.1}$$

for some fixed positive parameters h and $\mathcal{D}$ and any $\mathbf{x}, \mathbf{y} \in \mathbb{R}^n$. Then there exists the integral

$$u_{\mathbf{m}} := \int\limits_{\mathbb{R}^n} \mathrm{e}^{-\pi^2\mathcal{D}|\mathbf{y}|^2}\, u(h\mathbf{m} + i\pi\mathcal{D}h\mathbf{y})\, d\mathbf{y} \tag{7.2}$$

for any $\mathbf{m} \in \mathbb{Z}^n$.

THEOREM 7.1. *If the entire function u satisfies (7.1) and the semi-discrete convolution*

$$u_h(\mathbf{x}) = \sum_{\mathbf{m}\in\mathbb{Z}^n} u_{\mathbf{m}}\, \mathrm{e}^{-|\mathbf{x}-h\mathbf{m}|^2/\mathcal{D}h^2}$$

converges absolutely for given $\mathbf{x} \in \mathbb{R}^n$, then

$$u_h(\mathbf{x}) - u(\mathbf{x}) = \sum_{\boldsymbol{\nu}\in\mathbb{Z}^n\setminus\{\mathbf{0}\}} u(\mathbf{x}+i\pi\mathcal{D}h\boldsymbol{\nu})\, \mathrm{e}^{-\pi^2\mathcal{D}|\boldsymbol{\nu}|^2}\, \mathrm{e}^{\frac{2\pi i}{h}\langle\mathbf{x},\boldsymbol{\nu}\rangle}\,. \tag{7.3}$$

PROOF. We may write

$$u_{\mathbf{m}}\, \mathrm{e}^{-|\mathbf{x}-h\mathbf{m}|^2/\mathcal{D}h^2} = (\pi\mathcal{D}h)^{-n}\, \mathrm{e}^{-|\mathbf{x}-h\mathbf{m}|^2/\mathcal{D}h^2} \int_{\mathbb{R}^n} u(h\mathbf{m}+i\mathbf{y})\, \mathrm{e}^{-|\mathbf{y}|^2/\mathcal{D}h^2}\, d\mathbf{y}.$$

Applying Cauchy's Theorem several times we obtain

$$\begin{aligned} u_{\mathbf{m}}\, \mathrm{e}^{-|\mathbf{x}-h\mathbf{m}|^2/\mathcal{D}h^2} &= (\pi\mathcal{D}h)^{-n}\, \mathrm{e}^{-|\mathbf{x}|^2/\mathcal{D}h^2} \int_{\mathbb{R}^n} u(i\mathbf{y})\, \mathrm{e}^{-|\mathbf{y}|^2/\mathcal{D}h^2}\, \mathrm{e}^{2\langle\mathbf{x}-i\mathbf{y},\mathbf{m}\rangle/\mathcal{D}h} \\ &= (\pi\mathcal{D}h)^{-n} \int_{\mathbb{R}^n} u(\mathbf{x}+i\mathbf{y})\, \mathrm{e}^{-|\mathbf{y}|^2/\mathcal{D}h^2}\, \mathrm{e}^{2i\langle\mathbf{x}/h-\mathbf{m},\mathbf{y}\rangle/\mathcal{D}h}\, d\mathbf{y} \\ &= \int_{\mathbb{R}^n} u(\mathbf{x}+i\pi\mathcal{D}h\mathbf{y})\, \mathrm{e}^{-\pi^2\mathcal{D}|\mathbf{y}|^2}\, \mathrm{e}^{2\pi i\langle\mathbf{x}/h-\mathbf{m},\mathbf{y}\rangle}\, d\mathbf{y} \\ &= \int_{\mathbb{R}^n} f_{\mathbf{x}}(\mathbf{y})\, \mathrm{e}^{-2\pi i\langle\mathbf{m},\mathbf{y}\rangle}\, d\mathbf{y} = \mathcal{F}f_{\mathbf{x}}(\mathbf{m})\,, \end{aligned}$$

where we use the notation

$$f_{\mathbf{x}}(\mathbf{y}) = u(\mathbf{x}+i\pi\mathcal{D}h\mathbf{y})\, \mathrm{e}^{-\pi^2\mathcal{D}|\mathbf{y}|^2}\, \mathrm{e}^{2\pi i\langle\mathbf{x},\mathbf{y}\rangle/h}\,.$$

Now it remains to apply Poisson's summation formula (2.17) which gives

$$\sum_{\mathbf{m}\in\mathbb{Z}^n} \mathcal{F}f_{\mathbf{x}}(\mathbf{m}) = \sum_{\boldsymbol{\nu}\in\mathbb{Z}^n} f_{\mathbf{x}}(\boldsymbol{\nu}) = \sum_{\boldsymbol{\nu}\in\mathbb{Z}^n} u(\mathbf{x}+i\pi\mathcal{D}h\boldsymbol{\nu})\, \mathrm{e}^{-\pi^2\mathcal{D}|\boldsymbol{\nu}|^2}\, \mathrm{e}^{2\pi i\langle\mathbf{x},\boldsymbol{\nu}\rangle/h}$$

and is valid since the series converge absolutely in view of (7.2). □

Note that under the condition

$$\int_{\mathbb{R}^n} |\mathcal{F}u(\boldsymbol{\lambda})|\, \mathrm{e}^{\pi^2\mathcal{D}h^2|\boldsymbol{\lambda}|^2}\, d\boldsymbol{\lambda} < \infty\,, \tag{7.4}$$

there is an equivalent formula for the coefficient

$$u_{\mathbf{m}} = (\pi\mathcal{D})^{-n/2} \int_{\mathbb{R}^n} \mathcal{F}u(\boldsymbol{\lambda})\, \mathrm{e}^{\pi^2\mathcal{D}h^2|\boldsymbol{\lambda}|^2}\, \mathrm{e}^{2\pi i h\langle\mathbf{m},\boldsymbol{\lambda}\rangle}\, d\boldsymbol{\lambda}\,. \tag{7.5}$$

Condition (7.4) is satisfied, for example, if u is an entire function subject to

$$|\mathbf{x}^{\boldsymbol{\alpha}} u(\mathbf{x}+i\mathbf{y})| \le C_{\boldsymbol{\alpha}} \exp\Big(\sum_{j=1}^n a_j|y_j|\Big)\,.$$

Then its Fourier transform has compact support

$$\operatorname{supp}\mathcal{F}u \subseteq \{\boldsymbol{\lambda} : -a_j \le \lambda_j \le a_j\}\,.$$

Relation (7.3) shows that the semi-discrete convolution u_h can be a very precise approximant to entire functions $u(\mathbf{z})$ even for relatively large steps h, if $\mathcal{D}$ is properly chosen.

7.1.2. Examples.

7.1.2.1. *Polynomials.* From the relations

$$u_m = \int_{-\infty}^{\infty} (hm + i\pi\mathcal{D}hy)^j \, \mathrm{e}^{-\pi^2\mathcal{D}y^2} \, dy = \frac{d^j}{d\lambda^j} \int_{-\infty}^{\infty} \mathrm{e}^{-\pi^2\mathcal{D}y^2} \, \mathrm{e}^{(hm+i\pi\mathcal{D}hy)\lambda} \, dy \Big|_{\lambda=0}$$

and

$$\frac{d^j}{d\lambda^j} \int_{-\infty}^{\infty} \mathrm{e}^{-\pi^2\mathcal{D}y^2} \, \mathrm{e}^{(hm+i\pi\mathcal{D}hy)\lambda} \, dy = \frac{1}{\sqrt{\pi\mathcal{D}}} \frac{d^j}{d\lambda^j} \Big(\mathrm{e}^{hm\lambda} \, \mathrm{e}^{-\mathcal{D}h^2\lambda^2/4} \Big)$$

$$= \frac{\mathrm{e}^{m^2/\mathcal{D}}}{\sqrt{\pi\mathcal{D}}} \frac{d^j}{d\lambda^j} \, \mathrm{e}^{-(m/\sqrt{\mathcal{D}}-\sqrt{\mathcal{D}}h\lambda/2)^2}$$

and the definition of the Hermite polynomials (3.6), we obtain

$$u_m = (-1)^j \frac{\mathrm{e}^{m^2/\mathcal{D}}}{\sqrt{\pi\mathcal{D}}} \Big(\frac{\sqrt{\mathcal{D}}h}{2} \Big)^j H_j\Big(\frac{m}{\sqrt{\mathcal{D}}} - \frac{\sqrt{\mathcal{D}}h\lambda}{2} \Big) \, \mathrm{e}^{-(m/\sqrt{\mathcal{D}}-\sqrt{\mathcal{D}}h\lambda/2)^2} \Big|_{\lambda=0}$$

$$= \frac{1}{\sqrt{\pi\mathcal{D}}} \Big(\frac{\sqrt{\mathcal{D}}h}{2} \Big)^j H_j\Big(\frac{m}{\sqrt{\mathcal{D}}} \Big) .$$

Hence, in view of (7.3),

$$x^j = \frac{1}{\sqrt{\pi\mathcal{D}}} \Big(\frac{\sqrt{\mathcal{D}}h}{2} \Big)^j \sum_{m=-\infty}^{\infty} H_j\Big(\frac{m}{\sqrt{\mathcal{D}}} \Big) \, \mathrm{e}^{-(x-hm)^2/\mathcal{D}h^2} + R_j(x) ,$$

so that the monomial x^j is approximated by

$$\frac{1}{\sqrt{\pi\mathcal{D}}} \Big(\frac{\sqrt{\mathcal{D}}h}{2} \Big)^j \sum_{m=-\infty}^{\infty} H_j\Big(\frac{m}{\sqrt{\mathcal{D}}} \Big) \, \mathrm{e}^{-(x-hm)^2/\mathcal{D}h^2}$$

with the error estimated by

$$|R_j(x)| \le \sum_{m\in\mathbb{Z}\setminus\{0\}} |x + i\pi\mathcal{D}hm|^j \, \mathrm{e}^{-\pi^2\mathcal{D}m^2} .$$

In the case of n dimensions and with $h = 1$ we get, in particular,

$$\mathbf{x}^{\boldsymbol{\alpha}} = \frac{1}{(\pi\mathcal{D})^{n/2}} \Big(\frac{\sqrt{\mathcal{D}}h}{2} \Big)^{[\boldsymbol{\alpha}]} \sum_{\mathbf{m}\in\mathbb{Z}^n} H_{\boldsymbol{\alpha}}\Big(\frac{\mathbf{m}}{\sqrt{\mathcal{D}}} \Big) \, \mathrm{e}^{-|\mathbf{x}-\mathbf{m}|^2/\mathcal{D}} + R_{\boldsymbol{\alpha}}(\mathbf{x}) , \tag{7.6}$$

where $H_{\boldsymbol{\alpha}}$ denotes the Hermite polynomial of n variables

$$H_{\boldsymbol{\alpha}}(\mathbf{x}) = (-1)^{[\boldsymbol{\alpha}]} \, \mathrm{e}^{|\mathbf{x}|^2} \, \partial^{\boldsymbol{\alpha}} \, \mathrm{e}^{-|\mathbf{x}|^2} , \tag{7.7}$$

and

$$R_{\boldsymbol{\alpha}}(\mathbf{x}) = \sum_{\boldsymbol{\nu}\in\mathbb{Z}^n\setminus\{\mathbf{0}\}} (\mathbf{x} + i\pi\mathcal{D}\boldsymbol{\nu})^{\boldsymbol{\alpha}} \, \mathrm{e}^{-\pi^2\mathcal{D}|\boldsymbol{\nu}|^2} \, \mathrm{e}^{2\pi i\langle \mathbf{x},\boldsymbol{\nu}\rangle} .$$

Note that for any multi-index $\boldsymbol{\alpha} = (\alpha_1, \ldots, \alpha_n)$

$$H_{\boldsymbol{\alpha}}(\mathbf{x}) = \prod_{j=1}^{n} H_{\alpha_j}(x_j)\,,$$

with the univariate Hermite polynomial defined by (3.6). We see from (7.6) that any polynomial $P(\mathbf{x})$ can be approximated by linear combinations of the shifted Gaussians $\mathrm{e}^{-|\mathbf{x}-\mathbf{m}|^2/\mathcal{D}}$, $\mathbf{m} \in \mathbb{Z}^n$, with an arbitrarily small relative error $\epsilon > 0$ if $\mathcal{D}$ is chosen large enough.

7.1.2.2. *Exponential function.* Next, we consider the approximation of the function $u(\mathbf{x}) = \mathrm{e}^{\langle \mathbf{x},\mathbf{a}\rangle}$ with a fixed $\mathbf{a} \in \mathbb{C}^n$. Theorem 7.1 leads to

$$\begin{aligned} u_h(\mathbf{x}) &= (\pi\mathcal{D})^{-n/2}\, \mathrm{e}^{-\mathcal{D}h^2\mathbf{a}^2/4} \sum_{\mathbf{m}\in\mathbb{Z}^n} \mathrm{e}^{\langle h\mathbf{m},\mathbf{a}\rangle}\, \mathrm{e}^{-|\mathbf{x}-h\mathbf{m}|^2/\mathcal{D}h^2} \\ &= \mathrm{e}^{\langle \mathbf{x},\mathbf{a}\rangle} \Big(1 + \sum_{\boldsymbol{\nu}\in\mathbb{Z}^n\setminus\{\mathbf{0}\}} \mathrm{e}^{-\pi^2\mathcal{D}|\boldsymbol{\nu}|^2}\, \mathrm{e}^{2\pi i\langle \mathbf{x}/h+\mathcal{D}h\mathbf{a}/2,\boldsymbol{\nu}\rangle}\Big). \end{aligned}$$

Thus, if $\mathbf{a} \in \mathbb{R}^n$, then for any $h > 0$, the series u_h approximates the exponential function with the relative error less than

$$\sum_{\boldsymbol{\nu}\in\mathbb{Z}^n\setminus\{\mathbf{0}\}} \mathrm{e}^{-\pi^2\mathcal{D}|\boldsymbol{\nu}|^2}\,.$$

If $\mathbf{a} = \mathbf{u}+i\mathbf{v}$, $\mathbf{u},\mathbf{v} \in \mathbb{R}^n$ with $\mathbf{v} \neq \mathbf{0}$, and $h \le \pi/(2|\mathbf{v}|)$, then u_h approximates $\mathrm{e}^{\langle \mathbf{x},\mathbf{a}\rangle}$ with the relative error less than or equal to

$$\sum_{\boldsymbol{\nu}\in\mathbb{Z}^n\setminus\{\mathbf{0}\}} \mathrm{e}^{-\pi^2\mathcal{D}|\boldsymbol{\nu}|^2/2}\,.$$

If $n = 1$, then we get some special cases of the well-known transformation formula for Jacobi's Theta function (see (1.5))

$$\vartheta_3(\pi z|ia) = \sum_{m\in\mathbb{Z}} \mathrm{e}^{-m^2\pi a+2\pi imz} = a^{-1/2} \sum_{m\in\mathbb{Z}} \mathrm{e}^{-\pi(z-m)^2/a}\,, \tag{7.8}$$

for any $z \in \mathbb{C}$ and $\operatorname{Re} a > 0$. In particular, the substitution $z = 2x - ia/2$ provides the identity

$$\begin{aligned} \sum_{m\in\mathbb{Z}} (-1)^m\, \mathrm{e}^{-\pi(2x-m)^2/a} &= a^{1/2} \sum_{m\in\mathbb{Z}} \mathrm{e}^{-\pi a(2m+1)^2/4}\, \mathrm{e}^{2\pi i(2m+1)x} \\ &= 2\,a^{1/2} \sum_{m=0}^{\infty} \mathrm{e}^{-\pi a(2m+1)^2/4} \cos 2\pi(2m+1)x\,, \end{aligned} \tag{7.9}$$

which shows, for example, that

$$\begin{aligned} &\cos 2\pi x = \frac{\mathrm{e}^{\pi^2\mathcal{D}/4}}{2\sqrt{\pi\mathcal{D}}} \sum_{m=-\infty}^{\infty} (-1)^m\, \mathrm{e}^{-(2x-m)^2/\mathcal{D}} - R(x)\,, \\ &\text{with}\quad R(x) = \mathrm{e}^{\pi^2\mathcal{D}/4} \sum_{k=1}^{\infty} \cos\big(2\pi(2k+1)x\big)\, \mathrm{e}^{-\pi^2\mathcal{D}(2k+1)^2/4} = \mathcal{O}(\mathrm{e}^{-2\pi^2\mathcal{D}})\,. \end{aligned} \tag{7.10}$$

7.1.2.3. *Gaussian function.* Finally we apply Theorem 7.1 and formula (7.5) to $e^{-|\mathbf{x}|^2/\mathcal{D}_1}$, $\mathcal{D}_1 > \mathcal{D}h^2$, and arrive at the expansion

$$
\begin{aligned}
e^{-|\mathbf{x}|^2/\mathcal{D}_1} &= \Big(\frac{\mathcal{D}_1}{\pi\mathcal{D}(\mathcal{D}_1 - \mathcal{D}h^2)}\Big)^{n/2} \sum_{\mathbf{m}\in\mathbb{Z}^n} e^{-h^2|\mathbf{m}|^2/(\mathcal{D}_1-\mathcal{D}h^2)}\, e^{-|\mathbf{x}-h\mathbf{m}|^2/\mathcal{D}h^2} \\
&\quad - e^{-|\mathbf{x}|^2/\mathcal{D}_1} \sum_{\boldsymbol{\nu}\in\mathbb{Z}^n\setminus\{\mathbf{0}\}} e^{2\pi i(\mathcal{D}_1-\mathcal{D}h^2)\langle\mathbf{x},\boldsymbol{\nu}\rangle/\mathcal{D}_1 h}\, e^{-\pi^2\mathcal{D}(\mathcal{D}_1-\mathcal{D}h^2)|\boldsymbol{\nu}|^2/\mathcal{D}_1}
\end{aligned}
\tag{7.11}
$$

which shows that the Gaussian function can be expanded very accurately into a series of shifts of thinner Gaussians with the error

$$
\Big| e^{-|\mathbf{x}|^2/\mathcal{D}_1} - \Big(\frac{\mathcal{D}_1}{\pi\mathcal{D}(\mathcal{D}_1 - \mathcal{D}h^2)}\Big)^{n/2} \sum_{\mathbf{m}\in\mathbb{Z}^n} e^{-h^2|\mathbf{m}|^2/(\mathcal{D}_1-\mathcal{D}h^2)}\, e^{-|\mathbf{x}-h\mathbf{m}|^2/\mathcal{D}h^2} \Big|
$$
$$
= \mathcal{O}(e^{-\pi^2\mathcal{D}(\mathcal{D}_1-\mathcal{D}h^2)/\mathcal{D}_1})\, e^{-|\mathbf{x}|^2/\mathcal{D}_1} .
$$

7.2. High-order quasi-interpolation

Here, we apply the approach developed in Subsection 3.5.2 to study high-order quasi-interpolants of the form

$$
\mathcal{D}^{-n/2} \sum_{\mathbf{m}\in\mathbb{Z}^n} u_{\mathbf{m}}^h\, e^{-|\mathbf{x}-h\mathbf{m}|^2/\mathcal{D}h^2} .
$$

7.2.1. Generating functions. According to Subsection 3.5.2, the function

$$
\eta_{\mathcal{D}} = \pi^{n/2}\mathcal{F}^{-1}(P_M\, e^{-\pi^2\mathcal{D}|\cdot|^2})
$$

fulfills the moment Condition 2.15 with $N = 2M$, if the coefficients of the trigonometric polynomial

$$
P_M(\boldsymbol{\lambda}) = \sum_{[\boldsymbol{\beta}]<M} a_{\boldsymbol{\beta}} \prod_{j=1}^{n} \cos 2\pi\beta_j\lambda_j \,, \qquad \boldsymbol{\beta} = (\beta_1,\dots,\beta_n) \in \mathbb{Z}^n_{\geq 0},
$$

satisfy the equations

$$
\sum_{[\boldsymbol{\beta}]<M} a_{\boldsymbol{\beta}}\, \boldsymbol{\beta}^{2\boldsymbol{\alpha}} = \Big(-\frac{\mathcal{D}}{4\pi^2}\Big)^{[\boldsymbol{\alpha}]} \partial^{2\boldsymbol{\alpha}}(\mathcal{F}\eta)^{-1}(\mathbf{0})
\tag{7.12}
$$

for all $\boldsymbol{\alpha}$, $0 \leq [\boldsymbol{\alpha}] < M$, with $\eta(\mathbf{x}) = e^{-|\mathbf{x}|^2}$. By (3.1), the right-hand side has the form

$$
\partial^{2\boldsymbol{\alpha}}(\mathcal{F}\eta)^{-1}(\mathbf{0}) = \pi^{-n/2}\, \partial^{2\boldsymbol{\alpha}}\, e^{\pi^2|\boldsymbol{\lambda}|^2}\Big|_{\boldsymbol{\lambda}=\mathbf{0}} = \pi^{-n/2}\pi^{2[\boldsymbol{\alpha}]}\partial^{2\boldsymbol{\alpha}}\big(e^{|\cdot|^2}\big)(\mathbf{0}),
$$

and we obtain from (3.7) and (3.8)

$$
\partial^{2\boldsymbol{\alpha}}\big(e^{|\cdot|^2}\big)(\mathbf{0}) = \frac{(2\boldsymbol{\alpha})!}{\boldsymbol{\alpha}!} .
\tag{7.13}
$$

This leads to the linear system for the coefficients $a_{\boldsymbol{\beta}}$

$$
\sum_{[\boldsymbol{\beta}]<M} a_{\boldsymbol{\beta}}\, \boldsymbol{\beta}^{2\boldsymbol{\alpha}} = \Big(-\frac{\mathcal{D}}{4}\Big)^{[\boldsymbol{\alpha}]} \frac{(2\boldsymbol{\alpha})!}{\pi^{n/2}\boldsymbol{\alpha}!} \,, \qquad [\boldsymbol{\alpha}] < M ,
\tag{7.14}
$$

which is uniquely solvable by Theorem 3.17. As mentioned in Remark 3.18, owing to the radial structure of $e^{-|\mathbf{x}|^2}$, the values of $a_{\boldsymbol{\beta}}$ coincide for all permutations $\sigma(\boldsymbol{\beta})$

of the components of the multi-index $\boldsymbol{\beta}$, $a_{\sigma(\boldsymbol{\beta})} = a_{\boldsymbol{\beta}}$. Therefore, the solution of (7.14) can be obtained from the reduced system

$$\sum_{\substack{[\boldsymbol{\beta}]<M \\ \beta_1\geq\ldots\geq\beta_n}} a_{\boldsymbol{\beta}} \sum_{\boldsymbol{\gamma}=\sigma(\boldsymbol{\alpha})} \boldsymbol{\beta}^{2\boldsymbol{\gamma}} = \Big(-\frac{\mathcal{D}}{4}\Big)^{[\boldsymbol{\alpha}]} \frac{(2\boldsymbol{\alpha})!}{\pi^{n/2}\boldsymbol{\alpha}!} \tag{7.15}$$

for all multi-indices $0 \leq [\boldsymbol{\alpha}] < M$ with $\alpha_1 \geq \ldots \geq \alpha_n$, and the second sum extends over all different multi-indices $\boldsymbol{\gamma}$, which are permutations of the components of $\boldsymbol{\alpha}$ with $\alpha_1 \geq \ldots \geq \alpha_n$.

If we denote the number of non-zero components of $\boldsymbol{\beta}$ by $\kappa(\boldsymbol{\beta})$ and write

$$\prod_{j=1}^{n} \cos 2\pi\beta_j\lambda_j = 2^{-\kappa(\boldsymbol{\beta})} \sum_{\xi(\mathbf{k})=\boldsymbol{\beta}} \mathrm{e}^{2\pi i\langle\mathbf{k},\boldsymbol{\lambda}\rangle},$$

then we see that the unique solution of (7.15) provides the required generating function

$$\eta_{\mathcal{D}}(\mathbf{x}) = \pi^{n/2}\mathcal{F}^{-1}(P_M(\boldsymbol{\lambda})\,\mathrm{e}^{-\pi^2\mathcal{D}|\boldsymbol{\lambda}|^2})(\mathbf{x}) = \sum_{|\xi(\mathbf{k})|<M} c_{\mathbf{k}} e^{-|\mathbf{x}-\mathbf{k}|^2/\mathcal{D}}, \tag{7.16}$$

where $c_{\mathbf{k}} = 2^{-\kappa(\mathbf{k})}\, a_{\xi(\mathbf{k})}$ and as before $\xi(\mathbf{k}) = (|k_1|, \ldots, |k_n|)$.

LEMMA 7.2. *For any $M > 0$ and $\mathcal{D} > 0$, there exist uniquely determined coefficients $c_{\mathbf{k}}$ with $\sum_{j=1}^{n} |k_j| < M$, such that the function* (7.16) *is symmetric and satisfies the moment condition* (2.47) *with $N = 2M$.*

EXAMPLE 7.3. To construct the sixth-order approximate quasi-interpolant in $\mathbb{R}^3$, one has to solve the linear system (7.14) with 10 unknowns $a_{\boldsymbol{\beta}}$, $[\boldsymbol{\beta}] \leq 2$, which can be reduced to (7.15) with 4 unknows, corresponding to the multi-indices $(0,0,0)$, $(1,0,0)$, $(1,1,0)$, and $(2,0,0)$. Its solution determines 25 coefficients $c_{\mathbf{k}} = 2^{-\kappa(\mathbf{k})}\, a_{\xi(\mathbf{k})}$ with $|\xi(\mathbf{k})| \leq 2$. These coefficients are given by

$$c_{(0,0,0)} = \frac{1}{\pi^{3/2}}\Big(1 + \frac{15\mathcal{D}}{8} + \frac{21\mathcal{D}^2}{16}\Big),$$

$$c_{\mathbf{k}} = \begin{cases} -\dfrac{\mathcal{D}}{\pi^{3/2}}\Big(\dfrac{1}{3} + \dfrac{3\mathcal{D}}{8}\Big), & |\xi(\mathbf{k})| = 1, \\[2mm] \dfrac{\mathcal{D}}{\pi^{3/2}}\Big(\dfrac{1}{48} + \dfrac{\mathcal{D}}{32}\Big), & |\xi(\mathbf{k})| = 2,\ \kappa(\mathbf{k}) = 1, \\[2mm] \dfrac{\mathcal{D}^2}{\pi^{3/2}16}, & |\xi(\mathbf{k})| = 2,\ \kappa(\mathbf{k}) = 2. \end{cases}$$

7.2.2. Saturation error. Now, we study the saturation error of the quasi-interpolation operator with the generating function (7.16).

LEMMA 7.4. *For any multi-index $\boldsymbol{\alpha}$, $[\boldsymbol{\alpha}] < 2M$, and $\boldsymbol{\nu} \in \mathbb{Z}^n$ it holds that*

$$\partial^{\boldsymbol{\alpha}}\mathcal{F}\eta_{\mathcal{D}}(\boldsymbol{\nu}) = (-2\pi\sqrt{\mathcal{D}})^{[\boldsymbol{\alpha}]}\, \boldsymbol{\nu}^{\boldsymbol{\alpha}}\, \mathrm{e}^{-\pi^2\mathcal{D}|\boldsymbol{\nu}|^2}. \tag{7.17}$$

PROOF. Since

$$\mathcal{F}\eta_{\mathcal{D}}(\boldsymbol{\lambda}) = \pi^{n/2}\,\mathrm{e}^{-\pi^2\mathcal{D}|\boldsymbol{\lambda}|^2}\, P_M(\boldsymbol{\lambda}),$$

the periodicity of P_M and the equalities (7.12) imply

$$\begin{aligned}
\partial^{\boldsymbol{\alpha}} \mathcal{F}\eta_{\mathcal{D}}(\boldsymbol{\nu}) &= \pi^{n/2} \sum_{\boldsymbol{\beta}\leq\boldsymbol{\alpha}} \frac{\boldsymbol{\alpha}!\,(\pi\sqrt{\mathcal{D}})^{[\boldsymbol{\alpha}-\boldsymbol{\beta}]}}{\boldsymbol{\beta}!\,(\boldsymbol{\alpha}-\boldsymbol{\beta})!}\, \partial^{\boldsymbol{\beta}} P_M(\boldsymbol{\nu})\, \partial^{\boldsymbol{\alpha}-\boldsymbol{\beta}}\big(\mathrm{e}^{-|\cdot|^2}\big)(\pi\sqrt{\mathcal{D}}\boldsymbol{\nu}) \\
&= \pi^{n/2} \sum_{\boldsymbol{\beta}\leq\boldsymbol{\alpha}} \frac{\boldsymbol{\alpha}!\,(\pi\sqrt{\mathcal{D}})^{[\boldsymbol{\alpha}-\boldsymbol{\beta}]}}{\boldsymbol{\beta}!(\boldsymbol{\alpha}-\boldsymbol{\beta})!}\, \partial^{\boldsymbol{\beta}} P_M(\mathbf{0})\, \partial^{\boldsymbol{\alpha}-\boldsymbol{\beta}}\big(\mathrm{e}^{-|\cdot|^2}\big)(\pi\sqrt{\mathcal{D}}\boldsymbol{\nu}) \\
&= (\pi\sqrt{\mathcal{D}})^{[\boldsymbol{\alpha}]} \sum_{\boldsymbol{\beta}\leq\boldsymbol{\alpha}} \frac{\boldsymbol{\alpha}!\,\mathcal{D}^{[\boldsymbol{\beta}]/2}}{\boldsymbol{\beta}!\,(\boldsymbol{\alpha}-\boldsymbol{\beta})!}\, \partial^{\boldsymbol{\beta}}\big(\mathrm{e}^{|\cdot|^2}\big)(\mathbf{0})\, \partial^{\boldsymbol{\alpha}-\boldsymbol{\beta}}\big(\mathrm{e}^{-|\cdot|^2}\big)(\pi\sqrt{\mathcal{D}}\boldsymbol{\nu})\,.
\end{aligned}$$

Noting that by (7.13)

$$\partial^{\boldsymbol{\gamma}}\big(\mathrm{e}^{|\cdot|^2}\big)(\mathbf{0}) = \begin{cases} \dfrac{(2\boldsymbol{\beta})!}{\boldsymbol{\beta}!}, & \boldsymbol{\gamma} = 2\boldsymbol{\beta},\ \boldsymbol{\beta}\in\mathbb{Z}^n_{\geq 0}, \\ 0, & \text{otherwise}, \end{cases}$$

we obtain

$$\begin{aligned}
\partial^{\boldsymbol{\alpha}} \mathcal{F}\eta_{\mathcal{D}}(\sqrt{\mathcal{D}}\boldsymbol{\nu}) &= (\pi\sqrt{\mathcal{D}})^{[\boldsymbol{\alpha}]} \sum_{2\boldsymbol{\beta}\leq\boldsymbol{\alpha}} \frac{\boldsymbol{\alpha}!}{\boldsymbol{\beta}!\,(\boldsymbol{\alpha}-2\boldsymbol{\beta})!}\, \partial^{\boldsymbol{\alpha}-\boldsymbol{\beta}}\big(\mathrm{e}^{-|\cdot|^2}\big)(\pi\sqrt{\mathcal{D}}\boldsymbol{\nu}) \\
&= (-\pi\sqrt{\mathcal{D}})^{[\boldsymbol{\alpha}]}\, \mathrm{e}^{-\pi^2\mathcal{D}|\boldsymbol{\nu}|^2} \sum_{2\boldsymbol{\beta}\leq\boldsymbol{\alpha}} \frac{\boldsymbol{\alpha}!}{\boldsymbol{\beta}!\,(\boldsymbol{\alpha}-2\boldsymbol{\beta})!}\, H_{\boldsymbol{\alpha}-2\boldsymbol{\beta}}(\pi\sqrt{\mathcal{D}}\boldsymbol{\nu})\,,
\end{aligned}$$

where we use the Hermite polynomial in n variables (7.7). It can be easily seen that the expansion of the monomial $\mathbf{x}^{\boldsymbol{\alpha}}$ with respect to the Hermite polynomials has the form

$$\mathbf{x}^{\boldsymbol{\alpha}} = 2^{-[\boldsymbol{\alpha}]} \sum_{2\boldsymbol{\beta}\leq\boldsymbol{\alpha}} \frac{\boldsymbol{\alpha}!}{\boldsymbol{\beta}!\,(\boldsymbol{\alpha}-2\boldsymbol{\beta})!}\, H_{\boldsymbol{\alpha}-2\boldsymbol{\beta}}(\mathbf{x})\,,$$

which leads to

$$\partial^{\boldsymbol{\alpha}} \mathcal{F}\eta_{\mathcal{D}}(\sqrt{\mathcal{D}}\boldsymbol{\nu}) = (-2\pi\sqrt{\mathcal{D}}\boldsymbol{\nu})^{\boldsymbol{\alpha}}\, \mathrm{e}^{-\pi^2\mathcal{D}|\boldsymbol{\nu}|^2}$$

and establishes the assertion. □

THEOREM 7.5. *Let $M \in \mathbb{N}$, and let $\{a_{\boldsymbol{\beta}}\}_{[\boldsymbol{\beta}]<M}$ be the solution of the system (7.14). We define the point functionals*

$$u^h_{\mathbf{m}} := \sum_{|\xi(\mathbf{k})|<M} 2^{-\kappa(\mathbf{k})}\, a_{\xi(\mathbf{k})}\, u(h(\mathbf{m}-\mathbf{k}))\,. \tag{7.18}$$

Then for any $u \in W^N_\infty(\mathbb{R}^n)$ and all $h > 0$ the estimate

$$\begin{aligned}
&\Big| u(\mathbf{x}) - \mathcal{D}^{-n/2} \sum_{\mathbf{m}\in\mathbb{Z}^n} u^h_m\, \mathrm{e}^{-|\mathbf{x}-h\mathbf{m}|^2/\mathcal{D}h^2} \Big| \\
&\leq h^N \sum_{[\boldsymbol{\alpha}]=N} \frac{\rho_{\boldsymbol{\alpha}}}{\boldsymbol{\alpha}!} \|\partial^{\boldsymbol{\alpha}} u\|_{L_\infty} + \sum_{[\boldsymbol{\alpha}]=0}^{N-1} |\partial^{\boldsymbol{\alpha}} u(\mathbf{x})|\, \frac{(\sqrt{\mathcal{D}}h)^{[\boldsymbol{\alpha}]}}{\boldsymbol{\alpha}!} \sum_{\boldsymbol{\nu}\in\mathbb{Z}^n\setminus\{\mathbf{0}\}} |\boldsymbol{\nu}^{\boldsymbol{\alpha}}|\, \mathrm{e}^{-\pi^2\mathcal{D}|\boldsymbol{\nu}|^2}
\end{aligned}$$

is valid, where

$$\rho_{\boldsymbol{\alpha}} = \max_{\mathbf{x}\in\mathbb{R}^n} \mathcal{D}^{-n/2} \sum_{\mathbf{m}\in\mathbb{Z}^n} \Big| (\mathbf{x}-\mathbf{m})^{\boldsymbol{\alpha}} \sum_{|\xi(\mathbf{k})|<M} 2^{-\kappa(\mathbf{k})} a_{\xi(\mathbf{k})}\, \mathrm{e}^{-|\mathbf{x}-\mathbf{m}-\mathbf{k}|^2/\mathcal{D}} \Big|.$$

PROOF. From the representation (2.50) and Lemmas 7.2 and 7.4, we derive that the quasi-interpolant

$$
\begin{aligned}
u_h(\mathbf{x}) &= \sum_{\mathbf{m}\in\mathbb{Z}^n} u(h\mathbf{m})\, \eta_{\mathcal{D}}\Big(\frac{\mathbf{x}}{h}-\mathbf{m}\Big) \\
&= \mathcal{D}^{-n/2} \sum_{\mathbf{m}\in\mathbb{Z}^n} \mathrm{e}^{-|\mathbf{x}-h\mathbf{m}|^2/\mathcal{D}h^2} \sum_{|\xi(\mathbf{k})|<M} 2^{-\kappa(\mathbf{k})}\, a_{\xi(\mathbf{k})}\, u(h(\mathbf{m}-\mathbf{k}))
\end{aligned}
$$

can be written as

$$
\begin{aligned}
u_h(\mathbf{x}) &= u(\mathbf{x}) + R_{N,h}(\mathbf{x}) + \sum_{[\boldsymbol{\alpha}]=0}^{N-1} \Big(\frac{h}{2\pi i}\Big)^{[\boldsymbol{\alpha}]} \frac{\partial^{\boldsymbol{\alpha}} u(\mathbf{x})}{\boldsymbol{\alpha}!} \sum_{\boldsymbol{\nu}\in\mathbb{Z}^n\setminus\{\mathbf{0}\}} \partial^{\boldsymbol{\alpha}} \mathcal{F}\eta_{\mathcal{D}}(\boldsymbol{\nu})\, \mathrm{e}^{\frac{2\pi i}{h}\langle \mathbf{x},\boldsymbol{\nu}\rangle} \\
&= u(\mathbf{x}) + R_{N,h}(\mathbf{x}) + \sum_{[\boldsymbol{\alpha}]=0}^{N-1} (i\sqrt{\mathcal{D}}h)^{[\boldsymbol{\alpha}]} \frac{\partial^{\boldsymbol{\alpha}} u(\mathbf{x})}{\boldsymbol{\alpha}!} \sum_{\boldsymbol{\nu}\in\mathbb{Z}^n\setminus\{\mathbf{0}\}} \boldsymbol{\nu}^{\boldsymbol{\alpha}}\, \mathrm{e}^{-\pi^2\mathcal{D}|\boldsymbol{\nu}|^2}\, \mathrm{e}^{\frac{2\pi i}{h}\langle \mathbf{x},\boldsymbol{\nu}\rangle},
\end{aligned}
$$

and $R_{N,h}$ is bounded by Theorem 2.17. □

REMARK 7.6. Any trigonometric polynomial $P(\boldsymbol{\lambda})$ of period 1 satisfying the equations (7.12) can be used to construct a generating function by

$$
\eta(\mathbf{x}) = \pi^{n/2}\mathcal{F}^{-1}\big(\mathrm{e}^{-\pi^2\mathcal{D}|\cdot|^2}\, P\big)(\mathbf{x})
$$

such that the assertion of the previous theorem holds. But it follows from Lemma 7.2 that formula (7.18) gives the quasi-interpolant depending on the minimal number of function values $u(h(\mathbf{m}-\mathbf{k}))$, $|\xi(\mathbf{k})| < M$.

$[\boldsymbol{\alpha}]$	$\mathcal{D}=3$	$\mathcal{D}=4$
0	$8.30\cdot 10^{-13}$	$4.29\cdot 10^{-17}$
1	$2.61\cdot 10^{-12}$	$1.80\cdot 10^{-16}$
2	$1.23\cdot 10^{-11}$	$1.13\cdot 10^{-15}$
3	$3.86\cdot 10^{-11}$	$4.73\cdot 10^{-15}$
4	$9.10\cdot 10^{-11}$	$1.49\cdot 10^{-14}$
5	$1.71\cdot 10^{-10}$	$3.74\cdot 10^{-14}$
6	$2.69\cdot 10^{-10}$	$7.83\cdot 10^{-14}$
7	$3.63\cdot 10^{-10}$	$1.41\cdot 10^{-13}$

TABLE 7.1. Coefficients of the saturation error for $n=3$.

In Table 7.1, we give upper bounds for the coefficients in the saturation error

$$
\frac{(\mathcal{D}\pi)^{[\boldsymbol{\alpha}]}}{\boldsymbol{\alpha}!} \sum_{\boldsymbol{\nu}\in\mathbb{Z}^n\setminus\{\mathbf{0}\}} |\boldsymbol{\nu}^{\boldsymbol{\alpha}}|\, \mathrm{e}^{-\pi^2\mathcal{D}|\boldsymbol{\nu}|^2}
$$

for different $[\boldsymbol{\alpha}]$ and $\mathcal{D}$, in the case of approximate quasi-interpolation in $\mathbb{R}^3$. Note that these coefficients are multiplied by $h^{[\boldsymbol{\alpha}]}$ (see Theorem 7.5).

7.3. Interpolation with Gaussian kernels

In the following, we consider the interpolation with the semi-discrete convolution

$$(\pi\mathcal{D})^{-n/2}\sum_{\mathbf{m}\in\mathbb{Z}^n} u_{\mathbf{m}}\, \mathrm{e}^{-|\mathbf{x}-h\mathbf{m}|^2/\mathcal{D}h^2} \tag{7.19}$$

at the lattice $\{h\mathbf{m},\ \mathbf{m}\in\mathbb{Z}^n\}$, i.e., for a given function u, we look for a function $Q_h u$ of the form (7.19) such that

$$Q_h u(h\mathbf{m}) = u(h\mathbf{m}) \quad \text{for all} \quad \mathbf{m}\in\mathbb{Z}^n\,.$$

7.3.1. Lagrangian function. As mentioned in Subsection 1.2.3, the interpolant $Q_h u$ has the simple form

$$Q_h u(\mathbf{x}) = \sum_{\mathbf{m}\in\mathbb{Z}^n} u(h\mathbf{m})\,\chi_{\mathcal{D}}\Big(\frac{\mathbf{x}}{h}-\mathbf{m}\Big)$$

with the Lagrangian function

$$\chi_{\mathcal{D}}(\mathbf{x}) = (\pi\mathcal{D})^{-n/2}\sum_{\mathbf{m}\in\mathbb{Z}^n} \chi_{\mathbf{m}}\, \mathrm{e}^{-|\mathbf{x}-\mathbf{m}|^2/\mathcal{D}} \tag{7.20}$$

which satisfies $\chi_{\mathcal{D}}(\mathbf{k}) = \delta_{0|\mathbf{k}|}$ for all integer vectors $\mathbf{k}\in\mathbb{Z}^n$. In order to determine the Lagrangian function $\chi_{\mathcal{D}}$, we introduce

$$g_{\mathcal{D}}(\boldsymbol{\lambda}) := \sum_{\boldsymbol{\nu}\in\mathbb{Z}^n} \mathrm{e}^{-\pi^2\mathcal{D}|\boldsymbol{\lambda}-\boldsymbol{\nu}|^2}\,, \tag{7.21}$$

which can be written, in view of Poisson's summation formula (2.17), in the form

$$g_{\mathcal{D}}(\boldsymbol{\lambda}) = (\pi\mathcal{D})^{-n/2}\sum_{\mathbf{m}\in\mathbb{Z}^n} \mathrm{e}^{-|\mathbf{m}|^2/\mathcal{D}}\,\mathrm{e}^{2\pi i\langle\mathbf{m},\boldsymbol{\lambda}\rangle}\,.$$

LEMMA 7.7. *The Lagrangian function (7.20) is given by*

$$\chi_{\mathcal{D}}(\mathbf{x}) = \int_{\mathbb{R}^n} \mathrm{e}^{2\pi i\langle\mathbf{x},\boldsymbol{\lambda}\rangle}\,\frac{\mathrm{e}^{-\pi^2\mathcal{D}|\boldsymbol{\lambda}|^2}}{g_{\mathcal{D}}(\boldsymbol{\lambda})}\,d\boldsymbol{\lambda}\,, \tag{7.22}$$

and the coefficients $\chi_{\mathbf{m}}$ are the Fourier coefficients of the function $1/g_{\mathcal{D}}$, i.e.,

$$\chi_{\mathbf{m}} = \int_Q \frac{\mathrm{e}^{-2\pi i\langle\mathbf{m},\boldsymbol{\lambda}\rangle}}{g_{\mathcal{D}}(\boldsymbol{\lambda})}\,d\boldsymbol{\lambda}\,, \tag{7.23}$$

where Q is the cube $\big[-\frac{1}{2},\frac{1}{2}\big]^n$.

PROOF. We transform

$$\int_{\mathbb{R}^n} \mathrm{e}^{2\pi i\langle\mathbf{x},\boldsymbol{\lambda}\rangle}\,\frac{\mathrm{e}^{-\pi^2\mathcal{D}|\boldsymbol{\lambda}|^2}}{g_{\mathcal{D}}(\boldsymbol{\lambda})}\,d\boldsymbol{\lambda} = \sum_{\mathbf{m}\in\mathbb{Z}^n}\int_{\mathbf{m}+Q} \mathrm{e}^{2\pi i\langle\mathbf{x},\boldsymbol{\lambda}\rangle}\,\frac{\mathrm{e}^{-\pi^2\mathcal{D}|\boldsymbol{\lambda}|^2}}{g_{\mathcal{D}}(\boldsymbol{\lambda})}\,d\boldsymbol{\lambda}$$

$$= \int_Q \mathrm{e}^{2\pi i\langle\mathbf{x},\boldsymbol{\lambda}\rangle}\sum_{\mathbf{m}\in\mathbb{Z}^n} \mathrm{e}^{2\pi i\langle\mathbf{x},\mathbf{m}\rangle}\,\frac{\mathrm{e}^{-\pi^2\mathcal{D}|\boldsymbol{\lambda}+\mathbf{m}|^2}}{g_{\mathcal{D}}(\boldsymbol{\lambda})}\,d\boldsymbol{\lambda}\,.$$

For $\mathbf{x}=\mathbf{k}$, it follows from (7.21) that

$$\chi_{\mathcal{D}}(\mathbf{k}) = \int_Q \frac{\mathrm{e}^{2\pi i\langle\mathbf{k},\boldsymbol{\lambda}\rangle}}{g_{\mathcal{D}}(\boldsymbol{\lambda})}\sum_{\mathbf{m}\in\mathbb{Z}^n} \mathrm{e}^{-\pi^2\mathcal{D}|\boldsymbol{\lambda}+\mathbf{m}|^2}\,d\boldsymbol{\lambda} = \delta_{0|\mathbf{k}|}\,.$$

Thus, we obtain from (7.20) and (7.22) that

$$\mathcal{F}\chi_{\mathcal{D}}(\boldsymbol{\lambda}) = \frac{\mathrm{e}^{-\pi^2\mathcal{D}|\boldsymbol{\lambda}|^2}}{g_{\mathcal{D}}(\boldsymbol{\lambda})} = \mathrm{e}^{-\pi^2\mathcal{D}|\boldsymbol{\lambda}|^2} \sum_{\mathbf{m}\in\mathbb{Z}^n} \chi_{\mathbf{m}}\, \mathrm{e}^{2\pi i\langle \mathbf{m},\boldsymbol{\lambda}\rangle}\,,$$

which establishes (7.23). □

The coefficients $\chi_{\mathbf{m}}$ can be derived using the following lemma.

LEMMA 7.8. *Let $a > 0$. Then*

$$\Big(\sum_{m\in\mathbb{Z}} \mathrm{e}^{-am^2}\, \mathrm{e}^{2\pi imx}\Big)^{-1} = \sum_{m\in\mathbb{Z}} a_m\, \mathrm{e}^{2\pi imx}\,, \tag{7.24}$$

where

$$a_m = \frac{\mathrm{e}^{\,am^2}}{\kappa(a)} \sum_{r=|m|}^{\infty} (-1)^r\, \mathrm{e}^{\,-a(r+1/2)^2}\,,$$

and the constant

$$\kappa(a) = \sum_{r=-\infty}^{\infty} (4r+1)\, \mathrm{e}^{\,-a(2r+1/2)^2} = \Big(\frac{\pi}{a}\Big)^{3/2} \sum_{r=-\infty}^{\infty} (4r+1)\, \mathrm{e}^{-\pi^2(2r+1/2)^2/a}\,.$$

The formula for the coefficients a_m can be deduced by applying the Residue Theorem to the Theta function ϑ_3, which leads to certain recurrence relations for these coefficients. Here we verify formula (7.24) by checking that

$$\sum_{m\in\mathbb{Z}} \mathrm{e}^{-a(k-m)^2}\, a_m = \delta_{k0}\,. \tag{7.25}$$

We consider the sum

$$\begin{aligned}
&\sum_{m=-\infty}^{\infty} \mathrm{e}^{\,-a(k-m)^2}\, \mathrm{e}^{\,am^2} \sum_{r=|m|}^{\infty} (-1)^r\, \mathrm{e}^{\,-a(r+1/2)^2} \\
&= \mathrm{e}^{\,-ak^2-a/4}\Big(\sum_{r=0}^{\infty}(-1)^r\, \mathrm{e}^{\,-ar(r+1)} + \sum_{m=1}^{\infty}(\mathrm{e}^{\,2amk}+\mathrm{e}^{\,-2amk}) \sum_{r=m}^{\infty}(-1)^r\, \mathrm{e}^{\,-ar(r+1)}\Big) \\
&= \mathrm{e}^{\,-ak^2-a/4}\Big(\sum_{r=0}^{\infty}(-1)^r\, \mathrm{e}^{\,-ar(r+1)} + \sum_{r=1}^{\infty}(-1)^r\, \mathrm{e}^{\,-ar(r+1)} \sum_{m=1}^{r}(\mathrm{e}^{\,2amk}+\mathrm{e}^{\,-2amk})\Big).
\end{aligned}$$

If $k = 0$, then

$$\begin{aligned}
&\sum_{m=-\infty}^{\infty}\sum_{r=|m|}^{\infty} (-1)^r\, \mathrm{e}^{\,-a(r+1/2)^2} = \mathrm{e}^{\,-a/4}\Big(1+\sum_{r=1}^{\infty}(-1)^r\, \mathrm{e}^{\,-ar(r+1)}(1+2r)\Big) \\
&= \sum_{r=0}^{\infty}(-1)^r(2r+1)\, \mathrm{e}^{\,-a(r+1/2)^2} = \sum_{r=-\infty}^{\infty} (4r+1)\, \mathrm{e}^{\,-a(2r+1/2)^2}\,.
\end{aligned}$$

For $k \neq 0$, from

$$\begin{aligned}
&\sum_{r=0}^{\infty}(-1)^r \mathrm{e}^{-ar(r+1)} + \sum_{r=1}^{\infty}(-1)^r \mathrm{e}^{-ar(r+1)} \sum_{m=1}^{r}(\mathrm{e}^{2amk} + \mathrm{e}^{-2amk}) \\
&= 1 + \sum_{r=1}^{\infty}(-1)^r \mathrm{e}^{-ar(r+1)} \left(1 + \frac{\mathrm{e}^{2ak(r+1)} - \mathrm{e}^{2ak}}{\mathrm{e}^{2ak} - 1} + \frac{\mathrm{e}^{-2ak(r+1)} - \mathrm{e}^{-2ak}}{\mathrm{e}^{-2ak} - 1}\right) \\
&= 1 + \sum_{r=1}^{\infty}(-1)^r \mathrm{e}^{-ar(r+1)} \frac{\mathrm{e}^{2ak(r+1)} - \mathrm{e}^{-2akr}}{\mathrm{e}^{2ak} - 1} \\
&= \frac{1}{\mathrm{e}^{2ak} - 1}\left(\mathrm{e}^{2ak} - 1 + \sum_{r=1}^{\infty}(-1)^r \left(\mathrm{e}^{-a(r+1)(r-2k)} - \mathrm{e}^{-ar(r+1+2k)}\right)\right) \\
&= \frac{1}{\mathrm{e}^{2ak} - 1}\left(\sum_{r=0}^{\infty}(-1)^r \mathrm{e}^{-a(r+1)(r-2k)} - \sum_{r=0}^{\infty}(-1)^r \mathrm{e}^{-ar(r+1+2k)}\right) \\
&= \frac{1}{\mathrm{e}^{2ak} - 1}\left(\sum_{r=0}^{\infty}(-1)^r \mathrm{e}^{-a(r+1)(r-2k)} + \sum_{r=-\infty}^{-1}(-1)^r \mathrm{e}^{-a(r+1)(r-2k)}\right)
\end{aligned}$$

we obtain the identity

$$\begin{aligned}
&\sum_{m=-\infty}^{\infty} \mathrm{e}^{-a(k-m)^2} \mathrm{e}^{am^2} \sum_{r=|m|}^{\infty} (-1)^r \mathrm{e}^{-a(r+1/2)^2} \\
&= \frac{\mathrm{e}^{-ak^2-a/4}}{\mathrm{e}^{2ak} - 1} \sum_{r=-\infty}^{\infty} (-1)^r \mathrm{e}^{-a(r+1)(r-2k)} = \frac{1}{\mathrm{e}^{ak} - \mathrm{e}^{-ak}} \sum_{r=-\infty}^{\infty} (-1)^r \mathrm{e}^{-a(r-k+1/2)^2}.
\end{aligned}$$

Now, we note that

$$\begin{aligned}
&\sum_{r=-\infty}^{\infty} (-1)^r \mathrm{e}^{-a(r-k+1/2)^2} = \sum_{r=-\infty}^{\infty} \mathrm{e}^{-a(2r-k+1/2)^2} - \sum_{r=-\infty}^{\infty} \mathrm{e}^{-a(2r-k-1/2)^2} \\
&= \sum_{r=-\infty}^{\infty} \mathrm{e}^{-a(2r-k+1/2)^2} - \sum_{m=-\infty}^{\infty} \mathrm{e}^{-a(2(k-m)-k-1/2)^2} \\
&= \sum_{r=-\infty}^{\infty} \mathrm{e}^{-a(2r-k+1/2)^2} - \sum_{r=-\infty}^{\infty} \mathrm{e}^{-a(2r-k+1/2)^2} = 0,
\end{aligned}$$

which confirms the assertion of Lemma 7.8.

Owing to (7.23) and Lemma 7.8, the coefficients of the Lagrangian function (7.20) are given by

$$\chi_{\mathbf{m}} = (\pi\mathcal{D})^{n/2} \frac{\mathrm{e}^{|\mathbf{m}|^2/\mathcal{D}}}{\kappa(\mathcal{D}^{-1})^n} \prod_{j=1}^{n} \sum_{r=|m_j|}^{\infty} (-1)^r \mathrm{e}^{-(r+1/2)^2/cD}.$$

Thus, in the one-dimensional case the Lagrangian function has the form

$$\begin{aligned}
\chi_{\mathcal{D}}(x) &= \frac{1}{\kappa(\mathcal{D}^{-1})} \sum_{m=-\infty}^{\infty} \mathrm{e}^{-(x-m)^2/\mathcal{D}} \mathrm{e}^{m^2/\mathcal{D}} \sum_{r=|m|}^{\infty} (-1)^r \mathrm{e}^{-(r+1/2)^2/\mathcal{D}} \\
&= \frac{1}{\kappa(\mathcal{D}^{-1})} \frac{1}{\mathrm{e}^{x/\mathcal{D}} - \mathrm{e}^{-x/\mathcal{D}}} \sum_{r=-\infty}^{\infty} (-1)^r \mathrm{e}^{-(r-x+1/2)^2/\mathcal{D}}, \qquad (7.26)
\end{aligned}$$

where the last equation is obtained similarly to the verification of (7.25). Consequently,

$$\chi_{\mathcal{D}}(x) = \frac{2}{\kappa(\mathcal{D}^{-1}) \sinh \dfrac{x}{\mathcal{D}}} \sum_{r=-\infty}^{\infty} \left(e^{-(x-(2r+1/2))^2/\mathcal{D}} - e^{-(x+(2r+1/2))^2/\mathcal{D}} \right)$$

$$= \frac{e^{-x^2/\mathcal{D}}}{\sinh \dfrac{x}{\mathcal{D}}} \frac{\sum\limits_{r=-\infty}^{\infty} e^{-(2r+1/2)^2/\mathcal{D}} \sinh\left((4r+1)x/\mathcal{D}\right)}{\sum\limits_{r=-\infty}^{\infty} (4r+1)\, e^{-(2r+1/2)^2/\mathcal{D}}} .$$

Summarizing, we obtain

LEMMA 7.9. *The Lagrangian function of the span* $\left\{ e^{-|\cdot - \mathbf{m}|^2/\mathcal{D}}, \mathbf{m} \in \mathbb{Z}^n \right\}$ *is given by*

$$\chi_{\mathcal{D}}(\mathbf{x}) = e^{-|\mathbf{x}|^2/\mathcal{D}} \frac{\prod\limits_{j=1}^{n} \sum\limits_{r_j=-\infty}^{\infty} \sinh \dfrac{(4r_j+1)x_j}{\mathcal{D}}\, e^{-(2r_j+1/2)^2/\mathcal{D}}}{\left(\sum\limits_{r=-\infty}^{\infty} (4r+1)\, e^{-(2r+1/2)^2/\mathcal{D}} \right)^n \prod\limits_{j=1}^{n} \sinh \dfrac{x_j}{\mathcal{D}}} .$$

In the next part, we show that a small perturbation of $\chi_{\mathcal{D}}$ leads to another Lagrangian function with simple analytic representation and interesting approximation properties.

7.3.2. Simplification.

We use the expression (7.26)

$$\chi_{\mathcal{D}}(x) = \frac{1}{2\kappa(\mathcal{D}^{-1}) \sinh(x/\mathcal{D})} \sum_{r=-\infty}^{\infty} (-1)^r\, e^{-(r-x+1/2)^2/\mathcal{D}}$$

and the identity

$$\sum_{r=-\infty}^{\infty} (-1)^r\, e^{-(r-x+1/2)^2/\mathcal{D}} = 2\sqrt{\pi\mathcal{D}} \sum_{k=0}^{\infty} (-1)^k\, e^{-\pi^2\mathcal{D}(2k+1)^2/4} \sin(2k+1)\pi x$$
$$= 2\sqrt{\pi\mathcal{D}}\, e^{-\pi^2\mathcal{D}/4} \sin \pi x + \mathcal{O}(e^{-9\pi^2\mathcal{D}/4}) ,$$

which follows from (7.9). Hence, the Lagrangian function $\chi_{\mathcal{D}}(x)$ can be represented as

$$\chi_{\mathcal{D}}(x) = \frac{\sqrt{\pi\mathcal{D}}\, e^{-\pi^2\mathcal{D}/4} \sin \pi x}{\kappa(\mathcal{D}^{-1}) \sinh(x/\mathcal{D})} + \mathcal{O}(e^{-9\pi^2\mathcal{D}/4})$$

with

$$\kappa(\mathcal{D}^{-1}) = (\pi\mathcal{D})^{3/2}\, e^{-\pi^2\mathcal{D}/4} + \mathcal{O}(e^{-9\pi^2\mathcal{D}/4}) .$$

Thus, we can replace $\chi_{\mathcal{D}}$ with the function

$$\Psi_{\mathcal{D}}(x) = \frac{\sin \pi x}{\pi\mathcal{D} \sinh \dfrac{x}{\mathcal{D}}} , \tag{7.27}$$

which, obviously, satisfies $\Psi_{\mathcal{D}}(k) = \delta_{0k}$ for all $k \in \mathbb{Z}$ and is a small perturbation of $\chi_{\mathcal{D}}$. It can be easily seen that

$$|\Psi_{\mathcal{D}}(x) - \chi_{\mathcal{D}}(x)| \le c\,\frac{\mathrm{e}^{-9\pi^2\mathcal{D}/4}}{\cosh\dfrac{x}{\mathcal{D}}}\,. \tag{7.28}$$

In Figs. 7.1 and 7.2 we depict the graphs of

$$\operatorname{sinc} x = \frac{\sin \pi x}{\pi x} \quad \text{and} \quad \Psi_2(x) = \frac{\sin \pi x}{2\pi\,\sinh\dfrac{x}{2}}\,,$$

which indicate that the use of the Lagrangian function $\Psi_{\mathcal{D}}(x)$ reduces drastically the number of summations required to obtain a smooth interpolant to a given function. On the other hand, we will show that up to negligible saturation errors $\Psi_{\mathcal{D}}(x)$ provides similar approximation properties as the sinc approximants.

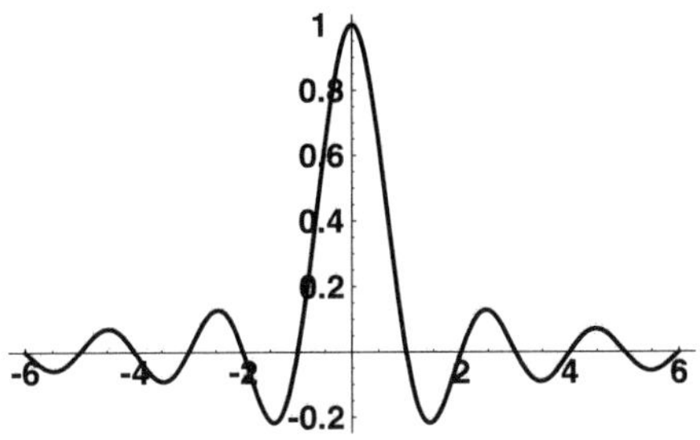

FIGURE 7.1. sinc x

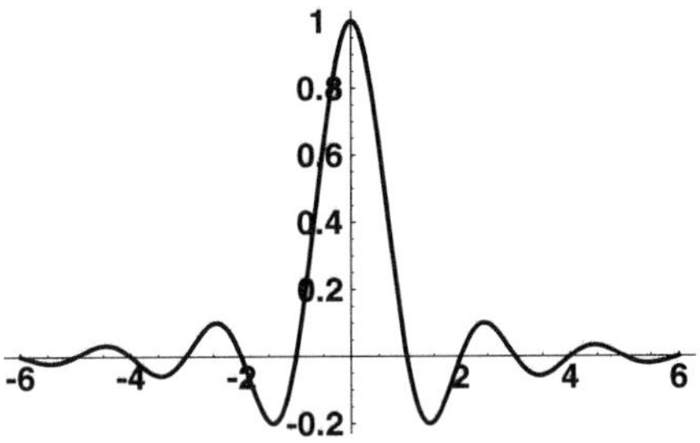

FIGURE 7.2. $\Psi_2(x)$

By the formula [**81**, 2.5.46.2]

$$\int_0^\infty \frac{\sin bx}{\sinh ax}\,dx = \frac{\pi}{2a}\,\tanh\frac{\pi b}{2a}\,,$$

the Fourier transform of $\Psi_{\mathcal{D}}$ computes to

$$\begin{aligned}\mathcal{F}\Psi_{\mathcal{D}}(\lambda) &= \frac{\sinh \pi^2\mathcal{D}}{2\cosh \pi^2\mathcal{D}(\lambda+1/2)\,\cosh \pi^2\mathcal{D}(\lambda-1/2)}\\ &= \frac{\sinh \pi^2\mathcal{D}}{\cosh 2\pi^2\mathcal{D}\lambda + \cosh \pi^2\mathcal{D}}\,.\end{aligned}$$

As in Subsection 7.3.1, we denote with the same symbol $\Psi_{\mathcal{D}}$ the Lagrangian function in $\mathbb{R}^n$ given by the tensor product

$$\Psi_{\mathcal{D}}(\mathbf{x}) = \frac{1}{(\pi\mathcal{D})^n}\prod_{j=1}^n \frac{\sin \pi x_j}{\sinh\dfrac{x_j}{\mathcal{D}}}\,, \qquad \mathbf{x} = (x_1, \ldots, x_n), \tag{7.29}$$

which has the Fourier transform

$$\mathcal{F}\Psi_{\mathcal{D}}(\boldsymbol{\lambda}) = \prod_{j=1}^n \frac{\sinh \pi^2\mathcal{D}}{\cosh 2\pi^2\mathcal{D}\lambda_j + \cosh \pi^2\mathcal{D}}\,.$$

7.3.3. Interpolation error. In the following, we study some approximation properties of the interpolant

$$\mathcal{Q}_h u(\mathbf{x}) = \sum_{\mathbf{m}\in\mathbb{Z}^n} u(h\mathbf{m})\Psi_{\mathcal{D}}\Big(\frac{\mathbf{x}}{h}-\mathbf{m}\Big). \tag{7.30}$$

Suppose that the continuous function u belongs to $L_1(\mathbb{R}^n)$ and $\mathcal{F}u \in L_1(\mathbb{R}^n)$. Then

$$\mathcal{Q}_h u(\mathbf{x}) = \sum_{\mathbf{m}\in\mathbb{Z}^n} \Psi_{\mathcal{D}}\Big(\frac{\mathbf{x}}{h}-\mathbf{m}\Big) \int_{\mathbb{R}^n} \mathcal{F}u(\boldsymbol{\lambda})\,\mathrm{e}^{2\pi i\langle h\mathbf{m},\boldsymbol{\lambda}\rangle}\,d\boldsymbol{\lambda}$$
$$= \int_{\mathbb{R}^n} \mathcal{F}u(\boldsymbol{\lambda}) \sum_{\mathbf{m}\in\mathbb{Z}^n} \Psi_{\mathcal{D}}\Big(\frac{\mathbf{x}}{h}-\mathbf{m}\Big)\,\mathrm{e}^{2\pi i\langle \mathbf{m},h\boldsymbol{\lambda}\rangle}\,d\boldsymbol{\lambda},$$

where the change of integration and summation is justified because the integrand is absolutely integrable and $\Psi_{\mathcal{D}} \in \mathcal{S}(\mathbb{R}^n)$. From Poisson's summation formula (2.21), we derive

$$\sum_{\mathbf{m}\in\mathbb{Z}^n} \Psi_{\mathcal{D}}\Big(\frac{\mathbf{x}}{h}-\mathbf{m}\Big)\,\mathrm{e}^{2\pi i h\langle \mathbf{m},\boldsymbol{\lambda}\rangle} = \mathrm{e}^{2\pi i\langle \mathbf{x},\boldsymbol{\lambda}\rangle} \sum_{\boldsymbol{\nu}\in\mathbb{Z}^n} \mathrm{e}^{\frac{2\pi i}{h}\langle \mathbf{x},\boldsymbol{\nu}\rangle}\,\mathcal{F}\Psi_{\mathcal{D}}(h\boldsymbol{\lambda}+\boldsymbol{\nu}).$$

Hence

$$\mathcal{Q}_h u(\mathbf{x}) = \int_{\mathbb{R}^n} \mathcal{F}u(\boldsymbol{\lambda})\,\mathrm{e}^{2\pi i\langle \mathbf{x},\boldsymbol{\lambda}\rangle} \sum_{\boldsymbol{\nu}\in\mathbb{Z}^n} \mathrm{e}^{\frac{2\pi i}{h}\langle \mathbf{x},\boldsymbol{\nu}\rangle}\,\mathcal{F}\Psi_{\mathcal{D}}(h\boldsymbol{\lambda}+\boldsymbol{\nu})\,d\boldsymbol{\lambda}.$$

Moreover, since

$$\sum_{\boldsymbol{\nu}\in\mathbb{Z}^n} \mathcal{F}\Psi_{\mathcal{D}}(h\boldsymbol{\lambda}+\boldsymbol{\nu}) = \sum_{\mathbf{m}\in\mathbb{Z}^n} \Psi_{\mathcal{D}}(\mathbf{m})\,\mathrm{e}^{-2\pi i h\langle \mathbf{m},\boldsymbol{\lambda}\rangle} = 1,$$

we obtain the representation of the interpolation error

$$u(\mathbf{x}) - \mathcal{Q}_h u(\mathbf{x}) = \int_{\mathbb{R}^n} \mathcal{F}u(\boldsymbol{\lambda})\,\mathrm{e}^{2\pi i\langle \mathbf{x},\boldsymbol{\lambda}\rangle} \Big(1 - \sum_{\boldsymbol{\nu}\in\mathbb{Z}^n} \mathrm{e}^{\frac{2\pi i}{h}\langle \mathbf{x},\boldsymbol{\nu}\rangle}\,\mathcal{F}\Psi_{\mathcal{D}}(h\boldsymbol{\lambda}+\boldsymbol{\nu})\Big)\,d\boldsymbol{\lambda}.$$

Let us introduce the function

$$\vartheta_{\mathcal{D}}(\mathbf{x},\boldsymbol{\lambda}) = \sum_{\boldsymbol{\nu}\in\mathbb{Z}^n} \mathrm{e}^{2\pi i\langle \mathbf{x},\boldsymbol{\nu}\rangle}\,\mathcal{F}\Psi_{\mathcal{D}}(\boldsymbol{\lambda}+\boldsymbol{\nu})$$
$$= \sum_{\boldsymbol{\nu}\in\mathbb{Z}^n} \prod_{j=1}^{n} \frac{\sinh \pi^2\mathcal{D}\,\mathrm{e}^{2\pi i x_j\nu_j}}{\cosh 2\pi^2\mathcal{D}(\lambda_j+\nu_j) + \cosh \pi^2\mathcal{D}},$$

such that

$$u(\mathbf{x}) - \mathcal{Q}_h u(\mathbf{x}) = \int_{\mathbb{R}^n} \mathcal{F}u(\boldsymbol{\lambda})\,\mathrm{e}^{2\pi i\langle \mathbf{x},\boldsymbol{\lambda}\rangle} \Big(1 - \vartheta_{\mathcal{D}}\Big(\frac{\mathbf{x}}{h}, h\boldsymbol{\lambda}\Big)\Big)\,d\boldsymbol{\lambda}.$$

Note that

$$|\vartheta_{\mathcal{D}}(\mathbf{x},\boldsymbol{\lambda})| \le 1$$

and for $\boldsymbol{\mu} \in \mathbb{Z}^n$

$$\vartheta_{\mathcal{D}}\Big(\frac{\mathbf{x}}{h}, h\boldsymbol{\lambda}+\boldsymbol{\mu}\Big) = \mathrm{e}^{-\frac{2\pi i}{h}\langle \mathbf{x},\boldsymbol{\mu}\rangle}\,\vartheta_{\mathcal{D}}\Big(\frac{\mathbf{x}}{h}, h\boldsymbol{\lambda}\Big). \tag{7.31}$$

Moreover, for fixed $\mathbf{x} \in \mathbb{R}^n$ the function $\vartheta_{\mathcal{D}}(\mathbf{x}, \mathbf{z})$ is meromorphic in $\mathbf{z} \in \mathbb{C}^n$ with simple poles at

$$\mathbf{k} + \frac{1}{2}\mathbf{e} + \frac{i}{\pi\mathcal{D}}\,(\mathbf{m} + \frac{1}{2}\mathbf{e}), \quad \mathbf{k}, \mathbf{m} \in \mathbb{Z}^n\,, \tag{7.32}$$

with the vector $\mathbf{e} = (1, \ldots, 1)$.

Using (7.31), the interpolation error can be transformed to

$$\begin{aligned} u(\mathbf{x}) - \mathcal{Q}_h u(\mathbf{x}) &= \int\limits_{\mathbb{R}^n} \mathrm{e}^{2\pi i \langle \mathbf{x}, \boldsymbol{\lambda} \rangle}\, \mathcal{F}u(\boldsymbol{\lambda}) \Big(1 - \vartheta_{\mathcal{D}}\Big(\frac{\mathbf{x}}{h}, h\boldsymbol{\lambda}\Big)\Big) d\boldsymbol{\lambda} \\ &= \sum_{\boldsymbol{\mu} \in \mathbb{Z}^n} \int\limits_{h^{-1}Q} \mathrm{e}^{2\pi i \langle \mathbf{x}, \boldsymbol{\lambda} + \boldsymbol{\mu}/h \rangle}\, \mathcal{F}u\big(\boldsymbol{\lambda} + \frac{\boldsymbol{\mu}}{h}\big) \Big(1 - \mathrm{e}^{-\frac{2\pi i}{h}\langle \mathbf{x}, \boldsymbol{\mu} \rangle}\, \vartheta_{\mathcal{D}}\Big(\frac{\mathbf{x}}{h}, h\boldsymbol{\lambda}\Big)\Big) d\boldsymbol{\lambda} \\ &= \int\limits_{h^{-1}Q} \mathrm{e}^{2\pi i \langle \mathbf{x}, \boldsymbol{\lambda} \rangle} \sum_{\boldsymbol{\mu} \in \mathbb{Z}^n} \mathcal{F}u\big(\boldsymbol{\lambda} + \frac{\boldsymbol{\mu}}{h}\big) \Big(\mathrm{e}^{\frac{2\pi i}{h}\langle \mathbf{x}, \boldsymbol{\mu} \rangle} - \vartheta_{\mathcal{D}}\Big(\frac{\mathbf{x}}{h}, h\boldsymbol{\lambda}\Big)\Big) d\boldsymbol{\lambda}\,, \end{aligned}$$

where $h^{-1}Q$ denotes the cube $[-\frac{1}{2h}, \frac{1}{2h}]^n$. Thus, we obtain the estimate

$$\begin{aligned} |u(\mathbf{x}) - \mathcal{Q}_h u(\mathbf{x})| &\le 2 \sum_{\boldsymbol{\mu} \in \mathbb{Z}^n \setminus \{\mathbf{0}\}} \int\limits_{h^{-1}Q} \big|\mathcal{F}u\big(\boldsymbol{\lambda} + \frac{\boldsymbol{\mu}}{h}\big)\big|\, d\boldsymbol{\lambda} \\ &\quad + \Big| \int\limits_{h^{-1}Q} \mathrm{e}^{2\pi i \langle \mathbf{x}, \boldsymbol{\lambda} \rangle}\, \mathcal{F}u(\boldsymbol{\lambda}) \Big(1 - \vartheta_{\mathcal{D}}\Big(\frac{\mathbf{x}}{h}, h\boldsymbol{\lambda}\Big)\Big) d\boldsymbol{\lambda} \Big| \\ &\le 2 \int\limits_{\mathbb{R}^n \setminus h^{-1}Q} |\mathcal{F}u(\boldsymbol{\lambda})|\, d\boldsymbol{\lambda} + \Big| \int\limits_{h^{-1}Q} \mathrm{e}^{2\pi i \langle \mathbf{x}, \boldsymbol{\lambda} \rangle}\, \mathcal{F}u(\boldsymbol{\lambda}) \Big(1 - \vartheta_{\mathcal{D}}\Big(\frac{\mathbf{x}}{h}, h\boldsymbol{\lambda}\Big)\Big) d\boldsymbol{\lambda} \Big|. \end{aligned} \tag{7.33}$$

7.3.4. Spectral convergence. Based on the inequality (7.33), one can derive several estimates for the interpolation error. Here and in the next subsection, we give two examples. We restrict ourselves to the case when the continuous function u belongs to $L_1(\mathbb{R}^n)$ and its Fourier transform satisfies certain integrability conditions.

THEOREM 7.10. *Suppose that the continuous function u is such that*

$$\|u\|'_N := \int\limits_{\mathbb{R}^n} |\mathcal{F}u(\boldsymbol{\lambda})|\, (1 + |\boldsymbol{\lambda}|)^N\, d\boldsymbol{\lambda} < \infty$$

for some $N \in \mathbb{N}$. Then the estimate

$$|u(\mathbf{x}) - \mathcal{Q}_h u(\mathbf{x})| \le c_N\, h^N\, \|u\|'_N + \sum_{[\boldsymbol{\alpha}]=0}^{N-1} h^{[\boldsymbol{\alpha}]}\, \frac{|a_{\boldsymbol{\alpha}}(\mathbf{x}/h)|}{\boldsymbol{\alpha}!}\, |\partial^{\boldsymbol{\alpha}} u(\mathbf{x})| \tag{7.34}$$

is valid, with the functions

$$a_{\boldsymbol{\alpha}}(\mathbf{x}) = \partial_{\boldsymbol{\lambda}}^{\boldsymbol{\alpha}} \Big(1 - \vartheta_{\mathcal{D}}(\mathbf{x}, \boldsymbol{\lambda})\Big)\Big|_{\boldsymbol{\lambda}=\mathbf{0}}\,, \tag{7.35}$$

and the constant c_N does not depend on u.

PROOF. Because

$$\int\limits_{\mathbb{R}^n \setminus h^{-1}Q} |\mathcal{F}u(\boldsymbol{\lambda})|\, d\boldsymbol{\lambda} < (2h)^N \|u\|'_N, \tag{7.36}$$

it remains to estimate the second term of the right-hand side of (7.33).

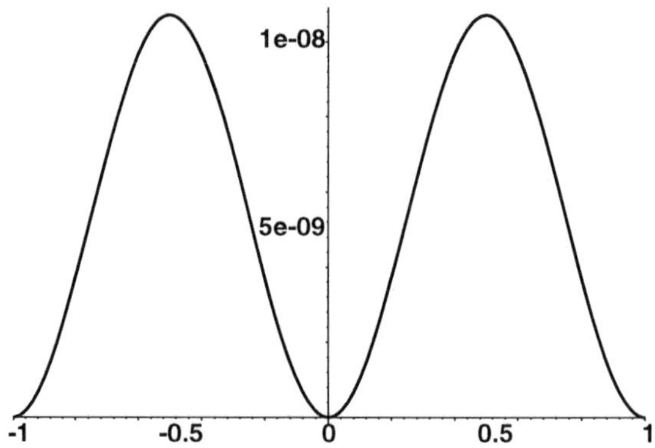

FIGURE 7.3. Plot of $a_0(x)$ for $\mathcal{D} = 2$

Note first that in view of (7.32) the series

$$1 - \vartheta_{\mathcal{D}}(\mathbf{x}, \mathbf{z}) = \sum_{[\boldsymbol{\alpha}]=0}^{\infty} a_{\boldsymbol{\alpha}}(\mathbf{x}) \frac{\mathbf{z}^{\boldsymbol{\alpha}}}{\boldsymbol{\alpha}!}$$

converges absolutely for all $\mathbf{z} = (z_1, \ldots, z_n) \in \mathbb{C}^n$ with

$$|z_j| \le \frac{1}{2\pi\mathcal{D}} \sqrt{1 + \pi^2\mathcal{D}^2} - \delta\,, \qquad \delta = \text{const} > 0\,,$$

uniformly for all $\mathbf{x} \in \mathbb{R}^n$. The functions $a_{\boldsymbol{\alpha}}(\mathbf{x})$ are given by formula (7.35) and they are smooth and periodic with period 1.

Since $\boldsymbol{\lambda} \in h^{-1}Q$ implies $|h\lambda_j| \le 1/2$, we can use the series expansion to split the second integral on the right-hand side of (7.33):

$$\begin{aligned}
&\int\limits_{h^{-1}Q} \mathrm{e}^{2\pi i \langle \mathbf{x}, \boldsymbol{\lambda} \rangle} \, \mathcal{F}u(\boldsymbol{\lambda}) \Big(1 - \vartheta_{\mathcal{D}}\Big(\frac{\mathbf{x}}{h}, h\boldsymbol{\lambda}\Big)\Big) d\boldsymbol{\lambda} \\
&= \sum_{[\boldsymbol{\alpha}]=0}^{\infty} \frac{a_{\boldsymbol{\alpha}}(\mathbf{x}/h)}{\boldsymbol{\alpha}!} h^{[\boldsymbol{\alpha}]} \int\limits_{h^{-1}Q} \mathrm{e}^{2\pi i \langle \mathbf{x}, \boldsymbol{\lambda} \rangle} \, \boldsymbol{\lambda}^{\boldsymbol{\alpha}} \, \mathcal{F}u(\boldsymbol{\lambda}) \, d\boldsymbol{\lambda} \\
&= \sum_{[\boldsymbol{\alpha}]=0}^{N} \frac{a_{\boldsymbol{\alpha}}(\mathbf{x}/h)}{\boldsymbol{\alpha}!} \bigg(\Big(\frac{h}{2\pi i}\Big)^{[\boldsymbol{\alpha}]} \partial^{\boldsymbol{\alpha}} u(\mathbf{x}) - h^{[\boldsymbol{\alpha}]} \int\limits_{\mathbb{R}^n \setminus h^{-1}Q} \mathrm{e}^{2\pi i \langle \mathbf{x}, \boldsymbol{\lambda} \rangle} \, \boldsymbol{\lambda}^{\boldsymbol{\alpha}} \, \mathcal{F}u(\boldsymbol{\lambda}) \, d\boldsymbol{\lambda} \bigg) \\
&\quad + \sum_{[\boldsymbol{\alpha}]=N}^{\infty} \frac{a_{\boldsymbol{\alpha}}(\mathbf{x}/h)}{\boldsymbol{\alpha}!} h^{[\boldsymbol{\alpha}]} \int\limits_{h^{-1}Q} \mathrm{e}^{2\pi i \langle \mathbf{x}, \boldsymbol{\lambda} \rangle} \, \boldsymbol{\lambda}^{\boldsymbol{\alpha}} \, \mathcal{F}u(\boldsymbol{\lambda}) \, d\boldsymbol{\lambda} \,.
\end{aligned}$$

If $[\boldsymbol{\alpha}] < N$, then

$$\begin{aligned}
h^{[\boldsymbol{\alpha}]} \int\limits_{\mathbb{R}^n \setminus h^{-1}Q} |\boldsymbol{\lambda}^{\boldsymbol{\alpha}}| \, |\mathcal{F}u(\boldsymbol{\lambda})| \, d\boldsymbol{\lambda} &\le h^{[\boldsymbol{\alpha}]} \, \|u\|'_N \max_{\boldsymbol{\lambda} \in \mathbb{R}^n \setminus h^{-1}Q} \frac{|\boldsymbol{\lambda}^{\boldsymbol{\alpha}}|}{(1 + |\boldsymbol{\lambda}|)^N} \\
&\le h^{[\boldsymbol{\alpha}]} \, \|u\|'_N \, (2h)^{N-[\boldsymbol{\alpha}]} = h^N 2^{N-[\boldsymbol{\alpha}]} \, \|u\|'_N \,,
\end{aligned}$$

whereas for $[\boldsymbol{\alpha}] \geq N$

$$\Big| \sum_{[\boldsymbol{\alpha}]=N}^{\infty} a_{\boldsymbol{\alpha}}(\mathbf{x}/h) \frac{h^{[\boldsymbol{\alpha}]}}{\boldsymbol{\alpha}!} \int\limits_{h^{-1}Q} \mathrm{e}^{2\pi i \langle \mathbf{x}, \boldsymbol{\lambda} \rangle} \boldsymbol{\lambda}^{\boldsymbol{\alpha}} \mathcal{F}u(\boldsymbol{\lambda})\, d\boldsymbol{\lambda} \Big|$$

$$\leq h^N \|u\|'_N \sum_{[\boldsymbol{\alpha}]=N}^{\infty} \frac{|a_{\boldsymbol{\alpha}}(\mathbf{x}/h)|}{\boldsymbol{\alpha}!} \max_{\boldsymbol{\lambda} \in h^{-1}Q} \frac{|(h\boldsymbol{\lambda})^{\boldsymbol{\alpha}}|}{(h + |h\boldsymbol{\lambda}|)^N} .$$

If we choose a multi-index $\boldsymbol{\beta} \leq \boldsymbol{\alpha}$ with $[\boldsymbol{\beta}] = N$, then for $\boldsymbol{\lambda} \in h^{-1}Q$

$$\frac{|(h\boldsymbol{\lambda})^{\boldsymbol{\alpha}}|}{(h + h|\boldsymbol{\lambda}|)^N} \leq |(h\boldsymbol{\lambda})^{\boldsymbol{\alpha}-\boldsymbol{\beta}}| \frac{|h\boldsymbol{\lambda}|^{\boldsymbol{N}}}{(h + h|\boldsymbol{\lambda}|)^N} \leq 2^{N-[\boldsymbol{\alpha}]} ,$$

which leads to

$$\Big| \sum_{[\boldsymbol{\alpha}]=N}^{\infty} \frac{a_{\boldsymbol{\alpha}}(\mathbf{x}/h)}{\boldsymbol{\alpha}!} h^{[\boldsymbol{\alpha}]} \int\limits_{h^{-1}Q} \mathrm{e}^{2\pi i \langle \mathbf{x}, \boldsymbol{\lambda} \rangle} \boldsymbol{\lambda}^{\boldsymbol{\alpha}} \mathcal{F}u(\boldsymbol{\lambda})\, d\boldsymbol{\lambda} \Big| \leq (2h)^N \|u\|'_N \sum_{[\boldsymbol{\alpha}]=N}^{\infty} \frac{|a_{\boldsymbol{\alpha}}(\mathbf{x}/h)|}{2^{[\boldsymbol{\alpha}]} \boldsymbol{\alpha}!} .$$

The last series is uniformly bounded, which implies together with (7.36) that the constant c_N can be estimated by

$$c_N < 2^N \Big(1 + \max_{\mathbf{x}} \sum_{[\boldsymbol{\alpha}]=0}^{\infty} \frac{|a_{\boldsymbol{\alpha}}(\mathbf{x})|}{2^{[\boldsymbol{\alpha}]} \boldsymbol{\alpha}!} \Big) . \qquad \square$$

In Table 7.2, we give some values of the coefficients $a_\alpha(x)/\alpha!$ of the saturation error in estimate (7.34) for different $\mathcal{D}$ and x in the one-dimensional case.

	$\mathcal{D} = 3$		$\mathcal{D} = 4$	
α	$x = 0.25$	$x = 0.5$	$x = 0.25$	$x = 0.5$
0	$2.767 \cdot 10^{-13}$	$5.535 \cdot 10^{-13}$	$1.431 \cdot 10^{-17}$	$2.862 \cdot 10^{-17}$
1	$1.639 \cdot 10^{-11}$	0.0	$1.130 \cdot 10^{-15}$	0.0
2	$4.852 \cdot 10^{-10}$	$9.704 \cdot 10^{-10}$	$4.462 \cdot 10^{-14}$	$8.934 \cdot 10^{-14}$
3	$9.578 \cdot 10^{-9}$	0.0	$1.174 \cdot 10^{-12}$	0.0
4	$1.418 \cdot 10^{-7}$	$2.836 \cdot 10^{-7}$	$2.318 \cdot 10^{-11}$	$4.636 \cdot 10^{-11}$
5	$1.679 \cdot 10^{-6}$	0.0	$3.661 \cdot 10^{-10}$	0.0
6	$1.657 \cdot 10^{-5}$	$3.315 \cdot 10^{-5}$	$4.817 \cdot 10^{-9}$	$9.634 \cdot 10^{-9}$
7	$1.402 \cdot 10^{-4}$	0.0	$5.433 \cdot 10^{-8}$	0.0
8	$1.038 \cdot 10^{-3}$	$2.076 \cdot 10^{-3}$	$5.362 \cdot 10^{-7}$	$1.072 \cdot 10^{-6}$

TABLE 7.2. Values of $a_\alpha(x)/\alpha!$

REMARK 7.11. Approximation properties of $\mathcal{Q}_h u$ could also be studied with the methods developed in Chapter 2. However, the generating function $\Psi_\mathcal{D}$ has a more complicated dependence on $\mathcal{D}$ than the generating functions considered there. Additionally, $\Psi_\mathcal{D}$ does not satisfy the moment Condition 2.15 exactly. In fact, the parameter $\mathcal{D}$ ensures that the moment condition is satisfied approximately for large N, if $\mathcal{D}$ is sufficiently large, which ensures high-order convergence up to small saturation errors.

7.3.5. Exponential convergence. Finally, we show that the interpolation with (7.30) converges exponentially up to a saturation error.

THEOREM 7.12. *If for some $a > 0$ the continuous function u satisfies*

$$\|u\|'_a := \int_{\mathbb{R}^n} |\mathcal{F}u(\boldsymbol{\lambda})| \, \mathrm{e}^{a|\boldsymbol{\lambda}|} \, d\boldsymbol{\lambda} < \infty ,$$

then the error of the interpolation with (7.30) can be estimated for all $\mathbf{x} \in \mathbb{R}^n$ by

$$|u(\mathbf{x}) - \mathcal{Q}_h u(\mathbf{x})| \le \left(\mathrm{e}^{-a/(2h)} + 8 \sum_{k=1}^{n} \binom{n}{k} c^k \left(\tanh \frac{\pi^2 \mathcal{D}}{2} \right)^{n-k} \right) \|u\|'_a ,$$

where

$$c = \begin{cases} \mathrm{e}^{-\pi^2 \mathcal{D}} , & a \ge 2h\pi^2 \mathcal{D} , \\ \mathrm{e}^{-a/(2h)} , & a \le 2h\pi^2 \mathcal{D} . \end{cases}$$

PROOF. We use the estimate (7.33) again. Obviously,

$$\int_{\mathbb{R}^n \setminus h^{-1}Q} |\mathcal{F}u(\boldsymbol{\lambda})| \, d\boldsymbol{\lambda} \le \mathrm{e}^{-a/2h} \int_{\mathbb{R}^n} |\mathcal{F}u(\boldsymbol{\lambda})| \, \mathrm{e}^{a|\boldsymbol{\lambda}|} \, d\boldsymbol{\lambda} .$$

To estimate the second integral of the right-hand side of (7.33), we note that

$$\begin{aligned} \Big| 1 - \vartheta_{\mathcal{D}}\Big(\frac{\mathbf{x}}{h}, h\boldsymbol{\lambda}\Big) \Big| &\le \sum_{\boldsymbol{\nu} \in \mathbb{Z}^n \setminus \{\mathbf{0}\}} \Big| 1 - \mathrm{e}^{\frac{2\pi i}{h} \langle \mathbf{x}, \boldsymbol{\nu} \rangle} \Big| \mathcal{F}\Psi_{\mathcal{D}}(h\boldsymbol{\lambda} + \boldsymbol{\nu}) \\ &\le 2 \sum_{\boldsymbol{\nu} \in \mathbb{Z}^n \setminus \{\mathbf{0}\}} \mathcal{F}\Psi_{\mathcal{D}}(h\boldsymbol{\lambda} + \boldsymbol{\nu}) , \end{aligned}$$

which leads to

$$\begin{aligned} &\Big| \int_{h^{-1}Q} \mathrm{e}^{2\pi i \langle \mathbf{x}, \boldsymbol{\lambda} \rangle} \, \mathcal{F}u(\boldsymbol{\lambda}) \Big(1 - \vartheta_{\mathcal{D}}\Big(\frac{\mathbf{x}}{h}, h\boldsymbol{\lambda}\Big) \Big) \, d\boldsymbol{\lambda} \Big| \\ &\qquad \le 2 \int_{h^{-1}Q} |\mathcal{F}u(\boldsymbol{\lambda})| \, \mathrm{e}^{a|\boldsymbol{\lambda}|} \, d\boldsymbol{\lambda} \max_{\boldsymbol{\lambda} \in h^{-1}Q} \Big(\mathrm{e}^{-a|\boldsymbol{\lambda}|} \sum_{\boldsymbol{\nu} \in \mathbb{Z}^n \setminus \{\mathbf{0}\}} \mathcal{F}\Psi_{\mathcal{D}}(h\boldsymbol{\lambda} + \boldsymbol{\nu}) \Big) . \end{aligned}$$

From

$$\begin{aligned} \mathrm{e}^{-a|\boldsymbol{\lambda}|} \sum_{\boldsymbol{\nu} \in \mathbb{Z}^n \setminus \{\mathbf{0}\}} \mathcal{F}\Psi_{\mathcal{D}}(h\boldsymbol{\lambda} + \boldsymbol{\nu}) &= \sum_{\boldsymbol{\nu} \in \mathbb{Z}^n \setminus \{\mathbf{0}\}} \prod_{j=1}^{n} \frac{\sinh \pi^2 \mathcal{D} \, \mathrm{e}^{-a|\lambda_j|}}{\cosh 2\pi^2 \mathcal{D}(h\lambda_j + \nu_j) + \cosh \pi^2 \mathcal{D}} \\ = \prod_{j=1}^{n} \Big(\sum_{\nu_j \ne 0} \frac{\sinh \pi^2 \mathcal{D} \, \mathrm{e}^{-a|\lambda_j|}}{\cosh 2\pi^2 \mathcal{D}(h\lambda_j + \nu_j) + \cosh \pi^2 \mathcal{D}} &+ \frac{\sinh \pi^2 \mathcal{D} \, \mathrm{e}^{-a|\lambda_j|}}{\cosh 2\pi^2 \mathcal{D} h\lambda_j + \cosh \pi^2 \mathcal{D}} \Big) \\ &- \prod_{j=1}^{n} \frac{\sinh \pi^2 \mathcal{D} \, \mathrm{e}^{-a|\lambda_j|}}{\cosh 2\pi^2 \mathcal{D} h\lambda_j + \cosh \pi^2 \mathcal{D}} , \end{aligned}$$

we conclude that

$$\max_{\boldsymbol{\lambda} \in h^{-1}Q} \Big(\mathrm{e}^{-a|\boldsymbol{\lambda}|} \sum_{\boldsymbol{\nu} \in \mathbb{Z}^n \setminus \{\mathbf{0}\}} \mathcal{F}\Psi_{\mathcal{D}}(h\boldsymbol{\lambda} + \boldsymbol{\nu}) \Big) \le (c_1 + c_2)^n - c_2^n ,$$

where

$$c_1 = \max_{|h\lambda| \le 1/2} \sum_{k \ne 0} \frac{\sinh \pi^2 \mathcal{D} \, \mathrm{e}^{-a|\lambda|}}{\cosh 2\pi^2 \mathcal{D}(h\lambda + k) + \cosh \pi^2 \mathcal{D}}$$

and

$$c_2 = \max_{|h\lambda| \le 1/2} \frac{\sinh \pi^2 \mathcal{D} \, \mathrm{e}^{-a|\lambda|}}{\cosh 2\pi^2 \mathcal{D} h\lambda + \cosh \pi^2 \mathcal{D}} = \tanh \frac{\pi^2 \mathcal{D}}{2} .$$

Therefore, we obtain

$$\max_{\boldsymbol{\lambda} \in h^{-1}Q} \Big(\mathrm{e}^{-a|\boldsymbol{\lambda}|} \sum_{\boldsymbol{\nu} \in \mathbb{Z}^n \setminus \{\mathbf{0}\}} \mathcal{F}\Psi_{\mathcal{D}}(h\boldsymbol{\lambda} + \boldsymbol{\nu}) \Big) \le \sum_{k=1}^{n} \binom{n}{k} c_1^k \Big(\tanh \frac{\pi^2 \mathcal{D}}{2} \Big)^{n-k} .$$

Hence, the second integral on the right-hand side of (7.33) can be estimated by

$$\Big| \int\limits_{h^{-1}Q} \mathrm{e}^{2\pi i \langle \mathbf{x}, \boldsymbol{\lambda} \rangle} \mathcal{F}u(\boldsymbol{\lambda}) \Big(1 - \vartheta_{\mathcal{D}} \Big(\frac{\mathbf{x}}{h}, h\boldsymbol{\lambda} \Big) \Big) \, d\boldsymbol{\lambda} \Big|$$
$$\le 2 \sum_{k=1}^{n} \binom{n}{k} c_1^k \Big(\tanh \frac{\pi^2 \mathcal{D}}{2} \Big)^{n-k} \|u\|_a' .$$

The value c_1 will be estimated in the next lemma. □

LEMMA 7.13. *Let $a > 0$. Then*

$$\max_{|h\lambda| \le 1/2} \mathrm{e}^{-a|\lambda|} \sum_{k \ne 0} \frac{\sinh \pi^2 \mathcal{D}}{\cosh 2\pi^2 \mathcal{D}(h\lambda + k) + \cosh \pi^2 \mathcal{D}} < \begin{cases} 4\,\mathrm{e}^{-\pi^2 \mathcal{D}}, & a \ge 2h\pi^2 \mathcal{D}, \\ 4\,\mathrm{e}^{-a/(2h)}, & a \le 2h\pi^2 \mathcal{D}. \end{cases}$$

PROOF. Since $|h\lambda| \le 1/2$, we have the estimate

$$\frac{\sinh \pi^2 \mathcal{D}}{\cosh 2\pi^2 \mathcal{D}(h\lambda + k) + \cosh \pi^2 \mathcal{D}} < \frac{2 \sinh \pi^2 \mathcal{D}}{\mathrm{e}^{2\pi^2 \mathcal{D}(h\lambda + k)} + \mathrm{e}^{\pi^2 \mathcal{D}}}$$
$$< 2 \sinh \pi^2 \mathcal{D} \, \mathrm{e}^{-2\pi^2 \mathcal{D}(h\lambda + k)}$$

for $k = 1, 2, \ldots$. If $k < 0$, then $h\lambda + k < 0$ and therefore,

$$\frac{\sinh \pi^2 \mathcal{D}}{\cosh 2\pi^2 \mathcal{D}(h\lambda + k) + \cosh \pi^2 \mathcal{D}} < 2 \sinh \pi^2 \mathcal{D} \, \mathrm{e}^{2\pi^2 \mathcal{D}(h\lambda + k)} .$$

Hence, if $|h\lambda| \le 1/2$, then

$$\sum_{k \ne 0} \frac{\sinh \pi^2 \mathcal{D}}{\cosh 2\pi^2 \mathcal{D}(h\lambda + k) + \cosh \pi^2 \mathcal{D}}$$
$$< 2 \sinh \pi^2 \mathcal{D} \Big(\sum_{k=1}^{\infty} \mathrm{e}^{-2\pi^2 \mathcal{D}(h\lambda)} + \sum_{k=-\infty}^{-1} \mathrm{e}^{2\pi^2 \mathcal{D}(h\lambda + k)} \Big)$$
$$= 4 \sinh \pi^2 \mathcal{D} \cosh 2\pi^2 \mathcal{D} h\lambda \sum_{k=1}^{\infty} \mathrm{e}^{-2\pi^2 \mathcal{D} k}$$
$$= 4 \sinh \pi^2 \mathcal{D} \cosh 2\pi^2 \mathcal{D} h\lambda \frac{\mathrm{e}^{-2\pi^2 \mathcal{D}}}{1 - \mathrm{e}^{-2\pi^2 \mathcal{D}}}$$
$$= 2\,\mathrm{e}^{-\pi^2 \mathcal{D}} \cosh 2\pi^2 \mathcal{D} h\lambda .$$

So, it remains to determine

$$2\,\mathrm{e}^{-\pi^2\mathcal{D}} \max_{0\le h\lambda\le 1/2} \mathrm{e}^{-a\lambda}\cosh 2\pi^2\mathcal{D}h\lambda = 4 \max_{0\le h\lambda\le 1/2} \mathrm{e}^{-a\lambda}\,\mathrm{e}^{-\pi^2\mathcal{D}(1-2h\lambda)}$$

$$= \begin{cases} 4\,\mathrm{e}^{-\pi^2\mathcal{D}}, & a \ge 2h\pi^2\mathcal{D}, \\ 4\,\mathrm{e}^{-a/(2h)}, & a \le 2h\pi^2\mathcal{D}. \end{cases}$$

□

REMARK 7.14. Note that, in view of (7.28), the error estimates of Theorems 7.10 and 7.12 are valid also for the interpolant of the form (7.19).

7.4. Orthogonal projection

The construction of the orthogonal (in L_2) projections onto the scaled and shifted Gaussians uses some well-known facts about principal shift invariant spaces (see [**11**]), which we recall briefly. Denote by $S(\varphi)$ the L_2-closure of finite linear combinations of the shifts $\varphi(\cdot - \mathbf{m})$, $\mathbf{m} \in \mathbb{Z}^n$, of a generating function φ, and suppose that the shifts form an L_2-stable basis of $S(\varphi)$, i.e., for all $\{a_\mathbf{m}\} \in \ell_2(\mathbb{Z}^n)$

$$c_1 \,\|\{a_\mathbf{m}\}\|_{\ell^2} \le \Big\| \sum_{\mathbf{m}\in\mathbb{Z}^n} a_\mathbf{m} \phi(\cdot - \mathbf{m})\Big\|_{L_2(\mathbb{R}^n)} \le c_2 \,\|\{a_\mathbf{m}\}\|_{\ell_2}\,. \tag{7.37}$$

Note that for any

$$\phi(\mathbf{x}) = \sum_{\mathbf{m}\in\mathbb{Z}^n} a_\mathbf{m}\, \varphi(\mathbf{x}-\mathbf{m}) \in S(\varphi)\,,$$

we have

$$\mathcal{F}\phi(\boldsymbol{\lambda}) = \tau(\boldsymbol{\lambda})\,\mathcal{F}\varphi(\boldsymbol{\lambda}) \quad \text{with } \tau(\boldsymbol{\lambda}) := \sum_{\mathbf{m}\in\mathbb{Z}^n} a_\mathbf{m}\, \mathrm{e}^{-2\pi i\langle \mathbf{m},\boldsymbol{\lambda}\rangle} \in L_2((0,1)^n)\,.$$

Furthermore, if $\mathcal{F}\phi = \tau\,\mathcal{F}\varphi \in L_2(\mathbb{R}^n)$ with a 1-periodic function τ, then

$$\|\phi\|_{L_2(\mathbb{R}^n)} = \|\tau[\mathcal{F}\varphi,\mathcal{F}\varphi]^{1/2}\|_{L_2((0,1)^n)}\,, \tag{7.38}$$

where the bracket product stands for the 1-periodic function

$$[f,g] := \sum_{\mathbf{m}\in\mathbb{Z}^n} f(\cdot-\mathbf{m})\,\overline{g(\cdot-\mathbf{m})}\,.$$

For each $f \in L_2(\mathbb{R}^n)$, the L_2-projection $Pf \in S(\varphi)$ is given by

$$\mathcal{F}(Pf) = \frac{[\mathcal{F}f,\mathcal{F}\varphi]}{[\mathcal{F}\varphi,\mathcal{F}\varphi]}\,\mathcal{F}\varphi\,. \tag{7.39}$$

Then the shifts of the function $\check{\varphi} \in S(\varphi)$, defined via

$$\mathcal{F}\check{\varphi} = \frac{\mathcal{F}\varphi}{[\mathcal{F}\varphi,\mathcal{F}\varphi]}\,, \tag{7.40}$$

form the corresponding biorthogonal basis, i.e., $(\varphi(\cdot-\mathbf{m}),\check{\varphi})_{L_2} = \delta_{0|\mathbf{m}|}$, $\mathbf{m}\in\mathbb{Z}^n$.

Therefore, the best L_2-approximant $P_h u$ of the form (7.19) to $u \in L_2(\mathbb{R}^n)$ has the Fourier transform

$$\mathcal{F}(P_h u)(\boldsymbol{\lambda}) = \mathrm{e}^{-\pi^2\mathcal{D}h^2|\boldsymbol{\lambda}|^2}\, \frac{\displaystyle\sum_{\boldsymbol{\nu}\in\mathbb{Z}^n} \mathcal{F}u(\boldsymbol{\lambda}+\frac{\boldsymbol{\nu}}{h})\,\mathrm{e}^{-\pi^2\mathcal{D}|h\boldsymbol{\lambda}+\boldsymbol{\nu}|^2}}{\displaystyle\sum_{\boldsymbol{\nu}\in\mathbb{Z}^n} \mathrm{e}^{-2\pi^2\mathcal{D}|h\boldsymbol{\lambda}+\boldsymbol{\nu}|^2}} \tag{7.41}$$

and the approximation error is the L_2-norm of the function

$$\mathcal{F}(P_h u)(\boldsymbol{\lambda}) - \mathcal{F}u(\boldsymbol{\lambda})$$
$$= \frac{1}{g(h\boldsymbol{\lambda})} \sum_{\boldsymbol{\nu}\in\mathbb{Z}^n} \Big(\mathrm{e}^{-\pi^2\mathcal{D}h^2|\boldsymbol{\lambda}|^2} \mathcal{F}u(\boldsymbol{\lambda} + \frac{\boldsymbol{\nu}}{h}) - \mathrm{e}^{-\pi^2\mathcal{D}|h\boldsymbol{\lambda}+\boldsymbol{\nu}|^2} \mathcal{F}u(\boldsymbol{\lambda}) \Big) \mathrm{e}^{-\pi^2\mathcal{D}|h\boldsymbol{\lambda}+\boldsymbol{\nu}|^2},$$

where

$$g(\boldsymbol{\lambda}) := \sum_{\boldsymbol{\nu}\in\mathbb{Z}^n} \mathrm{e}^{-2\pi^2\mathcal{D}|\boldsymbol{\lambda}-\boldsymbol{\nu}|^2}.$$

7.5. Notes

The results of this chapter are based on the article [**68**]. The approximation with Gaussian kernels $\mathrm{e}^{-|\mathbf{x}|^2}$ is often mentioned in the literature and has been studied by many authors in connection with radial-basis functions and principal shift-invariant spaces.

Buhmann found in [**14**] that the interpolation and quasi-interpolation from the stationary ladder $S^h(\mathrm{e}^{-a|\cdot|^2})_h$ cannot yield strict convergence results, since polynomial reproduction is absent. As mentioned in the Notes to Chapter 2, it is possible to obtain approximations with Gaussian kernels converging for all $h \to 0$, if one chooses the parameter $\mathcal{D}$ depending on h. This was studied by several authors. Madych and Nelson [**58**] and Wu and Schaback [**98**] study interpolants of the form

$$\sum_{\mathbf{m}} u_{\mathbf{m}}\, \mathrm{e}^{-|\mathbf{x}-\mathbf{x}_{\mathbf{m}}|^2/\mathcal{D}},$$

where the points $\mathbf{x}_{\mathbf{m}}$ are allowed to be irregularly distributed over a domain Ω. It is shown that the corresponding expressions approximate smooth functions with arbitrarily high order or even with exponential order. Quasi-interpolation with Gaussians on uniform grids with variable $\mathcal{D}$ was studied in [**91**, Section 5.8] and [**8**], which was already mentioned in Section 2.5. The tensor product construction of a quasi-interpolation formula with Gaussians studied in [**8**] converges with the rate $\mathcal{O}(h^N|\log h|^N)$ if $\mathcal{D}(h) = N\,|\log h|/\pi^2$, but it depends on the values of u at $(2M-1)^n$ grid points. Other results on the approximation power of the non-stationary ladders of principal shift-invariant spaces $S^h(\mathrm{e}^{h|\cdot|^2})_h$ were obtained in particular by de Boor and Ron ([**13**]) and Johnson ([**39**], [**40**]), who proved convergence of any order k in L_2- and L_p-spaces, respectively.

In all these approaches, the use of finer grids enlarges the number of summands necessary to compute the approximate value at a fixed point $\mathbf{x}$ within a given tolerance. This is in contrast to the case of fixed $\mathcal{D}$, which we prefer, because this is advantageous in numerical applications and reflects the local character of the quasi-interpolants constructed in Section 7.2.

The results of Section 7.3 are new. The Lagrangian function $\Psi_{\mathcal{D}}$ can be considered as the approximate counterpart of the sinc function. Owing to the exponential decay, $\Psi_{\mathcal{D}}$ offers certain computational advantages, and it would be interesting to examine numerical procedures, developed for sinc function approximation, for $\Psi_{\mathcal{D}}$ approximation also.

CHAPTER 8

Approximate wavelets

8.1. Introduction

In this chapter we introduce the so-called *approximate wavelet decompositions* of spaces of approximating functions, which are generated by quite arbitrary functions in the Schwartz class. These functions do not satisfy refinement equations exactly, but in some approximate sense, which allows us to decompose fine scale spaces within a given tolerance into a direct sum of wavelet spaces. As an example, we give a detailed construction for the Gaussian generating function. The approximate multi-resolution approach allows us to derive sparse representations of pseudodifferential and other integral operators of mathematical physics.

The application of wavelet based methods to the representation of integral and differential operators is an actual research topic in the numerical analysis of solution methods for the corresponding operator equations. The usual setting is based on multi-resolution analysis, introduced in [**74**], [**61**]. Starting from a finite sequence of nested closed subspaces

$$V_0 \subset V_1 \subset \ldots \subset V_k \subset L_2(\mathbb{R}^n)\,, \tag{8.1}$$

the space of approximating functions V_k, corresponding to the finest grid, is decomposed into the orthogonal sum

$$V_k = V_0 \bigoplus_{j=0}^{k-1} W_j\,, \tag{8.2}$$

where the wavelet space W_j is the orthogonal complement $W_j = V_{j+1} \ominus V_j$.

The chain (8.1) is called a (stationary) multi-resolution analysis of V_k if the spaces V_j have the properties

(i) $f(\mathbf{x}) \in V_0$ if and only if $f(\mathbf{x}-\mathbf{m}) \in V_0$ for any $\mathbf{m} \in \mathbb{Z}^n$;
(ii) $f(\mathbf{x}) \in V_j$ if and only if $f(2\mathbf{x}) \in V_{j+1}$ for any $j = 0, \ldots, k-1$;
(iii) there exists a function ϕ such that $\{\phi(\cdot-\mathbf{m})\}_{\mathbf{m}\in\mathbb{Z}^n}$ is an L_2-stable basis in V_0 (cf. (7.37)).

Then the spaces V_j are spanned by the dilated shifts $\phi(2^j\cdot-\mathbf{m})$, $\mathbf{m} \in \mathbb{Z}^n$, of the *scaling function* ϕ. The main goal of the multi-resolution is to determine a new basis of the space V_k, which is used in numerical procedures. In n dimensions, there exist $2^n - 1$ functions $\psi_v \in W_0$, called prewavelets, such that the shifts $\{\psi_v(\cdot-\mathbf{m}), \mathbf{m} \in \mathbb{Z}^n, v \in \mathcal{V}'\}$ form an L_2-stable basis in the space W_0. Here, the prewavelets ψ_v are indexed by the set $\mathcal{V}' = \mathcal{V}\backslash\{\mathbf{0}\}$ with $\mathcal{V}$ denoting the set of vertices of the cube $[0,1/2]^n$. Thus, one obtains an L_2-stable basis of the space V_k consisting of

$$\{\phi(\cdot-\mathbf{m}), \mathbf{m} \in \mathbb{Z}^n\} \quad \text{and} \quad \{2^{nj/2}\psi_v(2^j\cdot-\mathbf{m}), \mathbf{m} \in \mathbb{Z}^n, v \in \mathcal{V}', j = 0, \ldots, n-1\}\,.$$

Similarly to other transform methods, elements of V_k are now expanded into the new basis and the computations take place in this system of coordinates, where one hopes to achieve that the computation is faster than in the original system. Additionally, some features of wavelets, such as the localization in both space and frequency domains and vanishing moment properties, lead to a number of new and interesting properties of wavelet based numerical methods. The multi-resolution structure of the wavelet expansion leads to an effective organization of transformations. Furthermore, if the wavelets have a high number of vanishing moments, then pseudodifferential operators admit sparse matrix representations within a prescribed accuracy, which allows one to design fast numerical algorithms for these operators.

From the properties (i) and (ii), it follows that the scaling function ϕ has to satisfy the so-called *refinement equation*

$$\phi(\mathbf{x}) = \sum_{\mathbf{m}\in\mathbb{Z}^n} a_{\mathbf{m}}\, \phi(2\mathbf{x}-\mathbf{m})\,. \tag{8.3}$$

In the applications to integral operators, as a rule, the scaling functions are piecewise polynomials satisfying some smoothness and vanishing moment conditions. One drawback of these functions is that it is practically impossible to derive analytic formulas for the action of important integral operators of mathematical physics on these functions, especially in the multi-dimensional case. Thus, it is necessary to use cubatures for integral operators with singular kernel functions applied to piecewise polynomials.

A different approach to the cubature of integral operators of mathematical physics was developed in Chapter 4, where the density of the integral operator is approximated by a quasi-interpolant (2.23) generated by some smooth function η. Besides the decay and moment conditions, this function has the property that the integral $K\eta$ can be evaluated efficiently, either analytically or by simple one-dimensional quadrature. Here, we show that it is possible to find, similarly to the wavelet approach, a new basis in the linear span

$$\{\eta\big((\mathbf{x}-h\mathbf{m})/\sqrt{\mathcal{D}}h\big), \mathbf{m}\in\mathbb{Z}^n\}$$

such that the integral operators admit a sparse representation.

It is clear that the basic functions we have in mind do not satisfy a refinement equation of the form (8.3). But it turns out that for a wide class of functions, refinement equations are valid in an approximate sense. For example, the expansion (7.11) shows that the Gaussian $\varphi_{\mathcal{D}}(x) = \mathrm{e}^{-x^2/\mathcal{D}}$ satisfies the inequality

$$\begin{aligned}\Big| \mathrm{e}^{-x^2/\mathcal{D}} - \sum_{m\in\mathbb{Z}} \frac{\mathrm{e}^{-\mu^2 m^2/\mathcal{D}(1-\mu^2)}}{\sqrt{\pi\mathcal{D}(1-\mu^2)}}\, \mathrm{e}^{-(x-\mu m)^2/\mathcal{D}\mu^2} \Big| \\ \le \mathrm{e}^{-x^2/\mathcal{D}} \sum_{\nu\in\mathbb{Z}\setminus\{0\}} \mathrm{e}^{-\pi^2\mathcal{D}(1-\mu^2)\nu^2}\,.\end{aligned} \tag{8.4}$$

This means that if we set $\mu = 1/2$, then $\varphi_{\mathcal{D}}(x)$ can be expressed by a linear combination of the integer shifts of $\varphi_{\mathcal{D}}(2x)$ as in the usual refinement equation (8.3), but only modulo the error

$$\mathrm{e}^{-x^2/\mathcal{D}} \sum_{\nu\in\mathbb{Z}\setminus\{0\}} \mathrm{e}^{-3\pi^2\mathcal{D}\nu^2/4}\,,$$

which can be made smaller than any prescribed tolerance if $\mathcal{D}$ is chosen sufficiently large. This leads to the idea of performing an approximate multi-resolution analysis and wavelet construction analogously to the case where one has an exact refinement equation. This is the goal of the present chapter.

Note also that the Gaussian satisfies the aforementioned approximate refinement equation for any parameter $\mu < 1$. Hence in general one is not restricted to the particular choice of $\mu = 1/2$. However, in order to simplify notation, we will employ dyadic dilations.

In Section 8.2, we prove that there exists a large class of basic functions satisfying approximate refinement equations in the sense of (8.4), which is the basis for an approximate multi-resolution analysis. This is the topic of Section 8.3. We will show that, for any function η from this class, sufficiently large $\mathcal{D}$, and fixed $k \in \mathbb{N}$, any element of the L_2-closure of the linear span

$$\mathbf{V}_k := \Big\{\eta\Big(\frac{2^k \cdot -\mathbf{m}}{\sqrt{\mathcal{D}}}\Big), \mathbf{m} \in \mathbb{Z}^n\Big\} \tag{8.5}$$

can be approximated by elements of the direct sum

$$\mathbf{X}_k := \mathbf{V}_0 \dot{+} \mathbf{W}_0 \dot{+} \ldots \dot{+} \mathbf{W}_{k-1}\,, \tag{8.6}$$

with some small relative error. The wavelet spaces $\mathbf{W}_j$ are almost orthogonal so that the approximate decomposition (8.6) of $\mathbf{V}_k$ can be performed, using the orthogonal projections P_0 onto $\mathbf{V}_0$ and Q_j onto $\mathbf{W}_j$.

A central point is the construction of univariate approximate wavelets. For the example of the Gaussian function, this is given in Section 8.4. The wavelet spaces W_j are spanned by rapidly decaying analytic functions, plotted in Fig. 8.1, which are small perturbations of elements of V_{j+1} and which are orthogonal to all elements of V_j. This allows one to obtain simple analytic formulas for these approximate wavelets. More precisely, in the one-dimensional case, the space W_0 can be defined within the assumed tolerance $\varepsilon = \mathcal{O}(\mathrm{e}^{-\pi^2\mathcal{D}})$ as the L_2-span of the integer translates of the function

$$\psi_{\mathcal{D}}(x) = \mathrm{e}^{-(2x-1)^2/6\mathcal{D}} \cos\frac{5\pi}{6}(2x-1)\,. \tag{8.7}$$

Since the values of many integral operators applied to the wavelets can be given analytically, one can use approximants from $\mathbf{X}_k$ to derive the cubature of these operators. One assumes that the transformation to the basis in $\mathbf{X}_k$ leads to data compression, at least for sufficiently smooth densities u. Additionally, the moments of the prewavelets are very small and can be controlled by the parameter $\mathcal{D}$. This implies, for example, a fast decay of the integrals $\mathcal{K}\psi$ if the kernel $k(\mathbf{x},\mathbf{y})$ satisfies

$$|\partial_{\mathbf{y}}^{\boldsymbol{\alpha}} k(\mathbf{x},\mathbf{y})| \le c_{\boldsymbol{\alpha}}\, |\mathbf{x}-\mathbf{y}|^{-(\gamma+|\boldsymbol{\alpha}|)} \quad \text{for some } \gamma > 0\,.$$

In Section 8.5, we consider multi-variate approximate wavelets and the multi-resolution structure of the spaces spanned by the Gaussian radial function. Furthermore, we give explicit formulas for the ortho-projection P_0 onto $\mathbf{V}_0$ and almost orthogonal projections $\widetilde{Q}_j$ onto $\mathbf{W}_j$, which are proved in Section 8.6, so that for any $\varphi_k \in \mathbf{V}_k$, the estimate

$$\big\|\varphi_k - P_0\varphi_k - \sum_{j=0}^{k-1} \widetilde{Q}_j\varphi_k\big\|_{L_2} \le c\,\varepsilon\, \|\varphi_k\|_{L_2} \tag{8.8}$$

holds with some constant not depending on φ_k and $\mathcal{D}$. Furthermore, we show that this wavelet basis has the property that the action of important pseudodifferential operators can be obtained efficiently.

The proposed approximate multi-resolution analysis combines the advantages of well-established wavelet methods in numerical analysis with the use of simple approximating formulas based on smooth generating functions. The drawback of non-convergence and non-exact refinement equations can be overcome by an appropriate choice of parameters to force the errors within the round-off required.

8.2. Approximate refinement equations

In Section 7.1, we noticed the excellent approximation properties of the Gaussian to smooth functions. In particular, formula (7.11) applied to the Gaussian itself shows that for any $\mu < 1$

$$\begin{aligned} e^{-|\mathbf{x}|^2/\mathcal{D}} &= (\pi\mathcal{D}(1-\mu^2))^{-n/2} \sum_{\mathbf{m}\in\mathbb{Z}^n} e^{-\mu^2|\mathbf{m}|^2/\mathcal{D}(1-\mu^2)}\, e^{-|\mathbf{x}-\mu\mathbf{m}|^2/\mathcal{D}\mu^2} \\ &\quad - e^{-|\mathbf{x}|^2/\mathcal{D}} \sum_{\boldsymbol{\nu}\in\mathbb{Z}^n\setminus\{\mathbf{0}\}} e^{2\pi i(1-\mu^2)\langle\mathbf{x},\boldsymbol{\nu}\rangle/\mu}\, e^{-\pi^2\mathcal{D}(1-\mu^2)|\boldsymbol{\nu}|^2}. \end{aligned} \tag{8.9}$$

The second sum on the right-hand side is, for properly chosen μ and $\mathcal{D}$, a small remainder, which can be ignored under certain assumptions, discussed later on. Here, we show that equations of this type are valid for a large class of functions from the Schwartz class $\mathcal{S}(\mathbb{R}^n)$.

8.2.1. Approximate refinement equations for η. Equation (8.9) is a relation of the form

$$\eta\Big(\frac{\mathbf{x}}{\sqrt{\mathcal{D}}}\Big) = \mathcal{D}^{-n/2} \sum_{\mathbf{m}\in\mathbb{Z}^n} \tilde{\eta}\Big(\frac{\mu\mathbf{m}}{\sqrt{\mathcal{D}}}\Big)\, \eta\Big(\frac{\mathbf{x}-\mu\mathbf{m}}{\mu\sqrt{\mathcal{D}}}\Big) + \text{ small remainder term,} \tag{8.10}$$

which we call the *approximate refinement equation*. In the example (8.9), the function $\tilde{\eta}$ is given by

$$\tilde{\eta}(\mathbf{x}) = \frac{e^{-|\mathbf{x}|^2/(1-\mu^2)}}{(\pi(1-\mu^2))^{n/2}} \tag{8.11}$$

and the "small remainder term" means that for a fixed μ, the equality (8.10) holds up to an arbitrarily small error if $\mathcal{D}$ is sufficiently large.

Approximate refinement equations are valid for a large class of basic functions η as shown by the following assertion.

THEOREM 8.1. *Suppose that $\eta \in \mathcal{S}(\mathbb{R}^n)$ has a non-vanishing Fourier transform, $\mathcal{F}\eta \neq 0$, and for some positive $\mu < 1$, the function $\tilde{\eta}$ satisfies*

$$\mathcal{F}\tilde{\eta} = \frac{\mathcal{F}\eta}{\mathcal{F}\eta(\mu\cdot)} \in \mathcal{S}(\mathbb{R}^n). \tag{8.12}$$

Then

$$\eta\Big(\frac{\mathbf{x}}{\sqrt{\mathcal{D}}}\Big) = \mathcal{D}^{-n/2} \sum_{\mathbf{m}\in\mathbb{Z}^n} \tilde{\eta}\Big(\frac{\mu\mathbf{m}}{\sqrt{\mathcal{D}}}\Big)\eta\Big(\frac{\mathbf{x}-\mu\mathbf{m}}{\mu\sqrt{\mathcal{D}}}\Big) + R_{\eta,\mu,\mathcal{D}}(\mathbf{x}), \tag{8.13}$$

where $R_{\eta,\mu,\mathcal{D}} \in \mathcal{S}(\mathbb{R}^n)$*, and, moreover, for any* $\varepsilon > 0$ *and* $k > 0$ *there exists* $\mathcal{D}_0 = \mathcal{D}_0(\eta,\mu,k) > 0$ *such that* $(1+|\mathbf{x}|^k)|R_{\eta,\mu,\mathcal{D}}(\mathbf{x})| < \varepsilon$ *if* $\mathcal{D} \geq \mathcal{D}_0$.

PROOF. We apply Poisson's summation formula (2.17) to the sum

$$\mathcal{D}^{-n/2} \sum_{\mathbf{m}\in\mathbb{Z}^n} g\Big(\frac{\mu\mathbf{m}}{\sqrt{\mathcal{D}}}\Big)\,\eta\Big(\frac{\mathbf{x}-\mu\mathbf{m}}{\sqrt{\mathcal{D}}\mu}\Big) = \sum_{\boldsymbol{\nu}\in\mathbb{Z}^n} \int_{\mathbb{R}^n} \tilde{\eta}(\mu\mathbf{y})\eta\Big(\frac{\mathbf{x}}{\sqrt{\mathcal{D}}\mu}-\mathbf{y}\Big)\,\mathrm{e}^{-2\pi i\sqrt{\mathcal{D}}\langle\mathbf{y},\boldsymbol{\nu}\rangle}\,d\mathbf{y}$$

$$= \sum_{\boldsymbol{\nu}\in\mathbb{Z}^n} \int_{\mathbb{R}^n} \mathcal{F}\tilde{\eta}(\boldsymbol{\lambda}) \int_{\mathbb{R}^n} \eta(\mathbf{y})\,\mathrm{e}^{2\pi i\langle\mathbf{x}/\sqrt{\mathcal{D}}\mu-\mathbf{y},\mu\boldsymbol{\lambda}-\sqrt{\mathcal{D}}\boldsymbol{\nu}\rangle}\,d\mathbf{y}\,d\boldsymbol{\lambda}$$

$$= \sum_{\boldsymbol{\nu}\in\mathbb{Z}^n} \mathrm{e}^{2\pi i\langle\mathbf{x},\boldsymbol{\nu}\rangle/\mu} \int_{\mathbb{R}^n} \frac{\mathcal{F}\eta(\boldsymbol{\lambda})}{\mathcal{F}\eta(\mu\boldsymbol{\lambda})}\,\mathcal{F}\eta(\mu\boldsymbol{\lambda}+\sqrt{\mathcal{D}}\boldsymbol{\nu})\,\mathrm{e}^{2\pi i\langle\mathbf{x},\boldsymbol{\lambda}\rangle/\sqrt{\mathcal{D}}}\,d\boldsymbol{\lambda}$$

$$= \eta\Big(\frac{\mathbf{x}}{\sqrt{\mathcal{D}}}\Big) + \sum_{\boldsymbol{\nu}\in\mathbb{Z}^n\setminus\{\mathbf{0}\}} \mathrm{e}^{2\pi i\langle\mathbf{x},\boldsymbol{\nu}\rangle/\mu} \int_{\mathbb{R}^n} \frac{\mathcal{F}\eta(\boldsymbol{\lambda})}{\mathcal{F}\eta(\mu\boldsymbol{\lambda})}\,\mathcal{F}\eta(\mu\boldsymbol{\lambda}+\sqrt{\mathcal{D}}\boldsymbol{\nu})\,\mathrm{e}^{2\pi i\langle\mathbf{x},\boldsymbol{\lambda}\rangle/\sqrt{\mathcal{D}}}\,d\boldsymbol{\lambda}\,.$$

Using the notation

$$\boldsymbol{\xi}(\mathbf{x},\mathbf{y}) := \int_{\mathbb{R}^n} \mathcal{F}\tilde{\eta}(\boldsymbol{\lambda})\,\mathcal{F}\eta(\mu\boldsymbol{\lambda}+\mathbf{y})\,\mathrm{e}^{2\pi i\langle\mathbf{x},\boldsymbol{\lambda}\rangle}\,d\boldsymbol{\lambda}\,,$$

the assertion is proved if we show that

$$R_{\eta,\mu,\mathcal{D}}(\mathbf{x}) = -\sum_{\boldsymbol{\nu}\in\mathbb{Z}^n\setminus\{\mathbf{0}\}} \boldsymbol{\xi}\Big(\frac{\mathbf{x}}{\sqrt{\mathcal{D}}},\sqrt{\mathcal{D}}\boldsymbol{\nu}\Big)\,\mathrm{e}^{2\pi i\langle\mathbf{x},\boldsymbol{\nu}\rangle/\mu}$$

belongs to $\mathcal{S}(\mathbb{R}^n)$ and can be made arbitrarily small by choosing $\mathcal{D}$ large enough.

Because $\mathcal{F}\eta, \mathcal{F}\tilde{\eta} \in \mathcal{S}(\mathbb{R}^n)$ and

$$\int_{\mathbb{R}^n} \frac{d\boldsymbol{\lambda}}{(1+|\boldsymbol{\lambda}|)^j\,(1+|\mu\boldsymbol{\lambda}+\mathbf{y}|)^k}$$

$$= \int_{|\boldsymbol{\lambda}|<|\mathbf{y}|/2\mu} \frac{d\boldsymbol{\lambda}}{(1+|\boldsymbol{\lambda}|)^j\,(1+|\mu\boldsymbol{\lambda}+\mathbf{y}|)^k} + \int_{|\boldsymbol{\lambda}|>|\mathbf{y}|/2\mu} \frac{d\boldsymbol{\lambda}}{(1+|\boldsymbol{\lambda}|)^j\,(1+|\mu\boldsymbol{\lambda}+\mathbf{y}|)^k}$$

$$\leq \Big(1+\frac{|\mathbf{y}|}{2}\Big)^{-k} \int_{\mathbb{R}^n} \frac{d\boldsymbol{\lambda}}{(1+|\boldsymbol{\lambda}|)^j} + \Big(1+\frac{|\mathbf{y}|}{2\mu}\Big)^{-j} \int_{\mathbb{R}^n} \frac{d\boldsymbol{\lambda}}{(1+|\mu\boldsymbol{\lambda}+\mathbf{y}|)^k}$$

$$= \Big(1+\frac{|\mathbf{y}|}{2}\Big)^{-k} \int_{\mathbb{R}^n} \frac{d\boldsymbol{\lambda}}{(1+|\boldsymbol{\lambda}|)^j} + \mu^{-n}\Big(1+\frac{|\mathbf{y}|}{2\mu}\Big)^{-j} \int_{\mathbb{R}^n} \frac{d\boldsymbol{\lambda}}{(1+|\boldsymbol{\lambda}|)^k}\,,$$

the estimate

$$|\mathbf{y}^{\boldsymbol{\gamma}}\partial_{\mathbf{x}}^{\boldsymbol{\alpha}}\,\boldsymbol{\xi}(\mathbf{x},\mathbf{y})| \leq C_{\boldsymbol{\alpha\beta}}$$

holds uniformly in $\mathbf{x}$. This enables one to write

$$\mathbf{x}^{\boldsymbol{\alpha}}\partial_{\mathbf{x}}^{\boldsymbol{\beta}} R_{\eta,\mu,\mathcal{D}}(\mathbf{x}) = -\sum_{\boldsymbol{\nu}\in\mathbb{Z}^n\setminus\{\mathbf{0}\}} \mathbf{x}^{\boldsymbol{\alpha}}\partial_{\mathbf{x}}^{\boldsymbol{\beta}}\Big(\boldsymbol{\xi}\Big(\frac{\mathbf{x}}{\sqrt{\mathcal{D}}},\sqrt{\mathcal{D}}\boldsymbol{\nu}\Big)\,\mathrm{e}^{2\pi i\langle\mathbf{x},\boldsymbol{\nu}\rangle/\mu}\Big).$$

By the same argument,

$$|\mathbf{x}^{\boldsymbol{\alpha}}\mathbf{y}^{\boldsymbol{\gamma}}\partial_{\mathbf{x}}^{\boldsymbol{\beta}}\,\boldsymbol{\xi}(\mathbf{x},\mathbf{y})| \leq C\int_{\mathbb{R}^n} \big|\mathbf{y}^{\boldsymbol{\gamma}}\partial_{\boldsymbol{\lambda}}^{\boldsymbol{\alpha}}\big(\boldsymbol{\lambda}^{\boldsymbol{\beta}}\mathcal{F}\tilde{\eta}(\boldsymbol{\lambda})\,\mathcal{F}\eta(\mu\boldsymbol{\lambda}+\mathbf{y})\big)\big|\,d\boldsymbol{\lambda} \leq C_{\boldsymbol{\alpha\beta\gamma}}$$

for all $\mathbf{x}, \mathbf{y} \in \mathbb{R}^n$, which implies $R_{\eta,\mu,\mathcal{D}} \in \mathcal{S}(\mathbb{R}^n)$. Additionally, from

$$(1+|\mathbf{x}|^k)|\boldsymbol{\xi}(\mathbf{x},\mathbf{y})| \le C_{k,j}|\mathbf{y}|^{-j},$$

it follows that the sum

$$(1+|\mathbf{x}|^k) \sum_{\boldsymbol{\nu}\in\mathbb{Z}^n\setminus\{\mathbf{0}\}} \Big|\boldsymbol{\xi}\Big(\frac{\mathbf{x}}{\sqrt{\mathcal{D}}}, \sqrt{\mathcal{D}}\boldsymbol{\nu}\Big)\Big|$$

can be made arbitrarily small by choosing $\mathcal{D}$ large enough. □

REMARK 8.2. By (8.12), the functions η and $\tilde{\eta}$ are connected by the convolution

$$\eta(\mathbf{x}) = \int_{\mathbb{R}^n} \tilde{\eta}(\mathbf{x}-\mu\mathbf{y})\eta(\mathbf{y})\,d\mathbf{y} = \frac{1}{\mu}\int_{\mathbb{R}^n} \eta\Big(\frac{\mathbf{x}-\mathbf{y}}{\mu}\Big)\,\tilde{\eta}(\mathbf{y})\,d\mathbf{y}.$$

8.2.2. Examples of mask functions. In the following, the function $\tilde{\eta}$ which is defined by (8.12) will be referred to as the *mask function* corresponding to η. It will also play an important rule in Chapter 9 in developing quasi-interpolation formulas on graded meshes. In the context of the present chapter, only the validity of approximate refinement equations and effective estimates for the remainder are important.

Some of the generating functions considered in Chapter 3 allow analytic formulas for $\tilde{\eta}$. For example, the generating functions $\eta_{2M}(\mathbf{x}) = L_{M-1}^{(n/2)}(|\mathbf{x}|^2)\,\mathrm{e}^{-|\mathbf{x}|^2}$ from Theorem 3.5 satisfy the requirements of Theorem 8.1, since by (3.19) they possess positive Fourier transforms. By (8.11), the analytic expression of the mask to η_2 is

$$\tilde{\eta}_2(\mathbf{x}) = \frac{\mathrm{e}^{-|\mathbf{x}|^2/\alpha}}{(\pi\alpha)^{n/2}}, \quad \mathbf{x}\in\mathbb{R}^n,$$

where $\alpha = 1-\mu^2$. $\tilde{\eta}_4$ and $\tilde{\eta}_6$, in the case of one space dimension, are given by

$$\begin{aligned} \tilde{\eta}_4(t) &= \frac{1}{\mu^2}\Big[\tilde{\eta}_2(t) - \frac{\alpha}{\mu}\,\mathrm{W}\big(\frac{\sqrt{\alpha}}{\mu}, \frac{t}{\sqrt{\alpha}}\big)\Big], \\ \tilde{\eta}_6(t) &= \frac{1}{\mu^4}\Big\{\tilde{\eta}_2(t) - 2\frac{\alpha}{\mu}\,\mathrm{Re}\Big[\frac{1+i\mu^2}{\sqrt{1+i}}\,\mathrm{W}\big(\tfrac{\sqrt{\alpha(1+i)}}{\mu}, \frac{t}{\sqrt{\alpha}}\big)\Big]\Big\}. \end{aligned} \tag{8.14}$$

The function $\mathrm{W}(z,t)$ is defined by

$$\mathrm{W}(z,t) = \frac{\mathrm{e}^{-t^2}}{2}\big(\mathrm{w}(i(z+t)) + \mathrm{w}(i(z-t))\big) \tag{8.15}$$

with the Faddeeva function (5.7). Of course, these formulas allow one to obtain analytical representations for the mask functions in any space dimension when $\eta(\mathbf{x})$ is a product of one-dimensional functions:

$$\eta(\mathbf{x}) = \eta_{2M}(x_1)\ldots\eta_{2M}(x_n).$$

For $n=3$, the mask function to

$$\eta_4(\mathbf{x}) = \Big(\frac{5}{2} - |\mathbf{x}|^2\Big)\,\mathrm{e}^{-|\mathbf{x}|^2}, \quad \mathbf{x}\in\mathbb{R}^3,$$

has the representation

$$\tilde{\eta}_4(\mathbf{x}) = \frac{1}{\mu^2}\Big(\tilde{\eta}_2(\mathbf{x}) - \frac{\pi\alpha}{2\mu^2}\frac{\mathrm{e}^{-|\mathbf{x}|^2/\alpha}}{|\mathbf{x}|}\Big(\mathrm{w}\Big(\frac{i(\pi\alpha+|\mathbf{x}|)}{\mu\sqrt{\alpha}}\Big) - \mathrm{w}\Big(\frac{i(\pi\alpha-|\mathbf{x}|)}{\mu\sqrt{\alpha}}\Big)\Big)\Big).$$

8.3. Approximate multi-resolution analysis

In this section we perform an approximate multi-resolution analysis using the Gaussian in $\mathbb{R}^n$ as the scaling function.

8.3.1. Approximate chain of subspaces. We introduce the closed linear subspaces of $L_2(\mathbb{R}^n)$

$$\mathbf{V}_j := \Big\{ \sum_{\mathbf{m}\in\mathbb{Z}^n} a_{\mathbf{m}}\, \phi_{\mathcal{D}}(2^j \cdot - \mathbf{m}) \,,\; \{a_{\mathbf{m}}\} \in \ell_2(\mathbb{Z}^n) \Big\} \,,\; j \in \mathbb{Z} \,,$$

with the normed Gaussian function

$$\phi_{\mathcal{D}}(\mathbf{x}) = \Big(\frac{2}{\pi\mathcal{D}}\Big)^{n/4} \mathrm{e}^{-|\mathbf{x}|^2/\mathcal{D}} \,, \quad \mathbf{x} \in \mathbb{R}^n \,, \tag{8.16}$$

i.e., $\|\phi_{\mathcal{D}}\|_{L_2} = 1$. Since

$$\Big\| \sum_{\mathbf{m}\in\mathbb{Z}^n} a_{\mathbf{m}}\, \phi_{\mathcal{D}}(\cdot - \mathbf{m}) \Big\|_{L_2}^2 = (2\pi\mathcal{D})^{n/2} \int_{\mathbb{R}^n} \mathrm{e}^{-2\pi^2\mathcal{D}|\boldsymbol{\lambda}|^2} \Big| \sum_{\mathbf{m}\in\mathbb{Z}^n} a_{\mathbf{m}}\, \mathrm{e}^{2\pi i\langle \mathbf{m},\boldsymbol{\lambda}\rangle} \Big|^2 d\boldsymbol{\lambda}$$

$$= (2\pi\mathcal{D})^{n/2} \int_{[0,1]^n} \sum_{\mathbf{k}\in\mathbb{Z}^n} \mathrm{e}^{-2\pi^2\mathcal{D}|\boldsymbol{\lambda}-\mathbf{k}|^2} \Big| \sum_{\mathbf{m}\in\mathbb{Z}^n} a_{\mathbf{m}}\, \mathrm{e}^{2\pi i\langle \mathbf{m},\boldsymbol{\lambda}\rangle} \Big|^2 d\boldsymbol{\lambda}$$

and

$$\sum_{\mathbf{k}\in\mathbb{Z}^n} \mathrm{e}^{-2\pi^2\mathcal{D}|\boldsymbol{\lambda}-\mathbf{k}|^2} \geq c_{\mathcal{D}} > 0 \,,$$

the set $\{\phi_{\mathcal{D}}(\cdot - \mathbf{m})\}_{\mathbf{m}\in\mathbb{Z}^n}$ is an L_2-stable basis in $\mathbf{V}_0$.

Furthermore, with $\mu = 1/2$, from (8.9) we have the approximate refinement equation

$$\phi_{\mathcal{D}}(\mathbf{x}) = \frac{2^n}{(3\pi\mathcal{D})^{n/2}} \sum_{\mathbf{m}\in\mathbb{Z}^n} e^{-|\mathbf{m}|^2/3\mathcal{D}}\, \phi_{\mathcal{D}}(2\mathbf{x} - \mathbf{m}) + \mathcal{O}(\varepsilon_{\mathcal{D}})\phi_{\mathcal{D}}(\mathbf{x}) \,,$$

where

$$\varepsilon_{\mathcal{D}} = \sum_{\boldsymbol{\nu}\in\mathbb{Z}^n\setminus\{\mathbf{0}\}} \mathrm{e}^{-3\pi^2\mathcal{D}\,|\boldsymbol{\nu}|^2/4} \,.$$

Additionally, equation (8.9) shows that for any $l < j$ the space $\mathbf{V}_l$ is almost included in $\mathbf{V}_j$. For any element of $\varphi_l \in \mathbf{V}_l$ the small perturbation

$$\tilde{\varphi}_l(\mathbf{x}) = \varphi_l(\mathbf{x}) \Big(1 + \sum_{\boldsymbol{\nu}\in\mathbb{Z}^n\setminus\{\mathbf{0}\}} e^{-\pi^2\mathcal{D}(1-4^{l-j})\,|\boldsymbol{\nu}|^2}\, e^{2\pi i(2^{j-l}-2^{l-j})\langle \mathbf{x},\boldsymbol{\nu}\rangle} \Big)$$

admits the inclusion

$$\tilde{\varphi}_l \in \mathbf{V}_j \,. \tag{8.17}$$

Let us introduce a measure for the distance between subspaces. Let A, B be two closed linear manifolds of a Hilbert space X. The gap (cf. [**49**, §IV.2]) between the subspaces A and B is defined as

$$\hat{\delta}(A, B) = \max(\delta(A, B), \delta(A, B)) \,,$$

where

$$\delta(A, B) = \sup_{u\in S_A} \mathrm{dist}(u, B)$$

and S_A is the unit sphere of A (the set of all $u \in A$ with $\|u\| = 1$).

Since X is a Hilbert space, the gap can be determined from the relation $\hat{\delta}(A, B) = \|P_A - P_B\|$, where P_A and P_B denote the orthogonal projections onto A and B, respectively. Hence, $\hat{\delta}$ satisfies the triangle inequality and it can be used to define a distance between subspaces, and the set of all closed subspaces of A becomes a metric space.

Therefore, by (8.17)

$$\delta(\mathbf{V}_l, \mathbf{V}_j) \leq \sum_{\boldsymbol{\nu} \in \mathbb{Z}^n \setminus \{\mathbf{0}\}} \mathrm{e}^{-3\pi^2 \mathcal{D} |\boldsymbol{\nu}|^2/4} = \varepsilon_{\mathcal{D}} \tag{8.18}$$

for all $l < j$.

8.3.2. Almost orthogonal decomposition. We introduce the closed subspace $\widetilde{\mathbf{W}}_j \subset \mathbf{V}_{j+1}$ of all functions which are orthogonal to $\mathbf{V}_j$ and we choose subspaces $\mathbf{W}_j$ such that $\hat{\delta}(\mathbf{W}_j, \widetilde{\mathbf{W}}_j) \leq \varepsilon_{\mathcal{D}}$.

It can be easily seen that

$$|(\varphi_j, \varphi_l)| \leq \left((1+\varepsilon)^{|j-l|-1} - 1\right) \|\varphi_j\|_{L_2} \|\varphi_l\|_{L_2}$$

for $\varphi_j \in \mathbf{W}_j$. Thus, the situation is very similar to the case when exact refinement equations are valid, which was mentioned in Section 8.1.

Let us fix some integer $k > 0$, which determines the grid for the approximating functions. In the following, we show that any element of $\mathbf{V}_k$ can be represented within some prescribed tolerance as an element of the multi-resolution structure

$$\mathbf{X}_k := \mathbf{V}_0 \dotplus \mathbf{W}_0 \dotplus \ldots \dotplus \mathbf{W}_{k-1} \,. \tag{8.19}$$

To this end, introduce the orthogonal projections $P_j : L_2(\mathbb{R}^n) \to \mathbf{V}_j$, $j = 0, \ldots, k$, and $Q_j : L_2(\mathbb{R}^n) \to \mathbf{W}_j$, $\widetilde{Q}_j : L_2(\mathbb{R}^n) \to \widetilde{\mathbf{W}}_j$, $j = 0, \ldots, k-1$, and denote $Q_{-1} = P_0$.

THEOREM 8.3. *Any $\varphi_k \in \mathbf{V}_k$ can be approximatively represented as an element of the multi-resolution structure* (8.19) *and the estimate*

$$\Big\|\varphi_k - \sum_{j=-1}^{k-1} Q_j \varphi_k\Big\|_{L_2} \leq k \, \frac{3\varepsilon_{\mathcal{D}} - 2\varepsilon_{\mathcal{D}}^2}{1 - \varepsilon_{\mathcal{D}}} \, \|\varphi_k\|_{L_2}$$

holds.

PROOF. The result follows from the use of the telescopic series

$$\begin{aligned}
\varphi_k = P_k \varphi_k &= \sum_{j=1}^{k} (P_j - P_{j-1}) \varphi_k + P_0 \varphi_k \\
&= \sum_{j=1}^{k} (P_j - P_{j-1}) \varphi_k + Q_{-1} \varphi_k = \sum_{j=-1}^{k-1} Q_j \varphi_k + \sum_{j=0}^{k-1} (P_{j+1} - P_j - Q_j) \varphi_k
\end{aligned}$$

and the following auxiliary assertion.

LEMMA 8.4. *The estimate*

$$\|P_j + Q_j - P_{j+1}\|_{L_2} \leq \frac{3\varepsilon_{\mathcal{D}} - 2\varepsilon_{\mathcal{D}}^2}{1 - \varepsilon_{\mathcal{D}}} \,, \quad j = 0, \ldots, k-1 \,,$$

holds.

PROOF. We have

$$\|P_j + Q_j - P_{j+1}\|_{L_2} \le \|P_j + \widetilde{Q}_j - P_{j+1}\|_{L_2} + \|Q_j - \widetilde{Q}_j\|_{L_2}$$

with

$$\|Q_j - \widetilde{Q}_j\|_{L_2} = \hat{\delta}(\mathbf{W}_j, \widetilde{\mathbf{W}}_j) \le \varepsilon_{\mathcal{D}} .$$

Note that (8.18) implies the inequality

$$\|\varphi_j - P_{j+1}\varphi_j\|_{L_2} \le \varepsilon_{\mathcal{D}} \|\varphi_j\|_{L_2} , \quad \forall \varphi_j \in \mathbf{V}_j . \tag{8.20}$$

Since $P_{j+1}(\mathbf{V}_j) = \mathbf{V}_{j+1} \ominus \mathbf{W}_j$, any $\varphi_{j+1} \in \mathbf{V}_{j+1}$ can be written in the form

$$\varphi_{j+1} = P_{j+1}\varphi_j + Q_j\varphi_{j+1}$$

with some $\varphi_j \in \mathbf{V}_j$. From (8.20), we therefore derive

$$\|\varphi_{j+1} - (\varphi_j + Q_j\varphi_{j+1})\|_{L_2} = \|P_{j+1}\varphi_j - \varphi_j\|_{L_2} \le \varepsilon_{\mathcal{D}}\|\varphi_j\|_{L_2}$$

and

$$\|\varphi_j\|_{L_2} \le \frac{1}{1-\varepsilon_{\mathcal{D}}} \|P_{j+1}\varphi_j\|_{L_2} = \frac{1}{1-\varepsilon_{\mathcal{D}}} \|\varphi_{j+1} - Q_j\varphi_{j+1}\|_{L_2} \le \frac{\|\varphi_{j+1}\|_{L_2}}{1-\varepsilon_{\mathcal{D}}} .$$

Now we use that the sum $P_j + Q_j$ is the ortho-projection onto $\mathbf{V}_j \oplus \mathbf{W}_j$. Hence for any $u \in L_2(\mathbb{R}^n)$, we obtain the estimate

$$\|(I - (P_j + Q_j))P_{j+1}u\|_{L_2} = \inf_{v \in \mathbf{V}_j \oplus \mathbf{W}_j} \|P_{j+1}u - v\|_{L_2} \le \frac{\varepsilon_{\mathcal{D}}}{1-\varepsilon_{\mathcal{D}}} \|u\|_{L_2} ,$$

leading together with (8.20) to

$$\begin{aligned}
\|(P_j + Q_j - P_{j+1})u\|_{L_2} &\le \|(I - P_{j+1})(P_j + Q_j)u\|_{L_2} + \|P_{j+1}(I - P_j - Q_j)u\|_{L_2} \\
&\le \inf_{v \in \mathbf{V}_{j+1}} \|(P_j + Q_j)u - v\|_{L_2} + \|(I - P_j - Q_j)P_{j+1}\|_{L_2}\|u\|_{L_2} \\
&\le \big(\varepsilon_{\mathcal{D}} + \frac{\varepsilon_{\mathcal{D}}}{1-\varepsilon_{\mathcal{D}}}\big)\|u\|_{L_2} .
\end{aligned}$$

□

8.4. Approximate univariate wavelets

8.4.1. Gaussian as scaling function. To find an analytic expression of the wavelet, we apply some well-known constructions from wavelet theory. The scaling function is

$$\phi_{\mathcal{D}}(x) = \Big(\frac{2}{\pi\mathcal{D}}\Big)^{1/4} \mathrm{e}^{-x^2/\mathcal{D}} . \tag{8.21}$$

We denote the corresponding scaled and wavelet spaces by V_j and W_j, respectively. The function

$$\sum_{m\in\mathbb{Z}} (-1)^{m-1} \mu_{m-1} \phi_{\mathcal{D}}(2x - m) \quad \text{with} \quad \mu_m = \int_{\mathbb{R}} \phi_{\mathcal{D}}(x)\, \phi_{\mathcal{D}}(2x + m)\, dx \tag{8.22}$$

is orthogonal to all integer shifts of $\phi_{\mathcal{D}}$, modulo constants. It has the form

$$\sum_{m\in\mathbb{Z}} (-1)^{m-1} \mathrm{e}^{-(m-1)^2/5\mathcal{D}}\, \phi_{\mathcal{D}}(2x - m) \in \widetilde{W}_0 \subset V_0 .$$

Its Fourier transform is given by

$$\frac{\sqrt{5\pi\mathcal{D}}}{2}\, \mathrm{e}^{-\pi i\lambda}\, \mathrm{e}^{-\pi^2\mathcal{D}\lambda^2/4}\, \sigma_{5\mathcal{D}}\Big(\frac{\lambda+1}{2}\Big) , \tag{8.23}$$

where σ_α denotes the positive and 1-periodic function

$$\sigma_\alpha(\lambda) = \frac{1}{\sqrt{\alpha\pi}} \sum_{m\in\mathbb{Z}} \mathrm{e}^{-m^2/\alpha}\, \mathrm{e}^{2\pi i m\lambda} = \sum_{j\in\mathbb{Z}} \mathrm{e}^{-\alpha\pi^2(\lambda+j)^2} .$$

Hence, the integer shifts of this function form a Riesz basis in $\widetilde{W}_0$.

As indicated above, we want to use a very accurate approximation of this function by a simpler analytic representation. Using (7.10), we derive the formula

$$\begin{aligned} &\sum_{m\in\mathbb{Z}} (-1)^m \mathrm{e}^{-m^2/5\mathcal{D}}\, \phi_{\mathcal{D}}(2x-1-m) \\ &\quad = \mathrm{e}^{-(2x-1)^2/6\mathcal{D}} \sum_{m\in\mathbb{Z}} (-1)^m \exp\Big(-\frac{6}{5\mathcal{D}}\big(\frac{5}{6}(2x-1)-m\big)^2\Big) \\ &\quad = \sqrt{\frac{10\pi\mathcal{D}}{3}}\, \mathrm{e}^{-5\pi^2\mathcal{D}/24}\, \mathrm{e}^{-(2x-1)^2/6\mathcal{D}} \big(\cos\frac{5\pi}{6}(2x-1) + R_{\mathcal{D}}(x)\big) , \end{aligned} \tag{8.24}$$

with

$$R_{\mathcal{D}}(x) = \sum_{k=1}^{\infty} \cos\frac{5\pi}{6}(2k+1)(2x-1)\, \mathrm{e}^{-5\pi^2\mathcal{D}(k^2+k)/6} = \mathcal{O}(\mathrm{e}^{-5\pi^2\mathcal{D}/3}) .$$

The L_2-norm of the function $\mathrm{e}^{-(2x-1)^2/6\mathcal{D}} \cos\frac{5\pi}{6}(2x-1)$ is

$$\kappa_{\mathcal{D}}^{-1} = \Big\| \mathrm{e}^{-(2\cdot-1)^2/6\mathcal{D}} \cos\frac{5\pi}{6}(2\cdot-1) \Big\|_{L_2} = \frac{(3\pi\mathcal{D})^{1/4}}{2} \sqrt{1-\mathrm{e}^{-25\pi^2\mathcal{D}/12}} . \tag{8.25}$$

We introduce the normed *approximate prewavelet*

$$\psi_{\mathcal{D}}(x) := \kappa_{\mathcal{D}}\, \mathrm{e}^{-(2x-1)^2/6\mathcal{D}} \cos\frac{5\pi}{6}(2x-1) , \tag{8.26}$$

shown in Figs. 8.1 and 8.2 for two values of $\mathcal{D}$. It differs from the corresponding

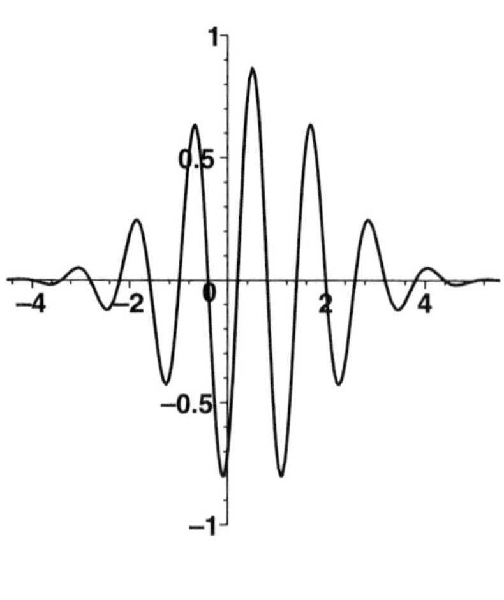

FIGURE 8.1. Wavelet $\psi_3(x)$

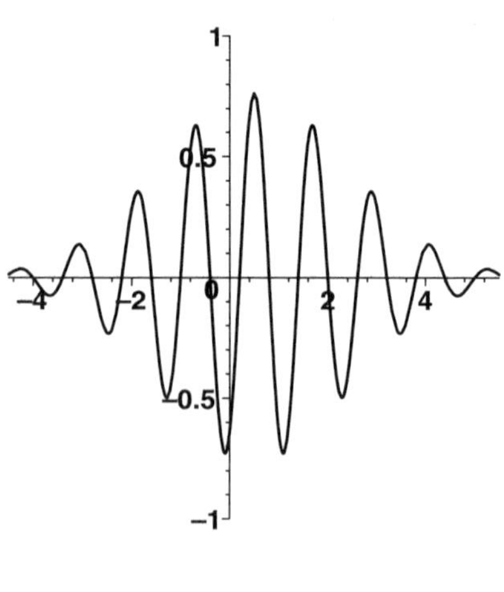

FIGURE 8.2. Wavelet $\psi_5(x)$

function in $\widetilde{W}_0$

$$\tilde{\psi}_{\mathcal{D}}(x) := \kappa_{\mathcal{D}} \sqrt{\frac{3}{10\pi\mathcal{D}}}\, \mathrm{e}^{5\pi^2\mathcal{D}/24} \sum_{m\in\mathbb{Z}} (-1)^m\, \mathrm{e}^{-m^2/5\mathcal{D}}\, \phi_{\mathcal{D}}(2x-1-m)$$

by

$$\begin{aligned}\tilde{\psi}_{\mathcal{D}}(x) - \psi_{\mathcal{D}}(x) &= \kappa_{\mathcal{D}}\, \mathrm{e}^{-(2x-1)^2/6\mathcal{D}} \sum_{k=1}^{\infty} \mathrm{e}^{-5\pi^2\mathcal{D}(k^2+k)/6} \cos\frac{5\pi(2k+1)}{6}(2x-1)\\ &= \psi_{\mathcal{D}}(x) \sum_{k=1}^{\infty} \mathrm{e}^{-5\pi^2\mathcal{D}(k^2+k)/6}\, \frac{\cos(5\pi(2k+1)(2x-1)/6)}{\cos(5\pi(2x-1)/6)}\,.\end{aligned}$$

Note that

$$\frac{\cos(5\pi(2k+1)(2x-1)/6)}{\cos(5\pi(2x-1)/6)} = (-1)^k \Big(1 + 2\sum_{j=1}^{k} (-1)^j \cos\frac{5\pi j}{3}(2x-1)\Big).$$

Hence,

$$\tilde{\psi}_{\mathcal{D}}(x) - \psi_{\mathcal{D}}(x) = \psi_{\mathcal{D}}(x) \sum_{k=1}^{\infty} (-1)^k\, \mathrm{e}^{-5\pi^2\mathcal{D}(k^2+k)/6} \Big(1 + 2\sum_{j=1}^{k} (-1)^j \cos\frac{5\pi j}{3}(2x-1)\Big)$$

and, consequently, the gap admits the estimate $\hat{\delta}(W_j, \widetilde{W}_j) < \varepsilon_{\mathcal{D}}$. Moreover, the functions $\psi_{\mathcal{D}}(\cdot - m)$, $m \in \mathbb{Z}$, form a Riesz basis in W_0.

In view of

$$\hat{\delta}(V_0 \dot{+} W_0, V_1) \le \frac{3\varepsilon_{\mathcal{D}} - 2\varepsilon_{\mathcal{D}}^2}{1 - \varepsilon_{\mathcal{D}}}$$

(see Lemma 8.4), the system $\{\phi_{\mathcal{D}}(\cdot - m), \psi_{\mathcal{D}}(\cdot - k), k, m \in \mathbb{Z}\}$ is a Riesz basis in $V_0 \dot{+} W_0$.

8.4.2. Moments. Besides the fast decay of $\psi_{\mathcal{D}}$, we are interested in the moments of the approximate wavelet $\psi_{\mathcal{D}}$. Since this function is almost orthogonal to the integer shifts of the Gaussian $\phi_{\mathcal{D}}$, which approximate polynomials very accurately as seen in Subsection 7.1.2, one can expect that even higher moments are very small and decrease if $\mathcal{D}$ increases. Using the Fourier transform of the approximate wavelet

$$\begin{aligned}\mathcal{F}\psi_{\mathcal{D}}(\lambda) &= \kappa_{\mathcal{D}} \frac{\sqrt{6\pi\mathcal{D}}}{2}\, \mathrm{e}^{-\pi i\lambda}\, \mathrm{e}^{-3\pi^2\mathcal{D}\lambda^2/2}\, \mathrm{e}^{-25\pi^2\mathcal{D}/24} \cosh\frac{5\pi^2\mathcal{D}\lambda}{2}\\ &= \frac{(3\pi\mathcal{D})^{1/4}\sqrt{2}}{\sqrt{\mathrm{e}^{25\pi^2\mathcal{D}/12} - 1}}\, \mathrm{e}^{-\pi i\lambda}\, \mathrm{e}^{-3\pi^2\mathcal{D}\lambda^2/2} \cosh\frac{5\pi^2\mathcal{D}\lambda}{2},\end{aligned} \tag{8.27}$$

one can compute the moments of $\psi_{\mathcal{D}}$ by the formula

$$\begin{aligned}\int_{\mathbb{R}} x^k \psi_{\mathcal{D}}(x)\, dx &= (-2\pi i)^{-k} \frac{d}{d\lambda} \mathcal{F}\psi_{\mathcal{D}}(\lambda)\Big|_{\lambda=0}\\ &= \frac{(3\pi\mathcal{D})^{1/4}\sqrt{2}}{(-2\pi i)^k \sqrt{\mathrm{e}^{25\pi^2\mathcal{D}/12} - 1}} \frac{d}{d\lambda}\Big(\mathrm{e}^{-\pi i\lambda}\, \mathrm{e}^{-3\pi^2\mathcal{D}\lambda^2/2} \cosh\frac{5\pi^2\mathcal{D}\lambda}{2}\Big)\Big|_{\lambda=0}.\end{aligned}$$

The first 10 moments of $\psi_{\mathcal{D}}$ for different values of $\mathcal{D}$ are given in the Table 8.1.

moment	$\mathcal{D}=3$	$\mathcal{D}=4$	$\mathcal{D}=5$
0	$1.31\cdot 10^{-13}$	$4.84\cdot 10^{-18}$	$1.75\cdot 10^{-22}$
1	$6.57\cdot 10^{-14}$	$2.42\cdot 10^{-18}$	$8.77\cdot 10^{-23}$
2	$-1.79\cdot 10^{-11}$	$-1.18\cdot 10^{-15}$	$-6.70\cdot 10^{-20}$
3	$-2.69\cdot 10^{-11}$	$-1.77\cdot 10^{-15}$	$-1.01\cdot 10^{-19}$
4	$2.26\cdot 10^{-9}$	$2.72\cdot 10^{-13}$	$2.45\cdot 10^{-17}$
5	$5.69\cdot 10^{-9}$	$6.82\cdot 10^{-13}$	$6.13\cdot 10^{-17}$
6	$-2.61\cdot 10^{-7}$	$-5.89\cdot 10^{-11}$	$-8.54\cdot 10^{-15}$
7	$-9.35\cdot 10^{-7}$	$-2.086\cdot 10^{-10}$	$-3.010\cdot 10^{-14}$
8	$2.73\cdot 10^{-5}$	$1.20\cdot 10^{-8}$	$2.84\cdot 10^{-12}$
9	$1.29\cdot 10^{-4}$	$5.51\cdot 10^{-8}$	$1.29\cdot 10^{-11}$
10	$-2.54\cdot 10^{-3}$	$-2.25\cdot 10^{-6}$	$-8.93\cdot 10^{-10}$

TABLE 8.1. Moments of $\psi_{\mathcal{D}}$

Owing to the simple form of $\psi_{\mathcal{D}}$, it is also easy to compute scalar products of elements of V_l and W_j. For example,

$$(8.28)\qquad \begin{aligned}&(\phi_{\mathcal{D}}, \psi_{\mathcal{D}}(2^k\cdot -m))_{L_2}\\ &= \frac{2\sqrt[4]{6}}{\sqrt{4^{k+1}+6}}\,\frac{\mathrm{e}^{-(2m+1)^2/(4^{k+1}+6)\mathcal{D}}\,\mathrm{e}^{-50\cdot 4^k\pi^2\mathcal{D}/3(4^{k+1}+6)}}{\sqrt{1-\mathrm{e}^{-25\pi^2\mathcal{D}/12}}}\cos\frac{5\pi(2m+1)}{4^{k+1}+6}\,.\end{aligned}$$

Hence, if $k=0$, then the approximate wavelets $\psi_{\mathcal{D}}$ are even orthogonal to the integer shifts of the scaling function $\phi_{\mathcal{D}}$, and therefore $V_j\perp W_j$ for all j.

The effect of the nearly vanishing moments is shown in Figs. 8.3 and 8.4 which depict the graph of the Hilbert transform of the wavelets ψ_3 and ψ_5. This function has the analytic representation

$$(8.29)\qquad \begin{aligned}\mathcal{H}\psi_{\mathcal{D}}(x) &= \frac{1}{\pi}\int_{-\infty}^{\infty}\frac{\psi_{\mathcal{D}}(y)}{y-x}\,dy\\ &= \kappa_{\mathcal{D}}\Big(\mathrm{e}^{-\frac{25\pi^2\mathcal{D}}{24}}\,\mathrm{Im}\,\mathrm{w}\Big(\frac{5\pi i\mathcal{D}+2(2x-1)}{2\sqrt{6\mathcal{D}}}\Big)-\mathrm{e}^{-\frac{(2x-1)^2}{6\mathcal{D}}}\sin\frac{5\pi}{6}(2x-1)\Big).\end{aligned}$$

Here we use the Faddeeva function $\mathrm{w}(z)$ defined by (5.7).

To analyze (8.29), we note that for $z=x+iy$

$$\mathrm{Im}\,\mathrm{w}(z)=\frac{2}{\sqrt{\pi}}\int_0^{\infty}\mathrm{e}^{-t^2}\,\mathrm{e}^{-2yt}\sin 2xt\,dt\,.$$

Hence

$$\Big|\mathrm{Im}\,\mathrm{w}\Big(\frac{5\pi i\mathcal{D}+2(2x-1)}{2\sqrt{6\mathcal{D}}}\Big)\Big|\le 1$$

and this function converges to zero as $|x|\to\infty$. Thus the Hilbert transform of the approximate wavelet

$$\psi_{\mathcal{D}}(x)=\kappa_{\mathcal{D}}\,\mathrm{e}^{-(2x-1)^2/6\mathcal{D}}\cos\frac{5\pi}{6}(2x-1)$$

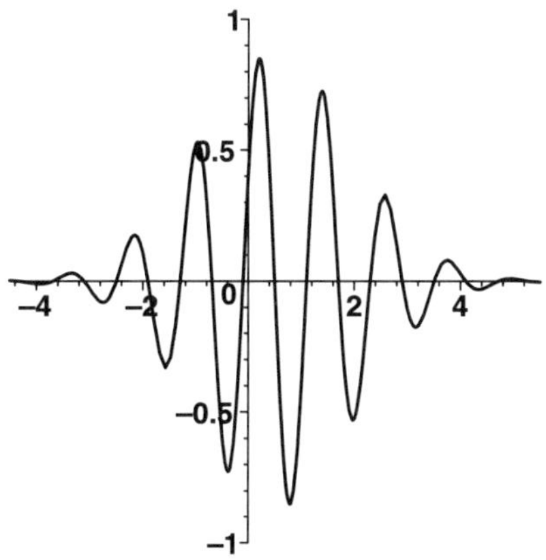

FIGURE 8.3. Hilbert transform $\mathcal{H}\psi_3(x)$

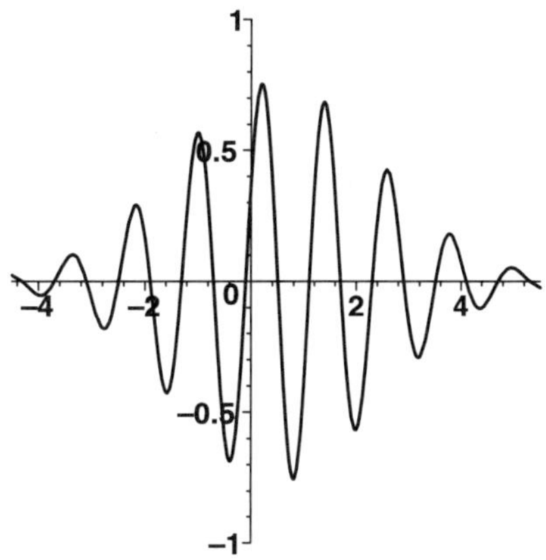

FIGURE 8.4. Hilbert transform $\mathcal{H}\psi_5(x)$

coincides within the tolerance $\kappa_{\mathcal{D}}\, \mathrm{e}^{-25\pi^2\mathcal{D}/24}$ with the function

$$-\kappa_{\mathcal{D}}\, \mathrm{e}^{-(2x-1)^2/6\mathcal{D}} \sin \frac{5\pi}{6}(2x-1)\,.$$

Since the essential supports of $\psi_{\mathcal{D}}$ and $\mathcal{H}\psi_{\mathcal{D}}$ are small, a high compression rate for the matrix representation of the Hilbert transform in the basis

$$\{\phi_{\mathcal{D}}(\cdot - m), m \in \mathbb{Z}\} \quad \text{and} \quad \{2^{j/2}\psi_{\mathcal{D}}(2^j \cdot - m), m \in \mathbb{Z}, j = 0, \dots, k-1\}$$

can be achieved.

8.4.3. Other examples. The above described construction of approximate wavelets can be performed for other generating functions of approximate quasi-interpolants. Here, we consider the scaling function

$$\phi_{\mathcal{D}}(x) = \mathrm{e}^{-x^2/\mathcal{D}} \cos \sqrt{\frac{2}{\mathcal{D}}}\, x\,, \quad x \in \mathbb{R}\,. \tag{8.30}$$

The quasi-interpolant generated by $\phi_{\mathcal{D}}$ is of the approximation order 4 (see the table in Subsection 3.2.1). We follow the wavelet construction indicated by (8.22). First one has to compute the integral

$$\mu_m = \int\limits_{\mathbb{R}} \phi_{\mathcal{D}}(x)\, \phi_{\mathcal{D}}(2x+m)\, dx = \sqrt{\mathcal{D}} \int\limits_{\mathbb{R}} \mathrm{e}^{-x^2} \cos \sqrt{2}x\; \mathrm{e}^{-(2x+y)^2} \cos \sqrt{2}(2x+y)\, dx$$

with $y = m/\sqrt{\mathcal{D}}$. Since

$$\mathrm{e}^{-x^2} \cos \sqrt{2}x = \mathrm{e}^{-1/2}\, \mathrm{Re}\, \mathrm{e}^{-(x+i/\sqrt{2})^2}$$

and

$$\mathrm{Re}\, u \cdot \mathrm{Re}\, v = \frac{1}{2}(\mathrm{Re}\, uv + \mathrm{Re}\, u\overline{v})\,,$$

we have to compute

$$\mu_m = \frac{\sqrt{\mathcal{D}}}{2\,\mathrm{e}} \operatorname{Re} \int_{\mathbb{R}} \Big(\mathrm{e}^{-(x+i/\sqrt{2})^2}\, \mathrm{e}^{-(2x+y+i/\sqrt{2})^2} + \mathrm{e}^{-(x+i/\sqrt{2})^2}\, \mathrm{e}^{-(2x+y-i/\sqrt{2})^2} \Big) dx\,.$$

From the relation

$$\int_{\mathbb{R}} \mathrm{e}^{-(x+a)^2}\, \mathrm{e}^{-(2x+b)^2}\, dx = \sqrt{\frac{\pi}{5}}\, \mathrm{e}^{-(2a-b)^2/5}\,,$$

we obtain

$$\mu_m = \frac{\sqrt{\pi\mathcal{D}}}{2\,\mathrm{e}\sqrt{5}} \operatorname{Re} \Big(\mathrm{e}^{-(m/\sqrt{\mathcal{D}}+3i/\sqrt{2})^2/5} + \mathrm{e}^{-(m/\sqrt{\mathcal{D}}+i/\sqrt{2})^2/5} \Big)\,.$$

Thus, by using (8.22), the element of the span of $\{\phi_{\mathcal{D}}(2\cdot+m)\}$ which is orthogonal to all integer shifts of $\phi_{\mathcal{D}}$ has the form

$$\sum_{m\in\mathbb{Z}} (-1)^m \operatorname{Re} \mathrm{e}^{-((2x-1-m)/\sqrt{\mathcal{D}}+i/\sqrt{2})^2} \times \operatorname{Re} \Big(\mathrm{e}^{-(m/\sqrt{\mathcal{D}}+3i/\sqrt{2})^2/5} + \mathrm{e}^{-(m/\sqrt{\mathcal{D}}+i/\sqrt{2})^2/5} \Big).$$

Therefore, in the sequel, we try to obtain simple analytic representations for the real part of the sums

$$\begin{aligned} &\sum_{m\in\mathbb{Z}} (-1)^m\, \mathrm{e}^{-((2x-1-m)/\sqrt{\mathcal{D}}+i/\sqrt{2})^2}\, \mathrm{e}^{-(m/\sqrt{\mathcal{D}}\pm i/\sqrt{2})^2/5}\,, \\ &\sum_{m\in\mathbb{Z}} (-1)^m\, \mathrm{e}^{-((2x-1-m)/\sqrt{\mathcal{D}}+i/\sqrt{2})^2}\, \mathrm{e}^{-(m/\sqrt{\mathcal{D}}\pm 3i/\sqrt{2})^2/5}\,. \end{aligned} \tag{8.31}$$

We start with

$$\sum_{m\in\mathbb{Z}} (-1)^m\, \mathrm{e}^{-((2x-1-m)/\sqrt{\mathcal{D}}+i/\sqrt{2})^2}\, \mathrm{e}^{-(m/\sqrt{\mathcal{D}}-3i/\sqrt{2})^2/5}\,. \tag{8.32}$$

Because

$$\begin{aligned} &\Big(\frac{2x-1-m}{\sqrt{\mathcal{D}}} + \frac{i}{\sqrt{2}}\Big)^2 + \frac{1}{5}\Big(\frac{m}{\sqrt{\mathcal{D}}} - \frac{3i}{\sqrt{2}}\Big)^2 \\ &\qquad = \Big(\frac{2x-1}{\sqrt{6\mathcal{D}}} - \frac{i}{\sqrt{3}}\Big)^2 + \frac{6}{5}\Big(\frac{m}{\sqrt{\mathcal{D}}} - \frac{4i}{3\sqrt{2}} - \frac{5(2x-1)}{6\sqrt{\mathcal{D}}}\Big)^2, \end{aligned}$$

(8.32) can be transformed to

$$\begin{aligned} &\mathrm{e}^{-((2x-1)/\sqrt{6\mathcal{D}}-i/\sqrt{3})^2} \sum_{m\in\mathbb{Z}} (-1)^m \exp\Big(-\frac{6}{5\mathcal{D}}\Big(m - \frac{4i\sqrt{\mathcal{D}}}{3\sqrt{2}} - \frac{5(2x-1)}{6}\Big)^2\Big) \\ &= \mathrm{e}^{-((2x-1)/\sqrt{6\mathcal{D}}-i/\sqrt{3})^2} \sqrt{\frac{10\pi\mathcal{D}}{3}} \\ &\quad\times \sum_{m=0}^{\infty} \mathrm{e}^{-5\pi^2\mathcal{D}(2m+1)^2/24} \cos(2m+1)\Big(\frac{5\pi}{6}(2x-1) + \frac{2\pi i\sqrt{2\mathcal{D}}}{3}\Big), \end{aligned}$$

where we use the special case (7.9) of Poisson's summation formula for the Theta function. The sum on the right-hand side can be approximated by its first term with high accuracy. Indeed, we have

$$\sum_{m=0}^{\infty} \mathrm{e}^{-5\pi^2\mathcal{D}(2m+1)^2/24} \cos(2m+1)\Big(\frac{5\pi}{6}y + \frac{2\pi i\sqrt{2\mathcal{D}}}{3}\Big)$$
$$= \mathrm{e}^{-5\pi^2\mathcal{D}/24}\Big(\cos\Big(\frac{5\pi}{6}y + \frac{2\pi i\sqrt{2\mathcal{D}}}{3}\Big) + R_{\mathcal{D}}(y)\Big)$$

with the sum

$$R_{\mathcal{D}}(y) = \sum_{m=1}^{\infty} \mathrm{e}^{-5\pi^2\mathcal{D}((2m+1)^2-1)/24} \cos(2m+1)\Big(\frac{5\pi}{6}y + \frac{2\pi i\sqrt{2\mathcal{D}}}{3}\Big).$$

The terms of $R_{\mathcal{D}}(y)$ can be estimated by

$$\Big| \mathrm{e}^{-5\pi^2\mathcal{D}((2m+1)^2-1)/24} \cos(2m+1)\Big(\frac{5\pi}{6}y + \frac{2\pi i\sqrt{2\mathcal{D}}}{3}\Big)\Big|$$
$$\leq 2\,\mathrm{e}^{16/15} \exp\Big(-\frac{5\pi^2\mathcal{D}}{24}\Big(\Big(2m+1-\frac{8\sqrt{2}}{5\pi\sqrt{\mathcal{D}}}\Big)^2 - 1\Big)\Big)$$
$$\leq 2\,\mathrm{e}^{16/15} \exp\Big(-\frac{5\pi^2\mathcal{D}}{6}\Big(\Big(m-\frac{4\sqrt{2}}{5\pi\sqrt{\mathcal{D}}}\Big)^2 + \Big(m-\frac{4\sqrt{2}}{5\pi\sqrt{\mathcal{D}}}\Big)\Big)\Big).$$

Therefore, the function

$$\rho_{\mathcal{D}}\, \mathrm{e}^{-((2x-1)/\sqrt{6\mathcal{D}}-i/\sqrt{3})^2} \cos\Big(\frac{5\pi}{6}(2x-1) + \frac{2\pi i\sqrt{2\mathcal{D}}}{3}\Big)$$
$$\text{with} \qquad \rho_{\mathcal{D}} := \mathrm{e}^{-5\pi^2\mathcal{D}/24}\sqrt{\frac{10\pi\mathcal{D}}{3}}$$

approximates (8.32) with the error

$$c\,\mathrm{e}^{-(2x-1)^2/6\mathcal{D}} \exp\Big(-\frac{5\pi^2\mathcal{D}}{6}\Big(1-\frac{4\sqrt{2}}{5\pi\sqrt{\mathcal{D}}}\Big)\Big(2-\frac{4\sqrt{2}}{5\pi\sqrt{\mathcal{D}}}\Big)\Big), \tag{8.33}$$

c being some constant independent of $\mathcal{D}$. Obviously, this error is in the range of saturation errors and can be ignored if $\mathcal{D}$ is large enough.

Thus, the real part of (8.32) can be approximated modulo saturation terms by the function

$$\operatorname{Re}\Big(\mathrm{e}^{-((2x-1)/\sqrt{6\mathcal{D}}-i/\sqrt{3})^2} \cos\Big(\frac{5\pi}{6}(2x-1) + \frac{2\pi i\sqrt{2\mathcal{D}}}{3}\Big)\Big).$$

The remaining three sums of (8.31) can be approximated by

$$\sum_{m\in\mathbb{Z}}(-1)^m\,\mathrm{e}^{-((2x-1-m)/\sqrt{\mathcal{D}}+i/\sqrt{2})^2}\,\mathrm{e}^{-(m/\sqrt{\mathcal{D}}+3i/\sqrt{2})^2/5} \asymp \rho_{\mathcal{D}}\,\mathrm{e}^{-((2x-1)/\sqrt{6\mathcal{D}}+2i/\sqrt{3})^2}\cos\left(\frac{5\pi}{6}(2x-1)+\frac{\pi i\sqrt{2\mathcal{D}}}{6}\right),$$

$$\sum_{m\in\mathbb{Z}}(-1)^m\,\mathrm{e}^{-((2x-1-m)/\sqrt{\mathcal{D}}+i/\sqrt{2})^2}\,\mathrm{e}^{-(m/\sqrt{\mathcal{D}}+i/\sqrt{2})^2/5} \asymp \rho_{\mathcal{D}}\,\mathrm{e}^{-((2x-1)/\sqrt{6\mathcal{D}}+i/\sqrt{3})^2}\cos\left(\frac{5\pi}{6}(2x-1)+\frac{\pi i\sqrt{2\mathcal{D}}}{3}\right),$$

$$\sum_{m\in\mathbb{Z}}(-1)^m\,\mathrm{e}^{-((2x-1-m)/\sqrt{\mathcal{D}}+i/\sqrt{2})^2}\,\mathrm{e}^{-(m/\sqrt{\mathcal{D}}-i/\sqrt{2})^2/5} \asymp \rho_{\mathcal{D}}\,\mathrm{e}^{-(2x-1)^2/6\mathcal{D}}\cos\left(\frac{5\pi}{6}(2x-1)+\frac{\pi i\sqrt{2\mathcal{D}}}{2}\right),$$

where the errors are less than that of (8.32). More precisely, the estimates corresponding to (8.33) are of the form

$$c\,\mathrm{e}^{-(2x-1)^2/6\mathcal{D}}\exp\left(-\frac{5\pi^2\mathcal{D}}{6}\left(1-\frac{j\sqrt{2}}{5\pi\sqrt{\mathcal{D}}}\right)\left(2-\frac{j\sqrt{2}}{5\pi\sqrt{\mathcal{D}}}\right)\right),\quad j=1,2,3.$$

So, we define the approximate wavelet $\psi_{\mathcal{D}}$, corresponding to the scaling function (8.30) by

$$\begin{aligned}\psi_{\mathcal{D}}(x) := \mathrm{e}^{-(2x-1)^2/6\mathcal{D}}\,\mathrm{Re}\Bigg(&\mathrm{e}^{4/3}\,\mathrm{e}^{-4i(2x-1)/\sqrt{18\mathcal{D}}}\cos\left(\frac{5\pi}{6}(2x-1)+\frac{\pi i\sqrt{2\mathcal{D}}}{6}\right)\\ &+\mathrm{e}^{1/3}\,\mathrm{e}^{-2i(2x-1)/\sqrt{18\mathcal{D}}}\Bigg(\cos\left(\frac{5\pi}{6}(2x-1)+\frac{\pi i\sqrt{2\mathcal{D}}}{3}\right)\\ &+\cos\left(\frac{5\pi}{6}(2x-1)-\frac{2\pi i\sqrt{2\mathcal{D}}}{3}\right)\Bigg)+\cos\left(\frac{5\pi}{6}(2x-1)+\frac{\pi i\sqrt{2\mathcal{D}}}{2}\right)\Bigg),\end{aligned}$$

and its distance to the elements of the span of $\{\phi_{\mathcal{D}}(2\cdot+m)\}$, which are orthogonal to all integer shifts of $\phi_{\mathcal{D}}$, is estimated by (8.33). Finally, in a more compact form, we get

$$\boxed{\psi_{\mathcal{D}}(x) = \mathrm{e}^{-(2x-1)^2/6\mathcal{D}}\left(\cos\frac{5\pi}{6}(2x-1)\,p(2x-1)+\sin\frac{5\pi}{6}(2x-1)\,q(2x-1)\right)}$$

with the trigonometric polynomials

$$\boxed{\begin{aligned}p(x) &= \cosh\frac{\sqrt{2\mathcal{D}}\pi}{2}+\mathrm{e}^{1/3}\left(\cosh\frac{\sqrt{2\mathcal{D}}\pi}{3}+\cosh\frac{2\sqrt{2\mathcal{D}}\pi}{3}\right)\cos\frac{\sqrt{2}x}{3\sqrt{\mathcal{D}}}\\ &\quad+\mathrm{e}^{4/3}\cosh\frac{\sqrt{2\mathcal{D}}\pi}{6}\cos\frac{2\sqrt{2}x}{3\sqrt{\mathcal{D}}},\\ q(x) &= \mathrm{e}^{1/3}\left(\sinh\frac{2\sqrt{2\mathcal{D}}\pi}{3}-\sinh\frac{\sqrt{2\mathcal{D}}\pi}{3}\right)\sin\frac{\sqrt{2}x}{3\sqrt{\mathcal{D}}}-\mathrm{e}^{4/3}\sinh\frac{\sqrt{2\mathcal{D}}\pi}{6}\sin\frac{2\sqrt{2}x}{3\sqrt{\mathcal{D}}}.\end{aligned}}$$

8.5. Approximate multi-variate wavelet decomposition

In the multi-variate case, first we have to define the approximate wavelet space $\mathbf{W}_0$. In accordance with Subsection 8.3.1, the elements of $\mathbf{V}_1$ which are orthogonal to $\mathbf{V}_0$ form the space $\widetilde{\mathbf{W}}_0$. To define the wavelet space, which is a perturbation of $\widetilde{\mathbf{W}}_0$, we denote

$$w_0(x) := \phi_{\mathcal{D}}(x) , \quad w_{1/2}(x) := \psi_{\mathcal{D}}(x) , \tag{8.34}$$

and we introduce the collection of 2^n functions given on $\mathbb{R}^n$ by

$$\Phi_{\mathbf{v}}(\mathbf{x}) = w_{v_1}(x_1) \cdots w_{v_n}(x_n) , \quad \mathbf{v} = (v_1, \dots, v_n) \in \mathcal{V} . \tag{8.35}$$

Here, $\mathcal{V}$ denotes the set of vertices of the cube $[0, 1/2]^n$. Furthermore, we define the principal shift invariant spaces

$$X_{\mathbf{v}} := \{\Phi_{\mathbf{v}}(\cdot - \mathbf{m}),\ \mathbf{m} \in \mathbb{Z}^n\} .$$

Note that $X_{\mathbf{0}} = \mathbf{V}_0$. Furthermore, we know from (8.28) that $(w_0(\cdot - m), w_{1/2})_{L_2} = 0$ for all $m \in \mathbb{Z}$. Hence, if $\mathbf{v} \neq \mathbf{v}'$, then also $(\Phi_{\mathbf{v}}(\cdot - \mathbf{m}), \Phi_{\mathbf{v}'})_{L_2} = 0$, for all $\mathbf{m} \in \mathbb{Z}^n$, which means that the principal shift invariant spaces $X_{\mathbf{v}}$, $\mathbf{v} \in \mathcal{V}$, are mutually orthogonal, $X_{\mathbf{v}} \perp X_{\mathbf{v}'}$. The approximate wavelet space $\mathbf{W}_0$ is defined as the orthogonal sum

$$\mathbf{W}_0 = \bigoplus_{\mathbf{v} \in \mathcal{V}'} X_{\mathbf{v}} ,$$

where $\mathcal{V}' = \mathcal{V} \backslash \{\mathbf{0}\}$. Obviously, $\hat{\delta}(\mathbf{V}_0 \dot{+} \mathbf{W}_0, \mathbf{V}_1) < \varepsilon$ and the functions $\Phi_{\mathbf{v}}(\mathbf{x} - \mathbf{m})$, $\mathbf{m} \in \mathbb{Z}^n$, $\mathbf{v} \in \mathcal{V}'$, form a Riesz basis in $\mathbf{W}_0$.

Now, we consider the problem of finding the approximate wavelet decomposition of a given element belonging to $\mathbf{V}_k$ and of proving estimate (8.8). Following Theorem 8.3, one has to determine the orthogonal projections onto $\mathbf{V}_0$ and $\mathbf{W}_j$.

8.5.1. Projections onto $\mathbf{V}_0$. Since $\phi_{\mathcal{D}}(\mathbf{x}) = \prod_{j=1}^n \phi_{\mathcal{D}}(x_j)$, the ortho-projection $\mathbf{P}_0$ onto $\mathbf{V}_0$ is the tensor product

$$\mathbf{P}_0 = R_0 \otimes \cdots \otimes R_0 \tag{8.36}$$

of n copies of the univariate L_2–projection R_0 onto V_0.

THEOREM 8.5. *The orthogonal projection R_0 onto V_0 has the form*

$$R_0 f = \sum_{k \in \mathbb{Z}} (f, \check{\phi}_{\mathcal{D}}(\cdot - m))_{L_2}\, \phi_{\mathcal{D}}(\cdot - m) , \tag{8.37}$$

where the function $\check{\phi}_{\mathcal{D}} \in V_0$ is given by the formula

$$\check{\phi}_{\mathcal{D}}(x) = \sum_{m \in \mathbb{Z}} a_m(\mathcal{D})\, \phi_{\mathcal{D}}(x - m)$$

with the coefficients

$$a_m(\mathcal{D}) = \frac{\mathrm{e}^{\,m^2/2\mathcal{D}}}{\rho(\mathcal{D})} \sum_{r=|m|}^{\infty} (-1)^r\ \mathrm{e}^{-(r+1/2)^2/2\mathcal{D}}$$

and the constant

$$\rho(\mathcal{D}) = (2\pi\mathcal{D})^{3/2} \sum_{r \in \mathbb{Z}} (4r+1)\ \mathrm{e}^{\,-2\pi^2\mathcal{D}(2r+1/2)^2} .$$

PROOF. The biorthogonal basis is given by the integer shifts of $\check{\phi}_{\mathcal{D}}$, whose Fourier transform is

$$\mathcal{F}\check{\phi}_{\mathcal{D}} = \frac{\mathcal{F}\phi_{\mathcal{D}}}{[\mathcal{F}\phi_{\mathcal{D}}, \mathcal{F}\phi_{\mathcal{D}}]}$$

(see (7.40)). Since

$$[\mathcal{F}\phi_{\mathcal{D}}, \mathcal{F}\phi_{\mathcal{D}}] = \sqrt{2\pi\mathcal{D}} \sum_{\nu\in\mathbb{Z}} \mathrm{e}^{-2\pi^2\mathcal{D}(\lambda-\nu)^2} = \sum_{m\in\mathbb{Z}} \mathrm{e}^{-m^2/2\mathcal{D}}\, \mathrm{e}^{2\pi i m\lambda},$$

the assertion follows immediately from Lemma 7.8. □

8.5.2. Projections onto $\mathbf{W}_0$. Because of the tensor product structure of the functions $\Phi_{\mathbf{v}}$, the ortho-projection onto $X_{\mathbf{v}}$ is the tensor product $R_{v_1} \otimes \cdots \otimes R_{v_n}$ with $\mathbf{v} = (v_1, \ldots, v_n) \in \mathcal{V}'$, where

$$R_0 : L_2(\mathbb{R}) \to V_0\,, \quad R_{1/2} : L_2(\mathbb{R}) \to W_0$$

denote the corresponding univariate ortho-projections. Since $\mathbf{W}_0$ is the orthogonal sum of the principal shift-invariant spaces $X_{\mathbf{v}}$, $\mathbf{v} \in \mathcal{V}'$, the orthogonal projection $\mathbf{Q}_0$ onto $\mathbf{W}_0$ is given by

$$\mathbf{Q}_0 := \sum_{\mathbf{v}\in\mathcal{V}'} R_{v_1} \otimes \cdots \otimes R_{v_d}\,. \tag{8.38}$$

In order to determine the approximate wavelet decomposition besides the ortho-projection R_0 onto the span of the Gaussians, we need a computable representation of the projection $R_{1/2}$ onto W_0. The following theorem gives an operator $\widetilde{R}_{1/2}$, which approximates $R_{1/2}$ in the operator norm within the required accuracy.

THEOREM 8.6. *The orthogonal projection $R_{1/2}$ onto W_0 can be approximated by the operator*

$$\widetilde{R}_{1/2} f = \sum_{m\in\mathbb{Z}} \left(f, \tilde{\psi}_{\mathcal{D}}(\cdot - m)\right)_2 \psi_{\mathcal{D}}(\cdot - m)\,,$$

where the function $\tilde{\psi}_{\mathcal{D}}$ is given as the sum

$$\tilde{\psi}_{\mathcal{D}}(x) := \sum_{m\in\mathbb{Z}} \tilde{a}_m\, \psi_{\mathcal{D}}(x-m) \in W_0\,,$$

with the coefficients

$$\tilde{a}_m = \sum_{j=0}^{\infty} (-1)^j \Big((-1)^m S_1\, \mathrm{e}^{-3(j+1/2)^2/4\mathcal{D}}\, \mathrm{e}^{-|m|(j+1/2)/\mathcal{D}} + S_0\, \mathrm{e}^{-3(j+1/2)^2/\mathcal{D}}\, \mathrm{e}^{-2|m|(j+1/2)/\mathcal{D}} \Big),$$

and the numbers S_0 and S_1, which depend on the parameter $\mathcal{D}$, are equal to

$$S_0 = \frac{2(1-\mathrm{e}^{-25\pi^2\mathcal{D}/12})}{(\pi\mathcal{D})^{3/2}\sqrt{3}} \Big(\sum_{j\in\mathbb{Z}} (-1)^j (6j+1)\, \mathrm{e}^{-\pi^2\mathcal{D}(6j+1)^2/12} \Big)^{-1}, \tag{8.39}$$

$$S_1 = \frac{1-\mathrm{e}^{-25\pi^2\mathcal{D}/12}}{(\pi\mathcal{D})^{3/2}\sqrt{3}} \Big(\sum_{j\in\mathbb{Z}} (-1)^j (6j+1)\, \mathrm{e}^{-\pi^2\mathcal{D}(6j+1)^2/3} \Big)^{-1}. \tag{8.40}$$

There exists a constant c such that

$$\|R_{1/2} f - \widetilde{R}_{1/2} f\|_{L_2} \le c\, \mathrm{e}^{-\pi^2\mathcal{D}}\, \|f\|_{L_2},\ \forall f \in L_2(\mathbb{R})\,. \tag{8.41}$$

The construction of an almost biorthogonal basis to $\{\psi(\cdot - m)\}_{m\in\mathbb{Z}}$ in the wavelet space W_0 is provided in Section 8.6.

Thus, instead of the ortho-projection Q_0 onto $\mathbf{W}_0$ defined by (8.38), we consider the operator

$$\widetilde{\mathbf{Q}}_0 := \sum_{\mathbf{v}\in\mathcal{V}'} \widetilde{R}_{v_1} \otimes \cdots \otimes \widetilde{R}_{v_n} \;:\; L_2(\mathbb{R}^n) \to \mathbf{W}_0\,, \tag{8.42}$$

where $\widetilde{R}_0 = R_0$ is given in Theorem 8.5 and $\widetilde{R}_{1/2}$ is described in Theorem 8.6. It is clear that we have

$$\|\widetilde{\mathbf{Q}}_0 - \mathbf{Q}_0\| \le c\,\mathrm{e}^{-\pi^2\mathcal{D}}\,, \tag{8.43}$$

with some constant c depending only on n. By Theorem 8.3, we obtain the following approximate wavelet decomposition of the space $\mathbf{V}_n$.

THEOREM 8.7. *There exists a constant c, depending on the space dimension n and on k, such that for any $\varphi_k \in \mathbf{V}_k$ the estimate*

$$\|\varphi_k - \sum_{j=-1}^{k-1} \widetilde{\mathbf{Q}}_j\varphi_k\|_{L_2} \le c\,\mathrm{e}^{-\pi^2\mathcal{D}}\;\|\varphi_k\|_{L_2}$$

holds, where $\widetilde{\mathbf{Q}}_{-1} = \mathbf{P}_0$ is defined in (8.36) and the mappings $\widetilde{\mathbf{Q}}_j$ onto $\mathbf{W}_j$ are obtained by scaling from $\widetilde{\mathbf{Q}}_0$ given in (8.42).

8.6. Proof of Theorem 8.6

By (7.40), the biorthogonal basis to the wavelet basis $\{\psi_\mathcal{D}(\cdot - m)\}$ is spanned by the function

$$\check{\psi}_\mathcal{D}(x) = \sum_{m\in\mathbb{Z}} w_m \psi_\mathcal{D}(x-m)\,, \tag{8.44}$$

where w_k are the Fourier coefficients of the reciprocal function of

$$G(\lambda) := \sum_{m\in\mathbb{Z}} |\mathcal{F}\psi_\mathcal{D}(\lambda+m)|^2 = [\mathcal{F}\psi_\mathcal{D}, \mathcal{F}\psi_\mathcal{D}](\lambda)\,. \tag{8.45}$$

From (8.27), we obtain

$$\mathcal{F}\psi_\mathcal{D}(\lambda) = \kappa_\mathcal{D}\frac{\sqrt{6\pi\mathcal{D}}}{4}\;\mathrm{e}^{-\pi i\lambda}\left(\mathrm{e}^{-3\pi^2\mathcal{D}(\lambda+5/6)^2/2} + \mathrm{e}^{-3\pi^2\mathcal{D}(\lambda-5/6)^2/2}\right), \tag{8.46}$$

and therefore,

$$G(\lambda) = \frac{3\pi\mathcal{D}\,\kappa_\mathcal{D}^2}{8}\sum_{m\in\mathbb{Z}}\Big(\mathrm{e}^{-3\pi^2\mathcal{D}(m+\lambda+5/6)^2} + \mathrm{e}^{-3\pi^2\mathcal{D}(m+\lambda-5/6)^2} + 2\,\mathrm{e}^{-25\pi^2\mathcal{D}/12}\;\mathrm{e}^{-3\pi^2\mathcal{D}(m+\lambda)^2}\Big)\,.$$

Obtaining a simple analytic expression of the Fourier coefficients of $1/G$ seems to be impossible. However, a very accurate approximation of these coefficients can be determined after some simplifications.

8.6.1. Simplification of $G(\lambda)$. Because

$$\begin{aligned}
\sum_{m\in\mathbb{Z}} \mathrm{e}^{-3\pi^2\mathcal{D}(m+\lambda+5/6)^2} &= \sum_{m\in\mathbb{Z}} \mathrm{e}^{-3\pi^2\mathcal{D}(m-\lambda+1/6)^2}, \\
\sum_{m\in\mathbb{Z}} \mathrm{e}^{-3\pi^2\mathcal{D}(m+\lambda-5/6)^2} &= \sum_{m\in\mathbb{Z}} \mathrm{e}^{-3\pi^2\mathcal{D}(m+\lambda+1/6)^2}, \\
2\sum_{m\in\mathbb{Z}} \mathrm{e}^{-3\pi^2\mathcal{D}(m+\lambda)^2} &= \sum_{m\in\mathbb{Z}} \left(\mathrm{e}^{-3\pi^2\mathcal{D}(m+\lambda)^2} + \mathrm{e}^{-3\pi^2\mathcal{D}(m-\lambda)^2}\right) \\
&= \mathrm{e}^{\pi^2\mathcal{D}/12} \sum_{m\in\mathbb{Z}} \frac{\mathrm{e}^{-3\pi^2\mathcal{D}(m+\lambda+1/6)^2} + \mathrm{e}^{-3\pi^2\mathcal{D}(m+\lambda-1/6)^2}}{2\cosh \pi^2\mathcal{D}(m+\lambda)} \\
&\quad + \mathrm{e}^{\pi^2\mathcal{D}/12} \sum_{m\in\mathbb{Z}} \frac{\mathrm{e}^{-3\pi^2\mathcal{D}(m-\lambda+1/6)^2} + \mathrm{e}^{-3\pi^2\mathcal{D}(m-\lambda-1/6)^2}}{2\cosh \pi^2\mathcal{D}(m-\lambda)} \\
&= \mathrm{e}^{\pi^2\mathcal{D}/12} \sum_{m\in\mathbb{Z}} \left(\frac{\mathrm{e}^{-3\pi^2\mathcal{D}(m+\lambda+1/6)^2}}{\cosh \pi^2\mathcal{D}(m+\lambda)} + \frac{\mathrm{e}^{-3\pi^2\mathcal{D}(m-\lambda+1/6)^2}}{\cosh \pi^2\mathcal{D}(m-\lambda)} \right),
\end{aligned}$$

$G(\lambda)$ can be written in the form

$$\begin{aligned}
G(\lambda) = \frac{3\pi\mathcal{D}\,\kappa_{\mathcal{D}}^2}{8} &\sum_{m\in\mathbb{Z}} \left(\mathrm{e}^{-3\pi^2\mathcal{D}(m+\lambda+1/6)^2} + \mathrm{e}^{-3\pi^2\mathcal{D}(m-\lambda+1/6)^2}\right) \\
+ \frac{3\pi\mathcal{D}\,\kappa_{\mathcal{D}}^2\,\mathrm{e}^{-2\pi^2\mathcal{D}}}{8} &\sum_{m\in\mathbb{Z}} \left(\frac{\mathrm{e}^{-3\pi^2\mathcal{D}(m+\lambda+1/6)^2}}{\cosh \pi^2\mathcal{D}(m+\lambda)} + \frac{\mathrm{e}^{-3\pi^2\mathcal{D}(m-\lambda+1/6)^2}}{\cosh \pi^2\mathcal{D}(m-\lambda)} \right).
\end{aligned}$$

Thus, $G(\lambda)$ is a small and smooth perturbation of the 1-periodic function

$$g(\lambda) = \frac{3\pi\mathcal{D}\,\kappa_{\mathcal{D}}^2}{8} \sum_{m\in\mathbb{Z}} \left(\mathrm{e}^{-3\pi^2\mathcal{D}(m+\lambda+1/6)^2} + \mathrm{e}^{-3\pi^2\mathcal{D}(m-\lambda+1/6)^2}\right), \tag{8.47}$$

and the perturbation is bounded by

$$0 < G(\lambda) - g(\lambda) < \mathrm{e}^{-2\pi^2\mathcal{D}}\, g(\lambda). \tag{8.48}$$

Instead of the Fourier coefficients w_k of $1/G(\lambda)$, we will determine the Fourier coefficients of $1/g(\lambda)$ in the following subsections.

8.6.2. Error of replacing $G(\lambda)$ by $g(\lambda)$. Let us estimate the error which is made by this simplification. The Fourier coefficients a_k of $1/g$ generate a function denoted by

$$\chi_{\mathcal{D}}(x) = \sum_{m\in\mathbb{Z}} a_m \psi_{\mathcal{D}}(\cdot - m) \tag{8.49}$$

and a linear operator $A : L_2 \to W_0$ defined by

$$Af = \sum_{m\in\mathbb{Z}} (f, \chi_{\mathcal{D}}(\cdot - m))_2\, \psi_{\mathcal{D}}(\cdot - m). \tag{8.50}$$

Owing to (7.39), the Fourier transform of $R_{1/2}f$, $f \in L_2(\mathbb{R})$, is given by

$$\mathcal{F}(R_{1/2}f)(\lambda) = \frac{[\mathcal{F}f, \mathcal{F}\psi_{\mathcal{D}}](\lambda)}{G(\lambda)}\, \mathcal{F}\psi_{\mathcal{D}}(\lambda),$$

whereas by (8.50), the Fourier transform of Af equals

$$\mathcal{F}(Af)(\lambda) = \frac{[\mathcal{F}f, \mathcal{F}\psi_{\mathcal{D}}](\lambda)}{g(\lambda)} \, \mathcal{F}\psi_{\mathcal{D}}(\lambda) .$$

Hence,

$$\begin{aligned}
\|(R_{1/2} - A)f\|_{L_2}^2 &= \int_{\mathbb{R}} \left| \frac{1}{G(\lambda)} - \frac{1}{g(\lambda)} \right|^2 |\mathcal{F}\psi_{\mathcal{D}}(\lambda)|^2 |[\mathcal{F}f, \mathcal{F}\psi_{\mathcal{D}}](\lambda)|^2 \, d\lambda \\
&\leq \int_{\mathbb{R}} \left| \frac{1}{G(\lambda)} - \frac{1}{g(\lambda)} \right|^2 |\mathcal{F}\psi_{\mathcal{D}}|^2 [\mathcal{F}f, \mathcal{F}f][\mathcal{F}\psi_{\mathcal{D}}, \mathcal{F}\psi_{\mathcal{D}}] \, d\lambda \\
&= \sum_{k \in \mathbb{Z}} \int_0^1 \left| \frac{1}{G(\lambda)} - \frac{1}{g(\lambda)} \right|^2 |\mathcal{F}\psi_{\mathcal{D}}(\lambda + k)|^2 [\mathcal{F}f, \mathcal{F}f][\mathcal{F}\psi_{\mathcal{D}}, \mathcal{F}\psi_{\mathcal{D}}] \, d\lambda \\
&= \int_0^1 \left| \frac{1}{G(\lambda)} - \frac{1}{g(\lambda)} \right|^2 [\mathcal{F}\psi_{\mathcal{D}}, \mathcal{F}\psi_{\mathcal{D}}]^2 \, [\mathcal{F}f, \mathcal{F}f] \, d\lambda \\
&= \int_0^1 \left| \frac{1}{G(\lambda)} - \frac{1}{g(\lambda)} \right|^2 G(\lambda)^2 \, [\mathcal{F}f, \mathcal{F}f] \, d\lambda ,
\end{aligned}$$

where we use (8.45). Thus, by (8.48),

$$\|(R_{1/2} - A)f\|_{L_2}^2 \leq \int_0^1 \left| 1 - \frac{G(\lambda)}{g(\lambda)} \right|^2 [\mathcal{F}f, \mathcal{F}f] \, d\lambda \leq \mathrm{e}^{-4\pi^2 \mathcal{D}} \int_{\mathbb{R}} |\mathcal{F}f|^2 \, d\lambda$$

and therefore, the operator A, based on the Fourier coefficients of $1/g$, differs from the ortho-projection onto W_0 by

$$\|R_{1/2} - A\| \leq \mathrm{e}^{-2\pi^2 \mathcal{D}} . \tag{8.51}$$

8.6.3. Equations for the Fourier coefficients of $1/g$. To find the Fourier coefficients

$$a_k = \int_0^1 \frac{\mathrm{e}^{-2\pi i k \lambda}}{g(\lambda)} \, d\lambda ,$$

we apply some elementary methods of complex function theory. The function $g(z)$, $z \in \mathbb{C}$, is analytic and quasi doubly-periodic with

$$g(z+1) = g(z) , \quad g\big(z + \frac{i}{\pi \mathcal{D}}\big) = -\, \mathrm{e}^{3/\mathcal{D}} \, \mathrm{e}^{-6\pi i z} \, g(z) . \tag{8.52}$$

Hence, if we take a rectangle $\mathcal{R} = (z, z+1) \times (z, z + i/\pi\mathcal{D})$, such that $g(z) \neq 0$ on $\partial\mathcal{R}$, then the change in $\arg g(z)$ equals 6π, if z traverses the boundary $\partial\mathcal{R}$ in a counterclockwise direction. Then by the Argument Principle, $g(z)$ has three zeros in $\mathcal{R}$. By the symmetry of $g(z)$ the points with $\operatorname{Re} z = j$ or $\operatorname{Re} z = j + 1/2$, $j \in \mathbb{Z}$, are candidates for zeros. Therefore, we choose the rectangular domain $\mathcal{R} = (-1/4, 3/4) \times (0, i/(\pi\mathcal{D}))$ as periodic cell.

It is easy to find the zero with $\operatorname{Re} z = 0$, since for $z_0 = i/(2\pi\mathcal{D})$

$$\begin{aligned} g(z_0) &= \frac{3\pi\mathcal{D}\,\kappa_{\mathcal{D}}^2\,\mathrm{e}^{3/4\mathcal{D}}}{8} \sum_{m\in\mathbb{Z}} \Big(\mathrm{e}^{-3\pi^2\mathcal{D}(m+1/6)^2}\,\mathrm{e}^{-3\pi i(m+1/6)} \\ &\qquad\qquad + \mathrm{e}^{-3\pi^2\mathcal{D}(m+1/6)^2}\,\mathrm{e}^{3\pi i(m+1/6)}\Big) \\ &= \frac{3\pi\mathcal{D}\,\kappa_{\mathcal{D}}^2\,\mathrm{e}^{3/4\mathcal{D}}}{8} \sum_{m\in\mathbb{Z}} (-1)^m\,\mathrm{e}^{-3\pi^2\mathcal{D}(m+1/6)^2}(\mathrm{e}^{-\pi i/2}+\mathrm{e}^{\pi i/2}) = 0\,. \end{aligned}$$

Next, we consider

$$\begin{aligned} \tilde g(y) &:= g\Big(\frac{1}{2}+iy\Big) = \frac{3\pi\mathcal{D}\kappa_{\mathcal{D}}^2}{8} \sum_{m\in\mathbb{Z}} \Big(\mathrm{e}^{-3\pi^2\mathcal{D}(m+2/3+iy)^2} + \mathrm{e}^{-3\pi^2\mathcal{D}(m-1/3-iy)^2}\Big) \\ &= \frac{3\pi\mathcal{D}\kappa_{\mathcal{D}}^2}{8}\,\mathrm{e}^{3\pi^2\mathcal{D}y^2} \sum_{m\in\mathbb{Z}} \mathrm{e}^{-3\pi^2\mathcal{D}(m+2/3)^2}(\mathrm{e}^{-6\pi^2\mathcal{D}i(m+2/3)y} + \mathrm{e}^{6\pi^2\mathcal{D}i(m+2/3)y}) \\ &= \frac{3\pi\mathcal{D}\kappa_{\mathcal{D}}^2}{4}\,\mathrm{e}^{3\pi^2\mathcal{D}y^2} \sum_{m=-\infty}^{\infty} \mathrm{e}^{-\pi^2\mathcal{D}(3m+2)^2/3}\cos 2\pi^2\mathcal{D}(3m+2)y \\ &= \frac{3\pi\mathcal{D}\kappa_{\mathcal{D}}^2}{4}\,\mathrm{e}^{3\pi^2\mathcal{D}y^2}\Big(\sum_{m=1}^{\infty} \mathrm{e}^{-\pi^2\mathcal{D}m^2/3}\cos 2\pi^2\mathcal{D}my - \sum_{m=1}^{\infty}\mathrm{e}^{-3\pi^2\mathcal{D}m^2}\cos 6\pi^2\mathcal{D}my\Big). \end{aligned}$$

The trigonometric series has the period $1/(\pi\mathcal{D})$ and

$$\tilde g\Big(\frac{1}{\pi\mathcal{D}} - y\Big) = \mathrm{e}^{3/\mathcal{D}}\,\mathrm{e}^{-6\pi y}\,\tilde g(y)\,.$$

Moreover, at least for $\mathcal{D} \geq 1$, $\tilde g$ is a small perturbation of

$$\frac{3\pi\mathcal{D}\kappa_{\mathcal{D}}^2}{4}\,\mathrm{e}^{-\pi^2\mathcal{D}/3}\,\mathrm{e}^{3\pi^2\mathcal{D}y^2}\,\cos 2\pi^2\mathcal{D}y\,,$$

such that its zeros y_1, y_2 are close to the zeros of $\cos 2\pi^2\mathcal{D}y$. In other words, we have $g(z) = 0$ at the points

$$z_j = \frac{1}{2} + iy_j\,, \quad j = 1, 2\,,$$

with

$$y_1 = \frac{1-\epsilon}{4\pi\mathcal{D}} \quad \text{and} \quad y_2 = \frac{3+\epsilon}{4\pi\mathcal{D}} \quad \text{with small} \quad \epsilon\,. \tag{8.53}$$

More precisely, ϵ has to satisfy the equation

$$\sum_{m=1}^{\infty}\Big(\mathrm{e}^{-\pi^2\mathcal{D}m^2/3}\cos\frac{\pi m(1-\epsilon)}{2} - \mathrm{e}^{-3\pi^2\mathcal{D}m^2}\cos\frac{\pi m(1+3\epsilon)}{2}\Big) = 0\,,$$

i.e.,

$$\sin\frac{\pi\epsilon}{2} = \mathrm{e}^{-\pi^2\mathcal{D}}\cos\pi\epsilon + \mathcal{O}(\mathrm{e}^{-5\pi^2\mathcal{D}})\,,$$

which shows that $0 < \epsilon < \mathrm{e}^{-\pi^2\mathcal{D}}$.

Knowing the poles of $1/g(z)$, we can apply the Residue Theorem and obtain

$$\int_{\partial\mathcal{R}} \frac{\mathrm{e}^{-2\pi ikz}}{g(z)}\,dz = r_0 + r_1 + r_2$$

with

$$r_j = 2\pi i \operatorname*{Res}_{z=z_j} \frac{e^{-2\pi i k z}}{g(z)} = 2\pi i \left.\frac{e^{-2\pi i k z}}{g'(z)}\right|_{z=z_j}.$$

Hence, the Fourier coefficients satisfy

$$a_k = \int_{-1/4}^{3/4} \frac{e^{-2\pi i k\lambda}}{g(\lambda)}\, d\lambda = \int_{-1/4+i/(\pi\mathcal{D})}^{3/4+i/(\pi\mathcal{D})} \frac{e^{-2\pi i k z}}{g(z)}\, dz + r_0 + r_1 + r_2\,,$$

which, together with (8.52), leads to the relation

$$a_k = -\,e^{(2k-3)/\mathcal{D}}\, a_{k-3} + r_0 + r_1 + r_2 \quad \text{for all} \quad k \in \mathbb{Z}\,. \tag{8.54}$$

Since

$$g'(z_0) = \frac{9\pi^3 i\mathcal{D}^2\kappa_{\mathcal{D}}^2}{2}\, e^{3/4\mathcal{D}} \sum_{m\in\mathbb{Z}} (-1)^m \big(m + \frac{1}{6}\big)\, e^{-3\pi^2\mathcal{D}(m+1/6)^2}\,,$$

we derive

$$r_0 = \frac{4\pi i\, e^{k/\mathcal{D}}}{9\pi^3 i\mathcal{D}^2\kappa_{\mathcal{D}}^2\, e^{3/4\mathcal{D}}} \Big(\sum_{m\in\mathbb{Z}} (-1)^m \big(m + \frac{1}{6}\big)\, e^{-3\pi^2\mathcal{D}(m+1/6)^2} \Big)^{-1} = C_0\, e^{k/\mathcal{D}}$$

with the constant

$$C_0 = \frac{8\, e^{-3/4\mathcal{D}}}{3\pi^2\mathcal{D}^2\kappa_{\mathcal{D}}^2} \Big(\sum_{m\in\mathbb{Z}} (-1)^m (6m+1)\, e^{-\pi^2\mathcal{D}(6m+1)^2/12} \Big)^{-1}. \tag{8.55}$$

To compute the residues r_1 and r_2, we note that

$$g'(z_j) = -i\,\tilde{g}'(y_j)\,, \quad j = 1, 2,$$

and that

$$\begin{aligned} \tilde{g}'(y_j) &= -\frac{3\pi^3\mathcal{D}^2\kappa_{\mathcal{D}}^2}{2}\, e^{3\pi^2\mathcal{D}y_j^2} \sum_{m=1}^{\infty} m\Big(e^{-\pi^2\mathcal{D}m^2/3} \sin 2\pi^2\mathcal{D}my_j \\ &\qquad\qquad - 3\, e^{-3\pi^2\mathcal{D}m^2} \sin 6\pi^2\mathcal{D}my_j \Big) \\ &= (-1)^{j-1}\, e^{3\pi^2\mathcal{D}y_j^2}\, b_\epsilon\,, \end{aligned}$$

where we denote

$$\begin{aligned} b_\epsilon &= \frac{3\pi^3\mathcal{D}^2\kappa_{\mathcal{D}}^2}{2} \\ &\quad \times \sum_{m=1}^{\infty} m\Big(e^{-\pi^2\mathcal{D}m^2/3} \sin \frac{\pi m(1-\epsilon)}{2} + 3\, e^{-3\pi^2\mathcal{D}m^2} \sin \frac{\pi m(1+3\epsilon)}{2} \Big). \end{aligned} \tag{8.56}$$

Therefore, we derive

$$\begin{aligned} r_1 + r_2 &= -2\pi\Big(\frac{e^{-2\pi i k z_1}}{\tilde{g}'(y_1)} + \frac{e^{-2\pi i k z_2}}{\tilde{g}'(y_2)} \Big) \\ &= 2\pi\Big(\frac{e^{-\pi i k}\, e^{(3+\epsilon)k/2\mathcal{D}}\, e^{-3(3+\epsilon)^2/16\mathcal{D}}}{b_\epsilon} - \frac{e^{-\pi i k}\, e^{(1-\epsilon)k/2\mathcal{D}}\, e^{-3(1-\epsilon)^2/16\mathcal{D}}}{b_\epsilon} \Big) \\ &= (-1)^k\, c_\epsilon\, e^{k/\mathcal{D}} \Big(e^{(2k-3)(1+\epsilon)/4\mathcal{D}} - e^{-(2k-3)(1+\epsilon)/4\mathcal{D}} \Big) \end{aligned}$$

with the abbreviation

$$c_\epsilon = \frac{2\pi\, \mathrm{e}^{-3/4\mathcal{D}}\, \mathrm{e}^{-3(1+\epsilon)^2/16\mathcal{D}}}{b_\epsilon}. \tag{8.57}$$

Finally, (8.54) takes the form

$$\begin{aligned} a_k = &- \mathrm{e}^{(2k-3)/\mathcal{D}}\, a_{k-3} + C_0\, \mathrm{e}^{k/\mathcal{D}} \\ &+ (-1)^k c_\epsilon\, \mathrm{e}^{k/\mathcal{D}} \left(\mathrm{e}^{(2k-3)(1+\epsilon)/4\mathcal{D}} - \mathrm{e}^{-(2k-3)(1+\epsilon)/4\mathcal{D}} \right), \end{aligned}$$

leading to the recurrence equations for the Fourier coefficients

$$\begin{aligned} \mathrm{e}^{-k/\mathcal{D}}\, a_k = &- \mathrm{e}^{(k-3)/\mathcal{D}}\, a_{k-3} + C_0 \\ &+ (-1)^k c_\epsilon \left(\mathrm{e}^{(2k-3)(1+\epsilon)/4\mathcal{D}} - \mathrm{e}^{-(2k-3)(1+\epsilon)/4\mathcal{D}} \right) \end{aligned} \tag{8.58}$$

which are to be solved.

8.6.4. Solution of (8.58). Since $g(\lambda)$ is an even function, it suffices to determine the Fourier coefficients a_k for $k \geq 0$. After some calculations, which are based on the representation of a_{3k+j} by a_j, $j = 0, 1, 2$, and which involve some tedious transformations, we obtain the series expansion

$$\begin{aligned} a_k = \mathrm{e}^{k^2/3\mathcal{D}} \Bigg(& \mathrm{e}^{3/4\mathcal{D}}\, C_0 \sum_{j=0}^{\infty} (-1)^j\, \mathrm{e}^{-(3(j+1/2)+k)^2/3\mathcal{D}} \\ &+ (-1)^k\, C_1(\epsilon) \sum_{j=0}^{\infty} \left(\mathrm{e}^{-(3(j+(1-\epsilon)/4)+k)^2/3\mathcal{D}} - \mathrm{e}^{-(3(j+(3+\epsilon)/4)+k)^2/3\mathcal{D}} \right) \Bigg) \end{aligned} \tag{8.59}$$

for $k \geq 0$, with a new constant $C_1(\epsilon) = 2\pi/b_\epsilon$, which, in view of (8.56), is given as

$$\begin{aligned} C_1(\epsilon) = &\frac{4}{3\pi^2\mathcal{D}^2\kappa_{\mathcal{D}}^2} \\ &\times \left(\sum_{m=1}^{\infty} m \left(\mathrm{e}^{-\pi^2\mathcal{D}m^2/3} \sin\frac{\pi m(1-\epsilon)}{2} + 3\,\mathrm{e}^{-3\pi^2\mathcal{D}m^2} \sin\frac{\pi m(1+3\epsilon)}{2} \right) \right)^{-1}. \end{aligned} \tag{8.60}$$

The exact value of ϵ is not known but we know that $\epsilon < \mathrm{e}^{-\pi^2\mathcal{D}}$. Therefore, in the following, we will use the formula for a_k with ϵ set to 0. We introduce the coefficients

$$\begin{aligned} \tilde{a}_k = \mathrm{e}^{k^2/3\mathcal{D}} \Bigg(& \mathrm{e}^{3/4\mathcal{D}}\, C_0 \sum_{j=0}^{\infty} (-1)^j\, \mathrm{e}^{-(3(j+1/2)+k)^2/3\mathcal{D}} \\ &+ (-1)^k\, C_1 \sum_{j=0}^{\infty} \left(\mathrm{e}^{-(3(j+1/4)+k)^2/3\mathcal{D}} - \mathrm{e}^{-(3(j+3/4)+k)^2/3\mathcal{D}} \right) \Bigg), \end{aligned}$$

where $C_1 = C_1(0)$. Taking into account (8.55) and (8.60), the coefficients transform to

$$\begin{aligned} \tilde{a}_k = &S_0 \sum_{j=0}^{\infty} (-1)^j\, \mathrm{e}^{-3(j+1/2)^2/\mathcal{D}}\, \mathrm{e}^{-2k(j+1/2)/\mathcal{D}} \\ &+(-1)^k S_1 \sum_{j=0}^{\infty} \left(\mathrm{e}^{-3(j+1/4)^2/\mathcal{D}}\, \mathrm{e}^{-2k(j+1/4)/\mathcal{D}} - \mathrm{e}^{-3(j+3/4)^2/\mathcal{D}}\, \mathrm{e}^{-2k(j+3/4)/\mathcal{D}} \right), \end{aligned} \tag{8.61}$$

where

$$S_0 = \frac{8}{3\pi^2\mathcal{D}^2\kappa_{\mathcal{D}}^2}\Big(\sum_{m\in\mathbb{Z}}(-1)^m(6m+1)\ \mathrm{e}^{-\pi^2\mathcal{D}(6m+1)^2/12}\Big)^{-1},$$

$$S_1 = \frac{4}{3\pi^2\mathcal{D}^2\kappa_{\mathcal{D}}^2}\Big(\sum_{m\in\mathbb{Z}}(-1)^m(6m+1)\ \mathrm{e}^{-\pi^2\mathcal{D}(6m+1)^2/3}\Big)^{-1}.$$

With the value of $\kappa_{\mathcal{D}}$ (see (8.25)), the expressions (8.39) and (8.40) of S_0 and S_1 follow. Noting that

$$\begin{aligned}
&\sum_{j=0}^{\infty}\Big(\mathrm{e}^{-3(j+1/4)^2/\mathcal{D}}\ \mathrm{e}^{-2k(j+1/4)/\mathcal{D}} - \mathrm{e}^{-3(j+3/4)^2/\mathcal{D}}\ \mathrm{e}^{-2k(j+3/4)/\mathcal{D}}\Big)\\
&=\sum_{j=0}^{\infty}\Big(\mathrm{e}^{-3(2j+1/2)^2/4\mathcal{D}}\ \mathrm{e}^{-k(2j+1/2)/\mathcal{D}} - \mathrm{e}^{-3(2j+1+1/2)^2/4\mathcal{D}}\ \mathrm{e}^{-k(2j+1+1/2)/\mathcal{D}}\Big)\\
&=\sum_{j=0}^{\infty}(-1)^j\ \mathrm{e}^{-3(j+1/2)^2/4\mathcal{D}}\ \mathrm{e}^{-k(j+1/2)/\mathcal{D}},
\end{aligned}$$

one can transform (8.61) to

$$\begin{aligned}
\tilde{a}_k &= S_0\sum_{j=0}^{\infty}(-1)^j\ \mathrm{e}^{-3(j+1/2)^2/\mathcal{D}}\ \mathrm{e}^{-2k(j+1/2)/\mathcal{D}}\\
&\quad + (-1)^k S_1\sum_{j=0}^{\infty}(-1)^j\ \mathrm{e}^{-3(j+1/2)^2/4\mathcal{D}}\ \mathrm{e}^{-k(j+1/2)/\mathcal{D}},
\end{aligned}$$

which gives all formulas mentioned in the formulation of Theorem 8.6.

8.6.5. Error of replacing a_k by $\tilde{a}_k$. It remains to estimate the difference between the operators A (see (8.50)) and $\tilde{R}_{1/2}$. The difference of the Fourier transforms of $\tilde{R}_{1/2}f$ and Af, $f \in L_2(\mathbb{R})$, can be written as

$$\mathcal{F}(\tilde{R}_{1/2}f)(\lambda) - \mathcal{F}(Af)(\lambda) = [\mathcal{F}f, \mathcal{F}\psi_{\mathcal{D}}](\lambda)\,\mathcal{F}\psi_{\mathcal{D}}(\lambda)\Big(\tau(\lambda) - \frac{1}{g(\lambda)}\Big)$$

with the periodic function

$$\tau(\lambda) = \sum_{k\in\mathbb{Z}}\tilde{a}_k\ \mathrm{e}^{2\pi ik\lambda}\,.$$

Similarly to the estimation of $\|R_{1/2} - A\|$, we have to find an upper bound to

$$\Big|\tau(\lambda) - \frac{1}{g(\lambda)}\Big|\,G(\lambda)\,.$$

Since

$$\frac{1}{g(\lambda)} = \sum_{k\in\mathbb{Z}} a_k\ \mathrm{e}^{2\pi ik\lambda},$$

one obtains from (8.59), after some elementary transformations,

$$\frac{1}{g(\lambda)} - \tau(\lambda) = F(\lambda,\epsilon) - F(\lambda,0)\,,$$

where the function $F(\lambda,\epsilon)$ is given by

$$F(\lambda,\epsilon)=C_1(\epsilon)\sum_{j=-\infty}^{\infty}\frac{\left(\mathrm{e}^{2(j+(1-\epsilon)/4)/\mathcal{D}}-\mathrm{e}^{-2(j+(1-\epsilon)/4)/\mathcal{D}}\right)\mathrm{e}^{-3(j+(1-\epsilon)/4)^2/\mathcal{D}}}{\mathrm{e}^{2(j+(1-\epsilon)/4)/\mathcal{D}}+\mathrm{e}^{-2(j+(1-\epsilon)/4)/\mathcal{D}}+2\cos 2\pi\lambda}\,.$$

Hence,

$$\left|\tau(\lambda)-\frac{1}{g(\lambda)}\right|G(\lambda)\le c\epsilon\,,$$

which completes the proof of Theorem 8.6.

8.7. Potentials of wavelet basis functions

This section is devoted to the cubature of various potentials of wavelet basis functions. We show that for any space dimension these potentials can be expressed as one-dimensional integrals with smooth integrands, which can be computed very efficiently. So, it is possible to combine the advantages of well-established wavelet methods in numerical analysis with the efficient computation of important integral operators. Indeed, let the function u be approximated by a quasi-interpolant (2.23) with $\eta(\mathbf{x})=\mathrm{e}^{-|\mathbf{x}|^2}$ and $h=2^{-k}$, i.e., with an element $\mathcal{M}_{2^{-k},\mathcal{D}}u\in\mathbf{V}_k$. We take the ortho-projection $\varphi_k=P(\mathcal{M}_{2^{-k},\mathcal{D}}u)\in\mathbf{X}_k$ and, in this way, we obtain a multivariate wavelet expansion of u, i.e.,

$$\varphi_k(\mathbf{x})=\sum_{\mathbf{m}\in\mathbb{Z}^n}a_{\mathbf{m}}\Phi_{\mathbf{0}}(\mathbf{x}-\mathbf{m})+\sum_{j=0}^{k-1}\sum_{\mathbf{m}\in\mathbb{Z}^n}\sum_{\mathbf{v}\in\mathcal{V}'}a_{j,\mathbf{m}}^{\mathbf{v}}\Phi_{\mathbf{v}}(2^j\mathbf{x}-\mathbf{m})\in\mathbf{X}_k\,,\tag{8.62}$$

where $\Phi_{\mathbf{v}}$ are the wavelet basis functions for the spaces spanned by the n-dimensional Gaussian.

In Subsection 8.7.1, we rewrite the wavelet basis functions as anisotropic Gaussians of complex arguments. In Subsection 8.7.2, the results of Section 6.3 are extended in order to treat the action of integral operators on those functions. As an application, we obtain formulas for harmonic, diffraction, and elastic potentials of these functions.

8.7.1. Representation of wavelet basis functions. We recall from Section 8.5 that the 2^n wavelet basis functions are given as products

$$\Phi_{\mathbf{v}}(\mathbf{x})=w_{v_1}(x_1)\cdots w_{v_n}(x_n)\,,\quad \mathbf{v}=(v_1,\dots,v_n)\in\mathcal{V}$$

with

$$w_0(x)=\phi_{\mathcal{D}}(x)=\rho\,\mathrm{e}^{-x^2/\mathcal{D}}\,,$$
$$w_{1/2}(x)=\psi_{\mathcal{D}}(x)=\kappa\,\mathrm{e}^{-(2x-1)^2/6\mathcal{D}}\cos\frac{5\pi}{6}(2x-1)$$

and the two norming factors

$$\rho=\rho_{\mathcal{D}}=\left(\frac{2}{\pi\mathcal{D}}\right)^{1/4},\qquad \kappa=\kappa_{\mathcal{D}}=\frac{2}{(3\pi\mathcal{D})^{1/4}\sqrt{1-\mathrm{e}^{-25\pi^2\mathcal{D}/12}}}\,.\tag{8.63}$$

Here, $\mathcal{V}$ denotes the set of vertices of the cube $[0,1/2]^n$ (see (8.35)). For the following computation of potentials, we write the wavelet basis functions $\Phi_{\mathbf{v}}$ as anisotropic Gaussians.

Let us denote the unit vectors in $\mathbb{R}^n$ by $e_j=(\delta_{jk})_{k=1}^n$, $j=1,\dots,n$, and set $\mathbf{e}=e_1+\ldots+e_n=(1,\dots,1)$. For given $\mathbf{v}\in\mathcal{V}$, we denote the projection matrix

$Q_{\mathbf{v}} = 2\,\mathrm{diag}(\mathbf{v})$ having the rank m, the number of non-zero components of $\mathbf{v}$. Let $P_{\mathbf{v}} = I - Q_{\mathbf{v}}$. Then definitions (8.16) and (8.25) of $\phi_{\mathcal{D}}$ and $\psi_{\mathcal{D}}$ imply

$$\Phi_{\mathbf{v}}(\mathbf{x}) = \rho^{n-m}\kappa^m \,\mathrm{e}^{-|P_{\mathbf{v}}\mathbf{x}|^2/\mathcal{D}}\,\mathrm{e}^{-|Q_{\mathbf{v}}(2\mathbf{x}-\mathbf{e})|^2/6\mathcal{D}} \prod_{j=1}^{n} \cos\frac{5\pi}{6}\langle Q_{\mathbf{v}}(2\mathbf{x}-\mathbf{e}), e_j\rangle. \tag{8.64}$$

Using the relation

$$\prod_{j=1}^{n} \cos\langle Q_{\mathbf{v}}\mathbf{y}, e_j\rangle = 2^{-m} \sum_{\mathbf{u}\in\mathcal{U}} \cos\langle Q_{\mathbf{v}}\mathbf{y}, \mathbf{u}\rangle, \tag{8.65}$$

where $\mathcal{U}$ denotes the set of the 2^n vectors $\mathbf{u} \in \mathbb{R}^n$ with components ± 1, we can write $\Phi_{\mathbf{v}}$ as the sum

$$\boxed{\Phi_{\mathbf{v}}(\mathbf{x}) = \frac{\rho^{n-m}\kappa^m}{2^m}\,\mathrm{e}^{-|P_{\mathbf{v}}\mathbf{x}|^2/\mathcal{D}}\,\mathrm{e}^{-|Q_{\mathbf{v}}(2\mathbf{x}-\mathbf{e})|^2/6\mathcal{D}} \sum_{\mathbf{u}\in\mathcal{U}} \cos\frac{5\pi}{6}\langle Q_{\mathbf{v}}(2\mathbf{x}-\mathbf{e}), \mathbf{u}\rangle.}$$

Note that

$$\begin{aligned}
&\mathrm{e}^{-|Q_{\mathbf{v}}(2\mathbf{x}-\mathbf{e})|^2/6\mathcal{D}} \cos\frac{5\pi}{6}\langle Q_{\mathbf{v}}(2\mathbf{x}-\mathbf{e}), \mathbf{u}\rangle \\
&= \frac{\mathrm{e}^{-|Q_{\mathbf{v}}(2\mathbf{x}-\mathbf{e})|^2/6\mathcal{D}}}{2}\Big(\mathrm{e}^{5\pi i\langle Q_{\mathbf{v}}(2\mathbf{x}-\mathbf{e}),\mathbf{u}\rangle/6} + \mathrm{e}^{-5\pi i\langle Q_{\mathbf{v}}(2\mathbf{x}-\mathbf{e}),\mathbf{u}\rangle/6}\Big) \\
&= \frac{\mathrm{e}^{-25\pi^2\mathcal{D}|Q_{\mathbf{v}}\mathbf{u}|^2/24}}{2}\Big(\mathrm{e}^{-|Q_{\mathbf{v}}(2\mathbf{x}-\mathbf{e}+5\pi i\mathcal{D}\mathbf{u}/2)|^2/6\mathcal{D}} + \mathrm{e}^{-|Q_{\mathbf{v}}(2\mathbf{x}+\mathbf{e}+5\pi i\mathcal{D}\mathbf{u}/2)|^2/6\mathcal{D}}\Big),
\end{aligned}$$

where, as in Section 5.2, we put

$$\langle \mathbf{y}, \mathbf{z}\rangle = \sum_{j=1}^{n} y_j z_j \quad \text{and} \quad |\mathbf{z}|^2 = \langle \mathbf{z}, \mathbf{z}\rangle\,, \tag{8.66}$$

for complex vectors $\mathbf{y}, \mathbf{z} \in \mathbb{C}^n$.

Using the abbreviation

$$f_{\mathbf{u}}(\mathbf{x}) := \mathrm{e}^{-|P_{\mathbf{v}}\mathbf{x}|^2/\mathcal{D}}\,\mathrm{e}^{-|Q_{\mathbf{v}}(2\mathbf{x}-\mathbf{e}+5\pi i\mathcal{D}\mathbf{u}/2)|^2/6\mathcal{D}} \tag{8.67}$$

and noting that rank $Q_{\mathbf{v}} = m$, we obtain

$$\Phi_{\mathbf{v}}(\mathbf{x}) = \frac{\rho^{n-m}\kappa^m\,\mathrm{e}^{-25\pi^2\mathcal{D}m/24}}{2^m} \sum_{\mathbf{u}\in\mathcal{U}} f_{\mathbf{u}}(\mathbf{x})\,. \tag{8.68}$$

Now, we note that the function $f_{\mathbf{u}}$ can be written as an anisotropic Gaussian function which was defined in Subsection 3.3.5. Introducing the matrix and the vector

$$A = \mathcal{D}P_{\mathbf{v}} + \frac{3\mathcal{D}}{2}Q_{\mathbf{v}}\,, \quad \mathbf{z} = Q_{\mathbf{v}}\Big(\frac{\mathbf{e}}{2} - \frac{5\pi i\mathcal{D}\mathbf{u}}{4}\Big) \in \mathbb{C}^n\,, \tag{8.69}$$

we can write $f_{\mathbf{u}}$ in the form

$$f_{\mathbf{u}}(\mathbf{x}) = \mathrm{e}^{-\langle A^{-1}(\mathbf{x}-\mathbf{z}),\mathbf{x}-\mathbf{z}\rangle}\,,$$

i.e., as an anisotropic Gaussian with complex vector-valued arguments.

8.7.2. Potentials of anisotropic Gaussians of a complex argument. Here, we compute potentials of shifts of the anisotropic Gaussians of the form

$$\mathrm{e}^{-\langle A^{-1}(\mathbf{x}-\mathbf{z}),\mathbf{x}-\mathbf{z}\rangle}\,, \tag{8.70}$$

where the matrix A has the same properties as in Section 6.3 and $\mathbf{z} \in \mathbb{C}^n$ is an arbitrary constant vector.

First, we remark that for $\mathbf{z} = \mathbf{u} + i\mathbf{v}$, $\mathbf{u}, \mathbf{v} \in \mathbb{R}^n$, and $A = A_R + iA_I$ with real symmetric matrices $A_R > 0$ and A_I, we have

$$\begin{aligned}\mathrm{Re}\langle A(\mathbf{x}+\mathbf{z}),\mathbf{x}+\mathbf{z}\rangle =& \langle A_R(\mathbf{x}+\mathbf{u}-A_R^{-1}A_I\mathbf{v}),\mathbf{x}+\mathbf{u}-A_R^{-1}A_I\mathbf{v}\rangle \\ &- \langle (A_R + A_I A_R^{-1} A_I)\mathbf{v},\mathbf{v}\rangle\,.\end{aligned}$$

Hence, for any given $\mathbf{z} \in \mathbb{C}^n$, the function (8.70) belongs to $L_2(\mathbb{R}^n)$. Thus, one can consider partial differential equations with the right-hand side (8.70), similarly to Section 6.3,

We note that in view of (8.66), the function $\mathrm{e}^{-\langle A^{-1}\mathbf{z},\mathbf{z}\rangle}$ is the analytic extension of the general Gaussian onto $\mathbb{C}^n$. Since the differential operators, occurring in Theorems 6.2, 6.4 and 6.5, have constant coefficients, the assertion of these theorems remains valid if the real vector $\mathbf{x} \in \mathbb{R}^n$ is replaced by $\mathbf{z} \in \mathbb{C}^n$. In particular, we obtain the following result.

COROLLARY 8.8. *Under the assumptions on the matrices A and B in Theorem 6.2, the following assertions are valid for any constant vector $\mathbf{z} \in \mathbb{C}^n$:*

(i) *If $n \geq 3$, then the function*

$$u(\mathbf{x}) = \frac{1}{4}\int_0^\infty \frac{\mathrm{e}^{-\langle (A+tB)^{-1}(\mathbf{x}+\mathbf{z}),\mathbf{x}+\mathbf{z}\rangle}}{\sqrt{\det(A+tB)}}\,dt$$

is a bounded solution of the equation

$$-\langle B\nabla,\nabla\rangle u(\mathbf{x}) = \frac{\mathrm{e}^{-\langle A^{-1}(\mathbf{x}+\mathbf{z}),\mathbf{x}+\mathbf{z}\rangle}}{\sqrt{\det A}}\,, \qquad \mathbf{x} \in \mathbb{R}^n\,.$$

(ii) *For $a \in \mathbb{C}$, $\mathrm{Re}\, a > 0$ and $n \geq 1$ or $\mathrm{Re}\, a = 0$ and $n \geq 3$, the function*

$$u(\mathbf{x}) = \frac{1}{4}\int_0^\infty \frac{\mathrm{e}^{-\langle (A+tB)^{-1}(\mathbf{x}+\mathbf{z}),\mathbf{x}+\mathbf{z}\rangle}}{\sqrt{\det(A+tB)}}\,\mathrm{e}^{-at/4}\,dt$$

is a bounded solution of the elliptic equation

$$-\langle B\nabla,\nabla\rangle u(\mathbf{x}) + au(\mathbf{x}) = \frac{\mathrm{e}^{-\langle A^{-1}(\mathbf{x}+\mathbf{z}),\mathbf{x}+\mathbf{z}\rangle}}{\sqrt{\det A}}\,, \qquad \mathbf{x} \in \mathbb{R}^n\,.$$

(iii) *If $n \geq 3$, then the function*

$$w_{kl}(\mathbf{x}) = -\frac{1}{16}\int_0^\infty t\,\frac{\partial^2}{\partial x_k \partial x_l}\frac{\mathrm{e}^{-\langle (A+tI)^{-1}(\mathbf{x}+\mathbf{z}),\mathbf{x}+\mathbf{z}\rangle}}{\sqrt{\det(A+tI)}}\,dt$$

is a solution of the bi-Laplace equation

$$-\Delta^2 w\,(\mathbf{x}) = \frac{\partial^2}{\partial x_k \partial x_l}\frac{\mathrm{e}^{-\langle A^{-1}(\mathbf{x}+\mathbf{z}),\mathbf{x}+\mathbf{z}\rangle}}{\sqrt{\det A}}\,, \qquad \mathbf{x} \in \mathbb{R}^n\,,$$

satisfying $w\,(\mathbf{x}) \to 0$ as $|\mathbf{x}| \to \infty$.

8.7.3. Harmonic potentials of approximate wavelets. Let $n \geq 3$. By (8.69), the matrix A has the form

$$A = \mathcal{D}P_{\mathbf{v}} + \frac{3\mathcal{D}}{2}Q_{\mathbf{v}}\,.$$

Thus, by Corollary 8.8(i), the harmonic potential of the function $f_{\mathbf{u}}$ defined by (8.67) equals

$$\mathcal{L}_n f_{\mathbf{u}}(\mathbf{x}) = \frac{\mathcal{D}}{4}\int_0^\infty \frac{\mathrm{e}^{-|P_{\mathbf{v}}\mathbf{x}|^2/\mathcal{D}(1+t)}\,\mathrm{e}^{-|Q_{\mathbf{v}}(2\mathbf{x}-\mathbf{e}+5\pi i\mathcal{D}\mathbf{u}/2)|^2/\mathcal{D}(6+4t)}}{(1+t)^{(n-m)/2}(1+2t/3)^{m/2}}\,dt\,,$$

which gives the following formula for the harmonic potential of the basis wavelet functions

$$\begin{aligned}\mathcal{L}_n\Phi_{\mathbf{v}}(\mathbf{x}) = {}& \frac{\mathcal{D}\rho^{n-m}\kappa^m\,\mathrm{e}^{-25\pi^2\mathcal{D}m/24}}{2^{m+2}}\\ &\times\sum_{\mathbf{u}\in\mathcal{U}}\int_0^\infty \frac{\mathrm{e}^{-|P_{\mathbf{v}}\mathbf{x}|^2/\mathcal{D}(1+t)}\,\mathrm{e}^{-|Q_{\mathbf{v}}(2\mathbf{x}-\mathbf{e}+5\pi i\mathcal{D}\mathbf{u}/2)|^2/\mathcal{D}(6+4t)}}{(1+t)^{(n-m)/2}(1+2t/3)^{m/2}}\,dt.\end{aligned}$$

Now we use that

$$\begin{aligned}&\mathrm{e}^{-25\pi^2\mathcal{D}m/24}\left(\mathrm{e}^{-|Q_{\mathbf{v}}(2\mathbf{x}-\mathbf{e}+5\pi i\mathcal{D}\mathbf{u}/2)|^2/\mathcal{D}(6+4t)}+\mathrm{e}^{-|Q_{\mathbf{v}}(2\mathbf{x}-\mathbf{e}-5\pi i\mathcal{D}\mathbf{u}/2)|^2/\mathcal{D}(6+4t)}\right)\\ &\quad= 2\,\mathrm{e}^{-25\pi^2\mathcal{D}mt/6(6+4t)}\,\mathrm{e}^{-|Q_{\mathbf{v}}(2\mathbf{x}-\mathbf{e})|^2/\mathcal{D}(6+4t)}\cos\frac{5\pi\langle Q_{\mathbf{v}}(2\mathbf{x}-\mathbf{e}),\mathbf{u}\rangle}{6+4t}\,,\end{aligned}$$

which gives, together with (8.65),

$$\begin{aligned}\mathcal{L}_n\Phi_{\mathbf{v}}(\mathbf{x}) = {}& \frac{\mathcal{D}\rho^{n-m}\kappa^m}{4}\\ &\times\int_0^\infty \frac{\mathrm{e}^{-|P_{\mathbf{v}}\mathbf{x}|^2/\mathcal{D}(1+t)}\,\mathrm{e}^{-|Q_{\mathbf{v}}(2\mathbf{x}-\mathbf{e})|^2/\mathcal{D}(6+4t)}}{(1+t)^{(n-m)/2}(1+2t/3)^{m/2}}\prod_{j=1}^n\cos\frac{5\pi\langle Q_{\mathbf{v}}(2\mathbf{x}-\mathbf{e}),e_j\rangle}{6+4t}\,dt.\end{aligned}$$

Introducing the two functions

$$\begin{aligned}&g_0(x,t) = \rho\frac{\mathrm{e}^{-x^2/\mathcal{D}(1+t)}}{\sqrt{1+t}}\,,\\ &g_{1/2}(x,t) = \tau\frac{\mathrm{e}^{-25\pi^2\mathcal{D}t/6(6+4t)}}{\sqrt{1+2t/3}}\,\mathrm{e}^{-(2x-1)^2/\mathcal{D}(6+4t)}\cos\frac{5\pi(2x-1)}{6+4t}\,,\end{aligned}\tag{8.71}$$

we can write the harmonic potential of the wavelet basis function in the form

$$\mathcal{L}_n\Phi_{\mathbf{v}}(\mathbf{x}) = \frac{\mathcal{D}}{4}\int_0^\infty g_{v_1}(x_1,t)\dots g_{v_n}(x_n,t)\,dt\,.\tag{8.72}$$

We conclude that the harmonic potential of a function u with the wavelet expansion (8.62) is approximated by

$$\begin{aligned}\mathcal{L}_{n,h}u(\mathbf{x}) &= \sum_{\mathbf{m}\in\mathbb{Z}^n} a_{\mathbf{m}} \frac{\mathcal{D}^{n/2}}{4|\mathbf{x}-\mathbf{m}|^{n-2}} \int\limits_0^{|\mathbf{x}-\mathbf{m}|^2/\mathcal{D}} t^{n/2-2}\,\mathrm{e}^{-t}\,dt \\ &+ \sum_{j=0}^{k-1} \frac{\mathcal{D}}{2^{2j+2}} \sum_{\mathbf{m}\in\mathbb{Z}^n}\sum_{\mathbf{v}\in\mathcal{V}'} a^{\mathbf{v}}_{j,\mathbf{m}} \int\limits_0^\infty g_{v_1}(2^j x_1 - m_1, t)\dots g_{v_n}(2^j x_n - m_n, t)\,dt\,. \end{aligned} \tag{8.73}$$

8.7.4. Diffraction potentials of approximate wavelets. Analogously to the previous subsection, one can determine the diffraction potential of the wavelet basis functions

$$\mathcal{S}_n \Phi_{\mathbf{v}}(\mathbf{x}) = \int\limits_{\mathbb{R}^n} \mathcal{E}_k(\mathbf{x}-\mathbf{y})\,\Phi_{\mathbf{v}}(\mathbf{y})\,d\mathbf{y}$$

with the fundamental solution $\mathcal{E}_k$ given by (5.2). By Corollary 8.8(ii), and using the arguments of Subsection 6.3.1, we derive

$$\mathcal{S}_n f_{\mathbf{u}}(\mathbf{x}) = \frac{i\rho^{n-m}\kappa^m\mathcal{D}}{4}\int\limits_0^\infty \frac{\mathrm{e}^{ik^2t/4}\,\mathrm{e}^{-|P_{\mathbf{v}}\mathbf{x}|^2/\mathcal{D}(1+it)}\,\mathrm{e}^{-|Q_{\mathbf{v}}(2\mathbf{x}-\mathbf{e}+5\pi i\mathcal{D}\mathbf{u}/2)|^2/\mathcal{D}(6+4it)}}{(1+it)^{(n-m)/2}(1+2it/3)^{m/2}}\,dt.$$

Because

$$\begin{aligned}&\mathrm{e}^{-25\pi^2\mathcal{D}m/24}\left(\mathrm{e}^{-|Q_{\mathbf{v}}(2\mathbf{x}-\mathbf{e}+5\pi i\mathcal{D}\mathbf{u}/2)|^2/\mathcal{D}(6+4it)} + \mathrm{e}^{-|Q_{\mathbf{v}}(2\mathbf{x}-\mathbf{e}-5\pi i\mathcal{D}\mathbf{u}/2)|^2/\mathcal{D}(6+4it)}\right) \\ &\quad = \mathrm{e}^{-25\pi^2 i\mathcal{D}mt/6(6+4it)}\,\mathrm{e}^{-|Q_{\mathbf{v}}(2\mathbf{x}-\mathbf{e})|^2/\mathcal{D}(6+4it)} \cos\frac{5\pi\langle Q_{\mathbf{v}}(2\mathbf{x}-\mathbf{e}),\mathbf{u}\rangle}{6+4it}\,,\end{aligned}$$

one obtains the compact form of the diffraction potentials

$$\boxed{\mathcal{S}_n\Phi_{\mathbf{v}}(\mathbf{x}) = \frac{i\mathcal{D}}{4}\int\limits_0^\infty \mathrm{e}^{ik^2t/4}\,g_{v_1}(x_1, it)\dots g_{v_n}(x_n, it)\,dt}$$

with g_0, g_1 defined by (8.71).

8.7.5. Elastic and hydrodynamic potentials of approximate wavelets. Here, we provide the formulas for the elastic and hydrodynamic potentials of the wavelet basis functions $\Phi_{\mathbf{v}}$ in $\mathbb{R}^3$. It follows from Corollary 8.8(iii) that the integrals

$$w_{kl}(\mathbf{x}) := \frac{1}{8\pi}\frac{\partial^2}{\partial x_k \partial x_l}\int\limits_{\mathbb{R}^3}|\mathbf{x}-\mathbf{y}|\,f_{\mathbf{u}}(\mathbf{y})\,d\mathbf{y} = \frac{1}{8\pi}\int\limits_{\mathbb{R}^3}|\mathbf{x}-\mathbf{y}|\,\frac{\partial^2}{\partial y_k\partial y_l}f_{\mathbf{u}}(\mathbf{y})\,d\mathbf{y}$$

with $f_{\mathbf{u}}$ defined by (8.67) can be written in the form

$$\begin{aligned} w_{kl}(\mathbf{x}) &= -\frac{\rho^{n-m}\kappa^m\mathcal{D}^2}{16} \\ &\quad\times\int\limits_0^\infty t\,\frac{\partial^2}{\partial x_k\partial x_l}\frac{\mathrm{e}^{-|P_{\mathbf{v}}\mathbf{x}|^2/\mathcal{D}(1+t)}\,\mathrm{e}^{-|Q_{\mathbf{v}}(2\mathbf{x}-\mathbf{e}+5\pi i\mathcal{D}\mathbf{u}/2)|^2/\mathcal{D}(6+4t)}}{(1+t)^{(3-m)/2}(1+2t/3)^{m/2}}\,dt.\end{aligned}$$

Then, together with formula (8.72), the one-dimensional integral representations of the elastic and hydrodynamic potentials of $\Phi_{\mathbf{v}}$ follow immediately. Since by (5.50)

$$\Gamma_{kl}(\mathbf{x}) = \frac{\lambda+\mu}{8\pi\mu(\lambda+2\mu)}\frac{\partial^2}{\partial x_k \partial x_l}|\mathbf{x}| - \frac{\delta_{kl}}{4\pi\mu|\mathbf{x}|},$$

the elastic potential of $\Phi_{\mathbf{v}}$ can be obtained from

$$\begin{aligned}
&\int_{\mathbb{R}^3} \Gamma_{kl}(\mathbf{x}-\mathbf{y})\Phi_{\mathbf{v}}(\mathbf{y})\,d\mathbf{y} \\
&= -\frac{\delta_{kl}}{\mu}\mathcal{L}_3(\Phi_{\mathbf{v}})(\mathbf{x}) + \frac{\lambda+\mu}{8\pi\mu(\lambda+2\mu)}\int_{\mathbb{R}^3} |\mathbf{x}-\mathbf{y}|\frac{\partial^2}{\partial y_k \partial y_l}\Phi_{\mathbf{v}}(\mathbf{y})\,d\mathbf{y} \\
&= -\frac{\mathcal{D}}{4\mu}\int_0^\infty \Big(\delta_{kl} + \frac{t\,\mathcal{D}(\lambda+\mu)}{4(\lambda+2\mu)}\frac{\partial^2}{\partial x_k \partial x_l}\Big) g_{v_1}(x_1,t)g_{v_2}(x_2,t)g_{v_3}(x_3,t)\,dt\,,
\end{aligned} \tag{8.74}$$

with the functions g_0, g_1 defined by (8.71).

Furthermore, by (5.49)

$$\Psi_{kl}(\mathbf{x}) = \frac{1}{8\pi\nu}\frac{\partial^2}{\partial x_k \partial x_l}|\mathbf{x}| - \frac{\delta_{kl}}{4\pi\nu|\mathbf{x}|},$$

which together with (5.50) leads to the hydrodynamic potential of the wavelet basis function

$$\begin{aligned}
&\int_{\mathbb{R}^3} \Psi_{kl}(\mathbf{x}-\mathbf{y})\Phi_{\mathbf{v}}(\mathbf{y})\,d\mathbf{y} \\
&= -\frac{\delta_{kl}}{\nu}\mathcal{L}_3(\Phi_{\mathbf{v}})(\mathbf{x}) + \frac{1}{8\pi\nu}\int_{\mathbb{R}^3} |\mathbf{x}-\mathbf{y}|\frac{\partial^2}{\partial y_k \partial y_l}\Phi_{\mathbf{v}}(\mathbf{y})\,d\mathbf{y} \\
&= -\frac{\mathcal{D}}{4\nu}\int_0^\infty \Big(\delta_{kl} + \frac{t\,\mathcal{D}}{4}\frac{\partial^2}{\partial x_k \partial x_l}\Big) g_{v_1}(x_1,t)\dots g_{v_n}(x_n,t)\,dt\,, \\
&-\frac{1}{4\pi}\int_{\mathbb{R}^3} \frac{x_k-y_k}{|\mathbf{x}-\mathbf{y}|^3}\Phi_{\mathbf{v}}(\mathbf{y})\,d\mathbf{y} = \frac{\mathcal{D}}{4}\frac{\partial}{\partial x_k}\int_0^\infty g_{v_1}(x_1,t)g_{v_2}(x_2,t)g_{v_3}(x_3,t)\,dt.
\end{aligned} \tag{8.75}$$

8.8. Numerical example

The decomposition described in Theorem 8.7 was implemented in the one-dimensional case to obtain compressed representations for density functions and it was applied to the computation of one-dimensional integral operators. Table 8.2 provides some numerical results concerning the evaluation of the Hilbert transform of different functions given on $(-500, 500)$. Originally, the functions are approximated by the quasi-interpolant

$$u_h(x) := \mathcal{D}^{-n/2} \sum_{m=-32000}^{32000} u(hm) \exp\Big(-\frac{(x-hm)^2}{\mathcal{D}h^2}\Big) \in V_6 \tag{8.76}$$

with $h = 1/64$, i.e., they are determined by 64001 point values. Table 8.2 lists the number of all basis functions in V_0 and W_j, $j = 0, \dots, 5$, necessary to compute the Hilbert transform at all grid points $\{mh\}$ with the prescribed accuracy $\varepsilon_{\mathcal{D}}$.

function	$\mathcal{D} = 2$	$\mathcal{D} = 3$	$\mathcal{D} = 4$
1	2563	3509	4689
$\lvert 500 - x\rvert$	2290	3174	4286
$\sin(\pi x)$	7624	7897	8116
$e^{-x^2/1000}$	811	897	1085
$e^{-x^2/1000}\sin(\pi x)$	3826	4140	4354

TABLE 8.2. Number of basis functions required to compute the Hilbert transform at all grid points with prescribed accuracy $\varepsilon_{\mathcal{D}}$

Except for $\exp(-x^2/1000)$, the density functions u under consideration are either non-smooth or oscillating, such that the support of the projections onto the spaces V_0 and W_j, $Q_j u$, is larger than the interval $(-500, 500)$. Nevertheless the table shows that it is possible to obtain significant compression rates for data if the density function behaves sufficiently well. The same applies to the computing time. For example, to evaluate one point value of the integral besides the Hilbert transform of all Gaussians $\exp(-(x-m)^2/\mathcal{D})$, which form $Q_{-1}u \in V_0$, one has to compute only the Hilbert transform of the wavelets with essential support near this point. Therefore, the number of the summands required to compute the Hilbert transform at one given point is essentially smaller than 64001, necessary if the representation (8.76) is used, and even much smaller than the number given in Table 8.2.

8.9. Notes

Approximate multi-resolution analysis and approximate wavelets were proposed by the authors in [**69**].

There is a large bibliography on wavelet theory. The basic material can be found, for example, in the books of Daubechies [**22**], Meyer [**74**], and Chui [**18**]. The theory of prewavelets is developed in [**74**] and by de Boor, DeVore, and Ron in [**10**]. Many interesting examples of scaling functions, which are used in numerical analysis and which satisfy certain smoothness and vanishing moment conditions, can be found in [**21**], [**22**], [**74**], [**19**], [**21**]. There exists a series of papers on the application of wavelet methods to the computation of integral operators and the solution of integral equations, where different types of scaling functions and wavelets are used (see [**2**], [**9**], [**21**], and [**78**] the references therein).

CHAPTER 9

Cubature over bounded domains

9.1. Introduction

In this chapter we extend the classes of cubature formulas introduced in Chapter 4 to integral operators over bounded domains.

It was mentioned there that the exact computation of volume potentials is an essential resource for the solution of boundary value problems with boundary integral methods. Even more important applications appear when one combines boundary integral methods with iteration procedures for linear problems with variable coefficients or for non-linear problems. Essentially, the approach for solving boundary problems for non-linear equations lumps the non-linearity into body forces and then solves the problem iteratively. This introduces domain integrals or volume potentials to the corresponding boundary integral equations.

If one wants to compute the integral

$$\mathcal{K}u(\mathbf{x}) = \int\limits_{\Omega} g(\mathbf{x}-\mathbf{y})\, u(\mathbf{y})\, d\mathbf{y} \tag{9.1}$$

by using known values of the integral $\mathcal{K}\eta$ applied to a generating function η, the approximation of the density u by a linear combination of dilated shifts of η has to take into account the following circumstances:

- The approximant should have a simple quasi-interpolation structure.
- Since one has to approximate u extended by zero outside Ω, the approximant must nearly vanish in $\Omega^c = \mathbb{R}^n \setminus \overline{\Omega}$.
- The error should be small in some integral norm over $\mathbb{R}^n$ since we are interested in the approximation of integral operators.
- Discontinuous functions can be approximated with smooth η centered on uniformly distributed nodes only with large errors; therefore mesh refinement near the boundary of the domain may prove useful.

Here, we develop an iteration scheme which satisfies these requirements. We obtain an approximation formula

$$\mathcal{B}_M u(\mathbf{x}) = \mathcal{D}^{-n/2} \sum_{\mathbf{m}\in\mathcal{Q}_0} u(h\mathbf{m})\eta\Big(\frac{\mathbf{x}-h\mathbf{m}}{h\sqrt{\mathcal{D}}}\Big) + \sum_{j=1}^{M}\sum_{\mathbf{m}\in\mathcal{Q}_j} c_{j,\mathbf{m}}\, \eta\Big(\frac{\mathbf{x}-h_j\mathbf{m}}{h_j\sqrt{\mathcal{D}}}\Big), \tag{9.2}$$

with $h_j = \mu^j h$, $0<\mu<1$ and coefficients $c_{j,\mathbf{m}}$, depending on point values of u near the nodes $h_j\mathbf{m}$.

The sets $\mathcal{Q}_j \subset \mathbb{Z}^n$ are such that hQ_0 consists of all nodes $h\mathbf{m}\in\Omega$ located at a certain distance to the boundary $\partial\Omega$ and such that $h_j\mathcal{Q}_j \subset \Omega$, $j=1,\dots,M$, lie in boundary layers of a width decreasing with j. We show that $\mathcal{B}_M u(\mathbf{x})$ approximates u on the whole $\mathbb{R}^n$ except for a small boundary layer of width decreasing exponentially with M, the number of iteration steps. This guarantees that $\mathcal{B}_M$ provides a similar approximation error in the L_p-norms as the quasi-interpolant $\mathcal{M}_{h,\mathcal{D}}$ generated by η, given in Theorem 2.28.

The operator $\mathcal{B}_M$ is the sum of the usual quasi-interpolation operator $\mathcal{M}_{h,\mathcal{D}}$ applied to the restriction of u to some interior subdomain of Ω and multi-resolution correction terms near the boundary. The construction, which will be discussed in Section 9.4, grants an easy computation of the coefficients $c_{j,\mathbf{m}}$, and the introduction of new higher-frequency terms in (9.2) does not require re-computation of the coefficients $c_{j,\mathbf{m}}$.

The accuracy provided by $\mathcal{B}_M$ for functions on domains gives high-order cubature formulas for (9.1) by setting

$$\mathcal{K}_h u(\mathbf{x}) = \mathcal{K}\mathcal{B}_M u(\mathbf{x}) = \sum_{j=0}^{M} \sum_{\mathbf{m}\in\mathcal{Q}_k} c_{j,\mathbf{m}}\, \mathcal{K}\eta\Big(\frac{\cdot - h_j\mathbf{m}}{h_j\sqrt{\mathcal{D}}}\Big)(\mathbf{x}) \tag{9.3}$$

with $c_{0,\mathbf{m}} = \mathcal{D}^{-n/2}u(h\mathbf{m})$. This will be discussed in Section 9.5.

In Section 9.6, we specialize this method for polyhedral domains and generating functions having tensor product structure. At the boundary layers, approximants on uniform meshes are constructed, but the meshes are refined only in the direction to the boundary. The anisotropic mesh refinement leads to a considerable reduction of data points and, which is most important, of the number of summands in $\mathcal{K}\mathcal{B}_M u$ required for the cubature of $\mathcal{K}u$. In Section 9.7, we consider an example showing how potentials of special anisotropic tensor product generating functions can be computed.

9.2. Simple approach

We start with a naive approach. By the local character of quasi-interpolation, which is described in Subsection 2.3.3, some of the above-mentioned requirements can be satisfied.

Recall that for arbitrarily small ε, there exist $\mathcal{D}$ and κ, such that for all $\mathbf{x} \in \Omega_{\kappa h} \subset \Omega$ the quasi-interpolant to any smooth function $u \in W_\infty^N(\Omega)$ provides the estimate

$$\begin{aligned} |u(\mathbf{x}) - \mathcal{M}_{h,\mathcal{D}}u(\mathbf{x})| \le\, & c(\sqrt{\mathcal{D}}h)^N \|\nabla_N u\|_{L_\infty(\Omega)} \\ & + \varepsilon\big(\|u\|_{L_\infty(\Omega)} + \sum_{k=1}^{N-1}(\sqrt{\mathcal{D}}h)^k |\nabla_k u(\mathbf{x})|\big), \end{aligned} \tag{9.4}$$

if $\eta \in \mathcal{S}(\mathbb{R}^n)$ is subject to the moment Condition 2.15 of order N (cf. Corollary 2.24). The subdomain $\Omega_{\kappa h}$ is defined by (2.67) and the constant c depends on the generating function η. Note further that by Lemma 2.22

$$|\mathcal{M}_{h,\mathcal{D}}(\chi_0 u)(\mathbf{x})| \le g_{\mathcal{D}}(h^{-1}\operatorname{dist}(\mathbf{x},\Omega) + \kappa, \eta) \sup_{\Omega}|u| \le \varepsilon \sup_{\Omega_{\kappa h}}|u|\,, \tag{9.5}$$

where χ_0 is the characteristic function of $\Omega_{\kappa h}$ and $\mathbf{x} \in \Omega^{\mathrm{c}} = \mathbb{R}^n \setminus \overline{\Omega}$.

Consequently, the function $\mathcal{M}_{h,\mathcal{D}}(\chi_0 u)$ is sufficiently small on Ω^{c}, and it approximates u in the subdomain $\Omega_{2\kappa h}$, in view of Corollary 2.24, with the estimate (9.4). So, we are left with the boundary layer $S_0 = \Omega \setminus \Omega_{2\kappa h}$, where the error

$$u_1(\mathbf{x}) := u(\mathbf{x}) - \mathcal{M}_{h,\mathcal{D}}(\chi_0 u)(\mathbf{x})$$

is large. In order to retain the simplicity of the approximation method, one could try to approximate this discrepancy by the quasi-interpolation operator $\mathcal{M}_{h_1,\mathcal{D}}$ on a finer mesh with step size $h_1 = \mu h$, $\mu < 1$. Since the approximant must be sufficiently small outside Ω, we introduce the characteristic function χ_1 of the boundary layer $S_1 = \Omega_{\kappa h_1} \setminus \Omega_{2\kappa h + \kappa h_1}$ and consider

$$\mathcal{M}_{h_1,\mathcal{D}}(\chi_1 u_1)(\mathbf{x}) = \mathcal{D}^{-n/2} \sum_{h_1\mathbf{m}\in S_1} u_1(h_1\mathbf{m})\eta\Big(\frac{\mathbf{x}-h_1\mathbf{m}}{\sqrt{\mathcal{D}}h_1}\Big), \tag{9.6}$$

which provides, in view of Corollary 2.24, the estimate

$$|u_1(\mathbf{x}) - \mathcal{M}_{h_1,\mathcal{D}}(\chi_1 u_1)(\mathbf{x})| \le c(\sqrt{\mathcal{D}}h_1)^N \|\nabla_N u_1\|_{L_\infty(S_1)} + \varepsilon\big(\|u\|_{L_\infty(\Omega)} + \sum_{k=1}^{N-1} (\sqrt{\mathcal{D}}h_1)^k |\nabla_k u(\mathbf{x})|\big)$$

for $\mathbf{x} \in \Omega_{2\kappa h_1} \setminus \Omega_{2\kappa h}$. However, since $\mathcal{M}_{h,\mathcal{D}}(\chi_0 u)$ nearly vanishes at $\partial\Omega$, one has

$$\max_{S_1} |\nabla_N \mathcal{M}_{h,\mathcal{D}}(\chi_0 u)| = \mathcal{O}((\sqrt{\mathcal{D}}h)^{-N}) \max_{\partial\Omega_{2\kappa h}} |u|\,,$$

so that

$$|u_1(\mathbf{x}) - \mathcal{M}_{h_1,\mathcal{D}}(\chi_1 u_1)(\mathbf{x})| \le c\,\big((\sqrt{\mathcal{D}}h_1)^N \|\nabla_N u\|_{L_\infty(\Omega)} + \mu^N \|u\|_{L_\infty(\Omega)}\big) + \varepsilon\big(\|u\|_{L_\infty(\Omega)} + \sum_{k=1}^{N-1} (\sqrt{\mathcal{D}}h_1)^k |\nabla_k u(\mathbf{x})|\big)\,. \tag{9.7}$$

Hence, the approximation error depends on the quotient $\mu = h_1/h$. Of course, in order to get approximation order $\mathcal{O}(h^N)$, one could choose $\mu = h$, but this is not practical because of the large number of summands in (9.6).

This procedure can be repeated iteratively as long as the remaining boundary layer is sufficiently small, but for each step the approximation error behaves like (9.7). This is confirmed in Fig. 9.1(a), where the Heaviside function is approximated with the above procedure, using 5 iteration steps, $\mu = 1/2$, and η is the Gaussian. Fortunately, there exists another approach to represent coarsely scaled generating

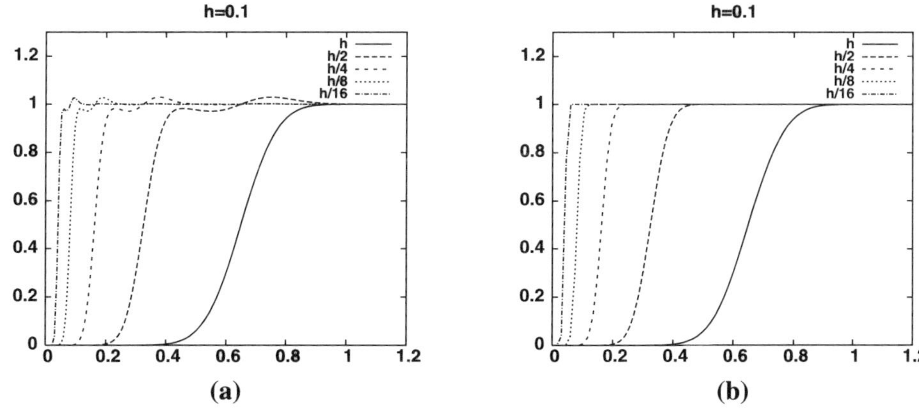

FIGURE 9.1. Multiscale approximation of the Heaviside function with Gaussians, based on **(a)** quasi-interpolation of the discrepancy and **(b)** the approximate refinement equation.

functions by linear combinations of finer scaled versions, which is based on the *approximate refinement equation* studied in Section 8.2. Using this idea, instead of the quasi-interpolation of the discrepancy, yields considerably better results, as shown in Fig. 9.1(b). In what follows, we utilize approximate refinement equations to construct multi-resolution approximants for functions given on domains.

9.3. Application of the approximate refinement equations

9.3.1. Approximate factorization of the quasi-interpolant. Theorem 8.1 states that identities of the form

$$\eta\Big(\frac{\mathbf{x}}{\sqrt{\mathcal{D}}}\Big) = \mathcal{D}^{-n/2} \sum_{\mathbf{m}\in\mathbb{Z}^n} \tilde{\eta}\Big(\frac{\mu\mathbf{m}}{\sqrt{\mathcal{D}}}\Big)\eta\Big(\frac{\mathbf{x}-\mu\mathbf{m}}{\mu\sqrt{\mathcal{D}}}\Big) + R_{\eta,\mu,\mathcal{D}}(\mathbf{x}) \tag{9.8}$$

with a small remainder term $R_{\eta,\mu,\mathcal{D}}$ are valid if η and the mask function

$$\tilde{\eta} = \mathcal{F}^{-1}(\mathcal{F}\eta/\mathcal{F}\eta(\mu\cdot))$$

belong to the Schwartz class $\mathcal{S}(\mathbb{R}^n)$ and

$$\mathcal{F}\eta \neq 0\,,$$

which is supposed throughout.

THEOREM 9.1. *If $\mu^{-1} \in \mathbb{N}$, then the approximate refinement equation* (9.8) *implies the approximate factorization of the quasi-interpolation operator* (2.23)

$$\mathcal{M}_{h,\mathcal{D}} = \mathcal{M}_{\mu h,\mathcal{D}}\,\widetilde{\mathcal{M}}_{h,\mathcal{D}} + \mathcal{R}_{h,\mathcal{D}} \tag{9.9}$$

with the quasi-interpolant

$$\widetilde{\mathcal{M}}_{h,\mathcal{D}}u(\mathbf{x}) := \mathcal{D}^{-n/2} \sum_{\mathbf{m}\in\mathbb{Z}^n} u(h\mathbf{m})\,\tilde{\eta}\Big(\frac{\mathbf{x}-h\mathbf{m}}{\sqrt{\mathcal{D}}h}\Big) \tag{9.10}$$

and the remainder

$$\mathcal{R}_{h,\mathcal{D}}u(\mathbf{x}) = \mathcal{D}^{-n/2} \sum_{\mathbf{m}\in\mathbb{Z}^n} u(h\mathbf{m})\,R_{\eta,\mu,\mathcal{D}}\Big(\frac{\mathbf{x}}{h}-\mathbf{m}\Big)\,. \tag{9.11}$$

PROOF. Using (9.8), one obtains

$$\begin{aligned}
\mathcal{D}^{-n/2} &\sum_{\mathbf{m}\in\mathbb{Z}^n} u(h\mathbf{m})\,\eta\Big(\frac{\mathbf{x}-h\mathbf{m}}{\sqrt{\mathcal{D}}\,h}\Big) \\
&= \mathcal{D}^{-n} \sum_{\boldsymbol{\nu},\mathbf{m}\in\mathbb{Z}^n} u(h\mathbf{m})\,\tilde{\eta}\Big(\frac{\mu\mathbf{m}}{\sqrt{\mathcal{D}}}\Big)\eta\Big(\frac{\mathbf{x}/(\mu h)-\mathbf{m}/\mu-\boldsymbol{\nu}}{\sqrt{\mathcal{D}}}\Big) \\
&\quad + \mathcal{D}^{-n/2} \sum_{\mathbf{m}\in\mathbb{Z}^n} u(h\mathbf{m})\,R_{\eta,\mu,\mathcal{D}}(\mathbf{x}/h-\mathbf{m}).
\end{aligned}$$

Since μ^{-1} is an integer, $\mathbf{k} = \boldsymbol{\nu} + \mu^{-1}\mathbf{m} \in \mathbb{Z}^n$. Thus, after re-indexing, one arrives at the representation

$$\begin{aligned}\mathcal{M}_{h,\mathcal{D}}u(\mathbf{x}) &= \mathcal{D}^{-n} \sum_{\mathbf{k},\mathbf{m}\in\mathbb{Z}^n} u(h\mathbf{m})\, \tilde{\eta}\Big(\frac{\mu\mathbf{k}-\mathbf{m}}{\sqrt{\mathcal{D}}}\Big)\eta\Big(\frac{\mathbf{x}-\mu h\mathbf{k}}{\mu h\sqrt{\mathcal{D}}}\Big) + \mathcal{R}_{h,\mathcal{D}}u(\mathbf{x}) \\ &= \mathcal{D}^{-n} \sum_{\mathbf{k},\mathbf{m}\in\mathbb{Z}^n} u(h\mathbf{m})\, \tilde{\eta}\Big(\frac{\mu h\mathbf{k}-h\mathbf{m}}{h\sqrt{\mathcal{D}}}\Big)\eta\Big(\frac{\mathbf{x}-\mu h\mathbf{k}}{\mu h\sqrt{\mathcal{D}}}\Big) + \mathcal{R}_{h,\mathcal{D}}u(\mathbf{x}) \\ &= \mathcal{D}^{-n/2} \sum_{\mathbf{k}\in\mathbb{Z}^n} \widetilde{\mathcal{M}}_{h,\mathcal{D}}u(\mu h\mathbf{k})\, \eta\Big(\frac{\mathbf{x}-\mu h\mathbf{k}}{\mu h\sqrt{\mathcal{D}}}\Big) + \mathcal{R}_{h,\mathcal{D}}u(\mathbf{x})\,. \qquad \square\end{aligned}$$

9.3.2. Properties of the mask function $\tilde{\eta}$. In Section 8.2, some analytic examples of mask functions $\tilde{\eta}$ were given. Here, we discuss some further properties of these functions. Let us note that for the computations in the algorithms described below, we do not need the analytic expression of the functions $\tilde{\eta}$. In the following section, we will show that for our purposes it suffices to precompute the values of $\tilde{\eta}$ just at several points, which is simple if η is a radial function. Otherwise, it can be done with some numerical method for computing the inverse Fourier transform.

Suppose that in addition to the requirements of Theorem 8.1, η is subject also to the moment Condition 2.15 of order N. Then, in view of (2.48), $\tilde{\eta}$ satisfies these conditions as well. Then, by Theorem 2.17 the quasi-interpolant $\widetilde{\mathcal{M}}_{h,\mathcal{D}}$ defined by (9.10) features the same rate of approximate convergence as $\mathcal{M}_{h,\mathcal{D}}$ (which is generated by η). Note that the terms of the saturation error can be estimated by

$$\|\varepsilon_{\boldsymbol{\alpha}}(\cdot,\tilde{\eta},\mathcal{D})\|_{L_\infty} \le \sum_{\boldsymbol{\nu}\in\mathbb{Z}^n\setminus\{\mathbf{0}\}} \left| \partial^{\boldsymbol{\alpha}}_{\boldsymbol{\lambda}} \frac{\mathcal{F}\eta(\boldsymbol{\lambda})}{\mathcal{F}\eta(\mu\boldsymbol{\lambda})}\Big|_{\boldsymbol{\lambda}=\sqrt{\mathcal{D}}\boldsymbol{\nu}} \right| .$$

For example, the quasi-interpolation operator $\widetilde{\mathcal{M}}_{h,\mathcal{D}}$ generated by the mask function $\tilde{\eta}_{2M}$ to

$$\eta_{2M}(\mathbf{x}) = \pi^{-n/2}\, L^{(n/2)}_{M-1}(|\mathbf{x}|^2)\ \mathrm{e}^{-|\mathbf{x}|^2}$$

(cf. (3.18)) provides the estimate

$$|u - \widetilde{\mathcal{M}}_{h,\mathcal{D}}u| = \mathcal{O}((\sqrt{\mathcal{D}}h)^{2M}) + \|\varepsilon_{\mathbf{0}}(\,\cdot\,,\tilde{\eta}_{2M},\mathcal{D})\|_{L_\infty}\,,$$

where

$$\|\varepsilon_{\mathbf{0}}(\,\cdot\,,\tilde{\eta}_{2M},\mathcal{D})\|_{L_\infty} \le \sum_{\boldsymbol{\nu}\in\mathbb{Z}^n\setminus\{\mathbf{0}\}} \frac{P_{M-1}(|\boldsymbol{\nu}|^2\pi^2\mathcal{D})}{P_{M-1}(\mu^2|\boldsymbol{\nu}|^2\pi^2\mathcal{D})}\ \mathrm{e}^{-(1-\mu^2)|\boldsymbol{\nu}|^2\pi^2\mathcal{D}}$$

with the polynomials

$$P_{M-1}(t) = \sum_{j=0}^{M-1} \frac{t^j}{j!}$$

(cf. (2.39), (3.19), (8.12)). Consequently

$$\|\varepsilon_{\mathbf{0}}(\,\cdot\,,\tilde{\eta}_{2M},\mathcal{D})\|_{L_\infty} = \mathcal{O}(n\mu^{2-2M}\ \mathrm{e}^{-(1-\mu^2)\pi^2\mathcal{D}})\,.$$

Similarly to (2.63), one can introduce the parameter $\tilde{\kappa}$ so that

$$\tilde{\kappa}^{[\boldsymbol{\alpha}]+1-N}\, r_{\mathcal{D}}(\tilde{\kappa},\tilde{\eta}) \le \mathcal{D}^{[\boldsymbol{\alpha}]/2}\|\varepsilon_{\boldsymbol{\alpha}}(\,\cdot\,,\tilde{\eta},\mathcal{D})\|_{L_\infty}\,,\ [\boldsymbol{\alpha}] < N\,, \tag{9.12}$$

where the function $r_{\mathcal{D}}(t,\tilde{\eta})$ is defined as in (2.61). Note that by (2.64)

$$g_{\mathcal{D}}(\tilde{\kappa},\tilde{\eta}) = \sup_{\mathbf{x}\in\mathbb{R}^n} \mathcal{D}^{-n/2} \sum_{|\mathbf{x}-\mathbf{m}|>\tilde{\kappa}} \Big|\tilde{\eta}\Big(\frac{\mathbf{x}-\mathbf{m}}{\sqrt{\mathcal{D}}}\Big)\Big| \le \|\varepsilon_{\mathbf{0}}(\,\cdot\,,\tilde{\eta},\mathcal{D})\|_{L_\infty}\,.$$

9.3.3. Convolutions based on the remainder term. In the following, we have to estimate semi-discrete convolutions $\mathcal{R}_{h,\mathcal{D}}u$ of the form (9.11) which are generated by the remainder term $R_{\eta,\mu,\mathcal{D}}(\mathbf{x})$ in Theorem 8.1. These sums are properly defined, since we have rapid decay in $\mathbf{x}$. Moreover, since $(1+|\mathbf{x}|^k)|R_{\eta,\mu,\mathcal{D}}(\mathbf{x})|$ can be made arbitrarily small by choosing $\mathcal{D}$ large enough, one can always find $\mathcal{D}$ such that

$$\|\mathcal{R}_{h,\mathcal{D}}u\|_{L_\infty} \le \varepsilon\|u\|_{L_\infty} \tag{9.13}$$

for any prescribed $\varepsilon > 0$.

Let us consider some examples. If η is the Gaussian ($\eta = \eta_2$), then by (8.11), the corresponding function $\tilde{\eta}_2$ is the scaled Gaussian

$$\tilde{\eta}_2(\frac{\mathbf{x}}{\sqrt{\mathcal{D}}}) = \eta_2(\frac{\mathbf{x}}{\sqrt{\mathcal{D}(1-\mu^2)}}) = \eta_2(\frac{\mathbf{x}}{\sqrt{\widetilde{\mathcal{D}}}}), \quad \widetilde{\mathcal{D}} = \mathcal{D}(1-\mu^2)\,.$$

Recall the approximate refinement equation (8.9) for this case

$$\begin{aligned}\mathrm{e}^{-|\mathbf{x}|^2/\mathcal{D}} =&(\pi\widetilde{\mathcal{D}})^{-n/2} \sum_{\mathbf{m}\in\mathbb{Z}^n} \mathrm{e}^{-\mu^2|\mathbf{m}|^2/\widetilde{\mathcal{D}}}\ \mathrm{e}^{-|\mathbf{x}-\mu\mathbf{m}|^2/\mathcal{D}\mu^2}\\ &- \mathrm{e}^{-|\mathbf{x}|^2/\mathcal{D}} \sum_{\boldsymbol{\nu}\in\mathbb{Z}^n\setminus\{\mathbf{0}\}} \mathrm{e}^{2\pi i(1-\mu^2)\langle\mathbf{x},\boldsymbol{\nu}\rangle/\mu}\ \mathrm{e}^{-\pi^2\widetilde{\mathcal{D}}|\boldsymbol{\nu}|^2}\,.\end{aligned}$$

Hence, the generating function of (9.11) is given by

$$R_{\eta_2,\mu,\mathcal{D}}(\mathbf{x}) = \eta_2\Big(\frac{\mathbf{x}}{\sqrt{\mathcal{D}}}\Big)\big[(I-\widetilde{\mathcal{M}}_{\mu,\mathcal{D}})1\,(\mathbf{x}_\mu)\big] = \eta_2\Big(\frac{\mathbf{x}}{\sqrt{\mathcal{D}}}\Big)\big[(I-\mathcal{M}_{\mu,\widetilde{\mathcal{D}}})1\,(\mathbf{x}_\mu)\big],$$

where $\mathcal{M}_{\mu,\widetilde{\mathcal{D}}}1$ is the quasi-interpolant $\mathcal{M}_{\mu,\widetilde{\mathcal{D}}}u$ for $u(\mathbf{x}) \equiv 1$ and $\mathbf{x}_\mu = (1-\mu^2)\mathbf{x}$. Thus by Theorem 2.17, $|R_{\eta_2,\mu,\mathcal{D}}(\mathbf{x})| \le \|\varepsilon_{\mathbf{0}}(\cdot\,,\eta_2,\widetilde{\mathcal{D}})\|_{L_\infty}$ and the semi-discrete convolution $\mathcal{R}_{\eta_2,h}u$ has the uniform bound

$$|\mathcal{R}_{\eta_2,h}u(\mathbf{x})| \le \|\varepsilon_{\mathbf{0}}(\cdot\,,\eta_2,\widetilde{\mathcal{D}})\|_{L_\infty}\|u\|_{L_\infty} = \|\varepsilon_{\mathbf{0}}(\cdot\,,\eta_2,\mathcal{D}(1-\mu^2))\|_{L_\infty}\|u\|_{L_\infty}.$$

The following lemma, which will be stated without proof, shows that the remainder terms $R_{\eta_{2M},\mu,\mathcal{D}}$ in the refinement equations, corresponding to the generating functions η_{2M} defined by (3.18), exhibit similar behavior as the remainder in the case of the Gaussian η_2:

LEMMA 9.2. *Suppose that the parameter $\mu \in (0,1)$ is fixed. Then there exist positive univariate polynomials Q_1 and Q_2 of degree $M-1$ such that for any sufficiently large $\mathcal{D}$*

$$|R_{\eta_{2M},\mu,\mathcal{D}}(\mathbf{x})| \le Q_1(|\mathbf{x}|^2/\mathcal{D})\ \mathrm{e}^{-|\mathbf{x}|^2/\mathcal{D}} \sum_{\boldsymbol{\nu}\in\mathbb{Z}^n\setminus\{\mathbf{0}\}} Q_2(\mathcal{D}|\boldsymbol{\nu}|^2)\ \mathrm{e}^{-\pi^2\mathcal{D}(1-\mu^2)|\boldsymbol{\nu}|^2}\,.$$

As a consequence, we obtain that the generating function of the convolution $\mathcal{R}_{\eta_{2M},h}$ has the amplitude of the same order as the saturation error.

9.4. Boundary layer quasi-interpolants

In this section, we use the approximate factorization (9.9) to construct a boundary layer quasi-interpolation operator $\mathcal{B}_M$. For a given sequence of step sizes $\{h_j\}_{j=0}^M$ with

$$h_j = \mu^j h, \quad 0 < h, \mu < 1, \quad \mu^{-1} \in \mathbb{Z},$$

we obtain an approximate multi-resolution decomposition of the quasi-interpolant on the highest resolution $\mathcal{M}_{h_M,\mathcal{D}}$, from which the desired boundary layer quasi-interpolation is derived after an appropriate truncation of the summation.

9.4.1. Multi-resolution decomposition. We use the notation

$$\mathcal{A}_j = \mathcal{M}_{\mu^j h,\mathcal{D}}, \qquad \widetilde{\mathcal{A}}_j = \widetilde{\mathcal{M}}_{\mu^j h,\mathcal{D}}, \qquad \mathcal{R}_j = \mathcal{R}_{\mu^j h,\mathcal{D}}, \qquad j = 0, 1, 2 \dots, \tag{9.14}$$

where $\mathcal{R}_{h,\mathcal{D}}$ is the semi-discrete convolution (9.11). By Theorem 9.1, we have

$$\mathcal{A}_j = \mathcal{A}_{j+1}\widetilde{\mathcal{A}}_j + \mathcal{R}_j, \qquad j = 0, 1, 2 \dots. \tag{9.15}$$

THEOREM 9.3. *Let $\{\chi_j\}_{j=0}^M$ be a set of linear operators. Then*

$$\mathcal{A}_M \chi_M = \mathcal{A}_0 \chi_0 + \sum_{j=1}^{M} \mathcal{A}_j(\chi_j - \widetilde{\mathcal{A}}_{j-1}\chi_{j-1}) - \sum_{j=0}^{M-1} \mathcal{R}_j \chi_j. \tag{9.16}$$

PROOF. By the approximate factorization identity (9.15), one has

$$\begin{aligned} \mathcal{A}_j \chi_j &= \mathcal{A}_{j-1}\chi_{j-1} + \mathcal{A}_j\chi_j - \mathcal{A}_{j-1}\chi_{j-1} \\ &= \mathcal{A}_{j-1}\chi_{j-1} + \mathcal{A}_j\chi_j - \mathcal{A}_j\widetilde{\mathcal{A}}_{-1}\chi_{j-1} - \mathcal{R}_{j-1}\chi_{j-1} \\ &= \mathcal{A}_{j-1}\chi_{j-1} + \mathcal{A}_j(\chi_j - \widetilde{\mathcal{A}}_{j-1}\chi_{j-1}) - \mathcal{R}_{j-1}\chi_{j-1}\,, \end{aligned}$$

and the assertion follows by induction. □

For the following we denote the distance of $\mathbf{x} \in \Omega$ to $\partial\Omega$ by

$$d(\mathbf{x}) := \operatorname{dist}(\mathbf{x}, \Omega^c) \tag{9.17}$$

and we define the set $\{\chi_j\}_{j=0}^M$ as the collection of operators of multiplication by characteristic functions of the domains $\Omega_{\tau h_j}$, where $\tau \geq 0$ is a free parameter, i.e.,

$$\chi_j u(\mathbf{x}) = \begin{cases} u(\mathbf{x}), & d(\mathbf{x}) > \tau h_j, \\ 0, & \text{otherwise}. \end{cases} \tag{9.18}$$

Denoting

$$\mathcal{B}_M^\star := \mathcal{A}_0\chi_0 + \sum_{j=1}^{M} \mathcal{A}_j(\chi_j - \widetilde{\mathcal{A}}_{j-1}\chi_{j-1})\,, \tag{9.19}$$

one can write

$$\mathcal{B}_M^\star u = \sum_{j=0}^{M} \mathcal{A}_j \tilde{u}_j \qquad \text{with} \quad \tilde{u}_j := \begin{cases} \chi_0 u, & j = 0, \\ \chi_j u - \widetilde{\mathcal{A}}_{j-1}(\chi_{j-1}u), & j \geq 1. \end{cases}$$

Hence the function $\mathcal{B}_M^\star u$ is the sum of quasi-interpolants applied to $\chi_0 u$ and to the discrepancy functions $\tilde{u}_j$, $j = 1, \dots, M$, computed on grids with different step sizes. Note that by (9.16)

$$u - \mathcal{B}_M^\star u = u - \mathcal{A}_0(\chi_0 u) - \sum_{j=1}^{M} \mathcal{A}_j \tilde{u}_j = u - \mathcal{A}_M(\chi_M u) - \sum_{j=0}^{M-1} \mathcal{R}_j(\chi_j u)\,. \tag{9.20}$$

Thus $\mathcal{B}^\star_M u$ approximates u like the quasi-interpolant on the finest grid modulo the small remainder term $\sum_{j=0}^{M-1} \mathcal{R}_j(\chi_j u)$.

9.4.2. Restriction of $\mathcal{A}_j$. Note that by properties of $\widetilde{\mathcal{M}}_{h_{j-1},\mathcal{D}}$, the function

$$\tilde{u}_j = \chi_j u - \widetilde{\mathcal{M}}_{h_{j-1},\mathcal{D}}(\chi_{j-1}u)\,,\ j \ge 1\,,$$

is small in an inner subdomain of Ω. This is used to restrict the application of $\widetilde{\mathcal{A}}_j = \widetilde{\mathcal{M}}_{h_j,\mathcal{D}}$ in (9.19) to a small boundary layer.

More precisely, in view of Lemma 2.23 and (9.4), one can choose $\mathcal{D}$ such that for any $\mathbf{x} \in \Omega_{(\tau+\tilde{\kappa})h_{j-1}}$, where $\tilde{\kappa}$ is determined from (9.12), the estimate

$$\begin{aligned} |\tilde{u}_j(\mathbf{x})| &\le c_{\tilde{\eta}}(\sqrt{\mathcal{D}}h_{j-1})^N \|\nabla_N u\|_{L_\infty(\Omega)} \\ &\quad + \varepsilon\Big(\|u\|_{L_\infty(\Omega)} + \sum_{k=1}^{N-1} (\sqrt{\mathcal{D}}h_{j-1})^k |\nabla_k u(\mathbf{x})|\Big) \end{aligned} \tag{9.21}$$

holds. Hence one can ignore the contribution of $\mathcal{A}_j \tilde{u}_j(\mathbf{x})$ to the sum $\mathcal{B}^\star_M u(\mathbf{x})$ if $d(\mathbf{x}) > (\tau + \tilde{\kappa})h_{j-1}$. Therefore, in the following definition, we introduce the operator $\mathcal{B}_M$ in which the summation is performed layer by layer with only minimal overlapping:

DEFINITION 9.4. Let $\{\widetilde{\chi}_j\}_{j=1}^M$ be the operator sequence

$$\widetilde{\chi}_j u(\mathbf{x}) = \begin{cases} u(\mathbf{x}), & \tau h_j \le d(\mathbf{x}) \le (\tau+\tilde{\kappa})h_{j-1}, \\ 0, & \text{otherwise.} \end{cases}$$

We define the boundary layer approximation operator by

$$\mathcal{B}_M := \mathcal{A}_0\chi_0 + \sum_{j=1}^{M} \mathcal{A}_j \widetilde{\chi}_j(\chi_j - \widetilde{\mathcal{A}}_{j-1}\chi_{j-1}). \tag{9.22}$$

Alternatively, as we indicated in the beginning of this chapter, we can rewrite (9.22) in the form

$$\mathcal{B}_M u(\mathbf{x}) = \sum_{j=0}^{M} \sum_{\mathbf{m}\in\mathcal{Q}_j} c_{j,\mathbf{m}}\, \eta\Big(\frac{\mathbf{x} - h_j\mathbf{m}}{h_j\sqrt{\mathcal{D}}}\Big), \tag{9.23}$$

with the coefficients

$$c_{j,\mathbf{m}} = \begin{cases} u(h_0\mathbf{m}), & j = 0\,, \\ u(h_j\mathbf{m}) - \widetilde{\mathcal{M}}_{h_{j-1},\mathcal{D}}(\chi_{j-1}u)(h_j\mathbf{m}), & j \ge 1\,, \end{cases} \tag{9.24}$$

and the set of indices

$$\mathcal{Q}_j = \begin{cases} \{\mathbf{m} \in \mathbb{Z}^n : \tau h_0 \le d(h_0\mathbf{m})\}, & j = 0, \\ \{\mathbf{m} \in \mathbb{Z}^n : \tau h_j \le d(h_j\mathbf{m}) \le (\tau+\tilde{\kappa})h_{j-1}\}, & j \ge 1. \end{cases}$$

Note that in view of the truncation, the values of $\widetilde{\mathcal{M}}_{h_{j-1},\mathcal{D}}(\chi_{j-1}u)(h_j\mathbf{m})$ for $\mathbf{m} \in \mathcal{Q}_j$ require only the point values $u(h_{j-1}\mathbf{m})$ for $\tau h_{j-1} \le d(h_{j-1}\mathbf{m}) \le (\tau + 2\tilde{\kappa})h_{j-1}$.

REMARK 9.5. The practical implementation of Theorem 9.3 does not require an explicit formula for $\tilde{\eta}$. Indeed, in order to calculate $\mathcal{B}_M u(\mathbf{x})$ by (9.23), one has to compute the coefficients $c_{j,\mathbf{m}}$, i.e., tabulate $(\chi_j - \widetilde{\mathcal{A}}_{j-1}\chi_{j-1})u$ at the points $h_j\mathbf{m}$ (cf. (9.24)). By Remark 2.21, the computation of $\widetilde{A}_{j-1}\chi_{j-1}u(h_j\mathbf{m})$ requires only the summation over the indices $\boldsymbol{\nu}$, for which

$$|h_j\mathbf{m}/h_{j-1} - \boldsymbol{\nu}| = |\mu\mathbf{m} - \boldsymbol{\nu}| \le \tilde{\kappa},$$

where $\tilde{\kappa}$ is such that (2.63) holds. These $(\mu^{-1}(2\kappa+1))^n$ values (or just $\mu^{-1}(2\kappa+1)$, if $\tilde{\eta}$ is a radial function) can be precomputed using the numerical Fourier inversion of (8.12).

9.4.3. Pointwise estimates. For a sufficiently smooth function u on Ω, we consider the approximation error for $\mathbf{x} \in \Omega$. Using (9.19), (9.20), and (9.22), one derives

$$u - \mathcal{B}_M u = u - \mathcal{A}_M(\chi_M u) - \sum_{j=0}^{M-1} \mathcal{R}_j(\chi_j u) + (\mathcal{B}_M^\star - \mathcal{B}_M)u\,. \tag{9.25}$$

Since the characteristic functions satisfy

$$(I - \widetilde{\chi}_j)\chi_j = (I - \widetilde{\chi}_j)\chi_{j-1}\,, \tag{9.26}$$

it follows from (9.19) and (9.22) that

$$\mathcal{B}_M^\star - \mathcal{B}_M = \sum_{j=1}^{M} \mathcal{A}_j(I - \widetilde{\chi}_j)(\chi_j - \widetilde{\mathcal{A}}_{j-1}\chi_{j-1}) = \sum_{j=1}^{M} \mathcal{A}_j(I - \widetilde{\chi}_j)(I - \widetilde{\mathcal{A}}_{j-1})\chi_{j-1}\,. \tag{9.27}$$

LEMMA 9.6. *Let κ and $\tilde{\kappa}$ be such that (9.4) and (9.21), respectively, hold and let $\tau \ge 0$ satisfy $\tau > \mu\kappa - \tilde{\kappa}$. If $d(\mathbf{x}) \ge (\tau + \tilde{\kappa})h_{j-1} - \kappa h_j$, then*

$$|\mathcal{A}_j(I - \widetilde{\chi}_j)(I - \widetilde{\mathcal{A}}_{j-1})(\chi_{j-1}u)(\mathbf{x})| \le \|\rho_0(\cdot, \eta, \mathcal{D})\|_{L_\infty}\big(c(\sqrt{\mathcal{D}}h_{j-1})^N \|\nabla_N u\|_{L_\infty(\Omega)}$$
$$+ 2\,\varepsilon \sum_{k=0}^{N-1} (\sqrt{\mathcal{D}}h_{j-1})^k \|\nabla_k u\|_{L_\infty(\Omega)}\big)$$

with a constant c, depending only on $\tilde{\eta}$, whereas for $d(\mathbf{x}) < (\tau + \tilde{\kappa})h_{j-1} - \kappa h_j$

$$|\mathcal{A}_j(I - \widetilde{\chi}_j)(I - \widetilde{\mathcal{A}}_{j-1})(\chi_{j-1}u)(\mathbf{x})|$$
$$\le \varepsilon\,(1 + \|\rho_0(\cdot, \eta, \mathcal{D})\|_{L_\infty} + \|\rho_0(\cdot, \tilde{\eta}, \mathcal{D})\|_{L_\infty})\|u\|_{L_\infty(\Omega)}.$$

PROOF. Split $(I - \widetilde{\chi}_j)(I - \widetilde{\mathcal{A}}_{j-1})(\chi_{j-1}u)(\mathbf{x}) = v_1(\mathbf{x}) + v_2(\mathbf{x})$ with

$$v_1(\mathbf{x}) := \begin{cases} (I - \widetilde{\mathcal{A}}_{j-1})(\chi_{j-1}u)(\mathbf{x})\,, & d(\mathbf{x}) \ge (\tau + \tilde{\kappa})h_{j-1}, \\ 0, & \textit{otherwise}, \end{cases}$$

$$v_2(\mathbf{x}) := \begin{cases} -\widetilde{\mathcal{A}}_{j-1}(\chi_{j-1}u)(\mathbf{x})\,, & d(\mathbf{x}) < \tau h_j, \\ 0, & \textit{otherwise}. \end{cases}$$

In view of (9.5), we have

$$|\mathcal{A}_j v_1(\mathbf{x})| \le \begin{cases} \|\rho_0(\cdot, \eta, \mathcal{D})\|_{L_\infty} \sup|v_1|, & d(\mathbf{x}) \ge (\tau + \tilde{\kappa})h_{j-1}\,, \\ g_{\mathcal{D}}(h_j^{-1}((\tau + \tilde{\kappa})h_{j-1} - d(\mathbf{x})), \eta) \sup|v_1|, & d(\mathbf{x}) < (\tau + \tilde{\kappa})h_{j-1}\,. \end{cases}$$

By definition, $g_{\mathcal{D}}(\kappa, \eta) \leq \varepsilon$, so that for $d(\mathbf{x}) \leq (\tau + \tilde{\kappa})h_{j-1} - \kappa h_j$

$$|\mathcal{A}_j v_1(\mathbf{x})| \leq \varepsilon(1 + \|\rho_0(\cdot, \tilde{\eta}, \mathcal{D})\|_{L_\infty})\|u\|_{L_\infty(\Omega)} .$$

Furthermore, by (9.21)

$$|v_1(\mathbf{x})| \leq c_{\tilde{\eta}}(\sqrt{\mathcal{D}}h_{j-1})^N \|\nabla_N u\|_{L_\infty(\Omega)} + \varepsilon\big(\|u\|_{L_\infty(\Omega)} + \sum_{k=1}^{N-1}(\sqrt{\mathcal{D}}h_{j-1})^k |\nabla_k u(\mathbf{x})|\big)$$

which implies obviously for the case $d(\mathbf{x}) > (\tau + \tilde{\kappa})h_{j-1} - \kappa h_j$

$$|\mathcal{A}_j v_1(\mathbf{x})|$$

$$\leq \|\rho_0(\cdot, \eta, \mathcal{D})\|_{L_\infty} \big(c_{\tilde{\eta}}(\sqrt{\mathcal{D}}h_{j-1})^N \|\nabla_N u\|_{L_\infty(\Omega)} + \varepsilon \sum_{k=0}^{N-1}(\sqrt{\mathcal{D}}h_{j-1})^k \|\nabla_k u\|_{L_\infty(\Omega)}\big) .$$

To estimate $v_2(\mathbf{x})$ in the case $d(\mathbf{x}) < \tau h_j$, we note that $\operatorname{dist}(\mathbf{x}, \operatorname{supp} \chi_{j-1}) \geq (\tau + \tilde{\kappa})h_{j-1} - \tau h_j$. Hence by Lemma 2.22

$$|v_2(\mathbf{x})| \leq g_{\mathcal{D}}(\tau(1 - \mu) + \tilde{\kappa}, \tilde{\eta}) \sup_{\Omega} |u| ,$$

which leads to

$$|\mathcal{A}_j v_2(\mathbf{x})| \leq \varepsilon \|\rho_0(\cdot, \eta, \mathcal{D})\|_{L_\infty} \|u\|_{L_\infty(\Omega)} . \qquad \square$$

THEOREM 9.7. *Suppose the generating function $\eta \in \mathcal{S}(R^n)$ satisfies the assumptions of Theorem 8.1 and the moment Condition 2.15 of order N. Let $u \in C^N(\overline{\Omega})$ and let $\varepsilon > 0$ be given. There exist positive $\mathcal{D} > 0$, κ, and $\tilde{\kappa}$ such that for any non-negative $\tau > \kappa\mu - \tilde{\kappa}$, the boundary layer quasi-interpolation operator $\mathcal{B}_M$ defined by (9.22) satisfies the estimate*

$$|(I - \mathcal{B}_M)u(\mathbf{x})| \leq c(\sqrt{\mathcal{D}}h_j)^N \|\nabla_N u\|_{L_\infty(\Omega)} + \varepsilon \sum_{k=0}^{N-1}(\sqrt{\mathcal{D}}h_j)^k \|\nabla_k u\|_{L_\infty(\Omega)} ,$$

where the index j can be determined by

$j = 0,$	*if*	$d(\mathbf{x}) > (\tau + \tilde{\kappa} - \kappa\mu)h_0 ,$
$j = 1, \dots, M-1,$	*if*	$(\tau + \tilde{\kappa} - \kappa\mu)h_j < d(\mathbf{x}) \leq (\tau + \tilde{\kappa} - \kappa\mu)h_{j-1} ,$
$j = M,$	*if*	$(\tau + \max(\tilde{\kappa} - \kappa\mu, \kappa))h_M < d(\mathbf{x}) \leq (\tau + \tilde{\kappa} - \kappa\mu)h_{M-1} ,$

where $d(\mathbf{x})$ is the distance from $\mathbf{x}$ to $\partial\Omega$.

PROOF. First, we choose $\mathcal{D}$ large enough and such that the saturation errors of the quasi-interpolants $\mathcal{M}_{h,\mathcal{D}}$ and $\widetilde{\mathcal{M}}_{h,\mathcal{D}}$ are less than a sufficiently small δ. Also let the sum of the remainders satisfy

$$\Big|\sum_{j=0}^{M-1} \mathcal{R}_j(\chi_j u)(\mathbf{x})\Big| \leq \delta \|u\|_{L_\infty(\Omega)} ,$$

which is possible, in view of (9.13). Then we choose values of the parameters κ and $\tilde{\kappa}$ such that corresponding quasi-interpolants satisfy the estimate (9.4) with the given δ.

It remains to estimate the terms of the decomposition (9.25). Since by Corollary 2.24,

$$|u - \mathcal{A}_M(\chi_M u)(\mathbf{x})| \le c_\eta(\sqrt{\mathcal{D}}h_M)^N \, \|\nabla_N u\|_{L_\infty(\Omega)} + \varepsilon \sum_{k=0}^{N-1} (\sqrt{\mathcal{D}}h_M)^k |\nabla_k u(\mathbf{x})|$$

if $d(\mathbf{x}) \ge (\tau + \kappa)h_M$, the application of Lemma 9.6 completes the proof. □

Thus, the behavior of $\mathcal{B}_M u(\mathbf{x})$ is actually very close to that of $\mathcal{A}_j u(\mathbf{x})$ for some positive $j \le M$, where j increases as the distance from $\mathbf{x}$ to the boundary decreases. This leads to the effect that the approximation becomes better in the points $\mathbf{x} \in \Omega_{(\tau+\kappa)h_M}$ which lie nearer the boundary $\partial\Omega$.

9.4.4. Numerical examples. We give some numerical examples to illustrate the overall approximation properties of the operator $\mathcal{B}_M$ defined by (9.22), and especially the behavior of the error near the boundary. We shall use the boundary layer quasi-interpolation (9.23) generated by the functions η_2, η_4, η_6 based on the Gaussian (see (3.12)), providing the second-, fourth-, and sixth-orders of approximate convergence. The corresponding mask functions $\tilde{\eta}_2$, $\tilde{\eta}_4$, $\tilde{\eta}_6$ are given by (8.14). In all cases, we use $\mathcal{D} = 3$, which assures saturation levels of magnitude 10^{-12}, 10^{-11}, and 10^{-10} for quasi-interpolants $\mathcal{M}_{h,\mathcal{D}}$ based on η_2, η_4, η_6, respectively. The step refinement ratio in all examples is $\mu^{-1} = 3$.

We recall that by Theorem 9.7, $\mathcal{B}_M$ performs approximately as $\mathcal{A}_j$ on the j-th boundary strip $(\tau + \tilde{\kappa} - \kappa\mu)h_j < d(\mathbf{x}) \le (\tau + \tilde{\kappa} - \kappa\mu)h_{j-1}$, i.e., the nearer the boundary, the better the approximation. The approximation results are plotted over the boundary layer

$$\{\mathbf{x} \in \Omega : (\tau + \kappa)h_{M+1} \le d(\mathbf{x}) \le (\tau + \tilde{\kappa} - \kappa\mu)h_0\}$$

in order to illustrate the interplay between the different quasi-interpolants building the operator $\mathcal{B}_M$. Since the step-size used by $\mathcal{B}_M$ is proportional to the distance from the boundary, one can determine the order of the formula used by the slope of the error plot $|(I - \mathcal{B}_M)u|$ against the distance to the boundary in logarithmic scales.

Consider the plot in Fig. 9.2 showing the error from the approximation of $\cos(1000x)$ near the boundary using the second-order formula based on the Gaussian. One can clearly see the stepwise increase of the accuracy towards the boundary until a saturation is reached. The error remains unchanged within a boundary strip, since the step does not change there. Observe also the slope of the "staircase" — it is approximately two.

In Fig. 9.3 the same function is approximated using the sixth-order formula based on η_6. Here, the slope is approximately 6 : 1, but the saturation error is higher.

The plot in Fig. 9.4 shows the results for the approximation of the function $\log(x)$ near the origin again using the formulas of $\mathcal{O}(h^2)$-, $\mathcal{O}(h^4)$-, and $\mathcal{O}(h^6)$-orders of approximate convergence. Note that in contrast to the previous examples the absolute error $|(I - \mathcal{B}_M)\log(x)|$ does not decrease as the mesh size becomes finer near the origin. This is due to the fact that the second, fourth and sixth derivatives of the logarithmic function grow as x^{-2}, x^{-4}, and x^{-6} as $x \to 0$.

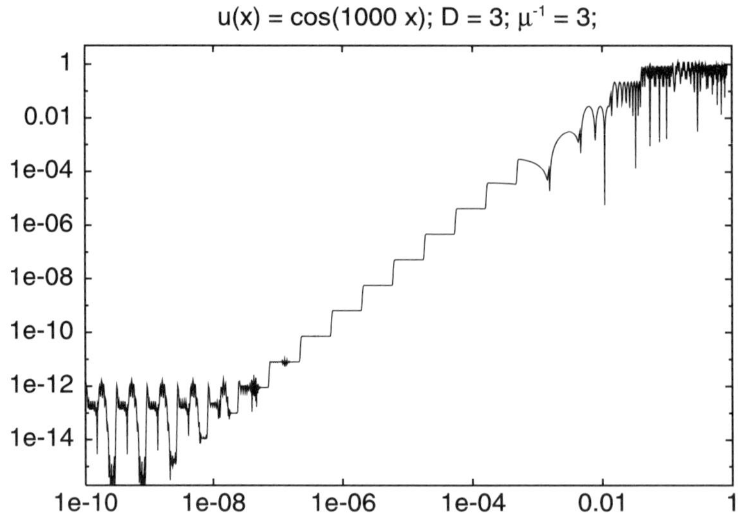

FIGURE 9.2. Error plot for $(I - \mathcal{B}_M)\cos(1000x)$ using boundary layer approximations of order $\mathcal{O}(h^2)$.

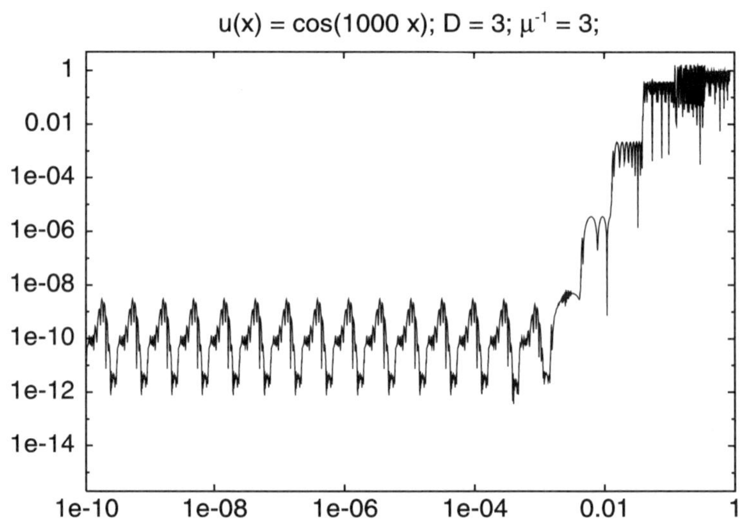

FIGURE 9.3. Error plot for $(I - \mathcal{B}_M)\cos(1000x)$ using boundary layer approximations of order $\mathcal{O}(h^6)$.

The second example represents boundary error plots for the approximation of the function

$$u(x_1, x_2) = \begin{cases} \cos(100\,|\mathbf{x}|^2), & x_1 > 0, x_2 > 0, \\ 0, & \text{otherwise}, \end{cases}$$

as an illustration for the action of a two-dimensional operator built as the tensor product of two one-dimensional operators $\mathcal{B}_M$ acting on the arguments of

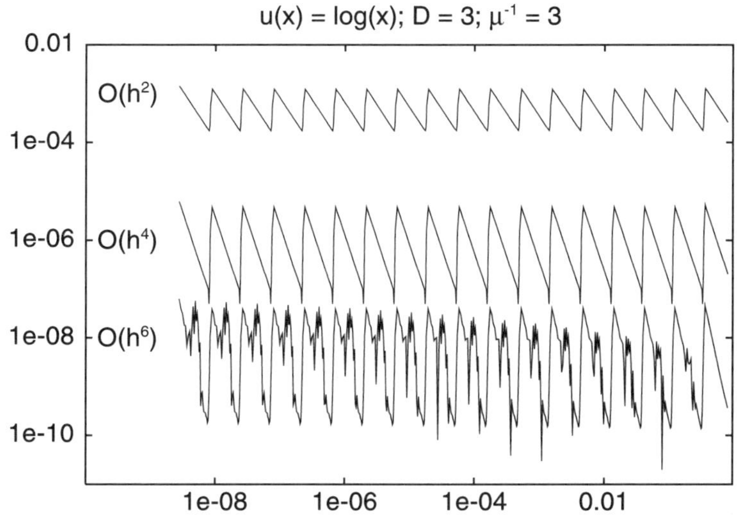

FIGURE 9.4. Boundary layer error plots using second-, fourth-, and sixth-order formulas for $(I - \mathcal{B}_M)\log x$.

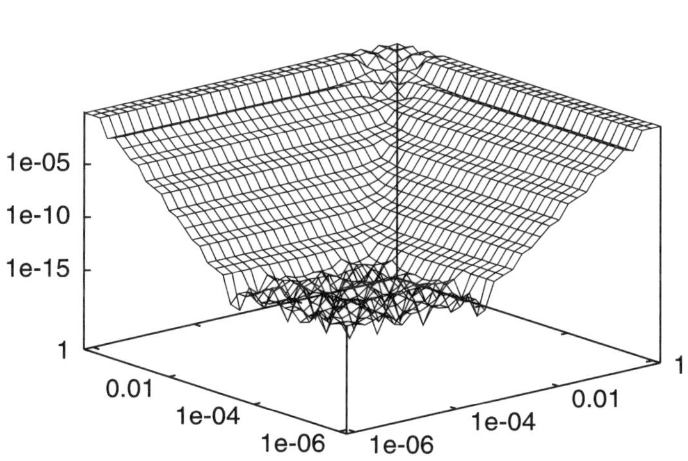

FIGURE 9.5. Boundary layer error plot for the function $\cos(100\,|\mathbf{x}|^2)$ with support on the first quadrant of $\mathbb{R}^2$. We use the tensor product of one-dimensional multi-resolution operators providing the order $\mathcal{O}(h^2)$.

$\mathbf{x} = (x_1, x_2)$. These one-dimensional operators are based on the generating functions η_2 and η_6, which provide the approximate order of convergence of $\mathcal{O}(h^2)$ and $\mathcal{O}(h^6)$, respectively. Similarly to the previous examples, we use $\mathcal{D} = 3$ and the step refinement ratio in all examples is $\mu^{-1} = 3$ in both the x_1- and x_2-directions. Again, the approximation results are plotted in logarithmic scales only in the interesting area near the vertex of the angle.

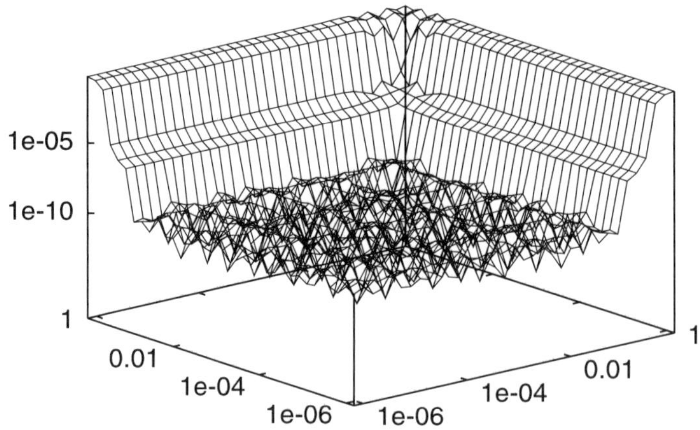

FIGURE 9.6. Boundary layer error plot for the function $\cos(100\,|\mathbf{x}|^2)$ with support on the first quadrant of $\mathbb{R}^2$. The tensor product of one-dimensional multi-resolution operators providing the order $\mathcal{O}(h^6)$ is used.

Precisely as in the one-dimensional examples, one can see the gradual increase of accuracy in the direction towards the boundary when the second-order formula is used (Fig. 9.5). The plot in Fig. 9.6 shows the approximation results, when the sixth-order formula is used. In this case, the saturation level is reached already after two iterations.

9.4.5. L_p-estimates for quasi-interpolants of functions in domains. We estimate the error of the boundary layer quasi-interpolation (9.22) in the L_p-norm. These estimates are the basis for the convergence studies of the cubature formulas (9.3); therefore we will need them on subdomains and on the whole space.

We start with some simple estimates:

LEMMA 9.8. *The multiplication operator with the characteristic function $\mathcal{X}_{S_h}$ of the boundary layer $S_h := \{\mathbf{x} \in \Omega : \operatorname{dist}(\mathbf{x}, \partial\Omega) < h\}$ can be estimated by*

$$\|\mathcal{X}_{S_h} u\|_{L_p(\Omega)} \leq c h^{(q-p)/pq} \|u\|_{L_q(\Omega)}, \qquad 1 \leq p \leq q \leq \infty\,, \tag{9.28}$$

$$\|\mathcal{X}_{S_h} u\|_{L_p(\Omega)} \leq c h^{r} \|u\|_{W_p^s(\Omega)}, \qquad 1 \leq p < \infty,\ 0 < r < s/n,\ r \leq 1/p\,, \tag{9.29}$$

$$\|\mathcal{X}_{S_h} u\|_{(W_p^s(\Omega))'} \leq c h^{r} \|u\|_{L_{p/(p-1)}(\Omega)}, \quad 1 \leq p < \infty,\ 0 < r < s/n,\ r \leq 1/p\,, \tag{9.30}$$

with constants depending only on Ω.

Here, $(W_p^s(\Omega))'$ denotes the dual space of $W_p^s(\Omega)$ with respect to the L_2 inner product.

PROOF. The first inequality follows with $q = pt$, $t \geq 1$, from

$$\int_\Omega |\mathcal{X}_{S_h} u|^p \, d\mathbf{x} \leq \big(\operatorname{meas} S_h\big)^{(t-1)/t} \, \|u\|^p_{L_{pt}(\Omega)}.$$

To prove (9.29), we note that $u \in W_p^t(\Omega)$, $t > n/p$, implies $u \in C(\overline{\Omega})$ by Sobolev's imbedding theorem. Hence,

$$\int_\Omega |\mathcal{X}_{S_h} u|^p \, d\mathbf{x} \leq \max_{\mathbf{x} \in S_h} |u(\mathbf{x})|^p \operatorname{meas} S_h \leq c \operatorname{meas} S_h \, \|u\|^p_{W_p^t(\Omega)} ,$$

so that

$$\|\mathcal{X}_{S_h} u\|_{L_p(\Omega)} \leq c h^{1/p} \, \|u\|_{W_p^t(\Omega)} ,$$

and by interpolation

$$\|\mathcal{X}_{S_h} u\|_{L_p(\Omega)} \leq c h^{\theta/p} \, \|u\|_{W_p^{\theta t}(\Omega)}, \qquad 0 \leq \theta \leq 1, \quad t > n/p.$$

Setting $r = \theta/p$ and $s = prt$ yields (9.29). Finally, since the operator $\mathcal{X}_{S_h}$ is symmetric, the estimate

$$\|\mathcal{X}_{S_h}\|_{L_{p/(p-1)}(\Omega) \to (W_p^s(\Omega))'} = \|\mathcal{X}_{S_h}\|_{W_p^s(\Omega) \mapsto L_p(\Omega)}$$

holds, which proves (9.30) and the lemma. □

Next, we consider the quasi-interpolant $\mathcal{M}_{h,\mathcal{D}} u$ where the function u is given in a domain Ω. In Lemma 2.30, we have considered the L_p-error in subdomains, so it remains to consider the behavior of $\mathcal{M}_{h,\mathcal{D}} u$ outside Ω and in the boundary layer.

To characterize the behavior of $\mathcal{M}_{h,\mathcal{D}} u$ outside $\operatorname{supp} u$, we use the functions

$$H\eta(t) := \int_{|\mathbf{x}|>t} |\eta(\mathbf{x})| \, d\mathbf{x}$$

and $g_\mathcal{D}(t, \eta)$, which was defined in (2.55). Since $\eta \in \mathcal{S}(\mathbb{R}^n)$, both functions decay to zero faster than any negative power of t.

LEMMA 9.9. *Consider two disjoint domains in Ω_1, $\Omega_2 \subset \mathbb{R}^n$ and let u be a bounded function with* $\operatorname{supp} u \subseteq \overline{\Omega_2}$. *Then*

$$\|\mathcal{M}_{h,\mathcal{D}} u\|_{L_p(\Omega_1)} \leq g_\mathcal{D}(h^{-1}\tau, \eta)^{1/q} \, H\eta((\sqrt{\mathcal{D}} h)^{-1}\tau)^{1/p} \, \|u\|_{p,h} \tag{9.31}$$

where $\tau := \operatorname{dist}(\Omega_1, \Omega_2)$ and $q = p/(p-1)$. If Ω_2 is bounded, then

$$\|\mathcal{M}_{h,\mathcal{D}} u\|_{L_p(\Omega_1)} \leq c \, g_\mathcal{D}(h^{-1}\tau, \eta)^{1/q} \, H\eta((\sqrt{\mathcal{D}} h)^{-1}\tau)^{1/p} \sup_{h\mathbf{m} \in \Omega_2} |u(h\mathbf{m})| \tag{9.32}$$

with a constant c depending on Ω_2.

PROOF. By Hölder's inequality,

$$|\mathcal{M}_{h,\mathcal{D}} u(\mathbf{x})|^p \leq \mathcal{D}^{-np/2} \Big(\sum_{h\mathbf{m} \in \Omega_2} \Big|\eta\Big(\frac{\mathbf{x} - h\mathbf{m}}{\sqrt{\mathcal{D}}\, h}\Big)\Big| \Big)^{p/q} \sum_{h\mathbf{m} \in \Omega_2} \Big|\eta\Big(\frac{\mathbf{x} - h\mathbf{m}}{\sqrt{\mathcal{D}}\, h}\Big)\Big| \, |u(h\mathbf{m})|^p ,$$

so that

$$\int_{\Omega_1} |\mathcal{M}_{h,\mathcal{D}} u(\mathbf{x})|^p$$

$$\le \mathcal{D}^{-np/2} \sup_{\mathbf{x}\in\Omega_1} \Big(\sum_{h\mathbf{m}\in\Omega_2} \Big|\eta\Big(\frac{\mathbf{x}-h\mathbf{m}}{\sqrt{\mathcal{D}}\,h}\Big)\Big|\Big)^{p/q} \int_{\Omega_1} \sum_{h\mathbf{m}\in\Omega_2} \Big|\eta\Big(\frac{\mathbf{x}-h\mathbf{m}}{\sqrt{\mathcal{D}}\,h}\Big)\Big|\, |u(h\mathbf{m})|^p \, d\mathbf{x}$$

$$\le \mathcal{D}^{-np(1-1/q)/2} g_{\mathcal{D}}(h^{-1}\tau,\eta)^{p/q} \sum_{h\mathbf{m}\in\Omega_2} |u(h\mathbf{m})|^p \int_{\Omega_1} \Big|\eta\Big(\frac{\mathbf{x}-h\mathbf{m}}{\sqrt{\mathcal{D}}\,h}\Big)\Big|\, d\mathbf{x}.$$

Since

$$\int_{\Omega_1} \Big|\eta\Big(\frac{\mathbf{x}-h\mathbf{m}}{\sqrt{\mathcal{D}}\,h}\Big)\Big|\, d\mathbf{x} \le \int_{|\mathbf{x}|>\mathrm{dist}(h\mathbf{m},\Omega_1)} \Big|\eta\Big(\frac{\mathbf{x}}{\sqrt{\mathcal{D}}\,h}\Big)\Big|\, d\mathbf{x} \le (\sqrt{\mathcal{D}}\,h)^n H\eta((\sqrt{\mathcal{D}}h)^{-1}\tau),$$

(9.31) follows. Note further that for bounded Ω_2

$$h^n \sum_{h\mathbf{m}\in\Omega_2} \longrightarrow \operatorname{meas}\Omega_2 \quad \text{as} \quad h\to 0.$$

Hence,

$$\|u\|_{p,h} = \Big(h^n \sum_{h\mathbf{m}\in\Omega_2} |u(h\mathbf{m})|^p\Big)^{1/p} \le c \operatorname{meas}\Omega_2 \sup_{h\mathbf{m}\in\Omega_2} |u(h\mathbf{m})|,$$

which proves (9.32). □

COROLLARY 9.10. *Let $\tau > 0$ and let $\chi_{\tau h}$ be the characteristic function of $\Omega_{\tau h}\subset\Omega$. Then*

$$\|\mathcal{M}_{h,\mathcal{D}}(\chi_{\tau h} u)\|_{L_p(\mathbb{R}^n\setminus\Omega)} \le c_\Omega\, g_{\mathcal{D}}(\tau,\eta)^{1/q}\, H\eta(\mathcal{D}^{-1/2}\tau)^{1/p} \|u\|_{L_\infty(\Omega)}.$$

LEMMA 9.11. *For a bounded domain Ω*

$$\|(I-\mathcal{M}_{h,\mathcal{D}})u\|_{L_p(\Omega\setminus\Omega_{\tau h})} \le c_\Omega(\tau h)^{1/p}\big(1+\|\rho_{\mathbf{0}}(\cdot,\eta,\mathcal{D})\|_{L_\infty(\mathbb{R}^n)}\big)\|u\|_{L_\infty(\Omega)},$$

where the constant c_Ω depends only on the domain Ω.

PROOF. With the notation $S_h = \Omega\setminus\Omega_{\kappa h}$,

$$\begin{aligned}\|(I-\mathcal{M}_{h,\mathcal{D}})u\|_{L_p(S_h)} &\le \|(I-\mathcal{M}_{h,\mathcal{D}})u\|_{L_\infty(S_h)}(\operatorname{meas} S_h)^{1/p}\\ &\le \big(1+\|\rho_{\mathbf{0}}(\cdot,\eta,\mathcal{D})\|_{L_\infty(\mathbb{R}^n)}\big)(\operatorname{meas} S_h)^{1/p}\sup_\Omega |u|.\end{aligned}$$ □

Lemmas 2.30 and 9.11 and Corollary 9.10 give L_p-estimates for the quasi-interpolation error on the whole of $\mathbb{R}^n$. Corollary 9.10 assesses the error accumulated outside Ω of the quasi-interpolant applied to the restriction of u to $\Omega_{\tau h}\subset\Omega$. Since η is in the Schwartz class $\mathcal{S}(\mathbb{R}^n)$, the product $g_{\mathcal{D}}(\tau,\eta)^{1/q} H\eta(\mathcal{D}^{-1/2}\tau)^{1/p}$ can be made of the same order of magnitude as the saturation error ε, by choosing τ larger. Note that ε is controlled by the parameter $\mathcal{D}$.

Then by Lemma 2.30, the quasi-interpolant $\mathcal{M}_{h,\mathcal{D}}(\chi_{\tau h}u)$ is a good approximation of u at internal points, lying at a distance larger than $(\tau+\kappa)h$ from the boundary. The error is then of order $\mathcal{O}\big((\sqrt{\mathcal{D}}h)^N\big)+\varepsilon$ and can be controlled effectively by a proper choice of the step size h.

Thus, the main contribution to the overall error comes from the boundary strip $\Omega\setminus\Omega_{(\tau+\kappa)h}$, where, by Lemma 9.11, the error is of order $\mathcal{O}(h^{1/p})$ if u does not vanish on $\partial\Omega$. Since it will be numerically very expensive to make this term small by choosing h smaller, especially in higher space dimensions, we again use the boundary layer quasi-interpolation operator $\mathcal{B}_M$ to reduce the size of the boundary strip.

9.4.6. L_p-estimates for boundary layer quasi-interpolation. We estimate $\|(I-\mathcal{B}_M)u\|_{L_p(\mathbb{R}^n)}$, where $\mathcal{B}_M$ is defined by (9.22). Similarly to Theorem 9.7, one can prove that

$$\|(I-\mathcal{B}_M)u(\mathbf{x})\|_{L_p(\Omega_M)} \le c\,(\sqrt{\mathcal{D}}h)^N\,\|\nabla_N u\|_{L_p(\Omega)} + \varepsilon \sum_{k=0}^{N-1} (\sqrt{\mathcal{D}}h)^k \|\nabla_k u\|_{L_p(\Omega)}\,,$$

where $\Omega_M = \{\mathbf{x}\in\Omega : d(\mathbf{x}) > (\tau + \max(\tilde\kappa - \kappa\mu, \kappa))h_M\}$. So it remains to estimate $\|\mathcal{B}_M u(\mathbf{x})\|_{L_p(\Omega^c)}$ and $\|(I-\mathcal{B}_M)u(\mathbf{x})\|_{L_p(\Omega\setminus\Omega_M)}$. From Corollary 9.10, the representation

$$\mathcal{B}_M u = \mathcal{A}_0(\chi_0 u) + \sum_{j=1}^{M} \mathcal{A}_j \widetilde{\chi}_j (\chi_j - \widetilde{\mathcal{A}}_{j-1}\chi_{j-1})u$$

and the definitions of χ_0 and $\widetilde{\chi}_j$ we obtain directly that

$$\|\mathcal{B}_M u\|_{L_p(\Omega^c)} \le c_\Omega\, M\, g_{\mathcal{D}}(\tau,\eta)^{1/q}\, H\eta(\mathcal{D}^{-1/2}\tau)^{1/p}\|u\|_{L_\infty(\Omega)}\,.$$

Furthermore, by (9.25), the difference $(I-\mathcal{B}_M)u(\mathbf{x})$ at $\Omega\setminus\Omega_M$ is the sum of $(I-\mathcal{A}_M)(\chi_M u)(\mathbf{x})$ and terms which are less than the saturation error; hence from Lemma 9.11, we conclude that

$$\begin{aligned}&\|(I-\mathcal{B}_M)u(\mathbf{x})\|_{L_p(\Omega\setminus\Omega_M)}\\ &\quad\le c_\Omega(\tau + \max(\tilde\kappa-\kappa\mu,\kappa)h_M)^{1/p}\left(1+\|\rho_{\mathbf{0}}(\cdot,\eta,\mathcal{D})\|_{L_\infty(\mathbb{R}^n)}\right)\|u\|_{L_\infty(\Omega)}\,.\end{aligned}$$

THEOREM 9.12. *Let the generating function $\eta\in\mathcal{S}(R^n)$ satisfy the assumptions of Theorem 8.1 and the moment Condition 2.15 of order N. Suppose that $\Omega\subset\mathbb{R}^n$ is bounded with Lipschitz boundary and let $u\in W_p^N(\Omega)$ with $N>n/p$. For any $\varepsilon>0$, there exist $\mathcal{D}>0$ and a boundary layer approximation $\mathcal{B}_M$ such that*

$$\|u-\mathcal{B}_M u\|_{L_p(\mathbb{R}^n)} \le c_1(\mathcal{D}h)^N\|\nabla_N u\|_{L_p(\Omega)} + c_2(\mu^M h)^{1/p}\|u\|_{L_\infty(\Omega)} + \varepsilon\|u\|_{W_p^{N-1}(\Omega)}\,.$$

Therefore, for a given μ, the choice

$$M \approx (Np-1)\frac{\log h}{\log\mu}$$

guarantees that $\|u-\mathcal{B}_M u\|_{L_p(\mathbb{R}^n)}$ behaves as the usual quasi-interpolant generated by η.

9.4.7. Estimates in weak norms. We saw in Subsection 4.4.2 that the quasi-interpolation on uniform meshes converges in weak norms because the saturation error, which is caused by fast oscillating functions, converges weakly to zero. The same property holds for the case of non-uniform meshes considered here. For the proof of Theorem 9.7, the approximation error $(I-\mathcal{B}_M)u$ is decomposed as follows:

$$(I-\mathcal{B}_M)u = (I-\mathcal{A}_M\chi_M)u - \sum_{j=0}^{M-1}\mathcal{R}_j(\chi_j u) + (\mathcal{B}_M^\star - \mathcal{B}_M)u\,.$$

The second term consists of convolutions with small oscillating functions, whereas $(\mathcal{B}_M^\star - \mathcal{B}_M)u$, by Lemma 9.6, consists of functions with L_p-norms which do not exceed $c(\mathcal{D}h)^N \|\nabla_N u\|_{L_p(\Omega)}$, plus small oscillating functions.

Therefore one can show similarly to Theorem 4.6, by using the results of Subsection 9.4.6, that for $s > 0$ the norm in the Bessel potential space H_p^{-s} is bounded by

$$\Big\| \sum_{j=0}^{M-1} \mathcal{R}_j(\chi_j u) - (\mathcal{B}_M^\star - \mathcal{B}_M)u \Big\|_{H_p^{-s}} \leq c_\eta (\sqrt{\mathcal{D}}h)^N \|\nabla_N u\|_{L_p(\Omega)}$$
$$+ c_s\, h^s \sum_{[\boldsymbol{\alpha}]=0}^{N-1} (\sqrt{\mathcal{D}}h)^{[\boldsymbol{\alpha}]} \frac{\varepsilon_{\boldsymbol{\alpha}}(\eta, \mathcal{D})}{\boldsymbol{\alpha}!} \|\partial^{\boldsymbol{\alpha}} u\|_{L_p(\Omega)} \,.$$

Thus, it remains to estimate $\|(I - \mathcal{A}_M \chi_M)u\|_{H_p^{-s}}$. For integer $s > 0$, we have

$$\|(I - \mathcal{A}_M \chi_M)u\|_{H_p^{-s}} \leq c\big(\|\mathcal{A}_M \chi_M u\|_{L_p(\Omega^c)} + \|(I - \mathcal{A}_M \chi_M)u\|_{(W_q^s(\Omega))'}\big)$$

with $q = p/(p-1)$. Denoting by χ_{M+1} the characteristic function of the boundary layer $\Omega \setminus \Omega_{(\tau+\kappa)h_M}$, one gets from Lemma 9.8 that for $0 < r < s/n$, $r \leq 1/q$,

$$\|\chi_{M+1}(I - \mathcal{A}_M \chi_M)u\|_{(W_q^s(\Omega))'} \leq c h_M^r \|\chi_{M+1}(I - \mathcal{A}_M \chi_M)u\|_{L_p(\Omega)}$$
$$\leq c h_M^{r+1/p} \|u\|_{W_p^N(\Omega)} \,.$$

Furthermore,

$$\|(I - \chi_{M+1})(I - \mathcal{A}_M \chi_M)u\|_{(W_q^s(\Omega))'} = \sup_{\|\varphi\|_{W_q^s(\Omega)}=1} \Big| \int_{\Omega \setminus S_{M+1}} (I - \mathcal{A}_M \chi_M)u\, \varphi\, d\mathbf{x} \Big|$$
$$\leq c_\eta (\sqrt{\mathcal{D}}h_M)^N \|\nabla_N u\|_{L_p(\Omega)} + c_s\, h_M^s \sum_{[\boldsymbol{\alpha}]=0}^{N-1} (\sqrt{\mathcal{D}}h_M)^{[\boldsymbol{\alpha}]} \frac{\varepsilon_{\boldsymbol{\alpha}}(\eta, \mathcal{D})}{\boldsymbol{\alpha}!} \|\partial^{\boldsymbol{\alpha}} u\|_{L_p(\Omega)} \,,$$

so that the following approximation result is valid.

THEOREM 9.13. *Under the assumptions of Theorem 9.12 for any $\varepsilon > 0$, there exist $\mathcal{D} > 0$ and a boundary layer quasi-interpolant $\mathcal{B}_M$ such that*

$$\|u - \mathcal{B}_M u\|_{H_p^{-s}(\mathbb{R}^n)} \leq (c_1(\sqrt{\mathcal{D}}h)^N + c_2(\mu^M h)^{1/p+r}) \|u\|_{W_p^N(\Omega)} + \varepsilon\, h^s\, \|u\|_{W_p^{N-1}(\Omega)} \,,$$

where $0 < r < s/n$ and $r \leq (p-1)/p$.

9.5. Cubature of potentials in domains

In this section, we derive some estimates for the cubature of integral operators that often appear in problems of mathematical physics. As mentioned in the beginning of this chapter, the cubature formula $\mathcal{K}_h u$ for the integral operator

$$\mathcal{K}u(\mathbf{x}) = \int_\Omega k(\mathbf{x} - \mathbf{y}) u(\mathbf{y}) d\mathbf{y}$$

is easily obtained from the boundary layer quasi-interpolation of the density

$$\mathcal{K}_h u(\mathbf{x}) = \mathcal{K}\mathcal{B}_M u(\mathbf{x}) = \sum_{k=0}^{M} \sum_{h_k \mathbf{m} \in \mathcal{Q}_k} c_{k,\mathbf{m}} \int_{\mathbb{R}^n} k(\mathbf{x} - \mathbf{y}) \eta\left(\frac{\mathbf{y} - h_k \mathbf{m}}{h_k \sqrt{D}}\right) d\mathbf{y} \,, \tag{9.33}$$

if η is chosen such that the integrals can be obtained analytically or by simple one-dimensional quadrature.

Many interesting operators are bounded mappings

$$\mathcal{K} : L_p(\Omega) \to W_p^m(\Omega_1)\,, \tag{9.34}$$

with some domains $\Omega,\ \Omega_1 \subset \mathbb{R}^n$, i.e., $\mathcal{K} \in \mathcal{L}\big(L_p(\Omega), W_p^m(\Omega_1)\big)$. Note that the case $m = 0$ corresponds to singular integral operators, whereas the volume potentials associated with elliptic partial differential equations satisfy relation (9.34) with $m > 0$. If the operator $\mathcal{K}$ is such that (9.34) holds with $\Omega = \Omega_1 = \mathbb{R}^n$, Theorems 9.12 and 9.13 imply

THEOREM 9.14. *Suppose that $\eta \in \mathcal{S}(R^n)$ satisfies the assumptions of Theorem 8.1 and the moment Condition 2.15 of order N. Let $u \in W_p^N(\Omega)$ with $N > n/p$ and $\mathcal{K} \in \mathcal{L}(L_p(\mathbb{R}^n), H_p^m(\mathbb{R}^n))$. For any $\varepsilon > 0$ there exists $\mathcal{D} > 0$ such that*

$$\begin{aligned}\|\mathcal{K}u - \mathcal{K}_h u\|_{H_p^m(\mathbb{R}^n)} \le{}& c_1(\sqrt{\mathcal{D}}h)^N \|\nabla_N u\|_{L_p(\Omega)} \\ &+ c_2(\mu^M h)^{1/p}\|u\|_{L_\infty(\Omega)} + \varepsilon\|u\|_{W_p^{N-1}(\Omega)}\,.\end{aligned}$$

If additionally $\mathcal{K} \in \mathcal{L}(H_p^{-m}(\mathbb{R}^n), L_p(\mathbb{R}^n))$, then

$$\begin{aligned}\|\mathcal{K}u - \mathcal{K}_h u\|_{L_p(\mathbb{R}^n)} \le{}& (c_1(\sqrt{\mathcal{D}}h)^N \\ &+ c_2(\mu^M h)^{1/p+r})\|u\|_{W_p^N(\Omega)} + \varepsilon\, h^m\, \|u\|_{W_p^{N-1}(\Omega)}\,,\end{aligned}$$

where $0 < r < m/n$, $r \le (p-1)/p$.

However, very often, the integral operator $\mathcal{K}$ satisfies (9.34) only for bounded domains $\Omega,\ \Omega_1 \subset \mathbb{R}^n$. Important examples are the harmonic or elastic potentials. In this case, we are interested in the estimation of $\mathcal{K}u - \mathcal{K}_h u$ on some bounded domain Ω_1. Since, in general, $\operatorname{supp}\mathcal{B}_M u = \mathbb{R}^n$, we have to consider integrals of the form

$$\int\limits_{\Omega^c} k(\mathbf{x}-\mathbf{y})\mathcal{B}_M u(\mathbf{y})d\mathbf{y}\,, \quad \mathbf{x} \in \Omega_1\,.$$

To this end, we choose a ball B_R with radius R centered at the origin, which contains Ω and Ω_1. We suppose that the kernel satisfies the estimate

$$|\partial^{\boldsymbol{\alpha}} k(\mathbf{x}-\mathbf{y})| \le r_{\boldsymbol{\alpha}}(|\mathbf{y}|)\,, \quad \mathbf{x} \in \Omega_1\,,\ \mathbf{y} \in \mathbb{R}^n \setminus B_R\,, \tag{9.35}$$

for all multi-indexes $0 \le [\boldsymbol{\alpha}] \le m$ and some functions $r_{\boldsymbol{\alpha}}(x)$ of at most polynomial growth.

LEMMA 9.15. *For any $N > 0$, there exist constants $c_{N,\boldsymbol{\alpha},R}$ such that*

$$\Big\| \int\limits_{\mathbb{R}^n\setminus B_R} \partial^{\boldsymbol{\alpha}} k(\cdot-\mathbf{y})\,\mathcal{B}_M u(\mathbf{y})\,d\mathbf{y}\Big\|_{L_p(\Omega_1)} \le c_{N,\boldsymbol{\alpha},R}\, h^N (\operatorname{meas}\Omega_1)^{1/p}\|u\|_{L_\infty(\Omega)}\,.$$

If $R \to \infty$, then $c_{N,\boldsymbol{\alpha},R} \to 0$.

PROOF. We estimate

$$\Big| \int\limits_{\mathbb{R}^n \setminus B_R} \partial^{\boldsymbol{\alpha}} k(\mathbf{x}-\mathbf{y}) \sum_{h_k \mathbf{m} \in \mathcal{Q}_k} c_{k,\mathbf{m}}\, \eta \left(\frac{\mathbf{y} - h_k \mathbf{m}}{h_k \sqrt{\mathcal{D}}} \right) d\mathbf{y} \Big|^p$$

$$\leq c \|u\|^p_{L_\infty(\Omega)} \Big(\int\limits_{\mathbb{R}^n \setminus B_R} r_{\boldsymbol{\alpha}}(|\mathbf{y}|) \sum_{h_k \mathbf{m} \in \mathcal{Q}_k} \Big| \eta \left(\frac{\mathbf{y} - h_k \mathbf{m}}{h_k \sqrt{\mathcal{D}}} \right) \Big| \, d\mathbf{y} \Big)^p$$

$$\leq c \|u\|^p_{L_\infty(\Omega)} \Big(\int\limits_{\mathbb{R}^n \setminus B_R} r_{\boldsymbol{\alpha}}(|\mathbf{y}|) \, g_{\mathcal{D}}({h_k}^{-1} \operatorname{dist}(\mathbf{y}, \mathcal{Q}_k), \eta) \, d\mathbf{y} \Big)^p .$$

Let $r_{\boldsymbol{\alpha}}(y) \leq c_j y^j$ for $y \to \infty$. The rapid decay of $g_{\mathcal{D}}$ implies

$$g_{\mathcal{D}}({h_k}^{-1} \operatorname{dist}(\mathbf{y}, \mathcal{Q}_k), \eta) = g_{\mathcal{D}}(\tau + h_k^{-1} \operatorname{dist}(\mathbf{y}, \Omega), \eta) \leq c_N h_k^N \operatorname{dist}(\mathbf{y}, \Omega)^{-N}$$

for any N, so that the inequality

$$\int\limits_{\mathbb{R}^n \setminus B_R} r_{\boldsymbol{\alpha}}(|\mathbf{y}|) \, g_{\mathcal{D}}(\operatorname{dist}({h_k}^{-1} \operatorname{dist}(\mathbf{y}, \mathcal{Q}_k), \eta)) \, d\mathbf{y} \leq c h_k^N \int\limits_{\mathbb{R}^n \setminus B_R} \frac{|\mathbf{y}|^j}{\operatorname{dist}(\mathbf{y}, \Omega)^N} \, d\mathbf{y}$$

with $N > n + j$ proves the assertion. □

THEOREM 9.16. *Let $u \in W_p^N(\Omega)$ with $N > n/p$ and $\mathcal{K} \in \mathcal{L}(L_p(\Omega), W_p^m(\Omega_1))$ satisfying* (9.35). *Then for any $\varepsilon > 0$, there exists $\mathcal{D} > 0$ such that*

$$\|\mathcal{K}u - \mathcal{K}_h u\|_{W_p^m(\Omega_1)} \leq c_1 (\sqrt{\mathcal{D}} h)^N \|u\|_{W_p^N(\Omega)} + c_2 (\mu^M h)^{1/p} \|u\|_{L_\infty(\Omega)} + \varepsilon \|u\|_{W_p^{N-1}(\Omega)} .$$

If, additionally, $\mathcal{K} \in \mathcal{L}((W^m_{p/(p-1)}(\Omega))', L_p(\Omega_1))$, then

$$\|\mathcal{K}u - \mathcal{K}_h u\|_{L_p(\Omega_1)} \leq \left(c_1 (\sqrt{\mathcal{D}} h)^N + c_2 (\mu^M h)^{1/p + r}\right) \|u\|_{W_p^N(\Omega)} + \varepsilon\, h^m \, \|u\|_{W_p^{N-1}(\Omega)} , \tag{9.36}$$

where $0 < r < m/n$, $r \leq (p-1)/p$.

PROOF. Fix the ball B_R and split

$$\mathcal{K}_h u(\mathbf{x}) = \mathcal{K} \mathcal{B}_M u(\mathbf{x}) = \mathcal{K} \chi_{B_R} \mathcal{B}_M u(\mathbf{x}) + \mathcal{K}(1 - \chi_{B_R}) \mathcal{B}_M u(\mathbf{x}) .$$

The $W_p^m(\Omega_1)$-norm of the difference

$$\|\mathcal{K}u - \mathcal{K} \chi_{B_R} \mathcal{B}_M u\|_{W_p^m(\Omega_1)} = \|\mathcal{K} \chi_{B_R} (u - \mathcal{B}_M u)\|_{W_p^m(\Omega_1)} \leq c_R \|u - \mathcal{B}_M u\|_{L_p(B_R)}$$

can be estimated using Theorem 9.12. It follows from Lemma 9.15 and Sobolev's imbedding theorem that the $W_p^m(\Omega_1)$-norm of the second term is bounded by

$$\|\mathcal{K}(1 - \chi_{B_R}) \mathcal{B}_M u\|_{W_p^m(\Omega_1)} \leq c_0 h^N \|u\|_{L_\infty(\Omega)} \leq c\, h^N \|u\|_{W_p^N(\Omega)} .$$

The same arguments also apply for (9.36) by using the inequality

$$\|\mathcal{K}u - \mathcal{K} \chi_{B_R} \mathcal{B}_M u\|_{L_p(\Omega_1)} \leq c_R \|u - \mathcal{B}_M u\|_{(W^m_{p/(p-1)}(B_R))'} \leq c \|u - \mathcal{B}_M u\|_{H_p^{-m}(\mathbb{R}^n)}$$

and Theorem 9.13. □

Summarizing, for a large class of domain integral operators with singular kernels, one can define cubature formulas retaining the order $\mathcal{O}(h^N)$ plus some small saturation error, if the boundary layer quasi-interpolation of the density is used with appropriate parameters μ and M.

REMARK 9.17. Note that the previous construction and results are applicable if instead of the integer points $\{\mathbf{m} \in \mathbb{Z}^n\}$, the set of lattice points $\{A\mathbf{m} : \mathbf{m} \in \mathbb{Z}^n\}$ with a non-singular $n \times n$ matrix A is taken as basis grid. This case can be transformed to that considered above by introducing the coordinates $\mathbf{y} = A\mathbf{x}$ and the new generating function $\eta_A := |\det A|^{-1}\eta(A\cdot)$, which satisfies the conditions of Theorem 9.14.

Let us consider two simple cubature examples:

EXAMPLE 9.18. Consider the logarithmic potential

$$\mathcal{L}_2 u(\mathbf{x}) = \frac{1}{2\pi} \int\limits_{\Omega} \log \frac{1}{|\mathbf{x}-\mathbf{y}|}\, u(\mathbf{y})\, d\mathbf{y}.$$

Note that the mapping

$$\mathcal{L} : L_p(\Omega) \mapsto W_p^2(\Omega), \quad 1 < p < \infty,$$

is bounded, if Ω is a bounded domain. Thus, Theorem 9.16 yields the estimate

$$\begin{aligned}\|\mathcal{L}_2 u - \mathcal{L}_{2,h} u\|_{W_p^2(\Omega)} &\le c_1(\sqrt{\mathcal{D}}h)^N \|\nabla_N u\|_{L_p(\Omega)} \\ &\quad + c_2(\mu^M h)^{1/p}\|u\|_{L_\infty(\Omega)} + \varepsilon\|u\|_{W_p^{N-1}(\Omega)}\,.\end{aligned}$$

Consequently, if the boundary layer approximations are such that μ^M is of the same order of magnitude as h^{Np-1}, we get the approximation order $\mathcal{O}(h^N)$ modulo a small saturation error. If we measure the error in a weaker norm than W_p^2, the saturation error tends to zero together with h. For example, if $u \in W_2^N(\Omega)$ with $N > 1$, then we obtain

$$\|\mathcal{L}_2 u - \mathcal{L}_{2,h} u\|_{L_2(\Omega)} \le (c_1(\sqrt{\mathcal{D}}h)^N + c_2\mu^M h)\|u\|_{W_2^N(\Omega)} + \varepsilon\, h^2\, \|u\|_{W_2^{N-1}(\Omega)}\,,$$

so that even the choice $\mu^M \asymp h^{N-1}$ leads to the order $\mathcal{O}(h^N)$ plus a small error term converging to zero with the rate $\mathcal{O}(h^2)$. Note that Sobolev's imbedding theorem can be used to prove the convergence of the cubature $\mathcal{L}_{2,h}$ with respect to the uniform norm.

EXAMPLE 9.19. The Poisson integral (6.12), i.e.,

$$\mathcal{P}_t\varphi(\mathbf{x}) = (4\pi a t)^{-n/2} \int\limits_{\Omega} \mathrm{e}^{-|\mathbf{x}-\mathbf{y}|^2/4at}\, \varphi(\mathbf{y})\, d\mathbf{y},$$

provides a solution of the homogeneous heat equation with initial value $u(\mathbf{x}, 0) = \varphi(\mathbf{x})$, $\mathbf{x} \in \Omega$. In Subsection 6.2.1, we pointed out that the Poisson integrals of the Gaussian or related functions have simple analytic expressions. Since the kernel function is smooth for any fixed $t > 0$, the Poisson integral generates a bounded mapping from Sobolev or Bessel potential spaces of arbitrary negative order into L_p-spaces. Therefore, from Theorem 9.16, it follows that

$$\|\mathcal{P}_t\varphi - \mathcal{P}_{t,h}\varphi\|_{L_2(\mathbb{R}^n)} \le (c_1(\sqrt{\mathcal{D}}h)^N + c_2\mu^M h)\|\varphi\|_{W_2^N(\Omega)}\,,$$

with constants depending on $t > 0$ but not on φ and h. Hence, $P_{t,h}\varphi = \mathcal{P}_t(\mathcal{B}_M\varphi)$ represents a semi-analytic cubature of order $\mathcal{O}(h^N)$ without saturation errors.

REMARK 9.20. Theorem 9.16 shows that for $p < \infty$ and $\mathcal{D}$ large enough, an N-th order approximation of the integral $\mathcal{K}u$ up to a prescribed accuracy can be obtained by fixing $\mu^{-1} \in \mathbb{N}$ and choosing M large enough, for example, $M = \mathcal{O}((pN-1)\log h/\log\mu)$. The computational costs are, of course, proportional to the number of grid points used for the construction of $\mathcal{B}_M$. The thickness of the layers $\mathcal{Q}_k$, $k = 1, \ldots, M$, is proportional to the size $\mu^k h$ of the mesh chosen there; hence the number of grid points at $\mathcal{Q}_k$ is $\mathcal{O}((\mu^k h)^{-(n-1)})$. Therefore, the computational costs for evaluating the sum over all nodes of the layers $\mathcal{Q}_k$, $k = 1, \ldots, M$, are of the order

$$\mathcal{O}\left(\left(\frac{(\mu - \mu^M)h}{1-\mu}\right)^{-(n-1)}\right).$$

Thus, the boundary layer approximation with isotropic meshes reduces the order of the computational costs by 1. In the next section, we consider the use of anisotropic meshes for special domains leading to a further reduction of the complexity of the cubature. Note that a more detailed analysis performed in [**35**] indicates that an optimal choice for scaling subsequent meshes is $\mu = 1/3$, if the generating functions (3.12) are used.

9.6. Anisotropic boundary layer approximate approximation

The method developed in the previous sections uses refined isotropic meshes on each boundary layer. This is the reason why it can be applied to arbitrary bounded domains. But a closer look shows that, in fact, only mesh refinement towards the direction of the boundary is necessary in order to achieve the same approximation results.

Here, we consider this modification of the boundary layer quasi-interpolation for the case of three-dimensional polyhedral domains. This leads to formulas of reduced complexity but having the same accuracy as before.

Consider a three-dimensional bounded polyhedral domain Ω. To approximate the integral operator (9.1), we divide the domain into simpler parts and process them separately. Using a partition of unity, the function u is decomposed into the sum

$$u = \Big(\varphi_{int} + \sum_{corners} \varphi_{c_k} + \sum_{edges} \varphi_{e_k} + \sum_{faces} \varphi_{f_k}\Big)u \tag{9.37}$$

where the cut-off functions $\varphi \in C_0^\infty$, $0 \le \varphi \le 1$, are different from zero only on special parts of Ω. So, $\varphi_{c_k} = 1$ at a neighborhood of the k-th corner point and vanishes outside a larger neighborhood. Correspondingly, $\varphi_{e_k} = 1$ at a neighborhood of the interior of the k-th edge, $\varphi_{f_k} = 1$ at some interior part of the k-th face, and $\varphi_{int} = 1$ on some interior part of Ω. The approximation of $\varphi_{int}\, u$ can be performed by the usual quasi-interpolant (2.23) with a suitable generating function, whereas the functions $\varphi_{c_k} u$ are approximated on the domain supp φ_{c_k} by applying the multi-scale quasi-interpolation operator $\mathcal{B}_M$ considered in Section 9.4.

We study the approximation of the functions $\varphi_{e_k} u$ and $\varphi_{f_k} u$ by using mesh refinement only in the direction normal to the boundary side.

To start with the approximation near the faces of Ω, we consider a sufficiently smooth function u given in $\mathbb{R}^3_+ = \{\mathbf{x} = (\mathbf{x}', x_3) \in \mathbb{R}^3 : x_3 \ge 0\}$, with bounded support and $u(\mathbf{x}', 0) \not\equiv 0$. More precisely, suppose that $G \subset \mathbb{R}^3_+$ is a bounded domain whose boundary contains a two-dimensional domain $\Gamma \subset \{(\mathbf{x}', 0) : \mathbf{x}' \in \mathbb{R}^2\}$.

In the following, $\mathring{W}_p^N(G)$ denotes the subspace of all functions from $W_p^N(G)$, which can be smoothly extended by zero through $\partial G \backslash \Gamma$.

We are interested in constructing approximants to $u \in \mathring{W}_p^N(G)$ with respect to the sequence of anisotropically distributed mesh points $\{(h\mathbf{m}', \mu^k h m_3) : (\mathbf{m}', m_3) \in \mathbb{Z}^3, m_3 > 0\}$. Therefore, we consider generating functions of tensor product form

$$\eta_3(\mathbf{x}) = \eta_3(\mathbf{x}', x_3) = \eta_2(\mathbf{x}')\, \eta_1(x_3)\,, \tag{9.38}$$

where η_2 and η_1 are generating functions of the Schwartz class of two or one variables, respectively, with positive Fourier transform.

The following assertion can be proved analogously to Lemma 2.30.

LEMMA 9.21. *Suppose that both $\eta_2 \in \mathcal{S}(\mathbb{R}^2)$ and $\eta_1 \in \mathcal{S}(\mathbb{R})$ satisfy the moment Condition* 2.15 *for some given N and let $\mu \in (0,1)$. There exists $\kappa > 0$ such that the quasi-interpolant*

$$\mathcal{N}_{h,\mu} u(\mathbf{x}) := \mathcal{D}^{-3/2} \sum_{(h\mathbf{m}', \mu h m_3) \in G} u(h\mathbf{m}', \mu h m_3)\, \eta_3\Big(\frac{\mathbf{x}' - h\mathbf{m}'}{h\sqrt{\mathcal{D}}}, \frac{x_3 - \mu h m_3}{\mu h \sqrt{\mathcal{D}}}\Big) \tag{9.39}$$

approximates $u \in \mathring{W}_p^N(G)$, $N > n/p$, in the subdomain $G_{\kappa\mu h} = \{x \in G : x_3 > \kappa\mu h\}$ with the estimate

$$\|\mathcal{N}_{h,\mu} u - u\|_{L_p(G_{\kappa\mu h})} \leq c\,(\sqrt{\mathcal{D}}h)^N \|\nabla_N u\|_{L_p(G)} + \varepsilon \sum_{k=0}^{N-1} (\sqrt{\mathcal{D}}h)^k \|\nabla_k u\|_{L_p(G)}\,.$$

Thus, one obtains approximations of order $\mathcal{O}((\sqrt{\mathcal{D}}h)^N)$ modulo saturation terms, for all $\mathbf{x} \in G$ with the exception of the strip $0 < x_3 < \delta$ for arbitraryily small $\delta > 0$, if μ is chosen sufficiently small. Therefore, it suffices to refine the mesh only in the x_3-direction, i.e., to use the refinement equation for the one-dimensional generating function η_1. If we denote the corresponding mask function defined by (8.12) by $\tilde{\eta}_1$, then, obviously, the assertion of Theorem 9.1 remains true for $\mathcal{N}_{h,\mu}$ and

$$\widetilde{\mathcal{N}}_{h,\mu} u(\mathbf{x}) := \mathcal{D}^{-3/2} \sum_{\mathbf{m} \in \mathbb{Z}^3} u(h\mathbf{m}', \mu h m_3)\, \eta_2\Big(\frac{\mathbf{x}' - h\mathbf{m}'}{h\sqrt{\mathcal{D}}}\Big)\, \tilde{\eta}_1\Big(\frac{x_3 - \mu h m_3}{\mu h\sqrt{\mathcal{D}}}\Big),$$

i.e.,

$$\mathcal{N}_{h,\mu} = \mathcal{N}_{h,\mu^2} \widetilde{\mathcal{N}}_{h,\mu} + \mathcal{R}_{h,\mu}\,, \qquad \|\mathcal{R}_{h,\mu}\| < \varepsilon\,.$$

Analogously to the isotropic case, we may define the face approximant as

$$\mathcal{B}_{M,f} u(\mathbf{x}) - \mathcal{D}^{-3/2} \sum_{j=0}^{M} \sum_{(h\mathbf{m}', \mu^j h m_3) \in Q_j} c_{j,\mathbf{m}}\, \eta_3\Big(\frac{\mathbf{x}' - h\mathbf{m}'}{h\sqrt{\mathcal{D}}}, \frac{x_3 - \mu^k h m_3}{\mu^j h\sqrt{\mathcal{D}}}\Big) \tag{9.40}$$

with the coefficients

$$c_{j,\mathbf{m}} = \begin{cases} u(h\mathbf{m}', h m_3), & j = 0, \\ u(h\mathbf{m}', \mu^j h m_3) - (\widetilde{\mathcal{N}}_{h,\mu^{j-1}} \chi_{j-1} u)(h\mathbf{m}', \mu^j h m_3), & j \geq 1, \end{cases}$$

where

$$\mathcal{Q}_j = \{\mathbf{x} \in G : \tau h \mu^j \leq x_3 \leq (\tau + \widetilde{\kappa}) h \mu^{j-1}\}\,,$$

$\chi_j(\mathbf{x}) = 1$ for $x_3 > \tau\mu^j$ and zero otherwise, and the parameters τ, $\widetilde{\kappa}$ are obtained from the one-dimensional functions η_1 and $\tilde{\eta}_1$.

Obviously, for the approximant on the anisotropic mesh (9.40), the assertions of Theorems 9.12 and 9.13 remain true. In contrast to the approximation (9.2), which uses isotropic mesh refinement, the pointwise estimate on the layer $(\tau+\tilde{\kappa}-\kappa\mu)h_j \le x_3 \le (\tau+\tilde{\kappa}-\kappa\mu)h_{j-1}$ is $\mathcal{O}((\sqrt{\mathcal{D}}h)^N)$, which is worse than $\mathcal{O}((\sqrt{\mathcal{D}}h_j)^N)$. However, this does not influence the estimate in integral norms and, which is more important, the approximation quality of the resulting cubature formula. But most importantly, anisotropic mesh refinement leads to a considerable reduction of the number of mesh points and, therefore, of terms in (9.40), required to derive the N-th order approximation of integral operators. Since any $\mathcal{Q}_j$ contains $\mathcal{O}(h^{-2})$ mesh points, the numerical costs to obtain the sum over these nodes are of the order $\mathcal{O}(Mh^{-2})$.

Quite similarly the approximation of edge functions $u\,\varphi_{e_k}$ can be performed with the order $\mathcal{O}((\sqrt{\mathcal{D}}h)^N)$ plus small saturation, by using anisotropic mesh refinement. Consider the bounded domain G obtained after intersecting some wedge $W = \{(x_1,x_2,x_3) \in \mathbb{R}^3 : x_2 > a\,|x_1|\}$, $0 < a < \infty$, with another domain. The common part of ∂G and the boundary of the wedge ∂W is denoted by Γ and again we denote by $\mathring{W}_p^N(G)$ the subspace of all functions from $W_p^N(G)$, which can be smoothly extended by zero through $\partial G \backslash \Gamma$.

Now, we are interested in constructing approximants to $u \in \mathring{W}_p^N(G)$ with respect to the sequence of meshes, which are refined in the two directions normal to Γ. As basic mesh points, we choose the lattice $\{(hA\mathbf{m}', hm_3) : \mathbf{m}' \in \mathbb{Z}^2_{++}, m_3 \in \mathbb{Z}\}$, where A is a 2×2 matrix, transforming the quarter plane $\mathbb{R}^2_{++}$ onto $\{\mathbf{x} \in W : x_3 = 0\}$ with $\det A = 1$.

In view of Remark 9.17, the following assertion for the tensor product generating function (9.38) is valid.

LEMMA 9.22. *Suppose that both $\eta_2(\mathbf{x}')$ and $\eta_1(x_3)$ satisfy the moment Condition 2.15 for some given $N > n/p$ and let $\mu \in (0,1)$. There exists $\kappa > 0$ such that the quasi-interpolant*

$$\mathcal{N}_{\mu,h}u(\mathbf{x}) := \mathcal{D}^{-3/2} \sum_{(\mu hA\mathbf{m}', hm_3)\in G} u(\mu hA\mathbf{m}', hm_3)\, \eta_3\Big(\frac{\mathbf{x}' - \mu hA\mathbf{m}'}{\mu h\sqrt{\mathcal{D}}}, \frac{x_3 - hm_3}{h\sqrt{\mathcal{D}}}\Big)$$

approximates $u \in \mathring{W}_p^N(G)$ in the subdomain $G_{\kappa\mu h} = \{\mathbf{x} \in G : \operatorname{dist}(\mathbf{x},\Gamma) > \kappa\mu h\}$ with the estimate

$$\|\mathcal{N}_{\mu,h}u - u\|_{L_p(G_{\kappa\mu h})} \le c\,(\sqrt{\mathcal{D}}h)^N \|\nabla_N u\|_{L_p(G)} + \varepsilon \sum_{k=0}^{N-1} (\sqrt{\mathcal{D}}h)^k \|\nabla_k u\|_{L_p(G)}\,.$$

The mesh refinement towards Γ without loss of approximation order is possible owing to the factorization

$$\mathcal{N}_{\mu,h} = \mathcal{N}_{\mu^2,h}\widetilde{\mathcal{N}}_{\mu,h} + \mathcal{R}_{h,\mu}\,, \qquad \|\mathcal{R}_{h,\mu}\| < \varepsilon\,,$$

where again $\mu^{-1} \in \mathbb{N}$,

$$\widetilde{\mathcal{N}}_{\mu,h}u(\mathbf{x}) := \mathcal{D}^{-3/2} \sum_{\mathbf{m}\in\mathbb{Z}^3} u(\mu hA\mathbf{m}', hm_3)\, \tilde{\eta}_2\Big(\frac{\mathbf{x}' - \mu hA\mathbf{m}'}{\mu h'\sqrt{\mathcal{D}}}\Big)\, \eta_1\Big(\frac{x_3 - hm_3}{h\sqrt{\mathcal{D}}}\Big)\,,$$

and $\tilde{\eta}_2$ is the mask function of η_2 defined by (8.12). This follows immediately from the fact that the quasi-interpolant with $\eta_{2,A} := \eta_2(A\,\cdot)$ is based on the nodes $h\mathbb{Z}^2$ and that $\mathcal{F}\eta_{2,A} = \mathcal{F}\eta_2(A^{-1}\,\cdot)$.

Now the edge approximant to $u \in \mathring{W}_p^N(G)$ is defined as

$$\mathcal{B}_{M,e}u(\mathbf{x}) = \mathcal{D}^{-3/2} \sum_{j=0}^{M} \sum_{(\mu^j hA\mathbf{m}', hm_3) \in \mathcal{Q}_j} c_{k,\mathbf{m}}\, \eta_3\Big(\frac{\mathbf{x}' - \mu^j hA\mathbf{m}'}{\mu^j h\sqrt{\mathcal{D}}}, \frac{x_3 - hm_3}{h\sqrt{\mathcal{D}}}\Big) \tag{9.41}$$

with the coefficients

$$c_{j,\mathbf{m}} = \begin{cases} u(hA\mathbf{m}', hm_3), & j = 0, \\ u(\mu^j hA\mathbf{m}', hm_3) - (\widetilde{\mathcal{N}}_{\mu^{j-1},h}\chi_{j-1}u)(\mu^j hA\mathbf{m}', hm_3), & j \geq 1, \end{cases}$$

where

$$\mathcal{Q}_j = \{\mathbf{x} \in G : \tau h\mu^j \leq \operatorname{dist}(\mathbf{x}, \Gamma) \leq (\tau + \tilde{\kappa})h\mu^{j-1}\},$$

$\chi_j(\mathbf{x}) = 1$ for $\operatorname{dist}(\mathbf{x}, \Gamma) > \tau\mu^j$ and zero otherwise, and the parameters τ, $\tilde{\kappa}$ are obtained from the functions η_2 and $\tilde{\eta}_2$. Again, for (9.41), the assertions of Theorems 9.12 and 9.13 remain true. We note that the numerical costs to evaluate $\mathcal{B}_{M,e}u$ are of the order $\mathcal{O}(M^2h^{-2})$.

After these preparations, we can define the anisotropic boundary layer approximant to the function u given in the polyhedral domain Ω. Turning to the decomposition (9.37) and using the different types of approximants defined in (2.23), (9.2), (9.40), and (9.41), we introduce

$$\widetilde{\mathcal{B}}_M u = \mathcal{M}_h^\kappa(u\varphi_{int}) + \sum_{corners} \mathcal{B}_M(u\varphi_{c_k}) + \sum_{edges} \mathcal{B}_{M,e}(u\varphi_{e_k}) + \sum_{faces} \mathcal{B}_{M,f}(u\varphi_{f_k}).$$

THEOREM 9.23. *For any $\varepsilon > 0$ there exist $\mathcal{D} > 0$ and a boundary layer approximation $\widetilde{\mathcal{B}}_M$, which is anisotropic near faces and edges of the polyhedral domain $\Omega \in \mathbb{R}^3$, such that for any $u \in W_p^N(\Omega)$, $N > 3/p$,*

$$\|u - \widetilde{\mathcal{B}}_M u\|_{L_p(\mathbb{R}^3)} \leq c_1(\sqrt{\mathcal{D}}h)^N \|\nabla_N u\|_{L_p(\Omega)} + c_2(\mu^M h)^{1/p}\|u\|_{L_\infty(\Omega)} + \varepsilon\|u\|_{W_p^{N-1}(\Omega)}$$

and

$$\|u - \widetilde{\mathcal{B}}_M u\|_{H_p^{-s}(\mathbb{R}^3)} \leq (c_1(\sqrt{\mathcal{D}}h)^N + c_2(\mu^M h)^{1/p+r} + \varepsilon\, h^s)\|u\|_{W_p^{N-1}(\Omega)},$$

where $0 < r < s/3$ and $r \leq (p-1)/p$.

9.7. Potentials of tensor product generating functions

From the results of Section 9.5 and Theorem 9.23, it follows that the assertions of Theorem 9.16 are valid also for the approximation $\mathcal{K}\widetilde{\mathcal{B}}_M$ of the integral operator $\mathcal{K}$ over a polyhedral domain $\Omega \subset \mathbb{R}^3$. For the practical application, it remains to study the efficient computation of integrals of generating functions occurring in the approximant $\widetilde{\mathcal{B}}_M u$. Here, we consider the tensor product function η_3 with factors of the class η_{2M} defined in (3.12), i.e., we set

$$\begin{aligned} \eta_3(\mathbf{x}) &= \pi^{-3/2} L_{M-1}^{(1)}(|\mathbf{x}'|^2)\, L_{M-1}^{(1/2)}(x_3^2)\, \mathrm{e}^{-|\mathbf{x}'|^2 - x_3^2} \\ &= \pi^{-3/2} \sum_{j,k=0}^{M-1} \frac{(-1)^{j+k}}{j!\,k!\,4^{j+k}} \Delta_{\mathbf{x}'}^j \frac{\partial^{2k}}{\partial x_3^{2k}}\, \mathrm{e}^{-|\mathbf{x}|^2}, \end{aligned}$$

where $\Delta_{\mathbf{x}'} = (\partial/\partial x_1)^2 + (\partial/\partial x_2)^2$. Therefore, the anisotropic generating functions can be written in the form

$$\eta_3(\mathbf{x}', ax_3) = \pi^{-3/2} \sum_{j,k=0}^{M-1} \frac{(-1)^{j+k}}{j!\,k!\,4^{j+k}a^{2k}} \Delta_{\mathbf{x}'}^j\, \mathrm{e}^{-|\mathbf{x}'|^2} \frac{\partial^{2k}}{\partial x_3^{2k}}\, \mathrm{e}^{-a^2x_3^2}.$$

Then the convolution integrals with $\eta_3(\mathbf{x}', ax_3)$, $a > 0$, have the form

$$\int_{\mathbb{R}^3} g(\mathbf{x}-\mathbf{y})\,\eta_3(\mathbf{y}', ay_3)\,d\mathbf{y}$$

$$(9.42)\qquad = \pi^{-3/2}\sum_{j,k=0}^{M-1}\frac{(-1)^{j+k}}{j!\,k!\,4^{j+k}a^{2k}}\int_{\mathbb{R}^3} g(\mathbf{x}-\mathbf{y})\,\Delta^j_{\mathbf{y}'}\,\mathrm{e}^{-|\mathbf{y}'|^2}\,\frac{\partial^{2k}}{\partial y_3^{2k}}\,\mathrm{e}^{-a^2y_3^2}\,d\mathbf{y}$$

$$= \pi^{-3/2}\sum_{j,k=0}^{M-1}\frac{(-1)^{j+k}}{j!\,k!\,4^{j+k}a^{2k}}\,\Delta^j_{\mathbf{x}'}\,\frac{\partial^{2k}}{\partial x_3^{2k}}\int_{\mathbb{R}^3} g(\mathbf{x}-\mathbf{y})\,\mathrm{e}^{-|\mathbf{y}'|^2-a^2y_3^2}\,d\mathbf{y}\,.$$

We see that higher-order cubature formulas are easily derived from the integrals

$$(9.43)\qquad \int_{\mathbb{R}^3} g(\mathbf{x}-\mathbf{y})\,\mathrm{e}^{-|\mathbf{y}'|^2-a^2y_3^2}\,d\mathbf{y}\,,$$

which have been considered in Section 6.3. For integral kernels g, often occurring in applications, the integral (9.43) can be reduced to a one-dimensional integral with a smooth integrand. In Section 6.4, this approach was extended to higher-order generating functions, which are not tensor products. Here we apply this approach to potentials of $\eta_3(\mathbf{x}', ax_3)$. Note that the case $a \neq 1$ covers both types (9.40) for the face function and (9.41) for the edge function approximation.

Consider, for example, the harmonic potential of $\eta_3(\mathbf{x}', ax_3)$. In Subsection 6.3.1, we treated the harmonic potential of anisotropic Gaussians in the general case. In particular, applying formula (6.38), we obtain the Newton potential of the anisotropic Gaussian $\mathrm{e}^{-|\mathbf{x}'|^2-a^2x_3^2}$ as the one-dimensional integral

$$\frac{1}{4\pi}\int_{\mathbb{R}^3}\frac{\mathrm{e}^{-|\mathbf{y}'|^2-a^2y_3^2}}{|\mathbf{x}-\mathbf{y}|}\,d\mathbf{y} = \frac{1}{4}\int_0^\infty \frac{\mathrm{e}^{-|\mathbf{x}'|^2/(1+t)}}{1+t}\,\frac{\mathrm{e}^{-a^2x_3^2/(1+a^2t)}}{\sqrt{1+a^2t}}\,dt\,,$$

such that, in view of (9.42),

$$\frac{1}{4\pi}\int_{\mathbb{R}^3}\frac{\eta_3(\mathbf{y}', ay_3)}{|\mathbf{x}-\mathbf{y}|}\,d\mathbf{y}$$

$$= \frac{1}{4\pi^{3/2}}\sum_{j,k=0}^{M-1}\frac{(-1)^{j+k}}{j!\,k!\,4^{j+k}a^{2k}}\,\Delta^j_{\mathbf{x}'}\,\frac{\partial^{2k}}{\partial x_3^{2k}}\int_0^\infty \frac{\mathrm{e}^{-|\mathbf{x}'|^2/(1+t)}}{1+t}\,\frac{\mathrm{e}^{-a^2x_3^2/(1+a^2t)}}{\sqrt{1+a^2t}}\,dt\,.$$

From (3.15), we conclude that

$$\Delta^j_{\mathbf{x}'}\,\frac{\mathrm{e}^{-|\mathbf{x}'|^2/(1+t)}}{1+t} = (-1)^j j!4^j\,\frac{\mathrm{e}^{-|\mathbf{x}'|^2/(1+t)}}{(1+t)^{j+1}}\,L_j^{(0)}\Big(\frac{|\mathbf{x}'|^2}{1+t}\Big)$$

and

$$\frac{\partial^{2k}}{\partial x_3^{2k}}\,\frac{\mathrm{e}^{-a^2x_3^2/(1+a^2t)}}{\sqrt{1+a^2t}} = (-1)^k k!4^k\,\frac{a^{2k}\,\mathrm{e}^{-a^2x_3^2/(1+a^2t)}}{(1+a^2t)^{k+1/2}}\,L_k^{(-1/2)}\Big(\frac{a^2x_3^2}{1+a^2t}\Big).$$

Thus, the Newton potential of $\eta_3(\mathbf{x}', ax_3)$ is given by the one-dimensional integral

$$\frac{1}{4\pi}\int\limits_{\mathbb{R}^3}\frac{\eta_3(\mathbf{y}',ay_3)}{|\mathbf{x}-\mathbf{y}|}\,d\mathbf{y}$$
$$=\frac{1}{4\pi^{3/2}}\int\limits_0^\infty \frac{\mathrm{e}^{-|\mathbf{x}'|^2/(1+t)}}{1+t}\,\frac{\mathrm{e}^{-a^2x_3^2/(1+a^2t)}}{\sqrt{1+a^2t}}\sum_{j,k=0}^{M-1}\frac{L_j^{(0)}\left(\frac{|\mathbf{x}'|^2}{1+t}\right)}{(1+t)^j}\,\frac{L_k^{(-1/2)}\left(\frac{a^2x_3^2}{1+a^2t}\right)}{(1+a^2t)^k}\,dt\,.$$

Obviously, an analogous formula can be derived for the other potentials considered in Section 6.3.

9.8. Notes

The poor performance of the simple quasi-interpolation method described in Section 9.2 was observed by Ivanov. He proposed in [**35**] the approximate factorization of the quasi-interpolation operator and the boundary layer quasi-interpolation based on the multi-resolution decomposition discussed in Sections 9.3 and 9.4. Rigorous results on the approximation error in L_p were obtained in [**36**]. In the present chapter, we presented a modification of the proofs given there. The results of Section 9.6 on the anisotropic boundary layer approximation are taken from [**37**].

CHAPTER 10

More general grids

In this chapter, we show that the construction of approximate quasi-interpolants (2.23) is not restricted to uniform cubic grids.

It is required sometimes to approximate functions on the nodes of orthotropic partitions or on uniform non-orthogonal grids. In Section 10.1, we consider some variants for the construction of approximate quasi-interpolants on those uniform grids, among them two-dimensional regular triangular and hexagonal grids. Then we study the more interesting case to approximate a function by its values on non-uniformly distributed nodes.

For the case when the grid can be interpreted as a perturbation of a uniform grid, in Section 10.2, we construct quasi-interpolants which involve the function values on the non-uniform grid, but the generating functions are centered at the nodes of the uniform grid. Therefore the quasi-interpolants have the well-known simple structure.

For other distributions of nodes, we require that the approximate quasi-interpolant have a similar semi-discrete convolutional form in order to obtain cubature formulas as in the uniform case. This can be achieved if the nodes are obtained by a smooth transformation of a uniform grid. In Section 10.3, we derive formulas for quasi-interpolants which approximate functions given either on an n-dimensional domain or a manifold. We show that the approximants exhibit the same properties as the approximate quasi-interpolants on uniform grids.

10.1. Uniform grids

10.1.1. Orthotropic grid. If the function data are given on a rectangular grid

$$\{(h_1 m_1, \dots, h_n m_n) : \mathbf{m} \in \mathbb{Z}^n\},$$

then it is useful to consider the more general quasi-interpolation formula

$$\frac{1}{\sqrt{\mathcal{D}_1 \dots \mathcal{D}_n}} \sum_{m \in \mathbb{Z}^n} u(h_1 m_1, \dots, h_n m_n)\, \eta\Big(\frac{x_1 - h_1 m_1}{\sqrt{\mathcal{D}_1} h_1}, \dots, \frac{x_n - h_n m_n}{\sqrt{\mathcal{D}_n} h_n}\Big).$$

Applying the one-dimensional Poisson summation formula (1.3) in each variable to a sufficiently smooth and rapidly decaying function η, one obtains the formula

$$\sqrt{\mathcal{D}_1 \dots \mathcal{D}_n} \sum_{m \in \mathbb{Z}^n} \Big(\frac{x_1 - h_1 m_1}{\sqrt{\mathcal{D}_1} h_1}\Big)^{\alpha_1} \dots \Big(\frac{x_n - h_n m_n}{\sqrt{\mathcal{D}_n} h_n}\Big)^{\alpha_n} \eta\Big(\frac{x_1 - h_1 m_1}{\sqrt{\mathcal{D}_1} h_1}, \dots, \frac{x_n - h_n m_n}{\sqrt{\mathcal{D}_n} h_n}\Big)$$

$$= \Big(\frac{i}{2\pi}\Big)^{[\boldsymbol{\alpha}]} \sum_{\boldsymbol{\nu} \in \mathbb{Z}^n} \partial^{\boldsymbol{\alpha}} \mathcal{F}\eta(\sqrt{\mathcal{D}_1}\nu_1, \dots, \sqrt{\mathcal{D}_n}\nu_n) \prod_{j=1}^{n} e^{2\pi i x_j \nu_j / h_j},$$

which is valid if both series converge absolutely. Then error estimations, similar to those of Chapter 1, can be obtained.

A further generalization of the quasi-interpolation formula is the "tensor product" case, which means that the generating function has the form

$$\eta(\mathbf{x}) = \eta_1(x_1)\dots\eta_n(x_n)\,,$$

where it is possible that the one-dimensional generating functions η_j have zero moments of different order in different directions.

10.1.2. Non-orthogonal grid. The quasi-interpolation formulas can be easily generalized to the case when the values of u are given on a lattice

$$\Lambda_h := \{hA\mathbf{m}\,,\ \mathbf{m}\in\mathbb{Z}^n\}$$

with a real non-singular $n\times n$ matrix A. We want to use the same generating functions as before, for example, radial functions η. This situation can occur, for instance, in cubature formulas for pseudodifferential operators, where analytic expressions are available only after affine transformations.

We define the quasi-interpolant

$$\mathcal{M}_{\Lambda_h}u(\mathbf{x}) = \frac{\det A}{\mathcal{D}^{n/2}}\sum_{\mathbf{m}\in\mathbb{Z}^n} u(hA\mathbf{m})\,\eta\Big(\frac{\mathbf{x}-hA\mathbf{m}}{\sqrt{\mathcal{D}}h}\Big)\,. \tag{10.1}$$

Using the notation $u_A = u(A\cdot)$, $\eta_A = \det A\,\eta(A\cdot)$, $\mathbf{t} = A^{-1}\mathbf{x}$, the sum (10.1) transforms to

$$\begin{aligned}\mathcal{M}_{\Lambda_h}u(\mathbf{x}) &= \frac{\det A}{\mathcal{D}^{n/2}}\sum_{\mathbf{m}\in\mathbb{Z}^n} u_A(h\mathbf{m})\,\eta\Big(A\frac{A^{-1}\mathbf{x}-h\mathbf{m}}{\sqrt{\mathcal{D}}h}\Big)\\ &= \mathcal{D}^{-n/2}\sum_{\mathbf{m}\in\mathbb{Z}^n} u_A(h\mathbf{m})\eta_A\Big(\frac{\mathbf{t}-h\mathbf{m}}{\sqrt{\mathcal{D}}h}\Big) = \mathcal{M}_{h,\mathcal{D}}u_A(\mathbf{t})\,.\end{aligned}$$

This is exactly the quasi-interpolation formula (2.23) with the transformed generating function η_A applied to the function u_A. Since

$$\int_{\mathbb{R}^n}\mathbf{x}^{\boldsymbol{\alpha}}\eta_A(\mathbf{x})\,d\mathbf{x} = \det A\int_{\mathbb{R}^n}\mathbf{x}^{\boldsymbol{\alpha}}\eta(A\mathbf{x})\,d\mathbf{x} = \int_{\mathbb{R}^n}(A^{-1}\mathbf{x})^{\boldsymbol{\alpha}}\eta(\mathbf{x})\,d\mathbf{x}\,,$$

the generating function η_A satisfies the decay and the moment Conditions 2.12 and 2.15 together with η. Therefore one can apply all estimates obtained in Sections 2.3 and 2.4 to the quasi-interpolation formula (10.1).

Denoting the j-th component of the vector $A\nabla$ by $(A\nabla)_j$ and using the notation

$$(A\nabla)^{\boldsymbol{\alpha}} = (A\nabla)_1^{\alpha_1}\dots(A\nabla)_n^{\alpha_n}\,,$$

we have

$$\partial^{\boldsymbol{\alpha}}u_A(\mathbf{t}) = (A^t\nabla)^{\boldsymbol{\alpha}}u(A\mathbf{t})\,,\quad \partial^{\boldsymbol{\alpha}}\mathcal{F}\eta_A(\boldsymbol{\lambda}) = ((A^t)^{-1}\nabla)^{\boldsymbol{\alpha}}\mathcal{F}\eta((A^t)^{-1}\boldsymbol{\lambda})\,,$$

where A^t denotes the transpose of the matrix A. If, for example, $u\in W_\infty^N(\mathbb{R}^n)$, then representation (2.50) takes the form

$$\begin{aligned}&\mathcal{M}_{\Lambda_h}u(\mathbf{x}) = u(\mathbf{x}) + R_{N,h}(\mathbf{x})\\ &\quad+\sum_{[\boldsymbol{\alpha}]=0}^{N-1}\frac{(A^t\nabla)^{\boldsymbol{\alpha}}u(\mathbf{x})}{\boldsymbol{\alpha}!}\Big(\frac{\sqrt{\mathcal{D}}h}{2\pi i}\Big)^{[\boldsymbol{\alpha}]}\sum_{\boldsymbol{\nu}\in\mathbb{Z}^n\setminus\{\mathbf{0}\}}((A^t)^{-1}\nabla)^{\boldsymbol{\alpha}}\mathcal{F}\eta(\sqrt{\mathcal{D}}(A^t)^{-1}\boldsymbol{\nu})\,e^{\frac{2\pi i}{h}\langle A^{-1}\mathbf{x},\boldsymbol{\nu}\rangle}\,,\end{aligned}$$

with a remainder $R_{N,h}$, which is bounded by

$$|R_{N,h}(\mathbf{x})| \le c_{A,\eta}\,(h\sqrt{\mathcal{D}})^N\|\nabla_N u\|_{L_\infty}\,,$$

and the constant $c_{A,\eta}$ is independent of u, h, and $\mathcal{D}$.

We see that it is always possible to choose $\mathcal{D}$ such that the quasi-interpolant $\mathcal{M}_{\Lambda_h}$ satisfies an estimate of the form (2.49) or a local estimate similar to (2.58) for any $\varepsilon > 0$.

Let us note that by (2.22), Poisson's summation formula on the affine lattice $\Lambda = \{A\mathbf{m}\}_{\mathbf{m}\in\mathbb{Z}^n}$ has the form

$$\frac{\det A}{\mathcal{D}^{n/2}} \sum_{\mathbf{m}\in\mathbb{Z}^n} \eta\Big(\frac{\mathbf{x} - A\mathbf{m}}{\sqrt{\mathcal{D}}}\Big) = \sum_{\boldsymbol{\nu}\in\mathbb{Z}^n} \mathcal{F}\eta(\sqrt{\mathcal{D}}(A^t)^{-1}\boldsymbol{\nu})\, e^{2\pi i\langle \mathbf{x}, A\boldsymbol{\nu}\rangle}\,. \tag{10.2}$$

10.1.3. Examples. We consider approximate partitions of unity and quasi-interpolants based on two regular grids in $\mathbb{R}^2$. For the first, the triangular grid, we can directly apply the results concerning non-orthogonal grids. The second example is a hexagonal grid, which cannot be represented as an affine lattice.

First, we consider quasi-interpolants on a regular triangular grid.

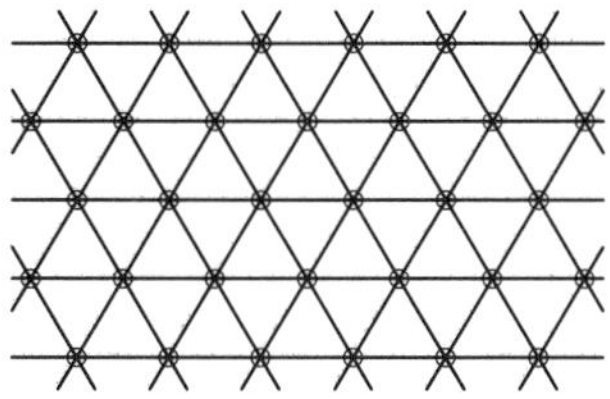

FIGURE 10.1. Tridiagonal grid

It is easy to check that the matrix

$$A = \begin{pmatrix} 1 & 1/2 \\ 0 & \sqrt{3}/2 \end{pmatrix}$$

maps the integer vectors $\mathbf{m} \in \mathbb{Z}^2$ onto the vertices $\mathbf{y}_{\mathbf{m}}^{\triangle} = A\mathbf{m}$ of a partition of $\mathbb{R}^2$ into equilateral triangles of side length 1 indicated in Fig. 10.1. From (10.1), we see that a quasi-interpolant on the nodes $\{h\mathbf{y}_{\mathbf{m}}^{\triangle} = hA\mathbf{m}\}_{\mathbf{m}\in\mathbb{Z}^2}$ of a regular tridiagonal partition of $\mathbb{R}^2$ can be given as

$$\mathcal{M}_h^{\triangle} u(\mathbf{x}) := \frac{\sqrt{3}}{2\mathcal{D}} \sum_{\mathbf{m}\in\mathbb{Z}^2} u(hA\mathbf{m})\, \eta\Big(\frac{\mathbf{x} - h\mathbf{y}_{\mathbf{m}}^{\triangle}}{\sqrt{\mathcal{D}}h}\Big)\,,$$

which has the approximation properties described in Chapter 1. In particular, the function system $\left\{\dfrac{\sqrt{3}}{2\mathcal{D}}\eta\Big(\dfrac{\mathbf{x} - \mathbf{y}_{\mathbf{m}}^{\triangle}}{\sqrt{\mathcal{D}}}\Big)\right\}_{\mathbf{m}\in\mathbb{Z}^2}$ forms an approximate partition of unity and

$$\Big|1 - \frac{\sqrt{3}}{2\mathcal{D}} \sum_{\mathbf{m}\in\mathbb{Z}^2} \eta\Big(\frac{\mathbf{x} - \mathbf{y}_{\mathbf{m}}^{\triangle}}{\sqrt{\mathcal{D}}}\Big)\Big| \le \sum_{\boldsymbol{\nu}\in\mathbb{Z}^2\setminus\{\mathbf{0}\}} \Big|\int_{\mathbb{R}^2} \eta(\mathbf{x})\, e^{-2\pi i\sqrt{\mathcal{D}}\langle A^{-1}\mathbf{x}, \boldsymbol{\nu}\rangle} d\mathbf{x}\Big|\,.$$

From the relation

$$A^{-1} = \begin{pmatrix} 1 & -1/\sqrt{3} \\ 0 & 2/\sqrt{3} \end{pmatrix}$$

and from (10.2), we obtain Poisson's summation formula for $\eta(\mathbf{x}) = \pi^{-1}\,\mathrm{e}^{-|\mathbf{x}|^2}$ on a triangular grid

$$
\begin{aligned}
&\frac{\sqrt{3}}{2\pi\mathcal{D}} \sum_{\mathbf{m}\in\mathbb{Z}^2} \mathrm{e}^{-|\mathbf{x}-\mathbf{y}^{\triangle}_{\mathbf{m}}|^2/\mathcal{D}} \\
&= \frac{1}{\pi} \sum_{\boldsymbol{\nu}\in\mathbb{Z}^2} \mathrm{e}^{2\pi i(x_1\nu_1 + x_2(2\nu_2-\nu_1)/\sqrt{3})} \int_{\mathbb{R}^2} \mathrm{e}^{-|\mathbf{y}|^2}\, \mathrm{e}^{-2\pi i\sqrt{\mathcal{D}}(y_1\nu_1+y_2(2\nu_2-\nu_1)/\sqrt{3})}\, d\mathbf{y} \qquad (10.3)\\
&= \sum_{\boldsymbol{\nu}\in\mathbb{Z}^2} \mathrm{e}^{-4\pi^2\mathcal{D}(\nu_1^2-\nu_1\nu_2+\nu_2^2)/3}\, \mathrm{e}^{2\pi i(x_1\nu_1+x_2(2\nu_2-\nu_1)/\sqrt{3})}\,.
\end{aligned}
$$

Hence, the main saturation error which corresponds to $\boldsymbol{\alpha} = (0,0)$ is bounded by

$$
\begin{aligned}
\left|1 - \frac{\sqrt{3}}{2\pi\mathcal{D}} \sum_{\mathbf{m}\in\mathbb{Z}^2} e^{-|\mathbf{x}-A\mathbf{m}|^2/\mathcal{D}}\right| &\le \sum_{(\nu_1,\nu_2)\neq(0,0)} \mathrm{e}^{-4\pi^2\mathcal{D}(\nu_1^2-\nu_1\nu_2+\nu_2^2)/3} \\
&= 6\,\mathrm{e}^{-4\pi^2\mathcal{D}/3} + \mathcal{O}(\mathrm{e}^{-4\pi^2\mathcal{D}})\,.
\end{aligned}
$$

Note that this difference is less than the single and double precision of floating point arithmetics of modern computers, if $\mathcal{D} \ge 1.5$ and $\mathcal{D} \ge 3.0$, respectively.

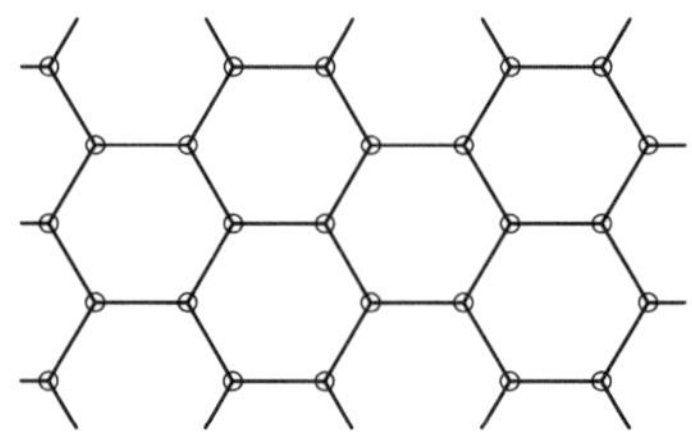

FIGURE 10.2. Hexagonal grid

Next, we consider a regular hexagonal grid of side length 1, depicted in Fig. 10.2. It can be obtained when the nodes of a triangular grid of side length $\sqrt{3}$ are removed from the nodes of a regular triangular grid of side length 1. This is indicated in Fig. 10.3, where the eliminated triangular grid is depicted with dashed lines. The

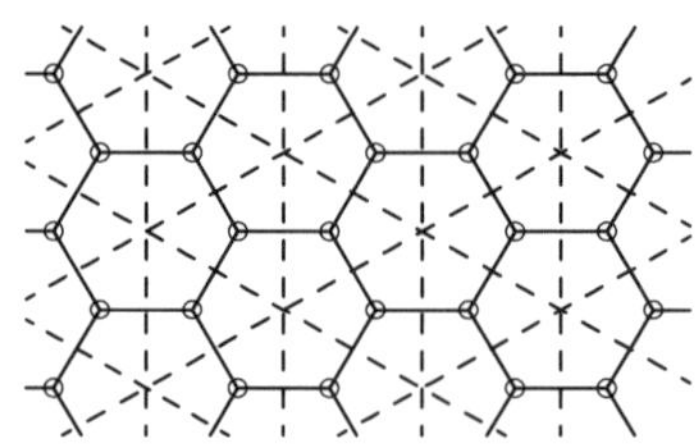

FIGURE 10.3. Nodes of a hexagonal grid

removed nodes can be written in the form $B\mathbf{m}$, $\mathbf{m} \in \mathbb{Z}^2$, with the matrix

$$
B = \begin{pmatrix} 3/2 & 0 \\ \sqrt{3}/2 & \sqrt{3} \end{pmatrix}.
$$

Hence, the set of nodes $\mathbf{X}^\diamond$ of the regular hexagonal grid are given by

$$\mathbf{X}^\diamond = \{A\mathbf{m}\}_{\mathbf{m}\in\mathbb{Z}^2} \setminus \{B\mathbf{m}\}_{\mathbf{m}\in\mathbb{Z}^2}\,,$$

and the sum of the shifted basis functions $\eta(\cdot/\sqrt{\mathcal{D}})$ centered at the nodes of $\mathbf{X}^\diamond$ can be written as

$$\sum_{\mathbf{y}^\diamond\in\mathbf{X}^\diamond} \eta\Big(\frac{\mathbf{x}-\mathbf{y}^\diamond}{\sqrt{\mathcal{D}}}\Big) = \sum_{\mathbf{m}\in\mathbb{Z}^2} \eta\Big(\frac{\mathbf{x}-A\mathbf{m}}{\sqrt{\mathcal{D}}}\Big) - \sum_{\mathbf{m}\in\mathbb{Z}^2} \eta\Big(\frac{\mathbf{x}-B\mathbf{m}}{\sqrt{\mathcal{D}}}\Big).$$

Under the condition $\mathcal{F}\eta(\mathbf{0}) = 1$, we have from (10.2) that

$$\sum_{\mathbf{m}\in\mathbb{Z}^2} \eta\Big(\frac{\mathbf{x}-A\mathbf{m}}{\sqrt{\mathcal{D}}}\Big) = \frac{\mathcal{D}}{\det A}\Big(1 + \sum_{\boldsymbol{\nu}\in\mathbb{Z}^2\setminus\{\mathbf{0}\}} \mathcal{F}\eta(\sqrt{\mathcal{D}}(A^t)^{-1}\boldsymbol{\nu})\, e^{2\pi i\langle\mathbf{x},(A^t)^{-1}\boldsymbol{\nu}\rangle}\Big).$$

Thus we obtain

$$\begin{aligned}\sum_{\mathbf{y}^\diamond\in\mathbf{X}^\diamond} \eta\Big(\frac{\mathbf{x}-\mathbf{y}^\diamond}{\sqrt{\mathcal{D}}}\Big) &= \frac{2\mathcal{D}}{\sqrt{3}} + \frac{2\mathcal{D}}{\sqrt{3}} \sum_{\boldsymbol{\nu}\in\mathbb{Z}^2\setminus\{\mathbf{0}\}} \mathcal{F}\eta(\sqrt{\mathcal{D}}(A^t)^{-1}\boldsymbol{\nu})\, e^{2\pi i\langle\mathbf{x},(A^t)^{-1}\boldsymbol{\nu}\rangle}\\ &\quad - \frac{2\mathcal{D}}{3\sqrt{3}} - \frac{2\mathcal{D}}{3\sqrt{3}} \sum_{\boldsymbol{\nu}\in\mathbb{Z}^2\setminus\{\mathbf{0}\}} \mathcal{F}\eta(\sqrt{\mathcal{D}}(B^t)^{-1}\boldsymbol{\nu})\, e^{2\pi i\langle\mathbf{x},(B^t)^{-1}\boldsymbol{\nu}\rangle}.\end{aligned}$$

Hence, an approximate partition of unity centered at the hexagonal grid is given by

$$\begin{aligned}\frac{3\sqrt{3}}{4\mathcal{D}} \sum_{\mathbf{y}^\diamond\in\mathbf{X}^\diamond} \eta\Big(\frac{\mathbf{x}-\mathbf{y}^\diamond}{\sqrt{\mathcal{D}}}\Big) &= 1 + \frac{3}{2} \sum_{\boldsymbol{\nu}\in\mathbb{Z}^2\setminus\{\mathbf{0}\}} \mathcal{F}\eta(\sqrt{\mathcal{D}}(A^t)^{-1}\boldsymbol{\nu})\, e^{2\pi i\langle\mathbf{x},(A^t)^{-1}\boldsymbol{\nu}\rangle}\\ &\quad - \frac{1}{2} \sum_{\boldsymbol{\nu}\in\mathbb{Z}^2\setminus\{\mathbf{0}\}} \mathcal{F}\eta(\sqrt{\mathcal{D}}(B^t)^{-1}\boldsymbol{\nu})\, e^{2\pi i\langle\mathbf{x},(B^t)^{-1}\boldsymbol{\nu}\rangle}.\end{aligned}$$

Now we define the quasi-interpolant on the h-scaled hexagonal grid

$$h\mathbf{X}^\diamond = \{hA\mathbf{m}\}_{\mathbf{m}\in\mathbb{Z}^2} \setminus \{hB\mathbf{m}\}_{\mathbf{m}\in\mathbb{Z}^2}$$

as

$$\mathcal{M}_h^\diamond u(\mathbf{x}) := \frac{3\sqrt{3}}{4\mathcal{D}} \sum_{\mathbf{y}^\diamond\in\mathbf{X}^\diamond} u(h\mathbf{y}^\diamond)\,\eta\Big(\frac{\mathbf{x}-h\mathbf{y}^\diamond}{h\sqrt{\mathcal{D}}}\Big). \tag{10.4}$$

Since it can be written in the form

$$\mathcal{M}_h^\diamond u(\mathbf{x}) = \frac{3\sqrt{3}}{4\mathcal{D}}\Big(\sum_{\mathbf{m}\in\mathbb{Z}^2} u(hA\mathbf{m})\eta\Big(\frac{\mathbf{x}-hA\mathbf{m}}{h\sqrt{\mathcal{D}}}\Big) - \sum_{\mathbf{m}\in\mathbb{Z}^2} u(hB\mathbf{m})\eta\Big(\frac{\mathbf{x}-hB\mathbf{m}}{h\sqrt{\mathcal{D}}}\Big)\Big),$$

we see that under conditions (2.45) and (2.47) the quasi-interpolant $\mathcal{M}_h^\diamond$ provides the usual estimates (2.49), (2.58) and (2.60) for sufficiently large $\mathcal{D}$.

Because

$$B^{-1} = \begin{pmatrix} 2/3 & 0 \\ -1/3 & \sqrt{3}/3 \end{pmatrix},$$

we obtain, by using (10.3), Poisson's summation formula for Gaussians on the hexagonal grid

$$\frac{3\sqrt{3}}{4\pi\mathcal{D}}\sum_{\mathbf{y}^\diamond\in\mathbf{X}^\diamond} \mathrm{e}^{-|\mathbf{x}-\mathbf{y}^\diamond|^2/\mathcal{D}} = \frac{3\sqrt{3}}{4\pi\mathcal{D}}\Big(\sum_{\mathbf{m}\in\mathbb{Z}^2} \mathrm{e}^{-|\mathbf{x}-A\mathbf{m}|^2/\mathcal{D}} - \sum_{\mathbf{m}\in\mathbb{Z}^2} \mathrm{e}^{-|\mathbf{x}-B\mathbf{m}|^2/\mathcal{D}}\Big)$$
$$= \frac{3}{2}\sum_{\boldsymbol{\nu}\in\mathbb{Z}^2} \mathrm{e}^{-4\pi^2\mathcal{D}(\nu_1^2-\nu_1\nu_2+\nu_2^2)/3}\,\mathrm{e}^{2\pi i(x_1\nu_1+x_2(2\nu_2-\nu_1)/\sqrt{3})}$$
$$-\frac{1}{2}\sum_{\boldsymbol{\nu}\in\mathbb{Z}^2} \mathrm{e}^{-4\pi^2\mathcal{D}(\nu_1^2-\nu_1\nu_2+\nu_2^2)/9}\,\mathrm{e}^{2\pi i(x_1(2\nu_1-\nu_2)/3+x_2\nu_2/\sqrt{3})}\,.$$

Hence, the main saturation error for $\eta(\mathbf{x}) = \pi^{-1}\,\mathrm{e}^{-|\mathbf{x}|^2}$ is bounded by

$$\Big|1 - \frac{3\sqrt{3}}{4\pi\mathcal{D}}\sum_{\mathbf{y}^\diamond\in\mathbf{X}^\diamond} \mathrm{e}^{-|\mathbf{x}-\mathbf{y}^\diamond|^2/\mathcal{D}}\Big|$$
$$\le \frac{1}{2}\sum_{(\nu_1,\nu_2)\neq(0,0)} 3\,\mathrm{e}^{-4\pi^2\mathcal{D}(\nu_1^2-\nu_1\nu_2+\nu_2^2)/3} + \mathrm{e}^{-4\pi^2\mathcal{D}(\nu_1^2-\nu_1\nu_2+\nu_2^2)/9}$$
$$= 3\,\mathrm{e}^{-4\pi^2\mathcal{D}/9} + \mathcal{O}(\mathrm{e}^{-4\pi^2\mathcal{D}/3})\,.$$

10.2. Quasi-interpolants for data on perturbed uniform grids

Here we give a simple extension of the quasi-interpolation operator on a uniform grid, defined by (2.23), to a quasi-interpolant which uses the values $u(\mathbf{x}_j)$ on a set of scattered nodes $\mathbf{X} = \{\mathbf{x}_j\}_{j\in J} \subset \mathbb{R}^n$, close to a uniform grid.

10.2.1. Perturbed grid. Consider the following example. Let $\{x_j\}$ be a sequence of points on $\mathbb{R}$, close to the uniform grid $\{hj\}_{j\in\mathbb{Z}}$ and such that $x_{j+1}-x_j \ge c\,h > 0$. Choose a rapidly decaying function η satisfying the conditions

$$\Big|1 - \sum_{j\in\mathbb{Z}}\eta(x-j)\Big| < \varepsilon\,, \quad \Big|\sum_{j\in\mathbb{Z}}(x-j)\eta(x-j)\Big| < \varepsilon.$$

One can easily see (and it is a special case of the results of this section) that the quasi-interpolant

$$M_h u(x) = \sum_{j\in\mathbb{Z}}\Big(\frac{x_{j+1}-hj}{x_{j+1}-x_j}u(x_j) + \frac{hj-x_j}{x_{j+1}-x_j}u(x_{j+1})\Big)\,\eta\Big(\frac{x}{h}-j\Big)$$

satisfies the estimate

$$|M_h u(x) - u(x)| \le C\,h^2\,\|u''\|_{L_\infty(\mathbb{R})} + \varepsilon(|u(x)| + h|u'(x)|)\,,$$

where the constant C depends only on the function η.

In the following, we suppose

CONDITION 10.1. There exists a uniform grid Λ such that the quasi-interpolants

$$\mathcal{M}_{h,\mathcal{D}}\,u(\mathbf{x}) = \mathcal{D}^{-n/2}\sum_{\mathbf{y}_j\in\Lambda} u(h\mathbf{y}_j)\,\eta\Big(\frac{\mathbf{x}-h\mathbf{y}_j}{h\sqrt{\mathcal{D}}}\Big) \tag{10.5}$$

approximate sufficiently smooth functions u with the error

$$|u(\mathbf{x}) - \mathcal{M}_{h,\mathcal{D}}u(\mathbf{x})| \le c_\eta\,(\sqrt{\mathcal{D}}h)^N\,\|\nabla_N u\|_{L_\infty} + \varepsilon\sum_{k=0}^{N-1}(\sqrt{\mathcal{D}}h)^k|\nabla_k u(\mathbf{x})|\,, \tag{10.6}$$

for any $\varepsilon > 0$ and sufficiently large $\mathcal{D}$. Let $\mathbf{X}_h$ be a sequence of grids with the property that for $\kappa_1 > 0$ not depending on h and for any $\mathbf{y}_j \in \Lambda$, the ball $B(h\mathbf{y}_j, h\kappa_1)$ contains nodes of $\mathbf{X}_h$.

10.2.2. Construction.

DEFINITION 10.2. *Let $\mathbf{x}_j \in \mathbf{X}_h$. A collection of* $m_N = \dfrac{(N-1+n)!}{n!(N-1)!} - 1$ *nodes* $\mathbf{x}_k \in \mathbf{X}_h$ *will be called a star of* $\mathbf{x}_j$, *denoted by* $\mathrm{st}(\mathbf{x}_j)$, *if the Vandermonde matrix*

$$V_{j,h} = \Big\{ \Big(\frac{\mathbf{x}_k - \mathbf{x}_j}{h} \Big)^{\boldsymbol{\alpha}} \Big\}, \ [\boldsymbol{\alpha}] = 1, ..., N-1, \tag{10.7}$$

is not singular.

CONDITION 10.3. Let $\mathbf{X}_h$ be a sequence of grids satisfying Condition 10.1. Denote by $\widetilde{\mathbf{x}}_j \in \mathbf{X}_h$ the node closest to $h\mathbf{y}_j \in h\Lambda$. There exists $\kappa_2 > 0$ such that for any $\mathbf{y}_j \in \Lambda$ there exists a star $\mathrm{st}(\widetilde{\mathbf{x}}_j) \subset B(\widetilde{\mathbf{x}}_j, h\kappa_2)$ with $|\det V_{j,h}| \geq c > 0$ uniformly in h.

To describe a construction of the quasi-interpolants which use the data at $\mathbf{X}_h$, we denote the elements of the inverse matrix of $V_{j,h}$ by $\{b^{(j)}_{\boldsymbol{\alpha},k}\}$, $[\boldsymbol{\alpha}] = 1, \dots, N-1$, $\mathbf{x}_k \in \mathrm{st}(\widetilde{\mathbf{x}}_j)$, and we define the functional

$$F_{j,h}(u) = u(\widetilde{\mathbf{x}}_j)\Big(1 - \sum_{[\boldsymbol{\alpha}]=1}^{N-1} \Big(\mathbf{y}_j - \frac{\widetilde{\mathbf{x}}_j}{h}\Big)^{\boldsymbol{\alpha}} \sum_{\mathbf{x}_k \in \mathrm{st}(\widetilde{\mathbf{x}}_j)} b^{(j)}_{\boldsymbol{\alpha},k}\Big)$$
$$+ \sum_{\mathbf{x}_k \in \mathrm{st}(\widetilde{\mathbf{x}}_j)} u(\mathbf{x}_k) \sum_{[\boldsymbol{\alpha}]=1}^{N-1} b^{(j)}_{\boldsymbol{\alpha},k} \Big(\mathbf{y}_j - \frac{\widetilde{\mathbf{x}}_j}{h}\Big)^{\boldsymbol{\alpha}}.$$

Then the quasi-interpolant is defined as the sum

$$\mathbb{M}_{h,\mathcal{D}} u(\mathbf{x}) = \mathcal{D}^{-n/2} \sum_{\mathbf{y}_j \in \Lambda} F_{j,h}(u) \eta\Big(\frac{\mathbf{x} - h\mathbf{y}_j}{h\sqrt{\mathcal{D}}}\Big), \tag{10.8}$$

i.e., the generating functions are centered at the nodes of the uniform grid $h\Lambda$. As in the case of uniform grids, this construction allows one to apply fast methods for the cubature of integral operators.

10.2.3. Error estimates.

THEOREM 10.4. *Under Conditions* 10.1 *and* 10.3, *for any* $\varepsilon > 0$, *there exists* $\mathcal{D}$ *such that the quasi-interpolant* (10.8) *approximates any* $u \in W^N_\infty(\mathbb{R}^n)$ *with the error estimate*

$$|\mathbb{M}_{h,\mathcal{D}} u(\mathbf{x}) - u(\mathbf{x})| \leq c_{N,\eta,\mathcal{D}}\, h^N \|\nabla_N u\|_{L_\infty} + \varepsilon \sum_{k=0}^{N-1} (\sqrt{\mathcal{D}} h)^k |\nabla_k u(\mathbf{x})|, \tag{10.9}$$

where $c_{N,\eta,\mathcal{D}}$ *does not depend on* u *and* h.

PROOF. For a given $u \in W^N_\infty(\mathbb{R}^n)$ and the grid $\mathbf{X}_h$, we consider the quasi-interpolant (10.5) on the uniform grid $\{h\Lambda\}$

$$\mathcal{M}_{h,\mathcal{D}} u(\mathbf{x}) = \mathcal{D}^{-n/2} \sum_{\mathbf{y}_j \in \Lambda} u(h\mathbf{y}_j)\, \eta\Big(\frac{\mathbf{x} - h\mathbf{y}_j}{h\sqrt{\mathcal{D}}}\Big).$$

According to Condition 10.1, we can find $\mathcal{D}$ such that $\mathcal{M}_{h,\mathcal{D}}u$ satisfies the inequality

$$|\mathcal{M}_{h,\mathcal{D}}u(\mathbf{x}) - u(\mathbf{x})| \le c_{N,\eta}\,(\sqrt{\mathcal{D}}h)^N \|\nabla_N u\|_{L_\infty} + \varepsilon \sum_{k=0}^{N-1} (\sqrt{\mathcal{D}}h)^k |\nabla_k u(\mathbf{x})|\,. \tag{10.10}$$

Hence, it remains to estimate $|\mathcal{M}_{h,\mathcal{D}}u - \mathbb{M}_{h,\mathcal{D}}u|$. From the Taylor expansion (2.1) of u around $\mathbf{t} \in \mathbb{R}^n$, we have

$$u(\mathbf{x}) = \sum_{[\boldsymbol{\alpha}]=0}^{N-1} \frac{\partial^{\boldsymbol{\alpha}} u(\mathbf{t})}{\boldsymbol{\alpha}!} (\mathbf{x}-\mathbf{t})^{\boldsymbol{\alpha}} + R_N(\mathbf{x},\mathbf{t}) \tag{10.11}$$

with the remainder

$$R_N(\mathbf{x},\mathbf{t}) = \sum_{[\boldsymbol{\alpha}]=N} \frac{(\mathbf{x}-\mathbf{t})^{\boldsymbol{\alpha}}}{\boldsymbol{\alpha}!} U_{\boldsymbol{\alpha}}(\mathbf{t},\mathbf{x})\,,$$

which obviously satisfies

$$|R_N(\mathbf{x},\mathbf{t})| \le c_N |\mathbf{x}-\mathbf{t}|^N \sup_{B(\mathbf{t},|\mathbf{x}-\mathbf{t}|)} |\nabla_N u|\,. \tag{10.12}$$

For $\mathbf{y}_j \in \Lambda$, we choose $\widetilde{\mathbf{x}}_j \in \mathbf{X}$ and use (10.11) with $\mathbf{t} = \widetilde{\mathbf{x}}_j$. We split

$$\mathcal{M}_{h,\mathcal{D}}u(\mathbf{x}) = M^{(1)}u(\mathbf{x}) + \mathcal{D}^{-n/2} \sum_{\mathbf{y}_j\in\Lambda} R_N(h\mathbf{y}_j, \widetilde{\mathbf{x}}_j)\, \eta\Big(\frac{\mathbf{x}-h\mathbf{y}_j}{h\sqrt{\mathcal{D}}}\Big)$$

with

$$M^{(1)}u(\mathbf{x}) = \mathcal{D}^{-n/2} \sum_{\mathbf{y}_j\in\Lambda} \sum_{\boldsymbol{\alpha}=0}^{N-1} \frac{\partial^{\boldsymbol{\alpha}} u(\widetilde{\mathbf{x}}_j)}{\boldsymbol{\alpha}!} (h\mathbf{y}_j - \widetilde{\mathbf{x}}_j)^{\boldsymbol{\alpha}}\, \eta\Big(\frac{\mathbf{x}-h\mathbf{y}_j}{h\sqrt{\mathcal{D}}}\Big)\,. \tag{10.13}$$

Because $|h\mathbf{y}_j - \widetilde{\mathbf{x}}_j| \le \kappa_1 h$ for any $\mathbf{y}_j$, we derive from (10.12) that

$$\begin{aligned} &|M^{(1)}u(\mathbf{x}) - \mathcal{M}_{h,\mathcal{D}}u(\mathbf{x})| \\ &\qquad \le c_N\,(\kappa_1 h)^N \mathcal{D}^{-n/2} \sum_{\mathbf{y}_j\in\Lambda} \Big|\eta\Big(\frac{\mathbf{x}-h\mathbf{y}_j}{h\sqrt{\mathcal{D}}}\Big)\Big| \sup_{B(\mathbf{x},h\kappa_1)} |\nabla_N u|\,. \end{aligned} \tag{10.14}$$

The next step is to approximate $\partial^{\boldsymbol{\alpha}} u(\widetilde{\mathbf{x}}_j)$, $1 \le [\boldsymbol{\alpha}] < N$, by a linear combination of $u(\mathbf{x}_k)$, $\mathbf{x}_k \in \mathrm{st}(\widetilde{\mathbf{x}}_j)$. Consider the linear system with m_N unknowns

$$\sum_{[\boldsymbol{\alpha}]=1}^{N-1} \frac{a_{\boldsymbol{\alpha}}^{(j)}}{\boldsymbol{\alpha}!} (\mathbf{x}_k - \widetilde{\mathbf{x}}_j)^{\boldsymbol{\alpha}} = u(\mathbf{x}_k) - u(\widetilde{\mathbf{x}}_j)\,, \quad \mathbf{x}_k \in \mathrm{st}(\widetilde{\mathbf{x}}_j)\,, \tag{10.15}$$

which, in view of Condition 10.3, has a unique solution $\{a_{\boldsymbol{\alpha}}^{(j)}\}_{1\le[\boldsymbol{\alpha}]\le N-1}$. It follows from (10.11) and (10.15) that

$$\sum_{[\boldsymbol{\alpha}]=1}^{N-1} \frac{h^{[\boldsymbol{\alpha}]}}{\boldsymbol{\alpha}!} (a_{\boldsymbol{\alpha}}^{(j)} - \partial^{\boldsymbol{\alpha}} u(\widetilde{\mathbf{x}}_j)) \Big(\frac{\mathbf{x}_k - \widetilde{\mathbf{x}}_j}{h}\Big)^{\boldsymbol{\alpha}} = R_N(\mathbf{x}_k, \widetilde{\mathbf{x}}_j)\,.$$

Again by Condition 10.3, the norms of $V_{j,h}^{-1}$ are bounded uniformly for all j and h. This, together with (10.12), leads to the inequalities

$$\frac{|a_{\boldsymbol{\alpha}}^{(j)} - \partial^{\boldsymbol{\alpha}} u(\widetilde{\mathbf{x}}_j)|}{\boldsymbol{\alpha}!} \le C_2\, h^{N-[\boldsymbol{\alpha}]} \sup_{B(\widetilde{\mathbf{x}}_j, h\kappa_2)} |\nabla_N u|\,,\ 0 \le [\boldsymbol{\alpha}] < N\,. \tag{10.16}$$

Therefore, if we replace the derivatives $\partial^{\boldsymbol{\alpha}} u(\widetilde{\mathbf{x}}_j)$ in (10.13) by $a_{\boldsymbol{\alpha}}^{(j)}$, then we get the sum

$$\mathcal{D}^{-n/2} \sum_{\mathbf{y}_j \in \Lambda} \Big(u(\widetilde{\mathbf{x}}_j) + \sum_{[\boldsymbol{\alpha}]=1}^{N-1} \frac{a_{\boldsymbol{\alpha}}^{(j)}}{\boldsymbol{\alpha}!} (h\mathbf{y}_j - \widetilde{\mathbf{x}}_j)^{\boldsymbol{\alpha}} \Big) \eta\Big(\frac{\mathbf{x} - h\,\mathbf{y}_j}{h\sqrt{\mathcal{D}}} \Big),$$

which, in view of

$$a_{\boldsymbol{\alpha}}^{(j)} = \frac{\boldsymbol{\alpha}!}{h^{[\boldsymbol{\alpha}]}} \sum_{\mathbf{x}_k \in \mathrm{st}(\widetilde{\mathbf{x}}_j)} b_{\boldsymbol{\alpha},\mathbf{k}}^{(j)} \left(u(\mathbf{x}_k) - u(\widetilde{\mathbf{x}}_j) \right),$$

coincides with the quasi-interpolant $\mathbb{M}_{h,\mathcal{D}} u$ defined by (10.8). Moreover, by (10.13) and (10.16),

$$\begin{aligned} &|\mathbb{M}\, u(\mathbf{x}) - M^{(1)} u(\mathbf{x})| \\ &\le C_2 h^N \sum_{[\boldsymbol{\alpha}]=1}^{N-1} \kappa_1^{[\boldsymbol{\alpha}]} \mathcal{D}^{-n/2} \sum_{\mathbf{y}_j \in \Lambda} \Big| \eta\Big(\frac{\mathbf{x} - h\,\mathbf{y}_j}{h\sqrt{\mathcal{D}}} \Big) \Big| \sup_{B(\mathbf{x}, h\kappa_2)} |\nabla_N u| . \end{aligned} \tag{10.17}$$

Now, we use Lemma 2.6, which implies the inequality

$$\sup_{\mathbf{x} \in \mathbb{R}^n} \mathcal{D}^{-n/2} \sum_{\mathbf{y}_j \in \Lambda} \Big| \eta\Big(\frac{\mathbf{x} - \mathbf{y}_j}{\sqrt{\mathcal{D}}} \Big) \Big| \le C_3$$

for all $\mathcal{D} \ge \mathcal{D}_0 > 0$. Hence, from (10.14) and (10.17), we obtain

$$|\mathcal{M}_{h,\mathcal{D}} u(\mathbf{x}) - \mathbb{M}_{h,\mathcal{D}} u(\mathbf{x})| \le C_4 h^N \sup_{B(\mathbf{x}, h\kappa_2)} |\nabla_N u| ,$$

which, together with (10.10), establishes the estimate (10.9). □

10.2.4. Numerical experiments with quasi-interpolants. The behavior of the quasi-interpolant $\mathbb{M}_{h,\mathcal{D}} u$ was tested by one- and two-dimensional experiments. In all cases, the scattered grid is chosen such that any ball $B(h\mathbf{j}, h/2)$, $\mathbf{j} \in \mathbb{Z}^n$, $n = 1$ or $n = 2$, contains one randomly chosen node, which is denoted by $\mathbf{x}_{\mathbf{j}}$.

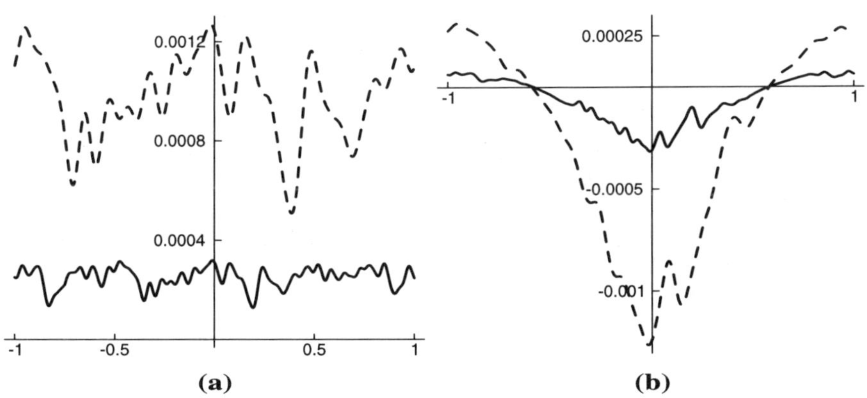

FIGURE 10.4. The graphs of $\mathbb{M}_{h,\mathcal{D}} u - u$ with $\eta(x) = \pi^{-1/2} e^{-x^2}$, $\mathcal{D} = 2$, $\mathrm{st}(x_j) = \{x_{j+1}\}$, when **(a)** $u(x) = x^2$ (on the left) and **(b)** $u(x) = (1 + x^2)^{-1}$. Dashed and solid lines correspond to $h = 1/32$ and $h = 1/64$, respectively.

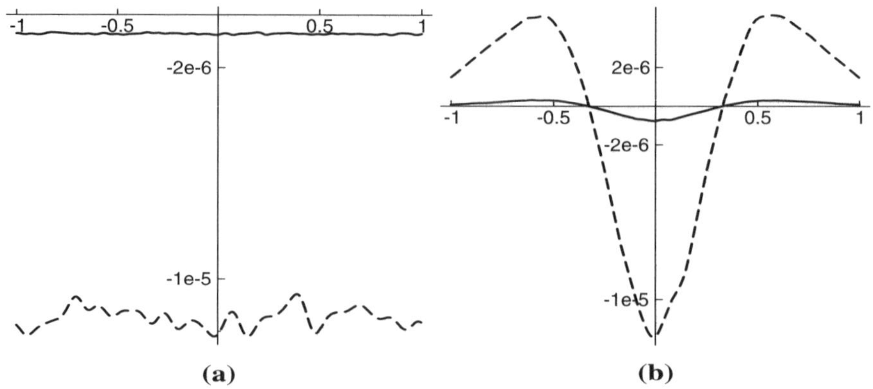

FIGURE 10.5. The graphs of $\mathbb{M}_{h,\mathcal{D}}u - u$ with $\eta(x) = \pi^{-1/2}(3/2 - x^2)e^{-x^2}$, $\mathcal{D} = 4$, $\mathrm{st}(x_j) = \{x_{j-2}, x_{j-1}, x_{j+1}\}$, when **(a)** $u(x) = x^4$ and **(b)** $u(x) = (1 + x^2)^{-1}$. Dashed and solid lines correspond to $h = 1/32$ and $h = 1/64$, respectively.

In the one-dimensional case, Figs. 10.4 and 10.5 show the graphs of $u - \mathbb{M}_{h,\mathcal{D}}u$ for different smooth functions u for $h = 1/32$ (dashed line) and $h = 1/64$ (solid line). We use $\pi^{-1/2}e^{-x^2}$ (Fig. 10.4) and $\pi^{-1/2}(3/2 - x^2)\,e^{-x^2}$ (Fig. 10.5) as generating functions.

As two-dimensional examples, in Figs. 10.6 and 10.7, we depict the quasi-interpolation error $\mathbb{M}_{h,\mathcal{D}}u - u$ for the function $u(\mathbf{x}) = (1 + |\mathbf{x}|^2)^{-1}$ and different h if generating functions of second- (with $\mathcal{D} = 2$) and fourth- (with $\mathcal{D} = 4$) orders of approximation are used. The h^2- and, respectively, h^4-convergence of the corresponding two-dimensional quasi-interpolants are confirmed by the L_∞-errors, which are given in Table 10.1.

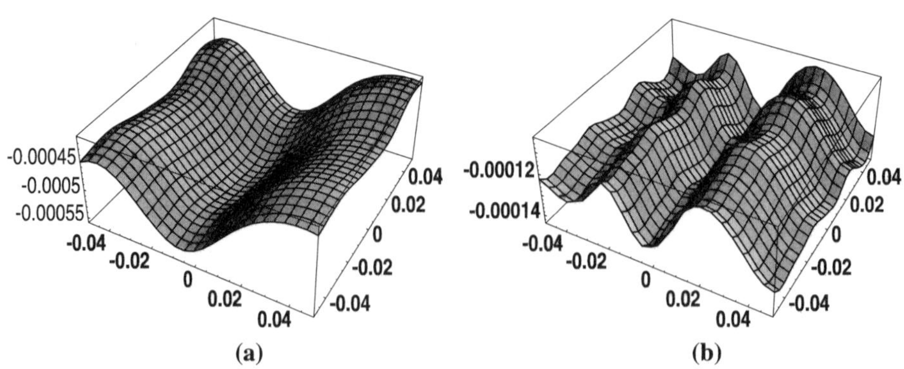

FIGURE 10.6. The graph of $\mathbb{M}_{h,\mathcal{D}}u - u$ with $u(\mathbf{x}) = (1 + |\mathbf{x}|^2)^{-1}$, $\mathcal{D} = 2$, $N = 2$, $\eta(\mathbf{x}) = \pi^{-1}e^{-|\mathbf{x}|^2}$, **(a)** $h = 2^{-6}$ and **(b)** $h = 2^{-7}$.

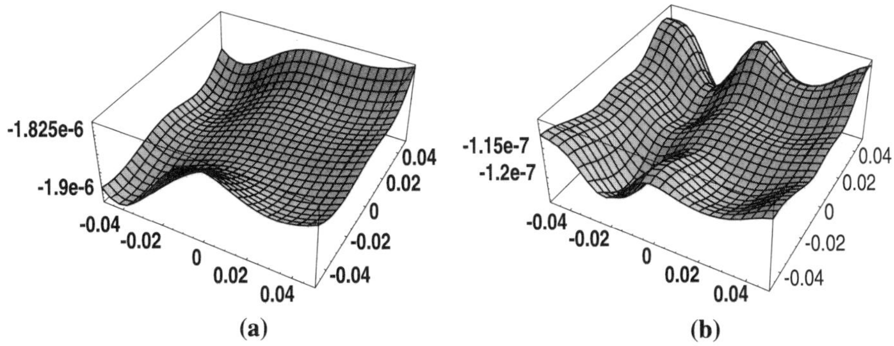

FIGURE 10.7. The graph of $\mathbb{M}_{h,\mathcal{D}}u - u$ with $u(\mathbf{x}) = (1 + |\mathbf{x}|^2)^{-1}$, $\mathcal{D} = 4$, $N = 4$, $\eta(\mathbf{x}) = \pi^{-1}(2 - |\mathbf{x}|^2)e^{-|\mathbf{x}|^2}$, **(a)** $h = 2^{-6}$ and **(b)** $h = 2^{-7}$.

h	$\mathcal{D} = 2$	$\mathcal{D} = 4$
2^{-4}	$8.75 \cdot 10^{-3}$	$1.57 \cdot 10^{-2}$
2^{-5}	$2.21 \cdot 10^{-3}$	$4.00 \cdot 10^{-3}$
2^{-6}	$5.51 \cdot 10^{-4}$	$1.01 \cdot 10^{-3}$
2^{-7}	$1.42 \cdot 10^{-4}$	$2.52 \cdot 10^{-4}$
2^{-8}	$3.56 \cdot 10^{-5}$	$6.50 \cdot 10^{-5}$

h	$\mathcal{D} = 4$	$\mathcal{D} = 6$
2^{-4}	$4.42 \cdot 10^{-4}$	$9.59 \cdot 10^{-4}$
2^{-5}	$2.95 \cdot 10^{-5}$	$6.61 \cdot 10^{-5}$
2^{-6}	$1.92 \cdot 10^{-6}$	$4.24 \cdot 10^{-6}$
2^{-7}	$1.24 \cdot 10^{-7}$	$2.68 \cdot 10^{-7}$
2^{-8}	$7.80 \cdot 10^{-9}$	$1.71 \cdot 10^{-8}$

TABLE 10.1. L_∞-approximation error for the function $u(\mathbf{x}) = (1+|\mathbf{x}|^2)^{-1}$ using $\mathbb{M}_{h,\mathcal{D}}u$ with $\eta(\mathbf{x}) = \pi^{-1}e^{-|\mathbf{x}|^2}$, $N = 2$ (on the left), and $\eta(\mathbf{x}) = \pi^{-1}(2 - |\mathbf{x}|^2)e^{-|\mathbf{x}|^2}$, $N = 4$ (on the right).

10.3. Non-uniformly distributed nodes

Here we study a case of non-uniform grids. To be more precise, we consider approximate quasi-interpolation for functions with compact support in some domain $\Omega \subset \mathbb{R}^n$ and for functions given on an n-dimensional manifold in the case when the nodes are images of a smooth mapping of uniformly distributed grid points. The approximant should have a simple semi-discrete convolutional form, similar to (2.23), in order to get effective methods for computing pseudodifferential operators.

10.3.1. Description of construction and error estimate. Suppose that $\omega \subset \mathbb{R}^n$ is a bounded domain and that $\boldsymbol{\phi} = (\varphi_1, \ldots, \varphi_s) : \omega \to \mathbb{R}^s$, $n \le s$, is a sufficiently smooth and non-singular mapping, defined in a neighborhood of ω. That means that

$$|\boldsymbol{\phi}'(\mathbf{y})| = \Big(\sum_{(i)} (\kappa_{(i)}(\mathbf{y}))^2\Big)^{1/2} \neq 0\,, \quad \mathbf{y} \in \omega\,, \tag{10.18}$$

where $\kappa^{(i)}$ denotes the minor of order n of the matrix

$$\phi'(\mathbf{y}) = \begin{Vmatrix} \partial\varphi_1/\partial y_1 & \cdots & \partial\varphi_1/\partial y_n \\ \vdots & \ddots & \vdots \\ \partial\varphi_s/\partial y_1 & \cdots & \partial\varphi_s/\partial y_n \end{Vmatrix},$$

corresponding to the rows with indices $i_1 < \ldots < i_n$. The sum is extended over all distinct $(i) = (i_1, \ldots, i_n)$, $1 \le i_p \le s$, of this kind. Then ϕ generates a one-to-one mapping between ω and $\Omega = \phi(\omega) \in \mathbb{R}^s$.

Let u be an N-times continuously differentiable function on Ω with compact support, i.e., the composition $u \circ \phi \in C_0^N(\omega)$. It the following, we study the approximation of u by the quasi-interpolant

$$u_h(\mathbf{x}) = \mathcal{D}^{-n/2} \sum_{h\mathbf{m} \in B} u(\phi(h\mathbf{m}))\, \eta\Big(\frac{\mathbf{x} - \phi(h\mathbf{m})}{\sqrt{\mathcal{D}}h\, |\phi'(h\mathbf{m})|^{1/n}}\Big), \quad \mathbf{x} = \phi(\mathbf{y}) \in \Omega, \tag{10.19}$$

where the summation extends over all $h\mathbf{m}$ belonging to some subset $B \subset \omega$, which depends on the point $\mathbf{x} \in \Omega$.

Note that formula (10.19) in the case $s = n$ corresponds to the quasi-interpolation on a domain $\Omega \in \mathbb{R}^n$ with respect to a set of data points $\{\mathbf{x_m}\}$, which can be represented as a sufficiently smooth image of a uniform lattice, $\mathbf{x_m} = \phi(h\mathbf{m})$, $h\mathbf{m} \in \omega$. If $s > n$, then $\Omega = \phi(\omega) \in \mathbb{R}^s$ can be considered as a part of an n-dimensional manifold parametrized by ϕ. The generating function η is given in $\mathbb{R}^s$; hence (10.19) defines a function in $\mathbb{R}^s$. We are interested in how the restriction of this linear combination to Ω approximates the function u on Ω. In the sequel, we suppose that in the case $s > n$ the generating function η is radial.

The following theorem states that the difference $u_h - u$ has a similar behavior as in the case of uniform grids provided that η satisfies the moment Condition 2.15 of order N and some additional smoothness and decay requirements.

THEOREM 10.5. *Assume, besides the moment Condition* 2.15 *of order* N, *that* η *is* $N+\mu-1$*-times continuously differentiable in* $\mathbb{R}^s$ *with* $\mu = [n/2]+1$, *the smallest integer greater than* $n/2$. *Additionally, all the derivatives* $\partial^{\boldsymbol\beta}\eta$, $0 \le [\boldsymbol\beta] \le N+\mu-1$, *have to satisfy the condition*

$$|\partial^{\boldsymbol\beta}\eta(\mathbf{x})| \le C_{\boldsymbol\beta} \begin{cases} (1+|\mathbf{x}|)^{-K-[\boldsymbol\beta]}, & 0 \le [\boldsymbol\beta] \le N, \\ (1+|\mathbf{x}|)^{-K-N}, & N < [\boldsymbol\beta] \le N+\mu-1, \end{cases} \tag{10.20}$$

for some number $K > N + n$. *Furthermore, we assume that* $\phi : \omega \to \Omega$ *is in the class* C^{N+1} *and we let* $u \in C_0^N(\Omega)$. *Then, for any* $\varepsilon > 0$, *there exists* $\mathcal{D} > 0$ *such that at any point* $\mathbf{x} \in \Omega$

$$|u_h(\mathbf{x}) - u(\mathbf{x})| \le c\, (\sqrt{\mathcal{D}}h)^N\, \|u\|_{C^N(\overline{\Omega})} + \varepsilon \sum_{k=0}^{N-1} c_k\, (\sqrt{\mathcal{D}}h)^k. \tag{10.21}$$

Here c *does not depend on* u, h, *and* $\mathcal{D}$ *and the numbers* c_k *can be obtained from the values* $\partial^{\boldsymbol\alpha}u(\mathbf{x})$, $[\boldsymbol\alpha] \le k$.

REMARK 10.6. For a given non-singular parametrization ϕ of Ω, the quasi-interpolant

$$\mathcal{D}^{-n/2} \sum_{m \in \mathbb{Z}^n} u(\phi(h\mathbf{m}))\, \eta\Big(\frac{\phi^{-1}(\mathbf{x}) - h\mathbf{m}}{\sqrt{\mathcal{D}}h}\Big)$$

is an approximation formula for a uniform grid on ω. It provides approximation errors similar to those established in Section 2.3. However, we want to have approximations with the property that, for example, the action of convolution operators can be determined efficiently. These integral operators have difference kernels with respect to the variable $\mathbf{x} \in \Omega$; therefore, formula (10.19) leads to semi-analytic approximations as soon as the convolution of η is known.

10.3.2. Quasi-interpolation on domains. The convergence proof is based on several lemmas. Let us fix a point $\mathbf{x} \in \Omega$, denote $\mathbf{y}_0 = \boldsymbol{\phi}^{-1}(\mathbf{x}) \in \omega$, and make the substitution

$$\boldsymbol{\xi}(\mathbf{y}) = \frac{\mathbf{x} - \boldsymbol{\phi}(\mathbf{y})}{|\boldsymbol{\phi}'(\mathbf{y})|^{1/n}} = \frac{\boldsymbol{\phi}(\mathbf{y}_0) - \boldsymbol{\phi}(\mathbf{y})}{|\boldsymbol{\phi}'(\mathbf{y})|^{1/n}}, \quad \mathbf{y} \in \omega . \tag{10.22}$$

LEMMA 10.7. *The mapping $\boldsymbol{\xi} : \omega \to \mathbb{R}^s$ can be represented in the form*

$$\boldsymbol{\xi}(\mathbf{y}) = A(\mathbf{y}_0 - \mathbf{y}) + |\mathbf{y}_0 - \mathbf{y}|^2 \, \widetilde{\boldsymbol{\xi}}(\mathbf{y}) , \quad \mathbf{y} \in \omega , \tag{10.23}$$

where $A : \mathbb{R}^n \to \mathbb{R}^s$ is a linear mapping with $|A| = 1$ and $\widetilde{\boldsymbol{\xi}} \in C^{N-1}(\omega)$. There exist a closed ball $B(\mathbf{y}_0, \kappa) \subset \omega$ around $\mathbf{y}_0$ with radius $\kappa > 0$ and positive constants C_1 and C_2 such that for any $\mathbf{y} \in B(\mathbf{y}_0, \kappa)$ and all real $s \in [0, 1]$

$$|\boldsymbol{\xi}'(\mathbf{y})| \geq C_1 \quad \textit{and} \quad \big|sA(\mathbf{y}_0 - \mathbf{y}) + (1-s)\,\boldsymbol{\xi}(\mathbf{y})\big| \geq C_2 |\mathbf{y}_0 - \mathbf{y}| . \tag{10.24}$$

PROOF. Since

$$\boldsymbol{\xi}'(\mathbf{y}) = -\frac{\boldsymbol{\phi}'(\mathbf{y})}{|\boldsymbol{\phi}'(\mathbf{y})|^{1/n}} - \frac{\boldsymbol{\xi}(\mathbf{y})(\nabla|\boldsymbol{\phi}'(\mathbf{y})|)^T}{n\,|\boldsymbol{\phi}'(\mathbf{y})|}$$

and $\boldsymbol{\xi}(\mathbf{y}_0) = \mathbf{0}$, we obtain $A = |\boldsymbol{\phi}'(\mathbf{y}_0)|^{-1/n}\,\boldsymbol{\phi}'(\mathbf{y}_0)$ which implies $|A| = 1$. Hence, the matrix A^*A is not singular and therefore $|A\mathbf{y}| \geq c|\mathbf{y}|$ with some positive constant c. Now, Taylor's formula

$$\begin{aligned} \boldsymbol{\xi}(\mathbf{y}) &= \boldsymbol{\xi}(\mathbf{y}_0) + \boldsymbol{\xi}'(\mathbf{y}_0)(\mathbf{y} - \mathbf{y}_0) + \mathcal{O}(|\mathbf{y} - \mathbf{y}_0|^2) \\ &= -\frac{\boldsymbol{\phi}'(\mathbf{y})}{|\boldsymbol{\phi}'(\mathbf{y})|^{1/n}}(\mathbf{y} - \mathbf{y}_0) + \mathcal{O}(|\mathbf{y} - \mathbf{y}_0|^2) \end{aligned}$$

leads to (10.23) and to

$$\begin{aligned} \big|sA(\mathbf{y}_0 - \mathbf{y}) + (1-s)\,\boldsymbol{\xi}(\mathbf{y})\big| &= \big|A(\mathbf{y}_0 - \mathbf{y}) + (1-s)\,|\mathbf{y}_0 - \mathbf{y}|^2\,\widetilde{\boldsymbol{\xi}}(\mathbf{y})\big| \\ &\geq \big(c - |\mathbf{y}_0 - \mathbf{y}||\widetilde{\boldsymbol{\xi}}(\mathbf{y})|\big)|\mathbf{y}_0 - \mathbf{y}| . \end{aligned}$$

□

After having fixed the ball $B(\mathbf{y}_0, \kappa)$, we will study the quasi-interpolant

$$u_h(\mathbf{x}) = \mathcal{D}^{-n/2} \sum_{h\mathbf{m} \in B(\mathbf{y}_0,\kappa)} u(\boldsymbol{\phi}(h\mathbf{m}))\,\eta\Big(\frac{\boldsymbol{\xi}(h\mathbf{m})}{\sqrt{\mathcal{D}}h}\Big) . \tag{10.25}$$

We give an asymptotic expansion of $u_h(\mathbf{x})$ for $h \to 0$ up to terms of the order $\mathcal{O}(h^N)$. By using (10.23) and the Taylor expansion of η, we split $u_h(\mathbf{x})$ into a finite sum of semi-discrete convolutions plus a remaining term. In the following, we denote the variables in $\mathbb{R}^s$ by $\mathbf{x}$, whereas the symbol $\mathbf{y}$ denotes variables in

$\omega \in \mathbb{R}^n$. Thus, the Taylor expansion (2.1) of the function η on $\mathbb{R}^s$ near the point $A(\mathbf{y}_0 - \mathbf{y})/\sqrt{\mathcal{D}}h$ yields, because $\boldsymbol{\xi}(\mathbf{y}) = A(\mathbf{y}_0 - \mathbf{y}) + |\mathbf{y}_0 - \mathbf{y}|^2 \widetilde{\boldsymbol{\xi}}(\mathbf{y})$,

$$\eta\Big(\frac{\boldsymbol{\xi}(\mathbf{y})}{\sqrt{\mathcal{D}}h}\Big) = \sum_{[\boldsymbol{\beta}]=0}^{N-1} \frac{\big(|\mathbf{y}_0 - \mathbf{y}|^2 \widetilde{\boldsymbol{\xi}}(\mathbf{y})\big)^{\boldsymbol{\beta}}}{\boldsymbol{\beta}!(\sqrt{\mathcal{D}}h)^{[\boldsymbol{\beta}]}} \partial^{\boldsymbol{\beta}} \eta\Big(\frac{A(\mathbf{y}_0 - \mathbf{y})}{\sqrt{\mathcal{D}}h}\Big)$$

$$+ \frac{N}{(\sqrt{\mathcal{D}}h)^N} \sum_{[\boldsymbol{\beta}]=N} \frac{\big(|\mathbf{y}_0 - \mathbf{y}|^2 \widetilde{\boldsymbol{\xi}}(\mathbf{y})\big)^{\boldsymbol{\beta}}}{\boldsymbol{\beta}!} \int_0^1 s^{N-1} \partial^{\boldsymbol{\beta}} \eta\Big(\frac{sA(\mathbf{y}_0 - \mathbf{y}) + (1-s)\,\boldsymbol{\xi}(\mathbf{y})}{\sqrt{\mathcal{D}}h}\Big)\, ds\,,$$

where $\partial^{\boldsymbol{\beta}}$ denotes the corresponding partial derivatives in $\mathbb{R}^s$. Denoting

$$\tilde{u}(\mathbf{y}) = u(\boldsymbol{\phi}(\mathbf{y})) \quad \text{and} \quad \tilde{\eta}_{A,\boldsymbol{\beta}}(\mathbf{y}) = |\mathbf{y}|^{2[\boldsymbol{\beta}]}\, \partial^{\boldsymbol{\beta}} \eta(A\mathbf{y})\,,$$

we obtain the splitting of the quasi-interpolant (10.25)

$$u_h(\mathbf{x}) = \sum_{[\boldsymbol{\beta}]=0}^{N-1} \frac{(\sqrt{\mathcal{D}}h)^{[\boldsymbol{\beta}]}}{\boldsymbol{\beta}!\mathcal{D}^{n/2}} \sum_{h\mathbf{m}\in B(\mathbf{y}_0,\kappa)} \tilde{u}(h\mathbf{m})\, \widetilde{\boldsymbol{\xi}}(h\mathbf{m})^{\boldsymbol{\beta}} \tilde{\eta}_{A,\boldsymbol{\beta}}\Big(\frac{\mathbf{y}_0 - h\mathbf{m}}{\sqrt{\mathcal{D}}h}\Big) + R_N(\mathbf{y}_0)\,, \tag{10.26}$$

where the remainder is of the form

$$R_N(\mathbf{y}_0) = N(\sqrt{\mathcal{D}}h)^N \sum_{[\boldsymbol{\beta}]=N} \frac{\mathcal{D}^{-n/2}}{\boldsymbol{\beta}!} \int_0^1 s^{N-1}$$

$$\times \sum_{h\mathbf{m}\in B(\mathbf{y}_0,\kappa)} \tilde{u}(h\mathbf{m})\, \widetilde{\boldsymbol{\xi}}(h\mathbf{m})^{\boldsymbol{\beta}} \frac{|\mathbf{y}_0 - h\mathbf{m}|^{2N}}{(\mathcal{D}h^2)^N}\, \partial^{\boldsymbol{\beta}} \eta\Big(\frac{sA(\mathbf{y}_0 - h\mathbf{m}) + (1-s)\,\boldsymbol{\xi}(h\mathbf{m})}{\sqrt{\mathcal{D}}h}\Big)\, ds\,.$$

LEMMA 10.8. *Suppose that η satisfies the decay condition* (10.20). *Then*

$$|R_N(\mathbf{y}_0)| \le c\,(\sqrt{\mathcal{D}}h)^N \|\tilde{u}\|_{C(B(\mathbf{y}_0,\kappa))}$$

with a constant c depending only on η and $\boldsymbol{\phi}$.

PROOF. If we use the abbreviation

$$c_u = \max_{\mathbf{y}\in B(\mathbf{y}_0,\kappa)} \big|\tilde{u}(\mathbf{y})\widetilde{\boldsymbol{\xi}}(\mathbf{y})^{\boldsymbol{\beta}}\big|\,,$$

then, in view of (10.20) and (10.24), we can estimate for $[\boldsymbol{\beta}] = N$

$$\Big|\mathcal{D}^{-n/2} \sum_{h\mathbf{m}\in B(\mathbf{y}_0,\kappa)} \tilde{u}(h\mathbf{m})\, \widetilde{\boldsymbol{\xi}}(h\mathbf{m})^{\boldsymbol{\beta}} \frac{|\mathbf{y}_0 - h\mathbf{m}|^{2N}}{(\mathcal{D}h^2)^N}\, \partial^{\boldsymbol{\beta}} \eta\Big(\frac{sA(\mathbf{y}_0 - h\mathbf{m}) + (1-s)\,\boldsymbol{\xi}(h\mathbf{m})}{\sqrt{\mathcal{D}}h}\Big)\Big|$$

$$\le c_u\, C_{\boldsymbol{\beta}}\, \mathcal{D}^{-n/2} \sum_{h\mathbf{m}\in B(\mathbf{y}_0,\kappa)} \frac{|\mathbf{y}_0 - h\mathbf{m}|^{2N}}{(\mathcal{D}h^2)^N} \Big(1 + \frac{|sA(\mathbf{y}_0 - h\mathbf{m}) + (1-s)\,\boldsymbol{\xi}(h\mathbf{m})|}{\sqrt{\mathcal{D}}h}\Big)^{-N-K}$$

$$\le c_u\, C_{\boldsymbol{\beta}}\, \mathcal{D}^{-n/2} \sum_{h\mathbf{m}\in B(\mathbf{y}_0,\kappa)} \frac{|\mathbf{y}_0 - h\mathbf{m}|^{2N}}{(\mathcal{D}h^2)^N} \Big(1 + \frac{C_2|\mathbf{y}_0 - h\mathbf{m}|}{\sqrt{\mathcal{D}}h}\Big)^{-N-K}$$

$$\le c_u\, C_{\boldsymbol{\beta}}\, C_2^{-2N}\, \mathcal{D}^{-n/2} \sum_{h\mathbf{m}\in B(\mathbf{y}_0,\kappa)} \Big(1 + \frac{C_2|\mathbf{y}_0 - h\mathbf{m}|}{\sqrt{\mathcal{D}}h}\Big)^{N-K}.$$

The last sum is uniformly bounded for $\mathcal{D} \geq \mathcal{D}_0 > 0$, since

$$\mathcal{D}^{-n/2} \sum_{h\mathbf{m} \in B(\mathbf{y}_0,\kappa)} \Big(1 + \frac{C_2|\mathbf{y}_0 - h\mathbf{m}|}{\sqrt{\mathcal{D}}h}\Big)^{N-K} \leq \sup_{\mathbb{R}^n} \mathcal{D}^{-n/2} \sum_{\mathbf{m}\in Z^n} \Big(1 + \frac{C_2|\mathbf{x} - \mathbf{m}|}{\sqrt{\mathcal{D}}}\Big)^{N-K} = C_2^{-n}\phi_{K-N-n}(\sqrt{\mathcal{D}}/C_2)$$

with the function ϕ_μ defined by (2.31), which proves the assertion. □

Next, we consider the semi-discrete convolutions in (10.26). It can be easily checked that, in view of (10.20), the generating function $\tilde{\eta}_{A,\boldsymbol{\beta}}$ satisfies

$$\big|\partial^{\boldsymbol{\alpha}}\tilde{\eta}_{A,\boldsymbol{\beta}}(\mathbf{y})\big| = |\mathbf{y}|^{2[\boldsymbol{\beta}]}\,\big|\partial^{\boldsymbol{\beta}}\eta(A\mathbf{y})\big| \leq C_{\boldsymbol{\beta}}(1+|\mathbf{y}|)^{-K+[\boldsymbol{\beta}]}$$

for any $0 \leq [\boldsymbol{\alpha}] \leq \mu$, i.e., it satisfies the extended decay Condition 2.12 with $K > N - [\boldsymbol{\beta}] + n$. Hence, we can expand the sum

$$\mathcal{D}^{-n/2} \sum_{h\mathbf{m}\in B(\mathbf{y}_0,\kappa)} \tilde{u}(h\mathbf{m})\,\widetilde{\boldsymbol{\xi}}(h\mathbf{m})^{\boldsymbol{\beta}}\,\tilde{\eta}_{A,\boldsymbol{\beta}}\Big(\frac{\mathbf{y}_0 - h\mathbf{m}}{\sqrt{\mathcal{D}}h}\Big)$$

for the function $\tilde{u}\,\widetilde{\boldsymbol{\xi}}^{\boldsymbol{\beta}} \in C^{N-[\boldsymbol{\beta}]}$, as in (2.36), up to the order $N - [\boldsymbol{\beta}]$ and we can apply Lemma 2.11 and Corollary 2.19. Then we obtain

LEMMA 10.9. *We have*

$$\frac{(\sqrt{\mathcal{D}}h)^{[\boldsymbol{\beta}]}}{\mathcal{D}^{n/2}} \sum_{h\mathbf{m}\in B(\mathbf{y}_0,\kappa)} \tilde{u}(h\mathbf{m})\,\widetilde{\boldsymbol{\xi}}(h\mathbf{m})^{\boldsymbol{\beta}}\,\tilde{\eta}_{A,\boldsymbol{\beta}}\Big(\frac{\mathbf{y}_0 - h\mathbf{m}}{\sqrt{\mathcal{D}}h}\Big)$$
$$= \sum_{[\boldsymbol{\alpha}]=0}^{N-[\boldsymbol{\beta}]-1} \frac{\partial^{\boldsymbol{\alpha}}\big(\tilde{u}\,\widetilde{\boldsymbol{\xi}}^{\boldsymbol{\beta}}\big)(\mathbf{y}_0)}{\boldsymbol{\alpha}!}\,\frac{(\sqrt{\mathcal{D}}h)^{[\boldsymbol{\alpha}]+[\boldsymbol{\beta}]}}{(2\pi i)^{[\boldsymbol{\alpha}]}} \sum_{\boldsymbol{\nu}\in\mathbb{Z}^n} \partial^{\boldsymbol{\alpha}}\mathcal{F}\tilde{\eta}_{A,\boldsymbol{\beta}}(\sqrt{\mathcal{D}}\boldsymbol{\nu})\,\mathrm{e}^{\frac{2\pi i}{h}\langle\mathbf{y}_0,\boldsymbol{\nu}\rangle}$$
$$+ R_{N-\boldsymbol{\beta}}(\mathbf{y}_0)\,,$$

with the remainder bounded by

$$|R_{N-\boldsymbol{\beta}}| \leq c\,(\sqrt{\mathcal{D}}h)^N\Big(\|\tilde{u}\,\widetilde{\boldsymbol{\xi}}^{\boldsymbol{\beta}}\|_{C(B(\mathbf{y}_0,\kappa))} + \|\nabla_{N-[\boldsymbol{\beta}]}\big(\tilde{u}\,\widetilde{\boldsymbol{\xi}}^{\boldsymbol{\beta}}\big)\|_{C(B(\mathbf{y}_0,\kappa))}\Big).$$

Now, (10.26) together with Lemma 10.9 leads to the representation of the quasi-interpolant (10.25)

$$u_h(\mathbf{x}) = \sum_{[\boldsymbol{\beta}]=0}^{N-1}\sum_{[\boldsymbol{\alpha}]=0}^{N-[\boldsymbol{\beta}]-1} \frac{\partial^{\boldsymbol{\alpha}}\big(\tilde{u}\,\widetilde{\boldsymbol{\xi}}^{\boldsymbol{\beta}}\big)(\mathbf{y}_0)}{\boldsymbol{\alpha}!\,\boldsymbol{\beta}!}\,\frac{(\sqrt{\mathcal{D}}h)^{[\boldsymbol{\alpha}]+[\boldsymbol{\beta}]}}{(2\pi i)^{[\boldsymbol{\alpha}]}} \sum_{\boldsymbol{\nu}\in\mathbb{Z}^n} \partial^{\boldsymbol{\alpha}}\mathcal{F}\tilde{\eta}_{A,\boldsymbol{\beta}}(\sqrt{\mathcal{D}}\boldsymbol{\nu})\,\mathrm{e}^{\frac{2\pi i}{h}\langle\mathbf{y}_0,\boldsymbol{\nu}\rangle}$$
$$+ \mathcal{O}((\sqrt{\mathcal{D}}h)^N)\,.$$

Hence, the behavior of u_h is determined by the values of the partial derivatives of the n-dimensional Fourier transforms

$$\partial^{\boldsymbol{\alpha}}\mathcal{F}\tilde{\eta}_{A,\boldsymbol{\beta}}(\sqrt{\mathcal{D}}\boldsymbol{\nu})\,,\quad 0 \leq [\boldsymbol{\beta}] \leq N-1\,,\quad 0 \leq [\boldsymbol{\alpha}] \leq N - [\boldsymbol{\beta}] - 1\,, \tag{10.27}$$

for $\boldsymbol{\nu} \in \mathbb{Z}^n$, which will be studied in the next lemmas.

First, we consider the case $s = n$, where the Fourier transform of the function $\tilde{\eta}_{A,\boldsymbol{\beta}}(\mathbf{y}) = |\mathbf{y}|^{2[\boldsymbol{\beta}]}\,\partial^{\boldsymbol{\beta}}\eta(A\mathbf{y})$ is easily found.

LEMMA 10.10. *Let $s = n$. Then*

$$\partial^{\boldsymbol\alpha} \mathcal{F}\tilde\eta_{A,\boldsymbol\beta}(\boldsymbol\lambda) = (2\pi i)^{-[\boldsymbol\beta]} \Delta^{[\boldsymbol\beta]}\, \partial^{\boldsymbol\alpha} \big(\boldsymbol\lambda^{\boldsymbol\beta} \mathcal{F}\eta(A^{-1}\boldsymbol\lambda)\big)\,.$$

If η satisfies the moment Condition 2.15 of order N, then, in particular,

$$\partial^{\boldsymbol\alpha} \mathcal{F}\tilde\eta_{A,\boldsymbol\beta}(\mathbf{0}) = \begin{cases} 1\,, & [\boldsymbol\alpha] = [\boldsymbol\beta] = 0\,,\\ 0\,, & 1 \le [\boldsymbol\beta] + [\boldsymbol\alpha] \le N-1\,. \end{cases}$$

Thus, the quasi-interpolant (10.25) defined in the domain $\Omega \subset \mathbb{R}^n$ has the representation

$$u_h(\mathbf{x}) = u(\mathbf{x}) + \sum_{[\boldsymbol\beta]=0}^{N-1} \sum_{[\boldsymbol\alpha]=0}^{N-[\boldsymbol\beta]-1} \frac{(\sqrt{\mathcal{D}}h)^{[\boldsymbol\alpha]+[\boldsymbol\beta]}}{\boldsymbol\alpha!\,\boldsymbol\beta!\,(2\pi i)^{[\boldsymbol\alpha]+[\boldsymbol\beta]}}\, \partial^{\boldsymbol\alpha}\big(\tilde u\, \widetilde{\boldsymbol\xi}^{\boldsymbol\beta}\big)(\mathbf{y}_0)$$

$$\times \sum_{\boldsymbol\nu \in \mathbb{Z}^n\setminus\{\mathbf{0}\}} \mathrm{e}^{\frac{2\pi i}{h}\langle \mathbf{y}_0,\boldsymbol\nu\rangle}\, \Delta^{[\boldsymbol\beta]}\, \partial^{\boldsymbol\alpha}\big(\boldsymbol\lambda^{\boldsymbol\beta} \mathcal{F}\eta(A^{-1}\boldsymbol\lambda)\big)\Big|_{\boldsymbol\lambda=\sqrt{\mathcal{D}}\boldsymbol\nu} + \mathcal{O}((\sqrt{\mathcal{D}}h)^N)\,,$$

which proves Theorem 10.5 in the case $s = n$.

10.3.3. Quasi-interpolation on manifolds. Now, we study the values of the Fourier transforms (10.27) in the case $s > n$, where the generating function η is radial and depends smoothly on the norm $|\mathbf{x}|$, $\mathbf{x} \in \mathbb{R}^s$.

LEMMA 10.11. *Let $A : \mathbb{R}^n \to \mathbb{R}^s$ be a linear mapping of rank n and let η be a smooth radial function in $\mathbb{R}^s$. Then, for any multi-indices $\boldsymbol\alpha \in \mathbb{Z}^n_{\ge 0}$ and $\boldsymbol\beta \in \mathbb{Z}^s_{\ge 0}$ with $[\boldsymbol\alpha] = [\boldsymbol\beta]$, there exist a collection of multi-indices $\boldsymbol\gamma, \boldsymbol\delta \in \mathbb{Z}^n_{\ge 0}$ and numbers $c_{\boldsymbol\gamma,\boldsymbol\delta}$ such that for all $\mathbf{y} \in \mathbb{R}^n$*

$$\mathbf{y}^{\boldsymbol\alpha}\, \partial^{\boldsymbol\beta}_{\mathbf{x}} \eta(A\mathbf{y}) = \sum_{[\boldsymbol\delta]=1}^{[\boldsymbol\beta]} \sum_{[\boldsymbol\gamma]=[\boldsymbol\delta]} c_{\boldsymbol\gamma,\boldsymbol\delta}\, \mathbf{y}^{\boldsymbol\gamma}\, \partial^{\boldsymbol\delta}_{\mathbf{y}} \eta(A\mathbf{y})\,. \tag{10.28}$$

PROOF. Since $A\mathbf{y} = \mathbf{x}$ with an $s\times n$ matrix A of rank n, there exist an invertible $n \times n$ matrix B and a subvector $\mathbf{x}'$ of length n, such that $\mathbf{y} = B\mathbf{x}'$. For definiteness suppose the ordering $\mathbf{x} = (\mathbf{x}', \mathbf{x}'')$ and $\partial^{\boldsymbol\beta}_{\mathbf{x}} = \partial^{\boldsymbol\beta'}_{\mathbf{x}'} \partial^{\boldsymbol\beta''}_{\mathbf{x}''}$. For any $[\boldsymbol\beta'] = 1$, the partial derivative $\partial^{\boldsymbol\beta'}_{\mathbf{x}'}$ is a linear combination of $\partial^{\boldsymbol\delta}_{\mathbf{y}}$ with $[\boldsymbol\delta] = 1$. Hence, if the multi-index $\boldsymbol\beta'' \in \mathbb{Z}^{s-n}_{\ge 0}$ satisfies $\boldsymbol\beta'' = \mathbf{0}$, then (10.28) is true with $c_{\boldsymbol\gamma,\boldsymbol\delta} = 0$ for all $[\boldsymbol\delta] < [\boldsymbol\beta]$.

Let $[\boldsymbol\beta''] > 0$. Since $\mathbf{y} = B\mathbf{x}'$ implies

$$\mathbf{y}^{\boldsymbol\alpha} = \sum_{[\boldsymbol\delta]=[\boldsymbol\alpha]} c_{\boldsymbol\delta}\, {\mathbf{x}'}^{\boldsymbol\delta}\,,$$

we have to transform ${\mathbf{x}'}^{\boldsymbol\delta} \partial^{\boldsymbol\beta'}_{\mathbf{x}'} \partial^{\boldsymbol\beta''}_{\mathbf{x}''} \eta$ with $[\boldsymbol\delta] = [\boldsymbol\beta'] + [\boldsymbol\beta'']$. Suppose that the variable x_j belongs to the first group $\mathbf{x}'$ with the corresponding multi-index $\mathbf{e}'_j \in \mathbb{Z}^n_{\ge 0}$ of norm 1. Similarly, a variable of $\mathbf{x}''$ will be denoted by x_k with the corresponding multi-index $\mathbf{e}''_k \in \mathbb{Z}^{s-n}_{\ge 0}$ of norm 1. Since, for any radial function η, the identity

$$x_j\, \partial_{x_k} \eta = x_k\, \partial_{x_j} \eta \tag{10.29}$$

holds, we have for $\boldsymbol\beta' = \mathbf{0}$

$${\mathbf{x}'}^{\boldsymbol\delta} \partial^{\boldsymbol\beta''}_{\mathbf{x}''} \eta = {\mathbf{x}''}^{\boldsymbol\beta''} \partial^{\boldsymbol\delta}_{\mathbf{x}'} \eta\,. \tag{10.30}$$

Moreover, as long as $\boldsymbol{\delta} - \mathbf{e}'_j \in \mathbb{Z}^n_{\geq 0}$ and $\boldsymbol{\beta}'' - \mathbf{e}''_k \in \mathbb{Z}^{s-n}_{\geq 0}$, we obtain from (10.29) that

$$\mathbf{x}'^{\,\boldsymbol{\delta}} \partial^{\boldsymbol{\beta}'}_{\mathbf{x}'} \partial^{\boldsymbol{\beta}''}_{\mathbf{x}''} \eta = \mathbf{x}'^{\,\boldsymbol{\delta}-\mathbf{e}'_j} x_j \, \partial^{\boldsymbol{\beta}'}_{\mathbf{x}'} \partial^{\boldsymbol{\beta}''}_{\mathbf{x}''} \eta$$

$$= \mathbf{x}'^{\,\boldsymbol{\delta}-\mathbf{e}'_j} \partial^{\boldsymbol{\beta}'}_{\mathbf{x}'} \partial^{\boldsymbol{\beta}''-\mathbf{e}''_k}_{\mathbf{x}''} x_j \, \partial_{x_k} \eta - \mathbf{x}'^{\,\boldsymbol{\delta}-\mathbf{e}'_j} \partial^{\boldsymbol{\beta}'-\mathbf{e}'_j}_{\mathbf{x}'} \partial^{\boldsymbol{\beta}''}_{\mathbf{x}''} \eta$$

$$= \mathbf{x}'^{\,\boldsymbol{\delta}-\mathbf{e}'_j} \partial^{\boldsymbol{\beta}'}_{\mathbf{x}'} \partial^{\boldsymbol{\beta}''-\mathbf{e}''_k}_{\mathbf{x}''} x_k \, \partial_{x_j} \eta - \mathbf{x}'^{\,\boldsymbol{\delta}-\mathbf{e}'_j} \partial^{\boldsymbol{\beta}'-\mathbf{e}'_j}_{\mathbf{x}'} \partial^{\boldsymbol{\beta}''}_{\mathbf{x}''} \eta$$

$$= \mathbf{x}'^{\,\boldsymbol{\delta}-\mathbf{e}'_j} x_k \, \partial^{\boldsymbol{\beta}'+\mathbf{e}'_j}_{\mathbf{x}'} \partial^{\boldsymbol{\beta}''-\mathbf{e}''_k}_{\mathbf{x}''} \eta + \mathbf{x}'^{\,\boldsymbol{\delta}-\mathbf{e}'_j} \partial^{\boldsymbol{\beta}'+\mathbf{e}'_j}_{\mathbf{x}'} \partial^{\boldsymbol{\beta}''-2\mathbf{e}''_k}_{\mathbf{x}''} \eta - \mathbf{x}'^{\,\boldsymbol{\delta}-\mathbf{e}'_j} \partial^{\boldsymbol{\beta}'-\mathbf{e}'_j}_{\mathbf{x}'} \partial^{\boldsymbol{\beta}''}_{\mathbf{x}''} \eta \,.$$

If the vectors $\boldsymbol{\beta}'' - 2\mathbf{e}''_k$ or $\boldsymbol{\beta}' - \mathbf{e}'_j$ have negative components, then the corresponding terms in the right-hand side are set to 0.

Obviously, repeating the last equations and using (10.30), we obtain expressions of the form

$$\mathbf{x}'^{\,\boldsymbol{\gamma}'} \mathbf{x}''^{\,\boldsymbol{\gamma}''} \partial^{\boldsymbol{\delta}}_{\mathbf{x}'} \eta \,, \quad 1 \leq [\boldsymbol{\gamma}'] + [\boldsymbol{\gamma}''] = [\boldsymbol{\delta}] \leq [\boldsymbol{\beta}] \,,$$

which, in view of $\mathbf{x} = A\mathbf{y}$ and $\mathbf{y} = B\mathbf{x}'$, completes the proof. □

Hence, the proof of Theorem 10.5 results from

LEMMA 10.12. *Let $A : \mathbb{R}^n \to \mathbb{R}^s$, $s > n$, be a linear mapping of rank n and let η be a smooth radial function in $\mathbb{R}^s$. Using the notation of Lemma 10.11, we have*

$$\mathcal{F}\tilde{\eta}_{A,\boldsymbol{\beta}}(\boldsymbol{\lambda}) = \sum_{[\boldsymbol{\alpha}]=[\boldsymbol{\beta}]} \sum_{[\boldsymbol{\delta}]=1}^{[\boldsymbol{\beta}]} \sum_{[\boldsymbol{\gamma}]=[\boldsymbol{\delta}]} c_{\boldsymbol{\gamma},\boldsymbol{\delta}} \frac{(-1)^{[\boldsymbol{\beta}]+[\boldsymbol{\gamma}]}[\boldsymbol{\beta}]!}{(2\pi i)^{[\boldsymbol{\beta}]}\boldsymbol{\alpha}!} \partial^{\boldsymbol{\alpha}+\boldsymbol{\gamma}} \big(\boldsymbol{\lambda}^{\boldsymbol{\delta}} \mathcal{F}\eta((A^*A)^{-1/2}\boldsymbol{\lambda})\big) \,.$$

In particular, under the moment Condition 2.15,

$$\partial^{\boldsymbol{\alpha}} \mathcal{F}\tilde{\eta}_{A,\boldsymbol{\beta}}(\mathbf{0}) = \begin{cases} 1 \,, & [\boldsymbol{\alpha}] = [\boldsymbol{\beta}] = 0 \,, \\ 0 \,, & 1 \leq [\boldsymbol{\beta}] + [\boldsymbol{\alpha}] \leq N-1 \,. \end{cases}$$

10.3.4. Quasi-interpolant of general form. Now, we apply the estimate for (10.25) to the quasi-interpolation formula

$$u_h(\mathbf{x}) = \mathcal{D}^{-n/2} \sum u(\mathbf{x_m}) \, \eta\Big(\frac{\mathbf{x}-\mathbf{x_m}}{\sqrt{\mathcal{D}}\, V_{\mathbf{m}}}\Big) \,, \tag{10.31}$$

containing numbers $V_{\mathbf{m}}$, which should be determined by using only nodes near $\mathbf{x_m}$. Of course, $V_{\mathbf{m}}$ will be an approximation of $h|\boldsymbol{\phi}'(h\mathbf{m})|^{1/n}$.

THEOREM 10.13. *Under the conditions of Theorem 10.5, the quasi-interpolant*

$$u_h(\mathbf{x}) = \mathcal{D}^{-n/2} \sum_{\mathbf{x_m} \in \boldsymbol{\phi}(B(\mathbf{y}_0,\kappa))} u(\mathbf{x_m}) \, \eta\Big(\frac{\mathbf{x}-\mathbf{x_m}}{\sqrt{\mathcal{D}}\, V_{\mathbf{m}}}\Big) \,, \quad \mathbf{x} \in \Omega \,,$$

approximates sufficiently smooth functions u with the estimate (10.21), if the numbers $V_{\mathbf{m}}$ satisfy

$$\big|(V_{\mathbf{m}})^n - h^n |\boldsymbol{\phi}'(h\mathbf{m})|\big| \leq c\, h^{N+n} \,.$$

PROOF. Evidently,

$$\sum_{\mathbf{x_m} \in \boldsymbol{\xi}(B(\mathbf{y}_0,\kappa))} u(\mathbf{x_m}) \Big(\eta\Big(\frac{\mathbf{x}-\mathbf{x_m}}{\sqrt{\mathcal{D}} h |\boldsymbol{\phi}'(h\mathbf{m})|^{1/n}}\Big) - \eta\Big(\frac{\mathbf{x}-\mathbf{x_m}}{\sqrt{\mathcal{D}}\, V_{\mathbf{m}}}\Big)\Big)$$

$$= \sum_{[\boldsymbol{\alpha}]=1} \sum_{\mathbf{x_m} \in \boldsymbol{\xi}(B(\mathbf{y}_0,\kappa))} u(\mathbf{x_m}) \frac{V_{\mathbf{m}} - h|\boldsymbol{\phi}'(h\mathbf{m})|^{1/n}}{V_{\mathbf{m}}} \eta_{\boldsymbol{\alpha}}\Big(\frac{\mathbf{x}-\mathbf{x_m}}{\sqrt{\mathcal{D}} h |\boldsymbol{\phi}'(h\mathbf{m})|^{1/n}}\Big) + R(\mathbf{x}) \,,$$

where we use the notation $\eta_{\boldsymbol{\alpha}}(\mathbf{x}) = \mathbf{x}^{\boldsymbol{\alpha}}\,\partial^{\boldsymbol{\alpha}}\eta(\mathbf{x})$ and the remainder satisfies

$$|R(\mathbf{x})| \le c\Big(\frac{V_{\mathbf{m}} - h|\boldsymbol{\phi}'(h\mathbf{m})|^{1/n}}{V_{\mathbf{m}}}\Big)^2 .$$

Since

$$\int\limits_{\mathbb{R}^n} \eta_{\boldsymbol{\alpha}}(\mathbf{y})\, d\mathbf{y} = -1\,, \quad [\boldsymbol{\alpha}] = 1\,,$$

it is clear from Theorem 10.5 and Lemma 10.7 that

$$\mathcal{D}^{-n/2} \sum_{\mathbf{x}_{\mathbf{m}} \in \boldsymbol{\xi}(B(\mathbf{y}_0,\kappa))} u(\mathbf{x}_{\mathbf{m}})\Big(\eta\Big(\frac{\mathbf{x}-\mathbf{x}_{\mathbf{m}}}{\sqrt{\mathcal{D}}h|\boldsymbol{\phi}'(h\mathbf{m})|^{1/n}}\Big) - \eta\Big(\frac{\mathbf{x}-\mathbf{x}_{\mathbf{m}}}{\sqrt{\mathcal{D}}\,V_{\mathbf{m}}}\Big)\Big)$$

is of the order $\mathcal{O}(h^N)$ if and only if

$$\Big|\frac{V_{\mathbf{m}} - h|\boldsymbol{\phi}'(h\mathbf{m})|^{1/n}}{V_{\mathbf{m}}}\Big| \le c\,h^N$$

for all data points $\mathbf{x}_{\mathbf{m}}$ in the neighborhood $\boldsymbol{\xi}(B(\mathbf{y}_0,\kappa))$ of $\mathbf{x}$. □

There exist different methods to find an N-th order approximations of $|\boldsymbol{\phi}'(h\mathbf{m})|$. The simplest way is to replace the partial derivatives $\partial\varphi_j/\partial y_\ell$ by difference quotient approximations Δ_ℓ^j of the order h^N which involve only the j-th coordinates $x_{\mathbf{k}}^j$ of the centers $\mathbf{x}_{\mathbf{k}} = \boldsymbol{\phi}(h\mathbf{k})$, $\mathbf{k} \in \mathbb{Z}^n$, near $\mathbf{x}_{\mathbf{m}}$. An example is given by

$$\Delta_\ell^j := \frac{2(x^j_{\mathbf{m}+\mathbf{e}_\ell} - x^j_{\mathbf{m}-\mathbf{e}_\ell})}{3h} - \frac{x^j_{\mathbf{m}+2\mathbf{e}_\ell} - x^j_{\mathbf{m}-2\mathbf{e}_\ell}}{12h}$$

approximating $\partial\varphi_j/\partial y_\ell$ with the order h^4. Obviously, the n-th root of

$$\big|\,\|h\Delta_\ell^j\|\,\big|\,, \quad j = 1,\dots,s\,,\ \ell = 1,\dots,n\,,$$

can be taken as $V_{\mathbf{m}}$ if $N = 4$.

Another method of defining $V_{\mathbf{m}}$, which uses the measure of grid patches near $\mathbf{x}_{\mathbf{m}}$, consists in the following. Consider a cube $Q \subset \mathbb{R}^n$ having all the corners at lattice points $\mathbf{k} \in \mathbb{Z}^n$ with $\mathbf{0} \in Q$. We denote its volume by $|Q|$ and introduce $Q_{\mathbf{m}} = h\mathbf{m} + hQ$. Then

$$\frac{1}{|Q|}\int\limits_{Q_{\mathbf{m}}} |\boldsymbol{\phi}'(\mathbf{y})|\, d\mathbf{y}$$

$$= h^n|\boldsymbol{\phi}'(h\mathbf{m})| + \sum_{[\boldsymbol{\alpha}]=1}^{N-1} \frac{\partial^{\boldsymbol{\alpha}}|\boldsymbol{\phi}'(h\mathbf{m})|}{\boldsymbol{\alpha}!}\,\frac{h^{n+[\boldsymbol{\alpha}]}}{|Q|}\int\limits_Q \mathbf{y}^{\boldsymbol{\alpha}}\, d\mathbf{y} + \mathcal{O}(h^{N+n})\,. \tag{10.32}$$

Therefore, by choosing different cubes Q^j of the above-mentioned type, one can form a linear combination of the equalities (10.32) such that the sum does not contain terms with $h^{n+[\boldsymbol{\alpha}]}$, $1 < [\boldsymbol{\alpha}] < N-1$. Thus, $h^n|\boldsymbol{\phi}'(h\mathbf{m})|$ can be approximated with the order $\mathcal{O}(h^{N+n})$ by linear combinations of the integrals

$$\int\limits_{Q^j_{\mathbf{m}}} |\boldsymbol{\phi}'(\mathbf{y})|\, d\mathbf{y} = \int\limits_{\boldsymbol{\phi}(Q^j_{\mathbf{m}})} d\mathbf{x} = |\boldsymbol{\phi}(Q^j_{\mathbf{m}})|$$

over the finite number of cubes $Q^j_{\mathbf{m}} = h\mathbf{m} + hQ^j$.

Consider, for example, a surface Γ in $\mathbb{R}^3$. We choose the squares $Q^1 = [-1,1]^2$ and Q^2 with corners at the points $(\pm 1, 0)$ and $(0, \pm 1)$. Then

$$\int\limits_{Q^1_{\mathbf{m}}} |\phi'(\mathbf{y})|\, d\mathbf{y} = 4h^2 |\phi'(h\mathbf{m})| + \frac{2h^4}{3}\big(\partial^{(2,0)}|\phi'(h\mathbf{m})| + \partial^{(0,2)}|\phi'(h\mathbf{m})|\big) + \mathcal{O}(h^6)\,,$$

$$\int\limits_{Q^2_{\mathbf{m}}} |\phi'(\mathbf{y})|\, d\mathbf{y} = 2h^2 |\phi'(h\mathbf{m})| + \frac{h^4}{6}\big(\partial^{(2,0)}|\phi'(h\mathbf{m})| + \partial^{(0,2)}|\phi'(h\mathbf{m})|\big) + \mathcal{O}(h^6)\,.$$

Consequently the quasi-interpolation formula (10.31) on Γ with

$$\eta(|\mathbf{x}|) = \pi^{-1} e^{-|\mathbf{x}|^2} \quad \text{and} \quad V_{\mathbf{m}} = \sqrt{|\phi(Q^1_{\mathbf{m}})|/2} \quad \text{or} \quad V_{\mathbf{m}} = \sqrt{|\phi(Q^2_{\mathbf{m}})|/2}$$

approximates with the order $\mathcal{O}(\mathcal{D}h^2)$ plus some saturation error. Similarly, approximate approximation with the order $\mathcal{O}(\mathcal{D}^2h^4)$ on Γ can be obtained with the generating function

$$\eta(|\mathbf{x}|) = \pi^{-1}(2 - |\mathbf{x}|)\, e^{-|\mathbf{x}|^2} \quad \text{and} \quad V_{\mathbf{m}} = \sqrt{|\phi(Q^2_{\mathbf{m}})| - |\phi(Q^1_{\mathbf{m}})|/4}\,.$$

10.4. Notes

In Sections 10.1 and 10.2, we followed the paper [**55**]. The approximation of scattered data by functions centered on a uniform grid has been considered in [**24**] as an auxiliary step for the construction of scattered data approximants with radial basis functions, which reproduce polynomials. It follows from the proof of Theorem 10.4 that if the quasi-interpolant $\mathcal{M}_{h,\mathcal{D}}u$ from Condition 10.1 approximates u without saturation error, then $\mathbb{M}_{h,\mathcal{D}}u$ has the same property.

The results of Section 10.3 are taken from [**71**]. In a series of papers (cf., e.g., [**26**], [**27**]), Fasshauer performed numerical tests for the quasi-interpolant (10.31) on a non-uniform grid.

CHAPTER 11

Scattered data approximate approximations

The aim of this chapter is to study quasi-interpolation for functions with values given on a rather general grid $\{\mathbf{x}_j\}_{j\in J}$. In order to treat this case, we will modify the approximating functions. More precisely, we consider approximations of the form

$$Mu(\mathbf{x}) = \sum_{j\in J} \sum_{\mathbf{x}_k \in \overline{\mathrm{st}}(\mathbf{x}_j)} u(\mathbf{x}_k)\, \mathcal{P}_{j,k}(\mathbf{x})\, \eta\Big(\frac{\mathbf{x}-\mathbf{x}_j}{h_j}\Big), \tag{11.1}$$

with some polynomials $\mathcal{P}_{j,k}$, where the set of nodes $\overline{\mathrm{st}}(\mathbf{x}_j)$ is an extension of the star $\mathrm{st}(\mathbf{x}_j)$ given in Definition 10.2. With such a quasi-interpolation, one can achieve high-order approximation of $u(\mathbf{x})$ with respect to the local mesh size, up to a small saturation error. Moreover, by a suitable choice of the basis functions η, one can obtain efficient approximation formulas for integral and pseudodifferential operators.

We give a simple example of a quasi-interpolation formula (11.1). Let $\{x_j\}$ be a sequence of points on $\mathbb{R}$ such that $0 < \kappa \le x_{j+1} - x_j \le 1$. Consider a sequence of functions ζ_j on $\mathbb{R}$ with support in a fixed neighborhood of the origin. Suppose that the sequence $\{\zeta_j(x - x_j)\}$ forms an approximate partition of unity on $\mathbb{R}$, which means that

$$|1 - \sum_j \zeta_j(x - x_j)| < \varepsilon$$

for a certain sufficiently small positive ε and for all $x \in \mathbb{R}$. Then

$$\Big|u(x) - \sum_j u(x)\zeta_j\Big(\frac{x}{h} - x_j\Big)\Big| < \varepsilon |u(x)|\,.$$

Now, we note that for $x \in x_j + h\,\mathrm{supp}\,\zeta_j$

$$\Big|u(x) - u(hx_j)\frac{x_{j+1} - x/h}{x_{j+1} - x_j} - u(hx_{j+1})\frac{x/h - x_j}{x_{j+1} - x_j}\Big| \le ch^2 \max_{t\in x_j + h\,\mathrm{supp}\,\zeta_j} |u''(t)|$$

with a constant depending on $\mathrm{diam}(\mathrm{supp}\,\zeta_j)$ and κ. Then, obviously, the quasi-interpolant

$$M_h u(x) = \sum_j u(hx_j)\left(\frac{x_{j+1} - x/h}{x_{j+1} - x_j}\zeta_j\Big(\frac{x}{h} - x_j\Big) + \frac{x/h - x_{j-1}}{x_j - x_{j-1}}\zeta_{j-1}\Big(\frac{x}{h} - x_{j-1}\Big)\right)$$

satisfies the estimate

$$|M_h u(x) - u(x)| \le c\,h^2\, \|u''\|_{L_\infty(\mathbb{R})} + \varepsilon\, |u(x)|\,,$$

where the constant c depends on the functions ζ_j.

11.1. Approximate partition of unity

We show in this section that an approximate partition of unity of $\mathbb{R}^n$ can be obtained from a given system of approximating functions centered at scattered nodes $\{\mathbf{x}_j\}_{j\in J}$, if these functions are multiplied by polynomials. Here, J denotes an infinite index set. We are mainly interested in rapidly decaying basis functions which are supported on the whole space, but we start with the simpler case of compactly supported basis functions.

11.1.1. Basis functions with compact support.

LEMMA 11.1. *Let $\{B(\mathbf{x}_j,h_j)\}_{j\in J}$ be an open locally finite covering of $\mathbb{R}^n$ by balls centered in $\mathbf{x}_j$ and with radii h_j. Suppose that the multiplicity of this covering does not exceed a positive constant μ_n and that there are positive constants c_1 and c_2 satisfying*

$$c_1 h_m \le h_j \le c_2 h_m \tag{11.2}$$

provided the balls $B(\mathbf{x}_j,h_j)$ and $B(\mathbf{x}_m,h_m)$ have common points. Furthermore, let $\{\eta_j\}$ be a bounded sequence of continuous functions on $\mathbb{R}^n$ such that $\operatorname{supp}\eta_j \subset B(\mathbf{x}_j,h_j)$. We assume that the functions $\mathbb{R}^n \ni \mathbf{y} \to \eta_j(h_j\,\mathbf{y})$ are continuous uniformly with respect to j and

$$s(\mathbf{x}) := \sum_{j\in J} \eta_j(\mathbf{x}) \ge c \quad \text{on } \mathbb{R}^n \tag{11.3}$$

where c is a positive constant. Then for any $\varepsilon > 0$ there exists a sequence of polynomials $\{\mathcal{P}_j\}$ with the following properties:

(i) *the function*

$$\Theta := \sum_{j\in J} \mathcal{P}_j\,\eta_j \tag{11.4}$$

satisfies

$$|\Theta(\mathbf{x}) - 1| < \varepsilon \quad \text{for all } \mathbf{x}\in\mathbb{R}^n\,; \tag{11.5}$$

(ii) *the degrees of all $\mathcal{P}_j$ are bounded (they depend on the least majorant of the continuity moduli of η_j and the constants ε, c, c_1, c_2, μ_n);*

(iii) *there is such a constant c_0 that*

$$|\mathcal{P}_j| < c_0 \quad \text{on } B(\mathbf{x}_j,\eta_j).$$

PROOF. Since the functions $B(\mathbf{x}_j,1) \ni \mathbf{y} \to s(h_j\,\mathbf{y})$ are continuous uniformly with respect to j, there exist polynomials $\mathcal{P}_j$ subject to

$$\Big|\mathcal{P}_j(\mathbf{x}) - \frac{1}{s(\mathbf{x})}\Big| < \delta \text{ on } B(\mathbf{x}_j,h_j)$$

for an arbitrary positive δ. The degrees $\deg\mathcal{P}_j$ are independent of j. Letting $\delta = \varepsilon\,(\mu_n\|\eta\|_{L_\infty})^{-1}$, we obtain

$$\Big|\eta_j(\mathbf{x})\Big(\mathcal{P}_j(\mathbf{x}) - \frac{1}{s(\mathbf{x})}\Big)\Big| \le \frac{\varepsilon}{\mu_n}\,.$$

Then

$$\sup_{\mathbb{R}^n}\sum_{j\in J}\Big|\eta_j(\mathbf{x})\Big(\mathcal{P}_j(\mathbf{x}) - \frac{1}{s(\mathbf{x})}\Big)\Big| \le \varepsilon\,, \tag{11.6}$$

since at most μ_n terms of this sum are different from zero. Furthermore,

$$\sum_{j\in J}\eta_j(\mathbf{x})\Big(\mathcal{P}_j(\mathbf{x})-\frac{1}{s(\mathbf{x})}\Big)=\sum_{j\in J}\eta_j(\mathbf{x})\mathcal{P}_j(\mathbf{x})-\frac{1}{s(\mathbf{x})}\sum_{j\in J}\eta_\mathbf{j}(\mathbf{x})$$
$$=\sum_{j\in J}\eta_j(\mathbf{x})\mathcal{P}_j(\mathbf{x})-1\,,$$

which proves (11.5). □

REMARK 11.2. Let the functions $\{\eta_j\}_{j\in J}$ in Lemma 11.1 satisfy the additional hypothesis $\eta_j\in C^r(\mathbb{R}^n)$. Then one can find a sequence of polynomials $\{\mathcal{P}_j\}$ of degrees L_j such that

$$\sup_{B(\mathbf{x}_j,h_\mathbf{j})}\Big|\mathcal{P}_j(\mathbf{x})-\frac{1}{s(\mathbf{x})}\Big|\le C(r)\,\frac{h_j^r}{L_j^r}\sup_{B(\mathbf{x}_j,h_j)}|\nabla_r s(\mathbf{x})|$$

(see, e.g., [**59**]). This shows that it suffices to take polynomials $\mathcal{P}_j$ of degrees greater than $c(r)\,\varepsilon^{-1/r}$ in order to achieve the error ε in (11.6).

11.1.2. Basis functions with non-compact support. Now, we consider the case when the approximating functions are supported on the whole $\mathbb{R}^n$. As in the previous chapters, we suppose that the functions η_j are scaled translates of a sufficiently smooth function η with rapid decay,

$$\eta_j(\mathbf{x})=\eta\Big(\frac{\mathbf{x}-\mathbf{x}_j}{h_j}\Big).$$

First, we formulate a result on weighted polynomial approximation which follows from [**60**, Thm. 4.2]. If we denote by $w_{\delta,p}$, $\delta>1$, $p>0$, the weight function

$$w_{\delta,p}(\mathbf{x})=\exp\Big(-p\sum_{k=1}^{n}|x_k|^\delta\Big)\,, \tag{11.7}$$

then, for any $g\in W_\infty^r(\mathbb{R}^n)$, there exists a polynomial $\mathcal{P}\in\Pi_{2N-1}$ such that

$$\|w_{\delta,p}(g-\mathcal{P})\|_{L_\infty}\le c\,N^{(1-\delta)r/\delta}\Big(\|w_{\delta,p}g\|_{L_\infty}+\sum_{k=1}^{n}\Big\|w_{\delta,p}\,\frac{\partial^r g}{\partial x_k^r}\Big\|_{L_\infty}\Big) \tag{11.8}$$

with a constant c depending only on the weight function. Here, Π_N denotes the set of polynomials which are of degree at most N in each variable $x_1,\dots,x_n$.

LEMMA 11.3. *Assume that the following assumptions on η, the nodes $\{\mathbf{x}_j\}_{j\in J}$, and the scaling parameters $\{h_j\}$ are satisfied:*

1. *There exists $K>0$ such that*

$$c_K:=\Big\|\sum_{j\in J}\big(1+h_j^{-1}|\cdot-\mathbf{x}_j|\big)^{-K}\Big\|_{L_\infty}<\infty\,. \tag{11.9}$$

2. *There exist $\delta>1$ and $p>0$ such that*

$$\Big\|\frac{(1+|\cdot|)^K}{w_{\delta,p}}\,\eta\Big\|_{L_\infty},\ \Big\|\frac{(1+|\cdot|)^K}{w_{\delta,p}}\,\nabla\eta\Big\|_{L_\infty}\le c_{\delta,p}<\infty \tag{11.10}$$

with the weight function $w_{\delta,p}$ defined in (11.7).

3. *There exists $C>0$ such that for all indices $j,m\in J$*

$$\frac{h_j}{h_m}\,w_{\delta,p}\Big(\frac{\mathbf{x}_j-\mathbf{x}_m}{h_j+h_m}\Big)\le C\,. \tag{11.11}$$

4. *The sum*

$$\sum_{j\in J} \eta_j(\mathbf{x}) ,$$

which, in view of (11.9) *and* (11.10), *converges absolutely to a smooth bounded function* s, *has a positive lower bound, i.e., fulfills* (11.3).

Then, for any $\varepsilon > 0$, *there exists* L_ε *and polynomials* $\{\mathcal{P}_j\}$ *of degree* $\deg \mathcal{P}_j \le L_\varepsilon$ *such that the function* Θ *defined by* (11.4) *satisfies*

$$|\Theta(\mathbf{x}) - 1| < \varepsilon \quad \textit{for all } \mathbf{x} \in \mathbb{R}^n .$$

PROOF. Suppose we have shown that for any $\varepsilon > 0$ and all indices j there exist polynomials $\mathcal{P}_j$ such that

$$\Big|\eta\Big(\frac{\mathbf{x}-\mathbf{x}_j}{h_j}\Big)\Big(\mathcal{P}_j\Big(\frac{\mathbf{x}-\mathbf{x}_j}{h_j}\Big) - \frac{1}{s(\mathbf{x})}\Big)\Big| \le \frac{\varepsilon}{c_K}\Big(1+\frac{|\mathbf{x}-\mathbf{x}_j|}{h_j}\Big)^{-K} \tag{11.12}$$

(c_K is defined in (11.9)) and $\deg \mathcal{P}_j \le L_\varepsilon$. Then

$$\sup_{\mathbb{R}^n} \sum_{j\in J} \Big|\eta_j\Big(\frac{\mathbf{x}-\mathbf{x}_j}{h_j}\Big)\Big(\mathcal{P}_j\Big(\frac{\mathbf{x}-\mathbf{x}_j}{h_j}\Big) - \frac{1}{s(\mathbf{x})}\Big)\Big| \le \varepsilon ,$$

and, as in the proof of Lemma 11.1, we conclude

$$\sup_{\mathbb{R}^n} \Big|\sum_{j\in J} \eta_j\Big(\frac{\mathbf{x}-\mathbf{x}_j}{h_j}\Big)\mathcal{P}_j\Big(\frac{\mathbf{x}-\mathbf{x}_j}{h_j}\Big) - 1\Big| \le \varepsilon .$$

Let us fix an index j and make the change of variables $\mathbf{y} = h_j{}^{-1}(\mathbf{x}-\mathbf{x}_j)$. Then (11.12) is proved if we show that there exists a polynomial $\mathcal{P}_j$ such that for all $\mathbf{y} \in \mathbb{R}^n$

$$\Big|\eta(\mathbf{y})\Big(\mathcal{P}_j(\mathbf{y}) - \frac{1}{\tilde{s}(\mathbf{y})}\Big)\Big| \le \frac{\varepsilon}{c_K}\big(1+|\mathbf{y}|\big)^{-K} \tag{11.13}$$

with $\tilde{s}(\mathbf{y}) = s(h_j\mathbf{y}+\mathbf{x}_j)$. Since $\tilde{s}^{-1} \in W^1_\infty(\mathbb{R}^n)$ according to (11.8), we can find a polynomial $\mathcal{P}_j$ satisfying

$$\sup_{\mathbb{R}^n} \Big|\mathcal{P}_j(\mathbf{y}) - \frac{1}{\tilde{s}(\mathbf{y})}\Big| w_{\delta,p}(\mathbf{y}) < \frac{\varepsilon}{c_{\delta,p}\, c_K}$$

with the constant $c_{\delta,p}$ in the decay condition (11.10). Now (11.13) follows directly from

$$|\eta(\mathbf{y})|\big(1+|\mathbf{y}|\big)^K \le c_{\delta,p}\, w_{\delta,p}(\mathbf{y}) .$$

By (11.8), the degree of $\mathcal{P}_j$ depends on the weighted norm

$$\begin{aligned}\sup_{\mathbb{R}^n} w_{\delta,p}(\mathbf{y})\Big|\nabla\frac{1}{\tilde{s}(\mathbf{y})}\Big| &= h_j \sup_{\mathbb{R}^n} w_{\delta,p}\Big(\frac{\mathbf{x}-\mathbf{x}_j}{h_j}\Big)\Big|\nabla\frac{1}{s(\mathbf{x})}\Big| \\ &\le \sup_{\mathbb{R}^n} \frac{1}{(s(\mathbf{x}))^2}\, w_{\delta,p}\Big(\frac{\mathbf{x}-\mathbf{x}_j}{h_j}\Big)\sum_{m\in J}\frac{h_j}{h_m}\Big|\nabla\eta\Big(\frac{\mathbf{x}-\mathbf{x}_m}{h_m}\Big)\Big| .\end{aligned} \tag{11.14}$$

Since by (11.10),

$$\Big|\nabla\eta\Big(\frac{\mathbf{x}-\mathbf{x}_m}{h_m}\Big)\Big| \le c_{\delta,p}\, w_{\delta,p}\Big(\frac{\mathbf{x}-\mathbf{x}_m}{h_m}\Big)\Big(1+\frac{|\mathbf{x}-\mathbf{x}_m|}{h_m}\Big)^{-K},$$

a uniform bound of (11.14) with respect to j can be established, if the sums

$$\sum_{m\in J}\frac{h_j}{h_m} w_{\delta,p}\Big(\frac{\mathbf{x}-\mathbf{x}_j}{h_j}\Big) w_{\delta,p}\Big(\frac{\mathbf{x}-\mathbf{x}_m}{h_m}\Big)\Big(1+\frac{|\mathbf{x}-\mathbf{x}_m|}{h_m}\Big)^{-K}$$

are uniformly bounded for all j. Consider the function

$$f(x) = \left|\frac{x-a}{\sigma}\right|^\delta + \left|\frac{x-b}{\tau}\right|^\delta$$

for $a < b$ and $\sigma, \tau > 0$, which attains its minimum at the point

$$x_0 = \frac{a\tau^{\delta/(\delta-1)} + b\sigma^{\delta/(\delta-1)}}{\sigma^{\delta/(\delta-1)} + \tau^{\delta/(\delta-1)}} \in (a, b)$$

with

$$f(x_0) = \frac{(b-a)^\delta}{(\sigma^{\delta/(\delta-1)} + \tau^{\delta/(\delta-1)})^{\delta-1}}\,.$$

Since for $\delta > 1$

$$(\sigma^{\delta/(\delta-1)} + \tau^{\delta/(\delta-1)}) \le (\sigma+\tau)^{\delta/(\delta-1)}\,,$$

we obtain the lower bound

$$f(x) \ge \frac{|a-b|^\delta}{(\sigma^{\delta/(\delta-1)} + \tau^{\delta/(\delta-1)})^{\delta-1}} \ge \left|\frac{a-b}{\sigma+\tau}\right|^\delta$$

for any $x \in \mathbb{R}$ and $\delta > 1$. Hence, we derive

$$w_{\delta,p}\Big(\frac{\mathbf{x}-\mathbf{x}_j}{h_j}\Big)\, w_{\delta,p}\Big(\frac{\mathbf{x}-\mathbf{x}_m}{h_m}\Big) \le w_{\delta,p}\Big(\frac{\mathbf{x}_j-\mathbf{x}_m}{h_j+h_m}\Big)$$

from (11.7). Therefore, the condition (11.11) on the nodes $\{\mathbf{x}_j\}$ and the corresponding parameters $\{h_j\}$ guarantees that the degrees of the polynomials $\mathcal{P}_j$ can be chosen independent of j. □

11.2. Quasi-interpolants of a general form

In this section, we study the approximation of functions $u \in W_\infty^N(\mathbb{R}^n)$ by the quasi-interpolant (11.1). We will show that within the class of generating functions of the form polynomial times compactly supported or rapidly decaying generating function, it suffices to have an approximate partition of unity in order to construct approximate quasi-interpolants of high-order accuracy up to some prescribed saturation error.

Recall the Definition 10.2 of the star of a node $\mathbf{x}_j \in \mathbf{X}$ as a collection of $m_N = \dfrac{(N-1+n)!}{n!(N-1)!} - 1$ nodes $\mathbf{x}_k \in \mathbf{X}$ such that the Vandermonde matrix

$$\{(\mathbf{x}_k - \mathbf{x}_j)^{\boldsymbol\alpha}\},\ [\boldsymbol\alpha] = 1, \ldots, N-1\,,\ \mathbf{x}_k \in \mathrm{st}(\mathbf{x}_j)\,,$$

is not singular. In the following, the union of the node $\mathbf{x}_j$ and its star $\mathrm{st}(\mathbf{x}_j)$ is denoted by $\overline{\mathrm{st}}(\mathbf{x}_j) = \mathbf{x}_j \cup \mathrm{st}(\mathbf{x}_j)$.

Let us assume the following hypothesis concerning the grid $\mathbf{X} = \{\mathbf{x}_j\}_{j\in J}$:

CONDITION 11.4. *For any $\mathbf{x}_j$, there exists a ball $B(\mathbf{x}_j, h_j)$ which contains m_N nodes $\mathbf{x}_k \in \mathrm{st}(\mathbf{x}_j)$ with*

$$|\det V_{j,h_j}| = \left|\det\left\{\Big(\frac{\mathbf{x}_k-\mathbf{x}_j}{h_\mathbf{j}}\Big)^{\boldsymbol\alpha}\right\}_{[\boldsymbol\alpha]=1,\mathbf{x}_k\in\mathrm{st}(\mathbf{x}_j)}^{N-1}\right| \ge c\,, \tag{11.15}$$

where c is positive and does not depend on $\mathbf{x}_j$.

11.2.1. Compactly supported basis functions.

THEOREM 11.5. *Suppose that the function system $\{\eta_j\}_{j\in J}$ satisfies the conditions of Lemma* 11.1 *and let $u \in W^N_\infty(\mathbb{R}^n)$ and $\varepsilon > 0$ be arbitrary. There exist polynomials $\mathcal{P}_{j,k}$, independent of u, whose degrees are uniformly bounded, such that the quasi-interpolant*

$$Mu(\mathbf{x}) = \sum_{k\in J} u(\mathbf{x}_k) \sum_{\overline{\mathrm{st}}(\mathbf{x}_j)\ni\mathbf{x}_k} \mathcal{P}_{j,k}(\mathbf{x})\eta_j(\mathbf{x}) \tag{11.16}$$

satisfies the estimate

$$|Mu(\mathbf{x}) - u(\mathbf{x})| \le C h_m^N \sup_{B(\mathbf{x}_m,\kappa\, h_m)} |\nabla_N\, u| + \varepsilon\, |u(\mathbf{x})|, \tag{11.17}$$

where $\mathbf{x}_m$ is an arbitrary node and $\mathbf{x}$ is any point of the ball $B(\mathbf{x}_m, h_m)$. By κ, we denote a constant greater than 1 *which depends on c_1 and c_2 in* (11.2). *The constant C does not depend on h_m, m, and ε.*

PROOF. For a given ε, we choose polynomials $\mathcal{P}_j(\mathbf{x})$ such that the function (11.4) satisfies

$$|\Theta(\mathbf{x}) - 1| < \varepsilon \quad \text{for all } \mathbf{x} \in \mathbb{R}^n$$

and we introduce the auxiliary quasi-interpolant

$$M^{(1)}u(\mathbf{x}) = \sum_{j\in J} \Big(\sum_{[\boldsymbol{\alpha}]=1}^{N-1} \frac{\partial^{\boldsymbol{\alpha}} u(\mathbf{x}_j)}{\boldsymbol{\alpha}!} (\mathbf{x} - \mathbf{x}_j)^{\boldsymbol{\alpha}} \Big) \mathcal{P}_j(\mathbf{x})\eta_j(\mathbf{x}) \,. \tag{11.18}$$

Using the Taylor expansion (10.11) with $\mathbf{y} = \mathbf{x}_j$, we write $M^{(1)}u(\mathbf{x})$ as

$$M^{(1)}u(\mathbf{x}) = u(\mathbf{x})\Theta(\mathbf{x}) - \sum_{j\in J} R_N(\mathbf{x}, \mathbf{x}_j)\mathcal{P}_j(\mathbf{x})\eta_j(\mathbf{x}) \,,$$

which gives

$$|M^{(1)}u(\mathbf{x}) - u(\mathbf{x})| \le \sum_{j\in J} |R_N(\mathbf{x}, \mathbf{x}_j)\mathcal{P}_j(\mathbf{x})\eta_j(\mathbf{x})| + |u(\mathbf{x})|\, |\Theta(\mathbf{x}) - 1| \,.$$

This, together with the estimate (10.12) for the remainder, shows that for $\mathbf{x} \in B(\mathbf{x}_m, h_m)$,

$$|M^{(1)}u(\mathbf{x}) - u(\mathbf{x})| \le C_1 h_m^N \sup_{B(\mathbf{x}_m,\kappa\, h_m)} |\nabla_N\, u| + \varepsilon |u(\mathbf{x})| \,, \tag{11.19}$$

where the ball $B(\mathbf{x}_m, \kappa\, h_m)$ contains all balls $B(\mathbf{x}_j,\, h_j)$ such that $B(\mathbf{x}_j,\, h_j)$ and $B(\mathbf{x}_m,\, h_m)$ intersect.

Similarly to the proof of Theorem 10.4, we approximate the values of the derivatives $\partial^{\boldsymbol{\alpha}} u(\mathbf{x}_j)$ in $M^{(1)}u$ by a linear combination of $u(\mathbf{x}_k)$, where $\mathbf{x}_k \in \mathrm{st}(\mathbf{x}_j)$. The solution of the algebraic system

$$\sum_{1\le[\boldsymbol{\alpha}]<N} \frac{a^{(j)}_{\boldsymbol{\alpha}}}{\boldsymbol{\alpha}!} (\mathbf{x}_k - \mathbf{x}_j)^{\boldsymbol{\alpha}} = u(\mathbf{x}_k) - u(\mathbf{x}_j) \,, \quad \mathbf{x}_k \in \mathrm{st}(\mathbf{x}_j) \,, \tag{11.20}$$

is given by

$$a^{(j)}_{\boldsymbol{\alpha}} = \frac{\boldsymbol{\alpha}!}{h_j^{[\boldsymbol{\alpha}]}} \sum_{\mathbf{x}_k\in\mathrm{st}(\mathbf{x}_j)} b^{(j)}_{\boldsymbol{\alpha},k}\, (u(\mathbf{x}_k) - u(\mathbf{x}_j)), \quad 1 \le [\boldsymbol{\alpha}] < N \,,$$

where $\{b^{(j)}_{\boldsymbol{\alpha},k}\}$ are the elements of the inverse of V_{j,h_j}. Replacing the derivatives $\{\partial^{\boldsymbol{\alpha}} u(\mathbf{x}_j)\}$ in (11.18) by $\{a^{(j)}_{\boldsymbol{\alpha}}\}$, we derive

$$\begin{aligned} Mu(\mathbf{x}) &= \sum_{j\in J}\Big\{u(\mathbf{x}_j)\Big(1-\sum_{\mathbf{x}_k\in \mathrm{st}(\mathbf{x}_j)}\sum_{[\boldsymbol{\alpha}]=1}^{N-1} b^{(j)}_{\boldsymbol{\alpha},k}\Big(\frac{\mathbf{x}-\mathbf{x}_j}{h_j}\Big)^{\boldsymbol{\alpha}}\Big) \\ &\qquad + \sum_{\mathbf{x}_k\in \mathrm{st}(\mathbf{x}_j)} u(\mathbf{x}_k)\sum_{[\boldsymbol{\alpha}]=1}^{N-1} b^{(j)}_{\boldsymbol{\alpha},k}\Big(\frac{\mathbf{x}-\mathbf{x}_j}{h_j}\Big)^{\boldsymbol{\alpha}}\Big\}\mathcal{P}_j(\mathbf{x})\eta_j(\mathbf{x}) \\ &= \sum_{j\in J}\sum_{\mathbf{x}_k\in \overline{\mathrm{st}}(\mathbf{x}_j)} u(\mathbf{x}_k)\,\mathcal{P}_{j,k}(\mathbf{x})\,\eta_j(\mathbf{x}) \end{aligned}$$

which can be rewritten in the form (11.16). Note that for $k\neq j$

$$\mathcal{P}_{j,k}(\mathbf{x}) = \sum_{[\boldsymbol{\alpha}]=1}^{N-1} b^{(j)}_{\boldsymbol{\alpha},k}\Big(\frac{\mathbf{x}-\mathbf{x}_j}{h_j}\Big)^{\boldsymbol{\alpha}}\mathcal{P}_j(\mathbf{x}),\quad \mathcal{P}_{j,j}(\mathbf{x}) = \mathcal{P}_j(\mathbf{x}) - \sum_{\mathbf{x}_k\in \mathrm{st}(\mathbf{x}_j)}\mathcal{P}_{j,k}(\mathbf{x});$$

hence the degree of the polynomials $\mathcal{P}_{j,k}$ is not larger than $\deg \mathcal{P}_j + N - 1$.

From (10.11) and (11.20), we obtain

$$\sum_{[\boldsymbol{\alpha}]=1}^{N-1}\frac{h_j^{[\boldsymbol{\alpha}]}}{\boldsymbol{\alpha}!}(a^{(j)}_{\boldsymbol{\alpha}} - \partial^{\boldsymbol{\alpha}}u(\mathbf{x}_j))\Big(\frac{\mathbf{x}_k-\mathbf{x}_j}{h_j}\Big)^{\boldsymbol{\alpha}} = R_N(\mathbf{x}_k,\mathbf{x}_j).$$

Hence the boundedness of $\|V^{-1}_{j,h_j}\|$ from Condition 11.4 and the estimate of the remainder (10.12) imply

$$|a^{(j)}_{\boldsymbol{\alpha}} - \partial^{\boldsymbol{\alpha}}u(\mathbf{x}_j)| \le \boldsymbol{\alpha}!\,C_2\,h_j^{N-[\boldsymbol{\alpha}]}\sup_{B(\mathbf{x}_j,h_j)}|\nabla_N u|.$$

Therefore, we have the inequality

$$|Mu(\mathbf{x}) - M^{(1)}u(\mathbf{x})| \le C_2\sum_{j\in J} h_j^N \sup_{B(\mathbf{x}_j,h_j)}|\nabla_N u|\sum_{[\boldsymbol{\alpha}]=1}^{N-1}\Big|\frac{\mathbf{x}-\mathbf{x}_j}{h_j}\Big|^{[\boldsymbol{\alpha}]}|\mathcal{P}_j(\mathbf{x})\eta_j(\mathbf{x})|$$

and, for any $\mathbf{x}\in B(\mathbf{x}_m,h_m)$,

$$|Mu(\mathbf{x}) - M^{(1)}u(\mathbf{x})| \le C_3 h_{\mathbf{m}}^N \sup_{B(\mathbf{x}_m,\kappa\, h_m)}|\nabla_N u|.$$

The last inequality and (11.19) lead to (11.17). □

11.2.2. Quasi-interpolants with non-compactly supported basis functions.

THEOREM 11.6. *Suppose that, in addition to the conditions of Lemma 11.3, the inequality*

$$\Big\|\sum_{j\in J}\big(1+h_j^{-1}|\cdot-\mathbf{x}_j|\big)^{N-K}\Big\|_{L_\infty} < \infty \tag{11.21}$$

holds and let $u\in W^N_\infty(\mathbb{R}^n)$ and $\varepsilon>0$ be arbitrary. There exist polynomials $\mathcal{P}_{j,k}$, independent on u, whose degrees are uniformly bounded, such that the quasi-interpolant

$$Mu(\mathbf{x}) = \sum_{k\in J} u(\mathbf{x}_k)\sum_{\overline{\mathrm{st}}(\mathbf{x_j})\ni \mathbf{x}_k}\mathcal{P}_{j,k}\Big(\frac{\mathbf{x}-\mathbf{x}_j}{h_j}\Big)\eta\Big(\frac{\mathbf{x}-\mathbf{x}_j}{h_j}\Big) \tag{11.22}$$

satisfies the estimate

$$|Mu(\mathbf{x}) - u(\mathbf{x})| \le C \sup_{m\in J} h_m^N \, \|\nabla_N \, u\|_{L_\infty} + \varepsilon \, |u(\mathbf{x})| \, . \tag{11.23}$$

The constant C does not depend on u and ε.

PROOF. Analogously to (11.18) in the proof of Theorem 11.5, we introduce the quasi-interpolant

$$M^{(1)}u(\mathbf{x}) = \sum_{j\in J} \Big(\sum_{[\boldsymbol{\alpha}]=1}^{N-1} \frac{\partial^{\boldsymbol{\alpha}} u(\mathbf{x}_j)}{\boldsymbol{\alpha}!} (\mathbf{x}-\mathbf{x}_j)^{\boldsymbol{\alpha}} \Big) \mathcal{P}_j\Big(\frac{\mathbf{x}-\mathbf{x}_j}{h_j}\Big) \eta\Big(\frac{\mathbf{x}-\mathbf{x}_j}{h_j}\Big)$$

and we obtain the estimate

$$|M^{(1)}u(\mathbf{x}) - u(\mathbf{x})| \le \sum_{j\in J} \Big| R_N(\mathbf{x},\mathbf{x}_j) \mathcal{P}_j\Big(\frac{\mathbf{x}-\mathbf{x}_j}{h_j}\Big) \eta\Big(\frac{\mathbf{x}-\mathbf{x}_j}{h_j}\Big)\Big| + |u(\mathbf{x})| \, |\Theta(\mathbf{x}) - 1| \, .$$

By (11.13), we have

$$\Big|\mathcal{P}_j\Big(\frac{\mathbf{x}-\mathbf{x}_j}{h_j}\Big) \eta\Big(\frac{\mathbf{x}-\mathbf{x}_j}{h_j}\Big)\Big| \le \frac{1}{c} \Big|\eta\Big(\frac{\mathbf{x}-\mathbf{x}_j}{h_j}\Big)\Big| + \frac{\varepsilon}{c_K} \Big(1 + \frac{|\mathbf{x}-\mathbf{x}_j|}{h_{\mathbf{j}}}\Big)^{-K}$$

with the lower bound c of $s(\mathbf{x})$ (see (11.3)). Together with (11.10) and (10.12), this leads to

$$\Big|R_N(\mathbf{x},\mathbf{x}_j) \mathcal{P}_j\Big(\frac{\mathbf{x}-\mathbf{x}_j}{h_j}\Big) \eta\Big(\frac{\mathbf{x}-\mathbf{x}_j}{h_j}\Big)\Big|$$
$$\le c_N h_j^N \|\nabla_N \, u\|_{L_\infty} \Big|\frac{\mathbf{x}-\mathbf{x}_j}{h_j}\Big|^N \Big(\frac{c_{\delta,p}}{c} \, w_{\delta,p}\Big(\frac{\mathbf{x}-\mathbf{x}_j}{h_j}\Big) + \frac{\varepsilon}{c_K} \Big) \Big(1 + \Big|\frac{\mathbf{x}-\mathbf{x}_{\mathbf{j}}}{h_{\mathbf{j}}}\Big|\Big)^{-K},$$

resulting in

$$|M^{(1)}u(\mathbf{x}) - u(\mathbf{x})| \le |u(\mathbf{x})| \, |\Theta(\mathbf{x}) - 1| + c_N \sup_{m\in J} h_m^N \|\nabla_N \, u\|_{L_\infty}$$
$$\times \Big(\frac{c_{\delta,p}}{c} \| w_{\delta,p}| \cdot |^N \|_{L_\infty} \sum_{j\in J} \Big(1 + \Big|\frac{\mathbf{x}-\mathbf{x}_j}{h_j}\Big|\Big)^{-K} + \frac{\varepsilon}{c_K} \sum_{j\in J} \Big(1 + \Big|\frac{\mathbf{x}-\mathbf{x}_j}{h_j}\Big|\Big)^{N-K} \Big).$$

Now, we can proceed as in the proof of Theorem 11.5. □

REMARK 11.7. Let the parameter $\kappa_{\mathbf{x}}$ be chosen for fixed $\mathbf{x}$ so that

$$\sum_{|\mathbf{x}_j - \mathbf{x}| > \kappa_{\mathbf{x}}} w_{\delta,p}\Big(\frac{\mathbf{x}-\mathbf{x}_j}{h_j}\Big) \Big|\frac{\mathbf{x}-\mathbf{x}_j}{h_j}\Big|^N \Big(1 + \Big|\frac{\mathbf{x}-\mathbf{x}_j}{h_j}\Big|\Big)^{-K} < \varepsilon \, .$$

Then the estimate (11.23) can be sharpened to

$$|Mu(\mathbf{x}) - u(\mathbf{x})| \le C \max_{|\mathbf{x}_j - \mathbf{x}| \le \kappa_{\mathbf{x}}} h_j^N \sup_{B(\mathbf{x},\kappa_{\mathbf{x}})} |\nabla_N \, u| + \varepsilon \, \big(|u(\mathbf{x})| + \|\nabla_N \, u\|_{L_\infty}\big) \, .$$

11.3. Computation of integral operators

Here, we discuss a direct application of the scattered data quasi-interpolant (11.22) for the Gaussian $\eta(\mathbf{x}) = e^{-|\mathbf{x}|^2}$. Suppose that the density of the integral operator with a radial kernel

$$\mathcal{K}u(\mathbf{x}) = \int_{\mathbb{R}^n} g(|\mathbf{x}-\mathbf{y}|) u(\mathbf{y}) \, d\mathbf{y} \tag{11.24}$$

is approximated by the quasi-interpolant

$$Mu(\mathbf{x}) = \sum_{j\in J} \sum_{\mathbf{x}_k \in \overline{\mathrm{st}}(\mathbf{x}_j)} u(\mathbf{x}_k)\, \mathcal{P}_{j,k}\Big(\frac{\mathbf{x}-\mathbf{x}_j}{h_j}\Big)\, e^{-|\mathbf{x}-\mathbf{x}_j|^2/h_j^2}\,. \tag{11.25}$$

Using the following lemma, it is easy to derive cubature formulas for (11.24).

LEMMA 11.8. *Any polynomial* $\mathcal{P}(\mathbf{x}) = \sum_{[\beta]=0}^{L} c_\beta\, \mathbf{x}^\beta$ *can be written as*

$$\mathcal{P}(\mathbf{x}) e^{-|\mathbf{x}|^2} = \sum_{[\beta]=0}^{L} c_\beta\, \mathcal{S}_\beta(\nabla_{\mathbf{x}})\, e^{-|\mathbf{x}|^2}$$

with the polynomials $\mathcal{S}_\beta(\mathbf{t})$ *being defined by*

$$\mathcal{S}_\beta(\mathbf{t}) = \Big(\frac{1}{2\,i}\Big)^{[\beta]} H_\beta\Big(\frac{\mathbf{t}}{2\,i}\Big)\,, \tag{11.26}$$

where H_β *denotes the Hermite polynomial in* n *variables introduced by* (7.7).

Here and in the following, by $\nabla_{\mathbf{x}}$ we denote the vector of partial differentiation with respect to $\mathbf{x}$,

$$\nabla_{\mathbf{x}} = \Big(\frac{\partial}{\partial x_1}, \dots, \frac{\partial}{\partial x_n}\Big).$$

PROOF. We are looking for the polynomial $\mathcal{S}_\beta(\mathbf{t})$ defined by the relation

$$\mathbf{x}^\beta e^{-|\mathbf{x}|^2} = \mathcal{S}_\beta(\nabla_{\mathbf{x}}) e^{-|\mathbf{x}|^2}, \quad \mathbf{x} \in \mathbb{R}^n. \tag{11.27}$$

Taking the Fourier transforms

$$\mathcal{F}(\mathcal{S}_\beta(\nabla_{\mathbf{x}}) e^{-|\mathbf{x}|^2})(\lambda) = \pi^{n/2} e^{-\pi^2|\lambda|^2} \mathcal{S}_\beta(2\pi i \lambda)$$

and

$$\mathcal{F}(\mathbf{x}^\beta e^{-|\mathbf{x}|^2})(\lambda) = \pi^{n/2}\Big(-\frac{\nabla_\lambda}{2\pi i}\Big)^\beta e^{-\pi^2|\lambda|^2}\,,$$

we obtain (11.26). □

Therefore, we can write $\mathcal{P}_{j,k}(\mathbf{x})\, e^{-|\mathbf{x}|^2} = \mathcal{T}_{j,k}(\nabla_{\mathbf{x}})\, e^{-|\mathbf{x}|^2}$ with some polynomials $\mathcal{T}_{j,k}(\mathbf{x})$. Then (11.25) can be rewritten as

$$Mu(\mathbf{x}) = \sum_{j\in J} \sum_{\mathbf{x}_k \in \overline{\mathrm{st}}(\mathbf{x}_j)} u(\mathbf{x}_k)\, \mathcal{T}_{j,k}(h_j\, \nabla_{\mathbf{x}})\, e^{-|\mathbf{x}-\mathbf{x}_j|^2/h_j^2}\,.$$

As in Chapter 4, the cubature formula for the integral $\mathcal{K}u$ is obtained by replacing u by its quasi-interpolant Mu, i.e.,

$$\begin{aligned} \widetilde{\mathcal{K}}u(\mathbf{x}) &= \mathcal{K}Mu(\mathbf{x}) \\ &= \sum_{j\in J} \sum_{\mathbf{x}_k \in \mathrm{st}(\mathbf{x}_j)} u(\mathbf{x}_k)\, \mathcal{T}_{j,k}(h_j\, \nabla_{\mathbf{x}}) \int_{\mathbb{R}^n} g(|\mathbf{x}-\mathbf{y}|)\, e^{-|\mathbf{y}-\mathbf{x}_j|^2/h_j^2}\, d\mathbf{y} \\ &= \sum_{j\in J} h_j^n \sum_{\mathbf{x}_k \in \mathrm{st}(\mathbf{x}_j)} u(\mathbf{x}_k)\, \mathcal{T}_{j,k}(h_j\, \nabla_{\mathbf{x}}) \int_{\mathbb{R}^n} g(h_j|\mathbf{y}|)\, e^{-|\mathbf{y}-\mathbf{t}_j|^2}\, d\mathbf{y}\,, \end{aligned} \tag{11.28}$$

where we set $\mathbf{t}_j = (\mathbf{x} - \mathbf{x}_j)/h_j$. Now we apply formula (5.15) for the convolution of radial kernels with Gaussians and derive

$$\int_{\mathbb{R}^n} g(h_j|\mathbf{y}|)e^{-|\mathbf{y}-\mathbf{t}_j|^2}\,d\mathbf{y} = \frac{2\pi^{n/2}\,\mathrm{e}^{-|\mathbf{t}_j|^2}}{|\mathbf{t}_j|^{n/2-1}}\int_0^\infty r^{n/2}\,e^{-r^2}g(h_j r)I_{n/2-1}(2|\mathbf{t}_j|r)\,dr\,.$$

Using the notation

$$\mathcal{L}_j(t) = \frac{2\,\pi^{n/2}\,e^{-t^2}}{t^{n/2-1}}\int_0^\infty r^{n/2}e^{-r^2}g(h_j r)I_{n/2-1}(2rt)dr\,,$$

relation (11.28) leads to the cubature formula

$$\tilde{\mathcal{K}}u(\mathbf{x}) = \sum_{j\in J} h_j^n \sum_{\mathbf{x}_k\in\overline{\mathrm{st}}(\mathbf{x}_j)} u(\mathbf{x}_k)\,\mathcal{T}_{j,k}(h_j\,\nabla_{\mathbf{x}})\,\mathcal{L}_j\Big(\frac{|\mathbf{x}-\mathbf{x}_j|}{h_j}\Big)$$

for the integral $\mathcal{K}u$.

11.4. Construction of the Θ-function with Gaussians

In Lemma 11.3 we gave conditions on a basis function η with non-compact support, on the nodes $\{\mathbf{x}_j\}$, and on the scalings h_j which ensure the existence of polynomials $\mathcal{P}_j$ such that the sum

$$\sum_{j\in J}\mathcal{P}_j(\mathbf{x})\,\eta\Big(\frac{\mathbf{x}-\mathbf{x}_j}{h_j}\Big)$$

is an approximate partition of unity. In this section, we describe a method for constructing this Θ-function if η is the Gaussian.

We start with an approximate partition of unity with Gaussians which are centered on a piecewise uniform grid. Then we approximate each of the Gaussians with functions of the form polynomial times Gaussian with scattered centers by solving some least squares problem. The advantage of this approach is its local character. The method does not require solving a large system of linear equations. Instead, in order to obtain a local representation of the partition of unity, one has to solve a small number of linear systems of moderate size.

11.4.1. Approximate partition of unity on a piecewise uniform grid. We point out that shifted Gaussians with different scalings can form an approximate partition of unity, provided that the centers $\mathbf{g}_j$ and scaling parameters $\mathcal{D}_j$ are chosen appropriately. More precisely, for a given bounded domain $\Omega \subset \mathbb{R}^n$ and any $\varepsilon > 0$, there exist a finite sequence of nodes $G = \{\mathbf{g}_j\}$ belonging to a piecewise uniform grid, parameters $\mathcal{D}_j$, and factors $a_j > 0$ such that

$$\Big|1 - \sum_{\mathbf{g}_j\in G} a_j\,\mathrm{e}^{-|\mathbf{x}-\mathbf{g}_j|^2/\mathcal{D}_j}\Big| < \varepsilon \qquad \text{for} \qquad \mathbf{x}\in\Omega\,.$$

Indeed, we start with a uniform grid $\{H\mathbf{m}+\mathbf{b}, \mathbf{m}\in\mathbb{Z}^n\}$ with $H > 0$ and $\mathbf{b}\in\mathbb{R}^n$. Omitting the Gaussians with centers far from Ω, we obtain from Poisson's summation formula that for any $\mathcal{D}$ there exists a finite subset $Z_1 \subset \mathbb{Z}^n$ such that

$$\Big|1 - \frac{1}{(\pi\mathcal{D})^{n/2}}\sum_{\mathbf{m}\in Z_1}\mathrm{e}^{-|\mathbf{x}-H\mathbf{m}-\mathbf{b}|^2/(\mathcal{D}H^2)}\Big| \le c_n\,\mathrm{e}^{-\pi^2\mathcal{D}}\,, \quad \mathbf{x}\in\Omega\,,$$

with a constant c_n depending only on the space dimension. By using the approximate refinement equation (7.11), we can expand Gaussians in this sum as a linear combinations of finer scaled Gaussians. For a given $\mu < 1/2$, we can find another finite subset $Z(\mu) \subset \mathbb{Z}^n$, depending on μ, so that

$$\Big| e^{-|\mathbf{x}-H\mathbf{m}-\mathbf{b}|^2/(\mathcal{D}H^2)} - \sum_{\mathbf{k}\in Z(\mu)} \frac{e^{-\mu^2|\mathbf{k}|^2/(\mathcal{D}(1-\mu^2))}}{(\pi\mathcal{D}(1-\mu^2))^{n/2}} e^{-|\mathbf{x}-\mathbf{b}-H(\mathbf{m}+\mu\mathbf{k})|^2/(\mathcal{D}H^2\mu^2)} \Big| \\ \le c_n\, e^{-3\pi^2\mathcal{D}/4}\, e^{-|\mathbf{x}-H\mathbf{m}-\mathbf{b}|^2/(\mathcal{D}H^2)}, \quad \mathbf{x}\in\Omega .$$

Hence, if we choose $Z_2 \subset Z_1$ and expand $e^{-|\mathbf{x}-H\mathbf{m}-\mathbf{b}|^2/(\mathcal{D}H^2)}$, $\mathbf{m}\in Z_2$, into scaled Gaussians with a scaling factor $\mu_{\mathbf{m}}$, then for $\mathbf{x}\in\Omega$

$$\Big| 1 - \frac{1}{(\pi\mathcal{D})^{n/2}} \sum_{\mathbf{m}\in Z_1\setminus Z_2} e^{-|\mathbf{x}-H\mathbf{m}-\mathbf{b}|^2/(\mathcal{D}H^2)} \\ - \frac{1}{(\pi\mathcal{D})^{n}} \sum_{\mathbf{m}\in Z_2}\sum_{\mathbf{k}\in Z(\mu_{\mathbf{m}})} \frac{e^{-\mu_{\mathbf{m}}^2|\mathbf{k}|^2/(\mathcal{D}(1-\mu_{\mathbf{m}}^2))}}{(1-\mu_{\mathbf{m}}^2)^{n/2}} e^{-|\mathbf{x}-\mathbf{b}-H(\mathbf{m}+\mu_{\mathbf{m}}\mathbf{k})|^2/(\mathcal{D}H^2\mu_{\mathbf{m}}^2)} \Big| \\ \le c_n\, e^{-3\pi^2\mathcal{D}/4}\Big(e^{-\pi^2\mathcal{D}/4} + \frac{1}{(\pi\mathcal{D})^{n/2}} \sum_{\mathbf{m}\in Z_2} e^{-|\mathbf{x}-H\mathbf{m}-\mathbf{b}|^2/(\mathcal{D}H^2)}\Big) \\ \le c_n\, e^{-3\pi^2\mathcal{D}/4}\Big(e^{-\pi^2\mathcal{D}/4} + 1 + c_n\, e^{-\pi^2\mathcal{D}}\Big).$$

Choosing $\mathcal{D}$ large enough, we obtain the approximate partition of unity

$$\Big\{ a_j\, e^{-|\mathbf{x}-\mathbf{g}_j|^2/\mathcal{D}_j} \Big\}_{\mathbf{g}_j\in G} \tag{11.29}$$

on the piecewise uniform partition

$$G = \{H\mathbf{m}+\mathbf{b}, \mathbf{m}\in Z_1\setminus Z_2\} \cup \bigcup_{\mathbf{m}\in Z_2} \{H(\mathbf{m}+\mu_{\mathbf{m}}\mathbf{k})+\mathbf{b}, \mathbf{k}\in Z(\mu_{\mathbf{m}})\}$$

with the factors

$$\mathcal{D}_j = \mathcal{D}H^2, \qquad a_j = \frac{1}{(\pi\mathcal{D})^{n/2}} \qquad \text{if } \mathbf{g}_j \in \{H\mathbf{m}+\mathbf{b} : \mathbf{m}\in Z_1\setminus Z_2\},$$

$$\mathcal{D}_j = \mathcal{D}(H\mu_{\mathbf{m}})^2, \; a_j = \frac{e^{-\mu_{\mathbf{m}}^2|\mathbf{k}|^2/(\mathcal{D}(1-\mu_{\mathbf{m}}^2))}}{(\pi\mathcal{D})^n(1-\mu_{\mathbf{m}}^2)^{n/2}} \\ \text{if } \mathbf{g}_j \in \{H(\mathbf{m}+\mu_{\mathbf{m}}\mathbf{k})+\mathbf{b} : \mathbf{m}\in Z_2, \mathbf{k}\in Z(\mu_{\mathbf{m}})\}.$$

This partition satisfies

$$\Big| 1 - \sum_{\mathbf{g}_j\in G} a_j\, e^{-|\mathbf{x}-\mathbf{g}_j|^2/\mathcal{D}_j} \Big| < \varepsilon \quad \text{for } \mathbf{x}\in\Omega .$$

Obviously, one can omit the grid points outside Ω, which appear during the refinement and have no influence on the value of the sum for $\mathbf{x}\in\Omega$. Moreover, the refinement procedure can be repeated for any of the Gaussian functions in (11.29) without violating the above estimate if the scaling factors $\mu < 1/2$. Then one obtains a new piecewise uniform grid G and explicitly given $\mathcal{D}_j$ and a_j.

We see that using only Poisson's summation formula and the approximate refinement equation, there is great flexibility in constructing approximate partitions of unity with Gaussians, which are centered on piecewise uniform grids and approximate the constant function 1 with any given precision.

11.4.2. Scattered nodes close to a piecewise uniform grid. The construction of the approximate partition of unity

$$\left\{\mathcal{P}_j(\mathbf{x})\,\mathrm{e}^{-|\mathbf{x}-\mathbf{x}_j|^2/h_j^2}\right\} \tag{11.30}$$

will be given for finite sets of scattered nodes $X = \{\mathbf{x}_j\}_{j\in J}$ which are close to piecewise uniform grids, described in the previous subsection. To be more precise, we suppose

CONDITION 11.9. Given a set of scattered nodes $X = \{\mathbf{x}_j\}_{j\in J}$. For any $\varepsilon > 0$, there exist a piecewise uniform grid G and an approximate partition of unity with Gaussians

$$\left\{a_k\,\mathrm{e}^{-|\mathbf{x}-\mathbf{g}_k|^2/\mathcal{D}_k} : \mathbf{g}_k \in G\right\}, \tag{11.31}$$

which satisfies

$$\Big|1 - \sum_{\mathbf{g}_k\in G} a_k\,\mathrm{e}^{-|\mathbf{x}-\mathbf{g}_k|^2/\mathcal{D}_k}\Big| < \varepsilon \qquad \text{for } \mathbf{x}\in\Omega\,. \tag{11.32}$$

For some fixed $\kappa > 1$ and for each grid point $\mathbf{g}_k \in G$, there exists a subset of scattered nodes $\Sigma(\mathbf{g}_k) \subset X$ with $|\mathbf{x}_j - \mathbf{g}_k| \le \kappa\sqrt{\mathcal{D}_k}$ if $\mathbf{x}_j \in \Sigma(\mathbf{g}_k)$ and

$$\bigcup_{\mathbf{g}_k\in G} \Sigma(\mathbf{g}_k) = X\,.$$

The main idea in constructing (11.30) is the following: We take the corresponding approximate partition of unity with Gaussians on a piecewise uniform grid (11.31). For given $\mathbf{g}_k \in G$ we fix a scaling parameter h_j for any $\mathbf{x}_j \in \Sigma(\mathbf{g}_k)$ and we determine, by using least squares approximation, a polynomial $\mathcal{P}_j^{(k)}$ of certain degree $L_j^{(k)}$ such that

$$\sum_{\mathbf{x}_j\in\Sigma(\mathbf{g}_k)} \mathcal{P}_j^{(k)}(\mathbf{x})\,\mathrm{e}^{-|\mathbf{x}-\mathbf{x}_j|^2/h_j^2} \tag{11.33}$$

approximates $\mathrm{e}^{-|\mathbf{x}-\mathbf{g}_k|^2/\mathcal{D}_k}$. For example, we require that the discrepancies

$$\omega_{\mathbf{g}_k}(\mathbf{x}) = \sum_{\mathbf{x}_j\in\Sigma(\mathbf{g}_k)} \mathcal{P}_j^{(k)}(\mathbf{x})\,\mathrm{e}^{-|\mathbf{x}-\mathbf{x}_j|^2/h_j^2} - \mathrm{e}^{-|\mathbf{x}-\mathbf{g}_k|^2/\mathcal{D}_k} \tag{11.34}$$

are subject to

$$\max_{\mathbf{x}} \sum_{\mathbf{g}_k\in G} a_k|\omega_{\mathbf{g}_k}(\mathbf{x})| < \delta\,. \tag{11.35}$$

Then, obviously,

$$\max_{\mathbf{x}}\Big|1 - \sum_{\mathbf{g}_k\in G} a_k \sum_{\mathbf{x}_j\in\Sigma(\mathbf{g}_k)} \mathcal{P}_j^{(k)}(\mathbf{x})\,\mathrm{e}^{-|\mathbf{x}-\mathbf{x}_j|^2/h_j^2}\Big| < \delta+\varepsilon\,,$$

where ε is the accuracy (11.32) of the approximate partition of unity on G. Thus, if δ is sufficiently small, then the function system

$$\Big\{\mathrm{e}^{-|\mathbf{x}-\mathbf{x}_j|^2/h_j^2} \sum_{\Sigma(\mathbf{g}_k)\ni\mathbf{x}_j} a_k\mathcal{P}_j^{(k)}(\mathbf{x})\Big\}$$

is the required partition of unity for the scattered nodes $\{\mathbf{x}_j\}$.

Note that if a node $\mathbf{x}_j$ belongs to different sets $\Sigma(\mathbf{g}_k)$, then the scaling parameter h_j has to take the same value in the approximations formulas (11.33) for the different Gaussians $\mathrm{e}^{-|\mathbf{x}-\mathbf{g}_k|^2/\mathcal{D}_k}$. Of course, the choice of the scaling parameters h_j in (11.33) is crucial. They should reflect the density or closeness of the nodes $\mathbf{x}_j \in \Sigma(\mathbf{g}_k)$ to preserve the local character of the resulting quasi-interpolants. Therefore, it is natural to assume that

$$h_j^2 \le \mathcal{D}_k \quad \text{for all} \quad \mathbf{x}_j \in \Sigma(\mathbf{g}_k)\,. \tag{11.36}$$

The choice of the scaling parameters h_j will be considered in Subsections 11.4.6 and 11.4.7 in more detail.

11.4.3. Discrepancy as convolution. We consider the prototype of the discrepancies $\omega_{\mathbf{g}_k}$ defined by (11.34)

$$\omega(\mathbf{x}) = \sum_{\mathbf{x}_j \in \Sigma} \mathcal{P}_j\Big(\frac{\mathbf{x}-\mathbf{x}_j}{h_j^2}\Big)\,\mathrm{e}^{-|\mathbf{x}-\mathbf{x}_j|^2/h_j^2} - \mathrm{e}^{-|\mathbf{x}|^2/\mathcal{D}}\,, \tag{11.37}$$

where the form of the argument of the polynomials $\mathcal{P}_j$ is chosen for technical reasons. Note that $\omega(\mathbf{x}) = a_k^{-1}\omega_{\mathbf{g}_k}(\mathbf{x}-\mathbf{g}_k)$ with the set of nodes $\Sigma = \Sigma(\mathbf{g}_k) - \mathbf{g}_k$. By Condition 11.9, the nodes $\mathbf{x}_j \in \Sigma$ satisfy $|\mathbf{x}_j| \le \kappa\sqrt{\mathcal{D}}$. Recall the assumption $h_j^2 \le \mathcal{D}$.

We will use a least squares method for constructing $\mathcal{P}_j \in \Pi_{L_j}$ such that for some $\tau > 0$ the discrepancy satisfies

$$\mathrm{e}^{\tau|\mathbf{x}|^2}\,|\omega(\mathbf{x})| < \varepsilon$$

for sufficiently large L_j. By Π_{L_j}, we denote the set of polynomials of degree L_j, and, in what follows, we use the representation

$$\mathcal{P}_j(\mathbf{x}) = \sum_{[\boldsymbol{\beta}]=0}^{L_j} c_{j,\boldsymbol{\beta}}\,\mathbf{x}^{\boldsymbol{\beta}}\,.$$

Then by Lemma 11.8,

$$\mathcal{P}_j\Big(\frac{\mathbf{x}-\mathbf{x}_j}{h_j^2}\Big)\,\mathrm{e}^{-|\mathbf{x}-\mathbf{x}_j|^2/h_j^2} = \sum_{[\boldsymbol{\beta}]=0}^{L_j} c_{j,\boldsymbol{\beta}}\,\mathcal{S}_{\boldsymbol{\beta}}(h_j\,\nabla_{\mathbf{x}})\,\mathrm{e}^{-|\mathbf{x}-\mathbf{x}_j|^2/h_j^2}\,,$$

and ω can be written as

$$\omega(\mathbf{x}) = \sum_{\mathbf{x}_j \in \Sigma} \sum_{[\boldsymbol{\beta}]=0}^{L_j} c_{j,\boldsymbol{\beta}}\,\mathcal{S}_{\boldsymbol{\beta}}(h_j\,\nabla_{\mathbf{x}})\,e^{-|\mathbf{x}-\mathbf{x}_j|^2/h_j^2} - e^{-|\mathbf{x}|^2/\mathcal{D}}\,. \tag{11.38}$$

To estimate the L_∞-norm of ω, we represent this function as a convolution in the following assertion.

LEMMA 11.10. *Let $\mathcal{P}(\mathbf{t})$ be a polynomial and let $0 < \mathcal{D}_0 < \mathcal{D}$. Then*

$$\mathcal{P}(\nabla_{\mathbf{x}})\,\mathrm{e}^{-|\mathbf{x}-\mathbf{y}|^2/\mathcal{D}} = c_1\,\mathrm{e}^{-|\mathbf{x}|^2/(\mathcal{D}-\mathcal{D}_0)} * \mathcal{P}(\nabla_{\mathbf{x}})\,\mathrm{e}^{-|\mathbf{x}-\mathbf{y}|^2/\mathcal{D}_0}$$

with the constant

$$c_1 = \Big(\frac{\mathcal{D}}{\pi\mathcal{D}_0(\mathcal{D}-\mathcal{D}_0)}\Big)^{n/2}$$

and $*$ standing for the convolution operator.

PROOF. From

$$\mathrm{e}^{-|\mathbf{x}-\mathbf{y}|^2/\mathcal{D}} = c_1 \int\limits_{\mathbb{R}^n} \mathrm{e}^{-|\mathbf{x}-\mathbf{t}|^2/(\mathcal{D}-\mathcal{D}_0)}\, \mathrm{e}^{-|\mathbf{t}-\mathbf{y}|^2/\mathcal{D}_0}\, d\mathbf{t}\,,$$

we obtain

$$\begin{aligned}\mathcal{P}(\nabla_{\mathbf{x}})\, \mathrm{e}^{-|\mathbf{x}-\mathbf{y}|^2/\mathcal{D}} &= \mathcal{P}(-\nabla_{\mathbf{y}})\, \mathrm{e}^{-|\mathbf{x}-\mathbf{y}|^2/\mathcal{D}} \\ &= c_1 \int\limits_{\mathbb{R}^n} \mathrm{e}^{-|\mathbf{x}-\mathbf{t}|^2/(\mathcal{D}-\mathcal{D}_0)}\, \mathcal{P}(\nabla_{\mathbf{t}})\, \mathrm{e}^{-|\mathbf{t}-\mathbf{y}|^2/\mathcal{D}_0}\, d\mathbf{t}\,.\end{aligned}$$

□

Now, we choose positive numbers δ_j and $\mathcal{D}_0$ such that

$$\sigma := \mathcal{D} - \mathcal{D}_0 = h_j^2 - \delta_j^2 > 0 \quad \text{for all } \mathbf{x}_j \in \Sigma\,,$$

and we write, using Lemma 11.10,

$$\begin{aligned}&\sum_{[\boldsymbol{\beta}]=0}^{L_j} c_{j,\boldsymbol{\beta}}\, \mathcal{S}_{\boldsymbol{\beta}}(h_j\, \nabla_{\mathbf{x}})\, \mathrm{e}^{-|\mathbf{x}-\mathbf{x}_j|^2/h_j^2} \\ &= \Big(\frac{h_j^2}{\pi \delta_j^2 \sigma}\Big)^{n/2} \int\limits_{\mathbb{R}^n} \mathrm{e}^{-|\mathbf{x}-\mathbf{t}|^2/\sigma} \sum_{\mathbf{x}_j \in \Sigma} \sum_{[\boldsymbol{\beta}]=0}^{L_j} c_{j,\boldsymbol{\beta}}\, \mathcal{S}_{\boldsymbol{\beta}}(h_j\, \nabla_{\mathbf{t}})\, \mathrm{e}^{-|\mathbf{t}-\mathbf{x}_j|^2/\delta_j^2}\, d\mathbf{t}\end{aligned}$$

and

$$\mathrm{e}^{-|\mathbf{x}|^2/\mathcal{D}} = \Big(\frac{\mathcal{D}}{\pi \mathcal{D}_0 \sigma}\Big)^{n/2} \int\limits_{\mathbb{R}^n} \mathrm{e}^{-|\mathbf{x}-\mathbf{t}|^2/\sigma}\, \mathrm{e}^{-|\mathbf{t}|^2/\mathcal{D}_0}\, d\mathbf{t}\,.$$

Thus, by (11.38),

$$\begin{aligned}\omega(\mathbf{x}) = \Big(\frac{\mathcal{D}}{\pi \mathcal{D}_0 \sigma}\Big)^{n/2} &\int\limits_{\mathbb{R}^n} \mathrm{e}^{-|\mathbf{x}-\mathbf{t}|^2/\sigma} \\ &\times \Big(\sum_{\mathbf{x}_j \in \Sigma} \Big(\frac{h_j^2 \mathcal{D}_0}{\delta_j^2 \mathcal{D}}\Big)^{n/2} \sum_{[\boldsymbol{\beta}]=0}^{L_j} c_{j,\boldsymbol{\beta}}\, \mathcal{S}_{\boldsymbol{\beta}}(h_j\, \nabla_{\mathbf{t}})\, \mathrm{e}^{-|\mathbf{t}-\mathbf{x}_j|^2/\delta_j^2} - \mathrm{e}^{-|\mathbf{t}|^2/\mathcal{D}_0} \Big)\, d\mathbf{t}\,.\end{aligned}$$

To simplify notation, we introduce

$$\chi_{\mathbf{c}}(\mathbf{t}) = \sum_{\mathbf{x}_j \in \Sigma} \Big(\frac{h_j^2 \mathcal{D}_0}{\delta_j^2 \mathcal{D}}\Big)^{n/2} \sum_{[\boldsymbol{\beta}]=0}^{L_j} c_{j,\boldsymbol{\beta}}\, \mathcal{S}_{\boldsymbol{\beta}}(h_j\, \nabla_{\mathbf{t}})\, \mathrm{e}^{-|\mathbf{t}-\mathbf{x}_j|^2/\delta_j^2} - \mathrm{e}^{-|\mathbf{t}|^2/\mathcal{D}_0} \tag{11.39}$$

with $\mathbf{c} = \{c_{j,\boldsymbol{\beta}}\}$ such that

$$\omega(\mathbf{x}) = \Big(\frac{\mathcal{D}}{\pi \mathcal{D}_0 \sigma}\Big)^{n/2} \int\limits_{\mathbb{R}^n} \mathrm{e}^{-|\mathbf{x}-\mathbf{t}|^2/\sigma}\, \chi_{\mathbf{c}}(\mathbf{t})\, d\mathbf{t}\,. \tag{11.40}$$

We obtain by Young's inequality (2.11) that

$$\|\omega\|_{L_\infty} \le \frac{\mathcal{D}^{n/2}}{\mathcal{D}_0^{n/2} (2\pi\sigma)^{n/4}} \|\chi_{\mathbf{c}}\|_{L_2}\,.$$

An estimate for the sum

$$\sum_{\mathbf{g}_k \in G} |\omega_{\mathbf{g}_k}(\mathbf{x})|$$

can be derived from the following.

LEMMA 11.11. *For given $h_j \le \sqrt{\mathcal{D}}$, $j \in \Sigma$, let the positive numbers δ_j, $\mathcal{D}_0$ be chosen such that $\mathcal{D} - \mathcal{D}_0 = h_j^2 - \delta_j^2 = \sigma > 0$. Denote $\tau = \dfrac{\sigma}{\sigma^2 + \mathcal{D}\mathcal{D}_0}$. Then*

$$|\omega(\mathbf{x})| \le c_2 \, \mathrm{e}^{-\tau|\mathbf{x}|^2} \sqrt{Q(\mathbf{c})}\,, \tag{11.41}$$

with the constant

$$c_2 = \Big(\frac{\mathcal{D}(\sigma^2 + \mathcal{D}\mathcal{D}_0)}{2\pi\mathcal{D}_0^3\sigma} \Big)^{n/4}.$$

Here $Q(\mathbf{c})$ is a quadratic form defined by

$$Q(\mathbf{c}) = \int\limits_{\mathbb{R}^n} \mathrm{e}^{2\sigma|\mathbf{t}|^2/(\mathcal{D}\mathcal{D}_0)} \, |\chi_{\mathbf{c}}(\mathbf{t})|^2 \, d\mathbf{t} \tag{11.42}$$

with the function $\chi_{\mathbf{c}}$ from (11.39).

PROOF. By using

$$|\mathbf{x} - \mathbf{t}|^2 = \Big| \sqrt{a}\mathbf{x} - \frac{\mathbf{t}}{\sqrt{a}} \Big|^2 + (1-a)|\mathbf{x}|^2 + \frac{a-1}{a}|\mathbf{t}|^2$$

for any $a > 0$, we derive the representation

$$\omega(\mathbf{x}) \, \mathrm{e}^{(1-a)|\mathbf{x}|^2/\sigma} = \Big(\frac{\mathcal{D}}{\pi\mathcal{D}_0\sigma} \Big)^{n/2} \int\limits_{\mathbb{R}^n} \mathrm{e}^{-|a\mathbf{x}-\mathbf{t}|^2/(a\sigma)} \, \mathrm{e}^{(1-a)|\mathbf{t}|^2/(a\sigma)} \, \chi_{\mathbf{c}}(\mathbf{t}) \, d\mathbf{t}$$

from (11.40). Then Young's inequality leads to

$$\Big| \omega(\mathbf{x}) \, \mathrm{e}^{(1-a)|\mathbf{x}|^2/\sigma} \Big| \le \Big(\frac{\mathcal{D}}{\pi\mathcal{D}_0\sigma} \Big)^{n/2} \big\| \mathrm{e}^{-a|\cdot|^2/\sigma} \big\|_{L_2} \big\| \mathrm{e}^{(1-a)|\cdot|^2/(a\sigma)} \, \chi_{\mathbf{c}} \big\|_{L_2}. \tag{11.43}$$

In view of (11.39), we have to choose the parameter a such that the functions

$$\mathrm{e}^{(1-a)|\mathbf{t}|^2/(a\sigma)} \, \mathrm{e}^{-|\mathbf{t}|^2/\mathcal{D}_0} \,, \ \mathrm{e}^{(1-a)|\mathbf{t}|^2/(a\sigma)} \, \mathrm{e}^{-|\mathbf{t}-\mathbf{x}_j|^2/\delta_j^2} \in L_2(\mathbb{R}^n)\,.$$

Since $\mathcal{D} - \mathcal{D}_0 = h_j^2 - \delta_j^2 = \sigma$ and $h_j^2 \le \mathcal{D}$, this can be achieved, for example, if a is chosen such that

$$\frac{(1-a)|\mathbf{t}|^2}{a\sigma} - \frac{|\mathbf{t}|^2}{\mathcal{D}_0} = -\frac{|\mathbf{t}|^2}{\mathcal{D}}\,, \quad \text{i.e.,} \quad a = \frac{\mathcal{D}\mathcal{D}_0}{\sigma^2 + \mathcal{D}\mathcal{D}_0}\,.$$

Then

$$\frac{(1-a)}{a\sigma} = \frac{\sigma}{\mathcal{D}\mathcal{D}_0}\,, \quad \tau = \frac{1-a}{\sigma} = \frac{\sigma}{\sigma^2 + \mathcal{D}\mathcal{D}_0}$$

and

$$\big\| \mathrm{e}^{-a|\cdot|^2/\sigma} \big\|_{L_2} = \Big(\frac{\pi\sigma}{2a} \Big)^{n/4} = \Big(\frac{\pi\sigma(\sigma^2 + \mathcal{D}\mathcal{D}_0)}{2\mathcal{D}\mathcal{D}_0} \Big)^{n/4}.$$

Hence, from (11.43), we derive

$$\Big| \omega(\mathbf{x}) \, \mathrm{e}^{\tau|\mathbf{x}|^2} \Big| \le \Big(\frac{\mathcal{D}(\sigma^2 + \mathcal{D}\mathcal{D}_0)}{2\pi\mathcal{D}_0^3\sigma} \Big)^{n/4} \big\| \mathrm{e}^{\sigma|\cdot|^2/(\mathcal{D}\mathcal{D}_0)} \, \chi_{\mathbf{c}} \big\|_{L_2}. \qquad \square$$

COROLLARY 11.12. *The following estimate holds:*

$$\min_{\mathcal{P}_j \in \Pi_{L_j}} \Big| \sum_{\mathbf{x}_j \in \Sigma} \mathcal{P}_j \, \mathrm{e}^{-|\cdot - \mathbf{x}_j|^2/h_j^2} - \mathrm{e}^{-|\cdot|^2/\mathcal{D}} \Big| \le c_2 \, \mathrm{e}^{-\tau|\mathbf{x}|^2} \min_{\mathbf{c}} \sqrt{Q(\mathbf{c})}\,. \tag{11.44}$$

In the next subsections, we describe a constructive method to find the vector $\mathbf{c} = \{c_{j,\boldsymbol{\beta}}\}$, which realizes

$$\min_{\mathbf{c}} Q(\mathbf{c})$$

and establish its uniqueness.

11.4.4. Construction of polynomials. Let us give an explicit expression of the quadratic form $Q(\mathbf{c})$. From (11.39) and (11.42), we obtain

$$\begin{aligned}Q(\mathbf{c}) = &\int\limits_{\mathbb{R}^n} \mathrm{e}^{2\sigma|\mathbf{t}|^2/(\mathcal{D}\mathcal{D}_0)}\, \mathrm{e}^{-2|\mathbf{t}|^2/\mathcal{D}_0}\, d\mathbf{t}\\
&-2\sum_{\mathbf{x}_j\in\Sigma}\Big(\frac{h_j^2\mathcal{D}_0}{\delta_j^2\mathcal{D}}\Big)^{n/2}\sum_{[\boldsymbol{\beta}]=0}^{L_j} c_{j,\boldsymbol{\beta}}\int\limits_{\mathbb{R}^n} \mathrm{e}^{2\sigma|\mathbf{t}|^2/(\mathcal{D}\mathcal{D}_0)}\,\mathrm{e}^{-|\mathbf{t}|^2/\mathcal{D}_0}\,\mathcal{S}_{\boldsymbol{\beta}}(h_j\nabla_{\mathbf{t}})\,\mathrm{e}^{-|\mathbf{t}-\mathbf{x}_j|^2/\delta_j^2}\,d\mathbf{t}\\
&+\sum_{\mathbf{x}_j,\mathbf{x}_k\in\Sigma}\Big(\frac{h_jh_k\mathcal{D}_0}{\delta_j\delta_k\mathcal{D}}\Big)^{n}\sum_{|\boldsymbol{\beta}|=0}^{L_j}\sum_{|\boldsymbol{\gamma}|=0}^{L_k} c_{j,\boldsymbol{\beta}}\,c_{k,\boldsymbol{\gamma}}\\
&\qquad\times\int\limits_{\mathbb{R}^n}\mathrm{e}^{2\sigma|\mathbf{t}|^2/(\mathcal{D}\mathcal{D}_0)}\,\mathcal{S}_{\boldsymbol{\beta}}(h_j\nabla_{\mathbf{t}})\,\mathrm{e}^{-|\mathbf{t}-\mathbf{x}_j|^2/\delta_j^2}\,\mathcal{S}_{\boldsymbol{\gamma}}(h_k\nabla_{\mathbf{t}})\,\mathrm{e}^{-|\mathbf{t}-\mathbf{x}_k|^2/\delta_k^2}\,d\mathbf{t}\,.\end{aligned}$$

Since

$$\mathcal{S}_{\boldsymbol{\beta}}(h_j\nabla_{\mathbf{x}})\,\mathrm{e}^{-|\mathbf{x}-\mathbf{y}|^2/\delta_j^2} = \mathcal{S}_{\boldsymbol{\beta}}(-h_j\nabla_{\mathbf{y}})\,\mathrm{e}^{-|\mathbf{x}-\mathbf{y}|^2/\delta_j^2}\,,$$

we introduce the functions

$$\begin{aligned}\mathcal{B}_{\boldsymbol{\beta}}^{(j)}(\mathbf{x}) &= \Big(\frac{h_j^2\mathcal{D}_0}{\delta_j^2\mathcal{D}}\Big)^{n/2}\mathcal{S}_{\boldsymbol{\beta}}(-h_j\nabla_{\mathbf{x}})\int\limits_{\mathbb{R}^n}\mathrm{e}^{2\sigma|\mathbf{t}|^2/(\mathcal{D}\mathcal{D}_0)}\,\mathrm{e}^{-|\mathbf{t}|^2/\mathcal{D}_0}\,\mathrm{e}^{-|\mathbf{t}-\mathbf{x}|^2/\delta_j^2}\,d\mathbf{t}\,,\\
\mathcal{C}_{\boldsymbol{\beta},\boldsymbol{\gamma}}^{(jk)}(\mathbf{x},\mathbf{y}) &= \Big(\frac{h_jh_k\mathcal{D}_0}{\delta_j\delta_k\mathcal{D}}\Big)^{n}\\
&\quad\times\mathcal{S}_{\boldsymbol{\beta}}(-h_j\nabla_{\mathbf{x}})\,\mathcal{S}_{\boldsymbol{\gamma}}(-h_k\nabla_{\mathbf{y}})\int\limits_{\mathbb{R}^n}\mathrm{e}^{2\sigma|\mathbf{t}|^2/(\mathcal{D}\mathcal{D}_0)}\,\mathrm{e}^{-|\mathbf{t}-\mathbf{x}|^2/\delta_j^2}\,\mathrm{e}^{-|\mathbf{t}-\mathbf{x}|^2/\delta_k^2}\,d\mathbf{t}\,,\end{aligned}$$

such that

$$\begin{aligned}Q(\mathbf{c}) = \Big(\frac{\pi\mathcal{D}}{2}\Big)^{n/2} &- 2\sum_{\mathbf{x}_j\in\Sigma}\sum_{[\boldsymbol{\beta}]=0}^{L_j} c_{j,\boldsymbol{\beta}}\mathcal{B}_{\boldsymbol{\beta}}^{(j)}(\mathbf{x}_j)\\
&+\sum_{\mathbf{x}_j,\mathbf{x}_k\in\Sigma}\sum_{[\boldsymbol{\beta}]=0}^{L_j}\sum_{[\boldsymbol{\gamma}]=0}^{L_k} c_{j,\boldsymbol{\beta}}\,c_{k,\boldsymbol{\gamma}}\,\mathcal{C}_{\boldsymbol{\beta},\boldsymbol{\gamma}}^{(jk)}(\mathbf{x}_j,\mathbf{x}_k).\end{aligned}$$

The minimum of the quadratic form $Q(\mathbf{c})$ is attained for the solution $\mathbf{c} = \{c_{j,\boldsymbol{\beta}}\}$ of the linear system

$$\sum_{\mathbf{x}_j\in\Sigma}\sum_{[\boldsymbol{\beta}]=0}^{L_j} c_{j,\boldsymbol{\beta}}\,\mathcal{C}_{\boldsymbol{\beta},\boldsymbol{\gamma}}^{(jk)}(\mathbf{x}_j,\mathbf{x}_k) = \mathcal{B}_{\boldsymbol{\gamma}}^{(j)}(\mathbf{x}_k)\,,\quad \mathbf{x}_k\in\Sigma,\ 0\le[\boldsymbol{\gamma}]\le L_k\,. \tag{11.45}$$

The integrals

$$\Big(\frac{h_j^2\mathcal{D}_0}{\delta_j^2\mathcal{D}}\Big)^{n/2}\int\limits_{\mathbb{R}^n} \mathrm{e}^{\,2\sigma|\mathbf{t}|^2/(\mathcal{D}\mathcal{D}_0)}\,\mathrm{e}^{\,-|\mathbf{t}|^2/\mathcal{D}_0}\,\mathrm{e}^{\,-|\mathbf{t}-\mathbf{x}|^2/\delta_j^2}\,d\mathbf{t}$$
$$= \Big(\frac{\pi\, h_j^2\, \mathcal{D}_0^2}{\delta_j^2(\mathcal{D}_0\mathcal{D}+\mathcal{D}-2\sigma)}\Big)^{n/2}\exp\Big(-\frac{(\mathcal{D}-2\sigma)|\mathbf{x}|^2}{\delta_j^2(\mathcal{D}_0\mathcal{D}+\mathcal{D}-2\sigma)}\Big)$$

and

$$\Big(\frac{h_jh_k\mathcal{D}_0}{\delta_j\delta_k\mathcal{D}}\Big)^{n}\int\limits_{\mathbb{R}^n} \mathrm{e}^{\,2\sigma|\mathbf{t}|^2/(\mathcal{D}\mathcal{D}_0)}\,\mathrm{e}^{\,-|\mathbf{t}-\mathbf{x}|^2/\delta_j^2}\,\mathrm{e}^{\,-|\mathbf{t}-\mathbf{y}|^2/\delta_k^2}\,d\mathbf{t}$$
$$= \frac{\pi^{n/2}h_j^nh_k^n\mathcal{D}_0^{3n/2}}{\mathcal{D}^{n/2}(\mathcal{D}_0\mathcal{D}(\delta_j^2+\delta_k^2)-2\sigma\delta_j^2\delta_k^2)^{n/2}}\exp\Big(\frac{2\sigma(\delta_k^2|\mathbf{x}|^2+\delta_j^2|\mathbf{y}|^2)-\mathcal{D}_0\mathcal{D}|\mathbf{x}-\mathbf{y}|^2}{\mathcal{D}_0\mathcal{D}(\delta_j^2+\delta_k^2)-2\sigma\delta_j^2\delta_k^2}\Big)$$

provide the explicit formulas for computing the coefficients $\mathcal{C}^{(jk)}_{\boldsymbol{\beta},\gamma}(\mathbf{x}_j,\mathbf{x}_k)$ and the right-hand side $\mathcal{B}^{(j)}_{\gamma}(\mathbf{x}_k)$ of the system (11.45):

$$\mathcal{B}^{(j)}_{\gamma}(\mathbf{x}) = \Big(\frac{\pi\, h_j^2\, \mathcal{D}_0^2}{\delta_j^2(\mathcal{D}_0\mathcal{D}+\mathcal{D}-2\sigma)}\Big)^{n/2}\mathcal{S}_{\boldsymbol{\beta}}(-h_j\nabla_{\mathbf{x}})\exp\Big(-\frac{(\mathcal{D}-2\sigma)|\mathbf{x}|^2}{\delta_j^2(\mathcal{D}_0\mathcal{D}+\mathcal{D}-2\sigma)}\Big),$$
$$\mathcal{C}^{(jk)}_{\boldsymbol{\beta},\gamma}(\mathbf{x},\mathbf{y}) = \frac{\pi^{n/2}h_j^nh_k^n\mathcal{D}_0^{3n/2}}{\mathcal{D}^{n/2}(\mathcal{D}_0\mathcal{D}(\delta_j^2+\delta_k^2)-2\sigma\delta_j^2\delta_k^2)^{n/2}}$$
$$\times\, \mathcal{S}_{\boldsymbol{\beta}}(-h_j\nabla_{\mathbf{x}})\mathcal{S}_{\gamma}(-h_k\nabla_{\mathbf{y}})\exp\Big(\frac{2\sigma(\delta_k^2|\mathbf{x}|^2+\delta_j^2|\mathbf{y}|^2)-\mathcal{D}_0\mathcal{D}|\mathbf{x}-\mathbf{y}|^2}{\mathcal{D}_0\mathcal{D}(\delta_j^2+\delta_k^2)-2\sigma\delta_j^2\delta_k^2}\Big).$$

In the next subsection, we show that (11.45) has a unique solution $\{c_{j,\boldsymbol{\beta}}\}$. Then by Corollary 11.12, the sum

$$\sum_{\mathbf{x}_j\in\Sigma}\mathcal{P}_j\Big(\frac{\mathbf{x}-\mathbf{x}_j}{h_j}\Big)\,\mathrm{e}^{\,-|\mathbf{x}-\mathbf{x}_j|^2/h_j^2} = \sum_{\mathbf{x}_j\in\Sigma}\sum_{[\boldsymbol{\beta}]=0}^{L_j} c_{j,\boldsymbol{\beta}}\Big(\frac{\mathbf{x}-\mathbf{x}_j}{h_j}\Big)^{\boldsymbol{\beta}}\,\mathrm{e}^{\,-|\mathbf{x}-\mathbf{x}_j|^2/h_j^2} \tag{11.46}$$

approximates $\mathrm{e}^{-|\mathbf{x}|^2/\mathcal{D}}$ with the error $c_2\,\mathrm{e}^{-\tau|\mathbf{x}|^2}\sqrt{Q(\mathbf{c})}$.

11.4.5. Existence and uniqueness. To establish the uniqueness of the minimizing vector $\mathbf{c}=\{c_{j,\boldsymbol{\beta}}\}$, we use another representation of the quadratic form $Q(\mathbf{c})$ defined by (11.42). If we introduce polynomials $T^{(j)}_{\boldsymbol{\beta}}$ of degree $[\boldsymbol{\beta}]$ by

$$T^{(j)}_{\boldsymbol{\beta}}(\mathbf{x}) = \Big(\frac{h_j^2\mathcal{D}_0}{\delta_j^2\mathcal{D}}\Big)^{n/2}\mathrm{e}^{\,|\mathbf{x}|^2/\delta_j^2}\,\mathcal{S}_{\boldsymbol{\beta}}(h_j\,\nabla)\,\mathrm{e}^{\,-|\mathbf{x}|^2/\delta_j^2}, \tag{11.47}$$

then by (11.39),

$$\chi_{\mathbf{c}}(\mathbf{t}) = e^{\,-|\mathbf{t}|^2/\mathcal{D}_0} - \sum_{\mathbf{x}_j\in\Sigma}\sum_{[\boldsymbol{\beta}]=0}^{L_j} c_{j,\boldsymbol{\beta}}T^{(j)}_{\boldsymbol{\beta}}(\mathbf{t}-\mathbf{x}_j)e^{\,-|\mathbf{t}-\mathbf{x}_j|^2/\delta_j^2}$$

and from (11.42), we conclude

$$Q(\mathbf{c}) = \int_{\mathbb{R}^n} \mathrm{e}^{2\sigma|\mathbf{t}|^2/(\mathcal{D}\mathcal{D}_0)} \Big(e^{-|\mathbf{t}|^2/\mathcal{D}_0} - \sum_{\mathbf{x}_j\in\Sigma} \sum_{[\boldsymbol{\beta}]=0}^{L_j} c_{j,\boldsymbol{\beta}} T^{(j)}_{\boldsymbol{\beta}}(\mathbf{t}-\mathbf{x}_j) e^{-|\mathbf{t}-\mathbf{x}_j|^2/\delta_j^2} \Big)^2 d\mathbf{t}$$

$$= \int_{\mathbb{R}^n} \Big(e^{-|\mathbf{t}|^2/\mathcal{D}} - \sum_{\mathbf{x}_j\in\Sigma} \sum_{[\boldsymbol{\beta}]=0}^{L_j} c_{j,\boldsymbol{\beta}} T^{(j)}_{\boldsymbol{\beta}}(\mathbf{t}-\mathbf{x}_j) e^{-|\mathbf{t}-\mathbf{x}_j|^2/\delta_j^2} \, \mathrm{e}^{\sigma|\mathbf{t}|^2/(\mathcal{D}\mathcal{D}_0)} \Big)^2 d\mathbf{t}\,.$$

Since

$$\frac{\sigma|\mathbf{t}|^2}{\mathcal{D}_0\mathcal{D}} - \frac{|\mathbf{t}-\mathbf{x}_j|^2}{\delta_j^2} = -\frac{\mathcal{D}_0\mathcal{D}-\delta_j^2\sigma}{\delta_j^2\mathcal{D}_0\mathcal{D}} \Big(\mathbf{t} - \frac{\mathcal{D}_0\mathcal{D}}{\mathcal{D}_0\mathcal{D}-\delta_j^2\sigma}\mathbf{x}_j\Big)^2 + \frac{\sigma\,|\mathbf{x}_j|^2}{\mathcal{D}_0\mathcal{D}-\delta_j^2\sigma}$$

and

$$\mathcal{D}_0\mathcal{D} - \delta_j^2\sigma = \mathcal{D}_0^2 + \sigma(\mathcal{D}_0 - \delta_j^2) > 0\,,$$

we can write

$$e^{-|\mathbf{t}-\mathbf{x}_j|^2/\delta_j^2}\, \mathrm{e}^{\sigma|\mathbf{t}|^2/(\mathcal{D}\mathcal{D}_0)} = \mathrm{e}^{\sigma|\mathbf{x}_j|^2/(\mathcal{D}_0\mathcal{D}-\delta_j^2\sigma)}\, \mathrm{e}^{-|\mathbf{t}-\mathbf{t}_j|^2/d_j}$$

with

$$d_j := \frac{\delta_j^2\mathcal{D}_0\mathcal{D}}{\mathcal{D}_0\mathcal{D}-\delta_j^2\sigma} > 0 \tag{11.48}$$

and the transformed points

$$\mathbf{t}_j = \frac{\mathcal{D}_0\mathcal{D}}{\mathcal{D}_0\mathcal{D}-\delta_j^2\sigma}\mathbf{x}_j = \frac{d_j}{\delta_j^2}\mathbf{x}_j\,, \quad \mathbf{x}_j\in\Sigma\,. \tag{11.49}$$

Note that $d_j = \mathcal{D}$ if $h_j^2 = \mathcal{D}$ and $d_j < \mathcal{D}$ otherwise.

Then $Q(\mathbf{c})$ can be written as

$$Q(\mathbf{c}) = \int_{\mathbb{R}^n} \Big(e^{-|\mathbf{t}|^2/\mathcal{D}} - \sum_{\mathbf{x}_j\in\Sigma} \sum_{[\boldsymbol{\beta}]=0}^{L_j} \widetilde{c}_{j,\boldsymbol{\beta}} T^{(j)}_{\boldsymbol{\beta}}(\mathbf{t}-\mathbf{x}_j)\, \mathrm{e}^{-|\mathbf{t}-\mathbf{t}_j|^2/d_j} \Big)^2 d\mathbf{t} \tag{11.50}$$

with the coefficients

$$\widetilde{c}_{j,\boldsymbol{\beta}} = c_{j,\boldsymbol{\beta}}\, \mathrm{e}^{\sigma|\mathbf{x}_j|^2/(\mathcal{D}_0\mathcal{D}-\delta_j^2\sigma)}\,.$$

Since $T^{(j)}_{\boldsymbol{\beta}}$ are polynomials of degree $[\boldsymbol{\beta}]$, the minimum problem for $Q(\mathbf{c})$ is equivalent, in view of (11.50), to the problem of finding the best L_2-approximation

$$\min_{b_{j,\boldsymbol{\beta}}} \int_{\mathbb{R}^n} \Big(e^{-|\mathbf{t}|^2/\mathcal{D}} - \sum_{\mathbf{x}_j\in\Sigma} \sum_{[\boldsymbol{\beta}]=0}^{L_j} b_{j,\boldsymbol{\beta}}(\mathbf{t}-\mathbf{t}_j)^{\boldsymbol{\beta}}\, \mathrm{e}^{-|\mathbf{t}-\mathbf{t}_j|^2/d_j} \Big)^2 d\mathbf{t}\,.$$

LEMMA 11.13. *Let $\Sigma = \{\mathbf{x}_j\}$ be a finite collection of nodes and let $d_j > 0$. For any $f \in L_2$ and all $L_j \ge 0$, the polynomials $\mathcal{P}_j \in \Pi_{L_j}$, which minimize*

$$\Big\| f - \sum_{\mathbf{x}_j\in\Sigma} \mathcal{P}_j(\cdot - \mathbf{x}_j)\, \mathrm{e}^{-|\cdot-\mathbf{x}_j|^2/d_j} \Big\|_{L_2}\,,$$

are uniquely determined.

PROOF. We write the polynomials $\mathcal{P}_j$ in the form

$$\mathcal{P}_j(\mathbf{x}) = \sum_{[\boldsymbol{\beta}]=0}^{L_j} b_{j,\boldsymbol{\beta}} \Big(\frac{\mathbf{x}}{\sqrt{d_j}}\Big)^{\boldsymbol{\beta}}.$$

Then the application of Lemma 11.8 gives

$$\Big\| f - \sum_{\mathbf{x}_j\in\Sigma} \sum_{[\boldsymbol{\beta}]=0}^{L_j} b_{j,\boldsymbol{\beta}} \Big(\frac{\cdot - \mathbf{x}_j}{\sqrt{d_j}}\Big)^{\boldsymbol{\beta}} \mathrm{e}^{-|\cdot - \mathbf{x}_j|^2/d_j} \Big\|^2_{L_2}$$

$$= \|f\|^2_{L_2} - 2\sum_{\mathbf{x}_j\in\Sigma} \sum_{[\boldsymbol{\beta}]=0}^{L_j} b_{j,\boldsymbol{\beta}}\, \mathcal{V}^{(j)}_{\boldsymbol{\beta}}(\mathbf{x}_j) + \sum_{\mathbf{x}_j,\mathbf{x}_k\in\Sigma} \sum_{[\boldsymbol{\beta}]=0}^{L_j} \sum_{[\boldsymbol{\gamma}]=0}^{L_k} b_{j,\boldsymbol{\beta}}\, b_{k,\boldsymbol{\gamma}}\, \mathcal{U}^{(jk)}_{\boldsymbol{\beta},\boldsymbol{\gamma}}(\mathbf{x}_j,\mathbf{y}_k)\,,$$

where the notation

$$\mathcal{V}^{(j)}_{\boldsymbol{\beta}}(\mathbf{x}) = \int_{\mathbb{R}^n} f(\mathbf{t})\, \mathcal{S}_{\boldsymbol{\beta}}(\sqrt{d_j}\, \nabla_{\mathbf{t}})\, \mathrm{e}^{-|\mathbf{t}-\mathbf{x}|^2/d_j}\, d\mathbf{x}$$

$$= \mathcal{S}_{\boldsymbol{\beta}}(-\sqrt{d_j}\nabla_{\mathbf{x}}) \int_{\mathbb{R}^n} f(\mathbf{t})\, \mathrm{e}^{-|\mathbf{t}-\mathbf{x}|^2/d_j}\, d\mathbf{x}\,,$$

$$\begin{aligned} \mathcal{U}^{(jk)}_{\boldsymbol{\beta},\boldsymbol{\gamma}}(\mathbf{x},\mathbf{y}) &= \int_{\mathbb{R}^n} \mathcal{S}_{\boldsymbol{\beta}}(\sqrt{d_j}\, \nabla_{\mathbf{t}})\, \mathrm{e}^{-|\mathbf{t}-\mathbf{x}|^2/d_j}\, \mathcal{S}_{\boldsymbol{\gamma}}(\sqrt{d_k}\, \nabla_{\mathbf{t}})\, \mathrm{e}^{-|\mathbf{t}-\mathbf{y}|^2/d_k}\, d\mathbf{t} \\ &= \Big(\frac{\pi d_j d_k}{d_j + d_k}\Big)^{n/2} \mathcal{S}_{\boldsymbol{\beta}}(-\sqrt{d_j}\, \nabla_{\mathbf{x}})\, \mathcal{S}_{\boldsymbol{\gamma}}(-\sqrt{d_k}\, \nabla_{\mathbf{y}})\, \mathrm{e}^{-|\mathbf{x}-\mathbf{y}|^2/(d_j+d_k)} \end{aligned} \tag{11.51}$$

is used. The norm is minimal if the coefficients $\{b_{j,\boldsymbol{\beta}}\}$ satisfy the linear system

$$\sum_{\mathbf{x}_j\in\Sigma} \sum_{[\boldsymbol{\beta}]=0}^{L_j} b_{j,\boldsymbol{\beta}}\, \mathcal{U}^{(jk)}_{\boldsymbol{\beta},\boldsymbol{\gamma}}(\mathbf{x}_j,\mathbf{x}_k) = \mathcal{V}^{(k)}_{\boldsymbol{\gamma}}(\mathbf{x}_k)\,, \quad \mathbf{x}_k \in \Sigma,\ 0 \le [\boldsymbol{\gamma}] \le L_k\,. \tag{11.52}$$

Hence, the uniqueness of the minimizing linear combination

$$\sum_{\mathbf{x}_j\in\Sigma} \sum_{[\boldsymbol{\beta}]=0}^{L_j} b_{j,\boldsymbol{\beta}} \Big(\frac{\mathbf{x}-\mathbf{x}_j}{\sqrt{d_j}}\Big)^{\boldsymbol{\beta}} \mathrm{e}^{-|\mathbf{x}-\mathbf{x}_j|^2/d_j}$$

is equivalent to the invertibility of the matrix $\|\mathcal{U}^{(jk)}_{\boldsymbol{\beta},\boldsymbol{\gamma}}(\mathbf{x}_j,\mathbf{x}_k)\|$ of the system (11.52). Thus, the assertion is proved by the following lemma. □

LEMMA 11.14. *The matrix $\|\mathcal{U}^{(jk)}_{\boldsymbol{\beta},\boldsymbol{\gamma}}(\mathbf{x}_j,\mathbf{x}_k)\|$ with the elements defined by* (11.51) *is positive definite, i.e., the sesquilinear form satisfies*

$$\sum_{\mathbf{x}_j,\mathbf{x}_k\in\Sigma} \sum_{[\boldsymbol{\beta}]=0}^{L_j} \sum_{[\boldsymbol{\gamma}]=0}^{L_k} \mathcal{U}^{(jk)}_{\boldsymbol{\beta},\boldsymbol{\gamma}}(\mathbf{x}_j,\mathbf{x}_k)\, v_{j,\boldsymbol{\beta}}\, \overline{v_{k,\boldsymbol{\gamma}}} > 0 \tag{11.53}$$

for any non-zero vector $\{v_{j,\boldsymbol{\beta}}\}$.

Here, as usual, $\overline{v_{k,\boldsymbol{\gamma}}}$ denotes the complex conjugate of $v_{k,\boldsymbol{\gamma}}$.

PROOF. To establish (11.53), we use the representation

$$\mathrm{e}^{-|\mathbf{x}-\mathbf{y}|^2/(d_j+d_k)} = \Big(\frac{d_j+d_k}{\pi}\Big)^{n/2} \int_{\mathbb{R}^n} \mathrm{e}^{-(d_j+d_k)|\mathbf{t}|^2}\, \mathrm{e}^{2i\langle \mathbf{t},\mathbf{x}\rangle}\, \mathrm{e}^{-2i\langle \mathbf{t},\mathbf{y}\rangle}\, d\mathbf{t}\,, \tag{11.54}$$

which implies, in view of (11.51),

$$\mathcal{U}^{(jk)}_{\boldsymbol{\beta},\boldsymbol{\gamma}}(\mathbf{x},\mathbf{y}) =(d_j d_k)^{n/2}\, \mathcal{S}_{\boldsymbol{\beta}}(-\sqrt{d_j}\nabla_{\mathbf{x}})\, \mathcal{S}_{\boldsymbol{\gamma}}(-\sqrt{d_k}\nabla_{\mathbf{y}}) \int\limits_{\mathbb{R}^n} \mathrm{e}^{-(d_j+d_k)|\mathbf{t}|^2+2i\langle t,\mathbf{x}-\mathbf{y}\rangle}\, d\mathbf{t}$$

$$=(d_j d_k)^{n/2} \int\limits_{\mathbb{R}^n} \mathcal{S}_{\boldsymbol{\beta}}(-2i\sqrt{d_j}\mathbf{t})\, \overline{\mathcal{S}_{\boldsymbol{\gamma}}(-2i\sqrt{d_k}\mathbf{t})}\, \mathrm{e}^{-(d_j+d_k)|\mathbf{t}|^2+2i\langle t,\mathbf{x}-\mathbf{y}\rangle}\, d\mathbf{t}\,.$$

Then for an arbitrary constant vector $\{v_{j,\boldsymbol{\beta}}\}$, the form in (11.53) can be rewritten as

$$\sum_{\mathbf{x}_j,\mathbf{x}_k\in\Sigma} \sum_{[\boldsymbol{\beta}]=0}^{L_j} \sum_{[\boldsymbol{\gamma}]=0}^{L_k} \mathcal{U}^{(jk)}_{\boldsymbol{\beta},\boldsymbol{\gamma}}(\mathbf{x}_j,\mathbf{x}_k)\, v_{j,\boldsymbol{\beta}}\, \overline{v_{k,\boldsymbol{\gamma}}}$$

$$= \sum_{\mathbf{x}_j,\mathbf{x}_k\in\Sigma} (d_j d_k)^{n/2} \sum_{[\boldsymbol{\beta}]=0}^{L_j} \sum_{[\boldsymbol{\gamma}]=0}^{L_k} v_{j,\boldsymbol{\beta}}\, \overline{v_{k,\boldsymbol{\gamma}}}$$

$$\times \int\limits_{\mathbb{R}^n} \mathcal{S}_{\boldsymbol{\beta}}(-2i\sqrt{d_j}\mathbf{t})\, \overline{\mathcal{S}_{\boldsymbol{\gamma}}(-2i\sqrt{d_k}\mathbf{t})}\, \mathrm{e}^{-(d_j+d_k)|\mathbf{t}|^2}\, \mathrm{e}^{2i\langle \mathbf{t},\mathbf{x}_j-\mathbf{x}_k\rangle}\, d\mathbf{t}$$

$$= \int\limits_{\mathbb{R}^n} \Big| \sum_{\mathbf{x}_j\in\Sigma} d_j^{n/2} \sum_{[\boldsymbol{\beta}]=0}^{L_j} v_{j,\boldsymbol{\beta}}\, \mathcal{S}_{\boldsymbol{\beta}}(-2i\sqrt{d_j}\mathbf{t})\, \mathrm{e}^{-d_j|\mathbf{t}|^2+2i\langle \mathbf{t},\mathbf{x}_j\rangle} \Big|^2 d\mathbf{t}\,,$$

which shows that

$$\sum_{\mathbf{x}_j,\mathbf{x}_k\in\Sigma} \sum_{[\boldsymbol{\beta}]=0}^{L_j} \sum_{[\boldsymbol{\gamma}]=0}^{L_k} \mathcal{U}^{(jk)}_{\boldsymbol{\beta},\boldsymbol{\gamma}}(\mathbf{x}_j,\mathbf{x}_k)\, v_{j,\boldsymbol{\beta}}\, \overline{v_{k,\boldsymbol{\gamma}}} \geq 0\,.$$

Here, the change of integration and summation is justified because the integrand is absolutely integrable and the sums are finite.

Next, we have to show that the inequality is strict, when $\{v_{j,\boldsymbol{\beta}}\} \neq 0$. This is equivalent to showing that

$$\sigma(\mathbf{t}) = \sum_{\mathbf{x}_j\in\Sigma} d_j^{n/2} \sum_{[\boldsymbol{\beta}]=0}^{L_j} v_{j,\boldsymbol{\beta}}\, \mathcal{S}_{\boldsymbol{\beta}}(-2i\sqrt{d_j}\mathbf{t})\, \mathrm{e}^{-d_j|\mathbf{t}|^2+2i\langle \mathbf{t},\mathbf{x}_j\rangle} = 0\,, \quad \mathbf{t}\in\mathbb{R}^n\,, \tag{11.55}$$

only if $v_{j,\boldsymbol{\beta}} = 0$ for all j and $\boldsymbol{\beta}$.

This will be established in the following way: Suppose that $\sigma(\mathbf{t}) = 0$ for all $\mathbf{t}\in\mathbb{R}^n$ and denote the minimal value of the scaling parameters d_j for all $\mathbf{x}_j\in\Sigma$ by $\kappa = \min d_j$. Then the function

$$f_\varepsilon(\mathbf{x}) = \int\limits_{\mathbb{R}^n} \mathrm{e}^{(\kappa-\varepsilon^2)|\mathbf{t}|^2}\, \sigma(\mathbf{t})\, \mathrm{e}^{-2i\langle \mathbf{t},\mathbf{x}\rangle}\, d\mathbf{t}$$

is identically zero for all $\mathbf{x}$ and $\varepsilon > 0$. Using the representation (11.55) of σ we have

$$
\begin{aligned}
f_\varepsilon(\mathbf{x}) &= \sum_{\mathbf{x}_j\in\Sigma} d_j^{n/2} \sum_{[\boldsymbol{\beta}]=0}^{L_j} v_{j,\boldsymbol{\beta}} \int\limits_{\mathbb{R}^n} \mathrm{e}^{-(d_j-\kappa+\varepsilon^2)|\mathbf{t}|^2}\, \mathcal{S}_{\boldsymbol{\beta}}(-2i\sqrt{d_j}\mathbf{t})\ \mathrm{e}^{2i\langle \mathbf{t},\mathbf{x}_j-\mathbf{x}\rangle}\, d\mathbf{t} \\
&= \sum_{\mathbf{x}_j\in\Sigma} d_j^{n/2} \sum_{[\boldsymbol{\beta}]=0}^{L_j} v_{j,\boldsymbol{\beta}}\, \mathcal{S}_{\boldsymbol{\beta}}(\sqrt{d_j}\,\nabla_{\mathbf{x}}) \int\limits_{\mathbb{R}^n} \mathrm{e}^{-(d_j-\kappa+\varepsilon^2)|\mathbf{t}|^2}\ \mathrm{e}^{2i\langle \mathbf{t},\mathbf{x}_j-\mathbf{x}\rangle}\, d\mathbf{t} \\
&= \sum_{\mathbf{x}_j\in\Sigma} \Big(\frac{\pi d_j}{d_j-\kappa+\varepsilon^2}\Big)^{n/2} \sum_{[\boldsymbol{\beta}]=0}^{L_j} v_{j,\boldsymbol{\beta}}\, \mathcal{S}_{\boldsymbol{\beta}}(\sqrt{d_j}\,\nabla_{\mathbf{x}})\, \mathrm{e}^{-|\mathbf{x}-\mathbf{x}_j|^2/(d_j-\kappa+\varepsilon^2)}\,.
\end{aligned}
$$

Let us denote the subset of nodes $\mathbf{x}_j$ for which $d_j = \kappa$ by Σ_1. We will show that $f_\varepsilon(\mathbf{x}) = 0$ implies $v_{j,\boldsymbol{\beta}} = 0$ for all $\mathbf{x}_j \in \Sigma_1$ and $\boldsymbol{\beta}$ with $0 \le [\boldsymbol{\beta}] \le L_j$.

The sum over these nodes can be written as

$$
\begin{aligned}
f_{1,\varepsilon}(\mathbf{x}) &= \sum_{\mathbf{x}_j\in\Sigma_1} \Big(\frac{\pi d_j}{d_j-\kappa+\varepsilon^2}\Big)^{n/2} \sum_{[\boldsymbol{\beta}]=0}^{L_j} v_{j,\boldsymbol{\beta}}\, \mathcal{S}_{\boldsymbol{\beta}}(\sqrt{d_j}\,\nabla_{\mathbf{x}})\, \mathrm{e}^{-|\mathbf{x}-\mathbf{x}_j|^2/(d_j-\kappa+\varepsilon^2)} \\
&= \Big(\frac{\pi\kappa}{\varepsilon^2}\Big)^{n/2} \sum_{\mathbf{x}_j\in\Sigma_1} \sum_{[\boldsymbol{\beta}]=0}^{L_j} v_{j,\boldsymbol{\beta}}\, \mathcal{S}_{\boldsymbol{\beta}}(\sqrt{d_j}\,\nabla_{\mathbf{x}})\, \mathrm{e}^{-|\mathbf{x}-\mathbf{x}_j|^2/\varepsilon^2}\,.
\end{aligned}
$$

We note that

$$
\mathcal{S}_{\boldsymbol{\beta}}(\sqrt{d_j}\,\nabla_{\mathbf{x}})\, \mathrm{e}^{-|\mathbf{x}-\mathbf{x}_j|^2/\varepsilon^2} \to 0 \quad \text{as} \quad \varepsilon \to 0
$$

uniformly outside any ball centered at $\mathbf{x}_j$, i.e., uniformly in $\mathbf{x} \in \mathbb{R}^n \setminus B(\mathbf{x}_j,\rho)$ for any radius ρ. Hence, for arbitrary $\rho > 0$

$$
\lim_{\varepsilon\to 0} f_{1,\varepsilon}(\mathbf{x}) = 0 \quad \text{uniformly in } \mathbf{x} \in \mathbb{X}_\rho = \mathbb{R}^n \setminus \bigcup_{\mathbf{x}_j\in\Sigma_1} B(\mathbf{x}_j,\rho)\,.
$$

Moreover, $d_j - \kappa > 0$ for all $\mathbf{x}_j \in \Sigma_2 = \Sigma \setminus \Sigma_1$, and therefore

$$
\Big(\frac{\pi d_j}{d_j-\kappa+\varepsilon^2}\Big)^{n/2} \mathcal{S}_{\boldsymbol{\beta}}(\sqrt{d_j}\,\nabla_{\mathbf{x}})\, \mathrm{e}^{-|\mathbf{x}-\mathbf{x}_j|^2/(d_j-\kappa+\varepsilon^2)}
$$

converge uniformly to

$$
\Big(\frac{\pi d_j}{d_j-\kappa}\Big)^{n/2} \mathcal{S}_{\boldsymbol{\beta}}(\sqrt{d_j}\,\nabla_{\mathbf{x}})\, \mathrm{e}^{-|\mathbf{x}-\mathbf{x}_j|^2/(d_j-\kappa)}
$$

if $\varepsilon \to 0$. Hence, $f_\varepsilon(\mathbf{x}) = 0$ implies

$$
\sum_{\mathbf{x}_j\in\Sigma_2} \Big(\frac{\pi d_j}{d_j-\kappa}\Big)^{n/2} \sum_{[\boldsymbol{\beta}]=0}^{L_k} v_{j,\boldsymbol{\beta}}\, \mathcal{S}_{\boldsymbol{\beta}}(\sqrt{d_j}\,\nabla_{\mathbf{x}})\, \mathrm{e}^{-|\mathbf{x}-\mathbf{x}_j|^2/(d_j-\kappa)} = 0
$$

for all $\mathbf{x} \in \mathbb{X}_\rho$. Since this function is real analytic, it vanishes for all $\mathbf{x}$.

We conclude that $f_{1,\varepsilon}(\mathbf{x}) \to 0$ uniformly for all $\mathbf{x} \in \mathbb{R}^n$ if ε tends to zero. Applying the subsequent Lemma 11.15, we derive that $v_{j,\boldsymbol{\beta}} = 0$ for all $\mathbf{x}_j \in \Sigma_1$ and $\boldsymbol{\beta}$ with $0 \le [\boldsymbol{\beta}] \le L_j$.

So we are left with the problem of showing that $v_{j,\boldsymbol{\beta}} = 0$ for all $\mathbf{x}_j \in \Sigma_2$ and $\boldsymbol{\beta}$ with $0 \le [\boldsymbol{\beta}] \le L_j$ if

$$\sigma(\mathbf{t}) = \sum_{\mathbf{x}_j \in \Sigma_2} d_j^{n/2} \sum_{[\boldsymbol{\beta}]=0}^{L_j} v_{j,\boldsymbol{\beta}}\, \mathcal{S}_{\boldsymbol{\beta}}(-2i\sqrt{d_j}\mathbf{t})\, \mathrm{e}^{-d_j|\mathbf{t}|^2+2i\langle \mathbf{t},\mathbf{x}_j\rangle} = 0$$

for all $\mathbf{t}$, which is solved by repeating the above procedure until all different scaling parameters d_j and nodes $\mathbf{x}_j$ are covered. □

LEMMA 11.15. *Let $\Sigma = \{\mathbf{x}_j\}$ be a finite collection of nodes. If*

$$\lim_{\varepsilon \to 0} \Big\| \sum_{\mathbf{x}_j \in \Sigma} \sum_{[\boldsymbol{\beta}]=0}^{L_j} v_{j,\boldsymbol{\beta}}\, \mathcal{S}_{\boldsymbol{\beta}}(\nabla_{\mathbf{x}})\, \mathrm{e}^{-|\mathbf{x}-\mathbf{x}_j|^2/\varepsilon^2} \Big\|_{L_\infty} = 0\,,$$

then $v_{j,\boldsymbol{\beta}} = 0$ for all $\mathbf{x}_j \in \Sigma$ and $\boldsymbol{\beta}$ with $0 \le [\boldsymbol{\beta}] < L_j$.

PROOF. Let us fix a node $\mathbf{x}_k \in \Sigma$ and consider

$$g_\varepsilon(\mathbf{x}) = \sum_{\mathbf{x}_j \in \Sigma} \sum_{[\boldsymbol{\beta}]=0}^{L_j} v_{j,\boldsymbol{\beta}}\, \mathcal{S}_{\boldsymbol{\beta}}(\nabla_{\mathbf{x}})\, \mathrm{e}^{-|\mathbf{x}-\mathbf{x}_j|^2/\varepsilon^2}$$

on the ball $B(\mathbf{x}_k, \varepsilon)$ for sufficiently small $\varepsilon > 0$. If there exists another node $\mathbf{x}_j \in \Sigma$, then obviously

$$\mathcal{S}_{\boldsymbol{\beta}}(\nabla_{\mathbf{x}})\, \mathrm{e}^{-|\mathbf{x}-\mathbf{x}_j|^2/\varepsilon^2} \to 0 \quad \text{as} \quad \varepsilon \to 0$$

uniformly on any sufficiently small ball around $\mathbf{x}_k$. Since $\|g_\varepsilon\|_{L_\infty} \to 0$, for any $\delta > 0$ there exists ε_0 such that for all $\varepsilon \in (0, \varepsilon_0)$ and $\mathbf{x} \in B(\mathbf{x}_k, \varepsilon)$

$$\Big| \sum_{[\boldsymbol{\beta}]=0}^{L_k} v_{k,\boldsymbol{\beta}}\, \mathcal{S}_{\boldsymbol{\beta}}(\nabla_{\mathbf{x}})\, \mathrm{e}^{-|\mathbf{x}-\mathbf{x}_k|^2/e^2} \Big| < \delta\,. \tag{11.56}$$

Setting $\mathbf{t} = (\mathbf{x} - \mathbf{x}_k)/\varepsilon$, the last inequality transforms to

$$\Big| \sum_{[\boldsymbol{\beta}]=0}^{L_k} v_{k,\boldsymbol{\beta}}\, \mathcal{S}_{\boldsymbol{\beta}}(\varepsilon^{-1}\nabla_{\mathbf{t}})\, \mathrm{e}^{-|\mathbf{t}|^2} \Big| = \mathrm{e}^{-|\mathbf{t}|^2} \Big| \sum_{[\boldsymbol{\beta}]=0}^{L_k} \varepsilon^{-[\boldsymbol{\beta}]}\, p_{\boldsymbol{\beta}}(\mathbf{t}) \Big| < \delta$$

for all $|\mathbf{t}| \le 1$ and $\varepsilon \in (0, \varepsilon_0)$, where $p_{\boldsymbol{\beta}}$ are certain polynomials of degree L_k not depending on ε. The inequality is valid for any $\delta > 0$ only, if these polynomials vanish, which implies for $\varepsilon = 1$ that

$$\sum_{[\boldsymbol{\beta}]=0}^{L_k} v_{k,\boldsymbol{\beta}}\, \mathcal{S}_{\boldsymbol{\beta}}(\nabla_{\mathbf{x}})\, \mathrm{e}^{-|\mathbf{x}-\mathbf{x}_k|^2/d_j} = 0\,.$$

Since by (11.27)

$$\mathcal{S}_{\boldsymbol{\beta}}(\nabla_{\mathbf{x}})\, \mathrm{e}^{-|\mathbf{x}-\mathbf{x}_k|^2} = (\mathbf{x} - \mathbf{x}_k)^{\boldsymbol{\beta}}\, \mathrm{e}^{-|\mathbf{x}-\mathbf{x}_k|^2}\,,$$

we deduce that $v_{k,\boldsymbol{\beta}} = 0$ for all $\boldsymbol{\beta}$. □

In the following two subsections, we show that under certain assumptions on the parameters $d_j \le \mathcal{D}$ and the given finite point set Σ in $\mathbb{R}^n$, the approximation error satisfies

$$\min_{\mathcal{P}_j \in \Pi_{L_j}} \Big\| \mathrm{e}^{-|\cdot|^2/\mathcal{D}} - \sum_{\mathbf{x}_j \in \Sigma} \mathcal{P}_j\, \mathrm{e}^{-|\cdot - \mathbf{x}_j|^2/d_j} \Big\|_{L_2} \to 0 \tag{11.57}$$

if the degrees L_j tend to infinity.

11.4.6. Approximation error in the case $d_j = \mathcal{D}$.

THEOREM 11.16. *Suppose that there exists* $\mathbf{x}_k \in \Sigma$ *with* $d_k = \mathcal{D}$. *Then*

$$\begin{aligned}\min_{\mathcal{P}_j \in \Pi_{L_j}} \Big\| \mathrm{e}^{-|\cdot|^2/\mathcal{D}} - \sum_{\mathbf{x}_j \in \Sigma} \mathcal{P}_j \, \mathrm{e}^{-|\cdot - \mathbf{x}_j|^2/d_j} \Big\|_{L_2}^2 \\ \le \Big(\frac{\pi \mathcal{D}}{2}\Big)^{n/2} \frac{|\mathbf{x}_k|^{2(L_k+1)}}{\mathcal{D}^{L_k+1}(L_k+1)!}\,.\end{aligned} \tag{11.58}$$

PROOF. Since

$$\sum_{\mathbf{x}_j \in \Sigma} \mathcal{P}_j \, \mathrm{e}^{-|\mathbf{x}-\mathbf{x}_j|^2/d_j} = \mathcal{P}_k \, \mathrm{e}^{-|\mathbf{x}-\mathbf{x}_k|^2/\mathcal{D}} + \sum_{\mathbf{x}_j \neq \mathbf{x}_k} \mathcal{P}_j \, \mathrm{e}^{-|\mathbf{x}-\mathbf{x}_j|^2/d_j}\,,$$

we have, obviously,

$$\begin{aligned}\min_{\mathcal{P}_j \in \Pi_{L_j}} \Big\| \mathrm{e}^{-|\cdot|^2/\mathcal{D}} - \sum_{\mathbf{x}_j \in \Sigma} \mathcal{P}_j \, \mathrm{e}^{-|\cdot - \mathbf{x}_j|^2/d_j} \Big\|_{L_2}^2 \\ \le \min_{\mathcal{P} \in \Pi_{L_k}} \int_{\mathbb{R}^n} \Big(\mathrm{e}^{-|\mathbf{x}|^2/\mathcal{D}} - \mathcal{P}(\mathbf{x})\, \mathrm{e}^{-|\mathbf{x}-\mathbf{x}_k|^2/\mathcal{D}} \Big)^2 d\mathbf{x}\,.\end{aligned}$$

The last integral can be transformed to

$$\begin{aligned}&\min_{\mathcal{P} \in \Pi_{L_k}} \int_{\mathbb{R}^n} \Big(\mathrm{e}^{-|\mathbf{x}|^2/\mathcal{D}} - \mathcal{P}(\mathbf{x})\, \mathrm{e}^{-|\mathbf{x}-\mathbf{x}_k|^2/\mathcal{D}} \Big)^2 d\mathbf{t} \\ &= \Big(\frac{\mathcal{D}}{2}\Big)^{n/2} \min_{\mathcal{P} \in \Pi_{L_k}} \int_{\mathbb{R}^n} \Big(\mathrm{e}^{-|\mathbf{t}|^2/2} - \mathcal{P}(\mathbf{t})\, \mathrm{e}^{-|\mathbf{t}-\mathbf{t}_k|^2/2} \Big)^2 d\mathbf{t} \\ &= \Big(\frac{\mathcal{D}}{2}\Big)^{n/2} \min_{\mathcal{P} \in \Pi_{L_k}} \int_{\mathbb{R}^n} \mathrm{e}^{-|\mathbf{t}|^2} \Big(\mathcal{P}(\mathbf{t}) - \mathrm{e}^{-|\mathbf{t}_k|^2} \mathrm{e}^{-\sqrt{2}\langle \mathbf{t}, \mathbf{t}_k \rangle} \Big)^2 d\mathbf{t}\end{aligned} \tag{11.59}$$

with $\mathbf{t}_k = \mathbf{x}_k/\sqrt{\mathcal{D}}$. It is well known that the Hermite polynomials

$$\Big\{ H_{\boldsymbol{\beta}}(\mathbf{x}),\ \boldsymbol{\beta} \in \mathbb{Z}^n_{\ge 0} \Big\}$$

form a closed orthogonal system in the weighted space $L_2(\mathbb{R}^n, w_{1,2})$ with the norm

$$\|f\|_{2,w_{1,2}} = \Big(\int_{\mathbb{R}^n} \mathrm{e}^{-|\mathbf{x}|^2} |f(\mathbf{x})|^2 \, d\mathbf{x} \Big)^{1/2}$$

(see (11.7)) and that

$$(H_{\boldsymbol{\alpha}}, H_{\boldsymbol{\beta}})_{2,w_{1,2}} = \int_{\mathbb{R}^n} \mathrm{e}^{-|\mathbf{x}|^2} H_{\boldsymbol{\alpha}}(\mathbf{x})\, H_{\boldsymbol{\beta}}(\mathbf{x})\, d\mathbf{x} = \begin{cases} 2^{[\boldsymbol{\beta}]}\, \boldsymbol{\beta}!\, \pi^{n/2}\,, & \boldsymbol{\alpha} = \boldsymbol{\beta}\,, \\ 0\,, & \text{otherwise}\,. \end{cases}$$

(See for example [**93**, Ch. 5.5].) Since $\mathrm{e}^{-\sqrt{2}\langle \cdot, \mathbf{t}_k \rangle} \in L_2(\mathbb{R}^n, w_{1,2})$, the minimum of (11.59) is attained, when

$$\mathcal{P}(\mathbf{t}) = \sum_{[\boldsymbol{\beta}]=0}^{L_k} \frac{a_{\boldsymbol{\beta}}}{\sqrt{2^{[\boldsymbol{\beta}]}\, \boldsymbol{\beta}!\, \pi^{n/2}}} H_{\boldsymbol{\beta}}(\mathbf{t})$$

with the coefficients

$$a_{\boldsymbol{\beta}} = \frac{\mathrm{e}^{-|\mathbf{t}_k|^2}}{\sqrt{2^{[\boldsymbol{\beta}]}\,\boldsymbol{\beta}!\,\pi^{n/2}}} \int\limits_{\mathbb{R}^n} \mathrm{e}^{-|\mathbf{t}|^2} H_{\boldsymbol{\beta}}(\mathbf{t})\,\mathrm{e}^{-\sqrt{2}\langle \mathbf{t},\mathbf{t}_k\rangle}\,d\mathbf{t} \tag{11.60}$$
$$= \frac{\mathrm{e}^{-|\mathbf{t}_k|^2}}{\sqrt{2^{[\boldsymbol{\beta}]}\,\boldsymbol{\beta}!\,\pi^{n/2}}} \int\limits_{\mathbb{R}^n} \mathrm{e}^{-\sqrt{2}\langle \mathbf{t},\mathbf{t}_k\rangle} (-\nabla_{\mathbf{t}})^{\boldsymbol{\beta}}\,\mathrm{e}^{-|\mathbf{t}|^2}\,d\mathbf{t}\,.$$

Integrating by parts, we obtain

$$a_{\boldsymbol{\beta}} = \frac{\pi^{n/4}\,(-\mathbf{t}_k)^{\boldsymbol{\beta}}}{\sqrt{\boldsymbol{\beta}!}}\,\mathrm{e}^{-|\mathbf{t}_k|^2/2}\,,$$

which together with

$$\sum_{[\boldsymbol{\beta}]=L_k+1}^{\infty} \frac{\mathbf{t}_k^{2\boldsymbol{\beta}}}{\boldsymbol{\beta}!} = \sum_{\ell=L_k+1}^{\infty} \frac{|\mathbf{t}_k|^{2\ell}}{\ell!} \le \frac{|\mathbf{t}_k|^{2(L_k+1)}}{(L_k+1)!}\,\mathrm{e}^{|\mathbf{t}_k|^2/2}$$

leads to the inequality

$$\sum_{[\boldsymbol{\beta}]=L_k+1}^{\infty} |a_{\boldsymbol{\beta}}|^2 \le \pi^{n/2} \frac{|\mathbf{t}_k|^{2(L_k+1)}}{(L_k+1)!}\,. \tag{11.61}$$

Since $\mathbf{t}_k = \mathbf{x}_k/\sqrt{\mathcal{D}}$, inequality (11.58) follows. Note that

$$\mathcal{P}(\mathbf{t}) = \sum_{[\boldsymbol{\beta}]=0}^{L_k} \frac{(-\mathbf{t}_k)^{\boldsymbol{\beta}}}{2^{[\boldsymbol{\beta}]/2}\,\boldsymbol{\beta}!}\,H_{\boldsymbol{\beta}}(\mathbf{t})\,.$$

□

11.4.7. Approximation error in the case $d_j < \mathcal{D}$. Now we assume that all scaling factors d_j in (11.57) satisfy $d_j < \mathcal{D}$ and we suppose first that $d_j = d$ for all $\mathbf{x}_j \in \Sigma$.

THEOREM 11.17. *Let, for given $\varepsilon > 0$, the scattered nodes $\{\mathbf{x}_j\}$ satisfy the following condition: There exists $0 < d < \mathcal{D}$ such that the ball B_R with the minimal radius R given by*

$$\Gamma\Big(\frac{n}{2}, \frac{R^2}{\mathcal{D}-d}\Big) < \frac{\varepsilon}{2}\Big(\frac{d}{\mathcal{D}}\Big)^{n/2}\Gamma\Big(\frac{n}{2}\Big) \tag{11.62}$$

can be partitioned into subdomains T_j, which contain at least one node $\mathbf{x}_j$ and satisfy $\max_{\mathbf{x}\in T_j} |\mathbf{x}-\mathbf{x}_j| \le \kappa\sqrt{d}$ with some positive constant κ. Then there exist polynomials of sufficiently large degree L such that the sum

$$\sum_j \mathcal{P}_j(\mathbf{x})\,\mathrm{e}^{-|\mathbf{x}-\mathbf{x}_j|^2/d}$$

approximates the Gaussian $\mathrm{e}^{-|\mathbf{x}|^2/\mathcal{D}}$ with

$$\Big|\mathrm{e}^{-|\mathbf{x}|^2/\mathcal{D}} - \sum_j \mathcal{P}_j(\mathbf{x})\,\mathrm{e}^{-|\mathbf{x}-\mathbf{x}_j|^2/d}\Big| < \min\big(\varepsilon, (1+\varepsilon/2)\,\mathrm{e}^{-|\mathbf{x}|^2/\mathcal{D}}\big)$$

for all $\mathbf{x} \in \mathbb{R}^n$.

Here, $\Gamma(a,x)$ denotes the *upper incomplete Gamma function*

$$\Gamma(a,x) = \Gamma(a) - \gamma(a,x) = \int\limits_x^\infty \tau^{a-1}\,\mathrm{e}^{-\tau}\,d\tau \tag{11.63}$$

with $\gamma(a,x)$ defined in (4.15).

The proof of Theorem 11.17 consists of several steps. It is based on a simple cubature formula for the integral

$$\mathrm{e}^{-|\mathbf{x}|^2/\mathcal{D}} = \Big(\frac{\mathcal{D}}{\pi d(\mathcal{D}-d)}\Big)^{n/2} \int\limits_{\mathbb{R}^n} \mathrm{e}^{-|\mathbf{x}-\mathbf{y}|^2/d}\,\mathrm{e}^{-|\mathbf{y}|^2/(\mathcal{D}-d)}\,d\mathbf{y}\,.$$

For a given $\varepsilon > 0$, we choose $R = R(\mathcal{D}, h)$ such that

$$\Big(\frac{\rho}{\pi}\Big)^{n/2} \int\limits_{|\mathbf{y}|>R} \mathrm{e}^{-|\mathbf{y}|^2/(\mathcal{D}-h^2)}\,d\mathbf{y} < \varepsilon/2\,,$$

where we set

$$\rho = \frac{\mathcal{D}}{d(\mathcal{D}-d)}\,.$$

Note that

$$\int\limits_{|\mathbf{y}|>R} \mathrm{e}^{-|\mathbf{y}|^2/(\mathcal{D}-d)}\,d\mathbf{y} = \frac{(\pi(D-d))^{n/2}}{\Gamma(n/2)}\,\Gamma\Big(\frac{n}{2},\frac{R^2}{\mathcal{D}-d}\Big),$$

so R has to be chosen such that

$$\Gamma\Big(\frac{n}{2},\frac{R^2}{\mathcal{D}-d}\Big) < \frac{\varepsilon}{2}\Big(\frac{d}{\mathcal{D}}\Big)^{n/2}\Gamma\Big(\frac{n}{2}\Big)\,.$$

Then the function

$$S_R(\mathbf{x}) = \Big(\frac{\rho}{\pi}\Big)^{n/2} \int\limits_{B_R} \mathrm{e}^{-|\mathbf{x}-\mathbf{y}|^2/d}\,\mathrm{e}^{-|\mathbf{y}|^2/(\mathcal{D}-d)}\,d\mathbf{y}$$

is subject to

$$0 < \mathrm{e}^{-|\mathbf{x}|^2/\mathcal{D}} - S_R(\mathbf{x}) < \varepsilon/2$$

for all $\mathbf{x} \in \mathbb{R}^n$. Here, we set $B_R = B(\mathbf{0}, R) = \{|\mathbf{y}| \le R\}$. Because

$$\mathrm{e}^{-|\mathbf{x}-\mathbf{y}|^2/d}\,\mathrm{e}^{-|\mathbf{y}|^2/(\mathcal{D}-d)} = \mathrm{e}^{-|\mathbf{x}|^2/\mathcal{D}}\,\mathrm{e}^{-\rho\,(\mathbf{x}/(\rho d)-\mathbf{y})^2}\,, \tag{11.64}$$

we can write

$$S_R(\mathbf{x}) = \Big(\frac{\rho}{\pi}\Big)^{n/2} \mathrm{e}^{-|\mathbf{x}|^2/\mathcal{D}} \int\limits_{B_R} \mathrm{e}^{-\rho\,(\mathbf{x}/(\rho d)-\mathbf{y})^2}\,d\mathbf{y}\,, \tag{11.65}$$

and therefore

$$\begin{aligned}\mathrm{e}^{-|\mathbf{x}|^2/\mathcal{D}} - S_R(\mathbf{x}) &= \Big(\frac{\rho}{\pi}\Big)^{n/2} \int\limits_{|\mathbf{y}|>R} \mathrm{e}^{-|\mathbf{x}-\mathbf{y}|^2/d}\,\mathrm{e}^{-|\mathbf{y}|^2/(\mathcal{D}-d)}\,d\mathbf{y} \\ &= \mathrm{e}^{-|\mathbf{x}|^2/\mathcal{D}} \Big(\frac{\rho}{\pi}\Big)^{n/2} \int\limits_{|\mathbf{y}|>R} \mathrm{e}^{-\rho\,(\mathbf{x}/(\rho d)-\mathbf{y})^2}\,d\mathbf{y} < \mathrm{e}^{-|\mathbf{x}|^2/\mathcal{D}}\,.\end{aligned}$$

This implies the estimate

$$0 < \mathrm{e}^{-|\mathbf{x}|^2/\mathcal{D}} - S_R(\mathbf{x}) < \min\big(\varepsilon/2, \mathrm{e}^{-|\mathbf{x}|^2/\mathcal{D}}\big)\,. \tag{11.66}$$

We subdivide the ball B_R into the subdomains T_j with the property that each T_j contains one node $\mathbf{x}_j$ and

$$\max_{\mathbf{x}\in T_j} |\mathbf{x}-\mathbf{x}_j| \le \kappa\sqrt{d}\,.$$

Then

$$\begin{aligned} S_R(\mathbf{x}) &= \Big(\frac{\rho}{\pi}\Big)^{n/2} \sum_j \int_{T_j} \mathrm{e}^{-|\mathbf{x}-\mathbf{y}|^2/d}\, \mathrm{e}^{-|\mathbf{y}|^2/(\mathcal{D}-d)}\, d\mathbf{y} \\ &= \Big(\frac{\rho}{\pi}\Big)^{n/2} \mathrm{e}^{-|\mathbf{x}|^2/\mathcal{D}} \sum_j \int_{T_j} \mathrm{e}^{-\rho\,(\mathbf{x}/(\rho d)-\mathbf{y})^2}\, d\mathbf{y}\,. \end{aligned}$$

LEMMA 11.18. *There exists a polynomial p_j of degree L such that*

$$\begin{aligned} &\Big| \int_{T_j} \mathrm{e}^{-\rho\,(\mathbf{x}/(\rho d)-\mathbf{y})^2}\, d\mathbf{y} - p_j(\mathbf{x})\, \mathrm{e}^{-\rho\,(\mathbf{x}/(\rho d)-\mathbf{x}_j)^2} \Big| \\ &\le c \operatorname{meas} T_j \max_{\mathbf{y}\in T_j} \mathrm{e}^{-\rho\,(\mathbf{x}/(\rho d)-\mathbf{y})^2/2} \Big(\frac{2\mathcal{D}\kappa^2}{\mathcal{D}-d}\Big)^{(L+1)/2} \sum_{[\boldsymbol{\alpha}]=L+1} \frac{1}{\sqrt{\boldsymbol{\alpha}!}} \end{aligned} \tag{11.67}$$

with a constant c depending only on n.

PROOF. Put

$$f_{\mathbf{x}}(\mathbf{y}) = \mathrm{e}^{-\rho\,(\mathbf{x}/(\rho d)-\mathbf{y})^2}$$

and expand this function into its Taylor series around $\mathbf{y}=\mathbf{x}_j$, i.e.,

$$f_{\mathbf{x}}(\mathbf{y}) = \sum_{[\boldsymbol{\alpha}]=0}^{L} \frac{\partial_{\mathbf{y}}^{\boldsymbol{\alpha}} f_{\mathbf{x}}(\mathbf{x}_j)}{\boldsymbol{\alpha}!}(\mathbf{y}-\mathbf{x}_j)^{\boldsymbol{\alpha}} + R_{L+1}(\mathbf{y},\mathbf{x}_j),$$

with the remainder

$$R_{L+1}(\mathbf{y},\mathbf{x}_j) := \sum_{[\boldsymbol{\alpha}]=L+1} \frac{(\mathbf{y}-\mathbf{x}_j)^{\boldsymbol{\alpha}}}{\boldsymbol{\alpha}!}\, \partial_{\mathbf{y}}^{\boldsymbol{\alpha}} f_{\mathbf{x}}(\mathbf{z})$$

for some $\mathbf{z}=\mathbf{x}_j+s(\mathbf{y}-\mathbf{x}_j)$, $s\in[0,1]$ (cf. (2.3)). Then

$$\int_{T_j} f_{\mathbf{x}}(\mathbf{y})\, d\mathbf{y} = \sum_{[\boldsymbol{\alpha}]=0}^{L} \frac{\partial_{\mathbf{y}}^{\boldsymbol{\alpha}} f_{\mathbf{x}}(\mathbf{x}_j)}{\boldsymbol{\alpha}!} \int_{T_j} (\mathbf{y}-\mathbf{x}_j)^{\boldsymbol{\alpha}}\, d\mathbf{y} + \int_{T_j} R_{L+1}(\mathbf{y},\mathbf{x}_j)\, d\mathbf{y}\,. \tag{11.68}$$

Note that by (7.7),

$$\partial_{\mathbf{y}}^{\boldsymbol{\alpha}} f_{\mathbf{x}}(\mathbf{y}) = \partial_{\mathbf{y}}^{\boldsymbol{\alpha}}\, \mathrm{e}^{-\rho\,(\mathbf{x}/(\rho d)-\mathbf{y})^2} = \rho^{[\boldsymbol{\alpha}]/2} H_{\boldsymbol{\alpha}}\Big(\sqrt{\rho}\big(\frac{\mathbf{x}}{\rho d}-\mathbf{y}\big)\Big)\, \mathrm{e}^{-\rho\,(\mathbf{x}/(\rho d)-\mathbf{y})^2}$$

with the multi-variate Hermite polynomial $H_{\boldsymbol{\alpha}}$. Hence, we set in (11.68)

$$\begin{aligned} &\frac{\partial_{\mathbf{y}}^{\boldsymbol{\alpha}} f_{\mathbf{x}}(\mathbf{x}_j)}{\boldsymbol{\alpha}!} \int_{T_j} (\mathbf{y}-\mathbf{x}_j)^{\boldsymbol{\alpha}}\, d\mathbf{y} \\ &= \frac{\rho^{[\boldsymbol{\alpha}]/2}}{\boldsymbol{\alpha}!} H_{\boldsymbol{\alpha}}\Big(\sqrt{\rho}\big(\frac{\mathbf{x}}{\rho d}-\mathbf{x}_j\big)\Big)\, \mathrm{e}^{-\rho\,(\mathbf{x}/(\rho d)-\mathbf{x}_j)^2} \int_{T_j} (\mathbf{y}-\mathbf{x}_j)^{\boldsymbol{\alpha}}\, d\mathbf{y}\,. \end{aligned}$$

If $[\boldsymbol{\alpha}] = L + 1$, then we have to estimate

$$|\partial_{\mathbf{y}}^{\boldsymbol{\alpha}} f_{\mathbf{x}}(\mathbf{y})| = \rho^{(L+1)/2} H_{\boldsymbol{\alpha}}\Big(\sqrt{\rho}\big(\frac{\mathbf{x}}{\rho d} - \mathbf{y}\big)\Big)\ \mathrm{e}^{-\rho\,(\mathbf{x}/(\rho d)-\mathbf{y})^2}$$

for $\mathbf{y} \in T_j$. We apply Cramer's inequality for Hermite polynomials [**1**, 22.14.17]

$$|H_j(t)| \le K\, 2^{j/2} \sqrt{j!}\ \mathrm{e}^{t^2/2} \tag{11.69}$$

with a constant $K \approx 1.086435$, which gives

$$H_{\boldsymbol{\alpha}}\Big(\sqrt{\rho}\big(\frac{\mathbf{x}}{\rho d} - \mathbf{y}\big)\Big) \le K^n 2^{[\boldsymbol{\alpha}]/2} \sqrt{\boldsymbol{\alpha}!}\ \mathrm{e}^{\rho\,(\mathbf{x}/(\rho d)-\mathbf{y})^2/2}\,.$$

Thus, we obtain

$$|\partial_{\mathbf{y}}^{\boldsymbol{\alpha}} f_{\mathbf{x}}(\mathbf{y})| \le (2\rho)^{(L+1)/2}\, K^n \sqrt{\boldsymbol{\alpha}!}\ \mathrm{e}^{-\rho\,(\mathbf{x}/(\rho d)-\mathbf{y})^2/2}\,,$$

implying

$$\big|R_{L+1}(\mathbf{y},\mathbf{x}_j)\big| \le (2\rho)^{(L+1)/2}\, K^n \max_{\mathbf{y}\in T_j} \mathrm{e}^{-\rho\,(\mathbf{x}/(\rho d)-\mathbf{y})^2/2} \sum_{[\boldsymbol{\alpha}]=L+1} \frac{\big|(\mathbf{y}-\mathbf{x}_j)^{\boldsymbol{\alpha}}\big|}{\sqrt{\boldsymbol{\alpha}!}}$$

and the estimate

$$\Big| \int_{T_j} f_{\mathbf{x}}(\mathbf{y})\, d\mathbf{y} - p_j(\mathbf{x})\ \mathrm{e}^{-\rho\,(\mathbf{x}/(\rho d)-\mathbf{x}_j)^2}\ \Big|$$
$$\le (2\rho)^{(L+1)/2}\, K^n \max_{\mathbf{y}\in T_j}\ \mathrm{e}^{-\rho\,(\mathbf{x}/(\rho d)-\mathbf{y})^2/2} \sum_{[\boldsymbol{\alpha}]=L+1} \frac{1}{\sqrt{\boldsymbol{\alpha}!}} \int_{T_j} \big|(\mathbf{y}-\mathbf{x}_j)^{\boldsymbol{\alpha}}\big|\, d\mathbf{y}$$

with the polynomial

$$p_j(\mathbf{x}) := \sum_{[\boldsymbol{\alpha}]=0}^{L} \frac{\rho^{[\boldsymbol{\alpha}]/2}}{\boldsymbol{\alpha}!} H_{\boldsymbol{\alpha}}\Big(\sqrt{\rho}\big(\frac{\mathbf{x}}{\rho d} - \mathbf{x}_j\big)\Big) \int_{T_j} (\mathbf{y}-\mathbf{x}_j)^{\boldsymbol{\alpha}}\, d\mathbf{y}\,.$$

Since

$$(2\rho)^{L+1/2} = d^{-(L+1)/2} \Big(\frac{2\mathcal{D}}{\mathcal{D}-d}\Big)^{(L+1)/2},$$

the assumption $\max\limits_{\mathbf{x}\in T_j} |\mathbf{x}-\mathbf{x}_j| \le \kappa\sqrt{d}$ implies

$$(2\rho)^{(L+1)/2} \sum_{[\boldsymbol{\alpha}]=L+1} \frac{1}{\sqrt{\boldsymbol{\alpha}!}} \int_{T_j} \big|(\mathbf{y}-\mathbf{x}_j)^{\boldsymbol{\alpha}}\big|\, d\mathbf{y}$$
$$\le \Big(\frac{2\mathcal{D}\kappa^2}{\mathcal{D}-d}\Big)^{(L+1)/2} \operatorname{meas} T_j \sum_{[\boldsymbol{\alpha}]=L+1} \frac{1}{\sqrt{\boldsymbol{\alpha}!}}\,,$$

and therefore, the estimate (11.67) is valid with the constant $c = K^n$. □

LEMMA 11.19. *For any $c > 0$, the sum*

$$c^{N/2} \sum_{[\boldsymbol{\alpha}]=N} \frac{1}{\sqrt{\boldsymbol{\alpha}!}} \to 0 \qquad \textit{as } N \to \infty\,.$$

PROOF. Since the number of multi-indices $\boldsymbol{\alpha} \in \mathbb{Z}^n_{\geq 0}$ of length $[\boldsymbol{\alpha}] = N$ is

$$\binom{N+n-1}{N} = \frac{(N+n-1)!}{N!\,(n-1)!}\,,$$

from

$$(x_1 + \cdots + x_n)^N = \sum_{[\boldsymbol{\alpha}]=N} \frac{N!}{\boldsymbol{\alpha}!} \mathbf{x}^{\boldsymbol{\alpha}}$$

and Cauchy's inequality

$$\sum_{[\boldsymbol{\alpha}]=N} \frac{1}{\sqrt{\boldsymbol{\alpha}!}} \leq \sqrt{\binom{N+n-1}{N}} \Big(\sum_{[\boldsymbol{\alpha}]=N} \frac{1}{\boldsymbol{\alpha}!} \Big)^{1/2} = \frac{n^{N/2}}{N!} \sqrt{\frac{(N+n-1)!}{(n-1)!}}$$

we obtain

$$c^{N/2} \sum_{[\boldsymbol{\alpha}]=N} \frac{1}{\sqrt{\boldsymbol{\alpha}!}} \leq \sqrt{\frac{c_1^N (N+n-1)!}{(N!)^2 (n-1)!}} = s_N$$

with $c_1 = c\,n$. From

$$\frac{s_{N+1}}{s_N} = \sqrt{\frac{c_1\,(N+n)}{(N+1)^2}} = \sqrt{\frac{c_1}{N+1}} \sqrt{1 + \frac{n-1}{N+1}}\,,$$

we see that s_N decays like

$$\sqrt{\frac{c_1^N}{(N+1)!}}$$

for large N. □

COMPLETION OF THE PROOF OF THEOREM 11.17. By Lemmas 11.18 and 11.19, there exist polynomials p_j of degree L such that

$$\Big| \int_{T_j} \mathrm{e}^{-\rho\,(\mathbf{x}/(\rho d) - \mathbf{y})^2}\, d\mathbf{y} - p_j(\mathbf{x})\, \mathrm{e}^{-\rho\,(\mathbf{x}/(\rho d) - \mathbf{x}_j)^2} \Big| < \delta_L \operatorname{meas} T_j$$

with a sequence $\delta_L \to 0$ if $L \to \infty$. Setting

$$P_R(\mathbf{x}) = \Big(\frac{\rho}{\pi}\Big)^{n/2} \mathrm{e}^{-|\mathbf{x}|^2/\mathcal{D}} \sum_j p_j(\mathbf{x})\, \mathrm{e}^{-\rho\,(\mathbf{x}/(\rho d) - \mathbf{x}_j)^2}\,, \tag{11.70}$$

we then obtain from (11.65) that

$$\begin{aligned} |S_R(\mathbf{x}) - P_R(\mathbf{x})| &\leq \mathrm{e}^{-|\mathbf{x}|^2/\mathcal{D}}\, \delta_L \Big(\frac{\rho}{\pi}\Big)^{n/2} \sum_j \operatorname{meas} T_j \\ &= \mathrm{e}^{-|\mathbf{x}|^2/\mathcal{D}}\, \delta_L \Big(\frac{\rho}{\pi}\Big)^{n/2} \operatorname{meas} B_R\,. \end{aligned}$$

Thus, one can choose the degree L of the polynomials p_j such that

$$|S_R(\mathbf{x}) - P_R(\mathbf{x})| < \mathrm{e}^{-|\mathbf{x}|^2/\mathcal{D}}\, \varepsilon/2\,.$$

Note that, in view of (11.64), we can rewrite

$$P_R(\mathbf{x}) = \sum_j \mathcal{P}_j(\mathbf{x})\; \mathrm{e}^{-|\mathbf{x} - \mathbf{x}_j|^2/d} \tag{11.71}$$

with the polynomials

$$\mathcal{P}_j(\mathbf{x}) = \Big(\frac{\rho}{\pi}\Big)^{n/2} \mathrm{e}^{-|\mathbf{x}_j|^2/(\mathcal{D}-d)} \, p_j(\mathbf{x})$$

$$= \frac{\mathrm{e}^{-|\mathbf{x}_j|^2/(\mathcal{D}-d)}}{\pi^{n/2}} \sum_{[\boldsymbol{\alpha}]=0}^{L} \frac{\rho^{(n+[\boldsymbol{\alpha}])/2}}{\boldsymbol{\alpha}!} H_{\boldsymbol{\alpha}}\Big(\sqrt{\rho}(\frac{\mathbf{x}}{\rho d} - \mathbf{x}_j)\Big) \int_{T_j} (\mathbf{y} - \mathbf{x}_j)^{\boldsymbol{\alpha}} \, d\mathbf{y} \,,$$

where $\rho = \mathcal{D}/d(\mathcal{D}-d)$. This together with (11.66) completes the proof of Theorem 11.17. □

COROLLARY 11.20. *Suppose that, for a given $\varepsilon > 0$, the assumptions of Theorem 11.17 are satisfied. There exist polynomials $\mathcal{P}_j$ of sufficiently large degree L_j such that*

$$\| \mathrm{e}^{-|\cdot|^2/\mathcal{D}} - \sum_j \mathcal{P}_j \, \mathrm{e}^{-|\cdot - \mathbf{x}_j|^2/d} \|_{L_p} < \varepsilon$$

for all $p \in [1, \infty]$.

11.4.8. Summary. Theorem 11.16 and Corollary 11.20 provide sufficient conditions under which

$$\min_{\mathbf{c}} Q(\mathbf{c}) = \min_{\mathcal{P}_j \in \Pi_{L_j}} \Big\| \mathrm{e}^{-|\cdot|^2/\mathcal{D}} - \sum_j \mathcal{P}_j \, \mathrm{e}^{-|\cdot - \mathbf{t}_j|^2/d_j} \Big\|_{L_2}^2$$

(see (11.50)) can be made arbitrary small if the degrees L_j of the polynomials are large enough. Applied to the discrepancy

$$\omega(\mathbf{x}) = \sum_{\mathbf{x}_j \in \Sigma} \mathcal{P}_j\Big(\frac{\mathbf{x} - \mathbf{x}_j}{h_j^2}\Big) \, \mathrm{e}^{-|\mathbf{x}-\mathbf{x}_j|^2/h_j^2} - \mathrm{e}^{-|\mathbf{x}|^2/\mathcal{D}},$$

we obtain the following conditions on the nodes $\mathbf{x}_j$ and parameters h_j.

To apply Theorem 11.16, we require $d_k = \mathcal{D}$, which is equivalent to $h_k^2 = \mathcal{D}$. Then

$$\min_{\mathbf{c}} Q(\mathbf{c}) \le \Big(\frac{\pi \mathcal{D}}{2}\Big)^{n/2} \frac{|\mathbf{t}_k|^{2(L_k+1)}}{\mathcal{D}^{L_k+1}(L_k+1)!} \,.$$

From (11.49) we have

$$\mathbf{t}_k = \frac{\mathcal{D}}{\mathcal{D}_0} \mathbf{x}_k$$

with $\mathbf{x}_k \in \Sigma$. Hence,

$$\min_{\mathbf{c}} Q(\mathbf{c}) \le \Big(\frac{\pi \mathcal{D}}{2}\Big)^{n/2} \Big(\frac{\mathcal{D}}{\mathcal{D}_0}\Big)^{L_k+1} \frac{|\mathbf{x}_k|^{2(L_k+1)}}{\mathcal{D}_0^{L_k+1}(L_k+1)!} \,.$$

Corollary 11.20 can be applied if we assume that $d_j = d < \mathcal{D}$, which means that $h_j = h$. Moreover, by (11.48),

$$d = \frac{(h^2 - \sigma)\mathcal{D}(\mathcal{D} - \sigma)}{\mathcal{D}(\mathcal{D} - \sigma) - (h^2 - \sigma)\sigma} \tag{11.72}$$

with $\sigma = \mathcal{D} - \mathcal{D}_0 < h^2$, which implies

$$h^2 = \frac{d\mathcal{D}(\mathcal{D} - \sigma)}{d\sigma + \mathcal{D}(\mathcal{D} - \sigma)} = d - \frac{d^2 \sigma}{d\sigma + \mathcal{D}(\mathcal{D} - \sigma)}$$

and, in particular, $h < \sqrt{\mathcal{D}}$. Additionally, it is required that there exists a partition of a sufficiently large ball B_R into subdomains T_j, which contain at least one node

$\mathbf{t}_j$ and satisfy $\max_{\mathbf{t}\in T_j} |\mathbf{t}-\mathbf{t}_j| \le \kappa\sqrt{d}$ with some positive constant κ. From (11.62) and (11.72), we derive that the radius R is given by the minimal value such that

$$\Gamma\Big(\frac{n}{2}, \frac{R^2(\mathcal{D}(\mathcal{D}-\sigma)-(h^2-\sigma)\sigma)}{\mathcal{D}^2(\mathcal{D}-h^2)}\Big) < \frac{\varepsilon}{2}\Big(\frac{(h^2-\sigma)(\mathcal{D}-\sigma)}{\mathcal{D}(\mathcal{D}-\sigma)-(h^2-\sigma)\sigma}\Big)^{n/2}\Gamma\Big(\frac{n}{2}\Big),$$

which is obviously satisfied if R is chosen to satisfy

$$\Gamma\Big(\frac{n}{2}, \frac{R^2}{\mathcal{D}-h^2}\Big) < \frac{\varepsilon}{2}\Big(\frac{h^2}{\mathcal{D}}\Big)^{n/2}\Gamma\Big(\frac{n}{2}\Big). \tag{11.73}$$

Furthermore, by (11.49),

$$\mathbf{t}_j = \frac{\mathcal{D}(\mathcal{D}-\sigma)}{\mathcal{D}(\mathcal{D}-\sigma)-(h^2-\sigma)\sigma}\mathbf{x}_j = \mathbf{x}_j + \frac{(h^2-\sigma)\sigma}{\mathcal{D}(\mathcal{D}-\sigma)-(h^2-\sigma)\sigma}\mathbf{x}_j$$

where $\mathbf{x}_j \in \Sigma$. Using (11.72), one can see that the condition

$$\mathbf{x}_j \in T_j \quad \text{and} \quad \max_{\mathbf{x}\in T_j} |\mathbf{x}-\mathbf{x}_j| \le \kappa h \tag{11.74}$$

ensures that the assumptions of Theorem 11.17 are fulfilled.

Thus, we obtain the following result:

THEOREM 11.21. *Let Σ be a finite set of points $\mathbf{x}_j$ in $\mathbb{R}^n$, and let $\mathcal{D}$ and ε be given positive numbers. For the positive parameters $h_j \le \sqrt{\mathcal{D}}$ and $\sigma < \min h_j^2$, there exist degrees L_j of the polynomials*

$$\mathcal{P}_j(\mathbf{x}) = \sum_{[\boldsymbol{\beta}]=0}^{L_j} c_{j,\boldsymbol{\beta}}\Big(\frac{\mathbf{x}-\mathbf{x}_j}{h_j}\Big)^{\boldsymbol{\beta}},$$

where the coefficients $c_{j,\boldsymbol{\beta}}$ satisfy the linear system (11.45), *such that*

$$\Big|\sum_{\mathbf{x}_j\in\Sigma} \mathcal{P}_j(\mathbf{x})\, e^{-|\mathbf{x}-\mathbf{x}_j|^2/h_j^2} - e^{-|\mathbf{x}|^2/\mathcal{D}}\Big| < \varepsilon\, e^{-\tau|\mathbf{x}|^2}$$

with

$$\tau = \frac{\sigma}{\mathcal{D}(\mathcal{D}-\sigma)+\sigma^2},$$

if one of the following two conditions is satisfied:

(i) $h_j = \sqrt{\mathcal{D}}$ *for at least one of the parameters,*

(ii) *there exists $h < \sqrt{\mathcal{D}}$ and $\Sigma' \subset \Sigma$ with $h_j = h$, $\mathbf{x}_j \in \Sigma'$, such that the ball B_R, with the radius given by* (11.73), *can be subdivided into subsets T_j which satisfy* (11.74).

Now, we are in a position to describe the construction method for the approximate partition of unity

$$\Big\{\mathcal{P}_j(\mathbf{x})\, e^{-|\mathbf{x}-\mathbf{x}_j|^2/h_j^2}\Big\}$$

for the set of scattered nodes $X = \{\mathbf{x}_j\}_{j\in J}$ close to a piecewise uniform grid G in the sense of Condition 11.9.

Assign a finite set of nodes $\Sigma(\mathbf{g}_k)$ to each grid point $\mathbf{g}_k \in G$ with $|\mathbf{x}_j - \mathbf{g}_k| \le \kappa_1\sqrt{\mathcal{D}_k}$, $\mathbf{x}_j \in \Sigma(\mathbf{g}_k)$. The scaling parameters $h_j > 0$ of the set of functions

$$\Big\{\mathcal{P}_j(\mathbf{x})\, e^{-|\mathbf{x}-\mathbf{x}_j|^2/h_j^2}\Big\}_{\mathbf{x}_j\in\Sigma(\mathbf{g}_k)} \tag{11.75}$$

which approximate $e^{-|\mathbf{x}-\mathbf{g}_k|^2/\mathcal{D}_k}$ should be chosen in the following way:

1. $h_j \le \sqrt{\mathcal{D}_k}$.
2. If there exists $h_k < \sqrt{\mathcal{D}_k}$ such that the ball $B(\mathbf{g}_k, R_k) = \{\mathbf{x} : \mathrm{e}^{-|\mathbf{x}-\mathbf{g}_k|^2/\mathcal{D}_k} > \varepsilon\}$ can be partitioned into subsets T_j such that

$$\mathbf{x}_j \in T_j \quad \text{and} \quad \max_{\mathbf{x}\in T_j} |\mathbf{x}-\mathbf{x}_j| \le \kappa_2 h_k \tag{11.76}$$

for $\mathbf{x}_j \in \Sigma'(\mathbf{g}_k) \subseteqq \Sigma(\mathbf{g}_k)$, then take $h_j = h_k$ if $\mathbf{x}_j \in \Sigma'(\mathbf{g}_k)$.
3. If (11.76) is not possible for $h_k < \sqrt{\mathcal{D}_k}$, then choose $h_j = \sqrt{\mathcal{D}_k}$ for at least one node $\mathbf{x}_j$.
4. If $\mathbf{x}_j$ belongs to $\Sigma(\mathbf{g}_k)$ for different grid points $\mathbf{g}_k$, then h_j has to be the same for all sets (11.75).

Then we fix positive $\sigma < h_j^2$, denote $\mathcal{D}_0 = \mathcal{D}_k - \sigma$ and $\delta_j^2 = h_j^2 - \sigma$, and solve, for a common degree L_k of the polynomials $\mathcal{P}_j$, the system

$$\sum_{\mathbf{x}_j\in\Sigma(\mathbf{g}_k)} \sum_{[\boldsymbol{\beta}]=0}^{L_k} c_{j,\boldsymbol{\beta}}\, \mathcal{C}^{(jl)}_{\boldsymbol{\beta},\boldsymbol{\gamma}}(\mathbf{x}_j, \mathbf{x}_l) = \mathcal{B}^{(j)}_{\boldsymbol{\gamma}}(\mathbf{x}_l)\,, \quad \mathbf{x}_l \in \Sigma(\mathbf{g}_k)\,,\ 0 \le [\boldsymbol{\gamma}] \le L_k\,, \tag{11.77}$$

where

$$\mathcal{B}^{(j)}_{\boldsymbol{\gamma}}(\mathbf{x}) = \Big(\frac{\pi\, h_j^2\, \mathcal{D}_0^2}{\delta_j^2(\mathcal{D}_0\mathcal{D}_k + \mathcal{D}_k - 2\sigma)}\Big)^{n/2} \mathcal{S}_{\boldsymbol{\beta}}(-h_j\nabla_{\mathbf{x}}) \exp\Big(-\frac{(\mathcal{D}_k - 2\sigma)|\mathbf{x}|^2}{\delta_j^2(\mathcal{D}_0\mathcal{D}_k + \mathcal{D}_k - 2\sigma)}\Big),$$

$$\begin{aligned}\mathcal{C}^{(jl)}_{\boldsymbol{\beta},\boldsymbol{\gamma}}(\mathbf{x},\mathbf{y}) &= \frac{\pi^{n/2} h_j^n h_l^n \mathcal{D}_0^{3n/2}}{\mathcal{D}_k^{n/2}(\mathcal{D}_0\mathcal{D}_k(\delta_j^2+\delta_l^2) - 2\sigma\delta_j^2\delta_l^2)^{n/2}}\\ &\times \mathcal{S}_{\boldsymbol{\beta}}(-h_j\nabla_{\mathbf{x}})\mathcal{S}_{\boldsymbol{\gamma}}(-h_l\nabla_{\mathbf{y}}) \exp\Big(\frac{2\sigma(\delta_l^2|\mathbf{x}|^2 + \delta_j^2|\mathbf{y}|^2) - \mathcal{D}_0\mathcal{D}_k|\mathbf{x}-\mathbf{y}|^2}{\mathcal{D}_0\mathcal{D}_k(\delta_j^2+\delta_l^2) - 2\sigma\delta_j^2\delta_l^2}\Big).\end{aligned}$$

Following (11.46), define the polynomials

$$\mathcal{P}^{(k)}_j(\mathbf{x}) = \sum_{[\boldsymbol{\beta}]=0}^{L_k} c_{j,\boldsymbol{\beta}} \Big(\frac{\mathbf{x}-\mathbf{x}_j}{h_j}\Big)^{\boldsymbol{\beta}}, \tag{11.78}$$

and, by Theorem 11.21, we have

$$\Big|\sum_{\mathbf{x}_j\in\Sigma} \mathcal{P}^{(k)}_j(\mathbf{x})\,\mathrm{e}^{-|\mathbf{x}-\mathbf{x}_j|^2/h_j^2} - \mathrm{e}^{-|\mathbf{x}-\mathbf{g}_k|^2/\mathcal{D}_k}\Big| < \delta\,\mathrm{e}^{-\tau_k|\mathbf{x}-\mathbf{g}_k|^2}$$

with

$$\tau_k = \frac{\sigma}{\mathcal{D}_k(\mathcal{D}_k - \sigma) + \sigma^2}\,,$$

if L_k is sufficiently large. Then

$$\Big|\sum_{\mathbf{g}_k\in G} a_k\Big(\mathrm{e}^{-|\mathbf{x}-\mathbf{g}_k|^2/\mathcal{D}_k} - \sum_{\mathbf{x}_j\in\Sigma(\mathbf{g}_k)} \mathcal{P}^{(k)}_j(\mathbf{x})\,\mathrm{e}^{-|\mathbf{x}-\mathbf{x}_j|^2/h_j^2}\Big)\Big| < \delta\, C\,,$$

where the constant

$$C = \sum_{\mathbf{g}_k \in G} a_k \, \mathrm{e}^{-\tau_k |\mathbf{x}-\mathbf{g}_k|^2}$$

does not depend on the scattered nodes $\mathbf{x}_j$ and the degrees L_k of the polynomials.

Hence, for sufficiently large L_k, the function

$$\sum_{\mathbf{g}_k \in G} a_k \sum_{\mathbf{x}_j \in \Sigma(\mathbf{g}_k)} \mathcal{P}_j^{(k)}(\mathbf{x}) \, \mathrm{e}^{-|\mathbf{x}-\mathbf{x}_j|^2/h_j^2} \tag{11.79}$$

is the required approximate partition of unity.

11.4.9. Numerical experiments. We have tested the construction given by (11.77), (11.78) in the one- and two-dimensional cases for randomly chosen nodes with the parameters $\mathcal{D} = 2$, $h = 1$, $\kappa_1 = 1/2$, and $\mathcal{D}_0 = 1$ and $\mathcal{D}_0 = 3/2$. To show the dependence of the approximation error on the number of nodes in $\Sigma(m)$, $m \in \mathbb{Z}$, and the degree of polynomials, we provide graphs of the difference to 1 for the following one-dimensional cases :

$\Sigma(m)$ consists of 1 point, $L = 3$ and $L = 4$ (Fig. 11.1);
$\Sigma(m)$ consists of 3 points, $L = 3$ and $L = 4$ (Fig. 11.2);
$\Sigma(m)$ consists of 5 points, $L = 2$ and $L = 3$ (Fig. 11.3).

In all cases, the choice $\mathcal{D}_0 = 3/2$ gives better results as can be seen from Fig. 11.1. All other figures correspond to $\mathcal{D}_0 = 3/2$.

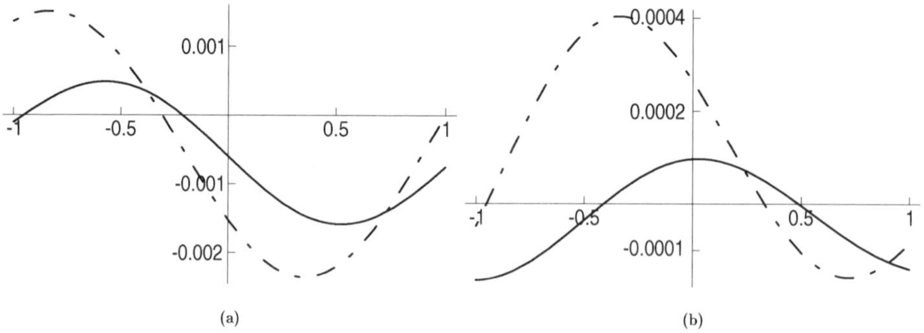

FIGURE 11.1. The graph of $\Theta(\mathbf{x}) - 1$ when $\Sigma(m)$ consists of 1 point, $D = 2$, **(a)** $L = 3$ and **(b)** $L = 4$. Solid and dot-dashed lines correspond to $\mathcal{D}_0 = 3/2$ and $\mathcal{D}_0 = 1$, respectively.

As expected, the approximation becomes better with increasing degree L and more points in the subsets $\Sigma(m)$. The use of only one node in $\Sigma(m)$ reduces the approximation error by a factor 10^{-1} if L increases by 1. The cases of 3 and 5 points indicate that enlarging the degree L of the polynomials by 1 gives a factor 10^{-2} for the approximation error.

Note that the plotted total error consists of two parts. Using (11.77), (11.78), we approximate the Θ-function

$$(2\pi)^{-1/2} \sum_{m \in \mathbb{Z}} \mathrm{e}^{-(x-m)^2/2} = 1 + 2 \sum_{j=1}^{\infty} \mathrm{e}^{-2\pi^2 j^2} \cos 2\pi j x \, . \tag{11.80}$$

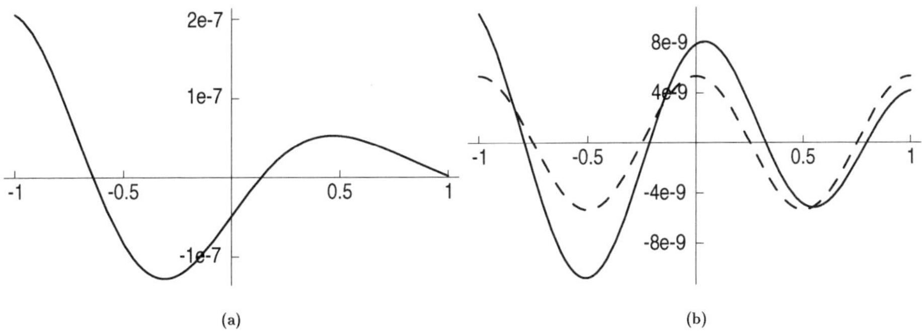

FIGURE 11.2. The graph of $\Theta(\mathbf{x}) - 1$ when $\Sigma(m)$ consists of 3 points, $D = 2$, $\mathcal{D}_0 = 3/2$, **(a)** $L = 3$ and **(b)** $L = 4$. The saturation term obtained on the uniform grid is depicted by dashed lines.

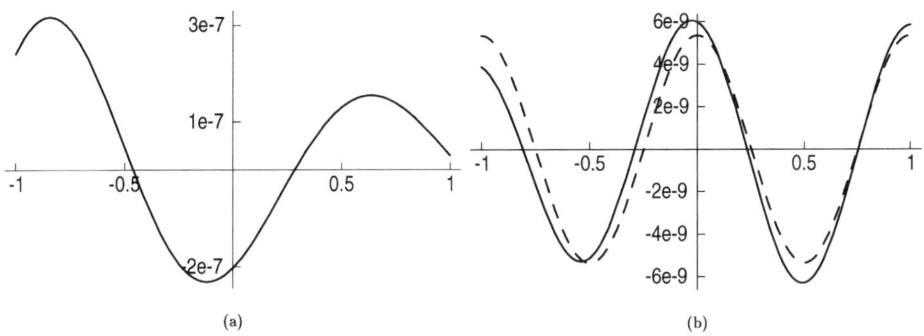

FIGURE 11.3. The graph of $\Theta(\mathbf{x}) - 1$ when $\Sigma(m)$ consists of 5 points, $D = 2$, $\mathcal{D}_0 = 3/2$, **(a)** $L = 2$ and **(b)** $L = 3$. The saturation term obtained on the uniform grid is depicted by dashed lines.

Hence, the plotted total error is the sum of the difference between (11.79) and (11.80) and the function

$$2 \sum_{j=1}^{\infty} \mathrm{e}^{-2\pi^2 j^2} \cos 2\pi j x \,, \tag{11.81}$$

which is the saturation term obtained for the uniform grid. The error plots in the right-hand side in Figs. 11.2 and 11.3 show that the total error is majorized by the saturation term (11.81), which is shown by dashed lines.

In Figs. 11.4 and 11.5, we depict the difference $Mu(x) - u(x)$ for the quasi-interpolation formula defined by (11.22) with Gaussian basis functions constructed via (11.46), (11.45) with $\Sigma(m)$ consisting of 5 points, and the approximation orders $N = 2$ and $N = 4$. For $N = 2$, we have used the parameters $L = 4$ (the degree of the polynomials $\mathcal{P}_j$), $\mathcal{D} = 2$, $\mathcal{D}_0 = 3/2$, and for $N = 4$, we have chosen $L = 6$, $\mathcal{D} = 4$, $\mathcal{D}_0 = 3$.

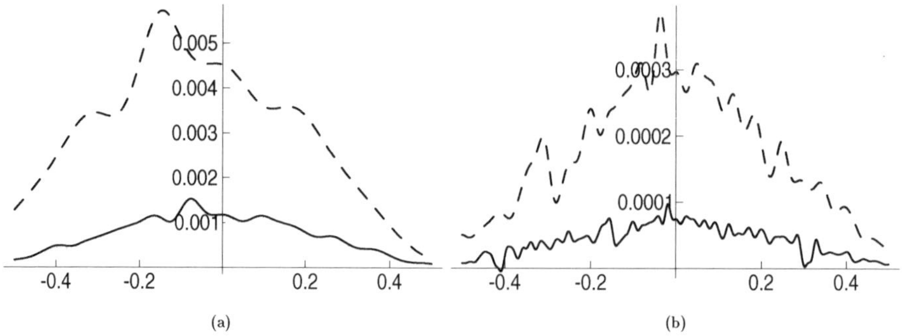

FIGURE 11.4. The graph of $Mu(x) - u(x)$ with $N = 2$, $u(x) = (1+x^2)^{-1}$. Dashed and solid lines correspond to **(a)** $h = 1/16$ and $h = 1/32$ and **(b)** $h = 1/64$ and $h = 1/128$.

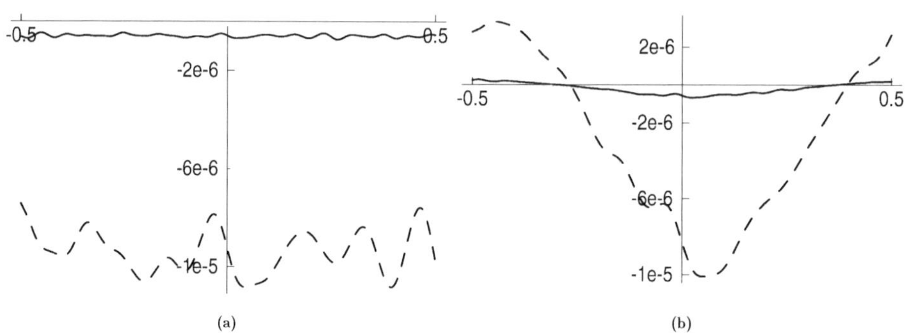

FIGURE 11.5. The graph of $Mu(x) - u(x)$ with $N = 4$, for **(a)** $u(x) = (1 + x^2)^{-1}$ and **(b)** $u(x) = x^4$. Dashed and solid lines correspond to $h = 1/32$ and $h = 1/64$, respectively.

The h^N-convergence of these one-dimensional quasi-interpolants is confirmed in Table 11.1, which contains the uniform error of $Mu - u$ on the interval $(-1/2, 1/2)$ for the function $u(x) = (1 + x^2)^{-1}$ with different values of h.

h	$N = 2$	$N = 4$
2^{-3}	$1.89 \cdot 10^{-2}$	$1.81 \cdot 10^{-3}$
2^{-4}	$5.72 \cdot 10^{-3}$	$1.38 \cdot 10^{-4}$
2^{-5}	$1.51 \cdot 10^{-3}$	$1.01 \cdot 10^{-5}$
2^{-6}	$3.81 \cdot 10^{-4}$	$6.65 \cdot 10^{-7}$
2^{-7}	$9.65 \cdot 10^{-5}$	$4.20 \cdot 10^{-8}$

TABLE 11.1. L_∞-approximation error for the function $u(x) = (1+x^2)^{-1}$ in the interval $(-1/2, 1/2)$ using Mu with $N = 2$ (on the left) and $N = 4$ (on the right).

Similar experiments have been performed for the two-dimensional case. Here we provide graphs of

$$1 - \sum_{\mathbf{x}_j \in X} \mathcal{P}_j(\mathbf{x})\, \mathrm{e}^{-|\mathbf{x}-\mathbf{x}_j|^2/\mathcal{D}}$$

for the following cases:

$\deg \mathcal{P}_j = 1$ and $\Sigma(\mathbf{m})$ consists of 1 or 5 points (Fig. 11.6);
$\deg \mathcal{P}_j = 4$ and $\Sigma(\mathbf{m})$ consists of 1 or 5 points (Fig. 11.7).

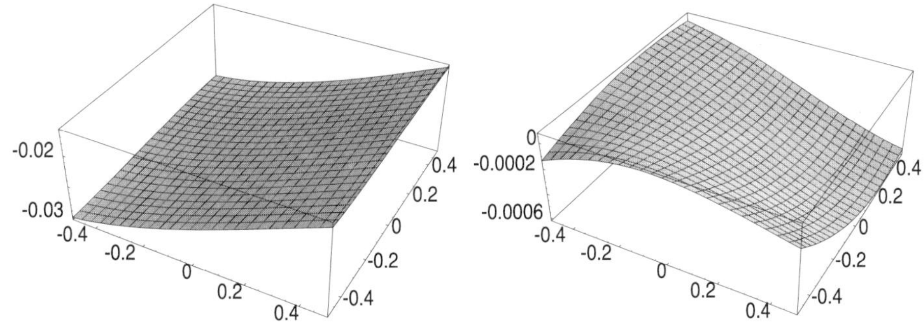

(a) **(b)**

FIGURE 11.6. The graph of $\Theta(\mathbf{x}) - 1$ when $L = 1$ and $\Sigma(\mathbf{m})$ consists of **(a)** 1 point and **(b)** 5 points.

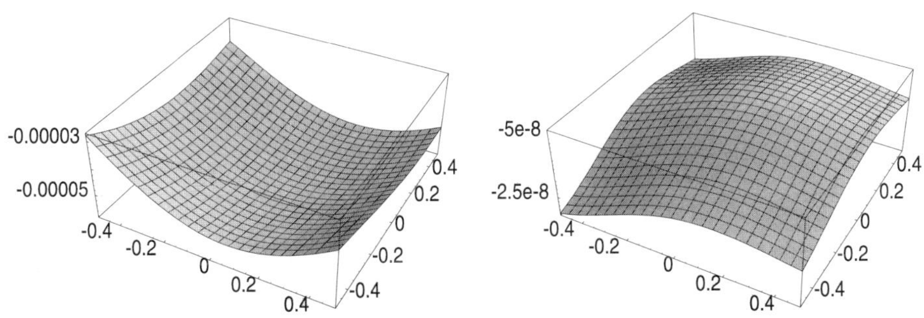

(a) **(b)**

FIGURE 11.7. The graph of $\Theta(\mathbf{x}) - 1$ when $L = 4$ and $\Sigma(\mathbf{m})$ consists of **(a)** 1 point and **(b)** 5 points.

11.5. Notes

Interpolation and quasi-interpolation by radial basis functions are promising methods for approximating multi-variate functions from scattered data. Various aspects of interpolation by radial functions are well developed; see, for example, the monograph by Buhmann [**15**] and the numerous papers referred to therein.

Since the construction of interpolants requires solving large systems of linear equations, it is very interesting to go over to quasi-interpolation. Scattered data quasi-interpolation by functions, which reproduce polynomials, has been studied by Buhmann, Dyn, Levin in [**16**] and Dyn, Ron in [**24**] (see also [**15**], [**97**] for further references).

Other methods for scattered data approximation include Moving Least Squares (see [**28**], [**57**]), which have attracted attention in the context of approximate solutions of partial differential equations as so-called meshless methods (see [**6**] and the references therein). As a rule, the methods reproduce polynomials, at least locally, but the shape functions η_j are not available analytically in simple forms. The computation of the approximant requires solving a linear algebraic system for each point $\mathbf{x} \in \mathbb{R}^n$.

In this chapter we gave a detailed presentation of the recent paper [**56**]; see also the preliminary version [**54**]. The aim of the approach is an analytic representation of the quasi-interpolant, which similarly to the case of uniformly distributed nodes can be used for the approximation of integral and pseudodifferential operators.

CHAPTER 12

Numerical algorithms based upon approximate approximations — linear problems

The real power of the approximate approximation is in the capability to treat multi-dimensional integral operators very efficiently. Therefore, it is natural to use it as an underlying approximation method in numerical algorithms for solving problems with integro-differential equations. Another very important application of approximate approximations is in the large field of integral equations methods for solving initial and boundary value problems for partial differential equations. For such problems, the method should be applied not directly to the partial differential formulation but to equations which involve potentials or other pseudodifferential operators.

In this chapter we describe three applications to the numerical solution of partial differential problems. Section 12.1 is devoted to the solution of Lippmann-Schwinger type equations, which involve volume integral operators and occur in scattering theory. We propose a collocation method which uses dilated shifts of the Gaussian as trial functions. Since the action of the volume potential can be given analytically, the computation time for the discrete system can be significantly reduced. We prove that the method provides spectral convergence order up to saturation errors.

As another example, in Section 12.2 we consider the boundary point method (BPM) as an application of approximate approximations to the solution of boundary integral equations, which are solved by collocation with dilated shifts of a rapidly decaying function. If the surface integrals are approximated by the integrals over the corresponding tangential plane, then the coefficients of the resulting discrete systems depend only on the coordinates of a finite number of points at the boundary and the direction of the normal at these points.

The accuracy of BPM is determined by the best approximation of the solution and by the approximation error of the surface integrals. The last problem is closely connected with cubature of integral operators over surfaces. The main idea is to combine the approximate quasi-interpolation of the surface density with the integration of the basis functions over the tangential plane by the use of appropriate asymptotic expansions. As an example, in Section 12.3, we discuss the computation of multi-dimensional single layer harmonic potentials and prove $\mathcal{O}(h^3|\log h|)$ approximation rate if the values of the normal and of the curvature of Γ at the nodes are used.

12.1. Numerical solution of the Lippmann-Schwinger equation by approximate approximations

Here, we apply an approximation method to the numerical solution of an integral equation of Lippmann-Schwinger type in diffraction theory

$$u(\mathbf{x}) + q(\mathbf{x}) \int_{\Omega} \mathcal{E}_k(\mathbf{x}-\mathbf{y})\, u(\mathbf{y})\, d\mathbf{y} = f(\mathbf{x}) \ , \ \Omega \subset \mathbb{R}^n, \tag{12.1}$$

where the kernel $\mathcal{E}_k$ is the fundamental solution of the Helmholtz equation defined in (5.2). Many scattering problems in inhomogeneous media can be transformed to volume integral equations of this form. Their numerical solution with standard cubature methods is very expensive because of the singularity of the kernel $\mathcal{E}_k$ and its fast oscillations (for large k), especially in the multi-dimensional case.

In our approach, the approximate solution is sought as a linear combination of scaled and shifted Gaussians. We saw in Section 5.1 that the diffraction potential of the Gaussian can be expressed by special functions; hence the discrete system can be computed very efficiently.

It turns out that our method provides spectral convergence up to some negligible saturation error. To be more precise, under the assumption that the solution u satisfies the smoothness condition

$$\int_{\mathbb{R}^n} |\mathcal{F}u(\boldsymbol{\lambda})|\, (1+|\boldsymbol{\lambda}|)^N \, d\boldsymbol{\lambda} < \infty \ ,$$

we show that

$$|u(\mathbf{x}) - u_h(\mathbf{x})| \le c_u\, h^N + c_1 \varepsilon h^2 \tag{12.2}$$

with small ε, negligible in numerical computations. This estimate does not depend on the wave number k, which is confirmed in numerical tests.

12.1.1. Problem. Consider, for example, the scattering problem

$$\Delta w + (k^2 - q) w = g \ , \ \mathbf{x} \in \mathbb{R}^n \ , \tag{12.3}$$

where $k > 0$ is a constant and the potential $q(\mathbf{x})$ and the right-hand side $g(\mathbf{x})$ are compactly supported complex-valued functions. The radiated field w has to satisfy Sommerfeld's radiation condition (5.4). The application of the diffraction potential

$$\mathcal{S}u(\mathbf{x}) = \mathcal{S}_n u(\mathbf{x}) = \int_{\mathbb{R}^n} \mathcal{E}_k(\mathbf{x}-\mathbf{y})\, u(\mathbf{y})\, d\mathbf{y} \tag{12.4}$$

leads to the integral equation for the radiated field

$$w(\mathbf{x}) + \mathcal{S}(qw)(\mathbf{x}) = -\mathcal{S}g(\mathbf{x}) \ . \tag{12.5}$$

In the following, we omit the index n, which indicates the space dimension of the diffraction potential $\mathcal{S}_n$.

In the special case when an incident field w^i, i.e., a given entire solution of the Helmholtz equation $\Delta w^i + k^2 w^i = 0$, is scattered by the potential $q(\mathbf{x})$, the right-hand side of (12.3) is given by $g = qw^i$, and equation (12.5) leads to the well-known Lippmann-Schwinger equation for the total field $w^{tot} = w + w^i$

$$w^{tot}(\mathbf{x}) + \mathcal{S}(q\, w^{tot})(\mathbf{x}) = w^i(\mathbf{x}) \ , \ \mathbf{x} \in \mathbb{R}^n. \tag{12.6}$$

We refer to [**20**] for more details concerning this equation.

In the following, we consider the equation (12.5). Multiplying both sides with the potential q, we derive an integral equation of the form

$$u(x) + q(\mathbf{x})\,\mathcal{S}u(\mathbf{x}) = -q(\mathbf{x})\mathcal{S}g(\mathbf{x}) \tag{12.7}$$

for the function $u = qw$. If a solution u of (12.7) is found, then from (12.5), one obtains the solution w of the original problem by the formula

$$w(\mathbf{x}) = -\int_{\mathbb{R}^n} \mathcal{E}_k(\mathbf{x}-\mathbf{y})\,(g(\mathbf{y}) + u(\mathbf{y}))\,d\mathbf{y}\,. \tag{12.8}$$

So the partial differential equation (12.3), given on the whole space $\mathbb{R}^n$, is transformed into an integral equation over a bounded domain Ω containing $\operatorname{supp} q$.

We propose a collocation method for solving (12.1), which is based on the direct computation of integrals of the basis functions. To solve (12.1), we choose as basis functions the elements of the set

$$X_h := \left\{ \mathrm{e}^{-|\mathbf{x}-h\mathbf{m}|^2/\mathcal{D}h^2},\ h\mathbf{m} \in \Omega_h, \mathbf{m} \in \mathbb{Z}^n \right\}, \tag{12.9}$$

where Ω_h is some domain containing the support of the potential q, $\Omega \subset \Omega_h$, the parameter $\mathcal{D}$ is a fixed positive number, and h is the discretization parameter. Thus the approximating functions are linear combinations of scaled Gaussians centered at the grid points $\{h\mathbf{m} \in \Omega_h\}$.

It was shown in Section 5.1 that in the one- and three-dimensional cases, we have analytic formulas for the diffraction potential of Gaussians (see (5.13) and (5.18)).

Other basis functions commonly used for solving three-dimensional problems, finite elements for example, do not give such simple formulas. Here special cubature formulas have to be utilized. The use of the Gaussian reduces the numerical expenses of discretizing the integral equation significantly. Moreover, since any point value of the integral operator applied the approximating functions can be computed exactly (of course, within the computer's precision), no cubature errors occur.

The integral equation (12.7) is solved by collocation: Find $u_h \in X_h$ such that

$$u_h(h\mathbf{m}) + q(h\mathbf{m})\,\mathcal{S}u_h(h\mathbf{m}) = -q(h\mathbf{m})\int_{\mathbb{R}^n} \mathcal{E}_k(h\mathbf{m}-\mathbf{y})\,g(\mathbf{y})\,d\mathbf{y} \tag{12.10}$$

for all grid points $h\mathbf{m} \in \Omega_h$. Hence, the coefficients $\{u_\mathbf{m}\}$ of the discrete solution

$$u_h(\mathbf{x}) = \sum_{h\mathbf{m}\in\Omega_h} u_\mathbf{m}\,\mathrm{e}^{-|\mathbf{x}-h\mathbf{m}|^2/\mathcal{D}h^2}$$

are determined from the linear system

$$u_\mathbf{m} + q(h\mathbf{m}) \sum_{h\mathbf{j}\in\Omega_h} a_{\mathbf{m}-\mathbf{j}} u_\mathbf{j} = -q(h\mathbf{m})\int_{\mathbb{R}^n} \mathcal{E}_k(h\mathbf{m}-\mathbf{y})\,g(\mathbf{y})\,d\mathbf{y}\,, \tag{12.11}$$

for $h\mathbf{m} \in \Omega_h$, where

$$a_\mathbf{j} = \chi(h\mathbf{j}) \quad \text{with} \quad \chi(\mathbf{x}) = \int_{\mathbb{R}^n} \mathcal{E}_k(\mathbf{x}-\mathbf{y})\,\mathrm{e}^{-|\mathbf{y}|^2/\mathcal{D}h^2}\,d\mathbf{y}\,. \tag{12.12}$$

In the one-dimensional case, (5.20) gives for $j \in \mathbb{Z}$

$$a_j = \frac{\sqrt{\pi\mathcal{D}}h}{2k}\left(\mathrm{e}^{-j^2/\mathcal{D}}\, L\Big(\frac{kh\sqrt{\mathcal{D}}}{2} + i\frac{|j|}{\sqrt{\mathcal{D}}}\Big) - i\,\mathrm{e}^{-k^2\mathcal{D}h^2/4}\,\mathrm{e}^{ikh|j|} \right),$$

whereas in $\mathbb{R}^3$ we obtain

$$a_{\mathbf{j}} = \frac{\sqrt{\pi \mathcal{D}^3}h^2}{4|\mathbf{j}|}\Big(\mathrm{e}^{-k^2\mathcal{D}h^2/4}\,\mathrm{e}^{ikh|\mathbf{j}|} - \mathrm{e}^{-|\mathbf{j}|^2/\mathcal{D}}\, K\Big(\frac{kh\sqrt{\mathcal{D}}}{2} + i\frac{|\mathbf{j}|}{\sqrt{\mathcal{D}}}\Big)\Big),$$

$$a_{(0,0,0)} = \frac{\mathcal{D}h^2}{2} + \frac{ik\sqrt{\pi\mathcal{D}^3}h^3}{4}\,\mathrm{w}\Big(\frac{kh\sqrt{\mathcal{D}}}{2}\Big).$$

Here $|\mathbf{j}|$ denotes the Euclidean norm of $\mathbf{j} \in \mathbb{Z}^3$.

Note that, in general, the right-hand side of the system (12.11) cannot be determined exactly. Hence, in order to obtain an estimate of the form (12.2), we use the high-order cubature formulas discussed in Section 5.1. We approximate $\mathcal{S}g$ by

$$\mathcal{S}_h g(\mathbf{x}) = \mathcal{D}^{-n/2} \sum_{h\mathbf{m}\in\Omega} g(h\mathbf{m})\chi_M(\mathbf{x} - h\mathbf{m})$$

with

$$\chi_M(\mathbf{x}) = (\sqrt{\mathcal{D}}h)^n \int_{\mathbb{R}^n} \mathcal{E}_k(\mathbf{x} - \sqrt{\mathcal{D}}h\mathbf{y})\,\eta_{2M}(\mathbf{y})\,d\mathbf{y}\,.$$

The following approximation theorem holds:

THEOREM 12.1. *Assume that $u \in W_\infty^N$, $N = 2M$, has compact support and let the mesh width satisfy $hk \le \kappa < 2\pi$ with k the wave number in* (12.3). *Then for any $\varepsilon > 0$, there exists $\mathcal{D} > 0$ such that*

$$|\mathcal{S}_h u(\mathbf{x}) - \mathcal{S}u(\mathbf{x})| \le c\,(\sqrt{\mathcal{D}}h)^N \sum_{[\boldsymbol{\alpha}]=N} \frac{\|\partial^{\boldsymbol{\alpha}} u\|_{L_\infty}}{\boldsymbol{\alpha}!} + h^2\varepsilon \|u\|_{W_\infty^{N-1}}\,.$$

PROOF. We use the expansion (2.50) of the quasi-interpolant

$$\mathcal{D}^{-n/2} \sum_{\mathbf{m}\in\mathbb{Z}^n} u(h\mathbf{m})\,\eta_{2M}\Big(\frac{\mathbf{x} - h\mathbf{m}}{\sqrt{\mathcal{D}}h}\Big),$$

which approximates the density u. Obviously, the remainder

$$R_{N,h}(\mathbf{x}) = (\sqrt{\mathcal{D}}h)^N \sum_{[\boldsymbol{\alpha}]=N} \mathcal{D}^{-n/2} \sum_{h\mathbf{m}\in \operatorname{supp} u} \Big(\frac{\mathbf{x}-h\mathbf{m}}{\sqrt{\mathcal{D}}h}\Big)^{\boldsymbol{\alpha}} \eta_{2M}\Big(\frac{\mathbf{x}-h\mathbf{m}}{\sqrt{\mathcal{D}}h}\Big)\,\frac{U_{\boldsymbol{\alpha}}(\mathbf{x}, h\mathbf{m})}{\boldsymbol{\alpha}!}$$

satisfies

$$\max_{\mathbf{x}} |\mathcal{S}R_{N,h}(\mathbf{x})| \le c\,(\sqrt{\mathcal{D}}h)^N \sum_{[\boldsymbol{\alpha}]=N} \frac{\|\partial^{\boldsymbol{\alpha}} u\|_{L_\infty}}{\boldsymbol{\alpha}!}\,.$$

Since the sums of the fast oscillating functions in (2.50), which represent the saturation error, converge absolutely, it remains to estimate the value of the diffraction potential applied to functions of the form $v(x)\,\mathrm{e}^{-2\pi i\langle \boldsymbol{\nu}, \mathbf{x}\rangle/h}$, $\boldsymbol{\nu} \in \mathbb{Z}^n \setminus \{\mathbf{0}\}$, with a compactly supported and sufficiently smooth function v. The pseudodifferential operator $\mathcal{S}$ has the symbol $(k^2 - 4\pi^2|\boldsymbol{\lambda}|^2)^{-1}$. Let us introduce the set

$$B = \{\boldsymbol{\lambda} \in \mathbb{R}^n : |4\pi^2|\boldsymbol{\lambda}|^2 - k^2| \le \rho\} \text{ with } \rho = \frac{(4\pi^2 - \kappa^2)k^2}{\kappa^2}$$

and a smooth function $a(\boldsymbol{\lambda})$ satisfying $a(\boldsymbol{\lambda}) = (k^2 - 4\pi^2|\boldsymbol{\lambda}|^2)^{-1}$ for $\boldsymbol{\lambda} \notin B$. Then

$$\mathcal{S}\big(v\,\mathrm{e}^{2\pi i\langle\cdot,\boldsymbol{\nu}\rangle/h}\big)(\mathbf{x}) = \iint\limits_{\mathbb{R}^n\mathbb{R}^n} \mathrm{e}^{2\pi i\langle\mathbf{x}-\mathbf{y},\boldsymbol{\lambda}\rangle}\, a(\boldsymbol{\lambda})\, v(\mathbf{y})\,\mathrm{e}^{-2\pi i\langle\mathbf{y},\boldsymbol{\nu}\rangle/h}\, d\boldsymbol{\lambda}\, d\mathbf{y}$$

$$+ \iint\limits_{\mathbb{R}^n B} \Big(\frac{1}{k^2 - 4\pi^2|\boldsymbol{\lambda}|^2} - a(\boldsymbol{\lambda})\Big)\, \mathrm{e}^{2\pi i\langle\mathbf{x},\boldsymbol{\lambda}\rangle}\,\mathrm{e}^{-2\pi i\langle\mathbf{y},\boldsymbol{\nu}/h-\boldsymbol{\lambda}\rangle}\, v(\mathbf{y})\, d\boldsymbol{\lambda}\, d\mathbf{y}$$

$$= \iint\limits_{\mathbb{R}^n\mathbb{R}^n} \mathrm{e}^{2\pi i\langle\mathbf{x}-\mathbf{y},\boldsymbol{\lambda}\rangle}\, a(\boldsymbol{\lambda})\, v(\mathbf{y})\,\mathrm{e}^{-2\pi i\langle\mathbf{y},\boldsymbol{\nu}\rangle/h}\, d\boldsymbol{\lambda}\, d\mathbf{y}$$

$$+ \int\limits_{B} \Big(\frac{1}{k^2 - 4\pi^2|\boldsymbol{\lambda}|^2} - a(\boldsymbol{\lambda})\Big)\, \mathrm{e}^{2\pi i\langle\mathbf{x},\boldsymbol{\lambda}\rangle}\, \mathcal{F}v(\boldsymbol{\lambda} - \boldsymbol{\nu}/h)\, d\boldsymbol{\lambda}\,.$$

The integral

$$I_1 := \iint\limits_{\mathbb{R}^n\mathbb{R}^n} \mathrm{e}^{2\pi i\langle\mathbf{x}-\mathbf{y},\boldsymbol{\lambda}\rangle}\, a(\boldsymbol{\lambda})\, v(\mathbf{y})\,\mathrm{e}^{-2\pi i\langle\mathbf{y},\boldsymbol{\nu}\rangle/h}\, d\boldsymbol{\lambda}\, d\mathbf{y}$$

is a pseudodifferential operator with smooth symbol $a \in S^{-2}(\Omega)$; hence the expansion

$$I_1 = \mathrm{e}^{2\pi i\langle x,\boldsymbol{\nu}\rangle/h} \sum_{[\boldsymbol{\alpha}]<N} \frac{\partial^{\boldsymbol{\alpha}} a(\boldsymbol{\nu}/h)\partial^{\boldsymbol{\alpha}} v(\mathbf{y})}{i^{[\boldsymbol{\alpha}]}\boldsymbol{\alpha}!} + R_N(\mathbf{x},h)$$

holds (see [**29**, Sect. 3.3.4]), where

$$|R_N(\mathbf{x},h)| \le C_{N,\Omega} h^{2+N}\,.$$

Since the number ρ is chosen so that $\boldsymbol{\nu}/h \notin B$ for all $\boldsymbol{\nu} \in \mathbb{Z}^n \setminus \{\mathbf{0}\}$, we obtain

$$\partial^{\boldsymbol{\alpha}} a(\boldsymbol{\nu}/h) = \partial^{\boldsymbol{\alpha}}_{\boldsymbol{\lambda}} \frac{1}{k^2 - 4\pi^2|\boldsymbol{\lambda}|^2}\Big|_{\boldsymbol{\lambda}=\boldsymbol{\nu}/h} = h^{2+[\boldsymbol{\alpha}]}\partial^{\boldsymbol{\alpha}}_{\boldsymbol{\nu}} \frac{1}{h^2k^2 - 4\pi^2|\boldsymbol{\nu}|^2}\,,$$

which leads to the asymptotics

$$I_1 = \mathrm{e}^{2\pi i\langle\mathbf{x},\boldsymbol{\nu}\rangle/h}\, h^2 \sum_{[\boldsymbol{\alpha}]<N} \frac{h^{[\boldsymbol{\alpha}]}\,\partial^{\boldsymbol{\alpha}} v(\mathbf{y})}{i^{[\boldsymbol{\alpha}]}\boldsymbol{\alpha}!}\,\partial^{\boldsymbol{\alpha}}_{\boldsymbol{\nu}} \frac{1}{h^2k^2 - 4\pi^2|\boldsymbol{\nu}|^2} + R_N(\mathbf{x},h)\,.$$

To estimate the second integral, we note that $|\boldsymbol{\lambda}|^N|\mathcal{F}v(\boldsymbol{\lambda})| \to 0$ as $|\boldsymbol{\lambda}| \to \infty$. Using spherical coordinates, one easily sees that the integral is a principal value integral of a compactly supported smooth function. Therefore

$$\Big|\int\limits_{B} \Big(\frac{1}{k^2 - 4\pi^2|\boldsymbol{\lambda}|^2} - a(\boldsymbol{\lambda})\Big)\, \mathrm{e}^{2\pi i\langle\mathbf{x},\boldsymbol{\lambda}\rangle}\, \mathcal{F}v(\boldsymbol{\lambda} - \boldsymbol{\nu}/h)\, d\boldsymbol{\lambda}\Big| \le c_v h^N\,. \qquad \square$$

12.1.2. Error analysis for the collocation method. To estimate the asymptotic error of the collocation method, we introduce an interpolation projection Q_h satisfying $Q_h f(h\mathbf{m}) = f(h\mathbf{m})$ for all $h\mathbf{m} \in \Omega_h$. Then the discretization (12.10) of the integral equation (12.5), which has the form

$$u + q\,\mathcal{S}u = -q\mathcal{S}g\,, \tag{12.13}$$

can be written as

$$Q_h u_h + Q_h q\,\mathcal{S}u_h = -Q_h q\mathcal{S}_h g\,, \tag{12.14}$$

where $\mathcal{S}_h$ is an appropriately chosen cubature for $\mathcal{S}$. The linear systems (12.14) are uniquely solvable for all sufficiently small h. This follows from the fact that $\mathcal{S}$ is a compact operator in any reasonable function space over bounded domains. Hence, under certain smoothness assumptions on q, the operator $q\mathcal{S}$ remains compact if the domain under consideration contains the support of q.

Furthermore, the interpolation problem with Gaussian functions is uniquely solvable (see [**79**] and Section 7.3). Moreover, by (11.54),

$$e^{-(\mathbf{j}-\mathbf{k})^2/\mathcal{D}} = (\pi\mathcal{D})^{n/2} \int_{\mathbb{R}^n} e^{-\pi^2\mathcal{D}|\mathbf{x}|^2}\, e^{2\pi i\langle \mathbf{j}-\mathbf{k},\mathbf{x}\rangle}\, d\mathbf{x}\,,$$

which shows that

$$(Av, v) = \sum_{h\mathbf{j},h\mathbf{k}\in\Omega_h} e^{-|\mathbf{j}-\mathbf{k}|^2/\mathcal{D}}\, v_{\mathbf{j}}\overline{v_{\mathbf{k}}} = (\pi\mathcal{D})^{n/2} \int_{\mathbb{R}^n} e^{-\pi^2\mathcal{D}|\mathbf{x}|^2} \Big| \sum_{h\mathbf{j}\in\Omega_h} v_{\mathbf{j}}\, e^{2\pi i\langle \mathbf{j},\mathbf{x}\rangle} \Big|^2 d\mathbf{x}\,,$$

for any vector $v = (v_{\mathbf{j}})_{h\mathbf{j}\in\Omega_h}$, $v_{\mathbf{j}} \in \mathbb{C}$. Therefore, we obtain the inequality

$$e^{-\pi^2\mathcal{D}n/4}\, \|v\|_{\ell^2}^2 < \frac{(Av, v)}{(\pi\mathcal{D})^{n/2}} < \max_{\mathbb{R}^n} \Big| \sum_{\mathbf{m}\in\mathbb{Z}^n} e^{-\pi^2\mathcal{D}|\mathbf{x}-\mathbf{m}|^2} \Big|\, \|v\|_{\ell^2}^2$$

with bounds not depending on the mesh size h and the number of unknowns of the interpolating sum of scaled and shifted Gaussians. Hence, the condition number of A can be estimated by $(\pi\mathcal{D})^{-n/2}\, e^{\pi^2\mathcal{D}\,n/4}$. This means that the numerical solution of the interpolation problem can cause stability problems for large $\mathcal{D}$. However, since the scaling of the trial functions coincides with the mesh width h, the condition number of A does not depend on the number of grid points and does not become worse for finer meshes. In practical calculations for three-dimensional problems, direct solvers from LAPACK are stable up to the parameter $\mathcal{D} = 4$.

Since the equations (12.14) are compact perturbations of the uniquely solvable interpolation problems, standard results for projection methods imply that these equations are solvable provided that (12.13) (or equivalently the original problem (12.3)) is uniquely solvable; see for example the monograph [**80**]. The approximate solution w_h to (12.3) is obtained then from the relation

$$w_h(\mathbf{x}) = -\mathcal{S}_h g(\mathbf{x}) - \mathcal{S} u_h(\mathbf{x})\,,\ \mathbf{x} \in \mathbb{R}^n\,, \tag{12.15}$$

where again the analytic formulas (5.13) or (5.18) for the diffraction operator can be used. Hence, by (12.14), one has

$$Q_h u_h - Q_h q(\mathcal{S}_h g + w_h) = -Q_h q \mathcal{S}_h g\,,$$

and therefore,

$$Q_h q w_h = Q_h u_h\,.$$

On the other hand, putting

$$\mathcal{S} u_h = \mathcal{S} Q_h u_h + \mathcal{S}(I - Q_h) u_h$$

into (12.15), we see that w_h solves the integral equation

$$w_h + \mathcal{S} Q_h q w_h = -\mathcal{S}_h g - \mathcal{S}(I - Q_h) u_h\,. \tag{12.16}$$

Since by (12.5)

$$w + \mathcal{S}(qw) = -\mathcal{S} g\,,$$

we obtain, under the assumption that the operator $(I+\mathcal{S}Q_h q)^{-1}$ exists,

$$\begin{aligned} w_h - w &= -(I+\mathcal{S}Q_h q)^{-1}\big(\mathcal{S}_h g + \mathcal{S}(I-Q_h)u_h\big) - w \\ &= -(I+\mathcal{S}Q_h q)^{-1}\big(\mathcal{S}_h g + w + \mathcal{S}Q_h q w + \mathcal{S}(I-Q_h)u_h\big) \\ &= (I+\mathcal{S}Q_h q)^{-1}\big((\mathcal{S}-\mathcal{S}_h)g + \mathcal{S}(I-Q_h)qw - \mathcal{S}(I-Q_h)u_h\big). \end{aligned} \tag{12.17}$$

Therefore, to estimate the error, we have to show that $(I+\mathcal{S}Q_h q)^{-1}$ exist and are uniformly bounded for sufficiently small h as well as finding upper bounds for the three terms inside the brackets. The first term $(\mathcal{S}-\mathcal{S}_h)g$ does not depend on the choice of the interpolation operator Q_h and has been estimated already in Theorem 12.1. The term $\mathcal{S}(I-Q_h)u_h$ vanishes if we choose Q_h to be the interpolation projection onto the space of Gaussians X_h defined in (12.9).

Before dealing with the second term and the operators $(I+\mathcal{S}Q_h q)^{-1}$, we briefly recall the result of Theorem 7.10. Roughly speaking, it was shown that the interpolant from the set of Gaussians X_h at the lattice $\{h\mathbf{m}\,,\ \mathbf{m}\in\mathbb{Z}^n\}$ approximates continuous functions with optimal order up to some saturation error. The saturation error of the interpolation can be represented in the form

$$\sum_{[\boldsymbol{\alpha}]=0}^{N} \frac{\partial^{\boldsymbol{\alpha}} u(\mathbf{x})}{\boldsymbol{\alpha}!}\Big(\frac{\pi\mathcal{D}h}{2}\Big)^{[\boldsymbol{\alpha}]} a_{\boldsymbol{\alpha}}(\mathbf{x})$$

with smooth h-periodic functions $a_{\boldsymbol{\alpha}}(\mathbf{x})$ given by (7.35). It was shown that the saturation error can be made arbitrarily small for sufficiently large $\mathcal{D}$ (for the exact formulation see Theorem 7.10).

Note that, obviously, these properties remain valid if we restrict the interpolation nodes to the set $\{h\mathbf{m}\in\Omega_h, \mathbf{m}\in\mathbb{Z}^n\}$, where the domain $\Omega_h\supset\Omega\supset\operatorname{supp} q$ is chosen so that the basis functions with centers at Ω are smaller than the saturation error outside Ω_h,

$$\mathrm{e}^{-|\mathbf{x}-h\mathbf{m}|^2/\mathcal{D}h^2} < \varepsilon \quad \text{for } \mathbf{x}\notin\Omega_h \text{ and } h\mathbf{m}\in\Omega\,.$$

Now, we are in a position to treat the remaining terms in (12.17). First, we show that the operators $I+\mathcal{S}Q_h q$ have uniformly bounded inverses for all sufficiently small h. Since the set $\|Q_h q\|$ is bounded and $\mathcal{S}$ is compact, the operators $\mathcal{S}Q_h q$ are collectively compact (cf., [**4**]). Since the saturation error of the interpolant consists of rapidly oscillating functions, we conclude as in the proof of Theorem 12.1 that

$$\|\mathcal{S}(I-Q_h)qu\| \to 0 \quad \text{as } h\to 0$$

for the dense subset of functions u with $\|u\|'_{N+\delta}<\infty$. Therefore, all conditions of the collectively compact operator theory are satisfied and we obtain, under the assumption that $I+\mathcal{S}q$ is invertible, that for sufficiently small h the inverse operators $(I+\mathcal{S}Q_h q)^{-1}$ exist and are uniformly bounded. On the other hand, if the solution w satisfies the condition

$$\|qw\|'_N = \int_{\mathbb{R}^n} |\mathcal{F}(qw)(\boldsymbol{\lambda})|\,(1+|\boldsymbol{\lambda}|)^N\,d\boldsymbol{\lambda} < \infty\,, \tag{12.18}$$

then the same arguments lead to the estimate

$$|\mathcal{S}(I-Q_h)q(\mathbf{x})w(\mathbf{x})| \le c\,(\sqrt{\mathcal{D}}h)^N\|qw\|'_N + c_1\,h^2\varepsilon\|qw\|_{C^{N-1}}\,,$$

which bounds the second term inside the brackets in (12.17).

Summarizing all arguments above, we obtain the following convergence result:

THEOREM 12.2. *Suppose that the solution w of equation* (12.3) *satisfies the smoothness condition* (12.18) *with $N = 2M$ and suppose the cubature formula for the right-hand side of* (12.7) *is generated by the function η_{2M}. Then the difference between w and the solutions w_h of the system* (12.11) *can be estimated as follows:*

$$|w(\mathbf{x}) - w_h(\mathbf{x})| \leq c_u \, (\sqrt{\mathcal{D}}h)^N + c_1 \varepsilon h^2 \, .$$

12.1.3. Numerical example. As a simple test, we consider the one-dimensional problem:

$$w'' + (k^2 + \mu \, q(x))w = -\delta(x) \, , \; x \in \mathbb{R} \, . \tag{12.19}$$

Here, μ is some constant parameter and the potential q is given as

$$q(x) = \frac{v(x)}{\mu \, \mathcal{S}v(x) + \mathrm{e}^{ik|x|} / 2ik}$$

with some function $v(x)$, $|x| \leq 1$, which is obviously a solution of the integral equation

$$u(x) - \frac{\mu \, q(x)}{2ik} \int_{-1}^{1} \mathrm{e}^{ik|x-y|} \; u(y) \, dy = \frac{q(x) \; \mathrm{e}^{ik|x|}}{2ik} \, . \tag{12.20}$$

Then (12.19) has the solution

$$w(x) = \mu \, \mathcal{S}v(x) + \frac{\mathrm{e}^{ik|x|}}{2ik} \, .$$

In Tables 12.1 and 12.2, we give convergence rates for $\max |w(x) - w_h(x)|$, where

$$w_h(x) = \mu \, \mathcal{S}u_h(x) + \frac{\mathrm{e}^{ik|x|}}{2ik} \, ,$$

with the approximate solution u_h of (12.20). In the tests, we tried different values of the parameter μ and the wave number k. The obtained convergence rates should correspond to the smoothness of v.

h^{-1}	$k=1$	$k=10$	$k=20$	$k=50$
20	5.1519	6.5723	8.1984	7.8392
30	1.1945	1.0650	0.8227	2.2741
40	1.0463	1.0665	0.9640	0.6128
50	1.0244	1.0482	0.9853	1.1107
60	1.0186	1.0462	0.9894	0.9759
70	1.0155	1.0417	0.9912	1.0305
80	1.0134	1.0292	0.9925	0.9321
90	1.0118	1.0303	0.9934	0.9793
100	1.0106	1.0263	0.9941	1.0000

TABLE 12.1. Convergence rates for $v \in C^1$

h^{-1}	$k=1$	$k=10$	$k=20$	$k=50$
20	16.5432	17.1947	21.2600	17.3789
30	4.4344	3.8187	3.3097	2.9959
40	4.0306	3.9816	6.3844	3.2325
50	4.0015	3.9991	4.1470	4.0169
60	3.9999	4.0020	3.4357	3.6604
70	3.9998	4.0029	4.7079	3.8351
80	3.9998	4.0033	4.0115	3.9777
90	3.9999	4.0000	4.0088	4.1684
100	3.9999	3.9982	4.0069	4.0862

TABLE 12.2. Convergence rates for $v \in C_0^4[-1,1]$

12.2. Applications to the solution of boundary integral equations

The possibility of obtaining explicit formulas for values of various integral and pseudodifferential operators of mathematical physics applied to the new classes of basis functions also makes approximate approximations attractive for the cubature of those integral operators over surfaces.

One important example is multi-dimensional surface potentials associated with elliptic differential operators. They are defined by surface integrals involving fundamental solutions of the differential operators which become singular when the observation point approaches the surface.

The numerical treatment of these integrals with singular kernels is an essential part of boundary integral methods, which have been established as efficient numerical procedures for solving boundary value problems for partial differential equations, which occur in mechanics, acoustics, electromagnetics, and other fields of mathematical physics.

In this section, we apply ideas of approximate approximations to the numerical solution of boundary integral equations. Owing to the rapid decay of the basis functions used in approximate approximations, the integration surface of the boundary integral operators for acting on these functions can be replaced by another surface, more suitable to the approximation of the boundary integral operator. The method, discussed below, replaces the integration surface by tangential planes, supported at certain boundary points. Therefore the coefficients of the resulting algebraic system depend only on the coordinates of a finite number of boundary points and the direction of the normal at these points; hence the name boundary point method seems quite natural ([**63**]).

12.2.1. Boundary integral equations. The methods of boundary integral equations reduce boundary value problems for partial differential equations with known fundamental solutions to equations with boundary integral operators with kernels involving the fundamental solution. Consider, for example, the second-order partial differential operator $L = -\Delta + c$ in $\mathbb{R}^n$, where Δ is the Laplacian and $c \in \mathbb{C}$, and denote by $\gamma(\mathbf{x})$ its fundamental solution. Using Green's formulas the Dirichlet and Neumann boundary value problems for the equation

$$Lu(\mathbf{x}) = 0 \tag{12.21}$$

in a bounded domain Ω or in the exterior $\mathbb{R}^n \setminus \overline{\Omega}$ can be transformed to equations on the boundary $\Gamma = \partial\Omega$ with integral operators of the form

$$\mathcal{V}\psi(\mathbf{x}) = \int_\Gamma \psi(\mathbf{y})\,\gamma(\mathbf{x}-\mathbf{y})\,d\sigma_{\mathbf{y}}\,, \tag{12.22}$$

$$\mathcal{K}\psi(\mathbf{x}) = \int_\Gamma \psi(\mathbf{y})\,\partial_n(\mathbf{y})\gamma(\mathbf{x}-\mathbf{y})\,d\sigma_{\mathbf{y}}\,, \tag{12.23}$$

$$\mathcal{K}'\psi(\mathbf{x}) = \int_\Gamma \psi(\mathbf{y})\,\partial_n(\mathbf{x})\gamma(\mathbf{x}-\mathbf{y})\,d\sigma_{\mathbf{y}}\,, \tag{12.24}$$

where $\mathbf{x} \in \Gamma$. Here $\partial_n(\mathbf{x})$ denotes the derivative in direction of the normal $\mathbf{n}(\mathbf{x})$ to Γ which points into the exterior of Ω.

The operator $\mathcal{V}$ has a weakly singular kernel, whereas $\mathcal{K}$ exists as a principal value singular integral, in general. The operator $\mathcal{K}'$ is the adjoint of (12.23) in $L_2(\Gamma)$. These operators appear as limits of the well-known single and double layer potentials

$$\int_\Gamma \psi(\mathbf{y})\,\gamma(\mathbf{z}-\mathbf{y})\,d\sigma_{\mathbf{y}}\,, \qquad \int_\Gamma \psi(\mathbf{y})\,\partial_n(\mathbf{y})\gamma(\mathbf{z}-\mathbf{y})\,d\sigma_{\mathbf{y}}\,,$$

if $\mathbf{z} \notin \Gamma$ approaches $\mathbf{x} \in \Gamma$. Note that both integrals are solutions of the differential equation outside Γ and the two-sided limits for $\mathbf{z} \to \mathbf{x} \in \Gamma$ satisfy well-known jump relations, which, for sufficiently smooth Γ, are as follows: If the density $\psi \in C(\Gamma)$, then the single layer potential

$$u(\mathbf{z}) = \int_\Gamma \psi(\mathbf{y})\,\gamma(\mathbf{z}-\mathbf{y})\,d\sigma_{\mathbf{y}}$$

is continuous in $\mathbb{R}^n$ and on Γ it holds that

$$u(\mathbf{x}) = \mathcal{V}\psi(\mathbf{x}) \tag{12.25}$$

and

$$\partial_n u^\pm(\mathbf{x}) = \mathcal{K}'\psi(\mathbf{x}) \mp \frac{1}{2}\psi(\mathbf{x}_0)\,, \tag{12.26}$$

where

$$\partial_n u^\pm(\mathbf{x}) = \lim_{h\to 0+} \langle \mathbf{n}(\mathbf{x}), \operatorname{grad} u(\mathbf{x} \pm h\mathbf{n}(\mathbf{x}))\rangle\,.$$

Since $h > 0$, we have $\mathbf{x} + h\mathbf{n}(\mathbf{x}) \in \mathbb{R}^n \setminus \overline{\Omega}$ and $\mathbf{x} - h\mathbf{n}(\mathbf{x}) \in \Omega$. The double layer potential

$$v(\mathbf{x}) = \int_\Gamma \psi(\mathbf{y})\,\partial_n(\mathbf{y})\gamma(\mathbf{x}-\mathbf{y})\,d\sigma_{\mathbf{y}}$$

with density $\psi \in C(\Gamma)$ can be continuously extended from Ω to $\overline{\Omega}$ and from $\mathbb{R}^n \setminus \overline{\Omega}$ to $\mathbb{R}^n \setminus \Omega$ with the limits

$$\lim_{h\to 0+} v(\mathbf{x} \pm h\mathbf{n}(\mathbf{x})) = \mathcal{K}\psi(\mathbf{x}) \pm \frac{1}{2}\psi(\mathbf{x}) \tag{12.27}$$

and the normal derivatives of v are continuous on Γ, i.e.,

$$\partial_n v^+(\mathbf{x}) = \partial_n v^-(\mathbf{x})\,. \tag{12.28}$$

Let us consider some boundary integral formulations of the Dirichlet problem

$$Lu = 0 \quad \text{in } \Omega\,, \quad u|_\Gamma = g\,. \tag{12.29}$$

The solution of the differential equation can be represented as

$$u(\mathbf{x}) = \int\limits_\Gamma \psi(\mathbf{y})\, \partial_n(\mathbf{y})\gamma(\mathbf{x}-\mathbf{y})\, d\sigma_\mathbf{y}\,, \quad \mathbf{x} \in \Omega\,,$$

with the density ψ which satisfies, because of (12.27), the integral equation of the second kind

$$\frac{1}{2}\psi(\mathbf{x}) - \mathcal{K}\psi(\mathbf{x}) = -g(\mathbf{x})\,, \quad \mathbf{x} \in \Gamma\,. \tag{12.30}$$

One can also use the representation

$$u(\mathbf{x}) = \int\limits_\Gamma \psi(\mathbf{y})\, \gamma(\mathbf{x}-\mathbf{y})\, d\sigma_\mathbf{y}\,, \quad \mathbf{x} \in \Omega\,,$$

which leads, in view of (12.25), to the integral equation of the first kind

$$\mathcal{V}\psi(\mathbf{x}) = g(\mathbf{x})\,, \quad \mathbf{x} \in \Gamma\,. \tag{12.31}$$

Moreover, from the representation formula

$$\begin{aligned} u(\mathbf{x}) = \int\limits_\Gamma \Big(\gamma(\mathbf{x}-\mathbf{y})\, \partial_n u(\mathbf{y}) - u(\mathbf{y})\, \partial_n(\mathbf{y})\gamma(\mathbf{x}-\mathbf{y})\Big) d\sigma_\mathbf{y} \\ + \int\limits_\Omega Lu(\mathbf{y})\, \gamma(\mathbf{x}-\mathbf{y})\, d\mathbf{y}\,, \end{aligned} \tag{12.32}$$

which follows from Green's formula and is valid for all reasonable functions u, domains Ω and all $\mathbf{x} \in \Omega$, one obtains the solution of (12.29) in the form

$$u(\mathbf{x}) = \int\limits_\Gamma \Big(\psi(\mathbf{y})\, \gamma(\mathbf{x}-\mathbf{y}) - g(\mathbf{y})\, \partial_n(\mathbf{y})\gamma(\mathbf{x}-\mathbf{y})\Big) d\sigma_\mathbf{y}\,.$$

Here, ψ is a solution of the integral equation of the first kind

$$\mathcal{V}\psi(\mathbf{x}) = \frac{1}{2}\, g(\mathbf{x}) + \mathcal{K}g(\mathbf{x})\,,\ \mathbf{x} \in \Gamma\,, \tag{12.33}$$

which is a consequence of the jump relations for the potentials (12.25) and (12.27).

Another classical example is given by the exterior Neumann problem

$$Lu = 0 \quad \text{in } \mathbb{R}^n \setminus \overline{\Omega}\,, \quad \partial_n u|_\Gamma = g\,, \tag{12.34}$$

with certain prescribed behavior at infinity. If the solution can be represented in the form of a single layer potential, then the density ψ can be found from the integral equation of the second kind

$$\frac{1}{2}\psi(\mathbf{x}) - \mathcal{K}'\psi(\mathbf{x}) = -g(\mathbf{x})\,, \quad \mathbf{x} \in \Gamma\,, \tag{12.35}$$

where the jump relation (12.26) is used.

Finally, we mention boundary integral formulations for the mixed boundary value problems

$$\begin{aligned} Lu = 0, \\ u|_{\Gamma_D} = g_0\,, \quad \partial_n u|_{\Gamma_N} = g_1 \end{aligned} \tag{12.36}$$

in Ω or the exterior $\mathbb{R}^n\backslash\overline{\Omega}$ and the boundary Γ is split into two parts, where Dirichlet or Neumann boundary conditions are imposed, $\overline{\Gamma_D}\cup\overline{\Gamma_N}=\Gamma$. For example, if the problem (12.36) is stated in Ω, then (12.32) together with the jump relations lead to the boundary integral equations for the unknown data $u|_{\Gamma_N}$ and $\partial_n u|_{\Gamma_D}$: for $\mathbf{x}\in\Gamma_N$

$$\begin{aligned}\frac{1}{2}u(\mathbf{x}) + \int\limits_{\Gamma_N} u(\mathbf{y})\,\partial_n(\mathbf{y})\gamma(\mathbf{x}-\mathbf{y})\,d\sigma_{\mathbf{y}} - \int\limits_{\Gamma_D} \gamma(\mathbf{x}-\mathbf{y})\,\partial_n u(\mathbf{y})\,d\sigma_{\mathbf{y}} \\ = -\int\limits_{\Gamma_D} g_0(\mathbf{y})\,\partial_n(\mathbf{y})\gamma(\mathbf{x}-\mathbf{y})\,d\sigma_{\mathbf{y}} + \int\limits_{\Gamma_N}\gamma(\mathbf{x}-\mathbf{y})\,g_1(\mathbf{y})\,d\sigma_{\mathbf{y}}\,,\end{aligned} \tag{12.37}$$

and for $\mathbf{x}\in\Gamma_D$

$$\begin{aligned}\int\limits_{\Gamma_D} \gamma(\mathbf{x}-\mathbf{y})\,\partial_n u(\mathbf{y})\,d\sigma_{\mathbf{y}} - \int\limits_{\Gamma_N} u(\mathbf{y})\,\partial_n(\mathbf{y})\gamma(\mathbf{x}-\mathbf{y})\,d\sigma_{\mathbf{y}} \\ = \frac{1}{2}g_0(\mathbf{x}) + \int\limits_{\Gamma_D} g_0(\mathbf{y})\partial_n(\mathbf{y})\gamma(\mathbf{x}-\mathbf{y})\,d\sigma_{\mathbf{y}} - \int\limits_{\Gamma_N}\gamma(\mathbf{x}-\mathbf{y})\,g_1(\mathbf{y})\,d\sigma_{\mathbf{y}}\,.\end{aligned} \tag{12.38}$$

If a solution of these integral equations is found, representation (12.32) can be used to determine the solution to (12.36) at any point $\mathbf{x}\in\Omega$.

We see that in using boundary integral equation techniques, different elliptic boundary value problems can be reduced to equations on the boundary of the domain, which involve integral operators with singular kernel functions. Here we do not consider the mathematical problems of these techniques, which include the equivalence of the integral equations to the boundary value problem, the solvability and uniqueness of these equations, and similar questions. Instead, in the next subsection we will discuss a certain application of approximate approximations to the numerical solution of boundary integral equations.

12.2.2. Boundary point method. Since boundary integral equations methods can treat both interior and exterior boundary value problems and reduce the dimension of the original problem by one, the numerical solution of the corresponding boundary integral equations became popular since the 1980s.

The widely used boundary element method (BEM), for example, is based on the application of finite element techniques, developed originally for solving partial differential equations. For the use of the BEM, the boundary surface is divided into a finite number of subareas and in every subarea the unknown functions are approximated by standard (as a rule polynomial) functions. After applying a collocation or Galerkin method, the boundary integral equations are reduced to the solution of a finite system of algebraic equations. Because it requires calculating only boundary values, rather than values throughout the domain, where the boundary value problem is formulated, the BEM is significantly more efficient in terms of computational resources for problems where there is a small surface/volume ratio. However, for many problems boundary element methods are less efficient than domain-based methods. This is caused by the need to construct a mesh of the surface and to compute the values of the integrals over the subareas at different points, distributed on the whole surface. In many cases, these integrals are singular and the complexity of their calculation depends on the type of approximating functions.

Since, especially in higher-dimensional problems, a great portion of computer time is spent in calculating the matrix elements, there is ongoing research to develop new effective algorithms. For example, compression techniques (e.g., multipole or wavelet expansions, panel clustering) can help to accelerate the BEM, though at the cost of added complexity.

Now, we describe an alternative approach for the discretization of boundary integral equations, which is not based upon the decomposition of the boundary into surface elements. The approximation on surfaces which have been studied in Section 10.3 and the possibility of obtaining explicit formulas for classical integral operators acting on generating functions of approximate approximations suggest using these functions also for the numerical solution of boundary integral equations. Then the coefficients of the resulting discrete systems depend only on the coordinates of a finite number of points at the boundary, which are the centers of the generating functions, and some surface characteristics near these points.

Let $\{\mathbf{x}_k\}_{k=1}^N$ be a collection of boundary points and let $\{h_k\}_{k=1}^N$ be a collection of positive constants, having the dimension of length. For example, h_k may be chosen as the average distance from $\mathbf{x}_k$ to the neighboring points of $\{\mathbf{x}_k\}_1^N$. The approximate solution of the boundary integral equation

$$A\phi(\mathbf{x}) = f(\mathbf{x}), \quad \mathbf{x} \in \Gamma, \tag{12.39}$$

is sought as a linear combination

$$\phi_N(\mathbf{x}) = \sum_{k=1}^{N} c_k \varphi_k(\mathbf{x}), \quad \mathbf{x} \in \Gamma,$$

with basis functions

$$\varphi_k(\mathbf{x}) = \eta\Big(\frac{|\mathbf{x} - \mathbf{x}_k|}{\sqrt{\mathcal{D}} h_k}\Big),$$

where η is a radial, rapidly decaying function used in approximate approximations and the parameter $\mathcal{D} > 0$ controls the saturation error.

The coefficients $\{c_k\}$ can be determined in different ways: for example, by the collocation method

$$A\phi_N(\mathbf{x}_j) = f(\mathbf{x}_j), \quad j = 1, \dots, N, \tag{12.40}$$

or by Galerkin's method

$$(A\phi_N, \varphi_j)_{L_2(\Gamma)} = (f, \varphi_j)_{L_2(\Gamma)}, \quad j = 1, \dots, N, \tag{12.41}$$

with the scalar product

$$(f, g)_{L_2(\Gamma)} = \int_\Gamma f(\mathbf{x})\, \overline{g(\mathbf{x})}\, d\sigma.$$

For the evaluation of $A\phi_N$, i.e., the action of the integral operator on the basis functions φ_k, we use the fact that φ_k decreases rapidly with the distance from the center $\mathbf{x}_k$. The simplest variant of the BPM replaces surface integration by integration over the tangent plane Γ_k to Γ at $\mathbf{x}_k$.

12.2.3. BPM for single layer potentials. Let A be the single layer potential $\mathcal{V}$ defined by (12.22). Then for the BPM discretization, we approximate

$$\mathcal{V}\varphi_k(\mathbf{x}) \approx \int\limits_{\Gamma_k} \gamma(\mathbf{x}-\mathbf{y})\,\varphi_k(\mathbf{y})\,d\sigma_{\mathbf{y}} = \int\limits_{\mathbb{R}^{n-1}} \gamma(\mathbf{x}-\mathbf{x}_k - O\mathbf{y})\,\eta\Big(\frac{|\mathbf{y}'|}{\sqrt{\mathcal{D}}h_k}\Big)\,d\mathbf{y}'\,, \tag{12.42}$$

where $\mathbf{y}=(\mathbf{y}',0)\in\mathbb{R}^n$ and O is the matrix of an orthogonal transformation of $\mathbb{R}^n$ that directs the normal $\mathbf{n}(\mathbf{x}_k)$ into the direction of the x_n-axis,

$$O\mathbf{n}(\mathbf{x}_k) = (0,\dots,0,1)\,.$$

Since the fundamental solution γ is a radial function, we can write

$$\gamma(\mathbf{x}-\mathbf{x}_k-O\mathbf{y}) = \gamma(O^{-1}(\mathbf{x}-\mathbf{x}_k)-\mathbf{y}) = g(|\mathbf{z}'-\mathbf{y}'|, z_n) \tag{12.43}$$

with the vector

$$(\mathbf{z}', z_n) = O^{-1}(\mathbf{x}-\mathbf{x}_k)\,.$$

Note that

$$z_n = \langle \mathbf{n}(\mathbf{x}_k), \mathbf{x}-\mathbf{x}_k\rangle \quad\text{and}\quad |\mathbf{z}'|^2 = |\mathbf{x}-\mathbf{x}_k|^2 - z_n^2\,. \tag{12.44}$$

For $n\ge 3$, formula (2.15) applied to convolutions in $\mathbb{R}^{n-1}$ has the form

$$\begin{aligned}&\int\limits_{\mathbb{R}^{n-1}} g(|\mathbf{z}'-\mathbf{y}'|, z_n)\,\eta(\mathbf{y}')\,d\mathbf{y}'\\ &\quad= \frac{2\pi}{|\mathbf{z}'|^{(n-3)/2}}\int\limits_0^\infty \mathcal{F}g(r,z_n)\,\mathcal{F}\eta(r)\,r^{(n-1)/2}\,J_{(n-3)/2}(2\pi r|\mathbf{z}'|)\,dr\,,\end{aligned} \tag{12.45}$$

where now $\mathcal{F}$ denotes the $(n-1)$-dimensional Fourier transform of radial functions, i.e.,

$$\mathcal{F}g(r,z_n) = \frac{2\pi}{r^{(n-3)/2}}\int\limits_0^\infty g(t,z_n)\,J_{(n-3)/2}(2\pi r t)\,t^{(n-1)/2}\,dt\,,$$

(see (2.12)). Thus, using (12.42) in the case $n\ge 3$, the value of $\mathcal{V}\varphi_k(\mathbf{x})$ is approximated by the one-dimensional integral

$$\begin{aligned}\mathcal{V}\varphi_k(\mathbf{x}) &\approx \int\limits_{\mathbb{R}^{n-1}} \gamma(\mathbf{x}-\mathbf{x}_k-O\mathbf{y})\,\eta\Big(\frac{|\mathbf{y}'|}{\sqrt{\mathcal{D}}h_k}\Big)\,d\mathbf{y}'\\ &= \frac{2\pi(\sqrt{\mathcal{D}}h_k)^{n-1}}{|\mathbf{z}'|^{(n-3)/2}}\int\limits_0^\infty \mathcal{F}g(r,z_n)\,\mathcal{F}\eta(\sqrt{\mathcal{D}}h_k r)\,J_{(n-3)/2}(2\pi r|\mathbf{z}'|)\,r^{(n-1)/2}\,dr\,,\end{aligned}$$

whereas for $n=2$

$$\mathcal{V}\varphi_k(\mathbf{x}) \approx \sqrt{\mathcal{D}}h_k\int\limits_{-\infty}^{\infty} g(z_1-\sqrt{\mathcal{D}}h_k t, z_2)\,\eta(t)\,dt$$

with $(z_1,z_2)=O^{-1}(\mathbf{x}-\mathbf{x}_k)$.

If the integral equation of the first kind (12.31) or (12.33) is solved by the collocation method (12.40), then the BPM approach approximates the elements of the collocation matrix

$$\Big(\mathcal{V}\varphi_k(\mathbf{x}_j)\Big)_{j,k=1}^N$$

by the values

$$\widehat{\mathcal{V}}_{jk} = \frac{2\pi(\sqrt{\mathcal{D}}h_k)^{n-1}}{Z_{jk}^{(n-3)/2}} \int_0^\infty \mathcal{F}g(r, z_{jk})\, \mathcal{F}\eta(\sqrt{\mathcal{D}}h_k r)\, J_{(n-3)/2}(2\pi r Z_{jk})\, r^{(n-1)/2}\, dr \quad \text{for } n \geq 3\,,$$

(12.46)

$$\widehat{\mathcal{V}}_{jk} = \sqrt{\mathcal{D}}h_k \int_{-\infty}^{\infty} g(Z_{jk} - \sqrt{\mathcal{D}}h_k t, z_{jk})\, \eta(t)\, dt\,, \qquad \text{for } n = 2\,,$$

where we denote, in accordance with (12.44),

$$z_{jk} = \langle \mathbf{n}(\mathbf{x}_k), \mathbf{x}_j - \mathbf{x}_k \rangle \quad \text{and} \quad Z_{jk} = \sqrt{|\mathbf{x}_j - \mathbf{x}_k|^2 - z_{jk}^2}\,. \tag{12.47}$$

12.2.4. BPM for double layer potentials. The BPM simplifies the collocation equations

$$\frac{1}{2}\phi_N(\mathbf{x}_j) - \mathcal{K}\phi_N(\mathbf{x}_j) = -g(\mathbf{x}_j)\,, \quad j = 1, \ldots, N\,, \tag{12.48}$$

for solving the integral equation of the second kind (12.30) by replacing the surface integrals $\mathcal{K}\varphi_k(\mathbf{x}_j)$ by integrals over the tangent planes at $\mathbf{x}_k$. In view of the jump relations (12.27) for the double layer potential, we make the substitution

$$\mathcal{K}\varphi_k(\mathbf{x}_j) \approx \int_{\Gamma_k} \varphi_k(\mathbf{y})\, \partial_n(\mathbf{y})\gamma(\mathbf{x}_j - \mathbf{y})\, d\sigma_{\mathbf{y}} - \frac{1}{2}\,\mathrm{sgn}(\langle \mathbf{n}(\mathbf{x}_k), \mathbf{x}_j - \mathbf{x}_k \rangle)\, \varphi_k(\mathbf{x}_j)\,, \tag{12.49}$$

where

$$\mathrm{sgn}(t) = \begin{cases} t/|t|\,, & t \neq 0\,, \\ 0\,, & t = 0\,, \end{cases}$$

which can be obtained by the following plausible argument.

Let $v^-(\mathbf{x}_j)$ and $v^+(\mathbf{x}_j)$ be the interior and exterior limit values of

$$v(\mathbf{x}_j) = \int_\Gamma \varphi_k(\mathbf{y})\, \partial_n(\mathbf{y})\gamma(\mathbf{x} - \mathbf{y})\, d\sigma_{\mathbf{y}}$$

as $\mathbf{x} \to \mathbf{x}_j$. By $\mathbf{x}_j^-$ and $\mathbf{x}_j^+$ we denote two points inside and outside Ω close to $\mathbf{x}_j$, placed on the normal $\mathbf{n}(\mathbf{x}_j)$. Then in the case $\langle \mathbf{n}(\mathbf{x}_k), \mathbf{x}_j - \mathbf{x}_k \rangle < 0$, we have

$$v^-(\mathbf{x}_j) \approx v(\mathbf{x}_j^-) \approx \int_{\Gamma_k} \varphi_k(\mathbf{y})\, \partial_n(\mathbf{y})\gamma(\mathbf{x}_j - \mathbf{y})\, d\sigma_{\mathbf{y}}$$

which along with the identity (12.27)

$$v^-(\mathbf{x}_j) = \mathcal{K}\varphi_k(\mathbf{x}_j) - \frac{1}{2}\varphi_k(\mathbf{x}_j),$$

leads to (12.49). If $\langle \mathbf{n}(\mathbf{x}_k), \mathbf{x}_j - \mathbf{x}_k \rangle > 0$, then

$$v^+(\mathbf{x}_j) \approx v(\mathbf{x}_j^+) \approx \int_{\Gamma_k} \varphi_k(\mathbf{y})\, \partial_n(\mathbf{y})\gamma(\mathbf{x}_j - \mathbf{y})\, d\sigma_{\mathbf{y}}$$

and by the relation

$$v^+(\mathbf{x}_j) = \mathcal{K}\varphi_k(\mathbf{x}_j) + \frac{1}{2}\varphi_k(\mathbf{x}_j)$$

we arrive again at (12.49). In the third case when $\langle \mathbf{n}(\mathbf{x}_k), \mathbf{x}_j - \mathbf{x}_k \rangle = 0$, we have

$$\begin{aligned} v^-(\mathbf{x}_j) &\approx \lim_{h\to 0+} \int\limits_{\Gamma_k} \varphi_k(\mathbf{y})\, \partial_n(\mathbf{y})\gamma(\mathbf{x}_j - h\mathbf{n}(\mathbf{x}_j) - \mathbf{y})\, d\sigma_{\mathbf{y}} \\ &= \int\limits_{\Gamma_k} \varphi_k(\mathbf{y})\, \partial_n(\mathbf{y})\gamma(\mathbf{x}_j - \mathbf{y})\, d\sigma_{\mathbf{y}} - \frac{1}{2}\varphi_k(x_j)\,. \end{aligned}$$

Since

$$v^-(\mathbf{x}_j) = \mathcal{K}\varphi_k(\mathbf{x}_j) - \frac{1}{2}\varphi_k(\mathbf{x}_j)\,,$$

we obtain relation (12.49) from

$$\mathcal{K}\varphi_k(\mathbf{x}_j) \approx \int\limits_{\Gamma_k} \varphi_k(\mathbf{y})\, \partial_n(\mathbf{y})\gamma(\mathbf{x}_j - \mathbf{y})\, d\sigma_{\mathbf{y}}\,.$$

The same argument applies to the approximation for the integral

$$\begin{aligned} &\int\limits_{\Gamma} \varphi_k(\mathbf{y})\, \partial_n(\mathbf{y})\gamma(\mathbf{x} - \mathbf{y})\, d\sigma_{\mathbf{y}} \\ &\quad\approx \int\limits_{\Gamma_k} \varphi_k(\mathbf{y})\, \partial_n(\mathbf{y})\gamma(\mathbf{x} - \mathbf{y})\, d\sigma_{\mathbf{y}} - \frac{1}{2}\big(1 + \operatorname{sgn}(\langle \mathbf{n}(\mathbf{x}_k), \mathbf{x} - \mathbf{x}_k \rangle)\big)\varphi_k(\mathbf{x}) \end{aligned} \tag{12.50}$$

if $\mathbf{x} \in \Omega$, which is used to determine the approximate solution

$$u_N(\mathbf{x}) = \int\limits_{\Gamma} \phi_N(\mathbf{y})\, \partial_n(\mathbf{y})\gamma(\mathbf{x} - \mathbf{y})\, d\sigma_{\mathbf{y}}$$

of the Dirichlet problem (12.29) in Ω if the solution ϕ_N of the discrete system (12.48) is found.

We note that in view of (12.43)

$$\int\limits_{\Gamma_k} \varphi_k(\mathbf{y})\, \partial_n(\mathbf{y})\gamma(\mathbf{x} - \mathbf{y})\, d\sigma_{\mathbf{y}} = - \int\limits_{\mathbb{R}^{n-1}} \varphi_k(\mathbf{z}')\, g_2(|\mathbf{z}' - \mathbf{y}'|, z_n)\, d\mathbf{y}'$$

with

$$g_2(t_1, t_2) = \frac{\partial g(t_1, t_2)}{\partial t_2}\,.$$

Hence, using (12.49) and (12.46), the BPM approximation of the elements

$$\Big(\big(\frac{1}{2} - \mathcal{K}\big)\varphi_k(\mathbf{x}_j)\Big)_{j,k=1}^{N}$$

of the linear system of the collocation method (12.48) is given by

(12.51)
$$\begin{aligned}\widehat{\mathcal{K}}_{jk} &= \frac{1+\operatorname{sgn}(\langle \mathbf{n}(\mathbf{x}_k), \mathbf{x}_j - \mathbf{x}_k\rangle)}{2}\,\varphi_k(\mathbf{x}_j) - \int\limits_{\Gamma_k} \varphi_k(\mathbf{y})\,\partial_n(\mathbf{y})\gamma(\mathbf{x}_j - \mathbf{y})\,d\sigma_{\mathbf{y}} \\ &= \frac{1+\operatorname{sgn}(z_{jk})}{2}\,\eta\Big(\frac{|\mathbf{x}_j - \mathbf{x}_k|}{\sqrt{\mathcal{D}}h_k}\Big) + W(Z_{jk}, z_{jk})\end{aligned}$$

with Z_{jk}, z_{jk} defined by (12.47) and for $n \geq 3$ the function $W(Z,z)$ is given by the one-dimensional integral

$$W(Z,z) = \frac{2\pi(\sqrt{\mathcal{D}}h_k)^{n-1}}{Z^{(n-3)/2}} \int\limits_0^\infty \mathcal{F}g_2(r,z)\,\mathcal{F}\eta(\sqrt{\mathcal{D}}h_k r)\,J_{(n-3)/2}(2\pi r Z)\,r^{(n-1)/2}\,dr\,.$$

In the case $n = 2$, we have

$$W(Z,z) = \sqrt{\mathcal{D}}h_k \int\limits_{-\infty}^{\infty} \eta(t)\,g_2((Z - \sqrt{\mathcal{D}}h_k t), z)\,dt\,.$$

For example, let η be the Gaussian and let $L = -\Delta$. Then

$$g_2(r,z) = \frac{\Gamma(\frac{n}{2})}{2\pi^{n/2}}\frac{z}{(r^2+z^2)^{n/2}}\,.$$

In the case $n = 2$, we get from (5.12)

(12.52)
$$W(Z_{jk}, z_{jk}) = \frac{1}{2}K\Big(\frac{Z_{jk}}{\sqrt{\mathcal{D}}h_k}, \frac{z_{jk}}{\sqrt{\mathcal{D}}h_k}\Big)$$

with the Voigt function $K(x,y)$. For $n = 3$ we obtain

(12.53)
$$W(Z_{jk}, z_{jk}) = \operatorname{sgn}(z_{jk}) \int\limits_0^\infty e^{-t^2 - 2|z_{jk}|t/(\sqrt{\mathcal{D}}h_k)}\,J_0\Big(\frac{2Z_{jk}t}{\sqrt{\mathcal{D}}h_k}\Big)\,t\,dt\,,$$

where J_0 is the Bessel function of the first kind and of order zero.

Finally, by the same arguments, the values of $\mathcal{K}'\psi_k(\mathbf{x}_j)$, which appear in the collocation method for solving the integral equation (12.35), can be approximated by

(12.54)
$$\mathcal{K}'\psi_k(\mathbf{x}_j) \approx \int\limits_{\Gamma_k} \psi_k(\mathbf{y})\,\partial_n(\mathbf{x})\gamma(\mathbf{x}_j - \mathbf{y})\,d\sigma_{\mathbf{y}} + \frac{1}{2}\operatorname{sgn}(\langle \mathbf{n}(\mathbf{x}_k), \mathbf{x}_j - \mathbf{x}_k\rangle)\,\psi_k(\mathbf{x}_j)\,.$$

12.2.5. Numerical experiments. Here, we provide the results of some numerical tests for solving the Dirichlet problem for the Laplace equation in two- and three-dimensional cases via the boundary integral equation of the second kind (12.30).

Let $\Omega \subset \mathbb{R}^2$ be an elliptic domain with the boundary

$$\mathbf{x}(t) = \begin{cases} x_1(t) = (1+c)\cos t, \\ x_2(t) = (1+c)\sin t, \end{cases} \quad t \in [0, 2\pi]\,,$$

with $0 \le c < 1$. Easy calculations show that (12.30) is given by

$$\psi(t) + \frac{1-c^2}{2\pi} \int_0^{2\pi} \frac{\psi(\tau)}{1+c^2-2c\cos(t-\tau)}\, d\tau = -2g(t)\,.$$

Letting

$$g_1(t) = -1\,,\ g_2(t) = -\frac{1+c}{2}\cos t\,,\ g_3(t) = -\frac{1+c}{2}\sin t\,,$$

the exact solutions are

$$\psi_1(t) = 1\,,\ \psi_2(t) = \cos t\,,\ \psi_3(t) = \sin t\,,$$

respectively.

As a basis for the collocation procedure the functions

$$\varphi_k(\mathbf{x}) = \mathrm{e}^{-|\mathbf{x}-\mathbf{x}_k|^2/\mathcal{D}h_k^2}\,,\ k = 1,\dots,N\,,$$

have been chosen, where the centers $\mathbf{x}_k$ are also the collocation points, chosen through a uniform subdivision of the parameter interval $\mathbf{x}_k = \mathbf{x}(t_k)$ with

$$t_k = \frac{2\pi(k-1)}{N}\,,\ k = 1,\dots,N\,.$$

The parameter $\mathcal{D} = 2$ and for each k, h_k is given as an approximation of the distance between neighboring points:

$$h_k = \frac{2\pi}{N}\sqrt{1+c^2-2c\cos\frac{2\pi(k-1)}{N}}\,,\ k = 1,\dots,N\,.$$

Table 12.3 provides the maximum errors between the exact solution ψ_i and the solution of the BPM equation

$$\sum_{k=1}^{N} c_k\varphi_k\,,$$

where $\{c_k\}$ is the solution of the linear system

$$\sum_{k=1}^{N} \widehat{\mathcal{K}}_{jk}c_k = -g_i(t_j)\,,\ j = 1,\dots,N\,,$$

with the coefficients $\widehat{\mathcal{K}}_{jk}$ given by (12.51) and (12.52).

	$c=0$			$c=1/3$		
N	g_1	g_2	g_3	g_1	g_2	g_3
16	0.1493	0.3012	0.3012	0.4761	0.5543	0.1624
32	0.0411	0.0923	0.0923	0.1738	0.2177	0.0812
64	0.0107	0.0247	0.0247	0.0443	0.0581	0.0229
128	0.0027	0.0063	0.0063	0.0109	0.0145	0.0059
256	0.0007	0.0016	0.0016	0.0027	0.0036	0.0015

TABLE 12.3. The maximum errors between the exact and the BPM solutions to (12.30)

In Table 12.4, we give the maximum errors in Ω for the BPM solution of the interior Dirichlet problem with the exact solution $u = x_1^2 - x_2^2$. This approximate solution is obtained, using the formula (12.50), in the form

$$u_N(\mathbf{x}) \approx \frac{1}{2}\sum_{k=1}^{N} c_k K\Big(\frac{Z_k}{\sqrt{\mathcal{D}}h_k}, \frac{z_k}{\sqrt{\mathcal{D}}h_k}\Big), \qquad \mathbf{x} \in \Omega,$$

where $z_k = \langle \mathbf{n}(\mathbf{x}_k), \mathbf{x} - \mathbf{x}_k \rangle$, $Z_k = \sqrt{|\mathbf{x} - \mathbf{x}_k|^2 - z_k^2}$, and $c_1, \dots, c_N$ solve the corresponding collocation equations.

N	$c = 0$	$c = 1/3$
16	$1.45 \cdot 10^{-3}$	$1.25 \cdot 10^{-2}$
32	$8.83 \cdot 10^{-5}$	$6.72 \cdot 10^{-4}$
64	$3.61 \cdot 10^{-6}$	$4.01 \cdot 10^{-5}$
128	$2.85 \cdot 10^{-7}$	$2.57 \cdot 10^{-6}$
256	$5.33 \cdot 10^{-8}$	$2.71 \cdot 10^{-7}$

TABLE 12.4. The maximum error for the interior Dirichlet problem with eccentricity c using N collocation points

In the present form, the BPM needs only the coordinates of boundary points and normal vectors associated with these points, which makes this method attractive for higher-dimensional problems. As a model problem, we consider ellipsoidal domains Ω in $\mathbb{R}^3$ with the boundary

$$\Gamma = \Big\{ \mathbf{x} \in \mathbb{R}^3 : \frac{x_1^2}{a} + \frac{x_2^2}{b} + \frac{x_3^2}{c} = 1 \Big\},$$

$a, b, c > 0$. The $N = 6M^2$ collocation points are obtained by projecting the midpoints of the M^2 rectangles on each side of the box with side lengths $(2a, 2b, 2c)$ onto Γ. The error from the exact solution

$$u(\mathbf{x}) = x_1^2 + x_2^2 - 2x_3^2$$

is calculated inside Ω. The parameters h_k are determined as the average distance to the neighboring points and $\mathcal{D} = 2$. In Table 12.5, numerical results are shown for the cases $a = b = c = 1$ and $a = 4/3$, $b = 2/3$, $c = 1$.

REMARK 12.3. If $\eta(\mathbf{x}) = \mathrm{e}^{-|\mathbf{x}|^2}$, then by using formula (5.15), the integrals over the tangent planes can be transformed to the one-dimensional integrals

$$\int\limits_{\mathbb{R}^{n-1}} g(|\mathbf{z}' - \mathbf{y}'|, z_n)\, \mathrm{e}^{-|\mathbf{y}'|^2/(\mathcal{D}h_k^2)}\, d\mathbf{y}'$$

$$= \mathrm{e}^{-|\mathbf{z}'|^2/(\mathcal{D}h_k^2)} \frac{2(\sqrt{\pi \mathcal{D}} h_k^2)^{n-1}}{|\mathbf{z}'|^{(n-3)/2}} \int\limits_0^\infty g(\mathcal{D}h_k^2 r, z_n)\, \mathrm{e}^{-\mathcal{D}h_k^2 r^2}\, I_{(n-3)/2}(2|\mathbf{z}'| r)\, r^{(n-1)/2}\, dr\,.$$

$a=b=c=1$			$a=4/3,\ b=2/3,\ c=1$		
M	N	Error	M	N	Error
3	54	$2.78\cdot 10^{-2}$	3	54	$7.43\cdot 10^{-2}$
5	150	$3.35\cdot 10^{-3}$	5	150	$1.83\cdot 10^{-2}$
7	294	$8.47\cdot 10^{-4}$	7	294	$1.03\cdot 10^{-2}$
9	486	$3.19\cdot 10^{-4}$	9	486	$9.37\cdot 10^{-3}$
11	726	$1.94\cdot 10^{-4}$	11	726	$8.18\cdot 10^{-3}$
13	1014	$9.73\cdot 10^{-5}$	13	1014	$7.08\cdot 10^{-3}$
15	1350	$1.12\cdot 10^{-4}$	15	1350	$5.14\cdot 10^{-3}$
17	1734	$4.51\cdot 10^{-5}$	17	1734	$4.31\cdot 10^{-3}$

TABLE 12.5. The maximum error for the interior Dirichlet problem inside the ellipsoid using $N = 6M^2$ collocation points

12.2.6. Stability analysis. We restrict ourselves to some partial results obtained for the integral equations (12.31), (12.33), and (12.30) in the case of a simply connected smooth boundary Γ. Even for this relatively simple situation the stability analysis of the BPM offers many unsolved problems. This is caused by the fact that the standard discretization method is collocation, which is theoretically less understood compared with the Galerkin method, especially if $n \geq 3$. Moreover, because of the choice of the approximating functions, we cannot expect the strong convergence of projection operators, required in the standard theory of projection methods.

The other peculiarity of BPM is the approximation of the integral operators. Whereas in the BEM the collocation or cubature points lie at the integration surface, either the boundary or some suitable approximation, in the BPM, the integration domain is approximated by a set of tangential planes supported at the collocation points. This leads to the additional terms in the formulas (12.49) and (12.54) for the approximation of $\mathcal{K}$ and $\mathcal{K}'$.

12.2.6.1. *Integral equations of the second kind.* Let us consider the equation

$$\frac{1}{2}u(\mathbf{x}) - \mathcal{K}u(\mathbf{x}) = f(\mathbf{x}), \quad \mathbf{x} \in \Gamma, \tag{12.55}$$

with the double layer potential $\mathcal{K}$ defined by (12.22),

$$\mathcal{K}u(\mathbf{x}) = \int_\Gamma \gamma'(|\mathbf{x}-\mathbf{y}|)\,\frac{\langle \mathbf{n}(\mathbf{y}), \mathbf{y}-\mathbf{x}\rangle}{|\mathbf{x}-\mathbf{y}|}\,u(\mathbf{y})\,d\mathbf{y}.$$

If Γ is a closed smooth surface, then this integral operator has a weakly singular kernel and therefore, it is compact in the space $C(\Gamma)$ of continuous functions on Γ.

It is well known that the solutions $u_N \in X_N$ of the collocation equations

$$\frac{1}{2}u_N(\mathbf{x}_j) - \mathcal{K}u_N(\mathbf{x}_j) = f(\mathbf{x}_j), \quad j = 1, \ldots, N, \tag{12.56}$$

exist for all sufficiently large N and converge to a solution of (12.55), if this equation is uniquely solvable and the N-dimensional subspaces $X_N \subset C(\Gamma)$ have the property that the interpolation projections

$$Q_N f(\mathbf{x}_j) = f(\mathbf{x}_j), \quad j = 1, \ldots, N,$$

with $Q_N f \in X_N$ exist and $\|(I - Q_N)\mathcal{K}\|_{C(\Gamma)} \to 0$ as $N \to \infty$.

As mentioned above, projection operators onto the linear span of basis functions of the BPM do not converge strongly. Instead, we make the following assumption on the function η and on the parameters $\{\mathbf{x}_k\}_1^N$, $\{h_k\}_1^N$, and $\mathcal{D}$, which define the finite-dimensional spaces

$$S_N(\eta, \mathcal{D}) = \Big\{\eta\Big(\frac{|\mathbf{x} - \mathbf{x}_k|}{\sqrt{\mathcal{D}}h_k}\Big), k = 1, \dots, N\Big\}. \tag{12.57}$$

CONDITION 12.4. For a given $\varepsilon > 0$, the parameters are chosen so that the interpolation operator $Q_N : C(\Gamma) \to \boldsymbol{S_N}(\eta, \mathcal{D})$ with $Q_N f(\mathbf{x}_j) = f(\mathbf{x}_j)$, $j = 1, \dots, N$, exists and

$$\lim_{N\to\infty} \|(I - Q_N)\mathcal{K}u\|_{C(\Gamma)} < \varepsilon\|u\|_{C(\Gamma)} \tag{12.58}$$

for any $u \in C(\Gamma)$.

For a given parametrization of Γ, one can apply Theorems 10.5 or 10.13 to choose sequences of nodes $\{\mathbf{x}_k\}_{k=1}^N$ and scaling parameters $\{h_k\}_{k=1}^N$ such that the quasi-interpolants

$$u_N(\mathbf{x}) = \mathcal{D}^{(1-n)/2} \sum_{k=1}^{N} u(\mathbf{x}_k)\, \eta\Big(\frac{|\mathbf{x} - \mathbf{x}_k|}{\sqrt{\mathcal{D}}h_k}\Big)$$

approximate Hölder continuous functions $u \in C^\alpha(\Gamma)$ with

$$|u(\mathbf{x}) - u_N(\mathbf{x})| \le c\, N^{-\alpha} + \varepsilon|u(\mathbf{x})|$$

and the saturation error ε is determined by $\mathcal{D}$. Since the operator $\mathcal{K}$ maps $C(\Gamma)$ into $C^\alpha(\Gamma)$, we see that $\mathcal{D}$ can be chosen so that (12.58) holds.

THEOREM 12.5. *Suppose that the operator* $\frac{1}{2}I - \mathcal{K}$ *is invertible in* $C(\Gamma)$ *and that the spaces* $S_N(\eta, \mathcal{D})$ *satisfy Condition* 12.4 *with* $\varepsilon < 1/\|(\frac{1}{2}I - \mathcal{K})^{-1}\|$. *Then the collocation equations* (12.56) *have unique solutions* $u_N \in S_N(\eta, \mathcal{D})$ *for sufficiently large* N, *which approach the solution* u *of* (12.55) *with*

$$\|u - u_N\|_{C(\Gamma)} \le c\,\|(I - Q_N)u\|_{C(\Gamma)}, \tag{12.59}$$

where the constant c *does not depend on* f.

PROOF. We write the collocation equations in the form

$$\Big(\frac{1}{2}I - Q_N\mathcal{K}\Big)u_N = Q_N f.$$

Because

$$\frac{1}{2}I - Q_N\mathcal{K} = \Big(\frac{1}{2}I - \mathcal{K}\Big)\Big(I - \Big(\frac{1}{2}I - \mathcal{K}\Big)^{-1}(I - Q_N)\mathcal{K}\Big),$$

the collocation equations are uniquely solvable if N is subject to

$$\Big\|\Big(\frac{1}{2}I - \mathcal{K}\Big)^{-1}(I - Q_N)\mathcal{K}\Big\|_{C(\Gamma)} < 1.$$

Since the solution of (12.55) satisfies

$$\frac{1}{2}u - Q_N\mathcal{K}u = \frac{1}{2}(I - Q_N)u + Q_N f,$$

we obtain

$$\Big(\frac{1}{2}I - Q_N\mathcal{K}\Big)(u - u_N) = \frac{1}{2}(I - Q_N)u,$$

and the estimate (12.59) follows. □

REMARK 12.6. A similar result is also valid for integral equations of the second kind with the adjoint of the double layer potential $\mathcal{K}'$.

The error estimate (12.59) indicates that similarly to Theorem 12.2 the collocation method can provide spectral convergence up to some saturation error if the spaces $S_N(\eta, \mathcal{D})$ are chosen properly. However, the interpolation with functions from $S_N(\eta, \mathcal{D})$ with parameters h_k, proportional to the average distance from $\mathbf{x}_k$ to the neighboring nodes of $\{\mathbf{x}_k\}_{k=1}^N$, has not been considered yet. Therefore, the error estimations of the BPM is still an open problem.

This problem becomes more difficult because of the approximation of the integrals. If the coefficients

$$\Big((\frac{1}{2} - \mathcal{K})\varphi_k(\mathbf{x}_j) \Big)_{j,k=1}^N$$

in the collocation matrix are replaced by $\widehat{\mathcal{K}}_{jk}$ given in (12.51), then we make an additional error, which has no influence on the solvability of the collocation equations, but on the approximation error of the BPM. Up to now only pessimistic estimates of the form $\mathcal{O}(h)$ can be proved, which do not explain the $\mathcal{O}(h^2)$ rate on the boundary and the order $\mathcal{O}(h^3)$ in the interior, which is obtained in the numerical experiments.

12.2.6.2. *Integral equations of the first kind.* For the stability analysis of the BPM for solving the equation

$$\mathcal{V}u(\mathbf{x}) = f(\mathbf{x}), \quad \mathbf{x} \in \Gamma, \tag{12.60}$$

we use the well-known representation

$$\mathcal{V} = \mathcal{V}_0 + K$$

with a bounded, symmetric operator $\mathcal{V}_0 : W_2^{-1/2}(\Gamma) \to W_2^{1/2}(\Gamma)$, which is positive definite, i.e.,

$$(\mathcal{V}_0 u, u)_{L_2(\Gamma)} \geq c_0 \|u\|^2_{W_2^{-1/2}(\Gamma)}, \tag{12.61}$$

and a compact operator $K : W_2^{-1/2}(\Gamma) \to W_2^{1/2}(\Gamma)$. The definition of the Sobolev spaces $W_2^{\pm 1/2}(\Gamma)$ has been given in Subsection 4.5.1.

It is a classical result that by (12.61) the Galerkin approximation for equation (12.60) with invertible $\mathcal{V}$ converges for any system of approximating functions X_N with the property

$$\lim_{N\to\infty} \operatorname{dist}\big(u, X_N; W_2^{-1/2}(\Gamma)\big) = 0 \quad \text{for any } u \in W_2^{-1/2}(\Gamma)$$

(cf., e.g., [**80**]). Let us show that the Galerkin equation

$$(\mathcal{V}u_N, v_N)_{L_2(\Gamma)} = (f, v_N)_{L_2(\Gamma)}, \quad \forall\, v_N \in S_N(\eta, \mathcal{D}), \tag{12.62}$$

has a unique solution for all sufficiently large N and that the approximate solutions provide quasi-optimal errors, if the following condition is satisfied:

CONDITION 12.7. For a given $\varepsilon > 0$, the parameters are chosen so that for any $u \in W_2^{-1/2}(\Gamma)$, the best approximation by elements of $S_N(\eta, \mathcal{D})$ satisfies

$$\lim_{N\to\infty} \operatorname{dist}\big(u, S_N(\eta, \mathcal{D}); W_2^{-1/2}(\Gamma)\big) < \varepsilon \|u\|_{W_2^{-1/2}(\Gamma)}.$$

LEMMA 12.8. *Suppose that the single layer potential $\mathcal{V} : W_2^{-1/2}(\Gamma) \to W_2^{1/2}(\Gamma)$ is invertible and suppose Condition 12.7 holds. There exist parameters for the spaces $S_N(\eta, \mathcal{D})$ such that for any sufficiently large N, the Galerkin equations (12.62) are uniquely solvable and the difference of the solutions $u_N \in S_N(\eta, \mathcal{D})$ and $u = \mathcal{V}^{-1}f$ is quasi-optimal, i.e.,*

$$\|u - u_N\|_{W_2^{-1/2}(\Gamma)} \le c \operatorname{dist}\left(u, S_N(\eta, \mathcal{D}); W_2^{-1/2}(\Gamma)\right).$$

PROOF. First, we show that for suitable parameters and N large enough

$$\sup_{v_N \in S_N(\eta,\mathcal{D})} \frac{\left|(\mathcal{V}u_N, v_N)_{L_2(\Gamma)}\right|}{\|v_N\|_{W_2^{-1/2}(\Gamma)}} \ge c_1 \|u_N\|_{W_2^{-1/2}(\Gamma)} \tag{12.63}$$

for any $u_N \in S_N(\eta, \mathcal{D})$. Indeed, suppose the existence of a sequence $u_N \in S_N(\eta, \mathcal{D})$, $\|u_N\|_{W_2^{-1/2}(\Gamma)} = 1$, with

$$\sup_{v_N \in S_N(\eta,\mathcal{D})} \frac{(\mathcal{V}u_N, v_N)_{L_2(\Gamma)}}{\|v_N\|_{W_2^{-1/2}(\Gamma)}} \to 0\,. \tag{12.64}$$

We choose a subsequence, again denoted by $\{u_N\}$, weakly converging in $W_2^{-1/2}(\Gamma)$ to some element u. There exists $v \in W_2^{-1/2}(\Gamma)$ with $\|v\|_{W_2^{-1/2}(\Gamma)} = 1$ such that

$$\|\mathcal{V}u\|_{W_2^{1/2}(\Gamma)} = (\mathcal{V}u, v)_{L_2(\Gamma)}\,,$$

and we choose a sequence $v_N \in S_N(\eta, \mathcal{D})$ with $\lim_{N\to\infty} \|v - v_N\|_{W_2^{-1/2}(\Gamma)} < \varepsilon$. Because

$$(\mathcal{V}u_N, v)_{L_2(\Gamma)} = (\mathcal{V}u_N, v_N)_{L_2(\Gamma)} + (\mathcal{V}u_N, v - v_N)_{L_2(\Gamma)}$$

and

$$(\mathcal{V}u_N, v)_{L_2(\Gamma)} \to (\mathcal{V}u, v)_{L_2(\Gamma)},$$

we derive $\|u\|_{W_2^{-1/2}(\Gamma)} \le \varepsilon c_{\mathcal{V}}$, where we denote

$$c_{\mathcal{V}} = \|\mathcal{V}\|_{W_2^{-1/2}(\Gamma)\to W_2^{1/2}(\Gamma)} \|\mathcal{V}^{-1}\|_{W_2^{1/2}(\Gamma)\to W_2^{-1/2}(\Gamma)}\,.$$

On the other hand,

$$(\mathcal{V}_0 u_N, u_N)_{L_2(\Gamma)} = (\mathcal{V}u_N, u_N)_{L_2(\Gamma)} - (Ku_N, u_N)_{L_2(\Gamma)}$$

and, because of K being a compact operator,

$$(Ku_N, u_N)_{L_2(\Gamma)} \to (Ku, u)_{L_2(\Gamma)}\,.$$

This implies, together with (12.64),

$$|(\mathcal{V}_0 u_N, u_N)_{L_2(\Gamma)}| \le 2\|K\|_{W_2^{-1/2}(\Gamma)\to W_2^{1/2}(\Gamma)} \|u\|^2_{W_2^{-1/2}(\Gamma)}$$

for all N large enough. Hence, if ε is chosen such that

$$2\varepsilon^2 \|K\|_{W_2^{-1/2}(\Gamma)\to W_2^{1/2}(\Gamma)} c_{\mathcal{V}}^2 < c_0,$$

we obtain the required contradiction to the coerciveness (12.61) of $\mathcal{V}_0$.

Thus, for suitably chosen parameters,

$$\left|(\mathcal{V}(u_N - w_N), v_N)_{L_2(\Gamma)}\right| \ge c_1 \|u_N - w_N\|_{W_2^{-1/2}(\Gamma)} \|v_N\|_{W_2^{-1/2}(\Gamma)}$$

for all $u_N, w_N, v_N \in S_N(\eta, \mathcal{D})$ and N large enough. Since the solutions u of (12.60) and $u_N \in S_N(\eta, \mathcal{D})$ of (12.62) satisfy

$$(\mathcal{V}(u_N - w_N), v_N)_{L_2(\Gamma)} = (\mathcal{V}(u - w_N), v_N)_{L_2(\Gamma)},$$

we obtain

$$c_1\|u_N - w_N\|_{W_2^{-1/2}(\Gamma)} \le \|\mathcal{V}\|_{W_2^{-1/2}(\Gamma)\to W_2^{1/2}(\Gamma)} \|u - w_N\|_{W_2^{-1/2}(\Gamma)},$$

which together with the triangle inequality

$$\|u - u_N\|_{W_2^{-1/2}(\Gamma)} \le \|u - w_N\|_{W_2^{-1/2}(\Gamma)} + \|u_N - w_N\|_{W_2^{-1/2}(\Gamma)}$$

establishes the error estimate. □

A similar result for the collocation method is not known in the general case. However, in some special situations, the collocation equations can be analyzed by using Lemma 12.8.

Let us consider the following problem on the unit circle in $\mathbb{R}^2$. We use the standard parametrization $\Gamma = \{\mathbf{x}(t) = (\cos 2\pi t, \sin 2\pi t)\}$, $0 \le t \le 1$, and the approximating functions

$$\varphi_k(\mathbf{x}(t)) = \sum_{m\in\mathbb{Z}} \mathrm{e}^{-(t-t_k-m)^2/\mathcal{D}h^2}, \quad k = 1, \dots, N,$$

where $t_k = (k-1)/N$ and $h = N^{-1}$. The approximation properties of the set

$$S_N(\mathcal{D}) = \Big\{\varphi_k(\mathbf{x}), k = 1, \dots, N\Big\}$$

coincide obviously with those of the Gaussians on a uniform grid in $\mathbb{R}$.

The single layer potential of these functions takes the form

$$\begin{aligned}\mathcal{V}\varphi_k(\mathbf{x}(t)) &= 2\pi \int_0^1 \gamma\big(|\mathbf{x}(t) - \mathbf{x}(s)|\big) \sum_{m\in\mathbb{Z}} \mathrm{e}^{-(s-t_k-m)^2/\mathcal{D}h^2}\, ds \\ &= 2\pi \int_{-\infty}^{\infty} \gamma\big(4\sin^2 \pi(t-s)\big)\, \mathrm{e}^{-(s-t_k)^2/\mathcal{D}h^2}\, ds.\end{aligned}$$

The collocation equations at the nodes $\mathbf{x}_j = \mathbf{x}(t_j)$, i.e.,

$$\mathcal{V}\phi_N(\mathbf{x}_j) = f(\mathbf{x}_j), \quad j = 1, \dots, N, \tag{12.65}$$

lead to a system of linear equations with the matrix

$$\Big(\mathcal{V}\varphi_k(\mathbf{x}(t_j))\Big)_{j,k=1}^{N}$$

with the elements

$$\mathcal{V}\varphi_k(\mathbf{x}(t_j)) = 2\pi \int_{-\infty}^{\infty} \gamma\big(4\sin^2 \pi s\big)\, \mathrm{e}^{-(s+t_j-t_k)^2/\mathcal{D}h^2}\, ds.$$

Since

$$\mathrm{e}^{-(s+t_j-t_k)^2/\mathcal{D}h^2} = \frac{2}{\sqrt{\pi\mathcal{D}}h} \int_{-\infty}^{\infty} \mathrm{e}^{-2(\tau-s-t_j)^2/\mathcal{D}h^2}\, \mathrm{e}^{-2(\tau-t_k)^2/\mathcal{D}h^2}\, d\tau,$$

we can write

$$\begin{aligned}\mathcal{V}\varphi_k(\mathbf{x}_j) &= \frac{4\sqrt{\pi}}{\sqrt{\mathcal{D}}h} \iint\limits_{-\infty}^{\infty} \gamma\big(4\sin^2 \pi s\big)\, \mathrm{e}^{-2(\tau-s-t_j)^2/\mathcal{D}h^2}\, \mathrm{e}^{-2(\tau-t_k)^2/\mathcal{D}h^2}\, d\tau\, ds \\ &= \frac{4\sqrt{\pi}}{\sqrt{\mathcal{D}}h} \int\limits_{-\infty}^{\infty} \mathrm{e}^{-2(\tau-t_k)^2/\mathcal{D}h^2}\, d\tau \int\limits_{-\infty}^{\infty} \gamma\big(4\sin^2 \pi(\tau-s)\big)\, \mathrm{e}^{-2(s-t_j)^2/\mathcal{D}h^2}\, ds \\ &= \frac{1}{\pi^{3/2}\sqrt{\mathcal{D}}h} \int\limits_{\Gamma} \widetilde{\varphi}_j(\mathbf{x}) \int\limits_{\Gamma} \gamma\big(|\mathbf{x}-\mathbf{y}|\big)\, \widetilde{\varphi}_k(\mathbf{y})\, d\sigma_{\mathbf{y}}\, d\sigma_{\mathbf{x}}\end{aligned}$$

with the functions

$$\widetilde{\varphi}_k(\mathbf{x}(t)) = \sum_{m\in\mathbb{Z}} \mathrm{e}^{-2(t-t_k-m)^2/\mathcal{D}h^2}, \quad k = 1, \dots, N.$$

Hence, except for a factor, the collocation matrix coincides with the matrix of the Galerkin method with the function system $\{\widetilde{\varphi}_k\}_{k=1}^N$ for the same integral equation of the first kind on the unit circle, i.e.,

$$\text{(12.66)} \qquad \Big(\mathcal{V}\varphi_k(\mathbf{x}_j)\Big)_{j,k=1}^N = \frac{1}{\pi^{3/2}\sqrt{\mathcal{D}}h} \Big(\big(\mathcal{V}\widetilde{\varphi}_k, \widetilde{\varphi}_j\big)_{L_2(\Gamma)}\Big)_{j,k=1}^N.$$

In view of

$$\int\limits_{\Gamma} f(\mathbf{x})\, \widetilde{\varphi}_j(\mathbf{x})\, d\sigma = 2\pi \int\limits_{-\infty}^{\infty} f(\mathbf{x}(t))\, \mathrm{e}^{-2(t-t_j)^2/\mathcal{D}h^2}\, dt \approx \pi^{3/2}\sqrt{2\mathcal{D}}h\, f(\mathbf{x}_j),$$

it is not hard to see that the collocation method with the set $S_N(\mathcal{D})$ of approximating functions has similar properties as the Galerkin method with $S_N(\mathcal{D}/2)$.

To this end, we introduce discrete Sobolev norms for $\mathbf{z} = (z_1, \dots, z_n) \in \mathbb{R}^N$,

$$\|\mathbf{z}\|_{s,N} = N^{-1}\Big(\sum_{j,k,l=1}^N \Big(1 + 4N^2 \sin^2 \frac{\pi l}{N}\Big)^s \mathrm{e}^{2\pi i l(j-k)/N}\, z_k z_j\Big)^{1/2}.$$

Since the nodes $\mathbf{x}_k$ are uniformly spaced, the discrete norm $\|\mathbf{z}\|_{s,N}$ is equivalent to

$$\Big\|\sum_{k=1}^N z_k \varphi_k\Big\|_{W_2^s(\Gamma)}.$$

Moreover, the stability of the Galerkin method

$$\text{(12.67)} \qquad \sup_{v_N\in S_N(\mathcal{D}/2)} \frac{\big|(\mathcal{V}u_N, v_N)_{L_2(\Gamma)}\big|}{\|v_N\|_{W_2^{-1/2}(\Gamma)}} \ge c\,\|u_N\|_{W_2^{-1/2}(\Gamma)}, \quad u_N \in S_N(\mathcal{D}/2),$$

is equivalent to the two-sided inequality for the Galerkin matrices

$$C_1\|\mathbf{z}\|_{-1/2,N} \le \Big\|N\Big(\sum_{k=1}^N \big(\mathcal{V}\widetilde{\varphi}_k, \widetilde{\varphi}_j\big)_{L_2(\Gamma)} z_k\Big)_{j=1}^N\Big\|_{1/2,N} \le C_2\|\mathbf{z}\|_{-1/2,N}$$

with constants independent of N and $\mathbf{z}$ (see [**80**]). Now, the circulant structure of the matrices implies that the last relation holds for arbitrary $s \in \mathbb{R}$; in particular, because of (12.66)

$$C_1\|\mathbf{z}\|_{0,N} \le \Big\| \Big(\sum_{k=1}^{N} \mathcal{V}\varphi_k(\mathbf{x}_j) z_k \Big)_{j=1}^{N} \Big\|_{1,N} \le C_2 \|\mathbf{z}\|_{0,N} \,.$$

Thus, writing the collocation equations (12.65) in the form

$$Q_N \mathcal{V} \phi_N = Q_N f \,, \quad \phi_N \in S_N(\mathcal{D}) \,,$$

with suitable interpolation operators Q_N onto $S_N(\mathcal{D})$, we see that the collocation operators $Q_N \mathcal{V}|_{S_N(\mathcal{D})}$ satisfy the stability estimate

$$\|Q_N \mathcal{V} u_N\|_{W_2^{\frac{1}{2}}(\Gamma)} \ge c\, \|u_N\|_{L_2(\Gamma)} \,, \quad u_N \in S_N(\mathcal{D}) \,,$$

if and only if (12.67) holds.

Since the approximation properties of the set $S_N(\mathcal{D})$ are controlled by the parameter $\mathcal{D}$, we obtain from Lemma 12.8 the convergence of the collocation method.

LEMMA 12.9. *Suppose that the single layer potential $\mathcal{V}$ is invertible. For any $\varepsilon > 0$, there exists $\mathcal{D}$ such that collocation equations (12.65) are uniquely solvable for all sufficiently large N and the approximate solutions u_N approach $u = \mathcal{V}^{-1} f$ with*

$$\|u - u_N\|_{L_2(\Gamma)} \le c \inf_{v_N \in S_N(\mathcal{D})} \|u - v_N\|_{L_2(\Gamma)} \,.$$

Similarly to the conclusions of Theorem 12.5, the collocation method for the single layer potential provides spectral convergence up to some saturation error. However, replacing the value of $\mathcal{V}\varphi_k(\mathbf{x}_j)$ by

$$\widehat{\mathcal{V}}_{jk} = \sqrt{\mathcal{D}} h \int_{-\infty}^{\infty} g(Z_{jk} - \sqrt{\mathcal{D}} h t, z_{jk})\, \mathrm{e}^{-t^2}\, dt$$

gives an approximation of the discrete operator $Q_N \mathcal{V}|_{S_N(\mathcal{D})}$ of the order $\mathcal{O}(h^2)$, which can deteriorate high convergence orders. Therefore, it is of interest to find higher-order approximations of the discrete operators by using the curvature of the boundary, which is considered in the next section.

12.3. Computation of multi-dimensional single layer harmonic potentials

In this section, we study higher-order cubature formulas for the single layer harmonic potential as an example of surface potentials. We combine the approximate approximation of the surface layer density with the integration of the basis functions over the tangential space by the use of appropriate asymptotic expansions. Our approach leads to cubature formulas involving only nodes of a regular grid. These formulas turn out to be efficient provided the saturation error of the approximate approximation is a priori chosen sufficiently small.

Consider the computation of multi-dimensional surface potentials of the form

$$\int_{\Gamma} Q(\mathbf{x} - \mathbf{y}) f(\mathbf{y})\, d\sigma_{\mathbf{y}} \,,$$

where Γ is a sufficiently smooth manifold in $\mathbb{R}^n$. It is well known that owing to the singularity of the kernel Q at $\mathbf{x} = \mathbf{y}$, the case when $\mathbf{x}$ is located on or close to the surface Γ requires special attention. This problem is usually addressed by sophisticated methods like special variable transformations or singularity subtraction combined with high-order cubature formulas and mesh refinement near to the singularity (see [**5**], [**31**], [**50**], [**84**], [**87**], and the references therein).

Here we propose cubature formulas for these singular or nearly singular integrals which use only the density values at the nodes of a regular grid and the corresponding surface parametrization. The underlying ideas are the following:

1. The density f is approximated by quasi-interpolation formulas using locally supported smooth radial functions which are centered at regularly distributed nodes on the surface. These approximations have been studied in Chapter 10, where we have shown that it is possible to construct formulas which provide arbitrarily high order approximations up to any prescribed accuracy.

2. The potentials of local basis functions over curved surfaces are approximated by a linear combination of integrals over the tangential space. This approximation is obtained from an asymptotic expansion of the potential by the use of the local parametrization of the surface at the center of the basis function. Again, arbitrarily high approximation orders can be achieved by taking into account the smoothness of the surface.

3. Since approximate approximations are very flexible concerning the choice of local basis functions, those are chosen so that the resulting integrals over the $(n-1)$-dimensional tangential space can be transformed to efficiently computable one-dimensional integrals. Thus, the proposed formulas are particularly well suited for the cubature of integral operators on high-dimensional surfaces.

Since, in principle, both the approximation of the density and the approximation of the potentials can be performed with arbitrarily high order, the proposed cubature formulas can provide very accurate approximations even for moderate grid sizes.

Let us consider this approach for the example of the single layer harmonic potential

$$\mathcal{V}f(\mathbf{x}) = \frac{\Gamma(\frac{n}{2}-1)}{4\pi^{n/2}} \int\limits_{\Gamma} \frac{f(\mathbf{y})}{|\mathbf{x}-\mathbf{y}|^{n-2}}\, d\sigma_{\mathbf{y}} = \omega_n \int\limits_{\Gamma} \frac{f(\mathbf{y})}{|\mathbf{x}-\mathbf{y}|^{n-2}}\, d\sigma_{\mathbf{y}} \,. \tag{12.68}$$

We derive a cubature formula which uses only the values of the normal and of the curvature of Γ at the nodes $\{\mathbf{y}_m\}$ of a regular grid. It is proved that this formula approximates the single layer potential uniformly with the order $\mathcal{O}(h^3|\log h|)$, where h denotes the grid size. It will be clear from the constructions given below how this approach can be applied to other types of potential operators and how higher-order formulas can be obtained by incorporating more smoothness data of Γ.

We may assume after applying a partition of the unity that the function f has compact support on Γ parametrized by $x_n = \varphi(\mathbf{x}')$, $\mathbf{x}' = (x_1, \dots, x_{n-1})$, with a sufficiently smooth function given on a bounded domain $\varphi : \gamma \subset \mathbb{R}^{n-1} \to \mathbb{R}$. Then (12.68) becomes

$$\mathcal{V}f(\mathbf{x}) = \omega_n \int\limits_{\gamma} \frac{f(\mathbf{y}', \varphi(\mathbf{y}))}{(|\mathbf{x}'-\mathbf{y}'|^2 + (x_n - \varphi(\mathbf{y}'))^2)^{n/2-1}} (1 + |\nabla\varphi(\mathbf{y}')|^2)^{1/2}\, d\mathbf{y}'$$

with $\mathbf{x} = (\mathbf{x}', x_n)$. To approximate the value of $\mathcal{V}f(\mathbf{x})$, we introduce a uniform grid $\{h\mathbf{m}' \in \gamma : \mathbf{m}' \in \mathbb{Z}^{n-1}\}$ and consider the cubature by the midpoint rule

$$\mathcal{V}_h f(\mathbf{x}) = \omega_n h^{n-1} \sum_{h\mathbf{m}' \in \gamma} \frac{f(h\mathbf{m}', \varphi(h\mathbf{m}'))(1 + |\nabla\varphi(h\mathbf{m}')|^2)^{1/2}}{(|\mathbf{x}' - h\mathbf{m}'|^2 + (x_n - \varphi(h\mathbf{m}'))^2)^{n/2-1}} . \tag{12.69}$$

Due to the well-known error estimate for the cubature of smooth integrands on uniform grids

$$\left| \int_\gamma g(\mathbf{y}')\, d\mathbf{y}' - h^{n-1} \sum_{h\mathbf{m}' \in \gamma} g(h\mathbf{m}') \right| \le c_\ell\, h^\ell \int_\gamma |\nabla_\ell g(\mathbf{y}')|\, d\mathbf{y}' , \ \ell = 1, 2, \dots$$

(see [**23**]), we have

$$|\mathcal{V}f(\mathbf{x}) - \mathcal{V}_h f(\mathbf{x})| \le c_\ell\, h^\ell (\operatorname{dist}(\mathbf{x}, \Gamma))^{1-\ell} \tag{12.70}$$

if $\operatorname{dist}(\mathbf{x}, \Gamma) > 0$ for sufficiently smooth f and φ. In the case when $\operatorname{dist}(\mathbf{x}, \Gamma)$ is small, formula (12.69) has to be modified. Usually, the cubature of potentials is based on special variable transformations or high-order cubature formulas and mesh refinement in the vicinity of the point $\mathbf{x}$.

To retain the grid and the simple structure of (12.69), we use the results of Section 10.3 on quasi-interpolation on surfaces. More precisely, we choose the quasi-interpolant (10.19) with a sufficiently smooth radial function $\eta(\mathbf{x}) = \psi(|\mathbf{x}|^2/2)$,

$$f_h(\mathbf{y}) = f_h(\mathbf{y}', \phi(\mathbf{y}')) = \mathcal{D}^{-n/2} \sum_{h\mathbf{m}' \in \gamma} f(\boldsymbol{\phi}(h\mathbf{m}'))\, \eta\Big(\frac{\mathbf{y}' - \boldsymbol{\phi}(h\mathbf{m}')}{\sqrt{\mathcal{D}}h\, |\boldsymbol{\phi}'(h\mathbf{m}')|^{1/(n-1)}} \Big) , \tag{12.71}$$

where, specified to the above-considered case, the mapping ϕ is of the form $\phi(\mathbf{x}') = (\mathbf{x}', \varphi(\mathbf{x}'))$ and thus $|\phi'(\mathbf{x}')| = \sqrt{1 + |\nabla\varphi(\mathbf{x}')|^2}$.

By Theorem 10.5, we have the following approximation result:

LEMMA 12.10. *Assume that the radial function $\eta \in \mathcal{S}(\mathbb{R}^{n-1})$ satisfies the moment conditions* (2.47) *and that $\varphi \in C^{N+1}(\gamma)$. If $f \in C_0^N(\Gamma)$, then for any $\varepsilon > 0$ there exists $\mathcal{D} > 0$ such that at any point $\mathbf{y} \in \Gamma$*

$$|f_h(\mathbf{y}) - f(\mathbf{y})| \le c\, (\sqrt{\mathcal{D}}h)^N\, \|f\|_{C^N(\overline{\Gamma})} + \varepsilon \sum_{k=0}^{N-1} c_k\, (\sqrt{\mathcal{D}}h)^k , \tag{12.72}$$

where c does not depend on f, h, and $\mathcal{D}$, and the numbers c_k depend on the values $\partial^{\boldsymbol{\alpha}} f(\mathbf{y})$ for $[\boldsymbol{\alpha}] \le k$.

Let us use the quasi-interpolant (12.71) of the density to obtain a cubature of the single layer potentials (12.68),

$$\mathcal{V}_h f(\mathbf{x}) = \frac{\omega_n}{\mathcal{D}^{(n-1)/2}} \sum_{h\mathbf{m}' \in \gamma} f(\phi(h\mathbf{m}')) \int_\Gamma \eta\Big(\frac{\mathbf{y} - \phi(h\mathbf{m}')}{\sqrt{\mathcal{D}}h\, |\phi'(h\mathbf{m}')|^{1/(n-1)}} \Big) \frac{d\sigma_{\mathbf{y}}}{|\mathbf{x} - \mathbf{y}|^{n-2}} . \tag{12.73}$$

Since $\mathcal{V}$ is a bounded mapping between suitable function spaces, the differences $\mathcal{V}f(\mathbf{x}) - \mathcal{V}_h f(\mathbf{x})$ behave like estimate (12.72). To obtain efficient methods for computing the integrals appearing in the sums over Γ, which, in general, has a small but curved integration domain, we approximate these by integrals over the tangential space at the points $\mathbf{y}_{\mathbf{m}'} = \phi(h\mathbf{m}')$. We are interested in the accuracy of this approximation if in addition to the first derivatives of ϕ, i.e., the direction of the normal, second derivatives also, i.e., the curvatures of Γ, are used to determine the integrals over the tangential space.

12.3.1. Asymptotic formulas. In the following, we derive asymptotic formulas of the single layer potential acting on local basis functions

$$\mathcal{V}_h\eta(\mathbf{x}) = \omega_n \int_\Gamma |\mathbf{x}-\mathbf{y}|^{2-n}\,\eta(\frac{\mathbf{y}-\mathbf{y}_0}{h})\,d\sigma_{\mathbf{y}} \tag{12.74}$$

as $h \to 0$. Given the normal $\mathbf{n}$ to Γ at $\mathbf{y}_0$, we choose a new coordinate system such that $\mathbf{y}_0$ becomes the origin $\mathbf{0}$ and the normal coincides with $e_n = (0,\dots,0,1)$.

Multiplying with a suitable cut-off function, we assume first that η is supported in the ball $B_\delta = \{|\mathbf{x}| \le \delta\}$, $\mathbf{x} \in \mathbb{R}^n$, $\mathbf{y} = (\mathbf{y}', y_n) \in \Gamma$. Without loss of generality we can assume that in a neighborhood U of the origin $\mathbf{0}$, the boundary Γ is given by

$$y_n = \varphi(\mathbf{y}') \text{ with } \varphi(\mathbf{0}) = 0 \text{ and } \nabla\varphi(\mathbf{0}) = 0\,. \tag{12.75}$$

We choose h such that $B_{h\delta} \cap \Gamma \subset U$ and we denote $U_h = \varphi^{-1}(U)$. Then (12.74) takes the form

$$\mathcal{V}_h\eta(\mathbf{x}) = \omega_n \int_{U_h} \frac{(1+|\nabla\varphi(\mathbf{y}')|^2)^{1/2}}{(|\mathbf{x}'-\mathbf{y}'|^2 + (x_n - \varphi(\mathbf{y}'))^2)^{n/2-1}}\,\eta(\frac{\mathbf{y}'}{h}, \frac{\varphi(\mathbf{y}')}{h})\,d\mathbf{y}'\,. \tag{12.76}$$

First, we show that $\mathcal{V}_h\eta(\mathbf{x})$ allows an asymptotic expansion in powers of h. Let the parametrization $\varphi(\mathbf{y}')$ be a real analytic function and denote the curvature tensor by $K = \|\partial_{jk}\varphi(\mathbf{0})\|_{j,k=1}^{n-1}$. Then, in view of (12.75),

$$\begin{aligned} \varphi(\mathbf{y}') &= \frac{1}{2}(K\mathbf{y}',\mathbf{y}') + \sum_{[\boldsymbol{\alpha}]\ge 3} \frac{(\mathbf{y}')^{\boldsymbol{\alpha}}}{\boldsymbol{\alpha}!}\,\partial^{\boldsymbol{\alpha}}\varphi(\mathbf{0})\,,\\ |\nabla\varphi(\mathbf{y}')|^2 &= |K\mathbf{y}'|^2 + \sum_{[\boldsymbol{\alpha}]\ge 3} \delta_{\boldsymbol{\alpha}}(\mathbf{y}')^{\boldsymbol{\alpha}}\,. \end{aligned} \tag{12.77}$$

Therefore, near $\mathbf{0}$ the area element is of the form

$$(1+|\nabla\varphi(\mathbf{y}')|^2)^{1/2} = 1 + \frac{1}{2}|K\mathbf{y}'|^2 + \sum_{k\ge 3}\pi_k(\mathbf{y}')\,,$$

where π_k are homogeneous polynomials of degree k. Hence, for $\mathbf{Y} = h^{-1}\mathbf{y}' \in \mathbb{R}^{n-1}$ we have

$$ds_{\mathbf{y}'} = h^{n-1}\Big(1 + \frac{1}{2}|K\mathbf{Y}|^2h^2 + \sum_{k\ge 3} h^3\pi_k(\mathbf{Y})\Big)d\mathbf{Y}\,.$$

Analogously, for the radial basis function $\eta(\mathbf{x}) = \psi(|\mathbf{x}|^2/2)$, one obtains from Taylor's expansion

$$\begin{aligned} \eta(\frac{\mathbf{y}'}{h}, \frac{\varphi(\mathbf{y}')}{h}) &= \sum_{j\ge 0} \frac{(2j-1)!!}{(2j)!}\psi^{(j)}(|\mathbf{Y}|^2/2)(h^{-1}\varphi(h\mathbf{Y}))^{2j}\\ &= \eta(\mathbf{Y},0) + h^2\frac{(K\mathbf{Y},\mathbf{Y})^2}{8}\,\psi'(|\mathbf{Y}|^2/2) + \sum_{j\ge 3} h^j p_j(\mathbf{Y},\frac{d}{dt})\,\psi(|\mathbf{Y}|^2/2) \end{aligned}$$

with certain differential operators $p_j(\mathbf{Y},\frac{d}{dt})$ of order $j/2$, having polynomial coefficients of degree $\le 2j$. Thus,

$$
\begin{aligned}
&\eta(\frac{\mathbf{y}'}{h},\frac{\varphi(\mathbf{y}')}{h})(1+|\nabla\varphi(\mathbf{y}')|^2)^{1/2}\\
&\quad=\psi(|\mathbf{Y}|^2/2)+\frac{h^2}{8}\Big(4\,\psi(|\mathbf{Y}|^2/2)\,|K\mathbf{Y}|^2+\psi'(|\mathbf{Y}|^2/2)\,(K\mathbf{Y},\mathbf{Y})^2\Big)\\
&\quad+\sum_{j\ge 3}h^j P_j(\mathbf{Y},\frac{d}{dt})\,\psi(|\mathbf{Y}|^2/2)
\end{aligned}
\tag{12.78}
$$

with differential operators $P_j(\mathbf{Y},\frac{d}{dt})$ of order $j/2$, having polynomial coefficients.

Similarly, the kernel function can be expanded in powers of h. We consider two zones, the far field $|\mathbf{x}|^2>4\delta^2h^2$ and the near field $|\mathbf{x}|^2<9\delta^2h^2$, where the kernel function is singular. In the far field we use the Taylor expansion

$$
(|\mathbf{x}'-\mathbf{y}'|^2+(x_n-\varphi(\mathbf{y}'))^2)^{1-n/2}=\sum_{j\ge 0}\frac{(-\varphi(\mathbf{y}'))^j}{j!}\,\partial_n^j\,\frac{1}{(|\mathbf{x}'-\mathbf{y}'|^2+x_n^2)^{n/2-1}}\,.
$$

Since $|\mathbf{x}-\mathbf{y}|\ge|\mathbf{x}|/2$, we obtain that

$$
\Big|\varphi(\mathbf{y}')^j\,\partial_n^j\,\frac{1}{(|\mathbf{x}'-\mathbf{y}'|^2+x_n^2)^{n/2-1}}\Big|\le\frac{c_j\,|\mathbf{y}'|^{2j}}{(|\mathbf{x}'|^2+x_n^2)^{(n-2+j)/2}}\le c_j\,h^{2-n+j}\,.
$$

If $|\mathbf{x}'|^2+|x_n|^2<9\delta^2h^2$, we expand the kernel with respect to $\varphi(\mathbf{x}')-\varphi(\mathbf{y}')$:

$$
\begin{aligned}
(|\mathbf{x}'-\mathbf{y}'|^2&+(x_n-\varphi(\mathbf{y}'))^2)^{1-n/2}\\
&=\sum_{j\ge 0}\frac{(\varphi(\mathbf{x}')-\varphi(\mathbf{y}'))^j}{j!}\,\partial_n^j\,\frac{1}{(|\mathbf{x}'-\mathbf{y}'|^2+(x_n-\varphi(\mathbf{x}'))^2)^{n/2-1}}\,.
\end{aligned}
$$

We denote $\tilde{x}_n:=x_n-\varphi(\mathbf{x}')$. Then

$$
\Big|(\varphi(\mathbf{x}')-\varphi(\mathbf{y}'))^j\,\partial_n^j\,\frac{1}{(|\mathbf{x}'-\mathbf{y}'|^2+\tilde{x}_n^2)^{n/2}}\Big|\le\frac{|\mathbf{x}'-\mathbf{y}'|^j\,\max|\nabla\varphi|^j}{(|\mathbf{x}'-\mathbf{y}'|^2+\tilde{x}_n^2)^{(n-2+j)/2}}\,,
$$

where the maximum of $|\nabla\varphi|$ is taken for $|\mathbf{x}'|\le 3\delta h$. Hence,

$$
\Big|(\varphi(\mathbf{x}')-\varphi(\mathbf{y}'))^j\,\partial_n^j\,\frac{1}{(|\mathbf{x}'-\mathbf{y}'|^2+\tilde{x}_n^2)^{n/2-1}}\Big|\le\frac{c_j\,h^j}{(|\mathbf{x}'-\mathbf{y}'|^2+\tilde{x}_n^2)^{n/2-1}}\,.
$$

Thus, $\mathcal{V}_h\eta(\mathbf{x})$ can be expanded, at least formally, into a power series with respect to h. The coefficients are given as integral operators over domains in $\mathbb{R}^{n-1}$.

In what follows, we determine the approximations of $\mathcal{V}_h\eta(\mathbf{x})$ by using the curvature tensor $K=\|\partial_{jk}\varphi(\mathbf{0})\|_{j,k=1}^{n-1}$ of Γ at $\mathbf{0}$. Due to (12.78), we obtain

$$
\Big(1+\frac{1}{2}|K\mathbf{y}'|^2\Big)\,\eta\Big(\frac{\mathbf{y}'}{h},\frac{\varphi(\mathbf{y}')}{h}\Big)=\sigma\Big(\frac{\mathbf{y}'}{h},h\Big)+\mathcal{O}(h^3)
$$

with the function

$$
\sigma(\mathbf{y}',h)=\psi(|\mathbf{y}'|^2/2)+\frac{h^2}{8}\Big(4\,\psi(|\mathbf{y}'|^2/2)\,|K\mathbf{y}'|^2+\psi'(|\mathbf{y}'|^2/2)\,(K\mathbf{y}',\mathbf{y}')^2\Big)\,,
\tag{12.79}
$$

and we have to analyze the integral

$$
\widetilde{\mathcal{V}}_h\eta(\mathbf{x})=\omega_n\int_{U_h}(|\mathbf{x}'-\mathbf{y}'|^2+(x_n-\varphi(\mathbf{y}'))^2)^{1-n/2}\,\sigma(\frac{\mathbf{y}'}{h},h)\,d\mathbf{y}'\,,
\tag{12.80}
$$

which differs from the original one by

$$|\mathcal{V}_h\eta(\mathbf{x}) - \widetilde{\mathcal{V}}_h\eta(\mathbf{x})| \le c\, h^3 \int\limits_{U_h} \frac{d\mathbf{y}'}{(|\mathbf{x}'-\mathbf{y}'|^2 + (x_n - \varphi(\mathbf{y}'))^2)^{n/2-1}}\,. \tag{12.81}$$

12.3.1.1. *Far field* $|\mathbf{x}'|^2 + |x_n|^2 > 4\delta^2 h^2$. Since in this area $|\mathbf{x}-\mathbf{y}| \ge |\mathbf{x}|/2$, we obtain

$$|\mathcal{V}_h\eta(\mathbf{x}) - \widetilde{\mathcal{V}}_h\eta(\mathbf{x})| \le c\, h^{2+n}\, |\mathbf{x}|^{2-n}\,. \tag{12.82}$$

The expansion of the kernel gives

$$\frac{1}{(|\mathbf{x}'-\mathbf{y}'|^2 + (x_n - \varphi(\mathbf{y}'))^2)^{n/2-1}} = \frac{1}{(|\mathbf{x}'-\mathbf{y}'|^2 + x_n^2)^{n/2-1}} + \frac{(n-2)\, x_n\, \varphi(\mathbf{y}')}{(|\mathbf{x}'-\mathbf{y}'|^2 + x_n^2)^{n/2}} + R_2(\mathbf{x},\mathbf{y})$$

with

$$R_2(\mathbf{x},\mathbf{y}) = \int\limits_0^1 (1-t)\mu''(t)\, dt\,, \quad \text{where} \quad \mu(t) = \frac{1}{(|\mathbf{x}'-\mathbf{y}'|^2 + (x_n - t\varphi(\mathbf{y}'))^2)^{n/2-1}}\,.$$

Note that

$$\mu''(t) = (2-n)\varphi(\mathbf{y}')^2\, \frac{|\mathbf{x}'-\mathbf{y}'|^2 + (1-n)(x_n - t\varphi(\mathbf{y}'))^2}{(|\mathbf{x}'-\mathbf{y}'|^2 + (x_n - t\varphi(\mathbf{y}'))^2)^{n/2+1}}$$

and therefore,

$$|\mu''(t)| \le \frac{|\varphi(\mathbf{y}')|^2}{|\mathbf{x}|^n}\,,$$

which implies in view of $|\mathbf{y}'| \le ch$ the estimate

$$\left| \omega_n \int\limits_\Gamma R_2(\mathbf{x},\mathbf{y})\, \sigma(\frac{\mathbf{y}'}{h}, h)\, d\sigma_{\mathbf{y}} \right| \le c\, \frac{h^{n+3}}{|\mathbf{x}|^n}\,.$$

Thus, it remains to consider the integrals

$$\omega_n \int\limits_{U_h} \frac{1}{(|\mathbf{x}'-\mathbf{y}'|^2 + x_n^2)^{n/2-1}}\, \sigma(\frac{\mathbf{y}'}{h}, h)\, d\mathbf{y}' + \frac{(n-2)\, \omega_n\, x_n}{2} \int\limits_{U_h} \frac{(K\mathbf{y}', \mathbf{y}')}{(|\mathbf{x}'-\mathbf{y}'|^2 + x_n^2)^{n/2}}\, \sigma(\frac{\mathbf{y}'}{h}, h)\, d\mathbf{y}' + (n-2)\, \omega_n\, x_n \int\limits_{U_h} \frac{\varphi_3(\mathbf{y}')}{(|\mathbf{x}'-\mathbf{y}'|^2 + x_n^2)^{n/2}}\, \sigma(\frac{\mathbf{y}'}{h}, h)\, d\mathbf{y}'\,,$$

with $\varphi_3(\mathbf{y}') = \sum_{[\boldsymbol{\alpha}]\ge 3} \delta_{\boldsymbol{\alpha}} (\mathbf{y}')^{\boldsymbol{\alpha}}$. In the new variables $\mathbf{X} = \dfrac{\mathbf{x}'}{h}$, $X_n = \dfrac{x_n}{h}$, $\mathbf{Y} = \dfrac{\mathbf{y}'}{h}$, the first two integrals transform to

$$\omega_n h \Big(\int\limits_{B'_\delta} \frac{\sigma(\mathbf{Y}, h)}{(|\mathbf{X}-\mathbf{Y}|^2 + X_n^2)^{n/2-1}}\, d\mathbf{Y} + \frac{(n-2)\, h X_n}{2} \int\limits_{B'_\delta} \frac{(K\mathbf{Y}, \mathbf{Y})\, \sigma(\mathbf{Y}, h)}{(|\mathbf{X}-\mathbf{Y}|^2 + X_n^2)^{n/2}}\, d\mathbf{Y} \Big),$$

whereas the third integral can be estimated by

$$\Big|(n-2)\,\omega_n\,x_n \int\limits_{U_h} \frac{\varphi_3(\mathbf{y}')}{(|\mathbf{x}'-\mathbf{y}'|^2+x_n^2)^{n/2}}\,\sigma(\frac{\mathbf{y}'}{h},h)\,d\mathbf{y}'\Big| \le c h^{n+2}|\mathbf{x}|^{1-n}\,. \tag{12.83}$$

Here $B'_\delta = B_\delta \cap \mathbb{R}^{n-1}$ denotes the support of the radial function η in $\mathbb{R}^{n-1}$. From (12.79) we obtain as an approximation to $\mathcal{V}_h\eta(\mathbf{x})$ in the far field

$$\begin{aligned}\widehat{\mathcal{V}}_h\eta(\mathbf{x}) &= \omega_n h \int\limits_{B'_\delta} \frac{\psi(|\mathbf{Y}|^2/2)}{(|\mathbf{X}-\mathbf{Y}|^2+X_n^2)^{n/2-1}}\,d\mathbf{Y}\\ &+ \frac{\omega_n\,(n-2)h^2X_n}{2}\int\limits_{B'_\delta}\frac{(K\mathbf{Y},\mathbf{Y})\,\psi(|\mathbf{Y}|^2/2)}{(|\mathbf{X}-\mathbf{Y}|^2+X_n^2)^{n/2}}\,d\mathbf{Y}\\ &+\frac{\omega_n h^3}{8}\int\limits_{B'_\delta}\frac{4\,|K\mathbf{Y}|^2\,\psi(|\mathbf{Y}|^2/2)+(K\mathbf{Y},\mathbf{Y})^2\,\psi'(|\mathbf{Y}|^2/2)}{(|\mathbf{X}-\mathbf{Y}|^2+X_n^2)^{n/2-1}}\,d\mathbf{Y}\,,\end{aligned} \tag{12.84}$$

which, in view of (12.82), approximates uniformly the far field $\mathcal{V}_h\eta(\mathbf{x})$ with the order $\mathcal{O}(h^{n+2}|\mathbf{x}|^{1-n})$. Note that the third integral in formula (12.84) is of the order $\mathcal{O}(h^{n+1}|\mathbf{x}|^{2-n})$, which for $|\mathbf{x}| = \mathcal{O}(h)$ is the same as for the error term (12.83).

12.3.1.2. *Near field* $|\mathbf{x}'|^2 + |x_n|^2 < 9\delta^2h^2$. Now, (12.81) leads to

$$|\mathcal{V}_h\eta(\mathbf{x}) - \widetilde{\mathcal{V}}_h\eta(\mathbf{x})| \le c\,h^3 \int\limits_{U_h}\frac{d\mathbf{y}'}{|\mathbf{x}'-\mathbf{y}'|^{n-2}} \le c\,h^4\,. \tag{12.85}$$

As mentioned before, we use the Taylor expansion of the kernel about the point $(\mathbf{x}'-\mathbf{y}',\tilde{x}_n)$, where $\tilde{x}_n = x_n - \varphi(\mathbf{x}')$. From (12.79), we obtain the integrals

$$\begin{aligned}\omega_n\int\limits_{U_h}\frac{1}{(|\mathbf{x}'-\mathbf{y}'|^2+\tilde{x}_n^2)^{n/2-1}}\,\sigma(\frac{\mathbf{y}'}{h},h)\,d\mathbf{y}' &= \omega_n h\int\limits_{B'_\delta}\frac{\psi(|\mathbf{Y}|^2/2)\,d\mathbf{y}}{(|\mathbf{X}-h\mathbf{Y}|^2+\widetilde{X}_n^2)^{n/2-1}}\\ &+\frac{\omega_n\,h^3}{8}\int\limits_{B'_\delta}\frac{4|K\mathbf{Y}|^2\psi(|\mathbf{Y}|^2/2)+(K\mathbf{Y},\mathbf{Y})^2\psi'(|\mathbf{Y}|^2/2)}{(|\mathbf{X}-h\mathbf{Y}|^2+\widetilde{X}_n^2)^{n/2-1}}\,d\mathbf{Y}\,,\end{aligned}$$

where $\mathbf{X} = h^{-1}\mathbf{x}'$, $\widetilde{X}_n = h^{-1}(x_n - \varphi(\mathbf{x}'))$. The succeeding term in Taylor's expansion gives

$$\omega_n(n-2)\,\tilde{x}_n\int\limits_{U_h}\frac{\varphi(\mathbf{y}')-\varphi(\mathbf{x}')}{(|\mathbf{x}'-\mathbf{y}'|^2+\tilde{x}_n^2)^{n/2}}\sigma(\frac{\mathbf{y}'}{h},h)\,d\mathbf{y}'\,.$$

If we replace $\varphi(\mathbf{y}')$ by $(K\mathbf{y}',\mathbf{y}')$, the error satisfies

$$\begin{aligned}\Big|\omega_n(n-2)\,\tilde{x}_n\int\limits_{U_h}\frac{\varphi_3(\mathbf{y}')-\varphi_3(\mathbf{x}')}{(|\mathbf{x}'-\mathbf{y}'|^2+\tilde{x}_n^2)^{n/2}}\sigma(\frac{\mathbf{y}'}{h},h)\,d\mathbf{y}'\Big|\\ \le c\int\limits_{U_h}\frac{(|\mathbf{y}'|^2+|\mathbf{x}'|^2)\,d\mathbf{y}'}{(|\mathbf{x}'-\mathbf{y}'|^2+\tilde{x}_n^2)^{n/2-1}} = \mathcal{O}(h^3)\,.\end{aligned}$$

Thus, for points $\mathbf{x} = (h\mathbf{X}, hX_n)$ in the near field we obtain the formula

$$\widehat{\mathcal{V}}_h\eta(\mathbf{x}) = \omega_n h \int\limits_{B'_\delta} \frac{\psi(|\mathbf{Y}|^2/2)}{(|\mathbf{X}-\mathbf{Y}|^2+\widetilde{X}_n^2)^{n/2-1}}\, d\mathbf{Y} + \frac{\omega_n(n-2)h^2\widetilde{X}_n}{2}\int\limits_{B'_\delta} \frac{((K\mathbf{Y},\mathbf{Y})-(K\mathbf{X},\mathbf{X}))\,\psi(|\mathbf{Y}|^2/2)}{(|\mathbf{X}-\mathbf{Y}|^2+\widetilde{X}_n^2)^{n/2}}\, d\mathbf{Y}\,, \tag{12.86}$$

which, by (12.85), provides uniform approximations of $\mathcal{V}_h\eta(\mathbf{x})$ of order h^3.

Thus, the single layer potential of the Laplacian is approximated by (12.84) and (12.86) for all $\mathbf{x}\in\mathbb{R}^n$ with the uniform error $\mathcal{O}(h^{n+2}/(|\mathbf{x}|+h)^{n-1})$.

12.3.1.3. *Matching in the area $2\delta h \le |\mathbf{x}| \le 3\delta h$.* According to the remark at the end of Subsection 12.3.1.1, we have to show that the sum of the first two integrals in (12.84) differs from (12.86) by higher-order terms, if the point $\mathbf{x}=(h\mathbf{X},hX_n)$ lies in the matching area. Since $X_n=\widetilde{X}_n+\varphi(\mathbf{x}')/h$ and $(|\mathbf{X}-\mathbf{Y}|^2+X_n^2)^{1-n/2}$ is smooth for $\mathbf{y}=(h\mathbf{Y},y_n)\in U_h$, we have

$$\frac{1}{(|\mathbf{X}-\mathbf{Y}|^2+X_n^2)^{n/2-1}} = \frac{1}{(|\mathbf{X}-\mathbf{Y}|^2+\widetilde{X}_n^2)^{n/2-1}} - \frac{\omega_n(n-2)\varphi(h\mathbf{X})\widetilde{X}_n}{h(|\mathbf{X}-\mathbf{Y}|^2+\widetilde{X}_n^2)^{-n/2}} + \mathcal{O}\Big(\frac{\varphi(h\mathbf{X})^2}{h^2}\Big)\,. \tag{12.87}$$

Thus, replacing $\varphi(h\mathbf{X})$ by $h^2(K\mathbf{X},\mathbf{X})/2$ in (12.87), we see that formula (12.84) differs from (12.86) by terms of the order $\mathcal{O}(h^3)$, i.e., in the overlapping region both formulas generate the same asymptotic error.

12.3.2. Approximation error. The approximation of $\mathcal{V}_h f(\mathbf{x})$ is now given by

$$\widetilde{\mathcal{V}}_h f(\mathbf{x}) = \mathcal{D}^{(1-n)/2}\sum_{h\mathbf{m}'\in\gamma} f(\phi(h\mathbf{m}'))\widehat{\mathcal{V}}_{h_{\mathbf{m}'}}\eta(\mathbf{x}-\phi(h\mathbf{m}'))\,, \tag{12.88}$$

where the parameter $h_{\mathbf{m}'}=\sqrt{\mathcal{D}}h\,|\phi'(h\mathbf{m}')|^{1/(n-1)}$ and the formulas for $\widehat{\mathcal{V}}_{h_{\mathbf{m}'}}\eta$ are determined by (12.84) or (12.86) in dependence on the value of $|\mathbf{x}-\phi(h\mathbf{m}')|$. By the uniform error estimate, the difference $|\mathcal{V}_h\eta(\mathbf{x})-\widehat{\mathcal{V}}_h\eta(\mathbf{x})|$ can be majorized by

$$\begin{aligned}
&\frac{\omega_n}{\mathcal{D}^{(n-1)/2}}\sum_{h\mathbf{m}'\in\gamma}|f(\phi(h\mathbf{m}'))|\,\frac{(\sqrt{\mathcal{D}}h\,|\phi'(h\mathbf{m}')|^{1/(n-1)})^{n+2}}{(|\mathbf{x}-\phi(h\mathbf{m}')|+\sqrt{\mathcal{D}}h\,|\phi'(h\mathbf{m}')|^{1/(n-1)})^{n-1}}\\
&\le c(\sqrt{\mathcal{D}}h)^3\int\limits_{\gamma}|f(\phi(\mathbf{y}'))|\frac{|\phi'(\mathbf{y}')|^{(n+2)/(n-1)}}{(|\mathbf{x}-\phi(\mathbf{y}')|+\sqrt{\mathcal{D}}h\,|\phi'(\mathbf{y}')|^{1/(n-1)})^{n-1}}\,d\mathbf{y}'\\
&\le c(\sqrt{\mathcal{D}}h)^3\int\limits_{\Gamma}\frac{|f(\mathbf{y})|}{(|\mathbf{x}-\mathbf{y}|+\sqrt{\mathcal{D}}h)^{n-1}}\,d\sigma_{\mathbf{y}}\\
&\le c\|f\|_{C(\overline{\Gamma})}(\sqrt{\mathcal{D}}h)^3|\log(\max(\sqrt{\mathcal{D}}h,\operatorname{dist}(\mathbf{x},\Gamma)))|\,.
\end{aligned} \tag{12.89}$$

Note that the integrals appearing in the formulas (12.84) or (12.86) are restricted to the domain B'_δ, which is the support of the basis function η in $\mathbb{R}^{n-1}$ after multiplication with a suitable cut-off function. Owing to the rapid decay of

η, one can obviously extend the integration domain to the whole of $\mathbb{R}^{n-1}$, making an error less than a prescribed tolerance ε.

Thus, let us fix $\delta' > 0$ such that

$$\int\limits_{\mathbb{R}^{n-1}\setminus B_{\delta'}} |\mathbf{y}'|^2 |\eta(\mathbf{y}')|\, d\mathbf{y}' \le \varepsilon,$$

with ε the saturation error from Lemma 12.10. To compute the approximation of (12.68), we choose a local coordinate system with the origin at the point $\phi(h\mathbf{m}')$ such that the x_n-axis is directed as the normal to Γ at this point. In the new coordinate system, the surface Γ is given locally by the mapping $x_n = \varphi(\mathbf{x}')$, $\mathbf{x}' \in \mathbb{R}^{n-1}$. We denote the corresponding curvature tensor by $K = \|\partial_{jk}\varphi(0)\|_{j,k=1}^{n-1}$. Let $\mathbf{x}-\phi(h\mathbf{m}') = (\mathbf{x}', x_n) = (h\mathbf{X}, hX_n)$, and consider the approximations of $\mathcal{V}_h\eta(\mathbf{x}-\phi(h\mathbf{m}))$:

1. If $|\mathbf{x} - \phi(h\mathbf{m})| \ge h\delta'$, then

$$\begin{aligned}
\widehat{\mathcal{V}}_h\eta(\mathbf{x} - \phi(h\mathbf{m}')) &= \omega_n h \int\limits_{\mathbb{R}^{n-1}} \frac{\psi(|\mathbf{Y}|^2/2)}{(|\mathbf{X}-\mathbf{Y}|^2 + X_n^2)^{n/2-1}}\, d\mathbf{Y} \\
&+ \frac{\omega_n\,(n-2)h^2 X_n}{2} \int\limits_{\mathbb{R}^{n-1}} \frac{(K\mathbf{Y},\mathbf{Y})\,\psi(|\mathbf{Y}|^2/2)}{(|\mathbf{X}-\mathbf{Y}|^2 + X_n^2)^{n/2}}\, d\mathbf{Y} \\
&+ \frac{\omega_n h^3}{8} \int\limits_{\mathbb{R}^{n-1}} \frac{4\,|K\mathbf{Y}|^2\,\psi(|\mathbf{Y}|^2/2) + (K\mathbf{Y},\mathbf{Y})^2\,\psi'(|\mathbf{Y}|^2/2)}{(|\mathbf{X}-\mathbf{Y}|^2 + X_n^2)^{n/2-1}}\, d\mathbf{Y}\,.
\end{aligned} \tag{12.90}$$

2. If $|\mathbf{x} - \phi(h\mathbf{m}')| < h\delta'$, then

$$\begin{aligned}
\widehat{\mathcal{V}}_h\eta(\mathbf{x} - \phi(h\mathbf{m}')) &= \omega_n h \int\limits_{\mathbb{R}^{n-1}} \frac{\psi(|\mathbf{Y}|^2/2)}{(|\mathbf{X}-\mathbf{Y}|^2 + \widetilde{X}_n^2)^{n/2-1}}\, d\mathbf{Y} \\
&+ \frac{\omega_n(n-2)h^2\,\widetilde{X}_n}{2} \int\limits_{\mathbb{R}^{n-1}} \frac{((K\mathbf{Y},\mathbf{Y}) - (K\mathbf{X},\mathbf{X}))\,\psi(|\mathbf{Y}|^2/2)}{(|\mathbf{X}-\mathbf{Y}|^2 + \widetilde{X}_n^2)^{n/2}}\, d\mathbf{Y}\,,
\end{aligned} \tag{12.91}$$

where $\widetilde{X}_n = X_n - h^{-1}\varphi(h\mathbf{X})$. Then, from Lemma 12.10 and (12.89), we derive

THEOREM 12.11. *Suppose that the radial function $\eta \in \mathcal{S}(\mathbb{R}^n)$ satisfies the moment condition* (2.47) *with $N = 4$. Then the single layer potential*

$$\mathcal{V}f(\mathbf{x}) = \omega_n \int\limits_{\Gamma} \frac{f(\mathbf{y})}{|\mathbf{x}-\mathbf{y}|^{n-2}}\, d\sigma_{\mathbf{y}}$$

is approximated by the sum (12.88) *with the order*

$$|\mathcal{V}f(\mathbf{x}) - \widetilde{\mathcal{V}}_h f(\mathbf{x})| = \mathcal{O}\Big((\sqrt{\mathcal{D}}h)^3 \big|\log(\max(\sqrt{\mathcal{D}}h, \operatorname{dist}(\mathbf{x},\Gamma)))\big| + \varepsilon\Big)\,, \tag{12.92}$$

provided the surface Γ has C^4-smoothness and $f \in C_0^3(\Gamma)$. The saturation term ε can be made negligibly small if $\mathcal{D}$ is large enough.

12.3.3. Cubature formula. Let us note that we use formula (12.90) only if $|\mathbf{x} - \phi(h\mathbf{m}')|$ is small; otherwise, we can take the simple midpoint rule (12.69). To give the corresponding bounds for $|\mathbf{x} - \phi(h\mathbf{m}')|$, we introduce a cut-off function χ_h with the property that $\chi_h(\mathbf{y}) = 1$ for $|\mathbf{y}| \le h^\beta$ and $\chi_h(\mathbf{y}) = 0$ for $|\mathbf{y}| \ge (h^\beta + h^{1/4})$

with some $\beta \in (0,1)$ to be specified later. We split the single layer potential into two integrals

$$\omega_n \int\limits_\Gamma \frac{f(\mathbf{y})\chi_h(\mathbf{x}-\mathbf{y})}{|\mathbf{x}-\mathbf{y}|^{n-2}}\, d\sigma_\mathbf{y} + \omega_n \int\limits_\Gamma \frac{f(\mathbf{y})(1-\chi_h(\mathbf{x}-\mathbf{y}))}{|\mathbf{x}-\mathbf{y}|^{n-2}}\, d\sigma'_\mathbf{y} \tag{12.93}$$

and we apply Theorem 12.11 to the first one. Note that $f(\mathbf{y})\chi_h(\mathbf{x}-\mathbf{y}) \neq 0$, $\mathbf{y} \in \Gamma$, only for $\operatorname{dist}(\mathbf{x},\Gamma) < (h^\beta + h^{1/4})$. Since $|\nabla \chi_h| \le c h^{-1/4}$, we have

$$\|f\, \chi_h(\mathbf{x}-\cdot)\|_{C^4(\overline{\Gamma})} \le c h^{-1} \|f\|_{C^4(\overline{\Gamma})}\,.$$

Thus, in view of Theorem 12.10, the function $f(\mathbf{y})\chi_h(\mathbf{x}-\mathbf{y})$ can be approximated on Γ by the quasi-interpolant

(12.94)

$$\mathcal{D}^{-(n-1)/2} \sum_{|\mathbf{x}-\phi(h\mathbf{m}')|<h^{\beta-1/4}} f(\phi(h\mathbf{m}'))\, \chi_h(\mathbf{x}-\phi(h\mathbf{m}'))\, \eta\Big(\frac{\mathbf{y}-\phi(h\mathbf{m}')}{\sqrt{\mathcal{D}}h\, |\phi'(h\mathbf{m}')|^{1/(n-1)}}\Big)$$

with the error

$$c\,(\sqrt{\mathcal{D}}h)^3\, \|f\|_{C^4(\overline{\Gamma})} + \varepsilon \sum_{k=0}^{3} c_k\, (\sqrt{\mathcal{D}}h)^{3k/4}\,.$$

Consequently, if $f \in C_0^4(\Gamma)$, then we can argue, as in Theorem 12.11, to derive the estimate

$$\Bigg| \mathcal{D}^{(1-n)/2} \sum_{|\mathbf{x}-\phi(h\mathbf{m}')|<h^\beta+h^{1/4}} f(\phi(h\mathbf{m}'))\chi_h(\mathbf{x}-\phi(h\mathbf{m}'))\widetilde{\mathcal{V}}_{h_{\mathbf{m}'}}\eta(\mathbf{x}-\phi(h\mathbf{m}'))$$
$$- \omega_n \int\limits_\Gamma \frac{f(\mathbf{y})\chi_h(\mathbf{x}-\mathbf{y})}{|\mathbf{x}-\mathbf{y}|^{n-2}}\, d\sigma_\mathbf{y} \Bigg| = \mathcal{O}((\sqrt{\mathcal{D}}h)^3 \big|\log(\sqrt{\mathcal{D}}h)\big| + \varepsilon)\,.$$

Thus, it remains to choose β such that the second integral in (12.93) is approximated with the order $\mathcal{O}(h^3)$ by

$$h^{n-1}\omega_n \sum_{h\mathbf{m}'\in\gamma} \frac{f(\phi(h\mathbf{m}'))(1-\chi_h(\mathbf{x}-\phi(h\mathbf{m}')))}{|\mathbf{x}-\phi(h\mathbf{m}')|^{n-2}}\, |\phi'(h\mathbf{m}')|\,.$$

If $\operatorname{dist}(\mathbf{x},\Gamma) \ge h^\beta + h^{1/4}$, then $\chi_h(\mathbf{x}-\phi(\mathbf{y}')) = 0$ and

$$\int\limits_\gamma \Big| \nabla_\ell \Big(\frac{f(\phi(\mathbf{y}'))}{|\mathbf{x}-\phi(\mathbf{y}')|^{n-2}} \sqrt{1+|\nabla\varphi(\mathbf{y}')|^2} \Big) \Big|\, d\mathbf{y}'$$
$$\le \sum_{j=0}^{\ell} c_j \Big| \nabla_{\ell-j}(f(\phi(\mathbf{y}'))\sqrt{1+|\nabla\varphi(\mathbf{y}')|^2}) \Big| \int\limits_\gamma \frac{d\mathbf{y}'}{|\mathbf{x}-\phi(\mathbf{y}')|^{n+j-2}}$$
$$\le \sum_{j=0}^{\ell} c_j \Big| \nabla_{\ell-j}(f(\phi(\mathbf{y}'))\sqrt{1+|\nabla\varphi(\mathbf{y}')|^2}) \Big| \int\limits_0^{\operatorname{diam}\gamma} \frac{r^{n-2}\, dr}{|r+h^\beta+h^{1/4}|^{n+j-2}}\,.$$

Therefore,

$$h^\ell \int\limits_\gamma \Big| \nabla_\ell \Big(\frac{f(\phi(\mathbf{y}'))}{|\mathbf{x}-\phi(\mathbf{y}')|^{n-2}} \sqrt{1+|\nabla\varphi(\mathbf{y}')|^2} \Big) \Big|\, d\mathbf{y}' \le c\, h^\ell (h^\beta + h^{1/4})^{1-\ell}\,.$$

If $\operatorname{dist}(\mathbf{x},\Gamma) < h^\beta + h^{1/4}$, we have

$$\left|\nabla_{\ell-j}\Big(f(\phi(\mathbf{y}'))(1-\chi_h(\mathbf{x}-\phi(\mathbf{y}')))\sqrt{1+|\nabla\varphi(\mathbf{y}')|^2}\Big)\right| \le c h^{(j-\ell)/4}$$

and

$$\int\limits_{|\mathbf{x}-\phi(\mathbf{y}')|\ge h^\beta} \frac{d\mathbf{y}'}{|\mathbf{x}-\phi(\mathbf{y}')|^{n+j-2}} \le c\, h^{\beta(1-j)},$$

so that

$$h^\ell \int\limits_\gamma \left|\nabla_\ell\Big(\frac{f(\phi(\mathbf{y}'))(1-\chi_h(\mathbf{x}-\phi(\mathbf{y}')))}{|\mathbf{x}-\phi(\mathbf{y}')|^{n-2}}\sqrt{1+|\nabla\varphi(\mathbf{y}')|^2}\Big)\right| d\mathbf{y}' \le c\, h^{\ell(1-\beta)} h^\beta .$$

Hence, depending on the smoothness $f \in C_0^\ell(\Gamma)$ with $\ell \ge 4$, the value of the parameter

$$\beta = 1 - 2/(\ell-1)$$

provides the following estimate of the cubature error:

$$\Bigg| h^{n-1}\omega_n \sum_{h\mathbf{m}'\in\gamma} \frac{f(\phi(h\mathbf{m}'))(1-\chi_h(\mathbf{x}-\phi(h\mathbf{m}')))}{|\mathbf{x}-\phi(h\mathbf{m}')|^{n-2}} |\phi'(h\mathbf{m}')|$$
$$- \omega_n \int\limits_\Gamma \frac{f(\mathbf{y})(1-\chi_h(\mathbf{x}-\mathbf{y}))}{|\mathbf{x}-\mathbf{y}|^{n-2}}\, d\sigma_{\mathbf{y}} \Bigg| \le c h^3 \|f\|_{C^\ell(\overline{\Gamma})} \cdot$$

Therefore, Theorem 12.11 remains valid if, for instance, $f \in C_0^5(\Gamma)$ and $\chi_h(\mathbf{y})$ is chosen such that $\chi_h(\mathbf{y}) = 1$ for $|\mathbf{y}| \le h^{1/2}$ and $\chi_h(\mathbf{y}) = 0$ for $|\mathbf{y}| \le h^{1/2} + h^{1/4}$. Then formula (12.90) is applied in the region $h\delta' \le |\mathbf{x}-\phi(h\mathbf{m}')| \le (h^{1/2}+h^{1/4})\delta'$ with the function values $f(h\mathbf{m}')\chi_h(\mathbf{x}-h\mathbf{m}')$. The midpoint rule with the values $f(h\mathbf{m}')(1-\chi_h(\mathbf{x}-h\mathbf{m}'))$ is applied in the region $|\mathbf{x}-\phi(h\mathbf{m}')| \ge h^{1/2}\delta'$.

Summarizing, we obtain the following result:

THEOREM 12.12. *Suppose that the surface Γ is $C^{\ell+1}$, $f \in C_0^\ell(\Gamma)$, $\ell \ge 4$, and set $\beta = 1-2/(\ell-1)$. Then the single layer potential*

$$\mathcal{V}f(\mathbf{x}) = \omega_n \int\limits_\Gamma \frac{f(\mathbf{y})}{|\mathbf{x}-\mathbf{y}|^{n-2}}\, d\sigma_{\mathbf{y}}$$

is approximated by the sum of

$$\mathcal{D}^{(1-n)/2} \sum_{|\mathbf{x}-\phi(h\mathbf{m}')|<h^\beta+h^{1/4}} f(\phi(h\mathbf{m}'))\chi_h(\mathbf{x}-\phi(h\mathbf{m}'))\widetilde{\mathcal{V}}_{h_{\mathbf{m}'}}(\mathbf{x}-\phi(h\mathbf{m}'))$$

and

$$h^{n-1}\omega_n \sum_{|\mathbf{x}-\phi(h\mathbf{m}')|>h^\beta} \frac{f(\phi(h\mathbf{m}'))(1-\chi_h(\mathbf{x}-\phi(h\mathbf{m}')))}{|\mathbf{x}-\phi(h\mathbf{m}')|^{n-2}} |\phi'(h\mathbf{m}')|$$

with the order

$$\mathcal{O}\Big((\sqrt{\mathcal{D}}h)^3 \big|\log(\max(\sqrt{\mathcal{D}}h, \operatorname{dist}(\mathbf{x},\Gamma)))\big| + \varepsilon\Big).$$

Here, $\chi_h(\mathbf{y})$ is a sufficiently smooth cut-off function in $\mathbb{R}^n$, vanishing outside the ball $|\mathbf{y}| > h^\beta + h^{1/4}$ and equal to 1 for $|\mathbf{y}| < h^\beta$. The saturation term ε can be made negligibly small if $\mathcal{D}$ is large enough.

12.3.4. Basis functions. We show here that the integrals appearing in the formulas (12.90) and (12.91) can be converted into one-dimensional integrals. Therefore, the proposed integration procedure for surface integration is well-suited for higher-dimensional cases. Since the basis function η is radial, one can use the formulas mentioned in Subsections 2.1.4 and 12.2.2 for the convolution of radial functions.

In the following, we use (12.45) to give explicit formulas for the integrals approximating the single layer harmonic potential, if the Gaussian is chosen as the local basis function. Since this function satisfies the moment condition (2.47) only with $N = 2$, one can take linear combinations of Gaussians as described in Subsection 3.4.1 in order to achieve the higher order for the quasi-interpolants as required in Theorem 12.11. In the numerical tests, we chose the generating function

$$\eta(\mathbf{x}) = \frac{2\,\mathrm{e}^{-|\mathbf{x}|^2}}{\pi^{(n-1)/2}} - \frac{\mathrm{e}^{-|\mathbf{x}|^2/2}}{(2\pi)^{(n-1)/2}}$$

which satisfies (2.47) in $\mathbb{R}^{n-1}$ with $N = 4$.

Next, we give the formulas based on (12.45) for the integrals appearing in (12.90) and (12.91) with $\psi(y) = \mathrm{e}^{-y}$. Since

$$\omega_n \int\limits_{\mathbb{R}^{n-1}} \frac{\mathrm{e}^{-2\pi i \langle \mathbf{x}', \boldsymbol{\lambda}' \rangle}}{(|\mathbf{x}'|^2 + x_n^2)^{n/2-1}}\, d\mathbf{x}' = \frac{\mathrm{e}^{-2\pi|\boldsymbol{\lambda}'||x_n|}}{4\pi|\boldsymbol{\lambda}'|}\,, \qquad \boldsymbol{\lambda}' \in \mathbb{R}^{n-1}\,,$$

formula (12.45) yields

$$\begin{aligned} \mathcal{I}_0(\mathbf{x}', x_n) &= \omega_n \int\limits_{\mathbb{R}^{n-1}} \frac{\mathrm{e}^{-|\mathbf{y}'|^2/2}}{(|\mathbf{x}'-\mathbf{y}'|^2 + x_n^2)^{n/2-1}}\, d\mathbf{y}' \\ &= \frac{(2\pi)^{(n-1)/2}}{|\mathbf{x}'|^{(n-3)/2}} \int\limits_0^\infty \mathrm{e}^{-2\pi r(\pi r + |x_n|)}\, r^{(n-3)/2} J_{(n-3)/2}(2\pi r|\mathbf{x}'|)\, dr\,. \end{aligned} \tag{12.95}$$

Then we consider the integral

$$\mathcal{I}_1(\mathbf{x}', x_n) = \omega_n\,(n-2) x_n \int\limits_{\mathbb{R}^{n-1}} \frac{(K\mathbf{y}', \mathbf{y}')\,\mathrm{e}^{-|\mathbf{y}'|^2/2}}{(|\mathbf{x}'-\mathbf{y}'|^2 + x_n^2)^{n/2}}\, d\mathbf{y}'\,.$$

Here, we use that

$$(K\mathbf{y}', \mathbf{y}')\,\mathrm{e}^{-|\mathbf{y}'|^2/2} = ((K\nabla, \nabla) + \mathrm{tr}K)\,\mathrm{e}^{-|\mathbf{y}'|^2/2}\,,$$

where $\nabla = (\partial_1, \dots, \partial_{n-1})$ and $\mathrm{tr}K = \Delta\varphi(\mathbf{0})$. Hence,

$$\mathcal{I}_1(\mathbf{x}', x_n) = \omega_n\,(n-2)((K\nabla, \nabla) + \mathrm{tr}K) \int\limits_{\mathbb{R}^{n-1}} \frac{x_n\,\mathrm{e}^{-|\mathbf{y}'|^2/2}}{(|\mathbf{x}'-\mathbf{y}'|^2 + x_n^2)^{n/2}}\, d\mathbf{y}'\,.$$

Since

$$\int\limits_{\mathbb{R}^{n-1}} \frac{x_n\,\mathrm{e}^{-2\pi i \langle \mathbf{x}', \boldsymbol{\lambda}' \rangle}}{(|\mathbf{x}'|^2 + x_n^2)^{n/2}}\, d\mathbf{x} = \frac{\pi^{n/2}\,\mathrm{sgn}(x_n)\,\mathrm{e}^{-2\pi|\lambda'||x_n|}}{\Gamma(n/2)}\,,$$

we derive

$$\mathcal{I}_1(\mathbf{x}', x_n) = \pi(2\pi)^{(n-1)/2}\,\mathrm{sgn}(x_n) \times \big((K\nabla,\nabla) + \mathrm{tr}K\big) \int_0^\infty e^{-2\pi r(\pi r+|x_n|)}\, r^{(n-1)/2}\, \frac{J_{(n-3)/2}(2\pi r|\mathbf{x}'|)}{|\mathbf{x}'|^{(n-3)/2}}\, dr\,.$$

The series expansion

$$J_n(2\pi r|\mathbf{x}'|) = (\pi r|\mathbf{x}'|)^n \sum_{j=0}^\infty \frac{(-1)^j(\pi r|\mathbf{x}'|)^{2j}}{j!\,\Gamma(j+n+1)}$$

shows that the function

$$(K\nabla,\nabla)\,\frac{J_{(n-3)/2}(2\pi r|\mathbf{x}'|)}{|\mathbf{x}'|^{(n-3)/2}} \tag{12.96}$$

is smooth. Let us consider a radial function $g(|\mathbf{x}'|)$ and denote $\omega = \mathbf{x}'/|\mathbf{x}'|$. Then

$$(K\nabla,\nabla)g(|\mathbf{x}'|) = (K\omega,\omega)\Big(g''(|\mathbf{x}'|) - \frac{g'(|\mathbf{x}'|)}{|\mathbf{x}'|}\Big) + \mathrm{tr}K\,\frac{g'(|\mathbf{x}'|)}{|\mathbf{x}'|}\,.$$

Therefore, (12.96) can be expressed by using the values of $(K\omega,\omega)$, $\mathrm{tr}K$, and either trigonometric functions (n even) or the Bessel functions J_0 and J_1 (n odd) . For example, if $n = 3$, then

$$(K\nabla,\nabla)J_0(2\pi r|\mathbf{x}'|) = 4\pi r(K\omega,\omega)\Big(\pi r J_0(2\pi r|\mathbf{x}'|) - \frac{J_1(2\pi r|\mathbf{x}'|)}{|\mathbf{x}'|}\Big) + 2\pi r\,\mathrm{tr}K\,\frac{J_1(2\pi r|\mathbf{x}'|)}{|\mathbf{x}'|}\,,$$

whereas for $n = 4$

$$\begin{aligned}(K\nabla,\nabla)\,\frac{J_{1/2}(2\pi r|\mathbf{x}'|)}{|\mathbf{x}'|^{1/2}} &= (K\nabla,\nabla)\,\frac{\sin(2\pi r|\mathbf{x}'|)}{\pi r^{1/2}|\mathbf{x}'|}\\ &= \mathrm{tr}K\,\frac{2\pi|x'|r\cos(2\pi r|\mathbf{x}'|) - \sin(2\pi r|\mathbf{x}'|)}{\pi r^{1/2}|\mathbf{x}'|^3}\\ &\quad - (K\omega,\omega)\,\frac{(4\pi^2r^2|\mathbf{x}'|^2-3)\sin(2\pi r|\mathbf{x}'|) + 6\pi r|\mathbf{x}'|\cos(2\pi r|\mathbf{x}'|)}{\pi r^{1/2}|\mathbf{x}'|^3}\,.\end{aligned}$$

Consider finally the integral

$$\omega_n \int_{\mathbb{R}^{n-1}} \frac{(4\,|K\mathbf{y}'|^2 - (K\mathbf{y}',\mathbf{y}')^2)\,e^{-|\mathbf{y}'|^2/2}}{(|\mathbf{x}'-\mathbf{y}'|^2 + x_n^2)^{n/2-1}}\, d\mathbf{y}'$$

appearing in formula (12.90). It is easy to see that

$$\begin{aligned}&\big(4\,|K\mathbf{y}'|^2 - (K\mathbf{y}',\mathbf{y}')^2\big)\,e^{-|\mathbf{y}'|^2/2}\\ &= -\big((K\nabla,\nabla)^2 + 2|K\nabla|^2 - (\mathrm{tr}K)^2 + 2\,\mathrm{det}K\,(\Delta+2)\big)\,e^{-|\mathbf{y}'|^2/2}\,.\end{aligned}$$

Therefore one has to determine

$$-\big((K\nabla,\nabla)^2 + 2|K\nabla|^2 - (\mathrm{tr}K)^2 + 2\,\mathrm{det}K\,(\Delta+2)\big)\,\frac{J_{(n-3)/2}(2\pi r|\mathbf{x}'|)}{|\mathbf{x}'|^{(n-3)/2}}\,.$$

For the radial function $g(|\mathbf{x}'|)$, we get

$$
\begin{aligned}
&-\Big((K\nabla,\nabla)^2+2|K\nabla|^2-(\mathrm{tr}K)^2+2\det K\,(\Delta+2)\Big)g(|\mathbf{x}'|)\\
&=(K\omega,\omega)^2\Big(-g^{(4)}(|\mathbf{x}'|)+6\frac{g^{(3)}(|\mathbf{x}'|)}{|\mathbf{x}'|}-15\frac{g''(|\mathbf{x}'|)}{|\mathbf{x}'|^2}+15\frac{g'(|\mathbf{x}'|)}{|\mathbf{x}'|^3}\Big)\\
&+|K\omega|^2\Big(-4\frac{g^{(3)}(|\mathbf{x}'|)}{|\mathbf{x}'|}+12\frac{g''(|\mathbf{x}'|)}{|\mathbf{x}'|^2}-12\frac{g'(|\mathbf{x}'|)}{|\mathbf{x}'|^3}-2g''(|\mathbf{x}'|)+2\frac{g'(|\mathbf{x}'|)}{|\mathbf{x}'|}\Big)\\
&+\mathrm{tr}K\,(K\omega,\omega)\Big(-2\frac{g^{(3)}(|\mathbf{x}'|)}{|\mathbf{x}'|}+6\frac{g''(|\mathbf{x}'|)}{|\mathbf{x}'|^2}-6\frac{g'(|\mathbf{x}'|)}{|\mathbf{x}'|^3}\Big)\\
&+(\mathrm{tr}K)^2\Big(-3\frac{g''(|\mathbf{x}'|)}{|\mathbf{x}'|^2}+3\frac{g'(|\mathbf{x}'|)}{|\mathbf{x}'|^3}-2\frac{g'(|\mathbf{x}'|)}{|\mathbf{x}'|}+g(|\mathbf{x}'|)\Big)\\
&+\det K\Big(4\frac{g''(|\mathbf{x}'|)}{|\mathbf{x}'|^2}-4\frac{g'(|\mathbf{x}'|)}{|\mathbf{x}'|^3}-2g''(|\mathbf{x}'|)+2\frac{g'(|\mathbf{x}'|)}{|\mathbf{x}'|}-4g(|\mathbf{x}'|)\Big).
\end{aligned}
$$

The differential expressions can easily be calculated by using computer programs for symbolic calculations.

12.3.5. Numerical examples. The proposed approach was tested numerically in the computation of single layer potentials for the three-dimensional Laplacian. We applied the combined formulas to obtain the integral

$$
\frac{1}{4\pi}\int\limits_{\Gamma}\frac{\mathrm{e}^{-|\mathbf{y}|^2}}{|\mathbf{x}-\mathbf{y}|}\,d\sigma_{\mathbf{y}} \tag{12.97}
$$

for a paraboloid Γ given by $x_3 = k_{11}x_1^2 + 2k_{12}x_1x_2 + k_{11}x_2^2$. Using the quasi-interpolation formula (12.94) with the local function

$$
\eta(\mathbf{y}) = \frac{2\,\mathrm{e}^{-|\mathbf{y}|^2}}{\pi} - \frac{\mathrm{e}^{-|\mathbf{y}|^2/2}}{2\pi},
$$

the approximation error of the density $\mathrm{e}^{-|\mathbf{y}|^2}$ is $\mathcal{O}((\sqrt{\mathcal{D}}h)^4+\varepsilon)$. The same rate is shown for the cubature of the potential for flat Γ, i.e. $k_{ij}=0$. In Table 12.6, we give the approximation order obtained by halving the step size h for a randomly chosen point $\mathbf{x}$ with prescribed distance from Γ. We have taken the parameter $\mathcal{D}=3.0$ in formula (12.71) in order to keep the saturation error less than 10^{-10}.

$\mathrm{dist}(\mathbf{x},\Gamma)$	$h=0.4$	$h=0.2$	$h=0.1$	$h=0.05$
2.0	12.55	12.47	7.48	0.0000
0.1	3.66	3.91	3.87	3.92
0.01	3.70	4.09	3.99	3.85
0.001	3.71	4.25	3.85	3.84
0.0001	3.71	4.37	3.74	3.86
0.00001	3.33	3.71	3.85	3.97
0.0	3.33	3.70	4.04	3.86

TABLE 12.6. Approximation order for different distances and the flat surface

The high orders for $\text{dist}(\mathbf{x}, \Gamma) = 2.0$ result from the fact that the simple midpoint is used for all mesh points. In the other cases, we approximate the density by a fourth-order quasi-interpolant. Since Γ is flat, the formulas (12.90), (12.91) provide the exact values of the potentials of the basis functions. Therefore, the single layer potential is approximated with the same order as the density.

In Table 12.7, we provide the approximation orders for the curved surface $x_3 = x_1^2 + 2x_1x_2 + 2x_2^2$, which are in agreement with the assertion $\mathcal{O}(h^3|\log h|)$ of Theorem 12.12.

$\text{dist}(\mathbf{x}, \Gamma)$	$h = 0.4$	$h = 0.2$	$h = 0.1$	$h = 0.05$
2.0	4.04	17.96	1.84	0.00
0.1	3.42	3.09	2.90	2.92
0.01	3.38	3.23	3.05	2.71
0.001	3.11	2.51	2.85	2.98
0.0001	3.18	2.82	3.18	2.82
0.00001	2.80	2.55	3.09	2.88
0.0	4.85	3.10	2.64	2.84

TABLE 12.7. Approximation order for different distances in the case of the paraboloid $x_3 = x_1^2 + 2x_1x_2 + 2x_2^2$

12.4. Notes

The material of Section 12.1 is taken from [**53**]. Two other approximation methods for solving the Lippmann-Schwinger equation were recently studied by Vainikko in [**95**]. The first one is a simple cubature method of second order, which can be applied in the case of piecewise smooth potentials. The second method is a sophisticated trigonometric collocation applied to periodized versions of the Lippmann-Schwinger equation. The values of the periodized diffraction operator on the trigonometric polynomials are computed via Fourier techniques and it is shown that this method provides optimal convergence orders if $q(\mathbf{x})$ is smooth on $\mathbb{R}^n$. Utilizing the convolution structure of the problems, Vainikko showed that by using FFT and two grid iterations, the discrete problems can be solved in $\mathcal{O}(N^n \log N)$ operations. The same approach can also be applied to the Gaussian collocation method considered in Section 12.1.

The BPM described in Section 12.2 was proposed in [**63**] and [**64**] for the numerical solution of the integral equations of the second kind for two- and three-dimensional potential problems with smooth boundary. Here we provide some of the numerical results obtained in the thesis [**3**] of H. Åkermark. Let us note that in a series of papers Kanaun and coworkers applied the BPM for solving problems from elasticity (cf., e.g., [**41**], [**42**], [**43**]).

In Section 12.3 we followed the paper [**73**]. The more accurate computation of boundary integrals by using the values of the normal and of the curvature of the boundary, which is considered here, improves the accuracy of the BPM. In [**3**] this is demonstrated by numerical tests for two-dimensional examples, where the surface integral of the basis function is replaced by the integral over the sphere which touches smoothly the boundary at the center of the basis function.

CHAPTER 13

Numerical algorithms based upon approximate approximations — non-linear problems

This chapter surveys some applications of the approximation method to the solution of evolution equations. In Section 13.1, we report on new semi-analytic time-marching algorithms for the numerical solution of quasi-linear parabolic equations. In Section 13.2 we extend these algorithms to obtain an explicit method for solving non-stationary Navier-Stokes equations. Section 13.3 gives an overview on the solution of Cauchy problems for non-linear evolution equations involving pseudodifferential operators.

13.1. Time-marching algorithms for non-linear parabolic equations

Some semi-analytic time-marching algorithms for the numerical solution of quasi-linear parabolic equations based upon approximate approximations are considered in this section. An important feature of the algorithms is that they are both explicit and stable under much milder restrictions to the time step, depending on the size of the grid, in comparison with the usual explicit difference schemes (cf., e.g., Richtmyer and Morton [**83**]). The algorithms give the time step approximation $u_h(\mathbf{x}, i\tau)$ in analytic form, which enables one to differentiate the approximate solution explicitly with respect to $\mathbf{x}$.

13.1.1. One-step time-marching algorithms. We start with the initial value problem for the semi-linear heat equation

$$
\begin{aligned}
u_t - \nu u_{xx} &= \frac{\partial}{\partial x} f(x,t,u) \quad \text{for} \quad x \in \mathbb{R}\,, \quad t > 0\,, \\
u(x,0) &= \varphi(x).
\end{aligned} \tag{13.1}
$$

For any $T > t \geq 0$, the equation (13.1) can be rewritten in the equivalent form

$$
\begin{aligned}
u(x,T) = u(x,t) &+ \int\limits_{\mathbb{R}} P(x-\xi, T-t)(u(\xi,t) - u(x,t))\, d\xi \\
&+ \frac{\partial}{\partial x} \int\limits_t^T \int\limits_{\mathbb{R}} P(x-\xi, T-\lambda)\, f(\xi,\lambda,u(\xi,\lambda))\, d\xi\, d\lambda\,,
\end{aligned} \tag{13.2}
$$

where P is the Poisson kernel

$$
P(x,t) = \frac{1}{\sqrt{4\nu\pi t}}\, \mathrm{e}^{-x^2/4\nu t}
$$

(cf., e.g., [**75**, Kapitel 25]). To derive a time-marching algorithm, we fix a time step τ and note that the solution $u(x,t)$ for $t = i\tau$, $i = 1, \ldots,$ can be obtained from

$$\begin{aligned} u(x,0) &= \varphi(x), \\ u(x,i\tau) &= u(x,(i-1)\tau) \\ &\quad + \int_{\mathbb{R}} P(x-\xi,\tau)\,(u(\xi,(i-1)\tau) - u(x,(i-1)\tau))\,d\xi \\ &\quad + \frac{\partial}{\partial x} \int_{(i-1)\tau}^{i\tau} \int_{\mathbb{R}} P(x-\xi, i\tau-\lambda) f(\xi,\lambda,u(\xi,\lambda))\,d\xi\,d\lambda\,. \end{aligned} \tag{13.3}$$

Representing the function $u(x,(i-1)\tau)$ by the simple Gaussian-based quasi-interpolant (1.7) and using (1.16), the first integral on the right-hand side can be approximated by

$$\begin{aligned} &\frac{h}{\sqrt{\pi\,(4\nu\tau + \mathcal{D}h^2)}} \sum_{m=-\infty}^{\infty} u(mh,(i-1)\tau)\ \mathrm{e}^{-(x-mh)^2/(4\nu\tau+\mathcal{D}h^2)} \\ &\qquad - \frac{1}{\sqrt{\pi\mathcal{D}}} \sum_{m=-\infty}^{\infty} u(mh,(i-1)\tau)\ \mathrm{e}^{-(x-mh)^2/\mathcal{D}h^2}\,. \end{aligned} \tag{13.4}$$

Using again the quasi-interpolant (1.7), we approximate the third term on the right-hand side of (13.3) by

$$\begin{aligned} &\frac{1}{\sqrt{\pi\mathcal{D}}} \frac{\partial}{\partial x} \int_{(i-1)\tau}^{i\tau} \int_{\mathbb{R}} \mathcal{P}(x-\xi, i\tau-\lambda) \\ &\qquad \times \sum_{m=-\infty}^{\infty} f(mh,\lambda,u(mh,\lambda))\ \mathrm{e}^{-(\xi-mh)^2/\mathcal{D}h^2}\ d\xi\,d\lambda\,, \end{aligned} \tag{13.5}$$

making an error $\mathcal{O}(\tau h^2)$ up to the saturation error. Here and in the following, we assume that the data and the solutions are sufficiently smooth. The one-step explicit time-marching procedure consists in replacing (13.5) by

$$\begin{aligned} I &= \frac{1}{\sqrt{\pi\mathcal{D}}} \sum_{m=-\infty}^{\infty} f(mh,(i-1)\tau,u(mh,(i-1)\tau)) \\ &\qquad \times \frac{\partial}{\partial x} \int_{(i-1)\tau}^{i\tau} \int_{\mathbb{R}} \mathcal{P}(x-\xi, i\tau-\lambda)\,\mathrm{e}^{-(\xi-mh)^2/\mathcal{D}h^2}\ d\xi\,d\lambda\,, \end{aligned} \tag{13.6}$$

which provides an error of order τ^2. The inner integral over $\mathbb{R}$ is equal to

$$\frac{h\sqrt{\mathcal{D}}}{\sqrt{4\nu(i\tau-\lambda)+\mathcal{D}h^2}}\ \mathrm{e}^{-(x-mh)^2/(4\nu(i\tau-\lambda)+\mathcal{D}h^2)}\,,$$

which along with the identity

$$\int_0^\tau \frac{(x-mh)e^{-(x-mh)^2/(4\nu\lambda+\mathcal{D}h^2)}}{(4\nu\lambda+\mathcal{D}h^2)^{3/2}}\,d\lambda = \frac{\sqrt{\pi}}{4\nu}\left(\operatorname{erf}\Big(\frac{x-mh}{\sqrt{\mathcal{D}}h}\Big) - \operatorname{erf}\Big(\frac{x-mh}{\sqrt{4\nu\tau+\mathcal{D}h^2}}\Big)\right)$$

leads to the explicit form of (13.6)

$$I = \frac{h}{2\nu}\sum_{m=-\infty}^{\infty} f(mh,(i-1)\tau,u(mh,(i-1)\tau)) \times \left(\operatorname{erf}\Big(\frac{x-mh}{\sqrt{4\nu\tau+\mathcal{D}h^2}}\Big) - \operatorname{erf}\Big(\frac{x-mh}{\sqrt{\mathcal{D}}h}\Big)\right).$$

Thus we have arrived at the step-by-step formula for determining the approximate solution $u_h(x,i\tau)$, $i=1,2,\ldots$,

$$\begin{aligned} u_h(x,i\tau) &= u_h(x,(i-1)\tau) \\ &+ \frac{h}{\sqrt{\pi}}\sum_{m=-\infty}^{\infty} u_h(mh,(i-1)\tau)\,\frac{\mathrm{e}^{-\Xi_m^2(\lambda)}}{(4\nu\lambda+\mathcal{D}h^2)^{1/2}}\Bigg|_{\lambda=0}^{\lambda=\tau} \\ &+ \frac{h}{2\nu}\sum_{m=-\infty}^{\infty} f(mh,(i-1)\tau,u_h(mh,(i-1)\tau))\operatorname{erf}(\Xi_m(\lambda))\Bigg|_{\lambda=0}^{\lambda=\tau}, \end{aligned} \tag{13.7}$$

with $u_h(x,0)=\varphi(x)$. Here and in the sequel, we use the notation

$$f(\lambda)\Big|_{\lambda=0}^{\lambda=\tau} = f(\tau)-f(0), \quad \Xi_m(\lambda) = \frac{x-mh}{\sqrt{4\nu\lambda+\mathcal{D}h^2}}.$$

By summarizing what was said on error estimates, we conclude that the algorithm just constructed gives the error $\mathcal{O}(\tau^2+\tau h^2)$ at each time step.

To obtain similar, but more precise, procedures, one can use the quasi-interpolants of higher approximation order. For example, the generating function

$$\eta_4(x) = \frac{1}{\sqrt{\pi}}\Big(\frac{3}{2}-x^2\Big)\,\mathrm{e}^{-x^2}$$

provides the explicit time-marching procedure

$$\begin{aligned} u_h(x,i\tau) &= u_h(x,(i-1)\tau) \\ &+ \frac{h}{\sqrt{\pi}}\sum_{m=-\infty}^{\infty} u_h(mh,(i-1)\tau)\left(4\nu\lambda + \frac{\mathcal{D}h^2(3-2\,\Xi_m^2(\lambda))}{2}\right)\frac{e^{-\Xi_m^2(\lambda)}}{(4\nu\lambda+\mathcal{D}h^2)^{3/2}}\Bigg|_{\lambda=0}^{\lambda=\tau} \\ &+ \frac{h}{2\nu}\sum_{m=-\infty}^{\infty} f(mh,(i-1)\tau,u_h(mh,(i-1)\tau)) \\ &\quad\times\left\{\frac{\mathcal{D}h^2}{\sqrt{\pi}}\,\Xi_m(\lambda)\,\frac{e^{-\Xi_m^2(\lambda)}}{4\nu\lambda+\mathcal{D}h^2} + \operatorname{erf}(\Xi_m(\lambda))\right\}\Bigg|_{\lambda=0}^{\lambda=\tau} \end{aligned} \tag{13.8}$$

with accuracy $\mathcal{O}(\tau^2+\tau h^4)$.

The above algorithms can be modified for the more general equation

$$u_t - \nu u_{xx} = \frac{\partial}{\partial x} f(x,t,u) + F(x,t,u) \quad \text{for} \quad x \in \mathbb{R}\,,\ t > 0. \tag{13.9}$$

The only difference in comparison with (13.7) is the appearance of the term

$$-\frac{h}{2\sqrt{\pi\nu}} \sum_{m=-\infty}^{\infty} F(mh,(i-1)\tau,u_h(mh,(i-1)\tau)) \\ \times \sqrt{4\nu\lambda + \mathcal{D}h^2}\left(e^{-\Xi_m^2(\lambda)} + \sqrt{\pi}\,\Xi_m(\lambda)\,\mathrm{erf}\,(\Xi_m(\lambda))\right)\Big|_{\lambda=0}^{\lambda=\tau}.$$

The analogous extra term for the algorithm (13.8) takes the form

$$-\frac{h}{4\sqrt{\pi\nu}} \sum_{m=-\infty}^{\infty} F(mh,(i-1)\tau,u_h(mh,(i-1)\tau)) \\ \times \sqrt{4\nu\lambda + \mathcal{D}h^2}\left(\frac{8\nu\lambda + \mathcal{D}h^2}{4\nu\lambda + \mathcal{D}h^2} e^{-\Xi_m^2(\lambda)} + 2\sqrt{\pi}\,\Xi_m(\lambda)\ \mathrm{erf}\,(\Xi_m(\lambda))\right)\Big|_{\lambda=0}^{\lambda=\tau}.$$

13.1.2. Higher-dimensional case. All the algorithms introduced above have direct analogs to the initial value problem for multi-dimensional equations

$$u_t - \nu\Delta u = \mathrm{div}\ \mathbf{f}(\mathbf{x},t,u)\,, \quad u(\mathbf{x},0) = \varphi(\mathbf{x})\,, \quad \mathbf{x} \in \mathbb{R}^n\,, \quad t > 0\,. \tag{13.10}$$

We rewrite the heat equation as

$$\begin{aligned} u(\mathbf{x},T) = u(\mathbf{x},t) &+ \int_{\mathbb{R}^n} \mathcal{P}(\mathbf{x}-\boldsymbol{\xi},T-t)(u(\boldsymbol{\xi},t) - u(\mathbf{x},t))\,d\boldsymbol{\xi} \\ &+ \mathrm{div}_{\mathbf{x}} \int_t^T \int_{\mathbb{R}^n} \mathcal{P}(\mathbf{x}-\boldsymbol{\xi},T-\lambda)\,\mathbf{f}(\boldsymbol{\xi},\lambda,u(\boldsymbol{\xi},\lambda))\,d\boldsymbol{\xi}\,d\lambda\,, \end{aligned} \tag{13.11}$$

with the Poisson kernel $\mathcal{P}$ (see (6.12)) and we employ the approximate quasi-interpolation with Gaussians in $\mathbb{R}^n$ to evaluate the integrals. We approximate $\mathbf{f}(\boldsymbol{\xi},\lambda,u(\boldsymbol{\xi},\lambda))$ by

$$(\pi\mathcal{D})^{-n/2} \sum_{\mathbf{m}\in\mathbb{Z}^n} \mathbf{f}(\mathbf{m}h,\lambda,u(\mathbf{m}h,\lambda))\ \mathrm{e}^{-|\boldsymbol{\xi}-\mathbf{m}h|^2/\mathcal{D}h^2}\,,$$

and we note that

$$
\begin{aligned}
(\pi\mathcal{D})^{-n/2}\operatorname{div}_{\mathbf{x}} \int\limits_{(i-1)\tau}^{i\tau} \int\limits_{\mathbb{R}^n} \mathcal{P}(\mathbf{x}-\boldsymbol{\xi}, i\tau-\lambda)\, \mathrm{e}^{-|\boldsymbol{\xi}-\mathbf{m}h|^2/\mathcal{D}h^2}\, d\boldsymbol{\xi}\, d\lambda \\
&= \frac{h^n}{\pi^{n/2}} \operatorname{div}_{\mathbf{x}} \int\limits_0^\tau \frac{\mathrm{e}^{-|\mathbf{x}-\mathbf{m}h|^2/(\mathcal{D}h^2+4\nu\lambda)}}{(\mathcal{D}h^2+4\nu\lambda)^{n/2}}\, d\lambda \\
&= -\frac{h^n(\mathbf{x}-\mathbf{m}h)}{2\nu\pi^{n/2}} \int\limits_{\mathcal{D}h^2}^{4\nu\tau+\mathcal{D}h^2} \frac{\mathrm{e}^{-|\mathbf{x}-\mathbf{m}h|^2/\lambda}}{\lambda^{1+n/2}}\, d\lambda \\
&= \frac{h^n}{2\nu\pi^{n/2}} \frac{\mathbf{x}-\mathbf{m}h}{|\mathbf{x}-\mathbf{m}h|^n} \int\limits_{z_{\mathbf{m}}(0)}^{z_{\mathbf{m}}(\tau)} \xi^{n/2-1}\, \mathrm{e}^{-\xi}\, d\xi
\end{aligned}
$$

with

$$
z_{\mathbf{m}}(\lambda) = \frac{|\mathbf{x}-h\mathbf{m}|^2}{4\nu\lambda+\mathcal{D}h^2}. \tag{13.12}
$$

Using the upper incomplete Gamma function defined by (11.63), we obtain

$$
\begin{aligned}
(\pi\mathcal{D})^{-n/2}\operatorname{div}_{\mathbf{x}} \int\limits_{(i-1)\tau}^{i\tau} \int\limits_{\mathbb{R}^n} \mathcal{P}(\mathbf{x}-\boldsymbol{\xi}, T-\lambda)\, \mathrm{e}^{-|\boldsymbol{\xi}-\mathbf{m}h|^2/\mathcal{D}h^2}\, d\boldsymbol{\xi}\, d\lambda \\
&= -\frac{h^n}{2\nu\pi^{n/2}} \frac{\mathbf{x}-\mathbf{m}h}{|\mathbf{x}-\mathbf{m}h|^n} \left(\Gamma\Big(\frac{n}{2}, z_{\mathbf{m}}(\tau)\Big) - \Gamma\Big(\frac{n}{2}, z_{\mathbf{m}}(0)\Big)\right).
\end{aligned}
$$

By the same arguments as in Subsection 13.1.1 and using formula (6.13) for the heat equation, we arrive at the computational formula

$$
\begin{aligned}
u_h(\mathbf{x}, i\tau) = {}& u_h(\mathbf{x}, (i-1)\tau) \\
&+ \frac{h^n}{\pi^{n/2}} \sum_{\mathbf{m}\in\mathbb{Z}^n} u_h(h\mathbf{m}, (i-1)\tau) \left.\frac{\mathrm{e}^{-z_{\mathbf{m}}(\lambda)}}{(4\nu\lambda+\mathcal{D}h^2)^{n/2}}\right|_{\lambda=0}^{\lambda=\tau} \\
&- \frac{h^n}{2\nu\pi^{n/2}} \sum_{\mathbf{m}\in\mathbb{Z}^n} \left\langle \mathbf{f}(h\mathbf{m}, (i-1)\tau, u_h(h\mathbf{m}, (i-1)\tau)), \frac{\mathbf{x}-h\mathbf{m}}{|\mathbf{x}-h\mathbf{m}|^n} \right\rangle \left.\Gamma\Big(\frac{n}{2}, z_{\mathbf{m}}(\lambda)\Big)\right|_{\lambda=0}^{\lambda=\tau}
\end{aligned} \tag{13.13}
$$

where $u_h(\mathbf{x}, 0) = \varphi(\mathbf{x})$. The accuracy is $\mathcal{O}(\tau^2 + \tau h^2)$ modulo saturation errors at each time step.

We note that in the case of even n the computational formulas (13.13) include only elementary functions, whereas for odd n the error function erf has to be evaluated (see Subsection 4.3.1).

For example, because $\Gamma(1,x) = \mathrm{e}^{-x}$, we obtain for $n=2$

$$\begin{aligned} u_h(\mathbf{x}, i\tau) = \ & u_h(\mathbf{x},(i-1)\tau) + \frac{h^2}{\pi} \sum_{\mathbf{m}\in\mathbb{Z}^2} u_h(h\mathbf{m},(i-1)\tau) \left. \frac{\mathrm{e}^{-z_{\mathbf{m}}(\lambda)}}{4\nu\lambda + \mathcal{D}h^2} \right|_{\lambda=0}^{\lambda=\tau} \\ & - \frac{h^2}{2\nu\pi} \sum_{\mathbf{m}\in\mathbb{Z}^2} \left\langle \mathbf{f}(h\mathbf{m},(i-1)\tau, u_h(h\mathbf{m},(i-1)\tau)), \frac{\mathbf{x}-h\mathbf{m}}{|\mathbf{x}-h\mathbf{m}|^2} \right\rangle \mathrm{e}^{-z_{\mathbf{m}}(\lambda)} \Big|_{\lambda=0}^{\lambda=\tau}. \end{aligned}$$

Since

$$\Gamma\Big(\frac{3}{2}, x\Big) = \mathrm{e}^{-x}\sqrt{x} + \frac{\sqrt{\pi}}{2}\operatorname{erfc}(\sqrt{x}),$$

in the case of $\mathbb{R}^3$, we derive the computational formula

$$\begin{aligned} u_h(\mathbf{x}, i\tau) = \ & u_h(\mathbf{x},(i-1)\tau) + \frac{h^3}{\pi^{3/2}} \sum_{\mathbf{m}\in\mathbb{Z}^3} u_h(h\mathbf{m},(i-1)\tau) \left. \frac{\mathrm{e}^{-z_{\mathbf{m}}(\lambda)}}{(4\nu\lambda + \mathcal{D}h^2)^{3/2}} \right|_{\lambda=0}^{\lambda=\tau} \\ & - \frac{h^3}{2\nu\pi^{3/2}} \sum_{\mathbf{m}\in\mathbb{Z}^3} \left\langle \mathbf{f}(h\mathbf{m},(i-1)\tau, u_h(h\mathbf{m},(i-1)\tau)), \frac{\mathbf{x}-h\mathbf{m}}{|\mathbf{x}-h\mathbf{m}|^3} \right\rangle \\ & \times \left(\mathrm{e}^{-z_{\mathbf{m}}(\lambda)} \frac{|\mathbf{x}-h\mathbf{m}|}{\sqrt{4\nu\lambda+\mathcal{D}h^2}} + \frac{\sqrt{\pi}}{2}\operatorname{erfc}(\sqrt{z_{\mathbf{m}}(\lambda)}) \right) \Big|_{\lambda=0}^{\lambda=\tau}. \end{aligned}$$

The higher-order approximate quasi-interpolants

$$\mathcal{M}_{h,\mathcal{D}}^{(2M)} u(\mathbf{x}) := \mathcal{D}^{-n/2} \sum_{\mathbf{m}\in\mathbb{Z}^n} u(h\mathbf{m})\, \eta_{2M}\Big(\frac{\mathbf{x}-h\mathbf{m}}{\sqrt{\mathcal{D}}\,h}\Big), \quad M = 2,3,\ldots,$$

with

$$\eta_{2M}(\mathbf{x}) = \pi^{-n/2} L_{M-1}^{(n/2)}(|\mathbf{x}|^2)\, \mathrm{e}^{-|\mathbf{x}|^2}$$

(cf. (3.18)) provide computational formulas of the order $\mathcal{O}(\tau^2 + \tau h^{2M})$.

We know from (6.15) that

$$\begin{aligned} & \int_{\mathbb{R}^n} \mathcal{P}(\mathbf{x}-\boldsymbol{\xi},\lambda)\, \eta_{2M}\Big(\frac{|\boldsymbol{\xi}-h\mathbf{m}|}{\sqrt{\mathcal{D}}h}\Big)\, d\boldsymbol{\xi} \\ & \quad = \frac{h^n\, \mathrm{e}^{-|\mathbf{x}-h\mathbf{m}|^2/(4\nu\lambda+\mathcal{D}h^2)}}{\pi^{n/2}(4\nu\lambda+\mathcal{D}h^2)^{n/2}} \sum_{j=0}^{M-1} \Big(\frac{\mathcal{D}h^2}{4\nu\lambda+\mathcal{D}h^2}\Big)^j L_j^{(n/2-1)}\Big(\frac{|\mathbf{x}-h\mathbf{m}|^2}{4\nu\lambda+\mathcal{D}h^2}\Big). \end{aligned}$$

One has only to determine

$$\begin{aligned} I_M & := \frac{\operatorname{div}_{\mathbf{x}}}{(\pi\mathcal{D})^{n/2}} \int_{(i-1)\tau}^{i\tau} \int_{\mathbb{R}^n} \mathcal{P}(\mathbf{x}-\boldsymbol{\xi}, i\tau-\lambda)\, L_{M-1}^{(n/2)}\Big(\frac{|\boldsymbol{\xi}|^2}{\mathcal{D}h^2}\Big)\, \mathrm{e}^{-|\boldsymbol{\xi}|^2/\mathcal{D}h^2}\, d\boldsymbol{\xi}\, d\lambda \\ & = \frac{h^n}{\pi^{n/2}} \operatorname{div}_{\mathbf{x}} \int_0^{\tau} \frac{\mathrm{e}^{-|\mathbf{x}|^2/(\mathcal{D}h^2+4\nu\lambda)}}{(\mathcal{D}h^2+4\nu\lambda)^{n/2}} \sum_{j=0}^{M-1} \Big(\frac{\mathcal{D}h^2}{4\nu\lambda+\mathcal{D}h^2}\Big)^j L_j^{(n/2-1)}\Big(\frac{|\mathbf{x}|^2}{4\nu\lambda+\mathcal{D}h^2}\Big)\, d\lambda \\ & = \frac{h^n}{4\nu\pi^{n/2}} \operatorname{div}_{\mathbf{x}} \int_{\mathcal{D}h^2}^{4\nu\tau+\mathcal{D}h^2} \frac{\mathrm{e}^{-|\mathbf{x}|^2/\lambda}}{\lambda^{n/2}} \sum_{j=0}^{M-1} \Big(\frac{\mathcal{D}h^2}{\lambda}\Big)^j L_j^{(n/2-1)}\Big(\frac{|\mathbf{x}|^2}{\lambda}\Big)\, d\lambda. \end{aligned}$$

By (4.38), we obtain

$$I_M = -\frac{h^n \mathbf{x}}{2\nu\pi^{n/2}} \int\limits_{\mathcal{D}h^2}^{4\nu\tau+\mathcal{D}h^2} \frac{\mathrm{e}^{-|\mathbf{x}|^2/\lambda}}{\lambda^{1+n/2}} \sum_{j=0}^{M-1} \Big(\frac{\mathcal{D}h^2}{\lambda}\Big)^j L_j^{(n/2)}\Big(\frac{|\mathbf{x}|^2}{\lambda}\Big)\, d\lambda$$

$$= \frac{h^n}{2\nu\pi^{n/2}} \frac{\mathbf{x}}{|\mathbf{x}|^n} \sum_{j=0}^{M-1} \frac{(\mathcal{D}h^2)^j}{|\mathbf{x}|^{2j}} \int\limits_{z(0)}^{z(\tau)} \mathrm{e}^{-\xi}\, \xi^{n/2-1+j}\, L_j^{(n/2)}(\xi)\, d\xi$$

$$= \frac{h^n}{2\nu\pi^{n/2}} \frac{\mathbf{x}}{|\mathbf{x}|^n} \Big(-\Gamma\Big(\frac{n}{2}, z(\lambda)\Big) + \mathrm{e}^{-z(\lambda)} \sum_{j=1}^{M-1} \frac{(\mathcal{D}h^2)^j}{j|\mathbf{x}|^{2j}} z(\lambda)^{j+n/2} L_{j-1}^{(n/2)}(z(\lambda))\Big)\Big|_{\lambda=0}^{\lambda=\tau}$$

with $z(\lambda) = |\mathbf{x}|^2/(4\nu\lambda + \mathcal{D}h^2)$, where in the last step, the formula

$$\int x^{j+\alpha-1}\, \mathrm{e}^{-x}\, L_j^{(\alpha)}(x)\, dx = \frac{1}{j}\, x^{j+\alpha}\, \mathrm{e}^{-x}\, L_{j-1}^{(\alpha)}(x)$$

(see [**82**, 1.14.2.9]) is used. After elementary transformations we derive

$$I_M = -\frac{h^n}{2\nu\pi^{n/2}} \frac{\mathbf{x}}{|\mathbf{x}|^n} \Gamma\Big(\frac{n}{2}, z(\lambda)\Big)\Big|_{\lambda=0}^{\lambda=\tau}$$
$$+ \frac{h^n}{2\nu\pi^{n/2}} \frac{\mathbf{x}\, \mathrm{e}^{-z(\lambda)}}{(4\nu\lambda+\mathcal{D}h^2)^{n/2}} \sum_{j=1}^{M-1} \frac{(\mathcal{D}h^2)^j}{j(4\nu\lambda+\mathcal{D}h^2)^j} L_{j-1}^{(n/2)}(z(\lambda))\Big|_{\lambda=0}^{\lambda=\tau}.$$

This leads to the time step algorithm

$$u_h(\mathbf{x}, i\tau) = u_h(\mathbf{x}, (i-1)\tau)$$
$$+ \frac{h^n}{\pi^{n/2}} \sum_{\mathbf{m}\in\mathbb{Z}^n} \Big\{ u_h(h\mathbf{m}, (i-1)\tau) \sum_{j=0}^{M-1} \frac{(\mathcal{D}h^2)^j\, \mathrm{e}^{-z_{\mathbf{m}}(\lambda)}}{(4\nu\lambda+\mathcal{D}h^2)^{n/2+j}} L_j^{(n/2-1)}(z_{\mathbf{m}}(\lambda))$$
$$(13.14)\qquad - \big\langle \mathbf{f}(h\mathbf{m}, (i-1)\tau, u_h(h\mathbf{m}, (i-1)\tau)), \mathbf{x} - h\mathbf{m}\big\rangle$$
$$\times \Big(|\mathbf{x}-h\mathbf{m}|^{-n}\, \Gamma\Big(\frac{n}{2}, z_{\mathbf{m}}(\lambda)\Big) - \sum_{j=1}^{M-1} \frac{(\mathcal{D}h^2)^j\, \mathrm{e}^{-z_{\mathbf{m}}(\lambda)}}{j(4\nu\lambda+\mathcal{D}h^2)^{n/2+j}} L_{j-1}^{(n/2)}(z_{\mathbf{m}}(\lambda))\Big)\Big\}\Big|_{\lambda=0}^{\lambda=\tau}$$

with $z_{\mathbf{m}}(\lambda)$ defined by (13.12). As mentioned above, this formula provides an approximate solution of (13.10) and has the order $\mathcal{O}(\tau^2 + \tau h^{2M})$ at each time step.

13.2. Application to the non-stationary Navier-Stokes equations

The Navier-Stokes equations describe the motion of a fluid in $\mathbb{R}^n$ ($n = 2$ or 3). They are to be solved for an unknown velocity vector $\mathbf{u}(\mathbf{x}, t) = (u_l(\mathbf{x}, t))_{1\le l\le n} \in \mathbb{R}^n$ and pressure $p(\mathbf{x}, t) \in \mathbb{R}$, defined at the point $\mathbf{x} \in \mathbb{R}^n$ and time $t \ge 0$. We restrict our attention to incompressible fluids filling all of $\mathbb{R}^n$. Then the equations are given by

$$\frac{\partial u_l}{\partial t} + \sum_{k=1}^{n} u_k \frac{\partial u_l}{\partial x_k} = \nu \Delta u_l - \frac{\partial p}{\partial x_l} + f_l(\mathbf{x}, t)\,, \qquad \mathbf{x} \in \mathbb{R}^n,\ t \geq 0\,, \tag{13.15}$$

$$\operatorname{div} \mathbf{u} = 0\,, \qquad \mathbf{x} \in \mathbb{R}^n,\ t \geq 0\,, \tag{13.16}$$

with the initial conditions

$$\mathbf{u}(\mathbf{x}, 0) = \boldsymbol{\varphi}(\mathbf{x})\,, \quad \mathbf{x} \in \mathbb{R}^n\,. \tag{13.17}$$

Here, $\boldsymbol{\varphi} = (\varphi_1, \ldots, \varphi_n)$ is a given smooth divergence-free vector field in $\mathbb{R}^n$, the viscosity ν is a positive coefficient, $f_l(\mathbf{x}, t)$ are the components of a given externally applied force.

13.2.1. Integral equation formulation. In the following we describe a time-marching algorithm for the numerical solution of this Cauchy problem for the Navier-Stokes equations. For simplicity, we assume that no external force is applied, i.e., $f_i = 0$. Using (13.16), which just says that the fluid is incompressible, equation (13.15) can be written in the vector form

$$\mathbf{u}_t - \nu \Delta \mathbf{u} = -\nabla p - \sum_{k=1}^{n} \frac{\partial}{\partial x_k}(u_k \mathbf{u})\,, \tag{13.18}$$

or, for any $0 \leq t < T$, analogously to (13.11), as

$$\begin{aligned} \mathbf{u}(\mathbf{x}, T) = \mathbf{u}(\mathbf{x}, t) &+ \int_{\mathbb{R}^n} \mathcal{P}(\mathbf{x} - \boldsymbol{\xi}, T - t)(\mathbf{u}(\boldsymbol{\xi}, t) - \mathbf{u}(\mathbf{x}, t))\, d\boldsymbol{\xi} \\ &- \nabla \int_t^T \int_{\mathbb{R}^n} \mathcal{P}(\mathbf{x} - \boldsymbol{\xi}, T - \lambda)\, p(\boldsymbol{\xi}, \lambda)\, d\boldsymbol{\xi}\, d\lambda \\ &- \sum_{k=1}^{n} \frac{\partial}{\partial x_k} \int_t^T \int_{\mathbb{R}^n} \mathcal{P}(\mathbf{x} - \boldsymbol{\xi}, T - \lambda)\, (u_k \mathbf{u})(\boldsymbol{\xi}, \lambda)\, d\boldsymbol{\xi}\, d\lambda\,. \end{aligned} \tag{13.19}$$

The second and third integrals on the right-hand side can be combined by the following observation: If we apply the divergence operator to (13.18), then we find by the continuity equation (13.16) that

$$\Delta p = -\sum_{j,k=1}^{n} \frac{\partial^2}{\partial x_j \partial x_k}(u_j u_k)\,.$$

Hence the pressure is expressed by

$$p = -\Delta^{-1} \sum_{j,k=1}^{n} \frac{\partial^2}{\partial x_j \partial x_k}(u_j u_k)\,. \tag{13.20}$$

Then the l-th component of the second integral in (13.19) transforms to

$$\begin{aligned}&\frac{\partial}{\partial x_l}\int_t^T\int_{\mathbb{R}^n}\mathcal{P}(\mathbf{x}-\boldsymbol{\xi},T-\lambda)\,p(\boldsymbol{\xi},\lambda)\,d\boldsymbol{\xi}\,d\lambda\\&\quad=-\Delta^{-1}\sum_{j=1}^n\frac{\partial^2}{\partial x_j\partial x_l}\sum_{k=1}^n\frac{\partial}{\partial x_k}\int_t^T\int_{\mathbb{R}^n}\mathcal{P}(\mathbf{x}-\boldsymbol{\xi},T-\lambda)(u_ju_k)(\boldsymbol{\xi},\lambda)\,d\boldsymbol{\xi}\,d\lambda\\&\quad=-\Delta^{-1}\sum_{j=1}^n\frac{\partial^2}{\partial x_j\partial x_l}\operatorname{div}\int_t^T\int_{\mathbb{R}^n}\mathcal{P}(\mathbf{x}-\boldsymbol{\xi},T-\lambda)(u_j\mathbf{u})(\boldsymbol{\xi},\lambda)\,d\boldsymbol{\xi}\,d\lambda\,.\end{aligned}$$

Moreover, the l-th component of the vector function

$$\sum_{k=1}^n\frac{\partial}{\partial x_k}\int_t^T\int_{\mathbb{R}^n}\mathcal{P}(\mathbf{x}-\boldsymbol{\xi},T-\lambda)\,(u_k\mathbf{u})(\boldsymbol{\xi},\lambda)\,d\boldsymbol{\xi}\,d\lambda$$

can be written in the form

$$\operatorname{div}\int_t^T\int_{\mathbb{R}^n}\mathcal{P}(\mathbf{x}-\boldsymbol{\xi},T-\lambda)(u_l\mathbf{u})(\boldsymbol{\xi},\lambda)\,d\boldsymbol{\xi}\,d\lambda\,.$$

This leads to

$$\begin{aligned}&\operatorname{div}\int_t^T\int_{\mathbb{R}^n}\mathcal{P}(\mathbf{x}-\boldsymbol{\xi},T-\lambda)(u_l\mathbf{u})(\boldsymbol{\xi},\lambda)\,d\boldsymbol{\xi}\,d\lambda\\&\qquad+\frac{\partial}{\partial x_l}\int_t^T\int_{\mathbb{R}^n}\mathcal{P}(\mathbf{x}-\boldsymbol{\xi},T-\lambda)\,p(\boldsymbol{\xi},\lambda)\,d\boldsymbol{\xi}\,d\lambda\\&=\sum_{j=1}^n\left(\delta_{jl}-\Delta_{\mathbf{x}}^{-1}\frac{\partial^2}{\partial x_j\partial x_l}\right)\operatorname{div}\int_t^T\int_{\mathbb{R}^n}\mathcal{P}(\mathbf{x}-\boldsymbol{\xi},T-\lambda)(u_j\mathbf{u})(\boldsymbol{\xi},\lambda)\,d\boldsymbol{\xi}\,d\lambda\,.\end{aligned}$$

Thus, by (13.19), the components u_l, $1\le l\le n$, of the solution $\mathbf{u}(\mathbf{x},t)$ satisfy the equations

$$\begin{aligned}u_l(\mathbf{x},T)=&u_l(\mathbf{x},t)+\int_{\mathbb{R}^n}\mathcal{P}(\mathbf{x}-\boldsymbol{\xi},T-t)(u_l(\boldsymbol{\xi},t)-u_l(\mathbf{x},t))\,d\boldsymbol{\xi}\\&-\sum_{j=1}^n\left(\delta_{jl}-\Delta_{\mathbf{x}}^{-1}\frac{\partial^2}{\partial x_j\partial x_l}\right)F_j\,,\end{aligned}\tag{13.21}$$

with the functions

$$F_j(\mathbf{x})=\operatorname{div}\int_t^T\int_{\mathbb{R}^n}\mathcal{P}(\mathbf{x}-\boldsymbol{\xi},T-\lambda)(u_j\mathbf{u})(\boldsymbol{\xi},\lambda)\,d\boldsymbol{\xi}\,d\lambda\,,\quad 1\le j\le n.$$

13.2.2. One-step time-marching algorithm. As before, we split the time interval into the subintervals $[(i-1)\tau, i\tau]$, $i = 1, 2, \dots$, and obtain

$$u_l(\mathbf{x}, i\tau) = u_l(\mathbf{x}, (i-1)\tau) + \int_{\mathbb{R}^n} \mathcal{P}(\mathbf{x}-\boldsymbol{\xi}, \tau)(u_l(\boldsymbol{\xi}, (i-1)\tau) - u_l(\mathbf{x}, (i-1)\tau))\, d\boldsymbol{\xi}$$

$$- \sum_{j=1}^{n} \Big(\delta_{jl} - \Delta_{\mathbf{x}}^{-1} \frac{\partial^2}{\partial x_j \partial x_l} \Big) \operatorname{div} \int_{(i-1)\tau}^{i\tau} \int_{\mathbb{R}^n} \mathcal{P}(\mathbf{x}-\boldsymbol{\xi}, i\tau - \lambda)(u_j \mathbf{u})(\boldsymbol{\xi}, \lambda)\, d\boldsymbol{\xi}\, d\lambda\,.$$

Now, we use (13.13) to approximate the functions

$$F_j^{(i)}(\mathbf{x}) = \operatorname{div} \int_{(i-1)\tau}^{i\tau} \int_{\mathbb{R}^n} \mathcal{P}(\mathbf{x}-\boldsymbol{\xi}, i\tau - \lambda)(u_j \mathbf{u})(\boldsymbol{\xi}, \lambda)\, d\boldsymbol{\xi}\, d\lambda$$

in the form
(13.22)

$$-\frac{h^n}{2\nu\pi^{n/2}} \sum_{\mathbf{m}\in\mathbb{Z}^n} \Big\langle (u_j \mathbf{u})(h\mathbf{m}, (i-1)\tau), \frac{\mathbf{x}-h\mathbf{m}}{|\mathbf{x}-h\mathbf{m}|^n} \Big\rangle \Gamma\Big(\frac{n}{2}, \frac{|\mathbf{x}-h\mathbf{m}|^2}{\kappa(\lambda)}\Big)\Big|_{\lambda=0}^{\lambda=\tau},$$

where $\kappa(\lambda) = (4\nu\lambda + \mathcal{D}h^2)$ and $\Gamma(a, x)$ is the incomplete Gamma function (11.63). Hence, we obtain the approximation

$$\Big(\delta_{jl} - \Delta_{\mathbf{x}}^{-1} \frac{\partial^2}{\partial x_j \partial x_l} \Big) F_j^{(i)}$$

$$\approx \frac{h^n}{2\nu\pi^{n/2}} \sum_{\mathbf{m}\in\mathbb{Z}^n} \sum_{k=1}^{n} (u_j u_k)(h\mathbf{m}, (i-1)\tau) \tag{13.23}$$

$$\times \Big(W_{jkl}(\lambda, \mathbf{x}-h\mathbf{m}) - \delta_{jl} \frac{x_k - hm_k}{|\mathbf{x}-h\mathbf{m}|^n} \Gamma\Big(\frac{n}{2}, \frac{|\mathbf{x}-h\mathbf{m}|^2}{\kappa(\lambda)}\Big)\Big)\Big|_{\lambda=0}^{\lambda=\tau},$$

where the abbreviation

$$W_{jkl}(\lambda, \mathbf{x}-h\mathbf{m}) = \Delta^{-1} \frac{\partial^2}{\partial x_j \partial x_l} \Big(\frac{x_k - hm_k}{|\mathbf{x}-h\mathbf{m}|^n} \Gamma\Big(\frac{n}{2}, \frac{|\mathbf{x}-h\mathbf{m}|^2}{\kappa(\lambda)}\Big)\Big) \tag{13.24}$$

is used. By setting $\mathbf{y} = \mathbf{x} - h\mathbf{m}$ and noting that

$$\frac{\partial}{\partial y_k} \int_{|\mathbf{y}|^2/\kappa(\lambda)}^{\infty} \Gamma\Big(\frac{n}{2}, t\Big) \frac{dt}{t^{n/2}} = -2\kappa(\lambda)^{n/2-1} \frac{y_k}{|\mathbf{y}|^n} \Gamma\Big(\frac{n}{2}, \frac{|\mathbf{y}|^2}{\kappa(\lambda)}\Big),$$

we get

$$W_{jkl}(\lambda, \mathbf{y}) = -\frac{1}{2\kappa(\lambda)^{n/2-1}} \Delta^{-1} \frac{\partial^3}{\partial y_j \partial y_k \partial y_l} \int_{|\mathbf{y}|^2/\kappa(\lambda)}^{\infty} \Gamma\Big(\frac{n}{2}, t\Big) \frac{dt}{t^{n/2}}$$

$$= \frac{1}{2\kappa(\lambda)^{n/2-2}} \frac{\partial^3}{\partial y_j \partial y_k \partial y_l} v\Big(\frac{|\mathbf{y}|^2}{\kappa(\lambda)}\Big),$$

with the function v defined by

$$v(|\mathbf{x}|^2) = -\Delta^{-1} \int_{|\mathbf{x}|^2}^{\infty} \Gamma\Big(\frac{n}{2}, t\Big) \frac{dt}{t^{n/2}}\,.$$

Turning to spherical coordinates, we obtain

$$\frac{1}{r^{n-1}}\frac{d}{dr}\Big(r^{n-1}\frac{d}{dr}v(r^2)\Big) = -\int_{r^2}^{\infty}\Gamma\Big(\frac{n}{2},t\Big)\frac{dt}{t^{n/2}},$$

which leads to the equation

$$4\tau v''(\tau) + 2nv'(\tau) = -\int_{\tau}^{\infty}\Gamma\Big(\frac{n}{2},t\Big)\frac{dt}{t^{n/2}},$$

or, equivalently, to

$$4(\tau^{n/2}v')' = -\tau^{n/2-1}\int_{\tau}^{\infty}\Gamma\Big(\frac{n}{2},t\Big)\frac{dt}{t^{n/2}}. \tag{13.25}$$

Note that the right-hand side of (13.25) can be given as

$$-\int_{\tau}^{\infty}\Gamma\Big(\frac{n}{2},t\Big)\frac{dt}{t^{n/2}} = \begin{cases} -\displaystyle\int_{\tau}^{\infty}\mathrm{e}^{-t}\frac{dt}{t} = -\Gamma(0,x), & n=2, \\ \dfrac{2}{n-2}\Big(\tau^{n/2-1}\,\mathrm{e}^{-\tau} - \Gamma\Big(\dfrac{n}{2},\tau\Big)\Big), & n>2. \end{cases}$$

Now we are in a position to derive analytic formulas for the functions W_{jkl}. Since

$$\begin{aligned} W_{jkl}(\lambda,\mathbf{y}) &= \frac{1}{2\kappa(\lambda)^{n/2-2}}\frac{\partial^3}{\partial y_j\partial y_k\partial y_l}v\Big(\frac{|\mathbf{y}|^2}{\kappa(\lambda)}\Big) \\ &= \frac{1}{\kappa(\lambda)^{n/2-1}}\frac{\partial^2}{\partial y_j\partial y_k}v'\Big(\frac{|\mathbf{y}|^2}{\kappa(\lambda)}\Big)y_l \\ &= \frac{2}{\kappa(\lambda)^{n/2}}v''\Big(\frac{|\mathbf{y}|^2}{\kappa(\lambda)}\Big)(\delta_{kl}y_j + \delta_{jl}y_k + \delta_{jk}y_l) \\ &\quad + \frac{4}{\kappa(\lambda)^{n/2+1}}v'''\Big(\frac{|\mathbf{y}|^2}{\kappa(\lambda)}\Big)y_jy_ky_l, \end{aligned} \tag{13.26}$$

it suffices to find the second and third derivatives of the solution of (13.25).

13.2.3. Formulas for the plane problem. Let us start with the case $n = 2$. The derivative of the solution of the equation

$$4\tau v'' + 4v' = -\int_{\tau}^{\infty}\mathrm{e}^{-t}\frac{dt}{t}$$

is given by

$$4\,v'(\tau) = \frac{\mathrm{e}^{-\tau}-1}{\tau} - \int_{\tau}^{\infty}\mathrm{e}^{-t}\frac{dt}{t}.$$

Hence one has only to insert

$$\frac{2}{\kappa(\lambda)}v''\Big(\frac{|\mathbf{y}|^2}{\kappa(\lambda)}\Big) = \kappa(\lambda)\,\frac{1-\mathrm{e}^{-|\mathbf{y}|^2/\kappa(\lambda)}}{2|\mathbf{y}|^4},$$

$$\frac{4}{\kappa(\lambda)^2}v'''\Big(\frac{|\mathbf{y}|^2}{\kappa(\lambda)}\Big) = 2\kappa(\lambda)\,\frac{\mathrm{e}^{-|\mathbf{y}|^2/\kappa(\lambda)}-1}{|\mathbf{y}|^6} + \frac{\mathrm{e}^{-|\mathbf{y}|^2/\kappa(\lambda)}}{|\mathbf{y}|^4}$$

into (13.26), which implies

$$\begin{aligned} W_{ikl}(\lambda,\mathbf{y}) =& \, y_i y_k y_l \frac{\mathrm{e}^{-|\mathbf{y}|^2/\kappa(\lambda)}}{|\mathbf{y}|^4} \\ &+ \kappa(\lambda)\,\frac{1-\mathrm{e}^{-|\mathbf{y}|^2/\kappa(\lambda)}}{2|\mathbf{y}|^4}\Big(\delta_{kl}y_i + \delta_{il}y_k + \delta_{ik}y_l - \frac{4y_iy_ky_l}{|\mathbf{y}|^2}\Big). \end{aligned}$$

Recall (13.23) and that we approximate

$$\begin{aligned} &\Big(\delta_{jl} - \Delta^{-1}\frac{\partial^2}{\partial x_j \partial x_l}\Big)F_j^{(i)} \\ &\qquad \approx \frac{h^2}{2\nu\pi}\sum_{\mathbf{m}\in\mathbb{Z}^2}\sum_{k=1}^{2}(u_ju_k)(h\mathbf{m},(i-1)\tau)Z_{jkl}^{(2)}(\lambda,\mathbf{x}-h\mathbf{m})\Big|_{\lambda=0}^{\lambda=\tau}, \end{aligned}$$

where the coefficients $Z_{jkl}^{(2)}$ are given by

$$\begin{aligned} Z_{jkl}^{(2)}(\lambda,\mathbf{y}) &= W_{jkl}(\lambda,\mathbf{y}) - \delta_{jl}\frac{y_k}{|\mathbf{y}|^2}\Gamma\Big(1,\frac{|\mathbf{y}|^2}{\kappa(\lambda)}\Big) \\ &= \Big(\frac{y_jy_l}{|\mathbf{y}|^2} - \delta_{jl}\Big)\frac{y_k\,\mathrm{e}^{-|\mathbf{y}|^2/\kappa(\lambda)}}{|\mathbf{y}|^2} \\ &\quad + \kappa(\lambda)\,\frac{1-\mathrm{e}^{-|\mathbf{y}|^2/\kappa(\lambda)}}{2|\mathbf{y}|^4}\Big(\delta_{kl}y_j + \delta_{jl}y_k + \delta_{jk}y_l - \frac{4y_jy_ky_l}{|\mathbf{y}|^2}\Big). \end{aligned}$$

It remains to notice that by (13.13) the first integral on the right-hand side of (13.19) can be approximated by the sum

$$\frac{h^2}{\pi}\sum_{\mathbf{m}\in\mathbb{Z}^2}\mathbf{u}_h(h\mathbf{m},(n-1)\tau)\,\frac{\mathrm{e}^{-|\mathbf{x}-h\mathbf{m}|^2/(4\nu\lambda+\mathcal{D}h^2)}}{4\pi\lambda+\mathcal{D}h^2}\Bigg|_{\lambda=0}^{\lambda=\tau}.$$

Combining this with (13.21), we arrive at the time-marching algorithm for the velocity components

$$\begin{aligned} u_{h,l}(\mathbf{x},i\tau) &= u_{h,l}(\mathbf{x},(i-1)\tau) \\ &\quad + \frac{h^2}{\pi}\sum_{\mathbf{m}\in\mathbb{Z}^2}u_{h,l}(h\mathbf{m},(i-1)\tau)\,\frac{\mathrm{e}^{-|\mathbf{x}-h\mathbf{m}|^2/(4\nu\lambda+\mathcal{D}h^2)}}{4\nu\lambda+\mathcal{D}h^2}\Bigg|_{\lambda=0}^{\tau} \\ &\quad - \frac{h^2}{2\nu\pi}\sum_{\mathbf{m}\in\mathbb{Z}^2}\sum_{j,k=1}^{2}(u_{h,j}u_{h,k})(h\mathbf{m},(i-1)\tau)Z_{jkl}^{(2)}(\lambda,\mathbf{x}-h\mathbf{m})\Big|_{\lambda=0}^{\lambda=\tau}, \end{aligned} \tag{13.27}$$

with $u_{h,l}(\mathbf{x},0) = \varphi_l(\mathbf{x})$, $l = 1, 2$,

$$Z^{(2)}_{jkl}(\lambda, \mathbf{x} - h\mathbf{m}) = \left(\frac{(x_j - hm_j)(x_l - hm_l)}{|\mathbf{x} - h\mathbf{m}|^2} - \delta_{jl} \right) \frac{x_k - hm_k}{|\mathbf{x} - h\mathbf{m}|^2} \,\mathrm{e}^{-z_{\mathbf{m}}(\lambda)}$$
$$+ \left\{ \delta_{kl} \frac{x_j - hm_j}{|\mathbf{x} - h\mathbf{m}|} + \delta_{jl} \frac{x_k - hm_k}{|\mathbf{x} - h\mathbf{m}|} + \delta_{jk} \frac{x_l - hm_l}{|\mathbf{x} - h\mathbf{m}|} \right.$$
$$\left. - 4 \frac{(x_j - hm_j)(x_k - hm_k)(x_l - hm_l)}{|\mathbf{x} - h\mathbf{m}|^3} \right\} \frac{1 - \mathrm{e}^{-z_{\mathbf{m}}(\lambda)}}{2|\mathbf{x} - h\mathbf{m}|\, z_{\mathbf{m}}(\lambda)}$$

and $z_{\mathbf{m}}(\lambda)$ is given by (13.12).

When the approximate velocity is found, we apply (13.20) to the quasi-interpolant

$$\mathcal{M}_{2,h}(u_{h,j}u_{h,k})(\mathbf{x}, i\tau) = \frac{1}{\pi \mathcal{D}} \sum_{\mathbf{m} \in \mathbb{Z}^2} (u_{h,i}u_{h,k})(h\mathbf{m}, i\tau) \,\mathrm{e}^{-|\mathbf{x}-h\mathbf{m}|^2/\mathcal{D}h^2},$$

i.e., the approximation of the pressure is computed from

$$p_h(\mathbf{x}, i\tau) = - \sum_{j,k=1}^{2} \frac{\partial^2}{\partial x_j \partial x_k} \Delta^{-1} \mathcal{M}_{2,h}(u_{h,j}u_{h,k})(\mathbf{x}, i\tau).$$

Now, we apply formula (6.4) stating that

$$\frac{\partial^2}{\partial x_j \partial x_k} \mathcal{L}_2(\mathrm{e}^{-|\cdot|^2})(\mathbf{x}) = \frac{2x_j x_k - \delta_{jk}|\mathbf{x}|^2}{2|\mathbf{x}|^4} (1 - \mathrm{e}^{-|\mathbf{x}|^2}) - \frac{x_j x_k}{|\mathbf{x}|^2} \,\mathrm{e}^{-|\mathbf{x}|^2},$$

which leads to the approximation formula for the pressure

$$(13.28) \quad p_h(\mathbf{x}, i\tau) = \frac{1}{\pi \mathcal{D}} \sum_{\mathbf{m} \in \mathbb{Z}^2} \sum_{j,k=1}^{2} (u_{h,j}u_{h,k})(h\mathbf{m}, i\tau) \left(\delta_{jk} \frac{\mathrm{e}^{-z_{\mathbf{m}}(0)} - 1}{2 z_{\mathbf{m}}(0)} + \frac{(x_j - hm_j)(x_k - hm_k)}{|\mathbf{x} - h\mathbf{m}|^2} \left(\frac{1 - \mathrm{e}^{-z_{\mathbf{m}}(0)}}{z_{\mathbf{m}}(0)} - \mathrm{e}^{-z_{\mathbf{m}}(0)} \right) \right).$$

Since $\mathbf{u}_h$ approximates the solution $\mathbf{u}$ with the order $\mathcal{O}(\tau^2 + \tau h^2)$, the pressure is approximated by (13.28) with the same order.

13.2.4. Numerical example. The two-dimensional algorithm (13.27), (13.28) was implemented and tested for different values of the viscosity ν, the discretization parameters τ and h, as well as for different initial functions φ.

In the Figs. 13.1 – 13.4 we depict the velocities $u_k(\mathbf{x}, t)$ and the pressure $p(\mathbf{x}, t)$ at the time stamps $t = 0.0$, 0.6, 1.2, 2.0 for the Navier-Stokes equation

$$\mathbf{u}_t - \nu \Delta \mathbf{u} = -\nabla p - \sum_{k=1}^{2} \frac{\partial}{\partial x_k}(u_k \mathbf{u}),$$

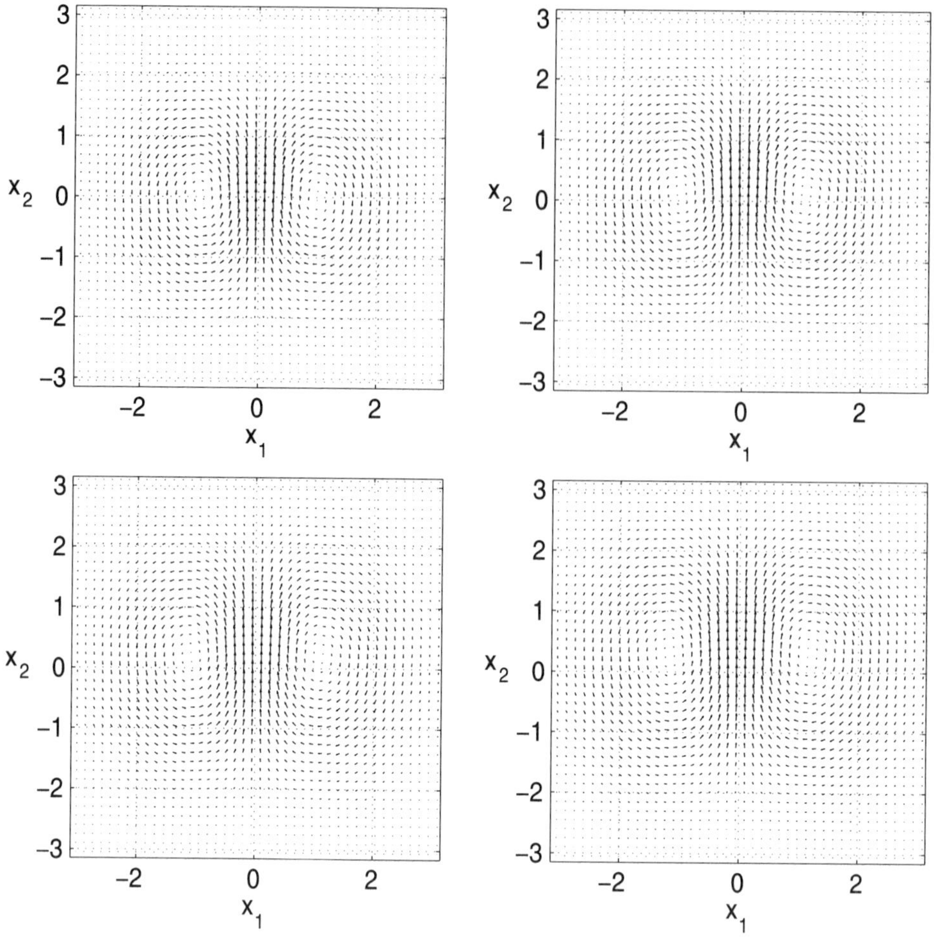

FIGURE 13.1. Evolution of the velocity field for $t = 0.0$ (upper left), $t = 0.6$ (upper right), $t = 1.2$ (lower left), and $t = 2.0$ (lower right)

with $\nu = 0.1$ and the initial conditions

$$u_1(\mathbf{x}, 0) = x_1 x_2 \, \mathrm{e}^{-|\mathbf{x}|^2/2}, \; u_2(\mathbf{x}, 0) = \left(1 - x_1^2\right) \mathrm{e}^{-|\mathbf{x}|^2/2}, \; \mathbf{x} \in \mathbb{R}^2 .$$

For this example, the time and spatial discretization parameters are $\tau = 0.01$ and $h = 0.05$.

13.2.5. Formulas for $\mathbb{R}^n$, $n > 2$. Using the relation

$$\Gamma(a+1, \tau) = a\Gamma(a, \tau) + \tau^a \, \mathrm{e}^{-\tau}, \tag{13.29}$$

one can easily check that the function

$$w(\tau) = \frac{1}{\tau^{n/2}} \left(\Gamma\left(\frac{n}{2} + 1, \tau\right) - (1 + \tau)\Gamma\left(\frac{n}{2}, \tau\right) \right)$$

is a solution of

$$(\tau^{n/2} w)' = -\left(\tau^{n/2-1} \, \mathrm{e}^{-\tau} - \Gamma\left(\frac{n}{2}, \tau\right) \right).$$

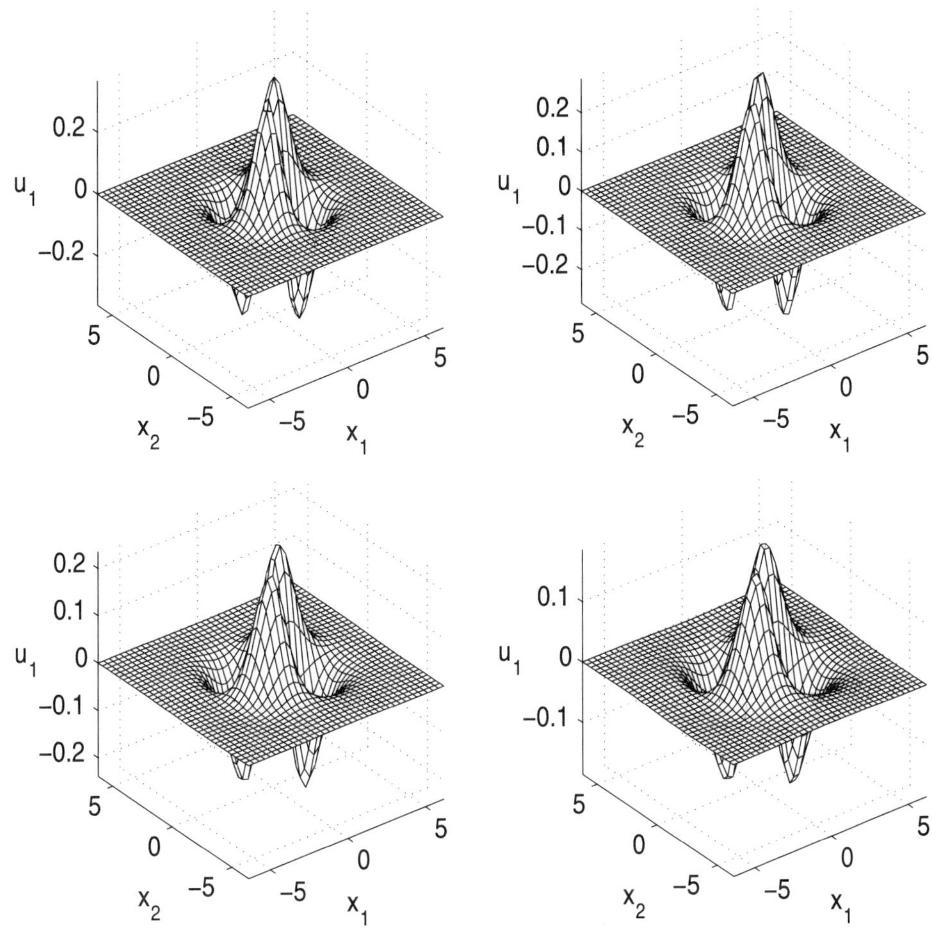

FIGURE 13.2. Evolution of u_1 for $t = 0.0$ (upper left), $t = 0.6$ (upper right), $t = 1.2$ (lower left), and $t = 2.0$ (lower right)

Hence, the derivative of the solution of (13.25) is equal to

$$4v'(\tau) = \frac{2}{(n-2)\tau^{n/2}}\Big(\Gamma\Big(\frac{n}{2}+1,\tau\Big) - (1+\tau)\Gamma\Big(\frac{n}{2},\tau\Big)\Big),$$

which gives after elementary transformations

$$\begin{aligned} 4v''(\tau) &= \frac{1}{\tau^{n/2}}\,\Gamma\Big(\frac{n}{2},\tau\Big) - \frac{1}{\tau^{n/2+1}}\,\Gamma\Big(\frac{n}{2}+1,\tau\Big), \\ 4v'''(\tau) &= \frac{n+2}{2\tau^{n/2+2}}\,\Gamma\Big(\frac{n}{2}+1,\tau\Big) - \frac{n}{2\tau^{n/2+1}}\,\Gamma\Big(\frac{n}{2},\tau\Big). \end{aligned}$$

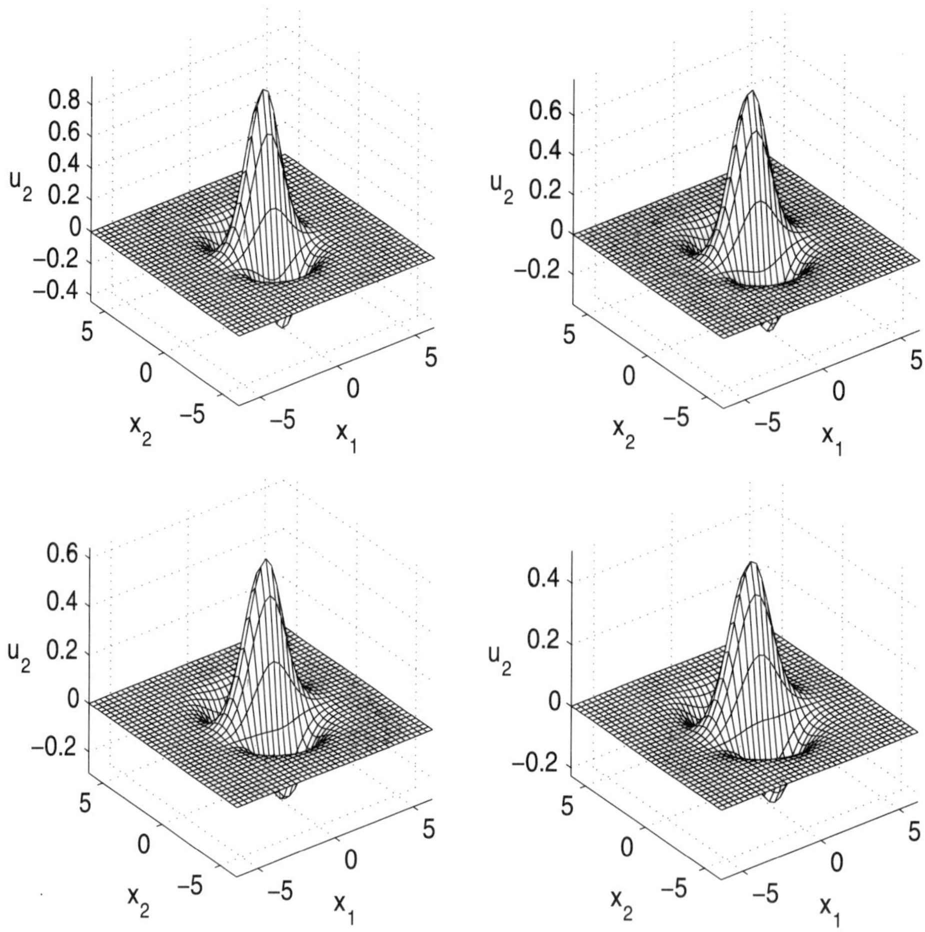

FIGURE 13.3. Evolution of u_2 for $t = 0.0$ (upper left), $t = 0.6$ (upper right), $t = 1.2$ (lower left), and $t = 2.0$ (lower right)

Inserting

$$\frac{2}{\kappa(\lambda)^{n/2}} v''\Big(\frac{|\mathbf{y}|^2}{\kappa(\lambda)}\Big) = \frac{1}{2|\mathbf{y}|^n}\,\Gamma\Big(\frac{n}{2}, \frac{|\mathbf{y}|^2}{\kappa(\lambda)}\Big) - \frac{\kappa(\lambda)}{2|\mathbf{y}|^{n+2}}\,\Gamma\Big(\frac{n}{2}+1, \frac{|\mathbf{y}|^2}{\kappa(\lambda)}\Big),$$

$$\frac{4}{\kappa(\lambda)^{n/2+1}} v'''\Big(\frac{|\mathbf{y}|^2}{\kappa(\lambda)}\Big) = \frac{(n+2)\kappa(\lambda)}{2|\mathbf{y}|^{n+4}}\,\Gamma\Big(\frac{n}{2}+1, \frac{|\mathbf{y}|^2}{\kappa(\lambda)}\Big) - \frac{n}{2|\mathbf{y}|^{n+2}}\,\Gamma\Big(\frac{n}{2}, \frac{|\mathbf{y}|^2}{\kappa(\lambda)}\Big)$$

into (13.26), we derive

$$\begin{aligned} W_{jkl}(\lambda, \mathbf{y}) = {} & \frac{\kappa(\lambda)}{2|\mathbf{y}|^{n+2}}\Gamma\Big(\frac{n}{2}+1, \frac{|\mathbf{y}|^2}{\kappa(\lambda)}\Big)\Big(\frac{(n+2)y_j y_k y_l}{|\mathbf{y}|^2} - (\delta_{kl}y_j + \delta_{jl}y_k + \delta_{jk}y_l)\Big) \\ & - \frac{1}{2|\mathbf{y}|^n}\Gamma\Big(\frac{n}{2}, \frac{|\mathbf{y}|^2}{\kappa(\lambda)}\Big)\Big(\frac{n\,y_j y_k y_l}{|\mathbf{y}|^2} - (\delta_{kl}y_j + \delta_{jl}y_k + \delta_{jk}y_l)\Big). \end{aligned}$$

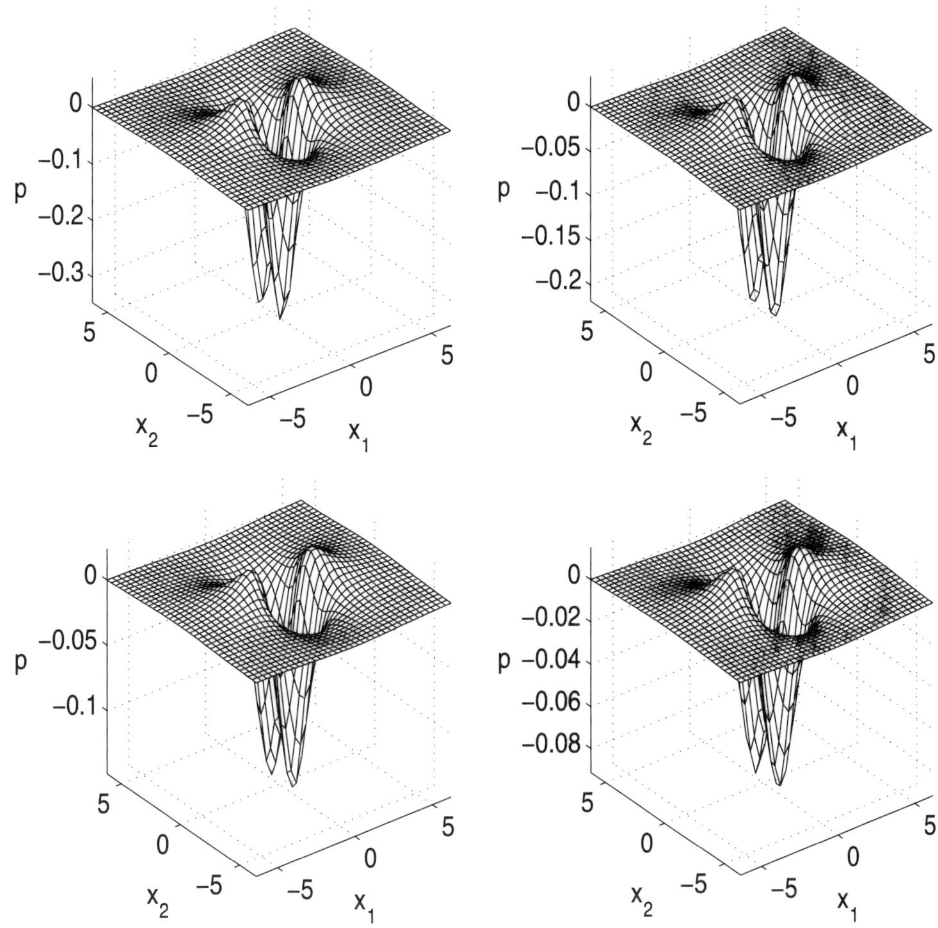

FIGURE 13.4. Evolution of pressure for $t = 0.0$ (upper left), $t = 0.6$ (upper right), $t = 1.2$ (lower left), and $t = 2.0$ (lower right)

Hence, in $\mathbb{R}^n$ we obtain the approximation

$$\begin{aligned}&\Big(\delta_{jl} - \Delta^{-1}\frac{\partial^2}{\partial x_j \partial x_l}\Big)F_j^{(i)}\\ &\qquad \approx \frac{h^n}{2\nu\pi^{n/2}}\sum_{\mathbf{m}\in\mathbb{Z}^n}\sum_{k=1}^{n}(u_j u_k)(h\mathbf{m},(i-1)\tau)Z_{jkl}^{(n)}(\lambda,\mathbf{x}-h\mathbf{m})\Big|_{\lambda=0}^{\lambda=\tau},\end{aligned}$$

where

$$\begin{aligned}Z_{jkl}^{(n)}(\lambda,\mathbf{y}) &= W_{jkl}(\lambda,\mathbf{y}) - \delta_{jl}\frac{y_k}{|\mathbf{y}|^n}\Gamma\Big(\frac{n}{2},\frac{|\mathbf{y}|^2}{\kappa(\lambda)}\Big)\\ &= \frac{\kappa(\lambda)}{2|\mathbf{y}|^{n+2}}\Gamma\Big(\frac{n}{2}+1,\frac{|\mathbf{y}|^2}{\kappa(\lambda)}\Big)\Big(\frac{(n+2)y_j y_k y_l}{|\mathbf{y}|^2} - (\delta_{kl}y_j + \delta_{jl}y_k + \delta_{jk}y_l)\Big) \qquad (13.30)\\ &\quad - \frac{1}{2|\mathbf{y}|^n}\Gamma\Big(\frac{n}{2},\frac{|\mathbf{y}|^2}{\kappa(\lambda)}\Big)\Big(\frac{n\,y_j y_k y_l}{|\mathbf{y}|^2} - (\delta_{kl}y_j + \delta_{jk}y_l - \delta_{jl}y_k)\Big).\end{aligned}$$

In particular, the formulas

$$\Gamma\Big(\frac{3}{2}, x\Big) = \sqrt{x}\,\mathrm{e}^{-x} + \frac{\sqrt{\pi}}{2}\,\mathrm{erfc}(\sqrt{x}),$$
$$\Gamma\Big(\frac{5}{2}, x\Big) = x^{3/2}\,\mathrm{e}^{-x} + \frac{3}{2}\sqrt{x}\,\mathrm{e}^{-x} + \frac{3\sqrt{\pi}}{4}\,\mathrm{erfc}(\sqrt{x}),$$

provide the following expression in the 3-dimensional case:

$$Z^{(3)}_{jkl}(\lambda, \mathbf{y}) = \frac{\mathrm{e}^{-|\mathbf{y}|^2/\kappa(\lambda)}}{|\mathbf{y}|^2}\bigg(\Big(\frac{y_j y_l}{|\mathbf{y}|^2} - \delta_{jl}\Big)\frac{y_k}{\sqrt{\kappa(\lambda)}}$$
$$+ \frac{3\sqrt{\kappa(\lambda)}}{4|\mathbf{y}|}\Big(\frac{5y_j y_k y_l}{|\mathbf{y}|^3} - \frac{\delta_{kl}y_j + \delta_{jl}y_k + \delta_{jk}y_l}{|\mathbf{y}|}\Big)\bigg)$$
$$+ \frac{\sqrt{\pi}}{8|\mathbf{y}|^4}\,\mathrm{erfc}\Big(\frac{|\mathbf{y}|}{\sqrt{\kappa(\lambda)}}\Big)\bigg((15\kappa(\lambda) - 6|\mathbf{y}|^2)\frac{y_j y_k y_l}{|\mathbf{y}|^3}$$
$$+ (2|\mathbf{y}|^2 - 3\kappa(\lambda))\frac{\delta_{kl}y_j + \delta_{jk}y_l}{|\mathbf{y}|} - (2|\mathbf{y}|^2 + 3\kappa(\lambda))\frac{\delta_{jl}y_k}{|\mathbf{y}|}\bigg).$$

By (13.13), the first integral on the right-hand side of (13.19) can be approximated as

$$\frac{h^n}{\pi^{n/2}} \sum_{\mathbf{m}\in\mathbb{Z}^n} u_h(h\mathbf{m}, (i-1)\tau)\, \frac{e^{-|\mathbf{x}-h\mathbf{m}|^2/(4\nu\lambda+\mathcal{D}h^2)}}{(4\nu\lambda + \mathcal{D}h^2)^{n/2}}\Bigg|_{\lambda=0}^{\lambda=\tau}.$$

Hence, we derive the time-marching algorithm for the velocity components in $\mathbb{R}^n$

$$\begin{aligned} u_{h,l}(\mathbf{x}, i\tau) &= u_{h,l}(\mathbf{x}, (i-1)\tau) \\ &+ \frac{h^n}{\pi^{n/2}} \sum_{\mathbf{m}\in\mathbb{Z}^n} u_{h,l}(h\mathbf{m}, (i-1)\tau)\frac{e^{-|\mathbf{x}-h\mathbf{m}|^2/(4\nu\lambda+\mathcal{D}h^2)}}{(4\nu\lambda + \mathcal{D}h^2)^{n/2}}\Bigg|_{\lambda=0}^{\tau} \\ &- \frac{h^n}{2\nu\pi^{n/2}} \sum_{\mathbf{m}\in\mathbb{Z}^n}\sum_{j,k=1}^{n} (u_{h,j}u_{h,k})(h\mathbf{m}, (i-1)\tau)Z^{(n)}_{jkl}(\lambda, \mathbf{x} - h\mathbf{m})\Bigg|_{\lambda=0}^{\lambda=\tau} \end{aligned} \tag{13.31}$$

and the coefficients $Z^{(n)}_{jkl}$ given in (13.30).

Finally, we give an approximation formula for the pressure p which is obtained in accordance with (13.20) from

$$p_h(\mathbf{x}, i\tau) = -\frac{1}{(\pi\mathcal{D})^{n/2}} \sum_{j,k=1}^{n} \frac{\partial^2}{\partial x_j \partial x_k}\Delta^{-1} \sum_{\mathbf{m}\in\mathbb{Z}^n} (u_{h,j}u_{h,k})(h\mathbf{m}, i\tau)e^{-|\mathbf{x}-h\mathbf{m}|^2/\mathcal{D}h^2}.$$

By (6.4) we have

$$\frac{\partial^2}{\partial x_i \partial x_k}\mathcal{L}_n(\mathrm{e}^{-|\cdot|^2})(\mathbf{x}) = \frac{n x_i x_k - \delta_{ik}|\mathbf{x}|^2}{2|\mathbf{x}|^{n+2}}\,\gamma\Big(\frac{n}{2}, |\mathbf{x}|^2\Big) - \frac{x_i x_k}{|\mathbf{x}|^2}\,\mathrm{e}^{-|\mathbf{x}|^2},$$

which implies the formula

$$
\boxed{
\begin{aligned}
&p_h(\mathbf{x}, i\tau) = \frac{1}{2(\pi\mathcal{D})^{n/2}} \sum_{\mathbf{m}\in\mathbb{Z}^n} \sum_{i,k=1}^{n} (u_{h,j}u_{h,k})(h\mathbf{m}, i\tau) \\
&\quad \Big\{ \Big(n\frac{(x_j - hm_j)(x_k - hm_k)}{|\mathbf{x} - h\mathbf{m}|^2} - \delta_{jk} \Big) \frac{1}{(z_{\mathbf{m}}(0))^{n/2}} \gamma\Big(\frac{n}{2}, z_{\mathbf{m}}(0)\Big) \\
&\quad - 2\frac{(x_j - hm_j)(x_k - hm_k)}{|\mathbf{x} - h\mathbf{m}|^2} \, \mathrm{e}^{-z_{\mathbf{m}}(0)} \Big\},
\end{aligned}
} \tag{13.32}
$$

approximating the pressure with the accuracy $\mathcal{O}(\tau^2 + \tau h^2)$.

13.3. Non-local evolution equations

Approximate approximations can be successfully applied for solving non-local evolution equations. Consider the Cauchy problem for equations of the general form

$$
\begin{aligned}
u_t - P_1(D_{\mathbf{x}})u &= P_2(D_{\mathbf{x}})\, F(\mathbf{x}, t, u, P_3(D_{\mathbf{x}})u) \,, \quad t > 0 \,,\ \mathbf{x} \in \mathbb{R}^n \,, \\
u(\mathbf{x}, 0) &= \varphi(\mathbf{x}) \,,
\end{aligned}
$$

where $D_{\mathbf{x}} = (-i\partial/\partial x_1, \dots, -i\partial/\partial x_n)$. We suppose that the operators $P_k(D_{\mathbf{x}})$ are convolutions with the symbol $P_k(2\pi\boldsymbol{\xi})$, $k = 1, 2, 3$, and that F is a smooth function. The equation is discretized in time by a two-parameter finite-difference approximation with a time step τ. Then $u(\mathbf{x}, t)$ is approximated by a sequence of functions $u_j(\mathbf{x}) = u(\mathbf{x}, j\tau)$, $j = 0, 1, 2, \dots$, satisfying

$$
\begin{aligned}
\tau^{-1}(u_j - u_{j-1}) - \theta_1 P_1(D_{\mathbf{x}})u_j &- (1 - \theta_1)P_1(D_{\mathbf{x}})u_{j-1} \\
&= P_2(D_{\mathbf{x}})\big((1 + \theta_2)F_{j-1}(\mathbf{x}) - \theta_2 F_{j-2}(\mathbf{x})\big) \,,
\end{aligned}
$$

where $F_j(\mathbf{x}) = F(\mathbf{x}, j\tau, u_j, P_3(D_{\mathbf{x}})u_j)$ and $0 < \theta_i \le 1$. With the notation

$$
\begin{aligned}
\mu &= \tau\theta_1 \,, \\
y &= u_j + (\theta_1^{-1} - 1)u_{j-1} \,, \\
f &= \theta_1^{-1} u_{j-1} \,, \\
g &= \tau\big((1 + \theta_2)F_{j-1} - \theta_2 F_{j-2}\big) \,,
\end{aligned}
$$

we derive the linear problem

$$
-\mu P_1(D_{\mathbf{x}})y + y = f + P_2(D_{\mathbf{x}})g \,, \quad \mathbf{x} \in \mathbb{R}^n \,,
$$

which has the solution

$$
y = f + (R - I)f + P_2 R g \qquad \text{with} \quad R = (I - \mu P_1)^{-1} \,.
$$

Replacing f and g by its approximate quasi-interpolants (2.23), one obtains the approximate solution

$$
\begin{aligned}
y_h &= f + (R - I)\mathcal{M}_{h,\mathcal{D}} f + P_2 R \mathcal{M}_{h,\mathcal{D}} g \\
&= f + \mathcal{D}^{-n/2} \sum_{\mathbf{m}\in\mathbb{Z}^n} f(h\mathbf{m}) \Big\{ (R\eta)\Big(\frac{\mathbf{x} - h\mathbf{m}}{\sqrt{\mathcal{D}}h}\Big) - \eta\Big(\frac{\mathbf{x} - h\mathbf{m}}{\sqrt{\mathcal{D}}h}\Big) \Big\} \\
&\quad + \mathcal{D}^{-n/2} \sum_{\mathbf{m}\in\mathbb{Z}^n} g(h\mathbf{m}) (P_2 R\eta)\Big(\frac{\mathbf{x} - h\mathbf{m}}{\sqrt{\mathcal{D}}h}\Big) \,.
\end{aligned}
$$

Thus, the values of y_h at the grid points $\mathbf{x_k} = h\mathbf{k}$ are linear combinations of

$$(R\eta)\Big(\frac{\mathbf{k}-\mathbf{m}}{\sqrt{\mathcal{D}}}\Big) \quad \text{and} \quad (P_2R\eta)\Big(\frac{\mathbf{k}-\mathbf{m}}{\sqrt{\mathcal{D}}}\Big),$$

which may be effectively computed for a suitably chosen generating function η. By this way one derives the following explicit scheme for computing the approximate solution $u_{j,h}$ of the Cauchy problem:

$$u_{j,h} = u_{j-1,h} + \theta_1^{-1}(R-I)\mathcal{M}_{h,\mathcal{D}}u_{j-1,h} + \tau P_2 R\mathcal{M}_{h,\mathcal{D}}\big((1+\theta_2)F_{j-1} - \theta_2 F_{j-2}\big).$$

For $\theta_1 = \theta_2 = 1/2$ the scheme is of second-order accuracy.

Compared with other explicit schemes for solving time-dependent problems, the proposed method is very robust with respect to variations of the ratio between time and spatial discretization. Numerous numerical tests for different equations have shown, that, of course, the non-linearity in the original equation imposes restrictions to the time step τ, but there exists no strict connection between τ and the mesh size h. In these tests the present method provides an accuracy of $\mathcal{O}(\tau^2+\tau h^N)$ at each time step, where N is the approximation order of the approximate quasi-interpolation with the generating function η. For $\theta_1 = \theta_2 = 1/2$ the numerical accuracy increases to $\mathcal{O}(\tau^3 + \tau h^N)$. However, the rigorous error analysis of this quite general method remains open. The estimation of the saturation errors, which occur at each time step, is rather involved and is understood at present only for some special equations. Here again for $\mathcal{D}$ sufficiently large, the saturation errors can be kept below a given error level, as expected also from the numerical experiments. In the following, we report on two examples of non-linear non-local evolution equations, to which the method was applied.

13.3.1. Joseph equation. The Joseph equation

$$u_t + \delta^{-1}u_x + (2\delta)^{-1}\int_{-\infty}^{\infty} u_{yy}\coth\frac{\pi(y-x)}{2\delta}\,dy = -(u^2)_x \tag{13.33}$$

describes the unidirectional propagation of small-amplitude, non-linear, dispersive, long waves in stratified fluids.

Note that the shallow water approximation ($\delta \ll 1$) of (13.33) is the Korteweg-de Vries equation, whereas the deep water approximation ($\delta \gg 1$) is the Benjamin-Ono equation

$$u_t + \mathcal{H}u_{xx} = -(u^2)_x$$

with the Hilbert transform $\mathcal{H}$. In Fig. 13.5 the computational results for Joseph's equation with the initial data $\varphi(x) = \delta\exp(-x^2)$ for $\delta = 0.1, 0.333, 1$, and 10, corresponding to shallow, intermediate, and deep water, are shown. In these computations η was the Gaussian function, $\mathcal{D} = 3$, $h = 0.1$, and $\tau = 0.001$.

13.3.2. Sivashinsky equation. Another interesting example which is difficult to solve by using finite-difference or finite-element methods represents the two-dimensional equation of flame front propagation

$$u_t + a\Delta^2 u + \epsilon\Delta u + b\Delta\int_{\mathbb{R}^2}\frac{u(\mathbf{y},t)}{|\mathbf{x}-\mathbf{y}|}\,d\mathbf{y} + cu = -\frac{1}{2}(\nabla u)^2\,,\ t>0\,,\ \mathbf{x}\in\mathbb{R}^2. \tag{13.34}$$

Here

$$P_1(\xi) = -a|\xi|^4 + \epsilon|\xi|^2 + b|\xi| - c,\ P_2(\xi) = 1,\ P_3(\xi) = i\xi,\ F(\mathbf{x},t,u,\mathbf{v}) = -|\mathbf{v}|^2/2\,.$$

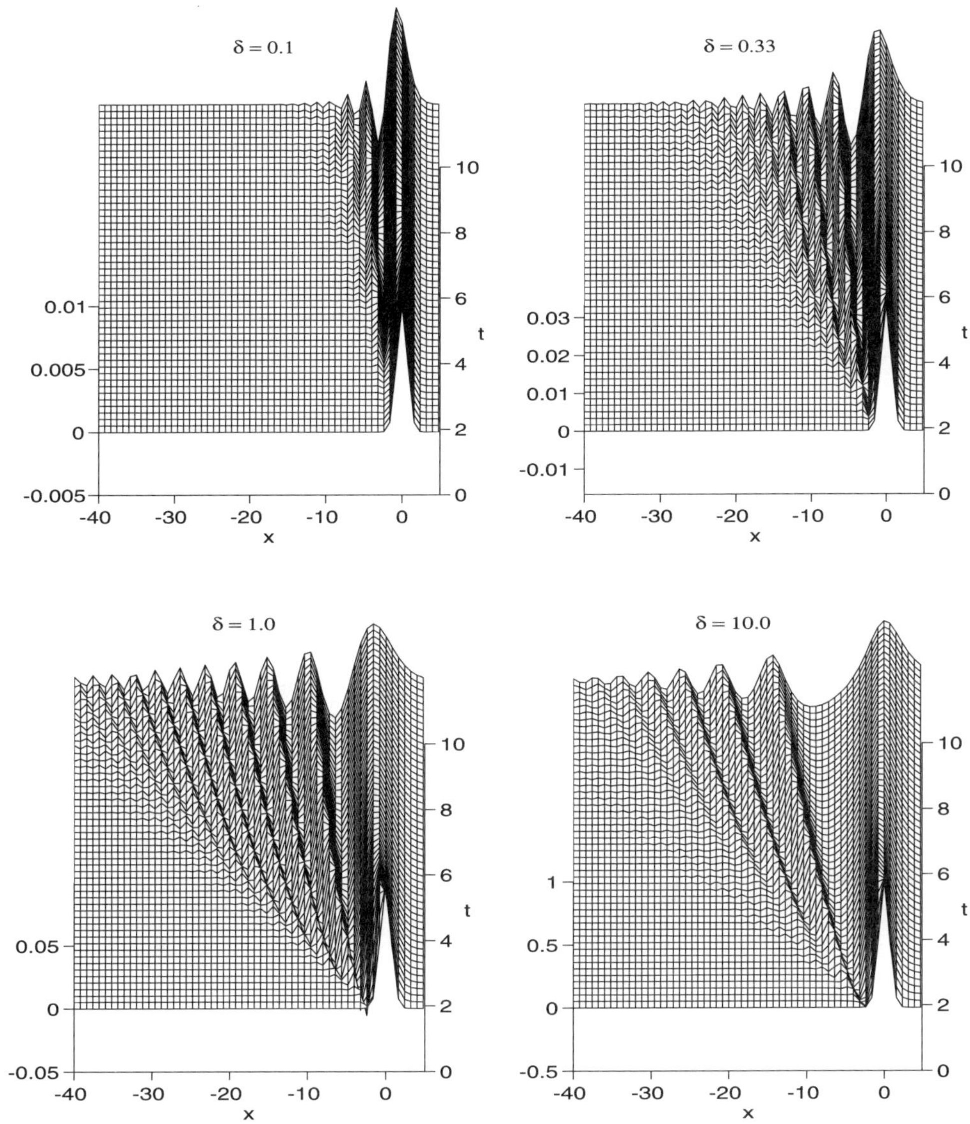

FIGURE 13.5. Solution of the Joseph equation for different δ

With the two-dimensional radial generating functions, the occurring integrals are transformed to the zero-order Hankel transform with smooth and rapidly decaying integrands and may be computed using standard quadrature procedure.

In Fig. 13.6, the results of the numerical solution of (13.34) with the initial data

$$\varphi(\mathbf{x}) = 1 + \frac{1}{2} \sin 2x_1 \, \sin 2x_2$$

FIGURE 13.6. Level lines of the surface of the flame front for $t = 0.0$ (upper left), $t = 1.0$ (upper right), $t = 5.0$ (lower left), and $t = 10.0$ (lower right)

and the parameters $a = 10^{-4}$, $\epsilon = 0.05$, $b = 0.005$, and $c = 1/6$ are given for different time values. The 2π-periodic solution was computed for the discretization parameters $h = \pi/32$ and $\tau = 10^{-3}$.

The given parameters correspond to a linear flame instability and the computations result in a corrugated flame front.

13.4. Notes

The material of Section 13.1 extends results from the papers [**65**], [**64**]. Note that we restrict ourselves to the simplest finite-difference approximation of the time derivative. In the papers mentioned, higher-order finite-difference approximations are considered as well, and numerical results for different model equations are given.

The approach presented in Subsection 13.1.1 is applied in [**45**] to the solution of initial boundary value problems for semi-linear heat and wave equations

$$u_t - u_{xx} = F(x,t,u)\,, \quad u(x,0) = \varphi_0(x)\,,\ u(\pm a,t) = \psi^{\pm}(t)\,,$$

and

$$u_{tt} - u_{xx} = F(x,t,u)\,, \quad u(x,0) = \varphi_0(x)\,,\ u_t(x,0) = \varphi_1(x)\,,\ u(\pm a,t) = \psi^{\pm}(t)\,,$$

where $x \in [-a,a]$, $t > 0$. Explicit analytic formulas for the approximate solution at each time step are obtained for the case of quasi-interpolation with the generating function η_4 and first- and second-order finite-difference time approximations. Numerical tests with smooth functions F, φ_j, and $\psi^{\pm}$ confirm that at each time step the accuracy of the algorithms is $\mathcal{O}(\tau^3 + \tau h^4)$.

The solution method for Navier-Stokes equations described in Section 13.2 was proposed in [**64**] for the two-dimensional case. The numerical tests were performed by V. Karlin.

The material of Section 13.3 is taken from [**46**]. Algorithmic and numerical aspects of solving the Sivashinsky equation, using approximate approximations, are discussed in more detail in [**48**].

In the dissertation [**76**] J. Niebsch studied error estimates for the algorithms considered in Sections 13.1 and 13.3. In particular, it is shown that the algorithm (13.14) for solving the semi-linear heat equation (13.10) provides the error estimate

$$\big\|u(\cdot,N\tau) - u_h(\cdot,N\tau)\big\|_{L_2} \le c_T\Big(\tau + (\sqrt{\mathcal{D}}h)^{2M} + \varepsilon\Big)\,\mathrm{e}^{TL_{\mathbf{f}}}\,, \tag{13.35}$$

where $T = N\tau$, $\mathbf{f}$ and φ are sufficiently smooth functions, and $L_{\mathbf{f}}$ is the Lipschitz constant

$$\big\|\mathbf{f}(\cdot,t,u) - \mathbf{f}(\cdot,t,v)\big\|_{L_2} \le L_{\mathbf{f}}\|u-v\|_{L_2}\,, \quad t \in (0,T)\,.$$

The saturation error ε is sufficiently small if the parameter $\mathcal{D}$ is chosen large enough. In special cases, for example linear problems or right-hand sides of (13.10) which ensure the existence of a global solution $u(\mathbf{x},t) \to 0$ if $t \to \infty$, the factor $\mathrm{e}^{TL_{\mathbf{f}}}$ in (13.35) can be omitted.

Further applications of approximate approximations are given in [**47**], where the method is used to solve hypersingular integral equations of the Peierls type

$$\int_{-\infty}^{\infty} K(x-y)\,u(y)\,dy = F(u(x))\,, \quad K(x) = -x^{-2} + \kappa(x)\,,$$

with κ being smooth and the integral is defined as Hadamard finite-part integral. Integral equations of this type occur in dislocation theory. By the efficient cubature formulas, critical Peierls stresses were calculated with very high accuracy for a variety of dislocations.

Bibliography

[1] M. Abramowitz and I. A. Stegun, *Handbook of Mathematical Functions*, Dover Publ., New York, 1968.

[2] B. Alpert, G. Beylkin, R. Coifman, and V. Rokhlin, *Waveletlike bases for the fast solution of second-kind integral equations*, SIAM J. Sci. Statist. Comp., 14 (1993), pp. 159–184.

[3] H. Åkermark, *Applications of the Boundary Point Method*, Theses No. 511, LIU-TEK-LIC-1995.

[4] P. M. Anselone, *Collectively compact operator theory and applications to integral equations*, Prentice Hall, Englewood Cliffs, New Jersey, 1971.

[5] K. Atkinson and D. Chien, *Piecewise polynomial collocation for boundary integral equations*, SIAM J. Sci. Comput., 16 (1995), pp. 651–681.

[6] I. Babuška, U. Banerjee, and J. E. Osborn, *On the approximability and the selection of particle shape functions*, Numer. Math. 96 (2004), pp. 601–640.

[7] H. Bateman and A. Erdélyi, *Tables of Integral Transforms*, Mc Graw-Hill, 1954.

[8] R. K. Beatson and W. A. Light, *Quasi-interpolation in the absence of polynomial reproduction*, in: Numerical Methods of Approximation Theory, Vol. 9, D. Braess and L. L. Schumaker (eds.), pp. 21-39, Birkhäuser, Basel, 1992.

[9] G. Beylkin, *Wavelets and fast numerical algorithms*, Proc. Symp. Appl. Math., 47 (1993), pp. 89–117.

[10] C. de Boor, R. A. DeVore, and A. Ron, *On the construction of multivariate (pre) wavelets*, Constr. Approx., 9 (1993), no. 2-3, pp. 123–166.

[11] ———, *Approximation from shift–invariant subspaces of* $L_2(\mathbb{R}^d)$, Trans. AMS, 341 (1994), no. 2, pp. 787–806.

[12] C. de Boor and R. Q. Jia, *Controlled approximation and a characterization of the local approximation order*, Proc. Amer. Math. Soc., 95 (1985), pp. 547–553.

[13] C. de Boor and A. Ron, *Fourier analysis of the approximation power of principal shift-invariant spaces*, Constr. Approx., 8 (1992), pp. 427–462.

[14] M. D. Buhmann, *Multivariate cardinal interpolation with radial-basis functions*, Constr. Approx., 6 (1990), pp. 225–255.

[15] ———, *Radial Basis Functions, Theory and Implementations*, Cambridge Monographs on Applied and Computational Mathematics, No. 12, Cambridge, 2003.

[16] M. D. Buhmann, N. Dyn, and D. Levin, *On quasi interpolation by radial basis functions with scattered data*, Constr. Appr., 11 (1995), pp. 239–254.

[17] H.-Q. Bui, and R. S. Laugesen, *Affine systems that span Lebesgue spaces*, J. Fourier Anal. Appl., 11 (2005), pp. 533–556.

[18] C. K. Chui, *An Introduction to Wavelets*, San Diego, Academic Press, 1992.

[19] C. K. Chui, J. Stöcker, and J. D. Ward, *Compactly supported box spline wavelets*, Proc. Amer. Math. Soc., 113 (1992), pp. 785–793.

[20] D. Colton and R. Kress, *Inverse acoustic and electromagnetic scattering theory*, Appl. Math. Sci., 93, Springer, Berlin, 1992.

[21] W. Dahmen, S. Prössdorf, and R. Schneider, *Wavelet approximation methods for pseudo-differential equations II*, Adv. Comp. Math., 1 (1993), pp. 259–335.

[22] I. Daubechies, *Ten lectures on wavelets*, vol. 61 of CBMS Conference Series in Applied Mathematics, SIAM, Philadelphia, 1992.

[23] P. J. Davis and P. Rabinowitz, *Methods of Numerical Integration*, Academic Press, Orlando, 1984.

[24] N. Dyn and A. Ron, *Radial basis function approximation: from gridded centers to scattered centers*, Proc. London Math. Soc., 71 (1995), pp. 76–108.

[25] D. M. Eidus, *On the principle of limiting absorption* (Russian), Mat. Sborn. (N.S.), 57 (99) (1962), pp. 13–44 .

[26] G. E. Fasshauer, *Approximate moving least-squares approximation with compactly supproted radial weights*, in: Lecture Notes in Computer Science and Engineering, Vol. 26, M. Griebel and M. A. Schweitzer (eds.), Springer, 2002, pp. 105–116.

[27] ———, *Toward approximate moving least squares approximation with irregularly spaced centers*, Comput. Methods Appl. Mech. Engrg., 193 (2004), pp. 1231–1243

[28] R. Farwig, *Multivariate interpolation of arbitrarily spaced data by moving least squares methods*, J. Comput. Appl. Math., 16 (1986), pp. 79–93.

[29] M. V. Fedoryuk, *Asymptotics: integrals and series* (Russian), Spravochnaya Matematicheskaya Biblioteka, Nauka, Moskva, 1987.

[30] P. Grisvard, *Elliptic Problems in Nonsmooth Domains*, Pitman, London, 1985.

[31] W. Hackbusch and S. Sauter, *On numerical cubatures of nearly singular surface integrals arising in bem collocation*, Computing, 52 (1994), pp. 139–159.

[32] H. Hakopian, *Integral remainder formula of the tensor product interpolation*, Bull. Ac. Pol. Math., 31 (1983), pp. 267–272.

[33] M. R. Hestenes, *Extension of the range of a differentiable function*, Duke Math. J., 8 (1941), pp. 183–192.

[34] L. Hörmander, *The Analysis of Linear Partial Differential Operators*, vol. I, Springer, Berlin, Heidelberg, New York, Tokyo, 1983.

[35] T. Ivanov, *Boundary layer approximate approximations and cubature of potentials in domains*, PhD thesis, Linköping University, 1997.

[36] T. Ivanov, V. Maz'ya, and G. Schmidt, *Boundary layer approximate approximations for the cubature of potentials in domains*, Adv. Comp. Math., 10 (1999), pp. 311–342.

[37] ———, *Boundary layer approximate approximations for the cubature of potentials*, in: Mathematical Aspects of Boundary Element Methods, M. Bonnet, A.-M. Sändig, and W. L. Wendland (eds.), vol. 10 of Research Notes in Mathematics, Chapman & Hall/CRC London, 1999, pp. 165–177.

[38] R. Q. Jia and J. Lei, *Approximation by multiinteger translates of functions having global support*, J. Approx. Th., 72 (1993), pp. 2–23.

[39] M. J. Johnson, *An upper bound on the approximation power of principal shift-invariant spaces*, Constructive Approximation, 13 (1997), pp. 155-176.

[40] ———, *On the approximation order of principal shift-invariant subspaces of $L_p(\mathbb{R}^d)$*, J. Approx. Th., 91 (1997), pp. 279–319.

[41] S. K. Kanaun, V. Romero, and J. Bernal, *A new numerical method for the solution of the second boundary value problem of elasticity for bodies with cracks*, Rev. Mexicana Fis., 47 (2001), 309–323.

[42] S. K. Kanaun, *A numerical method for the solution of electromagnetic wave diffraction problems on perfectly conducting screens*, J. Comp. Physics, 176 (2002), pp. 170–195.

[43] S. K. Kanaun and V. Romero, *Boundary point method in the dynamic and static problems of mathematical physics*, in: Domain Decomposition Methods in Science and Engineering, pp. 443–450, Natl. Auton. Univ. Mex., México, 2003.

[44] S. K. Kanaun and S. Kochekseraii, *A numerical method for the solution of thermo- and electro-static problems for a medium with isolated inclusions*, J. Comp. Physics, 192 (2003) pp. 471–493.

[45] V. Karlin and V. Maz'ya, *Time-marching algorithms for initial-boundary value problems for semi-linear evolution equations based upon approximate approximations*, BIT, 35 (1995), pp. 548–560.

[46] ———, *Time-marching algorithms for non-local evolution equations based upon "approximate approximations"*, SIAM J. Sci. Comput., 18 (1997), pp. 736–752.

[47] V. Karlin, V. Maz'ya, A. B. Movchan, J. R. Willis, and R. Bullogh, *Numerical solution of nonlinear hypersingular integral equations of the Peierls type in dislocation theory*, SIAM J. Appl. Math., 60 (2000), no. 2, pp. 664–678.

[48] V. Karlin, V. Maz'ya, and G. Schmidt, *High accuracy periodic solutions to the Sivashinsky equation*, J. Comput. Phys., 188 (2003), pp. 209–231.

[49] T. Kato, *Perturbation theory for linear operators*, vol. 132 of Grundlehren der mathematischen Wissenschaftem, Springer-Verlag, Berlin, Heidelberg, New York, 1966.

[50] R. Kieser, C. Schwab, and W. L. Wendland, *Numerical evaluation of singular and finite part integrals on curved surfaces using symbolic manipulation*, Computing, 49 (1992), no. 3, pp. 279–301.

[51] V. D. Kupradze, T. G. Gegelia, M. O. Basheleishvili, and T. V. Burchuladze, *Three-dimensional problems of the mathematical theory of elasticity and thermoelasticity*, North-Holland Publishing Co., Amsterdam-New York, 1979.

[52] O. A. Ladyzhenskaya, *The mathematical theory of viscous incompressible flow*, Gordon and Breach Science Publishers, New York, 1963.

[53] F. Lanzara, V. Maz'ya, and G. Schmidt, *Numerical Solution of the Lippmann–Schwinger Equation by Approximate Approximations*, J. Fourier Anal. Appl., 10 (2004), pp. 645-660.

[54] ———, *Approximate Approximations from Scattered Data*, WIAS Preprint No. 1058, 2005, Berlin.

[55] ———, *Approximate Approximations with Data on a Perturbed Uniform Grid*, to appear in ZAA (2007).

[56] ———, *Approximate Approximations from Scattered Data*, J. Approx. Th., 145 (2007), pp. 141–170.

[57] D. Levin, *The approximations power of moving least-squares*, Math. Comp., 67 (1998), pp. 1517–1531.

[58] W. R. Madych and S. A. Nelson, *Bounds on multivariate polynomials and exponential error estimates for multiquadric interpolation*, J. Approx. Th., 70 (1992), pp. 94-114.

[59] H. N. Mhaskar, *Approximation theory and neural networks*, in: Wavelet and Allied Topics, P. K. Jain et al. (eds.), Narosa Publishing House, New Dehli, 2001, pp. 247–289.

[60] ———, *On the degree of approximation in multivariate weighted approximation*, in: Advanced Problems in Constructive Approximation, M.D. Buhmann and D.H. Mache (eds.), ISNM **142**, Birkhäuser, Basel, 2002, pp. 129–141.

[61] S. G. Mallat, *Multiresolution approximations and wavelet orthonormal bases of* $L^2(\mathbb{R})$, Trans. Amer. Math. Soc., 315 (1989), pp. 69–87.

[62] V. Maz'ya, *A new approximation method and its applications to the calculation of volume potentials. Boundary point method,* in: 3. DFG-Kolloqium des DFG-Forschungsschwerpunktes "Randelementmethoden", 30 Sep–5 Oct 1991.

[63] ———, *Boundary point method*, Preprint LiTH-MAT-R-91-44, Linköping University, 1991.

[64] ———, *Approximate approximations*, in: The Mathematics of Finite Elements and Applications. Highlights 1993, J. R. Whiteman (ed.), Wiley & Sons, Chichester, 1994, pp. 77–104.

[65] V. Maz'ya and V. Karlin, *Semi-analytic time-marching algorithms for semi-linear parabolic equations*, BIT, 34 (1994), pp. 129–147.

[66] V. Maz'ya and G. Schmidt *On approximate approximations.* Preprint LiTH-MAT-R-94-12, Linköping University, 1994.

[67] ———, *"Approximate Approximations" and the cubature of potentials*, Rend. Mat. Acc. Lincei 9, 6 (1995), pp. 161–184.

[68] ———, *On approximate approximations using Gaussian kernels*, IMA J. Num. Anal., 16 (1996), pp. 13–29.

[69] ———, *Approximate wavelets and the approximation of pseudodifferential operators*, Appl. Comp. Harm. Anal., 6 (1999), pp. 287–313.

[70] ———, *Construction of basis functions for high order approximate approximations*, in: Mathematical Aspects of Boundary Element Methods, M. Bonnet, A.-M. Sändig, and W. L. Wendland (eds.), vol. 10 of Research Notes in Mathematics, Chapman & Hall/CRC London, 1999, pp. 191–202.

[71] ———, *On quasi–interpolation with non–uniformly distributed centers on domains and manifolds*, J. Approx. Th., 110 (2001), pp. 125–145.

[72] ———, *Potentials of Gaussians and approximate wavelets*, Mathematische Nachrichten, Vol. 280, No. 9-10 (2007).

[73] V. Maz'ya, G. Schmidt, and W. L. Wendland, *On the computation of multi-dimensional single layer harmonic potentials via approximate approximations*, Calcolo, 40 (2003), no. 1, pp. 33–53.

[74] Y. Meyer, *Wavelets and Operators*, University Press, Cambridge, 1992.

[75] S. G. Michlin, *Partielle Differentialgleichungen in der mathematischen Physik*, Mathematische Lehrbücher und Monographien. Bd. 30, Akademie-Verlag, Berlin, 1978.

[76] J. Niebsch, *Zur numerischen Lösung von Evolutionsgleichungen mit nichtlokalen Operatoren auf der Basis approximativer Approximationen*, PhD thesis, Logos Verlag, Berlin, 2000.

[77] P. W. Partridge, C. A. Brebbia, and L. C. Wrobel, *The Dual Reciprocity Boundary Element Method*, CMP, Southhampton, and Elsevier, London, 1991.

[78] T. von Petersdorf, R. Schneider, and C. Schwab, *Multi wavelet approximation for the double layer potential equation*, SIAM J. Numer. Anal., 34 (1997), no. 6, pp. 2212–2227.

[79] M. J. D. Powell, *The theory of radial basis functions in 1990*, in: Advances in Numerical Analysis. Vol. 2: Wavelets, Subdivision Algorithms, and Radial Basis Functions, W. Light (ed.), pp. 105–210, Clarendon Press, Oxford, 1992.

[80] S. Prössdorf and B. Silbermann, *Numerical analysis for integral and related operator equations*, Operator Theory: Advances and Applications, 52, Birkhäuser Verlag, Basel, 1991.

[81] A. P. Prudnikov, Yu. A. Brychkov, and O. I. Marichev, *Integrals and series. Vol. 1: Elementary functions*, Gordon & Breach Science Publishers, New York, 1986.

[82] ———, *Integrals and series. Vol. 2: Special functions*, Gordon & Breach Science Publishers, New York, 1988.

[83] R.D. Richtmyer and K.W. Morton, *Difference methods for initial-value problems*, Wiley & Sons, New York, London, Sydney, 1967.

[84] S. Sauter and C. Schwab, *Quadrature for hp-Galerkin BEM in* $\mathbb{R}^3$, Numer. Math., 78 (1997), pp. 211–258.

[85] R. Schaback and Z. Wu, *Construction techniques for highly accurate quasi–interpolation*, J. Approx. Th., 91 (1997), pp. 320–331.

[86] I. J. Schoenberg, *Contributions to the problem of approximation of equidistant data by analytic functions, Parts A & B*, Quarterly Appl. Math. IV (1946), pp. 45–99, pp. 112–141.

[87] C. Schwab and W. L. Wendland, *On numerical cubatures of singular surface integrals in the boundary element method*, Numer. Math., 62 (1992), pp. 343–369.

[88] S. L. Sobolev, *Introduction to the theory of cubature formulas*, Nauka, Moscow, 1974.

[89] E. M. Stein, *Singular Integrals and Differentiability Properties of Functions*, University Press, Princeton, 1970.

[90] E. M. Stein and G. Weiss, *Introduction to Fourier Analysis on Functions*, University Press, Princeton, 1971.

[91] F. Stenger, *Numerical Methods Based on Sinc and Analytic Functions*, Springer, Berlin, Heidelberg, New York, Tokyo, 1993.

[92] G. Strang and G. Fix, *A Fourier analysis of the finite element variational method*, in: Constructive Aspects of Functional Analysis, C.I.M.E. II Ciclo 1971, G. Geymonat (ed.), 1973, pp. 793–840.

[93] G. Szego, Orthogonal Polynomials, Amer. Math. Soc., Providence, RI, 1975 (fourth edition).

[94] H. Triebel, *Interpolation theory, function spaces, differential operators*, Deutscher Verlag der Wissenschaften, Berlin, 1978.

[95] G. Vainikko, *Fast solvers of the Lippmann-Schwinger equation*, in: Direct and Inverse Problems of Mathematical Physics, R. P. Gilbert et al. (eds.), Int. Soc. Anal. Appl. Comput., 5, pp. 423–440, Kluwer, Dordrecht, 2000.

[96] G. N. Watson, A treatise on the theory of Bessel functions, Cambridge University Press, 1922.

[97] Z. M. Wu and J. P. Liu, *Generalized Strang-Fix condition for scattered data quasi-interpolation*, Adv. Comp. Math., 23 (2005), pp. 201–214.

[98] Z. Wu and R. Schaback, *Local error estimates for radial basis function interpolation of scattered data*, IMA J. Numer. Anal., 13 (1993), pp. 13-27.

Index

Titles in This Series

141 **Vladimir Maz′ya and Gunther Schmidt,** Approximate approximations, 2007
140 **Elisabetta Barletta, Sorin Dragomir, and Krishan L. Duggal,** Foliations in Cauchy-Riemann geometry, 2007
139 **Michael Tsfasman, Serge Vlăduţ, and Dmitry Nogin,** Algebraic geometric codes: Basic notions, 2007
138 **Kehe Zhu,** Operator theory in function spaces, 2007
137 **Mikhail G. Katz,** Systolic geometry and topology, 2007
136 **Jean-Michel Coron,** Control and nonlinearity, 2007
135 **Bennett Chow, Sun-Chin Chu, David Glickenstein, Christine Guenther, James Isenberg, Tom Ivey, Dan Knopf, Peng Lu, Feng Luo, and Lei Ni,** The Ricci flow: Techniques and applications, Part I: Geometric aspects, 2007
134 **Dana P. Williams,** Crossed products of C^*-algebras, 2007
133 **Andrew Knightly and Charles Li,** Traces of Hecke operators, 2006
132 **J. P. May and J. Sigurdsson,** Parametrized homotopy theory, 2006
131 **Jin Feng and Thomas G. Kurtz,** Large deviations for stochastic processes, 2006
130 **Qing Han and Jia-Xing Hong,** Isometric embedding of Riemannian manifolds in Euclidean spaces, 2006
129 **William M. Singer,** Steenrod squares in spectral sequences, 2006
128 **Athanassios S. Fokas, Alexander R. Its, Andrei A. Kapaev, and Victor Yu. Novokshenov,** Painlevé transcendents, 2006
127 **Nikolai Chernov and Roberto Markarian,** Chaotic billiards, 2006
126 **Sen-Zhong Huang,** Gradient inequalities, 2006
125 **Joseph A. Cima, Alec L. Matheson, and William T. Ross,** The Cauchy Transform, 2006
124 **Ido Efrat, Editor,** Valuations, orderings, and Milnor K-Theory, 2006
123 **Barbara Fantechi, Lothar Göttsche, Luc Illusie, Steven L. Kleiman, Nitin Nitsure, and Angelo Vistoli,** Fundamental algebraic geometry: Grothendieck's FGA explained, 2005
122 **Antonio Giambruno and Mikhail Zaicev, Editors,** Polynomial identities and asymptotic methods, 2005
121 **Anton Zettl,** Sturm-Liouville theory, 2005
120 **Barry Simon,** Trace ideals and their applications, 2005
119 **Tian Ma and Shouhong Wang,** Geometric theory of incompressible flows with applications to fluid dynamics, 2005
118 **Alexandru Buium,** Arithmetic differential equations, 2005
117 **Volodymyr Nekrashevych,** Self-similar groups, 2005
116 **Alexander Koldobsky,** Fourier analysis in convex geometry, 2005
115 **Carlos Julio Moreno,** Advanced analytic number theory: L-functions, 2005
114 **Gregory F. Lawler,** Conformally invariant processes in the plane, 2005
113 **William G. Dwyer, Philip S. Hirschhorn, Daniel M. Kan, and Jeffrey H. Smith,** Homotopy limit functors on model categories and homotopical categories, 2004
112 **Michael Aschbacher and Stephen D. Smith,** The classification of quasithin groups II. Main theorems: The classification of simple QTKE-groups, 2004
111 **Michael Aschbacher and Stephen D. Smith,** The classification of quasithin groups I. Structure of strongly quasithin K-groups, 2004
110 **Bennett Chow and Dan Knopf,** The Ricci flow: An introduction, 2004
109 **Goro Shimura,** Arithmetic and analytic theories of quadratic forms and Clifford groups, 2004

TITLES IN THIS SERIES

108 **Michael Farber,** Topology of closed one-forms, 2004

107 **Jens Carsten Jantzen,** Representations of algebraic groups, 2003

106 **Hiroyuki Yoshida,** Absolute CM-periods, 2003

105 **Charalambos D. Aliprantis and Owen Burkinshaw,** Locally solid Riesz spaces with applications to economics, second edition, 2003

104 **Graham Everest, Alf van der Poorten, Igor Shparlinski, and Thomas Ward,** Recurrence sequences, 2003

103 **Octav Cornea, Gregory Lupton, John Oprea, and Daniel Tanré,** Lusternik-Schnirelmann category, 2003

102 **Linda Rass and John Radcliffe,** Spatial deterministic epidemics, 2003

101 **Eli Glasner,** Ergodic theory via joinings, 2003

100 **Peter Duren and Alexander Schuster,** Bergman spaces, 2004

99 **Philip S. Hirschhorn,** Model categories and their localizations, 2003

98 **Victor Guillemin, Viktor Ginzburg, and Yael Karshon,** Moment maps, cobordisms, and Hamiltonian group actions, 2002

97 **V. A. Vassiliev,** Applied Picard-Lefschetz theory, 2002

96 **Martin Markl, Steve Shnider, and Jim Stasheff,** Operads in algebra, topology and physics, 2002

95 **Seiichi Kamada,** Braid and knot theory in dimension four, 2002

94 **Mara D. Neusel and Larry Smith,** Invariant theory of finite groups, 2002

93 **Nikolai K. Nikolski,** Operators, functions, and systems: An easy reading. Volume 2: Model operators and systems, 2002

92 **Nikolai K. Nikolski,** Operators, functions, and systems: An easy reading. Volume 1: Hardy, Hankel, and Toeplitz, 2002

91 **Richard Montgomery,** A tour of subriemannian geometries, their geodesics and applications, 2002

90 **Christian Gérard and Izabella Łaba,** Multiparticle quantum scattering in constant magnetic fields, 2002

89 **Michel Ledoux,** The concentration of measure phenomenon, 2001

88 **Edward Frenkel and David Ben-Zvi,** Vertex algebras and algebraic curves, second edition, 2004

87 **Bruno Poizat,** Stable groups, 2001

86 **Stanley N. Burris,** Number theoretic density and logical limit laws, 2001

85 **V. A. Kozlov, V. G. Maz′ya, and J. Rossmann,** Spectral problems associated with corner singularities of solutions to elliptic equations, 2001

84 **László Fuchs and Luigi Salce,** Modules over non-Noetherian domains, 2001

83 **Sigurdur Helgason,** Groups and geometric analysis: Integral geometry, invariant differential operators, and spherical functions, 2000

82 **Goro Shimura,** Arithmeticity in the theory of automorphic forms, 2000

81 **Michael E. Taylor,** Tools for PDE: Pseudodifferential operators, paradifferential operators, and layer potentials, 2000

80 **Lindsay N. Childs,** Taming wild extensions: Hopf algebras and local Galois module theory, 2000

79 **Joseph A. Cima and William T. Ross,** The backward shift on the Hardy space, 2000

For a complete list of titles in this series, visit the AMS Bookstore at **www.ams.org/bookstore/**.